AF575454

Implementing Central Finance with SAP S/4HANA®

SAP PRESS is a joint initiative of SAP and Rheinwerk Publishing. The know-how offered by SAP specialists combined with the expertise of Rheinwerk Publishing offers the reader expert books in the field. SAP PRESS features first-hand information and expert advice, and provides useful skills for professional decision-making.

SAP PRESS offers a variety of books on technical and business-related topics for the SAP user. For further information, please visit our website: *www.sap-press.com*.

Anand Seetharaju

Implementing Central Finance with SAP S/4HANA®

Editor Rachel Gibson
Acquisitions Editor Emily Nicholls
Copyeditor Julie McNamee
Cover Design Graham Geary
Photo Credit iStockphoto: 174962301/© skodonnell
Layout Design Vera Brauner
Production Hannah Lane
Typesetting III-Satz, Germany
Printed and bound in the United States of America, on paper from sustainable sources

ISBN 978-1-4932-2607-8

1st edition 2025

Library of Congress Cataloging-in-Publication Data
Names: Seetharaju, Anand, author.
Title: Implementing central finance with SAP S/4HANA / by Anand Seetharaju.
Description: Bonn ; Boston : Rheinwerk Publishing, 2024. | Includes index.
Identifiers: LCCN 2024039388 | ISBN 9781493226078 (hardcover) | ISBN 9781493226085 (ebook)
Subjects: LCSH: SAP HANA (Electronic resource) | Business--Data processing--Management. | Enterprise resource planning. | Database management.
Classification: LCC HD30.213 .S366 2023 | DDC 650.0285--dc23/eng/20240906
LC record available at https://lccn.loc.gov/2024039388

Contents at a Glance

Contents

Preface

Welcome to *Implementing Central Finance with SAP S/4HANA*, a comprehensive guide to help you learn and master the Central Finance skills you need for successful Central Finance implementations. Central Finance, without a doubt, is the most feature-rich solution SAP has offered to customers in recent times that takes advantage of SAP S/4HANA's innovations. Central Finance integrates non-SAP systems with most SAP components, one way or the other, thus making Central Finance a powerful platform that streamlines processes and reports across an enterprise. However, with all its new features, it can be challenging to keep up, as new innovations are constantly announced in every major release cycle and through SAP Notes. Whether you're starting a new journey with Central Finance or you're already working in it, with this book, you'll gain the knowledge necessary and equip yourself with the tools to succeed in implementing Central Finance.

Objective of This Book

The main objective of this book is to provide you with a comprehensive guide covering both functional and technical aspects of Central Finance. After working on several Central Finance implementations, I observed there is a rising demand from the SAP community for a single source where readers can find all Central Finance-related topics with real-time insights. In this book, I tried to fulfill that goal. If you're new to Central Finance, this book provides you with the necessary foundational skills to jumpstart your career in Central Finance. Foundational SAP-related finance skills help you expedite your learning journey. If you're an experienced user, the book offers you the advanced skills necessary to perform customizing activities with step-by-step instructions and real-life examples, tips, and tricks. If you're a business user, you'll benefit from the functional knowledge provided in this book. The book was written based on the latest Central Finance features offered with SAP S/4HANA 2023.

Target Audience

To provide a comprehensive coverage of Central Finance topics, this book contains content relevant to various audiences. It offers both functional and technical knowledge with step-by-step instructions on the customizing activities necessary for functional or technical implementers to carry out on a new Central Finance project or financial transformation project. It was written with the following target audiences in mind:

- SAP functional finance (FI-CO) consultants involved in Central Finance implementation
- Finance managers, controllers, and chief financial officers
- Accounting professionals who practice and implement Central Finance
- IT analysts or business analysts
- Integrations consultants and specialists
- Other SAP cross-functional consultants who need a solid understanding of Central Finance, as it interacts with most other SAP components
- Other ERP specialists (e.g. Oracle Fusion, JD Edwards, and so on) who are integrating their ERP system with a Central Finance system

Structure of the Book

Each chapter of this book starts with a brief context, an overview on the subject, extensive coverage on the topic, customizing instructions, and then a summary of key points learned from the lesson. The book highlights tips, notes, and leading practices wherever applicable. This book is structured in a such way to provide readers with a logical sequence of each topic by building foundational knowledge in the early chapters and expanding to advanced topics in the later chapters. With that in mind, the book is organized into the following thirteen chapters:

- **Chapter 1, Introduction to Central Finance**
  The first chapter begins with an introduction to Central Finance and its product maturity, deployment options, and future roadmap. In addition, you'll review key use cases of Central Finance and understand the impact of process, technology, and organizational changes as a result of implementing Central Finance.
- **Chapter 2, Central Finance Processes**
  Central Finance enables customers to streamline processes based on harmonized and centralized data in the Universal Journal. This enhances reporting capabilities, as the data represents a single source of truth across an enterprise. In this chapter, you'll learn how centralized business processes integrate with SAP S/4HANA components with the help of process flow diagrams. In addition, the chapter covers Central Finance capabilities and explains how harmonized data helps organizations achieve maximum benefits when adopting centralized processes. Finally, we explain recent innovations in joint venture accounting (JVA) scenarios supported in Central Finance.
- **Chapter 3, Central Finance Architecture**
  A clear understanding of the Central Finance system landscape, architecture, and data flow are essential for a successful Central Finance implementation project. This chapter lays a solid foundation of Central Finance architecture and data flow from the source system to Central Finance (and integration with other systems, as applicable). You will learn various Central Finance terminology and tools that facilitate

the flow of data between source system and central system. The chapter concludes with information you need when integrating a third-party system with Central Finance.

- **Chapter 4, Master Data Management**
  Master data design and management are the most time-consuming activities in any SAP S/4HANA or Central Finance project. Poor quality data and incorrect mapping may lead to additional work or duplication of efforts. In this chapter, you'll learn about master data harmonization in Central Finance. Then the chapter transitions into data profiling in source systems, replication aspects, and master data mapping in Central Finance. Furthermore, the chapter explains how to configure enhanced mapping entities when adding additional fields in Central Finance.
- **Chapter 5, Configuring SAP Landscape Transformation Replication Server**
  SAP Landscape Transformation Replication Server enables real-time replication of data in Central Finance. It is essential that you understand how SAP Landscape Transformation Replication Server collects data from source systems and then routes to Central Finance. First, we give an overview of SAP Landscape Transformation Replication Server in Central Finance and then transition to the architecture of SAP Landscape Transformation Replication Server. Then the chapter explains how to configure SAP Landscape Transformation Replication Server with step-by-step instructions for data replication. The chapter also explains advanced settings such as filters, job settings, and monitoring.
- **Chapter 6, Configuring SAP Application Interface Framework**
  SAP Application Interface Framework is an error handling and monitoring tool used in Central Finance. This role-based tool offers several capabilities such as monitoring error messages, showing and confirming alerts, and sending email notifications to recipients. In this chapter, you'll learn about SAP Application Interface Framework terminology, runtime groups, and error handling. The chapter shows how alerts work with step-by-step instructions and concludes with miscellaneous topics in SAP Application Interface Framework.
- **Chapter 7, Configuring Central Finance and Source Systems**
  You'll learn important design elements and their considerations before starting Central Finance configuration. In addition, when implementing Central Finance, you need to understand prerequisites, security requirements, and configurations requirements. In this chapter, we lay a good foundation of the prerequisites to be followed, and then you'll learn how to configure the source system and Central Finance system with step-by-step instructions.
- **Chapter 8, Initial Load, Intermediate Data Retention, and Real-Time Replication**
  A Central Finance financial accounting initial load transfers general ledger balances, open items, and certain transactional data based on selection settings. You'll learn various initial load concepts, initial load prerequisites, sequences, validation, and initial load posting. You'll also review important considerations for the initial load and real-time replication.

- **Chapter 9, Central Payment**
  To emphasize the importance of central payments in Central Finance, this chapter goes in-depth into central payment considerations. First, you will learn prerequisites prior to considering central payments. Then you'll learn activating and configuring central payment functionality. Finally, the chapter explains dependent processes to be considered before activating central payment.
- **Chapter 10, Controlling**
  Central Finance offers additional replication scenarios in the Controlling space. This chapter dives deeper into those Controlling replication scenarios with step-by-step instructions. The chapter begins with project system master data before transitioning into other Controlling objects and transactional data replication.
- **Chapter 11, Accounting View of Logistics**
  In this chapter, you'll learn how a subset of logistics data transferred from source system to Central Finance system supports operational and statutory reporting in Central Finance. It starts with basic concepts and then transitions to step-by-step instructions on configuring accounting view of logistics (AVL) scenarios. The chapter further explains how to extract AVL data for third party systems.
- **Chapter 12, Reconciliation and Reporting in Central Finance**
  Reconciliation data at various stages of a Central Finance implementation plays a key role in making your overall SAP S/4HANA or Central Finance implementation project successful. Thus, SAP-delivered comparison reports are critical to reconcile transfer of postings from source system to Central Finance. In this chapter, you'll learn how to leverage SAP S/4HANA reports wherever applicable. We begin with an approach for Central Finance reconciliation. Then we transition into how to reconcile balances and line items with SAP-delivered Central Finance reports. The reconciliation reports provide different levels of analysis for financial accounting and Controlling data. Finally, we provide an overview of other SAP S/4HANA reports used to reconcile Central Finance data.
- **Chapter 13, Central Finance Project Management**
  In this chapter, you'll learn high-level recommendations for project management methodologies as well as the available tools for Central Finance. The chapter highlights important considerations from SAP activate methodology and discusses how to leverage predelivered Central Finance adoption content. Then you'll understand the change impact implementing Central Finance from a people, process, and technology perspective. The chapter concludes with Central Finance leading practices based on past implementation experience.

Note

This book uses screenshots taken from the Belize visual theme, which is used in SAP Fiori applications.

Throughout the book, we've also provided several elements that will help you access useful information.

[+]

Tips

Boxes marked with this icon provide specific recommendations that can make your work easier.

[«]

Notes

Boxes marked with this icon contain information about additional topics or important content that you should note.

Warnings

This icon refers to specifics that you should consider. It also warns about frequent errors or problems that can occur.

Acknowledgments

This work is dedicated to my daughter, Aditi, who has made me stronger, better, and more fulfilled than I could have ever imagined. Thank you, Aditi. I'm grateful for your love and unwavering support, especially your patience and understanding, which have been an anchor during the countless hours of writing the book during evenings and weekends over the past several months.

Building a fully functional Central Finance system involving multiple systems is an involved process. You need support and resources from areas such as technical, Basis, security, and so on. The content of this book wouldn't have been possible without the support of Amit Kumar Singh. I owe a great amount of gratitude to him for his time and effort, especially in building a complete working Central Finance environment. Amit is a highly experienced professional and technologist, and I truly appreciate his contributions across the book and his help with SAP Landscape Transformation Server. We have exchanged thoughts and learned from each other during our discussions over the last several months. Thank you, Amit, for providing your valuable feedback, and it's always great to hear a different perspective and present it to my readers. Amit is the go-to person for any complex Central Finance questions, and I always learn something new about Central Finance every time we talk!

A special thanks to Pardha Manyala and Bahram Maghsoudi for generously sharing their time and expertise. I value your insightful thoughts and perspectives on the subject. Next, I would like to thank Vaishnavi Ranganathan, Sridhar Baisa, Mazeen Naveed,

Anirudh Manikanta, and Srinivasa Rao Gunturu for their invaluable support. I can always count on their expert knowledge when I most need it.

Finally, I wanted to thank Emily Nicholls for the opportunity to write this book. I appreciate your help in formalizing ideas and aligning the content of the book by understanding the needs of the market. Next, I want to thank Rachel Gibson, the development editor, for motivating and mentoring me with guidance throughout the project. Your editing skills on organizing chapter contents were helpful in giving the manuscript the focus and clarity to logically present its topics. Lastly, I'd like to thank the copyeditor, Julie McNamee, for providing valuable feedback and proofing the copy to ensure accuracy. I look forward to collaborating with you all on future projects.

Anand Seetharaju, CPA, CMA

Chapter 1
Introduction to Central Finance

SAP provides multiple pathways to transform financial processes to SAP S/4HANA processes. Central Finance offers customers a consolidated view of taking advantage of SAP S/4HANA innovations. Customers with either a single ERP system or a distributed landscape with multiple ERP systems benefit from Central Finance transition without interrupting existing systems or processes. In this introductory chapter, you'll learn what Central Finance is all about, including adoption approaches, use cases, and road maps.

Large global organizations generally have multiple enterprise resource planning (ERP) systems to manage their day-to-day business operations. These disparate ERP systems are implemented during a company's organic growth or by integrating ERP systems through mergers and acquisitions over a period. In addition, companies often find difficulty in migrating to a newer ERP system, and sometimes it takes years to align on a common reporting structure. SAP's Central Finance solution offers such complex organizations the ability to perform consolidated reporting using the single source of truth—the Universal Journal—without interrupting existing processes or systems. In addition, the solution also enables companies to streamline financial processes across the enterprise, transform outdated processes with SAP S/4HANA leading processes, and centralize processes. Central Finance supports both SAP and non-SAP systems data replication. Let's understand how traditional systems function to achieve reporting from an integrated system landscape.

In this chapter, we'll begin with the typical challenges faced in a distributed landscape with multiple ERP systems. Then, we'll transition to discussing what Central Finance is about. Next, we'll cover SAP's clean core approach and deployment options. You'll also learn use cases to support potential business scenarios. Finally, the chapter concludes with the product road map of Central Finance.

1.1 Distributed Systems Landscape Challenges

In a distributed systems landscape, data integration is often challenging to get an integrated view of financial data across the enterprise. Figure 1.1 illustrates the traditional way of reporting data in a heterogeneous system landscape. Customers with multiple ERP instances typically use data warehousing systems (e.g., SAP Business Warehouse) to aggregate and report financial figures of the consolidated companies. This often requires companies extracting data from multiple ERP systems by running batch jobs overnight, which are scheduled periodically to extract and transform the data from multiple system landscapes or tables. From a reporting point of view, the data isn't real time or near real time. Therefore, often the reports aren't accurate at a given point in time. As a result of differences in data between the source and data in data warehousing systems, often reconciliation differences arise that result in a time-consuming activity to reconcile data between the systems. In addition, to transform data, companies build complex data models to transfer data through batch jobs that run at a regular interval to extract and transform the data from distributed landscapes to data warehouses. Often, batch jobs that transfer data aren't built with enough validation checks, which sometimes leads to receiving poor data in the reporting landscape, thereby impacting quality of reporting. This approach isn't real time and results in latency in reporting.

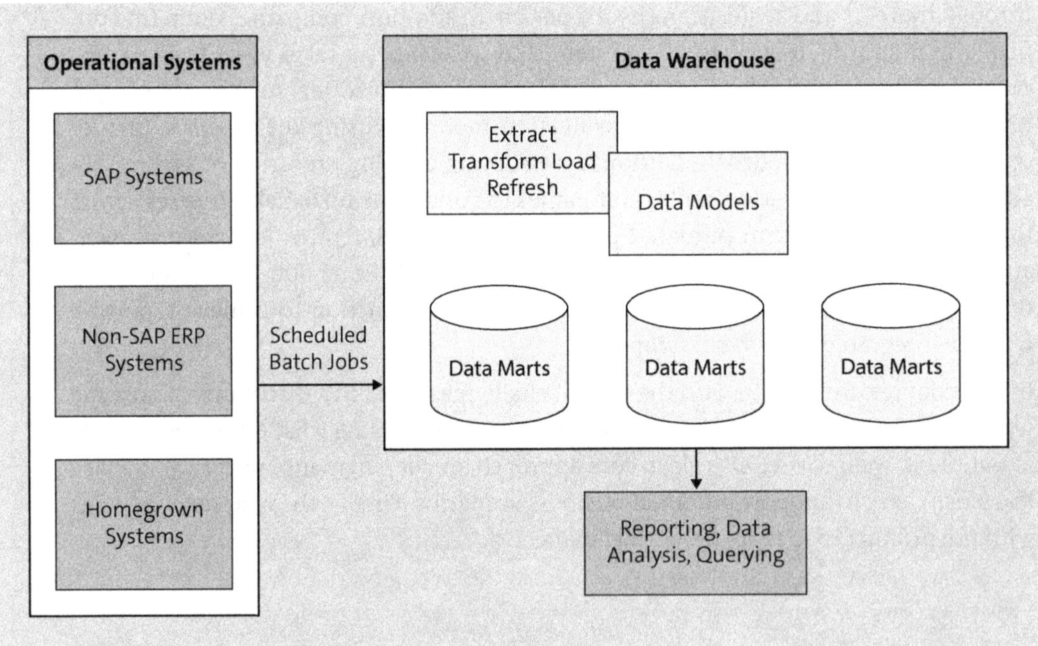

Figure 1.1 Distributed Landscape with Data Warehouses

As a result, companies with a distributed landscape with multiple ERP systems face challenges:

- Heterogeneous systems have homegrown systems or out of maintenance systems. In addition, as part of mergers and acquisitions, a growing system landscape has additional complexities.
- Systems integrated in a distributed landscape are in different release cycles and interdependencies on improvements or upgrades.
- No harmonization exists of organizational master data such as chart of accounts (CoA), profit center, and cost center. As a result, companies face duplicate data or issues with data quality.
- No common reporting structure limits unified reporting across the enterprise.
- There is a lack of transparency or limited transparency between systems in the distributed landscape.
- No real-time visibility of all company codes is available.
- Nonstandardized processes among systems create high duplication of effort or repeats of steps.
- Outdated tools or technology is in use.
- Complex models of extractions and transformation of data are developed between systems.
- Additional effort is involved with data reconciliation and data validation activities.

1.2 Getting Started with Central Finance

First, let's understand what Central Finance is about, followed by understanding its capabilities first to ensure a successful transition and then to its adoption. In this section, we'll explore paths to SAP S/4HANA and how Central Finance fits in the overall picture.

1.2.1 What Is Central Finance?

SAP offers Central Finance as a deployment option to streamline processes and consolidate data for unified reporting and centralized processes regardless of the source systems. In addition, Central Finance can be leveraged as a financial hub for processes outside of SAP S/4HANA. Central Finance has been available since the simple finance inception by the name of *Central Journal*. SAP rebranded Central Journal as Central Finance with several additional features. Nevertheless, there has been a lot of confusion among businesses and SAP user communities to truly understand what Central Finance is and its adoption strategies. Figure 1.2 shows how the different views of Central Finance can be perceived by business users and IT professionals:

❶ **Central Finance as a system**
Central Finance serves as a financial data hub by receiving replication data from both SAP and non-SAP source systems. The financial hub can serve to integrate other solutions beyond SAP S/4HANA. With Central Finance, customers with complex and distributed landscapes across geographies avert high costs of implementations, which come with risks of migrating and upgrading to a new system. Central Finance as a system offers customers a quick way to adopt SAP's architecturally advanced in-memory database and the latest reporting innovations with the Universal Journal without interrupting existing processes and systems with a sidecar approach. In this approach, Central Finance, in essence, replicates both financial and nonfinancial transactions from different source ERP systems into one single SAP S/4HANA system. While replicating data from source systems, existing processes and system landscape aren't interrupted. A nondisruptive solution means existing system processes remain untouched and current business operations aren't affected. Transaction-level data is replicated into Central Finance and leveraged with SAP's latest innovations—SAP HANA optimizations with built-in memory and cutting-edge user interface (UI) tools such as SAP Fiori, SAP Lumira, SAP Analytics Cloud, SAP Screen Personas, and so on. In addition, the system helps businesses transform mapping and harmonize master data, which leads to the SAP S/4HANA system of flexible data models. With harmonized master data and centralized data, you can perform planning, consolidating, and reporting activities.

❷ **Central Finance as a product**
This is delivered with the SAP S/4HANA suite as a product. To achieve goals of nondisruptive to existing processes and systems, Central Finance comes with its own tool set to support initial loads and replication: logging, mapping, error correction, initial load monitoring tools, reconciliation and comparison reports, and so on. Central Finance as a product is delivered with SAP Landscape Transformation Replication Server, SAP Application Interface Framework content, business configuration sets (BC sets), SAP Master Data Governance, and other supporting tools. Central Finance is scalable to adopt changes as business grows.

❸ **Central Finance as a solution**
You can streamline and centralize financial business processes in a single central location. Master data and configurations can be harmonized and transformed to support the streamlined and standardized processes. Centralized process execution capabilities can be introduced as an incremental step with the flexibility of business requirements to simplify business processes with SAP-delivered leading practices. Central Finance establishes the foundation for centralized financial processes, including central payment, central credit management, central collections and dispute management, central project reporting, central tax reporting, central closing, operational control, and so on.

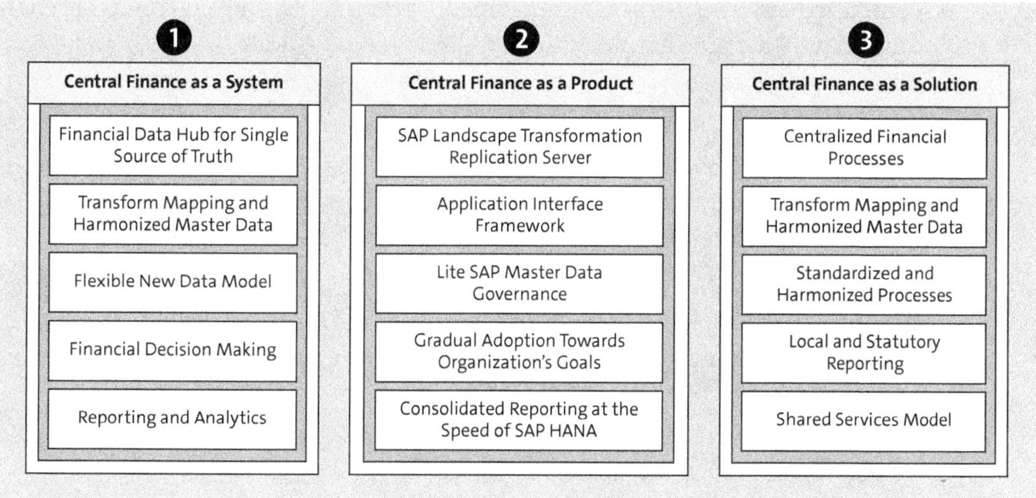

Figure 1.2 What's Central Finance?

The Central Finance system is built scalable, flexible, and adaptable for changing business requirements. With the implementation of Central Finance, the system becomes the single source of truth for financial data. Therefore, companies phase out source systems as they become out of support or outdated with numerous customizations. Figure 1.3 displays important benefits of using Central Finance. Here are a few additional general benefits obtained by implementing Central Finance:

- Real-time replication of source data without any interruption to existing business operations. The advantage of a nondisruptive approach enables customers to adopt SAP S/4HANA in a phased approach without touching the existing system immediately. In an incremental step, you'll phase out legacy systems at the customer's convenience.
- Centralized data in the Universal Journal, which provides a single source of truth for consolidated financial reporting. This enables faster consolidating and reporting and soft-closing capabilities.
- Group reporting capability with real-time data received from multiple source systems within an enterprise.
- Segment reporting capabilities to fulfill ASC 280 segment reporting.
- Harmonization of master data across multiple source systems to improve quality of master data and deduplication. For example, harmonized master data and reporting structures are available across various entities for CoA, cost center, profit center, vendors, customers, plants, and so on.
- Integrated business planning and forecasting for both short-term and long-term with centralized data.

- Centralized services such as central payment or central credit, dispute, and collections capabilities. Processes are simplified by eliminating duplicate processes and efforts, for example, central payment, central tax reporting, shared services, and so on.
- Personalized user experience with SAP Fiori and other presentation tools with SAP S/4HANA innovations.
- Improved productivity as a result of using cutting-edge UIs such as SAP Fiori, SAP Lumira, core data services (CDS) views, and so on.
- Centralized data for multiple system landscapes single source of truth via Central Finance's Universal Journal (multiple ERPs into one global SAP instance).
- Quickly integrate newly acquired companies into the system landscape via mergers and acquisitions.
- Replicate all open items to different systems with management and real-time liquidity reports, bank balances, and cash positions.

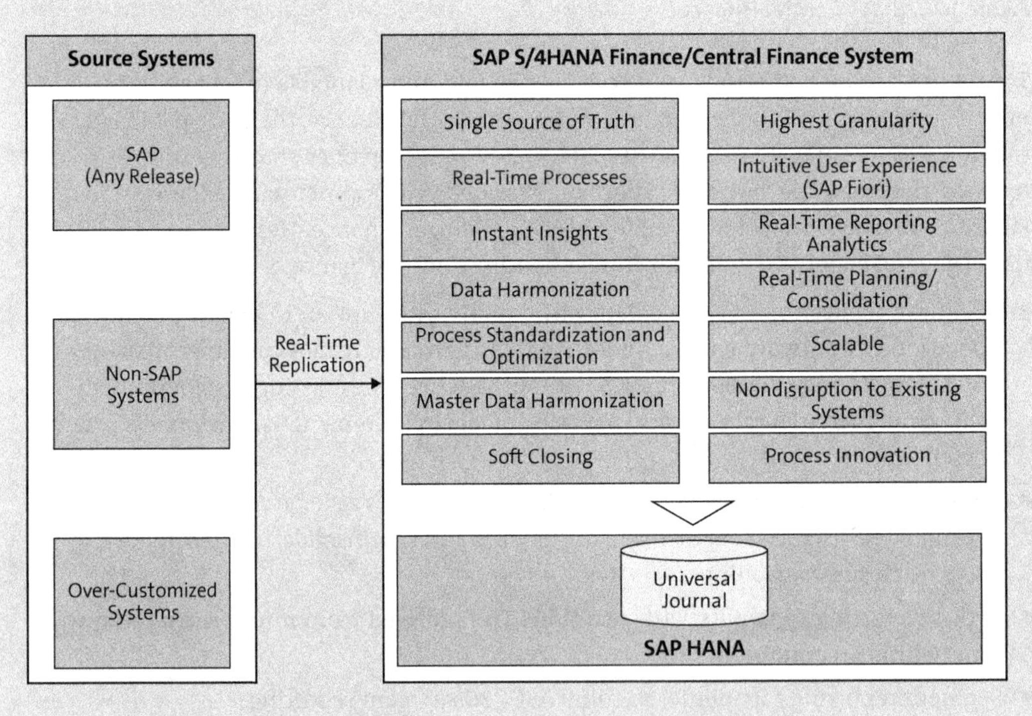

Figure 1.3 Central Finance Benefits

1.2.2 Paths to SAP S/4HANA

The opportunity to transition to SAP S/4HANA from a legacy ERP system presents customers with the possibility to modernize legacy infrastructure and applications. In addition, if a business sustains constant growth over a period of years, archaic hardware and software may not keep up with the growth and innovation required to run the business in the competitive world, resulting in companies sometimes losing a competitive edge in the market. When implementing SAP S/4HANA, options such as long-term vision of IT strategy, functional impact, cost of implementation, and business challenges must be considered carefully. SAP offers customers with different transition scenarios to leverage SAP S/4HANA innovations. Finance transformation also should be future proof and support the company's long-term goals. When implementing SAP S/4HANA, customers have a choice to pick a correct deployment strategy that fits the short- and long-term goals supporting their business strategy. Figure 1.4 illustrates paths to the SAP S/4HANA transformation journey. Let's understand each transition path's merits and demerits:

❶ **New system installation (greenfield approach)**
This is appropriate for customers starting a new implementation of SAP S/4HANA using SAP-delivered standard functionality and industry leading practices. The greenfield approach provides customers with an opportunity to leverage standard SAP functionality and redesign business processes with a best practices approach. The scenario is preferred where legacy systems have too much customer code or multiple controlling areas where cross-company code transactions aren't supported. In addition, the approach enables customers to redesign the organization structure such as CoA, operational hierarchies, and master data to fit new business requirements. On the other hand, depending on the complexity, significant effort may be required to deploy a greenfield approach. Although SAP offers conversion tools such as the Migrate Your Data app (app ID F3473) and Migration Object Modeler (Transaction LTMOM), transferring large volumes of data from legacy systems might consume a lot of effort and may extend project timelines. Besides, a lot of involvement from the business is required in designing business processes. SAP also provides best practices content to showcase the standard functionality to stakeholders without going through large exercise of configuration and testing the features.

❷ **System upgrade (brownfield approach)**
This allows customers to convert the existing SAP ERP system to SAP S/4HANA using a rapid conversion approach. After conversion, customers then take advantage of SAP S/4HANA's new functionalities such as introducing a new ledger or document splitting in an incremental step. The brownfield approach is suitable for systems with stable organizational structure and business processes such as CoA and cross-company code processes. The brownfield approach is accomplished with either a one-step or two-step approach. With the one-step approach, both database and system conversion happen at the same time; with the two-step approach, these upgrade events are done as separate projects. The brownfield implementation approach has the following benefits:

- SAP offers numerous conversion tools such as SAP Readiness Check, SAP Update Manager (SUM) prerequisites check, and simplification lists.
- Only delta configuration is required to support the conversion.

- All data is available from prior years, and no effort is involved in transferring legacy data to SAP S/4HANA.
- An easy to follow migration cockpit monitoring tool is provided with step-by-step instructions.
- Faster deployment is possible with proven tools and the activate methodology.

Although the brownfield approach is a faster approach to SAP S/4HANA transition, the system doesn't allow you to introduce a parallel ledger, currency, and document splitting functionality during conversion. The addition of such a feature requires subsequent implementation in a separate project after the system conversion. In addition, any errors or process gaps continue to exist in the new system.

❸ **Selective data transmission with Data Management and Landscape Transformation (DMLT)**

This helps customers to carve out or transform selected data from a complex data landscape with specific tools, methodologies, and leading industry approaches. SAP refers to the term *selective data* by applying restrictions to migrate certain master data, open items, and balances. The selected data transmission approach could be adopted in many situations where companies are involved in mergers and acquisitions or divestitures activities. The selective data transmission offers customers the benefits:

- Flexible carve-out rollouts per business requirements per organization units, for example, by company code
- Only pull information relevant by leaving unnecessary historical data behind
- Harmonization opportunities for data cleansing or master data
- Introduce new business processes as required by business requirements
- Low downtime or near-zero downtime

❹ **Central Finance**

This deployment approach offers customers the chance to take advantage of SAP S/4HANA features right away without interrupting existing ERP systems by replicating both financial and nonfinancial data to Central Finance. Alternatively, Central Finance can be used as an operating system with central business processes and SAP S/4HANA features. Central Finance provides multiple benefits to customers:

- Real-time advanced segment and group reporting without interrupting existing business processes
- Streamlined and harmonized live business processes in Central Finance
- Process centralization by eliminating duplication using the shared service model and economies of scale
- Landscape transformation capabilities toward building digital enterprise, digital core, Internet of Things (IoT), and cloud readiness
- Scalable as the companies grow with mergers, acquisitions, carve outs, and divestitures

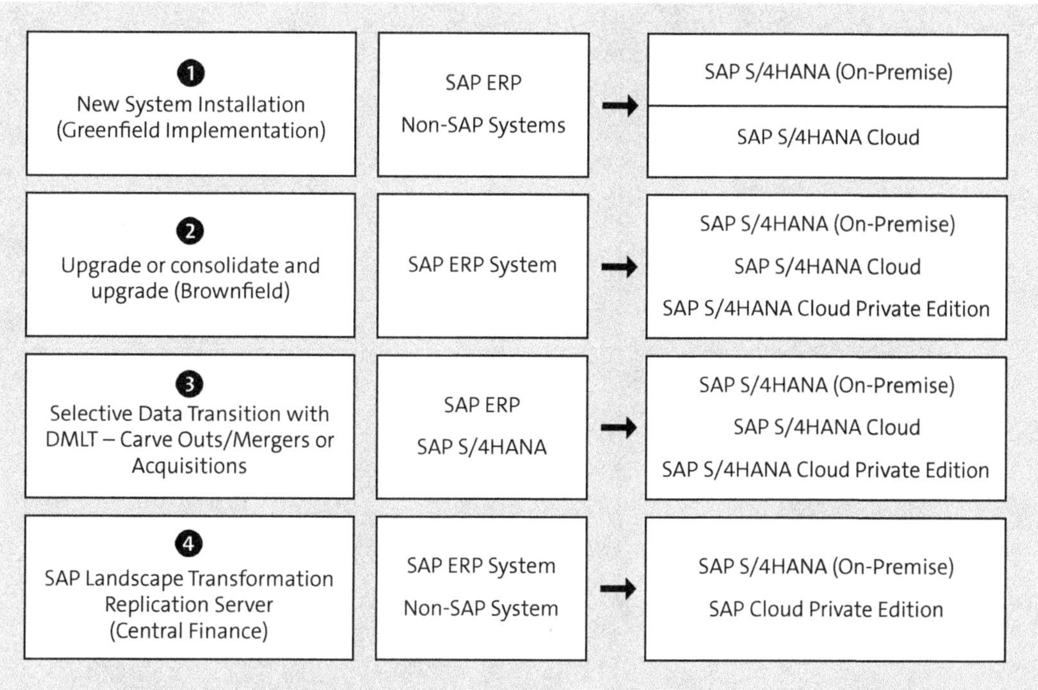

Figure 1.4 Paths to SAP S/4HANA

[«]

Note

The Central Finance sidecar approach referred to specifically application-based replication. This means transaction data received from multiple source systems are reposted in the Central Finance system for the various use cases we discussed in Section 1.4. Central Finance is implemented on the SAP S/4HANA system, which provides customers with the full capabilities of an ERP system. Businesses can gradually implement logistics and other business-relevant components in incremental steps or a phased approach per business goals.

1.2.3 Central Finance Real-Time Replication Capabilities

Figure 1.5 shows the real-time replication capabilities of Central Finance. The left side of the figure shows Central Finance replication capabilities for any SAP-based source systems ❶. The right side of the figure displays current replication capabilities as of writing this book ❷. We'll explain each replication capability briefly:

- **Replication of FI-CO documents**
  Replicate financial documents or journal entry postings (e.g., tables BSEG and BKPF in case of SAP ERP) and repost in Central Finance based on rederived account assignments, mapping, and configuration data.

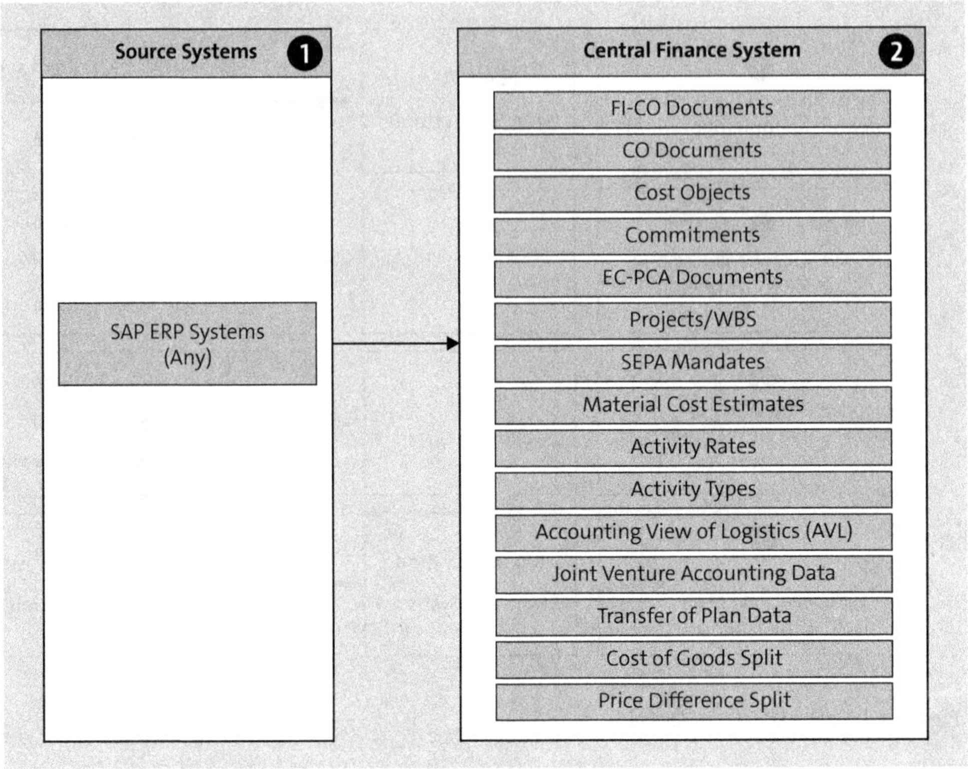

Figure 1.5 Central Finance Features

- **Replication of CO secondary documents**
  Replicate CO secondary postings from table COEP and dependent tables based on the rederived account assignments and configuration data.
- **Replication of cost objects**
  Replicate cost objects from table AUFK based on mapping and the cardinality rule.
- **Replication of commitments**
  Real-time replication of commitment from purchase requisitions or purchase orders.
- **Replication of plan data**
  Transfer financial plan data for cost centers, orders, and work breakdown structures (WBSs) to the Central Finance table (table ACDOCP).
- **Replication of EC-PCA postings**
  Replicate EC-PCA internal postings from the source system to the Central Finance system.
- **Material cost estimates**
  Replicate material cost estimates from the source system to the Central Finance system, which can be used for cost component split or other reporting scenarios.

- **Cost rates**
  Replicate prices of internal activities (table `cost`) from the source system to the Central Finance system.
- **Activity types**
  Replicate source system activity types used to classify the activities produced in a cost center within in a controlling area.
- **Cost of goods split**
  Based on configuration settings, split cost of goods sold (COGS) upon replication of financial documents.
- **Price difference split**
  Split price differences at reposting of financial documents in the Central Finance system based on variance categories transferred from the source system.
- **Project/WBS**
  Replicate project and WBS master data created in the source system to the Central Finance system.
- **Accounting view of logistics (AVL)**
  Replicate selected or a subset of logistics data from the source system to Central Finance on its own Central Finance tables.
- **Joint venture accounting (JVA) data**
  Replicate both master data and JVA documents from a source subsidiary system to the Central Finance system.
- **SEPA Mandate replication**
  Replicate SEPA mandates using ALE technologies.

1.3 Deployment Options

Deployment options for digital transformation vary with each project's requirements. Customers can take advantage of SAP's clean core methodology that primarily focuses on data, processes, integration, extensibility, and operations, as shown in Figure 1.6. SAP refers to the *clean core* as transparent, consistent, efficient, optimized, and innovative to drive organizations with legacy systems toward a faster digital transformation. In addition, the clean core provides customers a chance to standardize and streamline leading SAP processes and minimize customizations or modifications to the core SAP code. This approach especially helps customers who have highly customized legacy systems with significant custom code or complex interfaces benefit from the clean core Central Finance implementation approach. In addition, the strategic clean core approach provides customers with benefits from recent innovations: predictive accounting and SAP Intelligent Robotic Process Automation such as machine learning and artificial intelligence (AI) capabilities.

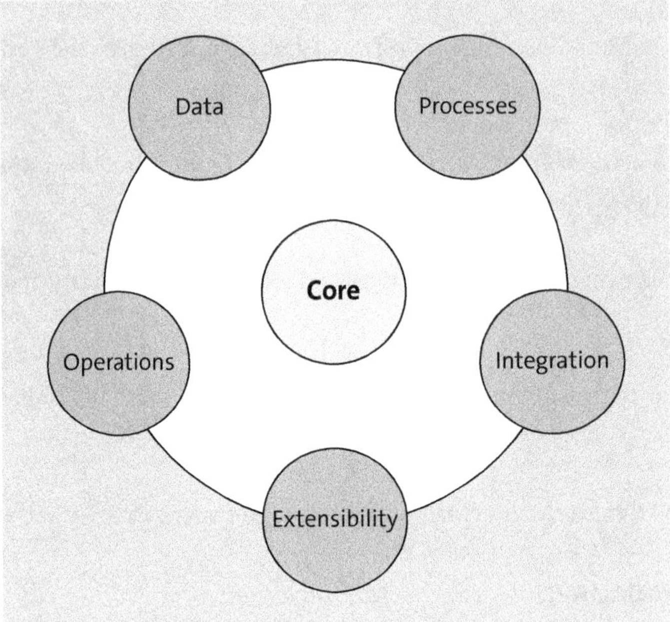

Figure 1.6 Clean Core Concept

Clean core advocates the *stepping stone* approach with incremental changes rather than going with a full-blown new implementation. The stepping stone approach enables customers to approach highly customized or complex systems with four simple steps: *simplify*, *innovate*, *standardize*, and *reduce customization*. This approach leads you to transition gradually from existing legacy systems and processes to SAP S/4HANA leading processes, thereby reducing customization. As companies are moving toward their digital journey, the clean core approach takes them the next step toward cloud computing.

Let's discuss the most common deployment options of Central Finance with SAP's clean core strategic approach.

1.3.1 Central Finance with On-Premise SAP S/4HANA

SAP offers on-premise models either managed by SAP or managed by SAP or in a hybrid model. This option provides customers with the ability to bring their own hardware and infrastructure to implement a Central Finance system with a tailored-to-fit model. Both models provide customers with full functionality of SAP S/4HANA, industry solutions, and local country-specific features. In addition, this option allows customers with customizations, modifications, and extensibility to provide full control on the system.

1.3.2 Central Finance with SAP S/4HANA Cloud Private Edition and RISE with SAP

For new customers, SAP offers Central Finance with SAP S/4HANA Cloud Private Edition. The first private cloud version was released in 2021 and comes with a full suite:

- Finance and procurement
- Manufacturing
- Product lifecycle
- Supply chain
- Assets

Central Finance can leverage all SAP S/4HANA innovations and tailor them to customer-specific requirements. SAP S/4HANA Cloud Private Edition is available for Central Finance implementations with the RISE subscription model. This option allows customers with customizations, modifications, and extensibility similar to on-premise edition.

1.3.3 SAP S/4HANA Cloud

Although SAP S/4HANA Cloud Public Edition is available as a ready-to-run solution, it's not supported as a target Central Finance system yet. On the other hand, besides SAP ERP and SAP S/4HANA on-premise systems as source systems, Central Finance also supports setting up an SAP S/4HANA Cloud Public Edition system as a source system in a Central Finance scenario. However, SAP recommends you don't have more than two years between the cloud release and SAP S/4HANA Central Finance. For example, Table 1.1 shows the releases and dependencies. Both master data and transactional data replication supported the cloud as a source system. However, classic initial load isn't supported; instead, intermediate data retention (IDR) is recommended for replication of initial load transactional data.

SAP S/4HANA Cloud Public Edition Release	Minimum SAP S/4HANA Central Finance Release
Public Cloud Edition 2402	SAP S/4HANA Central Finance 2021
Public Cloud Edition 2408	SAP S/4HANA Central Finance 2022 FPS0
Public Cloud Edition 2502	SAP S/4HANA Central Finance 2022 FPS1

Table 1.1 SAP S/4HANA Cloud Public Edition Release and Dependencies

Note

SAP Note 3398470 provides restrictions to be noted when SAP S/4HANA Cloud Public Edition is used as the source system.

1.4 Central Finance Use Cases

With centralized replicated data, Central Finance offers customers consolidated reporting with a single source of truth view of an enterprise. As a result, with a Central Finance implementation, companies can improve their decision-making, compliance, and operational processes across the enterprise significantly. Central Finance brings several use cases to customers, as shown in Figure 1.7. The use cases are divided into three broad categories: process transformation, technology transformation, and corporate strategy and planning.

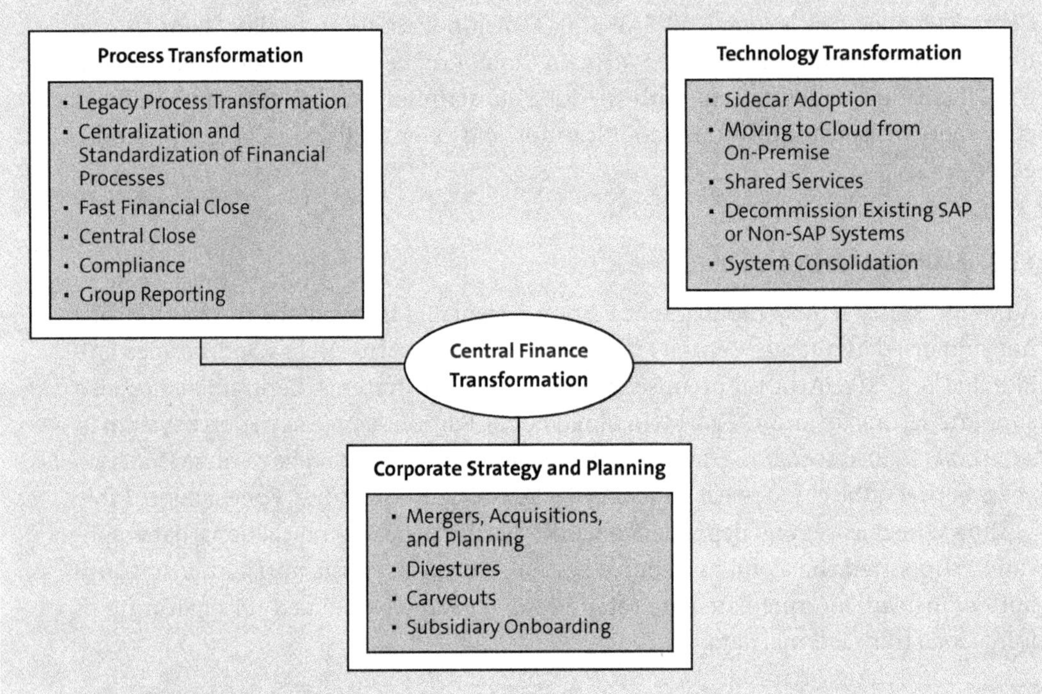

Figure 1.7 Central Finance Potential Use Cases

Let's explore each category of use case next.

1.4.1 Process Transformation

Organizations with heterogeneous ERP landscapes often end up with outdated, ineffective, and inefficient processes. With a Central Finance implementation, you can transform such outdated or inefficient processes by adopting SAP S/4HANA leading practices and optimizing processes. The streamlined financial processes in Central Finance result in improved reporting, better management decision-making, and compliance with regulatory requirements. Let's explore the important processes related to transformation use cases:

- **Legacy process consolidation**
  With Central Finance implementation, customers have an opportunity to streamline outdated and fragmented or duplicated financial processes into harmonized centralized processes with a single source of truth. Inefficient or outdated legacy processes cause companies to spend more time and money in manual processes, resulting in data silos and longer financial close cycles. To remain competitive in this emerging market, companies must evaluate legacy processes and consider consolidating or unifying to streamline financial processes.
- **Centralization and standardization of financial processes**
  For enterprises with heterogenous landscapes, to remain competitive, Central Finance provides opportunities to standardize and harmonize financial processes across the enterprise. The process of unification can begin with a harmonized CoA design in Central Finance that caters to current and future business processes. Standardization can then be expanded to other financial processes to centralize processes. For example, a few processes that can be centralized in Central Finance include the following:
  - Centralized payment
  - Centralized credit
  - Centralized dispute processes
  - Centralized collection processes
- **Fast financial close**
  With the Universal Journal in SAP S/4HANA, the advanced financial closing cockpit, and other innovative tools, customers will be able to accelerate closing by organizing, automating, and governing closing tasks. Fast financial close is achieved with improved processes as a result of the Central Finance centralized processes and following leading practices in SAP S/4HANA. Key benefits of fast close are as follows:
  - Visibility to pre-closing activities much early in the period
  - Quick access to data as a result of real-time data availability from various source systems
  - Accelerated closing with innovative tools offered in SAP S/4HANA
  - Real-time reconciliation due to the Universal Journal
  - Improved cash flow due to real-time insights into liquidity and forecasts for all ERP systems.
  - Support for local and statutory reporting
- **Central close**
  With Central Finance, companies can perform centralized close by consolidating and integrating financial processes across multiple source systems. With SAP-delivered closing tools such as intercompany reconciliation and advanced finance closing in SAP S/4HANA, you can organize and govern closing tasks to improve their

accuracy. In addition, central close helps organizations comply with regulatory requirements and provide timely insights to financial data for better decision-making:
  - Faster and efficient reconciliation process due to the Universal Journal
  - Availability of SAP S/4HANA closing tools
  - Enhanced workflow automation and approval processes related to closing
- **Compliance**
  Central Finance can leverage the SAP Document and Reporting Compliance tool, which is a next-generation legal and statutory reporting tool to support reporting and compliance needs for entities operating worldwide. This tool is supported in Central Finance to comply with country-specific reporting requirements such as electronic invoicing, real-time monitoring, and real-time visibility of processes to stakeholders. Central Finance coupled with SAP Document and Reporting Compliance brings customers the following:
  - Reduced risk of noncompliance penalties
  - Reduced errors as a result of process automation
  - Real-time insights into compliance processes and data
  - Mitigated audit risks in a distributed landscape with centralized data to facilitate audit processes
- **Group reporting**
  With Central Finance and group reporting, you have opportunities to standardize group reporting–related financial processes. Group reporting processes are often complex due to the nature of multiple processes unified in Central Finance. However, with financial data harmonization and group reporting integration with the Universal Journal, SAP made consolidated entity reporting seamless in Central Finance. Key benefits include the following:
  - Real-time integration of centralized data for consolidation and group reporting
  - Unified data model with the Universal Journal and reporting
  - Automated consolidation-related tasks
  - Intercompany reconciliation and elimination
  - Compliance and regulatory reporting
  - Quicker management decision-making with real-time insights

1.4.2 Technology Transformation

Next, we'll explore the technology-related use cases by transitioning to Central Finance:

- **Sidecar adoption**
  With Central Finance implemented with the sidecar approach, you can leverage SAP S/4HANA innovations without interrupting the core ERP systems. This approach

provides the immediate benefits of most SAP S/4HANA innovations in incremental ways while keeping legacy systems intact. Several benefits of a sidecar approach are listed here:

- Reap SAP S/4HANA benefits immediately in incremental steps with a nondisruptive approach to core ERP systems.
- Transition to SAP S/4HANA gradually or in a phased approach.
- Real-time replication of financial and nonfinancial data provides consolidated reporting capabilities.
- Increased scalability is provided compared to core ERP systems.
- Fallback to core ERP systems is possible for risk mitigation.

■ **Move to cloud**
Central Finance provides customers an opportunity to migrate from an on-premise system to a cloud environment. The cloud environment can be either an SAP cloud or other third-party clouds such as Microsoft Azure or Amazon AWS. the cloud environment helps customers centralize operations. In Central Finance view, a cloud environment provides customers with several benefits:
  - A scalable cloud environment infrastructure to support growing business needs
  - Flexibility in implementing the solution
  - Reduced capital infrastructure cost and maintenance cost
  - Less downtime due to service level agreements with cloud providers
  - Increased performance as a result of redundancy and failover infrastructure

■ **Shared services**
Central processes in a shared service model is one of the key use cases of Central Finance. Centralization of processes with the Universal Journal leads customers to leverage shared services opportunities in various departments such as accounts payable, accounts receivable, and payroll. Most important key benefits of Central Finance shared services include the following:
  - Unified data model to ensure shared services uses a single source of truth
  - Cost reduction due to simplification and reduced duplication of efforts across multiple source systems, resulting in reduced redundancies of work and reduced overhead cost
  - Economies of scale due to bulk processing and operations from Central Finance
  - Standardized and uniform processes across core ERP systems

1.4.3 Corporate Strategy and Planning

A company's corporate strategy and planning begins with a vision and mission statements that dictate the overall direction of a company and its key objectives. It's imperative Central Finance implementation is no exception when it comes to aligning both the short-term and long-term goals of a company. Corporate strategy and planning include mergers, acquisitions, consolidations, divestures, carveouts, and joint ventures. Let's discuss each corporate strategy next:

- **Mergers, acquisitions, and consolidations**
  Corporations evaluate mergers, acquisitions, and consolidations opportunities as part of their goals to keep a competitive edge with rivals or as part of overall strategic goals. This includes adding a new subsidiary or existing subsidiary to the system landscape. Central Finance provides companies with opportunities by integrating acquired or merged companies by simply replicating financial transactions to Central Finance without interrupting the existing acquisitions processes. In addition, Central Finance streamlines and centralizes processes from the acquired or merged company. This leads to a report on a unified view of mergers, acquisitions, and consolidated entities using consolidated data in the Universal Journal.
- **Divestitures**
  Companies divest or dispose of a business unit for multiple reasons such as strategic decisions or to streamline operations. Central Finance helps in such situations to divest data into a new instance by replicating data from a source system. You can identify divesture requirements and strategize how the Central Finance solution fits into those situations.
- **Carveouts**
  Carveouts lead to selling a subsidiary or spinning a division to form an independent legal entity. The Central Finance sidecar approach could be used to replicate both financial and nonfinancial data on the newly setup company code. In addition, selective data transmission is also possible based on the project requirements.
- **Joint ventures**
  Companies come forward to form joint ventures as a collaborate effort to achieve a specific goal or project. This is common in the oil and gas industries. Central Finance supports replicating joint venture accounting (JVA) data to Central Finance.

1.5 Central Finance Road Map Explorer

SAP offers customers the Road Map Explorer to navigate current, planned, and future innovations. As a leading practice, SAP recommends customers review the product road map during the discovery phase of the project to understand the following:

- **Current and planned innovations offered with the Central Finance Road Map Explorer**
  If required, perform a proof of concept in a sandbox environment.
- **Innovations to relevant stakeholders**
  Align with business requirements by explaining the innovations to the relevant stakeholders early in the project.

Figure 1.8 displays the **Road Map Explorer** initial screen. Upon entering the Road Map Explorer web page, you can use the tool in the following ways:

❶ Search by products, processes, and industries.

❷ Search directly by using a keyword search, for example, for "Central Finance".

❸ Explore roadmap priorities by product, for example, **Enterprise Resource Planning**, **Financial Management**, and so on.

❹ Explore roadmap priorities by industry.

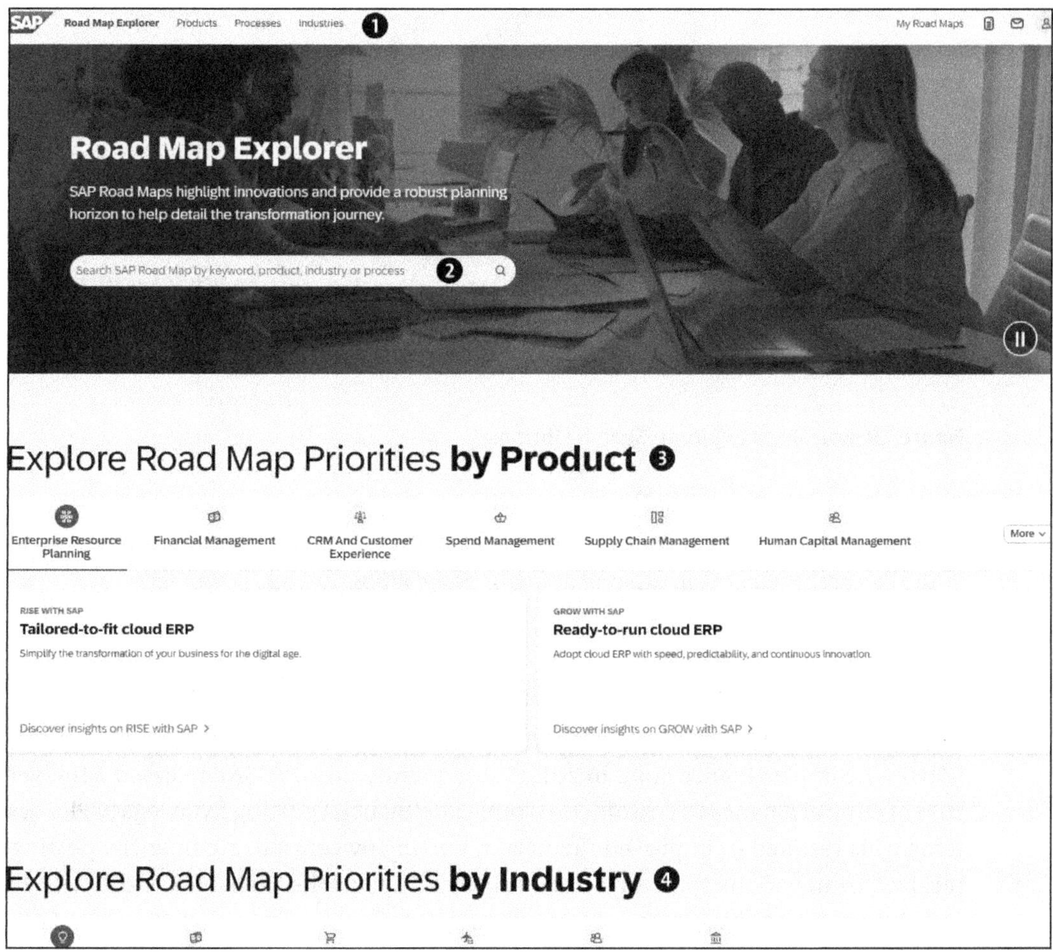

Figure 1.8 Road Map Explorer: Initial Screen

From Figure 1.8, when you search with a keyword, for example, "Central Finance", the web page takes you to the results page, as shown in Figure 1.9. You'll notice this page is loaded with **16 Road Map Items** ❶. You can further search by **Products** ❷, **Industries** ❸, and **Focus Topics** ❹. Furthermore, you can look up roadmap items by searching through the calendar tool ❺ for quarterly release updates.

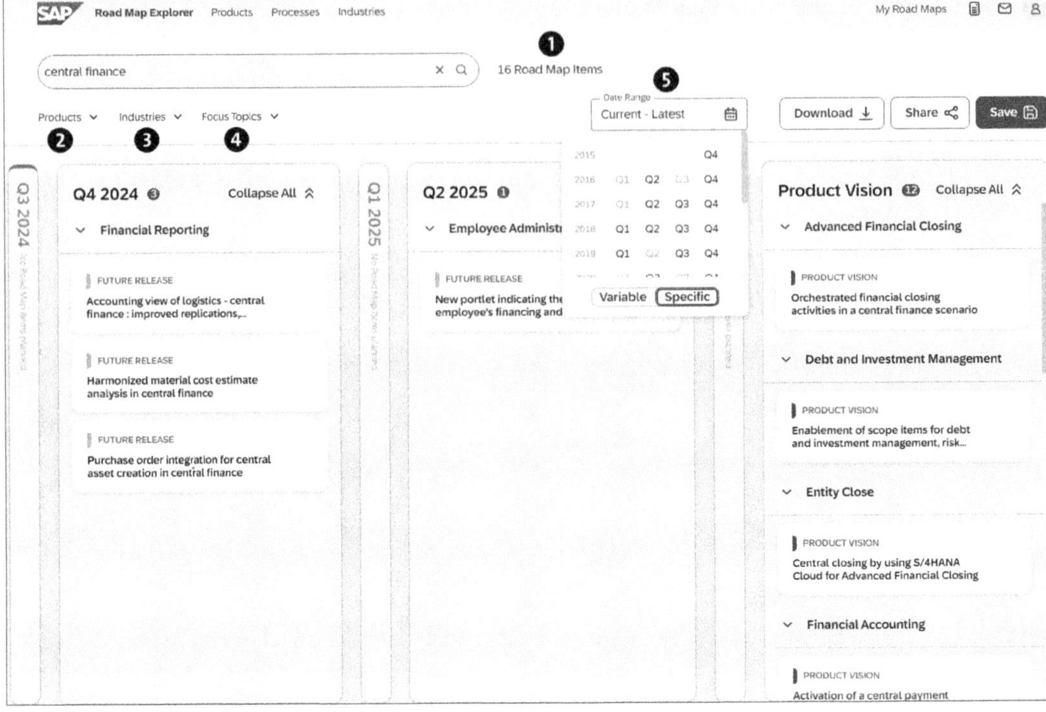

Figure 1.9 Road Map Explorer: Search Options

[»]

Note

You can access Road Map Explorer at *https://road maps.sap.com/welcome*.

1.5.1 Product Maturity

Central Finance has evolved with features since its introduction as part of SAP S/4HANA Simple Finance Suite in 2015. Subsequently, in 2016, SAP released a full version of Central Finance to customers to replicate financial posting from legacy ERP systems. This resulted in improved financial reporting by centralizing financial postings received from various source systems. Table 1.2 displays the available features in Central Finance released by SAP as of writing this book.

Infrastructure, Onboarding, and Operations	Central Reporting	Central Processing and Close
Initial load of financial data: Balances, open items, and transactions from a specific period	Group-wide general ledger and P&L reporting in real time	Central clearing and closing for AP and AR components

Table 1.2 Current Features of Central Finance

Infrastructure, Onboarding, and Operations	Central Reporting	Central Processing and Close
Real-time replication of financial documents	Advanced segment, entity, and group reporting	Central credit management
Third-party system support	Extended CO posting scenarios	Central dispute and collections management
Master data replication: SAP Master Data Governance, SAP Landscape Transformation, IDocs, SAP Data Services, and other tools	Replication of AP/AR clearing and document changes to Central Finance	Intercompany processing with advanced matching and reconciliation
Cost object framework to replicate internal orders, production orders, and product cost collectors	Central cash position and liquidity forecasts based on AP/AR open item replication	Down payments in sales and procurement processes with Central Finance
Harmonization of master data and configurations using key mapping and value mapping tools	Costing-based profitability analysis characteristics and value fields mapping	Handling historical payments in Central Finance
Drill-back functionality from the Central Finance system to the source system and SAP Fiori apps	Central projects (WBS)	
SAP S/4HANA Cloud Public Edition as a source system	Central asset accounting	
Key mapping change logs and audit trail	Central profit center reporting	
Retrigger individual documents: reversal and repost	Central reporting on commitments from procurement	
Consistency checks	SAP Document and Reporting Compliance capability with Central Finance	
Initial load with document splitting information	Statistical conditions, cost components, and price difference splits for central profitability analysis	

Table 1.2 Current Features of Central Finance (Cont.)

Infrastructure, Onboarding, and Operations	Central Reporting	Central Processing and Close
Central payment enhancements with cross-system process control (CSPC)	Group cost planning for distributed value chains with material cost estimates and activity rates	
SAP Application Interface Framework monitoring	Central general ledger open item management reporting	

Table 1.2 Current Features of Central Finance (Cont.)

Note

Besides current product features, SAP also provides Central Finance adoption content gathered from past Central Finance implementations.

1.5.2 Recent Innovations

To provide comprehensive financial reporting capabilities, Central Finance continues to evolve with new features. The recent innovations not only improve product features but also provide customers with features that enable them to have an enhanced experience with Central Finance. These recent innovations are available for customers to use. Table 1.3 displays recent innovations delivered with Central Finance as of SAP S/4HANA 2023. Some of these features are included with subsequent quarters after the SAP S/4HANA 2023 release. We recommend checking the Road Map Explorer to understand quarterly updates.

Area	Recently Released Functionality
Central Finance Features	
Accounting view of logistics (AVL)	▪ Comprehensive comparison reports for AVL support
Consistency checks	▪ Customer-defined consistency checks by project requirements ▪ Permanent checks and health checks
SAP Activate methodology	Availability of Central Finance–specific SAP Activate methodology content
Initial load	Initial load using the IDR approach
Document display	Displays the Central Finance document and other reference information in the source system

Table 1.3 Central Finance Recent Innovations

Area	Recently Released Functionality
Temporary posting	Temporary postings for documents in error to support month-end close
Process automation	Automatic mappings error resolution using SAP Intelligent Robotic Process Automation Bot Store
Third-party system support	■ Transfer of group currency, clearing information, profitability data, and document changes ■ Extended amount fields support for the third-party interface replication
Accounting principle	Support of multiple accounting principles with mapping account solution to parallel ledger approach.
Document splitting	Support of central document splitting for source systems with classic general ledger
Activity rates	Replication of activity rates in Central Finance
Plan data transfer	Transfer plan data of cost center and internal orders from the source system to the Central Finance system
Central Reporting	
Tax reporting	Statutory and local reporting with accounting view of logistics and SAP Document and Reporting Compliance
Margin analysis	Deriving profitability segments with COGS posting
Central Processes	
Central payment	■ Enhanced features in CSPC for central payment ■ Support for SAP S/4HANA Cloud Public Edition as the source system
Accruals and deferrals	Central accruals and deferrals for purchase orders
Central budgeting and availability control	Central budgeting and availability control for internal orders
Local statutory reporting	SAP Document and Reporting Compliance support for Central Finance
Central general ledger open item management	Manage general ledger open item clearing activities in Central Finance
Central asset accounting	Central asset accounting and central asset settlement for fixed assets
Closing	
Period close	Advanced financial closing for Central Finance

Table 1.3 Central Finance Recent Innovations (Cont.)

1.5.3 Future Roadmaps

Table 1.4 displays future innovations planned in the Central Finance product road map. We recommend checking the Road Map Explorer to understand quarterly updates, as discussed earlier in this section.

Area	Future Road Map for Central Finance
Central Finance Features	
Central Finance adoption content	Continue to enhance adoption content accelerators and guidelines
Stacking of Central Finance systems	▪ Support of connecting multiple regional Central Finance systems to a global Central Finance system ▪ Reporting and centralized processes in a regional Central Finance system ▪ Support of deviating CoAs in the global Central Finance system ▪ Support of higher aggregation
Third-party interface support	Additional functional enhancements and extensibility support
Focused run	Additional functionality for focused-run monitoring support
Asset accounting	▪ Improved integration of central asset accounting with other components such as project systems and plant maintenance ▪ Create asset based on accounting view of a purchase order in Central Finance ▪ Automating asset acquisition posting in Central Finance
AVL	▪ Additional sales order categories support: service contract and scheduling agreement ▪ Deleting AVL data by posting date ▪ Improved replications (support of AVL billing document with tax amount) and enhanced reporting
Service objects and service-accounting document replication	▪ Support of replicating service orders ▪ Support of service order–related financial postings and reposting the same on the Central Finance side of the corresponding service order
Parallel valuation	Multiple parallel valuations for material movements and activity confirmations

Table 1.4 Central Finance Future Roadmaps

Area	Future Road Map for Central Finance
Material ledger	Central material ledger processes
Cost analysis	Cost analysis with a lean central cost object: support of replicating production orders to a lean cost object in Central Finance for variance analysis of primary and secondary costs of a production order
Overhead costing	■ Harmonized material cost estimates analysis with one controlling area currency in Central Finance due to differing currency amounts as a result of multiple controlling areas with different currencies setup in the source system ■ Support of real-time replication of activity types ■ Support of activity rate replication from Central Finance to source systems
Product and portfolio management	Additional enhancements in the central project and portfolio management components
Central Reporting	
Predictive finance	Predictive accounting scenarios in Central Finance and related reporting
Tax reporting	Enhanced localized tax reporting with SAP Document and Reporting Compliance
Revenue accounting reporting	Central revenue accounting and reporting
Group reporting	Additional innovations with group reporting
Central Processing	
Central payment	■ Separation of central payment functionality for flexibility, for example, enabling accounts payable managed in Central Finance whereas accounts receivable is managed in source system
Central closing	■ Improved mapping and mapping of company codes as one entity between source systems and Central Finance
GR/IR reconciliation	■ Central GR/IR reconciliation and clearing
Invoice verification	Central invoice and verification

Table 1.4 Central Finance Future Roadmaps (Cont.)

1.6 Summary

In this introductory chapter, you've learned what Central Finance is and its capabilities. We covered the various pathways to benefit SAP S/4HANA innovations and how Central Finance helps customers achieve finance transformation goals without disturbing existing processes or systems. In addition, Central Finance use cases are divided into three major broad categories: process, technology, and corporate strategy. Finally, the chapter delved into the Central Finance Road Map Explorer with current, planned, and future innovations in the Central Finance space. In the next chapter, you'll learn about Central Finance processes.

Chapter 2
Central Finance Processes

Central Finance can be established as a platform to perform consolidated financial reporting using the Universal Journal by collecting financial data from various ERP systems via a sidecar approach. Additionally, Central Finance can be leveraged to streamline and centralize processes. This chapter highlights important Central Finance processes and how these processes can be harmonized and centralized to take advantage of SAP S/4HANA finance innovations.

Central Finance is one of the most important components of SAP's financial management software offered to customers in recent times. Central Finance can be implemented in multiple ways, and use cases vary with every business requirement. In general, a common recommendation is for customers to begin with consolidating reporting using replicated centralized data from different source systems across the enterprise. With this step, customers reap the benefits of SAP S/4HANA's Universal Journal features right away without interrupting existing processes. Next, you'll simplify and standardize by transforming finance processes from both financial and management accounting processes to Central Finance. In addition, this gives customers an opportunity to harmonize processes across business units in an enterprise.

In this chapter, Central Finance processes are divided into the following:

- **General ledger processes**
  These processes are essential for maintaining accurate accounting records. With the help of the Universal Journal, you'll achieve a single source truth for consolidated reporting and centralized processes.
- **Accounts receivables (AR) and accounts payable (AP) processes**
  These processes with centralized payments and intercompany reconciliation, companies can manage closing activities in Central Finance.
- **Financial supply chain management (FSCM)**
  In this section, you'll understand important centralized processes in financial supply chain management: credit management, dispute management, collections management, cash and liquidity management, and in-house cases.
- **Projects, fixed assets, and internal orders processes**
  Central Finance offers central project reporting, central budgeting for internal

orders, central budgeting for projects, and central asset settlement scenarios to manage these processes in a centralized approach.

- **Shared services framework**
  This framework provides customers a comprehensive service management solution with unique values propositions for customers with service centralization.
- **Joint venture accounting (JVA)**
  JVA in the Central Finance scenario offers customers the chance to replicate JVA master data and internal JVA documents to the Central Finance system.

Let's dive deeper into these processes next.

2.1 General Ledger

In this section, you'll learn how centralized general ledger processes help customers streamline processes based on harmonized and centralized data in Central Finance.

2.1.1 Central Open Item Management

In SAP, open item–managed general ledger accounts are balance sheet accounts created with the **Open Item Management** checkbox on the **Control Data** tab of general ledger account master data. During financial accounting (FI) transactional posting on an open item basis, the system marks line items with open item–managed general ledger accounts as either open or cleared (XOPVW).

Central Finance provides central open item management to manage open items centrally for the open item–managed general ledger account's postings in the source system. During the evaluation phase, you'll identify the appropriate use cases for the central open item management functionality. For example, you won't be able to centrally manage goods receipt/invoice receipt (GR/IR) accounts as these accounts are dependent on logistics processes, which isn't desirable for clearing centrally in Central Finance. However, there are scenarios where general ledger open items can be managed in Central Finance. For example, you can manage certain accruals or deferrals posted in the source system, but there is a desire to manage those postings in Central Finance. As a prerequisite, to manage general ledger open items centrally in Central Finance, you'll need to activate central clearing general ledger open items as discussed in Chapter 7, Section 7.4.6. In addition, you'll **Define G/L Accounts for Technical Clearing** in V_CFIN_CGL_ACC as discussed in Section 7.3.5.using Transaction SM30.

Figure 2.1 illustrates the data flow showing how general ledger open items are posted in the source system and then cleared centrally using the central open item management functionality in Central Finance:

❶ First, line items with open item–managed general ledger accounts from the documents posted in the source system are technically cleared (ALE-extern) in the source system. Once the open items are technically cleared, these open items are no longer cleared in the source system.

❷ Next, the open line items replicated to Central Finance are left with **Open** status to manage them from the Central Finance system.

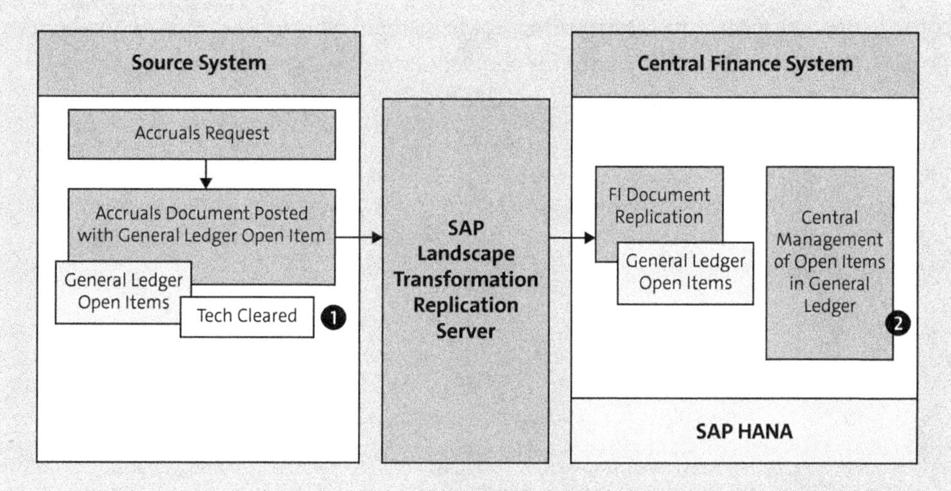

Figure 2.1 Central Open Items Management in General Ledger

Note

SAP provides Transaction CFIN_CGL_TECH_CLR (program FINS_CFIN_CGL_HISTL_OI_MASTER) to manage historical items posted before enabling the central open item management functionality.

Consult SAP Note 2985251 to understand prerequisite SAP Notes and configuration requirements for central clearing of general ledger open items.

2.1.2 Central Tax Reporting

Central Finance replicates tax data along with FI documents from the source systems to the Central Finance system. Harmonized and centralized data enables central tax reporting in Central Finance. If central payment isn't activated, all clearing activities happen in the source system, so you'll perform tax reporting in the source system. In addition, the replication of data transferred to Central Finance is enabled for company codes with the plants abroad setting active and company codes with Registration for Independent Taxation Abroad (RITA) enabled. However, replication tax data for a scenario where the plants abroad settings is not active in Central Finance, but is active in the source system, is not supported out of the box. On the other hand, when central payment is activated, you can perform tax reporting in Central Finance. However, Central Finance doesn't support official document numbering (ODN) to uniquely and consistently identify various business transactional documents for compliance purposes. You'll enable the ODN through a series of configurations in respective SAP components. As a result of this limitation, for countries requiring ODN as a legal requirement, Central Finance currently doesn't support that country-specific tax reporting.

When central payment is activated, to perform tax reporting, Central Finance supports core tax reports with some restrictions. Table 2.1 displays core tax reports and transactions supported in Central Finance for central tax reporting.

Program/Transaction	Description	Central Reporting Usage and Restrictions
Program RFUMSV00/ Transaction S_ALR_ 87012357/Transaction IDTW_VAT	Advance return for tax on sales/purchases	▪ The report generates tax for sales and purchases/value-added taxes in either XML-based format or some other form format.
Program RFUVDE00	Print program: Advance return for tax on sales/purchases (Germany)	▪ The report creates entries for the German form related to advance return for tax on sales and purchases. ▪ Additional testing may be required to understand restrictions.
Program RFUMSV10	Additional list for advance return for tax on sales/purchases	▪ SAP recommends implementing SAP Note 2489200 to access cross-company code documents (BKPF-BVORG) posted from the initial load. ▪ Country specific data relevant to purchasing account management documents, for example, countries Italy or France, isn't displayed. ▪ The report may not support displaying data from contract accounting (FI-CA) and HR Payroll (HRPAY).
Program RFID_ PTVPRADPRC00/ Transaction S_RFID_ PTVPRADPRC00	Pro-rata adjustments due to PR calculation	▪ This report is used for periodic VAT declarations for VAT transactions. ▪ Additional testing may be required to understand restrictions.
Program RFID_ PTVPRADPRV00/ Transaction S_RFID_ PTVPRADPRV00	Pro-rata adjustments due to PR variation	▪ This report is primarily used for preparation of VAT returns to comply with local reporting requirements. ▪ Additional testing may be required to understand restrictions.

Table 2.1 Central Finance Supported Core Tax Reports

Program/Transaction	Description	Central Reporting Usage and Restrictions
Program RFUMS-RVG00/Transaction FBTR	VAT refund	■ This report is used to prepare advance VAT returns in compliance with local tax reporting requirements. ■ Additional testing may be required to understand restrictions.
Program RFUMSV35	Tax adjustment	■ The program is used to reconcile VAT tax data to confirm that data sent to tax authorities is accurate and complies with tax reporting requirements. ■ Additional testing may be required to understand restrictions.
Program RFASLM00	EC sales list	■ This report may not display certain business partner information. ■ The report may not support displaying data from contract accounting (FI-CA) and HR Payroll (HRPAY).
Program RFASLD20	EC sales list in Data Medium Exchange (DME) format	■ This report supports creating necessary reporting files for tax authorities. ■ Additional testing may be required to understand restrictions.
Program RFUMSV50/ Transaction S_AC0_ 52000644	Deferred tax transfer (new)	■ This report supports tax adjustments and transfers. ■ Additional testing may be required to understand restrictions from the Central Finance aspect. ■ Consult SAP Note 1800344 for additional guidance on this report.

Table 2.1 Central Finance Supported Core Tax Reports (Cont.)

To perform tax reporting in Central Finance, you ensure that FI documents with tax data from the source systems are transferred consistently to Central Finance. To achieve consistent data transfer between systems, it's necessary that the tax-relevant configuration settings are identical between the source and Central Finance systems. By activating the **Activate Tax Consistency for Company Codes** configuration in Central Finance, the system checks tax-relevant inconsistencies between the source system and Central Finance configurating during tax data transfer in the initial load and online replication. SAP leading practice suggests running tax consistency checks to prevent tax errors during the initial load and online replication. If your source system uses an external bolt-on tax engine to calculate taxes, SAP recommends the same tax engine in

Central Finance for tax calculation. In addition, SAP recommends the tax engine certified for Central Finance scenario. SAP provides the certified solutions directory to validate whether the product is certified at *http://s-prs.co/v590106*.

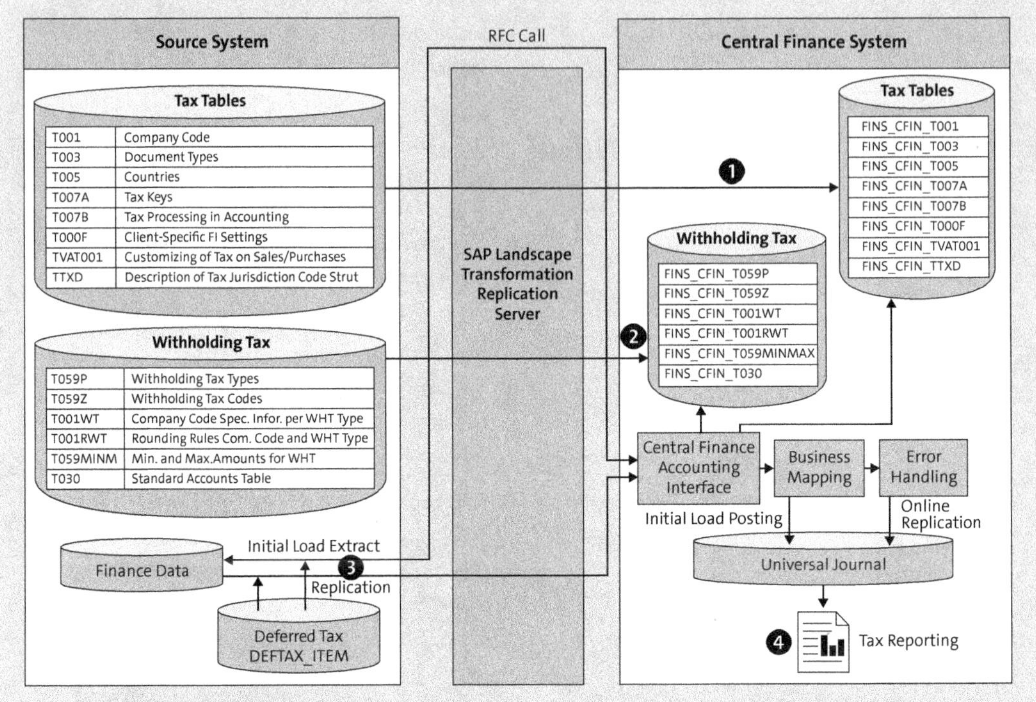

Figure 2.2 Central Finance Tax Reporting

Figure 2.2 shows the tax data flow architecture in Central Finance for central tax reporting:

❶ First, you'll set up SAP Landscape Transformation Replication Server trigger-based replication for tax-relevant tables in SAP Landscape Transformation Replication Server. This is covered in Chapter 5, Section 5.2.1. After configuring and enabling real-time replication, you'll activate the **Activate Tax Consistency Check for Company Codes** configuration in Central Finance. In addition, you'll carry out tax configuration consistency checks to avoid errors in the initial load or online replication. During the initial load, simulation tools are used to identify and fix any tax errors. Likewise, if errors occur during online replication, you'll correct tax configuration and repost SAP Application Interface Framework messages.

❷ To support withholding tax data replication, you'll set up trigger-based replication for withholding tax tables in SAP Landscape Transformation Replication Server. Central Finance supports extended withholding tax and doesn't support classic withholding tax. As a result, if your source system runs in classic withholding tax, you'll migrate to extended withholding tax. Consult SAP Note 2623514 for additional guidance on this topic.

During financial document replication from the source system to the Central Finance system, the system follows this sequence:

- *Step 1*: The system calculates tax values based on the Central Finance system configuration.
- *Step 2*: The system compares the expected outcome to the replicated value from the source system. In addition, the system compares customizing tables for any inconsistencies. Besides other things, the system checks tax code properties such as the **Error Message for Invalid Tax Amount** indicator in Transaction FTXP (field label **CheckID** and set T007A-PRUEF = X).
- *Step 3*: If significant differences are found, the system displays tax-related errors. You'll also configure an appropriate tolerance percentage for tax calculation. SAP recommends setting up the tolerance value as close to zero as possible to prevent errors and for reconciliation purposes.

Tax data replication works mostly, but there are cases where the system excludes certain conditions, and you must be aware of those situations. Table 2.2 displays conditions in which tax recalculation excluded during online replication or initial load.

Restrictions on Postings	Conditions That System Checks
Contract accounting (FI-CA) not supported	BKPF-AWTYP = FKKSU
Sales and distribution (SD) component	BKPF-AWTYP = VBRK
CRM-billing	BKPF-AWTYP = BEBD
Special general ledger transaction types down payment request and with tax code	BKPF-UMSKS = A; BKPF-MWSKZ <> (blank)
Cross-company code transactions	BSEG-KTOSL = BUV
Summarization over tax jurisdiction code	Both structure of tax jurisdiction code identified with schema of the country and accounting document header field BSEG-TAXPS are initial.
HR data split due to technical reasons, for example, lines exceeding 999 lines	BKPF-AWTYP = HRPAY and BSEG-KTOSL = HRA
Documents split with classical split logic	BKPF-AWTYP not in (MKPF, RMRP, VBRK, WBRK, BEBD, and BKPF-xsplit = X)

Table 2.2 Central Finance Tax Recalculation Exclusions during Online Replication

Restrictions on Postings	Conditions That System Checks
Initial load	Transaction Type = 0 ■ In case of the initial load, open items in prior periods extracted and transferred without the offsetting expense or revenue lines. ■ For periods identified in table VCFIN_SOURCE_SET for balances and open items in the source system, Central Finance doesn't support tax reporting. SAP recommends carrying out the tax reporting in the source system before you begin tax reporting in Central Finance. In addition, ensure you haven't activated central payment until you've completed tax reporting for the balances and historical open items. ■ SAP also recommends not posting any documents retrospectively in periods of initial load of open items and balances after you've decided to carry out tax reporting with BKPF-BUDAT or tax reporting date (BKPF-VATDATE).

Table 2.2 Central Finance Tax Recalculation Exclusions during Online Replication (Cont.)

❸ To perform central tax reporting in Central Finance, if deferred tax is in scope, all deferred tax data must be available. Central Finance supports deferred taxes replication (from DEFTAX_ITEM) from the source system to the Central Finance system with certain restrictions. Deferred taxes are included with incoming or outgoing payments, but they haven't been recognized in the books yet. Both initial load open items and online replication items include deferred tax codes replicated to Central Finance for central tax reporting. For countries where deferred tax processes are mandatory, you'll activate deferred tax replication along with central payment activation. SAP supports deferred tax variants for the initial load and online replication with some restrictions:

- During the initial load of open items and balance transfer, the system transfers deferred tax as a complete journal entry rather than open items. Due to limitations of transferring tax information in the initial load, especially for deferred taxes, you'll run the correction Transaction FINS_CFIN_DEFTAX (report RFINS_CFIN_CORR_DEFTAX_ITEM) subsequent to the initial load in Central Finance. This report makes corrections required in table DEFTAX_ITEM in Central Finance.

- During online replication, not all deferred tax items are created in table DEFTAX_ITEM. SAP recommends running the deferred tax transfer Transaction S_ACO_52000644 (report RFUMSV50) in Central Finance, which is used for transfer of deferred taxes from invoices, payments, down payments, tax transfer documents, and any other documents. This report creates missing deferred tax line items in Central Finance. The old deferred tax transfer Transaction S_P00_07000119 (report RFUMSV25) is still supported, but no longer added with additional functionality. Moreover, report RFUMSV25 uses data from BSEG and BSET tables. On the other hand, the report RFUMSV50 uses the table DEFTAX_ITEM. From a Central Finance perspective, SAP recommends using report RFUMSV50.
- Likewise, Transaction RFUMSV53 is used to resolve inconsistencies that can be run prior to closing the books. The report can be used to create additional missing deferred tax lines, correcting amounts, and removing any incorrect entries in table DEFTAX_ITEM.

[+]

Tip

Consult SAP Notes 3424979 and 2074813 for additional guidance on Transactions S_ACO_52000644 (report RFUMSV50), S_P00_07000119 (report RFUMSV25), and RFUMSV53 (program RFUMSV53).

[+]

Tip

Consult SAP Note 2787790 for additional guidance on activating deferred tax in the source and Central Finance systems.

❹ Once tax data is reposted to the Universal Journal in Central Finance, you can perform tax reporting based on posting date (BKPF-BUDAT) and tax reporting date (BKPF-VATDATE). You'll use core tax reports as referred to in Table 2.1, accounting view of logistics (AVL) data, and with SAP Document and Reporting Compliance to carry out tax reporting in Central Finance for local and statutory reporting.

[«]

Note

Consult SAP Notes 2494127 and 2668261 for the latest list of VAT tax and withholding tax tables, respectively.

2.1.3 Central Accruals/Deferrals

From an accounting point of view, if an economic event such as offering or receiving goods or services occurs first and then the cash transaction happens later, an accrual asset or liability is created, as shown in Figure 2.3 ❶. On the other hand, if the cash event happens first, and economic event occurs later, you'll create a deferral account to recognize expense or revenue as an asset or liability ❷. For example, AR and AP are part of accrual accounting, and they represent future cash inflows and outflows in financial statements. In addition, accrual and deferrals keep revenues and expenses on the balance sheet until they can be recognized on the profit and loss statement. Therefore, it's important to determine expenses and revenue accrued correctly in the correct period to reflect correct balances on the accounts in financial statements.

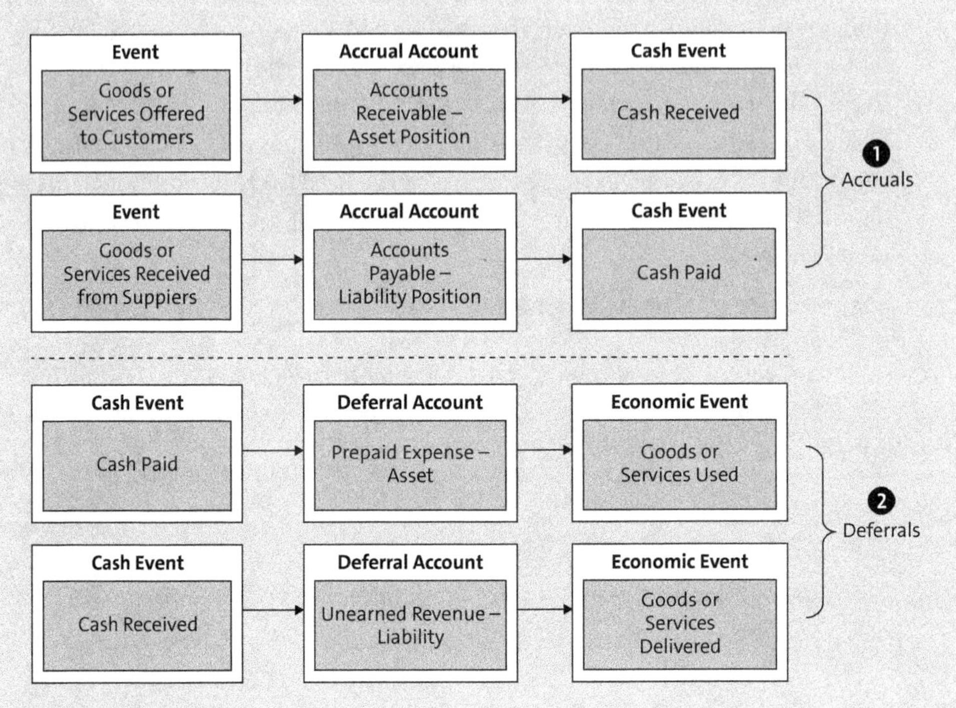

Figure 2.3 Accruals and Deferrals Concept

Central Finance offers a central accruals management functionality to manage accruals processes centrally. Thus, customers can streamline and standardize accruals and deferrals processes enterprise-wide to maintain consistency and to comply with regulatory requirements. In addition, by centralizing accrual processes, you'll improve efficiency of accrual process by reducing redundant accrual processes across multiple source systems. Central accruals management offers the following key benefits:

- Consistent accrual and deferral processes across multiple source systems leads to centralized control of accrual and deferral processes.
- Real-time insights with SAP S/4HANA reporting tools provide advanced reporting and analytics capabilities.
- By combining AVL data and accrual data, Central Finance supports local and statutory requirements.
- You can automate accruals and deferrals with SAP S/4HANA's accrual engine.

Figure 2.4 illustrates central accrual management using AVL and the accrual engine functionality for purchase orders created in the source system:

❶ First, purchase order information is replicated to Central Finance using AVL from the source systems and stored in AVL tables in Central Finance. After AVL replication, SAP recommends transferring purchase orders to the accrual engine either in manual mode with Transaction POAC_MM2ACE_TRANSFER or automatic mode by activating the **Online Integ.Active** flag in the company code in scope under IMG customizing **Financial Accounting • General Ledger Accounting • Business Transactions • Accruals Management • Applications of the Accrual Engine • Purchase Order Accruals • Transfer of Purchase Orders from MM to the Accrual Engine • Define Accrual Item Types and Methods for MM Purchase Order Items.**

❷ Next, with replicated purchase orders, the accrual engine is configured to calculate and post accruals and deferrals automatically. With the SAP accrual engine, you can do the following:

- Propose periodic amounts.
- Review manually adjusted periodic amounts.
- Approve periodic amounts.
- Perform periodic posting runs.

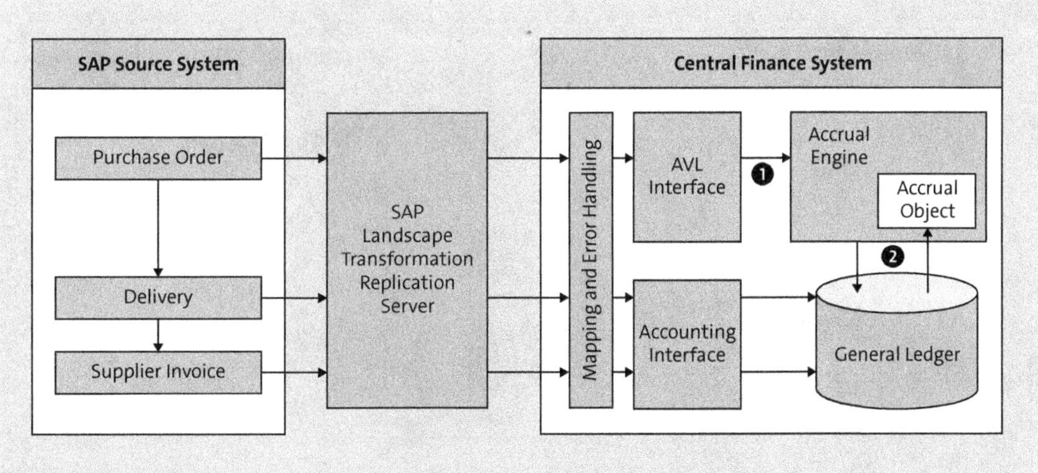

Figure 2.4 Central Accrual Management

Note

Accruals in Central Finance are supported since SAP S/4HANA 1909, and deferrals are supported from SAP S/4HANA 2022 onward. For SAP S/4HANA 1909 or higher releases, SAP recommends applying SAP Note 2882950.

SAP provides Transaction FINS_CFIN_INI_POSYS to manage historical data.

2.1.4 Central Reporting

Central Finance offers comprehensive reporting features, including detailed and analytical reports with the help of SAP S/4HANA reporting tools. In addition, with accounting views of logistics data and financial replicated documents, you can support country-specific reporting requirements. Figure 2.5 illustrates how AVL data coupled with replicated accounting documents from the Universal Journal can be used to support statutory reporting in Central Finance. Here's the flow of data:

❶ Logistics data, especially from sales order and purchasing data, is replicated through AVL functionality to AVL tables in Central Finance. However, the replicated data isn't visible via standard logistics transactions, for example, Transaction VA03, view a sales order. However, you'll access this data through SAP delivered comparison reports, or using core data services (CDS), or any other ad hoc reporting tools. You can build dashboards with key metrics.

❷ Financial data along with any country-specific data replicated through financial accounting (FI) and Controlling (CO) replication from the source systems. Replicated data is posted to the Universal Journal in Central Finance.

❸ SAP Document and Reporting Compliance can combine both AVL data with replicated financial data to support statutory reporting in Central Finance.

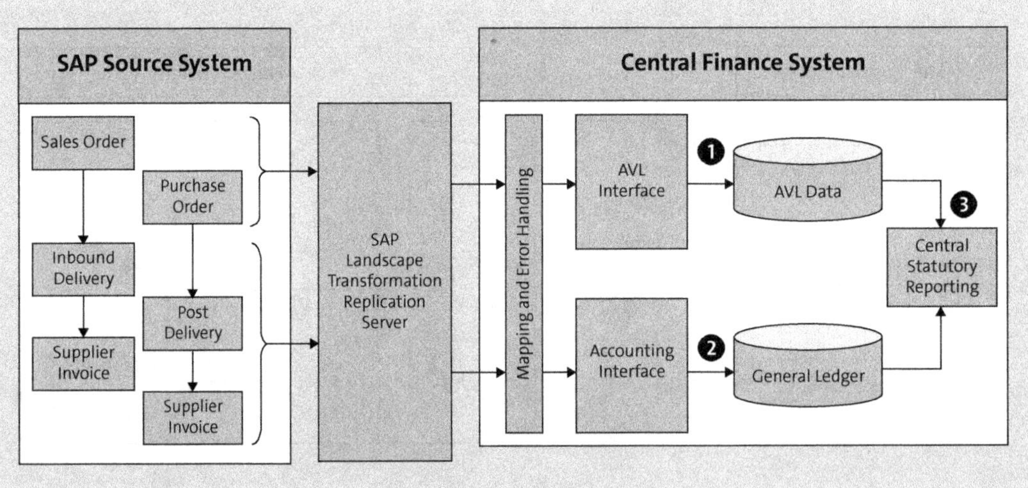

Figure 2.5 Central Statutory Reporting

[»]

Note

You can combine SAP Document and Reporting Compliance with replicated data AVL and data from the Universal Journal to cater to country-specific legal reporting requirements.

2.1.5 Central Closing

With the availability of financial data in the Universal Journal from the source systems, you can carry out both line-item level reporting and centralized consolidated reporting in Central Finance. In addition, as you expand Central Finance centralized features such as central payment, central credit management, and dispute and collections management, you'll carry out central closing activities in the Central Finance system.

Figure 2.6 illustrates central closing with a value flow diagram.

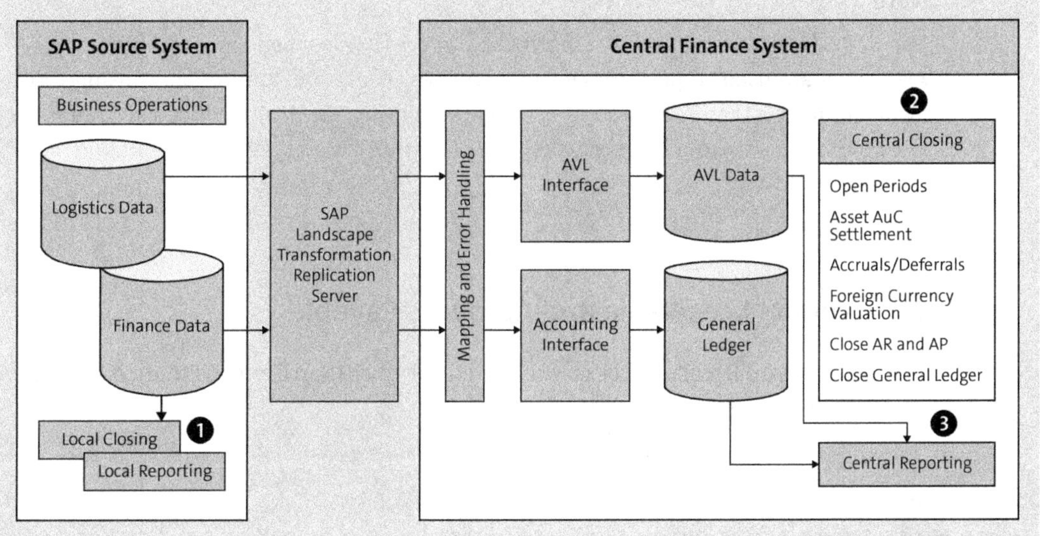

Figure 2.6 Central Closing

Let's discuss the flow of data between the source system and Central Finance, which helps you to achieve centralized closing in Central Finance:

❶ First, both financial data and logistics data from the source system should be sent to Central Finance. The centralized data is collected from source systems posted in the Universal Journal, which represents a single source of truth for consolidated reporting and central processes.

❷ In Central Finance, you'll carry out central closing activities. The following list isn't an exhaustive list as the closing activities vary with every company and reporting requirements. In general, you'll begin with the following closing activities:

- Open financial periods.
- Perform AP/AR clearing activities, and close A/P and A/R books.
- Run intercompany reconciliations.
- Carry out accruals and deferrals activities.
- Perform foreign currency valuation.

- Execute asset under construction (AuC) settlement and asset closing process.
- Perform asset accounting close activities.
- Perform universal allocations.
- Close general ledger books.

❸ Using centralized data in the Universal Journal, you'll perform central reporting activities in Central Finance.

[»]

Note

The SAP S/4HANA Financial Closing cockpit and SAP Advanced Financial Closing (SAP AFC) tools are used to perform closing activities. These are additional licensing tools available for customers exploring accelerated closing. Central Finance can integrate with these tools to offer customers with automation capabilities and streamline manual tasks in closing.

2.2 Accounts Receivable and Accounts Payable

In this section, you'll learn processes related to AP and AR in Central Finance. Both components have processes such as these:

- Posting invoices and clearing open items
- Payment processing
- Monitoring cash collections and dispute managements
- Bad debt management
- Intercompany reconciliation
- Reporting

In this section, let's explore how central payment plays a key role in fulfilling process centralization requirements.

2.2.1 Central Payment

With central payment activation scenario, core logistics processes are processed in the source system without interrupting existing processes, but the relevant financial open items are technically cleared (`ALE-extern`) in the source system. The corresponding financial document is then replicated to Central Finance to process payment activities centrally in the Central Finance system. Payment clearing dependent-related activities are moved to Central Finance as part of process centralization. Central payment is activated by company code (an entry in table `FINS_CFIN_CPCTL`) in Central Finance. Once central payment is activated, SAP S/4HANA becomes a system of record for financial transactions.

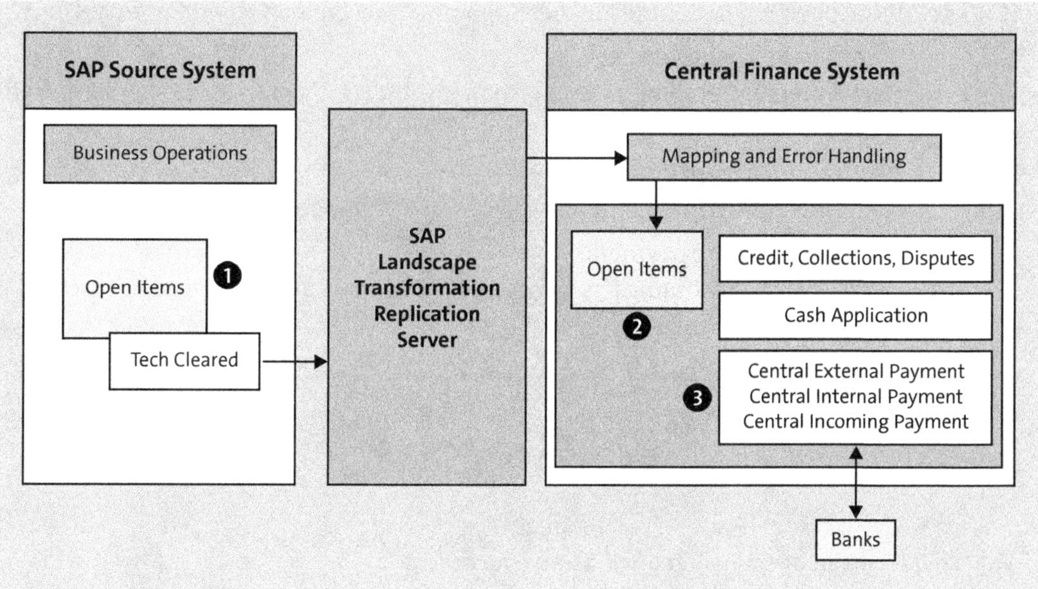

Figure 2.7 Central Payment

Figure 2.7 illustrates the central payment activation scenario in Central Finance:

❶ Upon activation of central payment, both AP/AR open items are technically cleared upon posting in the source system.

❷ Corresponding open items are replicated from the source system to the Central Finance system. Clearing activities of open items and relevant dependent activities are carried out in Central Finance. You'll perform the following dependent activities in Central Finance:

- Bank communication and bank statement processes
- Dunning
- Foreign exchange revaluation
- Interest calculations
- Customer and vendor correspondence
- Regrouping of AP/AR
- Provision for doubtful receivables
- Calculate and post balance sheet adjustments
- Intercompany reconciliations
- Month-end reporting
- Foreign trade report

❸ Central payment in Central Finance supports three types of payment processing:

- **Central external payment**
  You'll process outgoing external payments especially for outside suppliers, creditors, and so on.
- **Central internal payment**
  This is related to intercompany customers and vendors.
- **Central incoming payment**
  You'll process incoming payments from customers and other parties involved in the operation of the business.

Note

Central payment is extensively covered in Chapter 9.

2.2.2 Intercompany Reconciliation Processes

SAP enables customers to reconcile three types of reconciliation processes:

- General ledger account open items
- AR and AP open items (customer and vendor open items)
- Individual document reconciliation for specific general ledger accounts

With central payment and centralizing relevant processes in Central Finance, you'll run the intercompany processes, which can be streamlined and enhanced across entity levels in real time. To leverage the accelerated intercompany process in Central Finance, company codes involved in intercompany reconciliation must be activated for central payment. The streamlined intercompany reconciliation process supports preparing consolidated financial statements and to comply with legal statutory requirements.

Figure 2.8 illustrates the intercompany reconciliation process in a Central Finance scenario with central payment activation:

❶ Upon activation of central payment, AP/AR open items are technically cleared (`ALE-extern`) upon posting open items in the source system.

❷ Corresponding open items are then replicated from the source system to the Central Finance system. Clearing activities of open items are then carried out only in Central Finance.

❸ You'll run the intercompany matching and reconciliation (ICMR) application, an SAP NetWeaver–based interface, or SAP Fiori apps to manage intercompany reconciliation as part of period-end closing in Central Finance. The ICMR application accesses both AP and AR open items directly in Central Finance in real time without any latency, thereby providing real-time visibility of intercompany transactions.

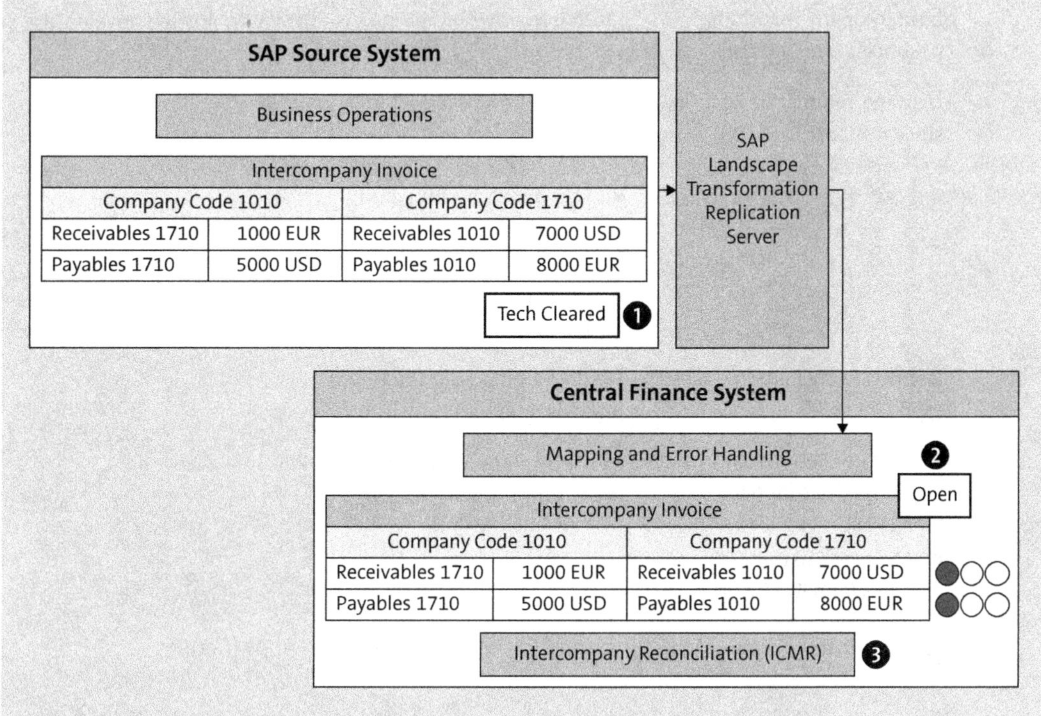

Figure 2.8 Central Finance Intercompany Reconciliation

SAP Fiori apps are provided to perform matching, reconciliation, and adjustment postings. Table 2.3 provides ICMR-related SAP Fiori apps.

Intercompany Matching/ Reconciliation Settings	Intercompany Matching Operation	Intercompany Reconciliation
Maintain Data Sources (app ID ICADS)	Define Matching Methods (app ID F3862)	Reconciliation Status Overview (app ID F3865)
Maintain Reason Codes (app ID ICARC)	Manage Matching Assignments (app ID F3870)	Reconciliation Balances (app ID F1819)
■ Manage Workflows - Intercompany Variance Adjustment (app ID F5766) ■ Manage Workflows - Reconciliation Close Requests (app ID F4993)	Manage Matching Documents (app ID F3868)	Display Matching Items (app ID F3869)

Table 2.3 Intercompany Matching and Reconciliation: SAP Fiori Apps

Intercompany Matching/ Reconciliation Settings	Intercompany Matching Operation	Intercompany Reconciliation
Manage Teams and Responsibilities (app ID F6783)	▪ Schedule Matching Run Jobs - By Company (app ID F7669) ▪ Schedule Matching Run Jobs - By Consolidation Unit (app ID F7669) ▪ Schedule Matching Run Jobs (app ID F5011)	
Define Reconciliation Cases (app ID F3863)	Flexible Upload of Matching Items (app ID ICAFU)	
▪ Mange Global Accounting Hierarchies (app ID F2918) ▪ Manage Customer Hierarchies (app ID F5794)	Verify Intercompany Variance Adjmt (app ID F5764)	

Table 2.3 Intercompany Matching and Reconciliation: SAP Fiori Apps (Cont.)

When reconciliation is performed on a system that isn't based on SAP HANA, three-step processes are followed. First, data selection and updating is done in the intercompany reconciliation ICR ledger. Then, you'll apply customized predefined rules to select and assign documents. Finally, in the last step, you'll reconcile intercompany documents manually. However, on the other hand, with the Universal Journal and SAP HANA–based systems, reconciliation is done in real time as data is accessed directly.

[»]

Note

The ICMR tool is delivered with SAP S/4HANA and easily integrated with Central Finance for intercompany reconciliation. There is no separate license required to use the ICMR tool in SAP S/4HANA. However, if you're planning to use machine learning features offered along with ICMR, you'll need a license for using this feature. To understand ICMR features offered with SAP S/4HANA compared to SAP ERP, consult SAP Note 3114271.

2.3 Financial Supply Chain Management

SAP offers FSCM, a comprehensive set of financial supply chain management components, to support financial processes across an enterprise's supply chain. The overall goal of these components is to optimize the cash flow cycle and thereby improve working capital and cash flows, as well as reduce operational risks. From the Central Finance perspective, you can standardize the centralized financial supply chain processes using replicated data collected from various systems in Central Finance. As shown in Figure 2.9, Central Finance supports the following financial supply chain processes:

- **Credit management**
  This component provides customers to manage credit limits and determine creditworthiness.
- **Dispute management**
  This enables you to manage dispute resolution, tracking, and subsequent reporting for management decisions.
- **Collections management**
  This functionality offers customers the chance to collect outstanding accounts receivable receives as part of the cash flow cycle.
- **Cash and liquidity management/treasury**
  These enable you to manage cash position and liquidity forecasting based on centralized data. In addition, you can improve bank communication with Bank Communication Management. In addition, SAP Treasury and Risk Management enables you to manage treasury operations and risk management.

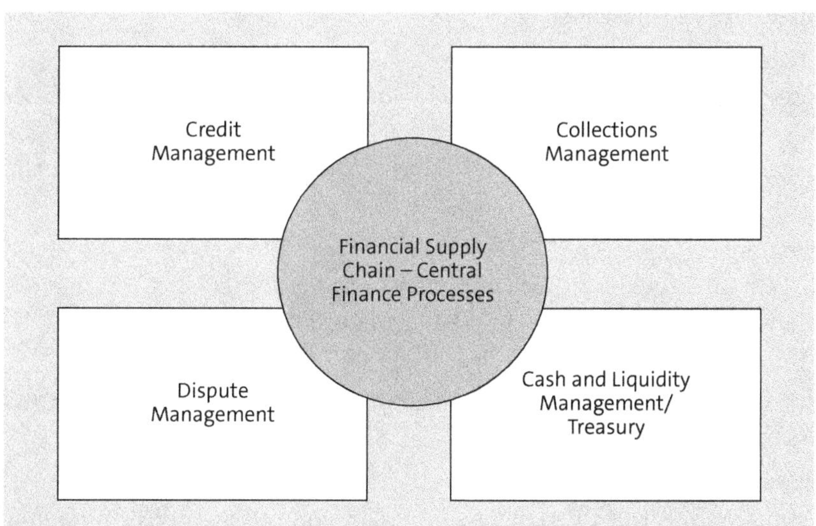

Figure 2.9 Central Finance Financial Supply Chain Management Components Processes

Let's explore the integration of each of the financial supply chain management components with Central Finance next.

2.3.1 Credit Management

In SAP, companies lower the risk of credit allocations and monitoring with the credit management component. Often companies face slow credit evaluation that could hamper the credit application process. This may sometimes lead to lost sales, delayed sales, or even cancellation of sales orders. In addition, inconsistent credit policies and guidelines may also expose the risk of bad debt and write-offs. Companies are seeking robust credit management functionality to fulfill growing business demands. SAP credit management alleviates most challenges companies facing today. By implementing SAP credit management, companies benefit from the following:

- Timely credit decisions via automated credit check process
- Automate credit application and credit approval processes
- Viewing documented credit decision and release blocked orders instantly
- Cut down payment delays, resulting improved days sales outstanding (DSO)
- Reduced bad debt
- Reduced credit refusal
- Improved customer relationships by concentrating on profitable customers

Credit management comes with the following features:

- **Define credit policy**
  Companies are able to implement an enterprise-wide credit management especially in a distributed system landscape.
- **Business partner**
  Business partner master data by master data management (MDM) is used in credit management, which houses all credit relevant ratings and analysis. To support credit management, you'll activate the business partner role for credit management.
- **Case management**
  Credit management provides automated credit checks and creates documented credit decisions (DCD) to manage credit cases in Central Finance.
- **Worklist**
  This is essential for managing credit-related activities such as using the credit rule engine, categorizing customers by scoring rules, and assigning credit limits.
- **Credit check rules**
  These are configuration rules that help companies control credit risk, as well as determine timing and how credit checks are performed for customers.
- **Credit information**
  This is obtained from external credit agencies with a connected interface. In addition, you'll obtain input parameters for scoring rules.
- **Credit manager portal**
  This provides credit management information data for analysis. Predefined reports are available to strategic credit analysts and credit management information is provided for salespeople.

[«]

Note

The credit management role of business partner in SAP S/4HANA stores all credit-relevant data within individual credit segments. The system uses the current credit limit, external rating value, and risk classes to determine credit limits. The business partner role has an audit trail to track changes made to business partner credit-related information. You can establish business partner relationships with global customer hierarchy features. In addition, a global credit segment can serve as the master data for all entities using the same relationship hierarchy.

With this basic understanding of credit management, let's explore how credit management functions in a Central Finance environment. In Central Finance, credit management is used as the repository for customer credit information for multiple source systems. You can collect, manage, and analyze credit information from those source systems centrally in Central Finance. Integration of credit management provides expedited credit information and helps you make better decisions. From a Central Finance context, let's explore credit management with the most common use cases:

- Central credit management with local processing
- Credit management with central payment
- Central Finance and credit management with mixed scenarios

We'll discuss each scenario next.

Central Credit Management with Local Processing

In this scenario, as shown in Figure 2.10, all payments are processed in the source system. Along with financial documents, credit exposure information is sent to the Central Finance system. In this case, credit management integration with Central Finance happens with web services or SAP Process Integration interfaces:

❶ Central credit management resides within the Central Finance system. Credit limit checks happen via application link enabling (ALE) or web services. AR updates the business partner master data with credit management information. When sales orders are created in the source SAP system, credit checks are performed in Central Finance. The credit request is evaluated based on credit data in the business partner, and a decision is sent back to the core system. Typically, a credit check is performed based on check rule:

- Credit exposure versus credit limit
- Past due
- Maximum order value

In addition, the risk class on the credit profile and the time horizon are used to determine the credit check. A credit analyst can use either internal credit information or external credit information to determine the customer score. For external credit information, large enterprises typically have interfaces to credit agencies. For internal credit, you'll potentially use the following:

- Information from business partner master data
- Payment behavior using past average days in arrears
- Dunning notices sent in the past
- Dispute management history

❷ Next, credit exposures updates happen via ALE or web services for order, delivery, and billing updates.

❸ In the Central Finance system, you'll calculate and manage credit limits for core systems. In addition, central credit reporting is available based on real-time data available from the source systems connected to Central Finance.

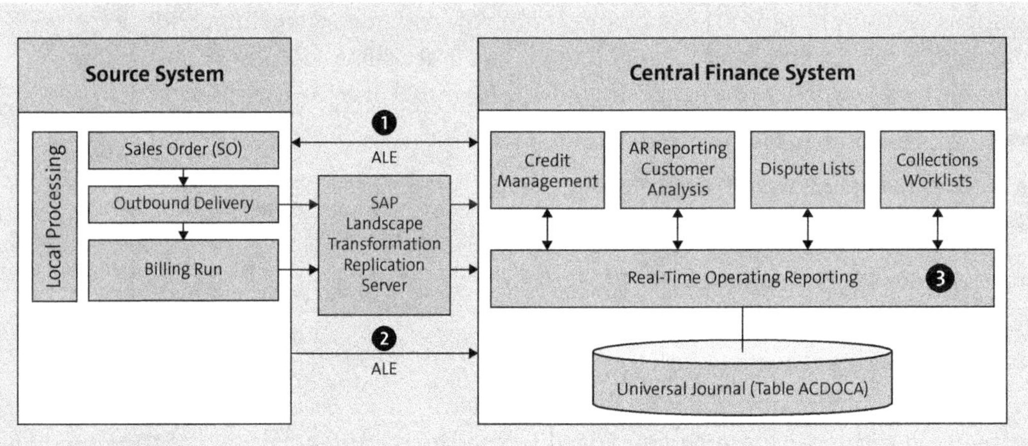

Figure 2.10 Credit Management with Central Finance

Credit Management with Central Payment

In this scenario, all payments are processed from the Central Finance system for the company code that is in scope for central payment activation. Figure 2.11 illustrates this scenario:

❶ Credit checks for SD objects are initiated from the local system and checked in the central credit management system. The credit management role is maintained on business partner in SAP S/4HANA. When sales orders are created in the source systems, the customer credit checks are carried out in Central Finance with credit information maintained in the business partner.

❷ AR open items are replicated to Central Finance, and corresponding open items are technically cleared in the source system.

❸ Credit exposure information from logistics processes, sales orders, and deliveries are transferred to the Central Finance system.

❹ AR open items which are transferred to Central Finance to consider risk ratings that determine credit limits assigned to customers in Central Finance. The system takes into account several factors such as order and payment behavior, dispute history, dunning history, industry, and legal structure for calculation of credit ratings based on a scoreboard.

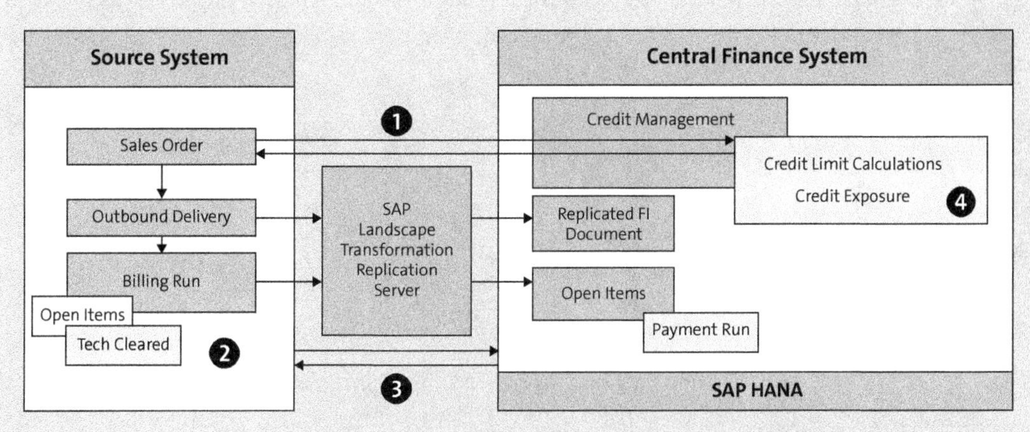

Figure 2.11 Credit Management with Central Payment

[«]

Note

SAP recommends that your credit management for central payment scenario is configured as follows:

- Credit management, business partner, and other relevant functionalities are configured in Central Finance.
- Consider exposures from open sales orders from the source system.
- Accounts receivable open items are considered for exposures from Central Finance.
- Credit decisions are performed in Central Finance using DCD for sales orders created in the source system.
- Take account of payment behavior key figures from Central Finance.

Credit Management with Mixed Scenarios

SAP also supports mixed migration options where one company is activated with central payment and the other company code's payments are processed in the source system. Although this scenario is allowed, this is considered a custom option and requires further evaluation of impacted processes. Figure 2.12 illustrates the mixed credit migration scenario:

❶ In the first scenario, central payment is activated for the company code in scope. As a result, all open items are technically cleared in the source system. Central credit management is active in Central Finance, so local systems use ALE or web services to perform credit limit and exposure updates.

❷ In the second scenario, payments are handled in the source system. FI documents are sent to Central Finance. In addition, credit limit calculations and credit exposure are captured in central credit management in Central Finance.

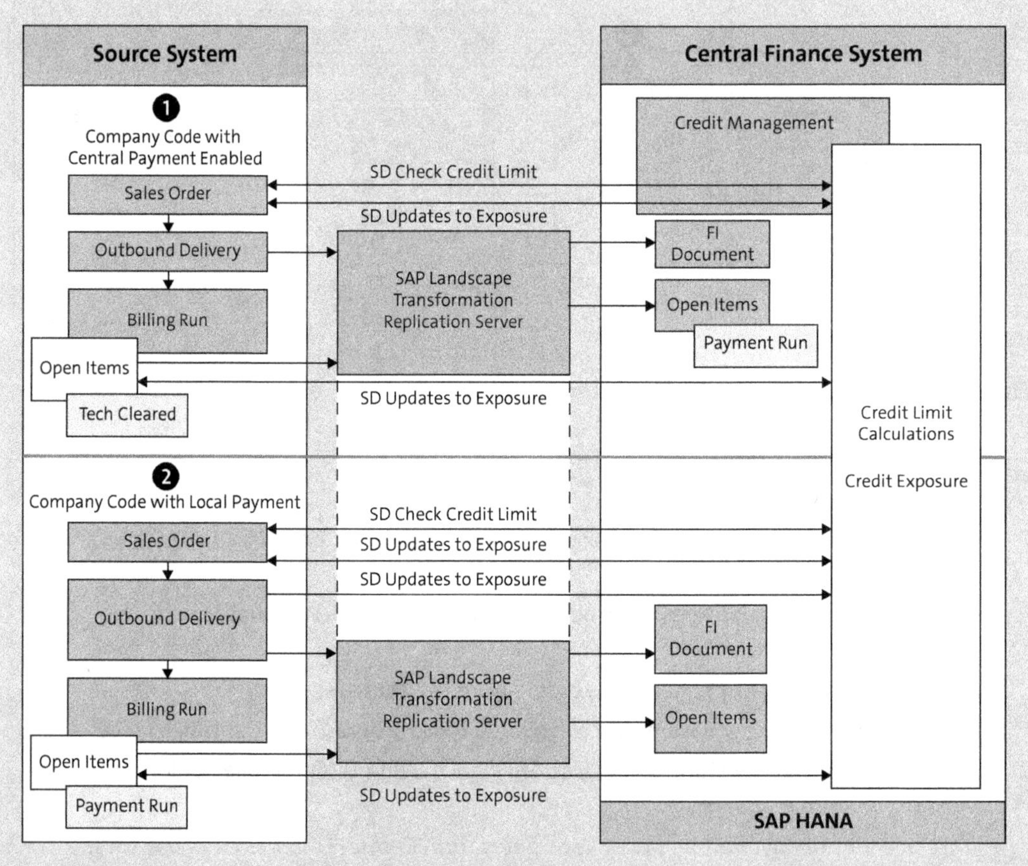

Figure 2.12 Credit Management with Central Payment: Multiple Scenarios

[»]

Note

For the second scenario to work correctly, you must implement SAP recommended BAdI `UKM_FILL~FILL_FIELDS` to filter out the company code as suggested in SAP Notes

2839326 and 2833145. Otherwise, the system doubles the credit exposure information for open invoices as the FSCM processes are run from the Central Finance system.

SAP also recommends rebuilding credit exposure with type 200 open invoices to fix any inconsistencies in credit exposure values. Transaction UKM_COMMITMENTS can be used to delete any inconsistencies in credit exposure values. Then, you'll run report UKM_TRANSFER_IEMS to reconstruct the credit exposures. Consult SAP Notes 2576445 and 1124271 for additional guidance on this topic.

2.3.2 Dispute Management

In SAP, dispute management offers companies the opportunity to process, escalate, and resolve AR disputes during the normal course of business. Disputes arise in cases where the seller and buyer disagree on the amount charged on invoices or discrepancies found on the invoice. The most common reasons for AR disputes are listed here:

- Goods or services not delivered, or goods damaged during transit
- Incorrect price stated on invoice
- Unacceptable delay in delivery
- Quantities discrepancies arising with differences between delivered quantity and quantity invoiced
- Unauthorized changes or unapproved charges
- Duplicate invoices submitted for the same order
- Tax discrepancies due to incorrect tax calculation or charges for a nontaxable item
- Discount or contractual agreements-related disputes
- Disputes arising when service level agreements (SLA) aren't followed
- Miscommunication or missing documentation
- Credit memo adjustments due to timing or credit memo not adjusted with the invoice

Central Finance offers companies an integrated dispute management platform to handle disputes effectively by streamlining the dispute management processes. The integrated dispute management functionality in Central Finance provides customers with the following:

- Streamlined dispute case process that provides full visibility to dispute cases within accounting transactions
- Dispute case resolution across departments or in an enterprise-wide view
- Documentation of dispute-related information and approval via delegation of authority
- Advanced reporting capabilities to detect quality issues, control workload, and track disputes, deductions per customer, and so on

Dispute cases are business objects in FSCM, which serves as central repository to resolve dispute cases. The central repository is like a folder organized in a structured way where it contains dispute-relevant information such as documentation and related objects. For example, a dispute case contains information related to invoices, delivery notes, correspondence, credit notes, supporting documentation, statuses, reason codes, and so on.

We'll explore how dispute management functions when central payment is active in the following sections.

Central Finance with Dispute Management

Figure 2.13 illustrates the integrated dispute management data flow when central is payment activated for a company code that is in scope. As shown in the figure, AR open items are technically cleared in the source system upon posting. The corresponding invoice open items are then replicated to Central Finance where central payment clearing activities are carried out. Let's explore how dispute management integrates with Central Finance:

❶ With integrated dispute management with AR, a dispute case is created and tied to the replicated open item in Central Finance.

❷ Centralized dispute management integration with AR provides the following:

- Log disputes at business partner level without a requirement to assign the dispute against an open item receivable
- Dispute resubmissions
- Customer contacts
- Automatic creation of correspondence to send dispute case to stakeholders to take further action
- Streamlined dispute processes
- Cross-department resolution of disputes

❸ Integrated dispute management offers customers centralized dispute handling capabilities with financial data replicated from various source systems. Besides robust integration with AR, centralized dispute management offers advanced reporting analytics with the following:

- Key performance indicators (KPIs) to monitor case status by following investigation and resolution process
- Dispute aging reporting to track time spent by responsible personnel tracking or resolving dispute cases
- Overdue receivables in dispute
- Role-based SAP Fiori apps to manage dispute cases: Manage Dispute Cases (app ID F0702A), Dispute Management Reporting (app ID F4370), Schedule Dispute

Management Jobs (app ID F3578), Display Dispute Management Log (app ID F6395), Display Dispute Case, and Edit Dispute Case (app ID /SCMTMS/INV_DISP), as well as other HTML GUID-based apps.

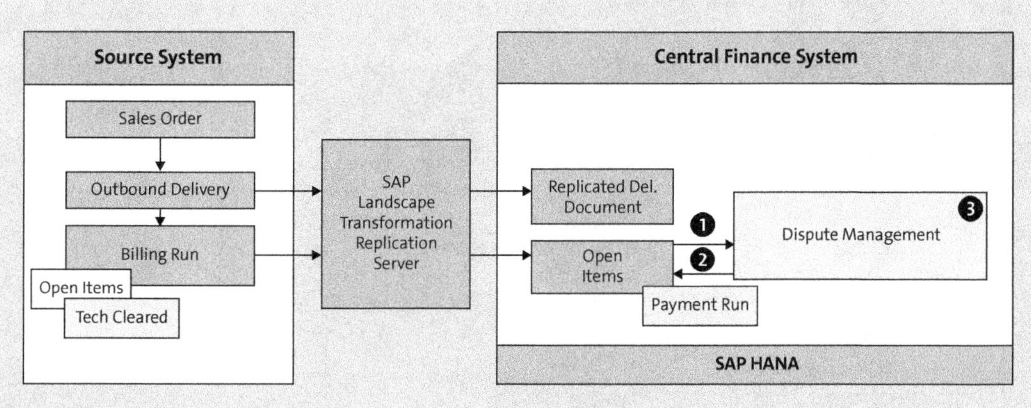

Figure 2.13 Dispute Management with Central Payment

[«]

Note

SAP recommends dispute management be implemented in Central Finance if central payment is activated.

Central Dispute Management Scenarios with and without Central Payment Activation

Figure 2.14 illustrates a mixed scenario where one company code is activated with central payment and the other processes payments locally. All disputes are centrally managed in the Central Finance system for both company codes. The hybrid approach is considered a custom approach, and you must evaluate all pros and cons of going with this approach. Let's understand how the mixed scenario functions for centralized dispute management:

❶ In the first scenario, central payment is activated for the company code in scope. As a result of central payment activation, all open items are technically cleared in the source system. Disputes are created for open items replicated to Central Finance.

❷ In the second scenario, payments are handled in the source system. FI documents are replicated to Central Finance. Although open items are managed in the source system, disputes can be created in the integrated dispute management component in Central Finance with classic ALE or the web SAP Process Integration interface.

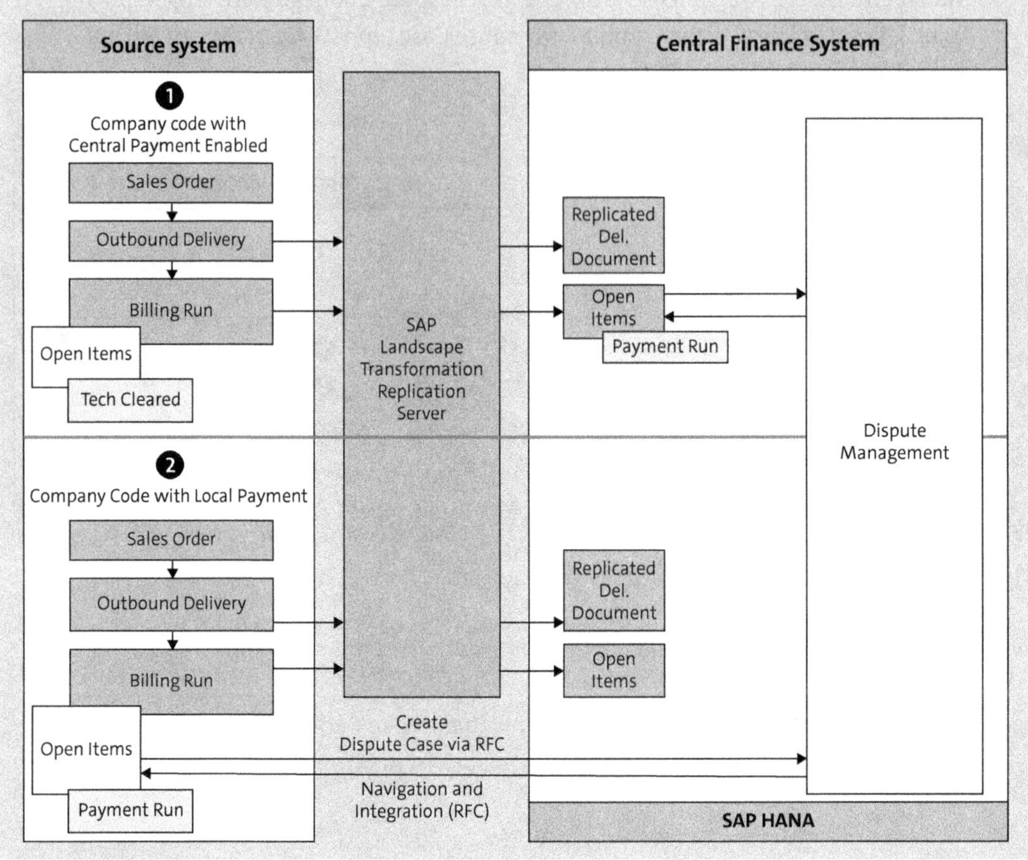

Figure 2.14 Dispute Management with Central Payment Multiple Scenarios

2.3.3 Collections Management

As companies grow in business, the customer base also increases, which makes managing AR more complex. This happens with most business models. With technological growth, the classic way of sending dunning notices isn't effective and often results in delays in managing collections. SAP collection management provides companies with strategies to manage AR collection activities. The collection management supports customers with the following:

- Creates collection strategies that offer streamline collection processes by allowing prioritization of collections
- Automatic creation of daily worklists for collection specialist
- Promise to pay agreements for invoices customer promises to pay
- Integration with dispute management with dispute cases
- Defines and aligns strategies

- Maintains collection groups
- Reassigns worklist items
- Monitors worklist processing
- Provides collection manager's ability to control collection receivables
- Sends payments reminders
- Follow-ups with customers

Central Finance supports central collection processes. With central collections management, you can perform collections from Central Finance for AR open items replicated from multiple source systems across the enterprise. This will often include multiple business units, products, services or geographies. With central collections management, you can do the following:

- Perform centralized collection activities in the Central Finance system.
- Generate worklist and processing views.
- Use the collection workbench, which captures an enterprise-wide view of customer invoice, payment, and dispute data.
- Perform advanced reporting capabilities with daily/weekly collection reports, as well as operating reporting analysis views with the help of SAP Fiori apps.
- Engage in centralized team management based on customer contact results.
- Use full integration of promises to pay, disputes, and correspondence with financial components.
- Drill back to the source system and non-SAP source systems for sales orders or other logistics documents with ALE.

[«]

Collections Management Terminology

Let's understand key terminology used in collections management in Central Finance:

- **Worklists**
  This contains past due open items in order of priority. The prioritization is determined by the collection strategy you've defined based on multiple parameters.
- **Collection strategy**
  The collection strategy is a grouping and prioritization of collection rules to expedite payments from customers in a timely manner.
- **Collection groups**
  Collection groups are used to streamline collection processes by grouping customers based on certain criteria such as customer industry, geography, and so on.
- **Collection segments**
  Collection segments group company codes for customers based on their risks and collection behavior for open items, dispute cases, and so on.

- **Collection profile**
  A collection profile is a group of collection segments used for collection management.
- **Collection specialist**
  A collection specialist is a person from the AR department involved in collection-related activities.

With the introduction of collection management concepts and terminology, let's dive deeper into the integration of collections management with Central Finance. We'll begin with collection management when central payment is active for a company code.

Collections Management with Central Payment

Figure 2.15 shows the collections management data flow when central payment is activated for a company code. AR open items are technically cleared in the source system. The corresponding invoice open items are replicated to Central Finance where central payment clearing activities are carried out as follows:

❶ Once open items are replicated over to Central Finance, first you'll create a worklist in the Central Finance system for open items that are overdue. The worklists can be created automatically based on collection strategies and prioritized worklists, and then a collection specialist is assigned. The collection manager assigns worklists to the collection specialist both manually or automatically.

❷ You'll begin invoice resolution processes with the Process Receivables app (app ID F0106A). The app enables the collection analyst to do the following:

- Process customer open receivables, credit memos, credit information. and contact history.
- View overall status of customer receivables, including invoice details and total amounts.
- Review business partner information, including head office or branch office relationships.
- Send invoices to customers via email or a print medium.
- Navigate to other apps to view correspondence.
- View aging of the receivables in a due date range.
- Create promises to pay for overdue receivables.
- Create dispute cases for overdue receivables.
- Create resubmissions for specific invoices or for customers.
- Create notes at the customer level.

In addition, besides SAP HTML GUI-based apps, SAP provides several HTML5-based SAP Fiori apps to support centralized collections management in Central Finance. The following list provides the most important SAP Fiori apps in collections management in SAP S/4HANA:

- Create List with Collection Items (app ID FPO3L)
- Manage Collection Strategies (app ID F2946)
- Cash Collection Tracker (app ID F3182)
- Manage Collection Contact Persons (app ID F4708)
- Collection Progress (app ID F1738)
- Collection Management Reporting (app ID F4495)
- Manage Collection Emails (app ID F6120)

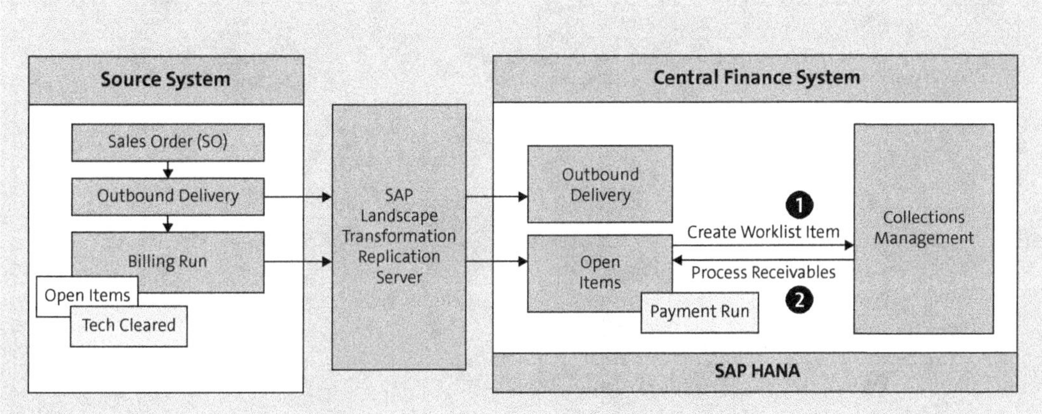

Figure 2.15 Collections Management with Central Payment

[+]

Note

Collections management must be implemented in Central Finance once central payment is activated. In order to run collections management in Central Finance, SAP recommends you assign one collection segment to one company code that is activated for central payment. As a result, you'll be able to run collection processes in the source system for the other company codes that are not activated for a central payment scenario.

Collection Management with and without Central Payment Activation

Figure 2.16 illustrates a mixed scenario where one company code is activated with central payment and the other company code processes payments locally. In both circumstances, all collection activities are centrally managed in the Central Finance system for both company codes. The hybrid approach is considered a custom approach, so you must evaluate all restrictions or limitations of going with this approach. Let's look at both scenarios:

❶ In the first scenario, central payment is activated for the company code in scope. As a result, all open items are technically cleared in the source system. As you've seen in the previous scenario, you'll create a worklist and process receivables within Central Finance.

❷ In the second scenario, payments are handled in the source system. FI documents are sent to Central Finance. In this scenario, worklist items and processing receivable open items are handled with web-based services or an ALE interface from the source system to the Central Finance system.

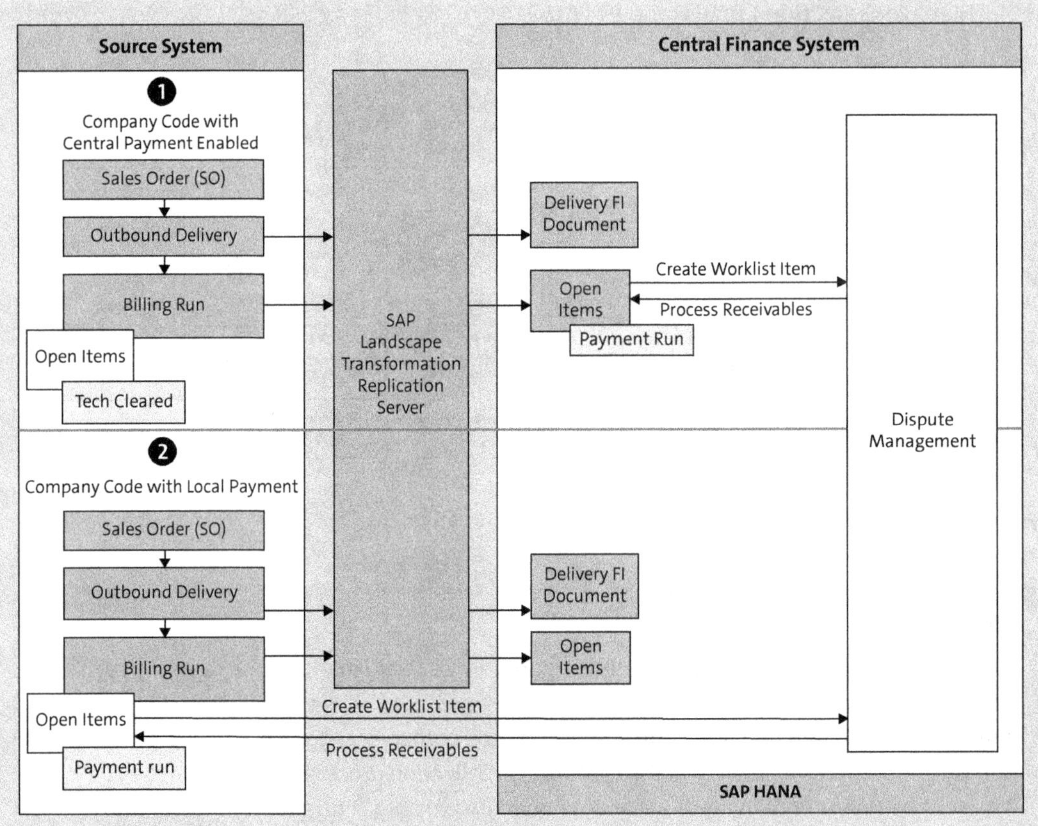

Figure 2.16 Collections Management with Central Payment Multiple Scenarios

2.3.4 Cash and Liquidity Management

Cash management evolved with new features in SAP S/4HANA. Most of the features are available and supported in Central Finance. With centralized data and central processes in Central Finance, you can achieve improved cash flows across an organization. In addition, with centralized data available in the Universal Journal, managers can gain access to cash positions, forecast future cash flows, and manage liquidity risk in real time. Figure 2.17 illustrates central cash management from a Central Finance view:

❶ A replicated transaction data set and harmonized master data set from the source system are available in the Central Finance system. With both data sets, you can obtain real-time data anytime from the Universal Journal, which represents a single source of truth.

❷ Most SAP S/4HANA components such as bank communication management, in-house cash, and treasury risk management features are available in the Central Finance system. In addition, you'll be able to implement bank account management as a hub for central bank account management, and then this information can be distributed to other systems via IDocs.

❸ One exposure in SAP S/4HANA provides stakeholders with a unified view of a company's cash and liquidity position. It serves as a central repository for replicated AR and AP, and other treasury-related data coming from multiple source systems. In addition, you'll be able to retrieve logistics data from SD and MM with ALE IDoc interfaces. Cash management personnel can get real-time insights into cash-related data to help them perform comprehensive cash flow analysis. Centralized data from One Exposure serves as a single source truth for managing any financial risks.

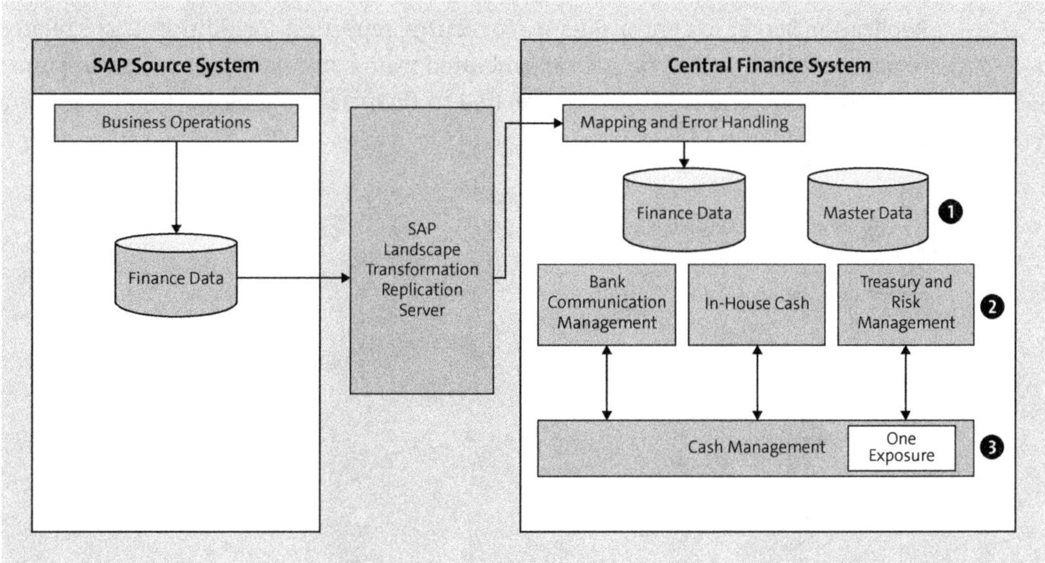

Figure 2.17 Central Cash Management

2.3.5 SAP In-House Cash

Customers can leverage SAP In-House Cash to automate internal and external payment transactions within a company or group of companies. SAP In-House Cash is especially useful when companies transact at a high volume involving banking transactions across the enterprise. By centralizing these transactions using SAP In-House Cash, customers reduce their number of bank accounts, thereby allowing companies to reduce operational costs, enhance liquidity, and improve cashflow. SAP supports internal payments, centralized payments, local payments, and centralized incoming payments processes.

From a Central Finance perspective, use cases for SAP In-House Cash processes can vary with business requirements. You can implement SAP In-House Cash as a sidecar approach or directly in your Central Finance system. We'll walk through each approach in the following sections.

SAP In-House Cash Implemented as a Sidecar Approach

Figure 2.18 illustrates SAP In-House Cash deployed as a sidecar approach. In this case, SAP In-House Cash resides in a separate system outside of the source system or Central Finance system for a Central Finance reporting or central payment scenario, respectively:

❶ SAP In-House Cash can be connected to the source system, as the source system is the book of record in this case. As shown in Figure 2.18, AP and AR open items and relevant clearings from subsidiary are replicated via SAP Landscape Transformation Replication Server to Central Finance for central reporting. In addition, IDocs related to payment files (PAYEXT or DIRDEB) generated from subsidiary sent to SAP In-House Cash. At period end closing, IDocs related to financial statements (FINSTA) are sent back to subsidiary.

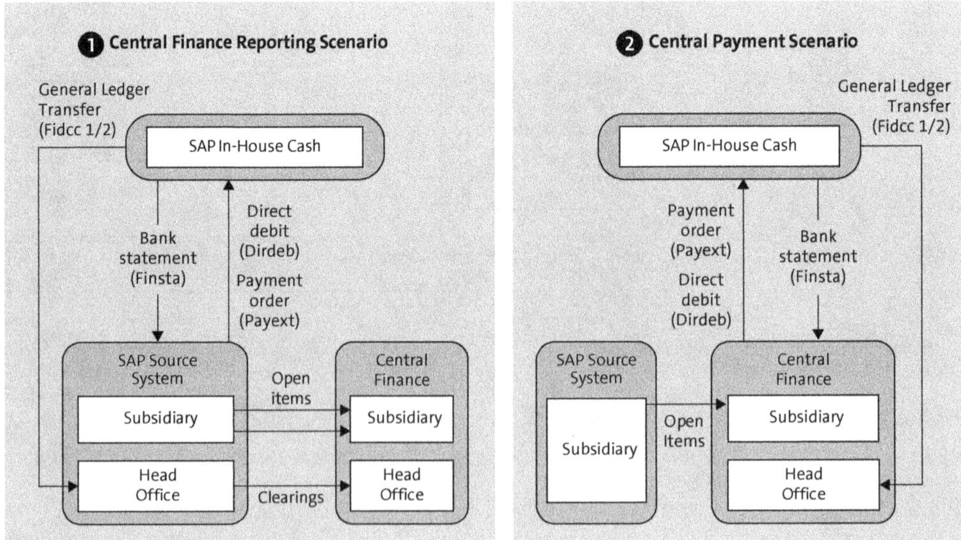

Figure 2.18 SAP In-House Cash as a Sidecar Approach

❷ The second scenario involves SAP In-House Cash connected as a sidecar approach from Central Finance. In this case, central payment is activated, and hence Central Finance is the book of record as a result of handling payments from various source systems. Open items are replicated from a subsidiary from the source system via SAP Landscape Transformation Replication Server. All payments are processed by the subsidiary from the Central Finance system. IDocs related to payment files (PAYEXT or DIRDEB) generated from subsidiary sent to SAP In-House Cash. At period-end closing, IDocs related to financial statements (FINSTA) are sent back to the subsidiary.

SAP In-House Cash Implemented on Central Finance

Figure 2.19 shows SAP In-House Cash implemented in Central Finance for both a Central Finance reporting scenario and a central payment scenario:

❶ In the central reporting scenario, both payment order- and bank statement-related IDocs from the source system subsidiary are sent and received from SAP In-House Cash in Central Finance. As shown in Figure 2.19, general ledger transfer from SAP In-House Cash is sent to the head office of the source system.

❷ Conversely, in a central payment scenario, both payment order- and bank statement-related IDocs from the Central Finance subsidiary are sent and received directly from SAP In-House Cash in Central Finance. In case of central payment scenario, the general ledger transfer occurs directly between SAP In-House Cash and the head office.

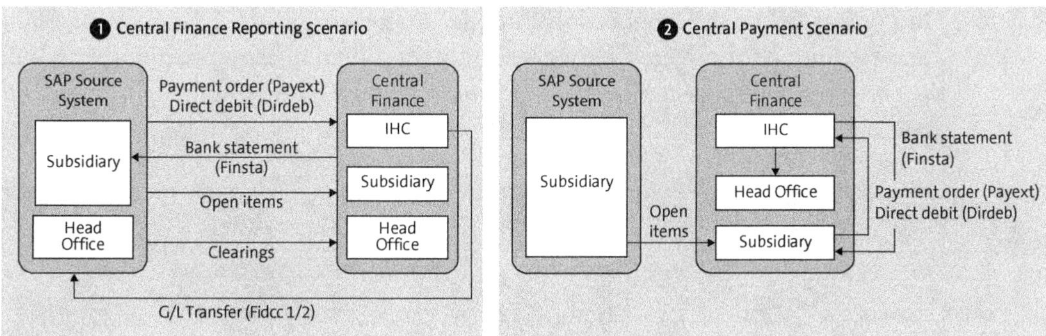

Figure 2.19 Inhouse cash on central finance system

[+]

Note

SAP recommends implementing in-house cash in central finance along with central payment and central processes as a long-term approach.

2.4 Projects, Fixed Assets, and Internal Orders

In this section, we'll explore centralized processes in the project systems, capital asset management, and internal order components.

2.4.1 Central Projects Reporting Scenario

Central Finance supports replicating projects and work breakdown structure (WBS) master data from the source system to the Central Finance system. Project replication doesn't use SAP Landscape Transformation technology; instead, it uses ALE and intermediate documents (IDocs) technologies to replicate master data. As a prerequisite, you'll configure ALE inbound and outbound configurations in both the source system and the Central Finance system.

Let's deep dive into how project replication helps users facilitate central projects reporting scenarios in Central Finance. Figure 2.20 illustrates the flow of project master data and subsequent financial posting in Central Finance:

❶ First, projects and WBSs are created in the source system. After releasing a project or WBS in the source system, the interface triggers an IDoc carrying master data information and sends it to Central Finance. These project system–related cost objects are then used in the financial transactional posting in the source system.

❷ The project systems master data is then replicated via the ALE IDoc into Central Finance where it goes through SAP Master Data Governance business mapping before the master data is created. The system maps the project or WBS ID and other important fields. SAP Application Interface Framework is used for error handling and resolution for projects and WBS replication. When financial postings with WBS occur in the source system, the corresponding WBS is mapped and reposted in Central Finance.

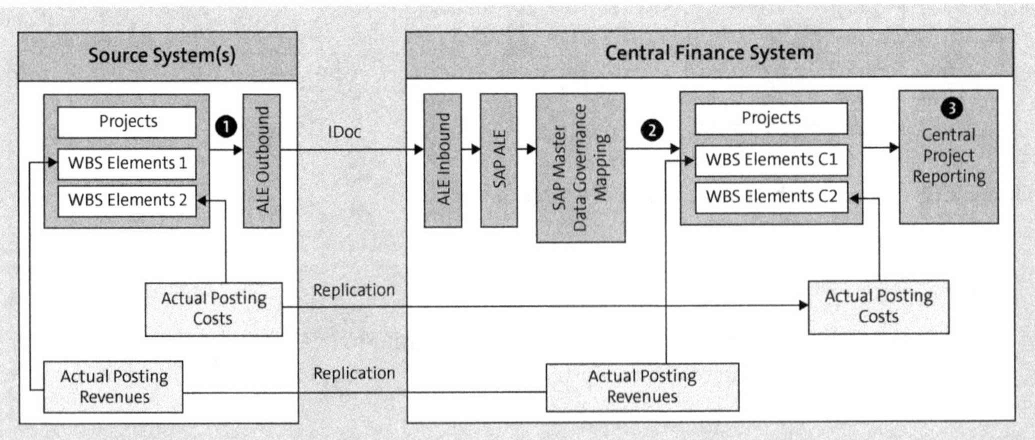

Figure 2.20 Central Project Reporting

❸ Financial postings involving both costs and revenues are replicated to Central Finance. Central project master data along with financial postings from the Universal Journal are used for centralized project reporting.

[+]

Tip

SAP provides report FINS_CFIN_PS_CHECK to check configuration inconsistencies prior to starting project and WBS replication.

2.4.2 Asset Settlement Scenario

With the Central Finance asset settlement scenario, the system creates AuCs automatically, settles cost to the AuC, and then finally settles cost to final fixed assets in Central Finance. This is useful especially when fixed assets processes are moved to Central Finance. You'll activate the central asset settlement by company code in Central Finance configuration. Figure 2.21 illustrates the data flow between the source system and the Central Finance system for the central projects asset settlement scenario:

❶ **Master data replication**
First, projects and WBS information is created in the source system and replicated to Central Finance. Project costs are collected in the source system on WBS via ongoing financial postings. These financial transactions are then replicated to Central Finance along with WBS master data.

❷ **Investment profile and project profiles**
To facilitate the asset settlement scenario in Central Finance, select the **Manage AuC** checkbox on the investment profile configuration in Central Finance only. As a result of this configuration, the system generates the AuC automatically in the Central Finance system. No AuC is generated in the source system as this flag is turned off on the investment profile in the source system. In addition, SAP recommends setting new project profiles for asset settlement scenarios.

❸ **Transactional postings in the source system**
Any financial transaction postings that use project or WBS master data as cost objects will replicate to Central Finance through SAP Landscape Transformation Replication Server.

❹ **Periodic settlement**
Once cost is accumulated on an AuC, you'll run periodic settlement in the Central Finance system. Periodic settlement moves cost from the WBS to the AuC generated from ❷.

❺ **Final settlement**
Once you're ready to capitalize the final asset per the asset accounting policy, you'll carry out final settlement in Central Finance. Full settlement moves costs from the AuC to the final asset.

Central Finance allows you to close a project or WBS element in Central Finance.

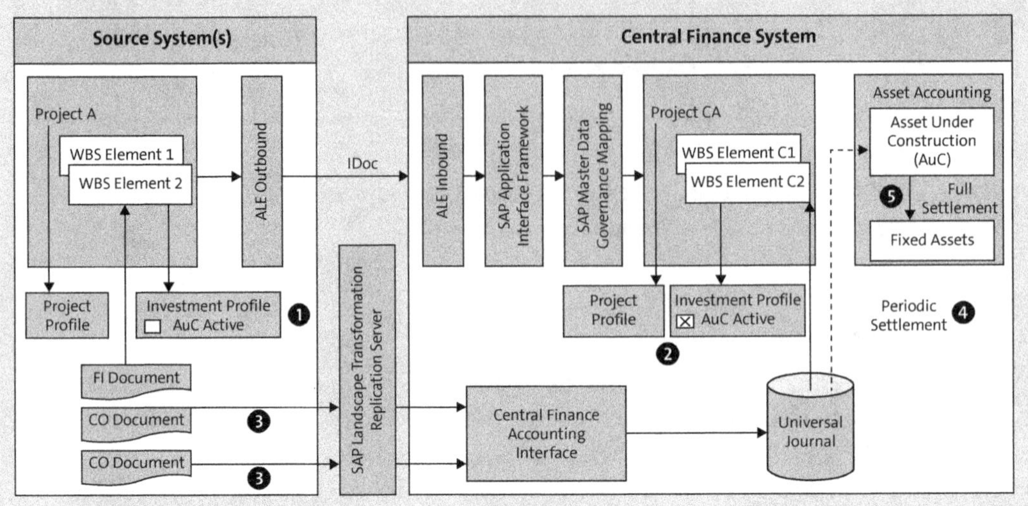

Figure 2.21 Central Asset Settlement

[!]

Restrictions

Although project asset settlement works with configured rules, there are several restrictions you should consider prior to implementing the central asset settlement scenario:

- If SAP S/4HANA Cloud Public Edition is configured as the source system, central projects asset settlement isn't supported in such situations.
- There is no write-back to the source system.
- Project network replication isn't supported; SAP recommends settling networks when projects contain networks.
- Plan data replication isn't supported.
- SAP Application Interface Framework doesn't support the reposting with user changes feature.
- After releasing the project or WBS, master data is replicated to Central Finance. You won't be able to delete WBS elements after that. However, the system allows you to set the deletion flag provided the deletion flag is set first in Central Finance for the corresponding WBS element.
- The system doesn't support changing IDs of the project and WBS in either system.
- Most project and WBS changes are done in the source system with the exception of **Close** and **Delete** system statuses for the asset settlement scenario.
- On WBS, settlement rules are restricted to only fixed assets and profitability segments.

- You'll set the **Close** status first in Central Finance prior to changing the status in the source system.
- The system allows you to delete investment profiles in the source system only if the AuC isn't capitalized yet. SAP provides program RFINS_CFIN_PS_AUC_DEL_IMPRF to delete AuCs in Central Finance.
- By clearing accounting assignment element indicator in the source system, the system doesn't delete the AuC. You can execute program RFINS_CFIN_PS_AUC_DEL_IMPRF to delete the investment profile and the corresponding AuC in the Central Finance system.
- The statistical indicator change on the WBS in the source system is allowed only if no AuC exists in the Central Finance system. SAP recommends you execute program RFINS_CFIN_PS_AUC_DEL_IMPRF to delete the investment profile and the corresponding AuC in the Central Finance system first.

2.4.3 Central Asset Accounting

In Central Finance, asset accounting information, such as asset number and transactions posted with financial line items, in the source system is removed. In addition, asset reconciliation accounts are mapped to the general ledger account in Central Finance. Conversely, SAP introduced the central accounting functionality with the SAP S/4HANA 2021 release to manage asset accounting in Central Finance. With this new functionality, you can manage fixed asset processes centrally in the Central Finance system while inactivating asset accounting in the source system. Figure 2.22 illustrates the data flow of central asset accounting in Central Finance. Let's explore the flow diagram further:

❶ After central asset accounting is activated, all asset-related transactions are no longer executed, or transactions are blocked in the source system. You'll apply appropriate security authorizations, organizational changes, or configurational changes to prevent posting asset-related transactions in the source systems

❷ All fixed transactions are executed in the Central Finance system. During cutover to new asset accounting, you'll establish baseline asset accounting configuration under IMG customizing **Financial Accounting • Asset Accounting**. Next, you'll bring in legacy asset master data and corresponding balances from the source system by following legacy data transfers available for new asset accounting. Finally, you'll need to repost asset-related postings from the general ledger account to the appropriate adjustment account to avoid duplicate postings from the legacy data transfer. Once new asset accounting is activated and relevant cutover activities are carried out, you'll perform all asset-related transactions in the Central Finance system.

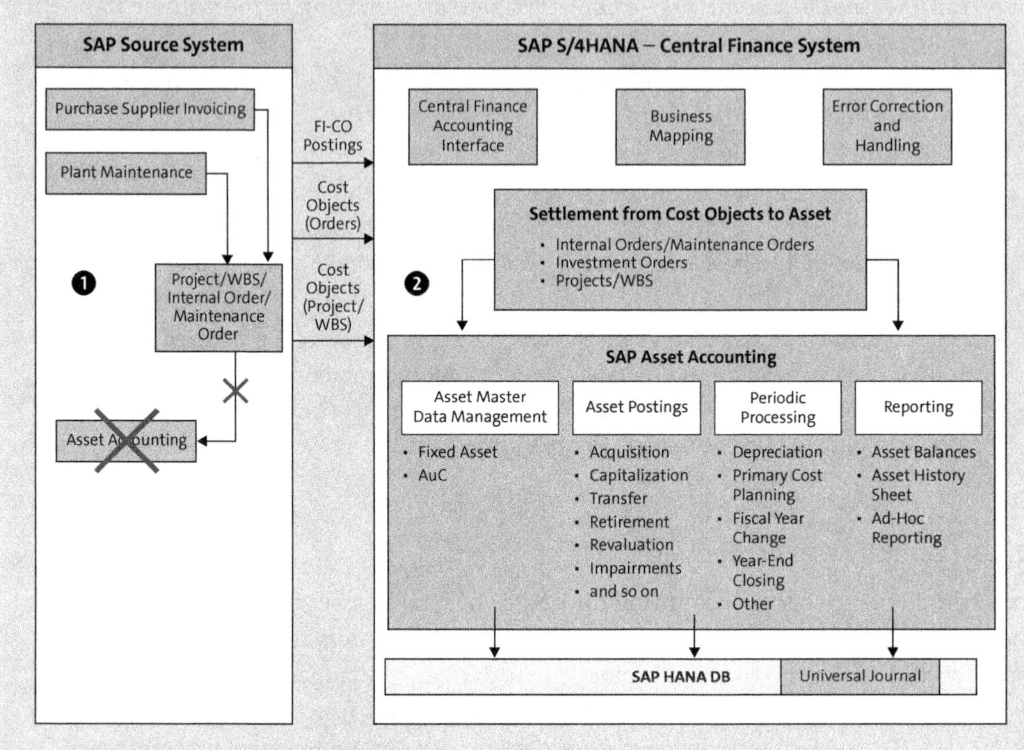

Figure 2.22 Central Asset Accounting

Tip

SAP strongly recommends using central asset accounting from the beginning of a fiscal year. This means you must activate central asset accounting at the end of a fiscal year once you've completed the year-end close for asset accounting. Consult SAP Note 3127262 for additional guidance on central asset accounting. This SAP Note provides a list of other SAP Notes for central asset accounting.

2.4.4 Central Budgeting for Projects

With central budgeting for the project scenario, you'll maintain budgets centrally for all source systems in a single place, that is, in Central Finance. During finance transactions posting in the source systems, the system checks the budget in the Central Finance system via RFC. If the actual amount posted is less than the budgeted amount plus defined tolerance limits, the system allows you to post the financial transaction in the source system. By centralizing the budget in a single system, you'll benefit from reduced customizing activities and maintenance activities across different source systems. Figure 2.23 displays central budgeting for the projects scenario:

1. First, the system checks whether the company code and project profile are configured for central budgeting for projects in the Central Finance system. If the configuration is enabled, the system initiates an RFC to Central Finance to perform an availability check in Central Finance.
2. The availability check is carried out in the Central Finance system per the budget profile and tolerance limit configuration parameters. As a prerequisite, the WBS system status must be in the availability control check (**AVC**) and budgeted (**BUDG**) status prior to posting the document with the relevant WBS.
3. If the availability check is successful, then financial transactions are allowed to be posted in the source system. If the availability check isn't successful for reasons such as insufficient funds as the budget amount is less than the actual amount being posted, the system displays a budget exceeded error. A successfully posted financial document is replicated through SAP Landscape Transformation Replication Server.

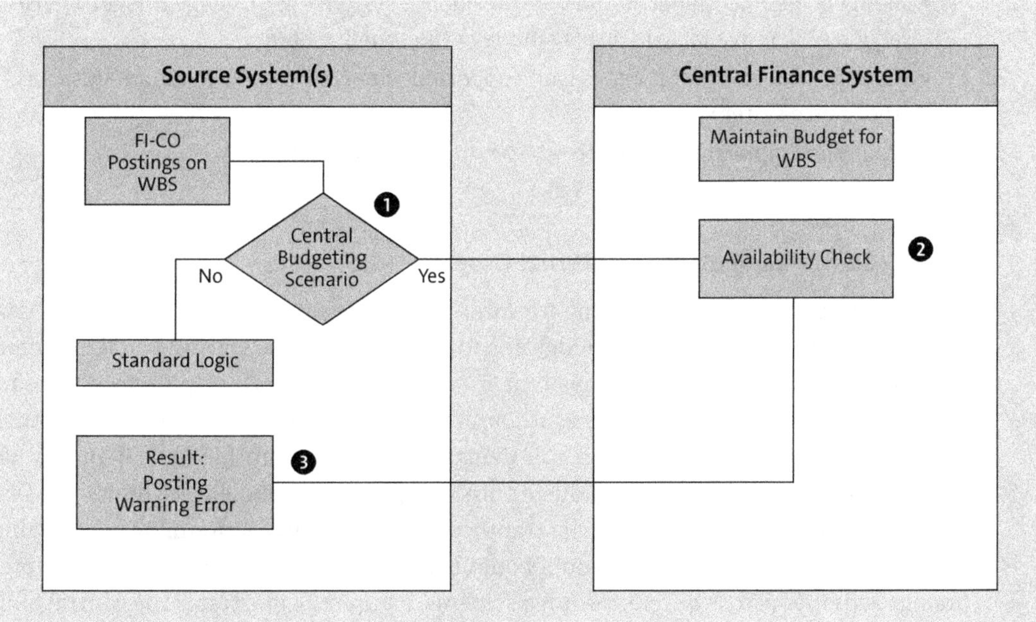

Figure 2.23 Central Budgeting: Projects

SAP recommends performing project reporting in Central Finance for the central budgeting projects.

[!]

Restrictions

SAP recommends reviewing the following constraints when considering the central budgeting for projects scenario:

- SAP doesn't provide any conversion tools for the existing projects.

- The Central Finance consistency check doesn't include central budgeting for projects to check any inconsistencies.
- If the relevant financial document gets stuck in SAP Application Interface Framework due to some other error and not replicated to Central Finance, the remote budget check will treat the budget as not yet consumed.
- The system checks availability control active (**AVAC**) or budgeted (**BUDG**) status in Central Finance during replication of a change in hierarchy for the WBS in the source system. In such situations, the change isn't replicated over to Central Finance, and an error is displayed in SAP Application Interface Framework.
- Budget maintenance is done when no RFC check is carried out on the WBS element.
- Project networks aren't supported.
- N:1 mappings aren't supported; Central Finance projects support only 1:1 mapping between the source system and the Central Finance system.
- Budgets are maintained in the Central Finance system only. Change budget Transaction CJ30 is available in display mode in the source system.
- Central budgeting for projects isn't supported when SAP S/4HANA Cloud is used as the source system.
- You must implement SAP Note 3025235 in the source system.

2.4.5 Central Budgeting for Internal Orders

SAP released the central budgeting for internal orders feature with the SAP S/4HANA 2020 FPS0 release. With central budgeting for internal order scenarios, you'll maintain budgets centrally for all source systems in a single location. During finance transactions posting with internal orders as account assignment objects in the source systems, the system checks the budget in the Central Finance system via RFC. If the actual amount posted is less than the budgeted amount plus defined tolerance limits, the system allows you to post the financial transaction in the source system. By centralizing the budget in a single system, you'll benefit from reduced customizing and maintenance activities across different source systems. Figure 2.24 illustrates the central budgeting for internal orders scenario. As you can see, similarities exist between central budgeting for WBS and central budgeting for internal orders:

❶ First, the system checks whether the company code and internal order are configured for central budgeting for the internal order scenario in the Central Finance system. If the configuration is enabled, the system initiates an RFC to Central Finance to perform an availability check.

❷ The availability check is carried out in the Central Finance system per the budget profile and tolerance limit configuration parameters tied to the order type.

❸ If the availability check is successful, then financial transactions are allowed to post in the source system. Subsequently, a successfully posted financial document replicated through SAP Landscape Transformation Replication Server.

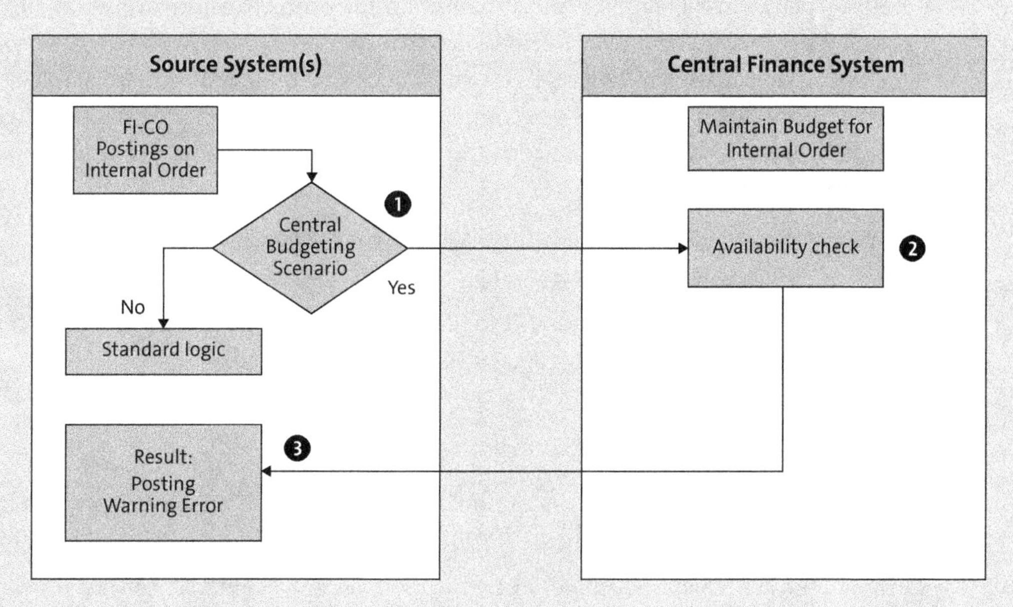

Figure 2.24 Central Budgeting: Orders

[!]

Constraints

SAP recommends reviewing the following constraints when considering central budgeting for internal order scenarios:

- Only internal orders replicated using the cost object framework are considered for the central budgeting for internal order scenario.
- Once central budgeting for internal orders is enabled, you'll manage budgets only in Central Finance. As a result, Transaction KO22 is used in display mode in the source system. Budgets created in the source system aren't considered anymore.
- SAP supports internal order to internal scenario for central budgeting. Other scenarios aren't supported.
- If the relevant financial document with internal order posting gets stuck in SAP Application Interface Framework due to some other error, the remote budget check will treat the budget as not yet consumed.
- You must implement SAP Note 2914565 in the source system considered for central budgeting for internal orders.

2.5 SAP Shared Services Framework

With the Central Finance implementation, the most common benefits include consolidated reporting and process centralization. The SAP Shared Service Framework is offered as a platform for customers to streamline and optimize central processes enterprise-wide, especially for global business service organizations. SAP Shared Service Framework is a licensed product offered to organizations to optimize their shared services operations. In addition, you'll accomplish process orchestration with centralized service delivery. With Central Finance as the shared services anchored solution, you'll accomplish the following:

- Centralization of all service-related transactions in Central Finance
- Efficient orchestration of shared services in Central Finance
- Integration with SAP processes and logical integration with components such as receivables management, SAP Ariba, and OpenText.
- Automated processes
- Centralized service type activities in a single place

Out-of-the-box integration components and backend processes enabled to support the Central Finance service

Figure 2.25 shows the shared services framework in a Central Finance context. Central Finance serves as both the corporate reporting and centralization system of the business process.

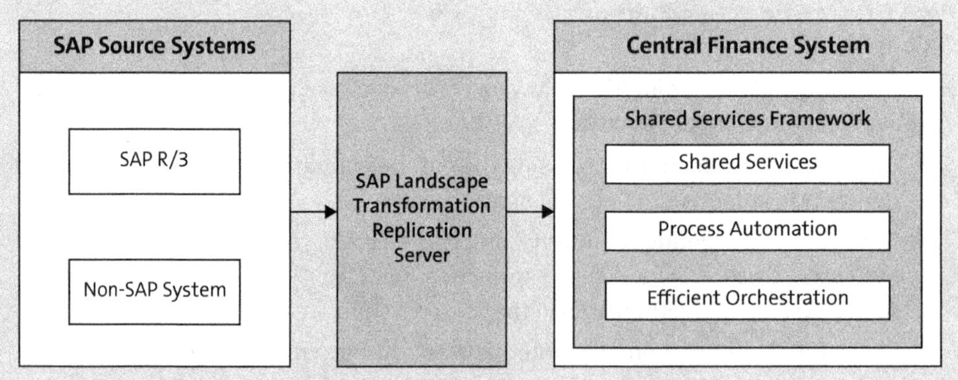

Figure 2.25 Central Finance Shared Services Framework

2.6 Joint Venture Accounting

SAP delivered the initial shipment of joint venture accounting (JVA) document replication in the on-premise SAP S/4HANA version 2020 and onward. With this JVA replication in Central Finance, Central Finance integrates JVA with the following features:

- JVA split results are included with FI document replication.
- Replicated internal JVA documents are posted as the result of multiple JVA period-end processes.
- Depending on the JVA setup in the source system, Central Finance supports internal JVA documents that are posted to classic JVA ledgers such as 4A, 4B, 4C, and 4D, or to the Universal Journal only. Documents aren't posted to FI or CO ledgers.
- Master data replication of JVA, such as joint operating agreements, ventures, equity groups, and master data, is supported.
- The initial load for JVA is supported with Transaction CJCFIL. Only JVA documents and master data are replicated over to the Central Finance system; balance transfer isn't supported.
- Central Finance offers comparison reports for JVA reporting replication.

In this section, we'll discuss the JVA replication scenarios in Central Finance. We'll begin with Central Finance supported scenarios for JVA and then transition into replication of JVA data. At the end, we'll discuss limitations and SAP-delivered comparison reports.

2.6.1 Supported Source System Joint Venture Accounting Scenarios

Central Finance supports the following JVA source scenarios using the Data Replication Framework (DRF), as shown in Figure 2.26:

❶ If the source system is running with SAP ERP with classic general ledger and classic JVA, the system enriches with JVA split information as part of FI documents.

❷ If the source system is running with SAP ERP with new general ledger and new general ledger integrated JVA, the system enriches with JVA information as part of FI documents by reading document splitting table FAGL_SPLINFO.

❸ If the source system is running with SAP S/4HANA release 101 with the general ledger splitter and new general ledger integrated classic JVA, the system enriches with JVA information as part of FI documents by reading document splitting table FAGL_SPLINFO.

❹ If the source system is running with SAP S/4HANA release 104 with general ledger splitter and JVA on the Universal Journal (table ACDOCA), the system enriches with JVA information as part of FI documents by reading document splitting table FAGL_SPLINFO.

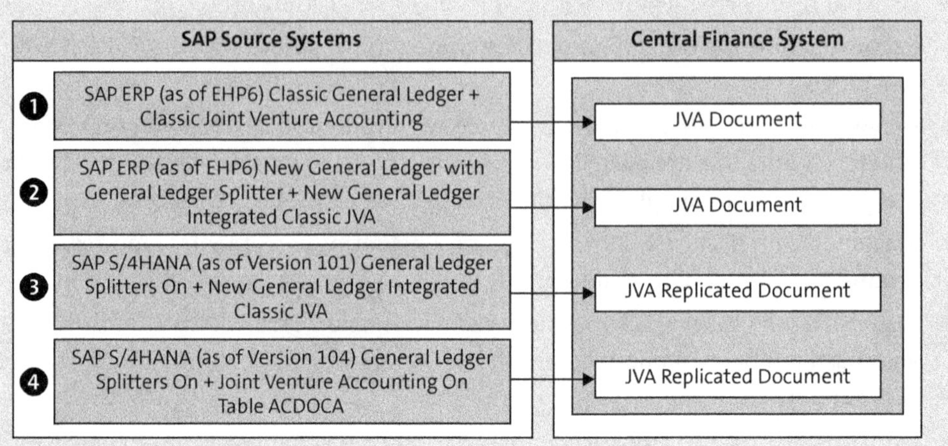

Figure 2.26 Central Finance JVA Replication

2.6.2 Replication of Joint Venture Accounting Processes

Let's understand replication of JVA data in Central Finance:

- **Master data replication**
  To replicate master data, Central Finance uses the DRF Transaction DRFOUT to replicate JVA master data from the source system to the Central Finance system. Master data replication includes both the initial load and ongoing replication. While performing the initial load due to dependencies, SAP recommends you follow a specific sequence. Consult SAP Note 292518 for additional guidance on sequencing.
- **Initial load of JVA replicated documents**
  You'll initiate the initial load of JVA data with Transaction GJCFIL With this transaction, you'll replicate the following:
  - Equity change
  - Equity adjustment
  - Suspense
  - Unsuspense
  - Farm in/out
  - Cutback documents

Tip

Central Finance doesn't support the initial load of balances as of writing this book.

- **Online replication of JVA documents**
  You'll activate online replication of JVA data from Central Finance by configuring the company code. Prior to enabling online replication, you'll replicate master data replication, perform data mapping, and activate JVA in Central Finance.

2.6.3 Limitations

As SAP released the initial shipment of JVA integration with Central Finance and the product is still maturing, SAP recommends reviewing several restrictions prior to considering JVA functionality:

- Central Finance doesn't support *network* cost objects assigned in FI-CO replication. SAP recommends mapping networks to some other cost objects through SAP Master Data Governance mapping or through BAdI enhancements.
- If additional currencies are defined in JVA but not as part of FI, Central Finance doesn't support such scenarios. You need to evaluate for the feasibility of enhancement in Central Finance.
- There is no write-back to the source system for any of the clearing activities performed in Central Finance. This may lead to incorrect cash-based billing.

2.6.4 Comparison Reports

Central Finance provides comparison reports that can be used to reconcile JVA data for any differences between the source system and Central Finance. The comparison reports offer different levels of reports to compare data between the source system and the Central Finance system. Table 2.4 lists the Central Finance to JVA comparison reports.

Area	Comparison Report Transaction Code	Transaction Name	Usage
Master Data	GJCF_V1	DFV Master Data Comparison JOA	To compare joint operating agreement fields
	GJCF_V2	DFV Master Data Comparison Venture	To display differences in venture fields
	GJCF_V3	DFV Master Data Comparison Partner	To display differences in partner fields
Online Document Replication	GJCF_V4	Comparison of JVA Document Count (Ledger 4A)	To show JVA document count difference
	GJCF_V5	Comparison of JVA Document Details (Ledger 4A)	To show differences in JVA document details
	GJCF_V6	Comparison of JVA Billing Count (Ledger 4B)	To show differences in JVA billing count
	GJCF_V7	Comparison of JVA Billing Details (Ledger 4B)	To show differences in JVA billing document details

Table 2.4 JVA Comparison Reports

2.7 Summary

In this chapter you've learned how Central Finance is used as a reporting solution to help businesses implement key central processes by application components. Customers have both choices and flexibility. In general, most projects introduce additional business processes in an incremental way. It's important to understand centralized processes to realize the value it brings to enterprises. We discussed general ledger processes first and then transitioned into other application components such as AR, AP, financial supply chain processes, and so on. Finally, the chapter concluded with an overview of JVA. In the next chapter, you'll learn about Central Finance architecture.

3

Chapter 3
Central Finance Architecture

In the Central Finance landscape, often there is a misconception among users on how data is flowing between systems and which tables are updated during the Central Finance initial load and/or during online replication. In addition, having a good understanding of the replication mechanism and its underlying system architecture is helpful to stakeholders when designing to achieve consolidated reporting and centralized processes. In this chapter, you'll learn what components or systems make up Central Finance for both SAP and non-SAP source systems.

In general, corporations operating in a heterogeneous landscape comprise complex enterprise resource planning (ERP) systems. This often includes both SAP-based and non-SAP-based ERP source systems. In addition, Central Finance integrates various application components within SAP S/4HANA and outside of SAP. Central Finance receives data from various source systems to get a consolidated view of a single source of truth to facilitate group reporting. While integrating these source systems, a solid understanding of Central Finance components is required: SAP Landscape Transformation Replication Server, SAP Application Interface Framework, and the Central Finance accounting interface. A heterogenous system architecture brings numerous challenges to organizations:

- Data integration challenges in getting a unified view from a complex distributed system architecture
- ERP systems in a heterogenous landscape with different release levels which may cause version compatibility
- Challenges in upgrading due to dependencies and complex interfaces
- Standardizing processes when dealing with heavily customized systems
- Integration issues due to incompatible data structures
- Homegrown systems
- Limited transparency between systems
- Long process execution or tasks
- Complex automation and efficiency problems

Central Finance addresses most of these challenges and provides customers with immediate tangible benefits with SAP S/4HANA innovations. A well-built Central Finance architecture provides customer with immediate benefits such as these:

- Real-time replication enables you to achieve harmonized master data, consolidated reporting, and centralized financial processes.
- Achieve a high level of transparency with a single source of truth as a result of harmonized financial data from multiple source systems.
- Central Finance offers customers the chance to quickly adopt SAP S/4HANA features by providing flexibility to introduce additional features in an incremental step or in a rapid approach as requirements demand.
- By streamlining and centralizing business processes, you can achieve greater efficiency in operations and reporting in Central Finance.
- Increased service levels and accuracy of data with timely reporting needs for stakeholders makes for better decision-making.
- By consolidating data into a single instance and process centralization, you can eliminate legacy systems and reduce the cost of maintaining those systems.
- Working capital and related ratios are improved by simplifying cash management processes, collections, and receivables management in Central Finance.
- Most importantly, you'll build Central Finance as a sidecar approach without interrupting existing processes and systems. This provides customers with business continuity while they continue to transform existing processes and systems into Central Finance.

Now that you understand the benefits of a well-architected Central Finance landscape, let's explore Central Finance architectural components next.

3.1 Central Finance Landscape Components

By replicating both financial and nonfinancial data from several source systems, the Central Finance landscape strives to create a centralized financial system to cater to various requirements. The Central Finance system can be configured in multiple ways with different systems. Most common Central Finance implementations contain five components:

- Source systems: SAP-based ERP systems
- Source Systems: Non-SAP-based ERP systems
- SAP Landscape Transformation Replication Server
- SAP Master Data Governance
- Central Finance on SAP S/4HANA

Figure 3.1 illustrates the Central Finance architecture with these components. Let's explore each component and how it helps to achieve logging, replication, error handling, and posting of data in Central Finance from source systems:

❶ **SAP-based source systems**
One of the main systems in the Central Finance architecture is the source system where ongoing day-to-day business activities are carried out in an enterprise. Central Finance promises nondisruption to the source systems is one of the key goals that businesses continue without any disruption or with minimal impact on existing business processes. This gives the opportunity to customers to continue to use the existing system while transitioning to SAP S/4HANA gradually or in a phased approach. To enable Central Finance–relevant functionality in the source systems, first you apply the necessary SAP Notes and relevant configurations based on the release level. As a result of applying SAP Notes, SAP source system tables are equipped with database triggers that ensure real-time data is sent to staging tables for Central Finance replication. In the case of SAP S/4HANA as a source system, Central Finance features are predelivered.

❷ **Non-SAP-based source systems**
Central Finance supports replicating both financial and nonfinancial data from external third-party systems to the Central Finance system. To facilitate replication of third-party systems data, SAP Landscape Transformation Replication Server is delivered with the necessary staging tables. SAP Landscape Transformation Replication Server identifies data on the header table within SAP Landscape Transformation Replication Server for the trigger to replicate third-party data into SAP.

❸ **SAP Landscape Transformation Replication Server**
This is a critical component in the Central Finance landscape that reads data from logging tables in the source systems and transforms that data to the Central Finance accounting interface for further processing. In the case of third-party systems, SAP Landscape Transformation Replication Server reads data from the header staging table within SAP Landscape Transformation Replication Server.

❹ **SAP Master Data Governance**
Central Finance is integrated with SAP Master Data Governance to access mapping tables without getting a fully licensed SAP Master Data Governance suite. The streamlined SAP Master Data Governance version is light in functionality compared to a fully licensed SAP Master Data Governance. With this version, SAP Master Data Governance mapping tables are available for Central Finance replication scenarios without installing the full suite of SAP Master Data Governance. On the other hand, SAP offers a fully featured SAP Master Data Governance tool (requires a separate license) as an option to manage and replicate master data across the enterprise. Alternatively, customers have an option to adopt other master data solutions to leverage full-scale benefits of master data functions per their business requirements.

❺ **Central Finance system**
This component within SAP S/4HANA is installed on the SAP HANA database (high-performance analytics appliance). SAP HANA is built on an in-memory database architecture that offers customers robust database performance and advanced ana-

lytics capabilities. Transferred data from the source systems goes through a series of business mapping and transformations in SAP Landscape Transformation Replication Server. Next, business mapping is performed with SAP Master Data Governance mapping tools. Central Finance uses the SAP Application Interface Framework, which handles errors and corrections prior to processing and posting data. Finally, for financial data, the internal accounting interface is called to post financial data to the Universal Journal (table ACDOCA). Central Finance as a system can be leveraged to serve as a financial hub to integrate solutions not only within the SAP S/4HANA system but also outside the system. In addition, Central Finance benefits from analytic and planning platform tools such as SAP Analytics Cloud and SAP Business Warehouse (SAP BW), which are offered with SAP S/4HANA.

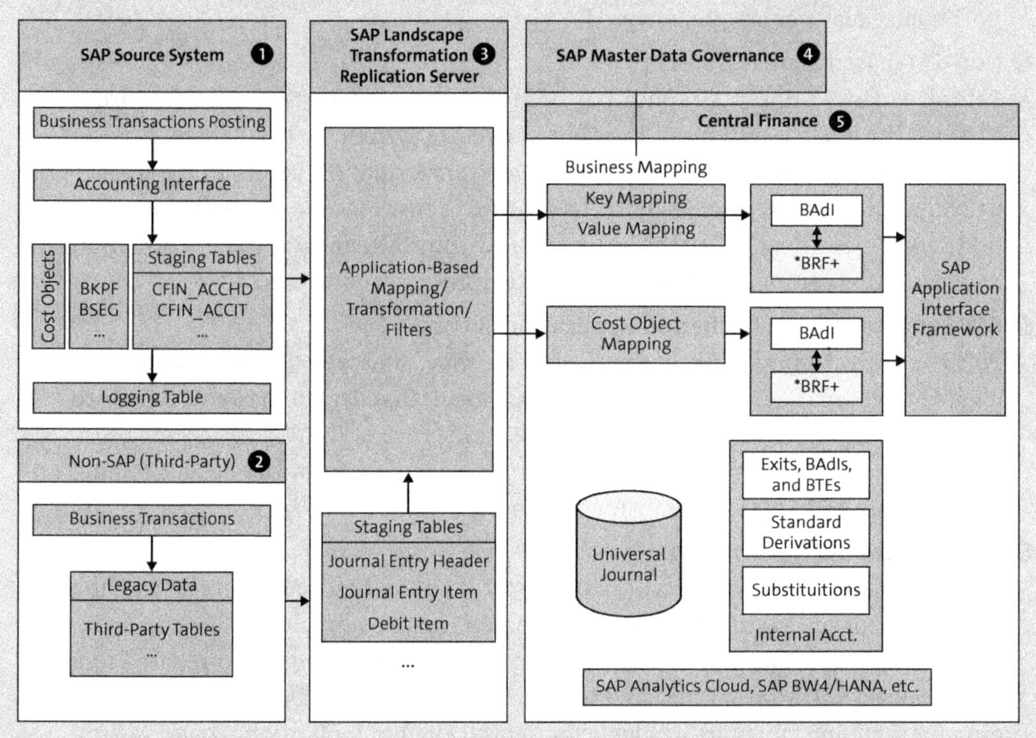

Figure 3.1 Central Finance Architecture

[»]

Note

The Central Finance architecture in Figure 3.1 is illustrated with SAP Landscape Transformation Replication Server. Besides SAP Landscape Transformation Replication Server, Central Finance also supports replicating using application link enabling (ALE) IDocs technology, for example, work breakdown structure (WBS) and project structures. Note that journal entries based on ALE business suite IDoc types FIDCCP01, FIDC-

CP02, FAGLDT01, or CODCMT01 from SAP ERP source systems aren't supported and aren't compatible with SAP S/4HANA. Consult SAP Note 2313879 for additional guidance on this topic. In Central Finance, replication of journal entries goes through SAP Landscape Transformation Replication Server as part of the standard Central Finance product scope.

Now that you've learned the high-level architecture of the Central Finance landscape, let's explore these components further.

3.1.1 Source Systems

Central Finance supports most of the SAP releases to replicate both financial and non-financial data to Central Finance. In certain releases, you'll carry out prerequisites to fulfill Central Finance integration requirements. Figure 3.2 illustrates Central Finance–supported SAP-based source systems and non-SAP-based source systems. Let's explore each of the SAP-supported releases with additional details:

❶ **SAP S/4HANA as a source system**
Central Finance supports on-premise SAP S/4HANA, SAP S/4HANA Cloud Private Edition, and SAP S/4HANA Cloud Public Edition as the source systems. The SAP S/4HANA source systems come with Central Finance integration components preinstalled. You still need to review any prerequisites that are missing for the required functionality you're evaluating for the Central Finance scenario.

❷ **SAP ERP 6.0 as a source system**
When SAP ERP 6.0 is the source system to support Central Finance, you'll identify missing SAP Notes and dependent SAP Notes from the SAP Support Portal and apply them with relevant manual activities to bring the system to support Central Finance. Depending on the service pack level, missing SAP Notes take significant time to implement. Alternatively, a service pack upgrade may bring the missing SAP Notes to a minimum. On the other hand, a service pack upgrade may introduce new features that require extensive regression testing and may add additional time and effort to both the business and IT staff. Evaluate the options by considering project schedule and other factors.

[«]

Note
Consult SAP Note 2323494 to understand prerequisite SAP Notes for SAP ERP 6.0.

❸ **SAP 4.6, 4.7, 5.0**
For lower SAP releases from SAP R/3 4.6c to SAP ERP 5.0 as the source systems, SAP recommends you create an incident under the FI-CF component. In these situations, you'll carry out additional project work to integrate lower releases of SAP source sys-

tems to the Central Finance system by *downporting*. Essentially, new features and fixes are applied to lower versions to bring the system to support Central Finance. This is normally done with the SAP consulting service. This is helpful for customers who can't upgrade to the latest version for some reason, yet they want to benefit from Central Finance innovations.

Note

SAP provides consulting SAP Note 2279674. However, this SAP Note isn't released to all customers. You need to create an SAP incident to get it released to the customer.

❹ **SAP Business ByDesign, SAP Business One, and non-SAP systems**
As shown in Figure 3.2, for all three systems, Central Finance replication is supported using an SAP Landscape Transformation Replication Server staging table. You'll build migration tools to bring data from these systems into the SAP Landscape Transformation Replication Server staging table.

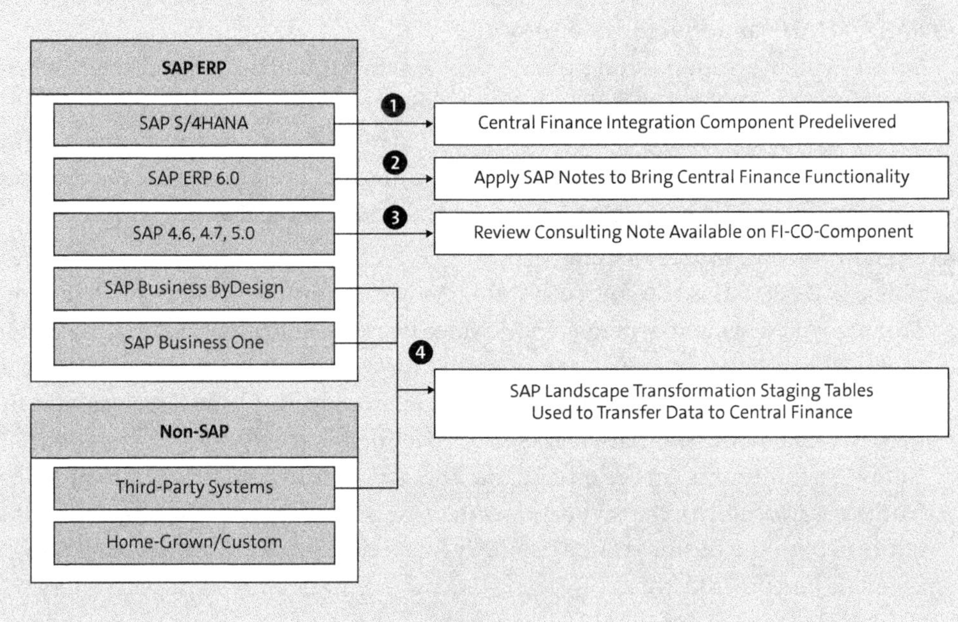

Figure 3.2 Supported SAP-Based Source Systems

Note

Likewise, consulting SAP Notes 2610660, 2713300, and 2300443 are available to understand SAP Landscape Transformation Replication Server staging tables and prerequisite SAP Notes for these scenarios.

3.1.2 SAP Landscape Transformation Replication Server

An essential and important component in the Central Finance landscape is SAP Landscape Transformation Replication Server. From a Central Finance context, application data from the source system transferred to Central Finance with SAP Landscape Transformation Replication Server's mapping and transformation engine. Figure 3.3 illustrates SAP Landscape Transformation Replication Server capabilities considering a Central Finance implementation:

❶ **Integration with the source systems (initial load and real-time data transfer from multiple source systems)**
SAP Landscape Transformation Replication Server is involved in the initial load and real-time transfer of data from the source systems. First, SAP Landscape Transformation Replication Server triggers the access plan calculation jobs in the source system. These calculation jobs pull the data from the source system depending upon the advanced filter setting in Transaction LTRS and transfer the data to the SAP Landscape Transformation Replication Server. In addition, SAP Landscape Transformation Replication Server creates the insert, update, and delete database triggers on application tables at the source system. SAP Landscape Transformation Replication Server also generates the SAP Landscape Transformation Replication Server logging table for each application table in the source system, which captures any insert and update done on the application table at the source system. The collected data from various source systems is then transferred to Central Finance using SAP Landscape Transformation Replication Server.

❷ **Mapping and transformation**
SAP Landscape Transformation Replication Server comes with rich features of mapping and transformation capabilities to transform source data. You'll apply the necessary mapping and transformation rules to convert data, populating additional information, combining fields, and filtering out unnecessary data based on project requirements. Custom logic can be written in SAP Landscape Transformation Replication Server prior to sending data to Central Finance by SAP Landscape Transformation Replication Server's write engine.

❸ **Continuous replication**
SAP Landscape Transformation Replication Server not only supports initial transfer of data but also delta replication and any changes to original data. This will result in data being captured in Central Finance accurately and with up-to-date information.

❹ **Integration with Central Finance**
SAP Landscape Transformation Replication Server integrates with Central Finance interfaces to transfer both financial and nonfinancial source system data. There are standard application interfaces for each Central Finance object that transfer the data to the Central Finance system through SAP Landscape Transformation Replication Server.

❺ **Monitoring**
SAP Landscape Transformation Replication Server offers monitoring and troubleshooting capabilities and improves the data replication process performance.

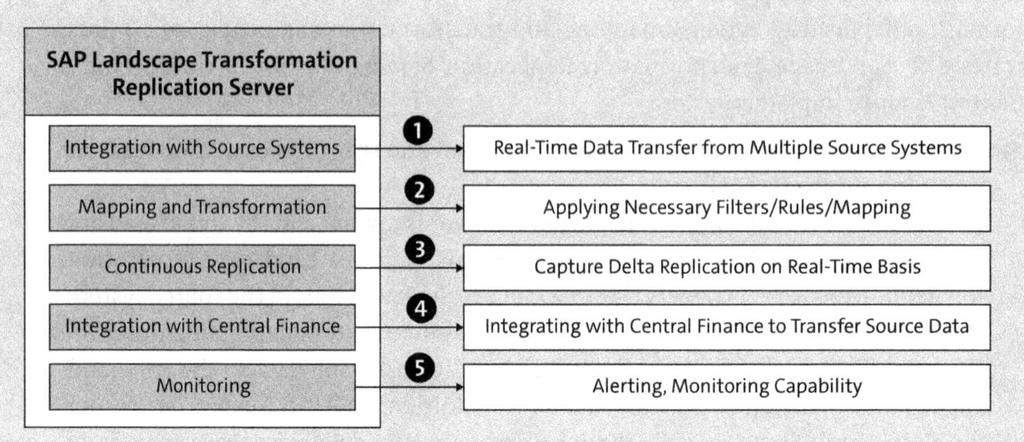

Figure 3.3 Functions of SAP Landscape Transformation Replication Server

Table 3.1 shows supported replication scenarios in SAP Landscape Transformation Replication Server for Central Finance scenarios.

Area	Replication Scenarios
Financial Accounting	FI-CO postings, VAT tables, withholding tax tables, intermediate data retention (IDR)
Controlling	CO internal postings
	Cost objects
	Commitments
	EC-PCA postings
	Material cost estimates
	Activity rates
Logistics	Accounting view of logistics (AVL)

Table 3.1 SAP Landscape Transformation Supported Replication Scenarios for Central Finance

[»]

Note
Consult SAP Note 2154420 for comprehensive information on trigger-based replication for Central Finance.

3.1.3 Central Finance System

Central Finance delivered on the SAP S/4HANA system uses the SAP HANA database. As shown in Figure 3.4, the Central Finance system comprises several subcomponents to support replication scenarios:

❶ **Business mapping**
Central Finance requires a system landscape directory (SLD) to integrate both the source system and target system information. Business mapping for key and value mapping requires an SLD assigned in Central Finance. In Central Finance, SAP foundation functions are called to perform business mapping. Optionally, custom rules can be mapped with business-rule mapping offered with the Business Rules Framework plus (BRFplus) tool.

❷ **SAP Application Interface Framework**
In Central Finance, IT support and key business users use SAP Application Interface Framework to monitor and perform error resolution prior to posting replicated data from the source systems. SAP Application Interface Framework interfaces and versions are specific to Central Finance for various replication scenarios.

❸ **Cost object framework interface**
The cost object framework transfers cost object master data from the source system to Central Finance. When cost objects are transferred from the write engine from SAP Landscape Transformation, the cost object framework interface is called to perform business mapping and eventually create master data in table AUFK in Central Finance.

❹ **FI and CO interface**
For financial data, the system calls the FI and CO interfaces to post transferred data to Central Finance.

❺ **Universal Journal**
FI-CO replicated data resides in the Universal Journal (table ACDOCA) and serves as a single source of truth for consolidated financial reporting and to execute financial processes centrally.

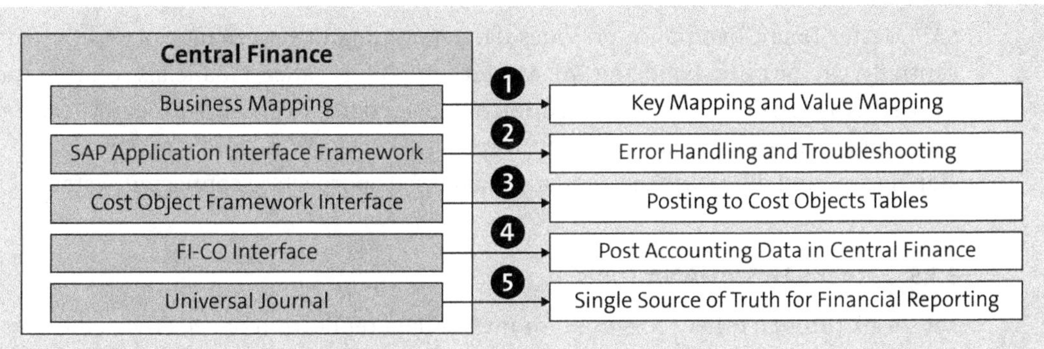

Figure 3.4 Central Finance Components

3.1.4 SAP Master Data Governance

Central Finance is integrated with SAP Master Data Governance for business mapping prior to posting replicated data in Central Finance. SAP Master Data Governance tables store mapping of master data for key mapping, value mapping, and cost object framework mapping information. You'll maintain master data mapping per project requirements in Central Finance. Central Finance uses the SAP Master Data Governance foundation function to check and maintain business mapping. Figure 3.5 illustrates how key master data can be harmonized and distributed with SAP Master Data Governance. In our example, first, the customer master data is created with SAP Master Data Governance and distributed in Central Finance ❶, and then it's created in the source ERP system ❷. Once the master data is available in Central Finance and the source system, key mapping is used to maintain customer master mapping in the Central Finance system ❸. During financial postings, the system reads key mapping and posts financial documents in the Central Finance system ❹.

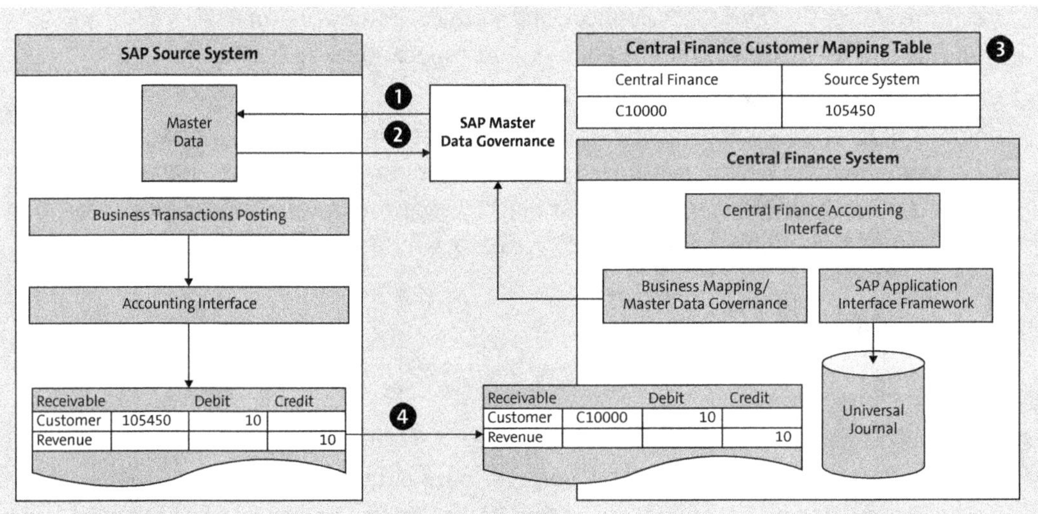

Figure 3.5 Master Data Key Mapping Architecture

SAP Master Data Governance provides data creation and governance of master data centrally. On the other hand, the SAP Master Data Governance, Financials solution can be used as a local or centralized execution of master data governance tasks. SAP Master Data Governance, Financials allows customers to create and maintain master data in a staging area and distribute master data in a phased approach to connected systems.

3.1.5 Other SAP S/4HANA Tools

Let's walk through other tools used for master data replication or selective data transmission:

- **IDocs**
  SAP's IDocs solution can be used to exchange data between different systems using the ALE technology. In Central Finance scenario, IDocs can transfer master data, such as material master data, business partners, general ledger accounts, and so on, from the source systems to the target Central Finance system. One advantage of IDocs is its own built-in error-handling feature with standard delivered functionality. SAP offers various message types for different master data objects. Moreover, IDocs can be enhanced for customer-specific requirements and be automated on a scheduler.
- **SAP Data Services**
  SAP Data Services offers data transformation services for transferring data from the source systems to a Central Finance system. It's an extract, transform, and load (ETL) tool to facilitate data transformation services. In addition, SAP Data Services requires a separate license. Most companies use SAP Data Services to migrate data from an old system to a new system. As mentioned earlier, Central Finance only replicates cost objects and project systems master data. Other master data such as customers, vendors, and cost centers aren't supported out of the box. For large volumes of data, you must build interfaces using the SAP Data Services technology, which comprises repositories and transformation rules to transfer master data to the Central Finance system or any other system. The relevancy rules vary by requirement. Some key benefits of SAP Data Services are listed here:
  - Supports SAP Application Interface Framework error monitoring
  - Can be used as an alternative to SAP Master Data Governance
  - Supports non-SAP ERP systems, including third-party applications
  - Automates data transformation from different source systems
  - Provides data profiling and data enriching capabilities
  - Dedupes master data to eliminate duplicate data from various source systems
  - Provides dashboards to view data from various sources
- **Migrate Your Data app**
  SAP offers the Migrate Your Data app (app ID F3473), which is a migration cockpit, to migrate data from the source systems or staging tables to the Central Finance system. Transaction LTMC (Legacy Transfer Migration Cockpit) used to create master data and transaction data in SAP S/4HANA is depreciated from SAP S/4HANA 20221 onward. Consult SAP Note 3355075. The app comes with templates and walks you through a sequence of steps to upload master data in SAP. This process requires manual intervention to migrate master data.
- **Other third-party tools**
  You can choose any third-party master data management hub, such as IBM Infosphere or insightsoftware (formerly Magnitude) SourceConnect. These products offer solutions for governing master data, including harmonizing and replicating

master data. insightsoftware SourceConnect provides customers with a master data replication functionality from both source SAP and non-SAP systems to a Central Finance system.

3.2 Data Flow in Central Finance for SAP-Based Source Systems

There is often confusion among the Central Finance user community on when Central Finance uses SAP Landscape Transformation Replication Server or remote function calls (RFCs) to transfer data. Table 3.2 displays important replication objects and methods used during the following:

- Smoke test
- Initial load
- Real-time replication

Object	Smoke Test		Initial Load		Real-Time Replication	
	Connection	Error Monitoring	Connection	Error Monitoring	Connection	Error Monitoring
AUFK	RFC	Application log	SAP Landscape Transformation Replication Server	SAP Application Interface Framework	SAP Landscape Transformation Replication Server	SAP Application Interface Framework
FI-CO			RFC	Application Log	SAP Landscape Transformation Replication Server	SAP Application Interface Framework
COBK	RFC	Application Log	SAP Landscape Replication Server	SAP Application Interface Framework	SAP Landscape Transformation Replication Server	SAP Application Interface Framework
Non-SAP			SAP Landscape Transformation Replication Server	SAP Application Interface Framework	SAP Landscape Transformation Replication Server	SAP Application Interface Framework

Table 3.2 Architecture Overview for Data Flow

[«]

Note

SAP Central Finance adoption content comes with architecture diagrams that will help you understand how data flows for different replication objects.

Now, that you've learned the basic concepts behind SAP Landscape Transformation Replication Server and RFC methods, let's discuss the FI initial load data flow.

3.2.1 Financial Accounting-Controlling Initial Load

The Central Finance initial load for FI works differently than online replication. For the classic FI initial load, you begin the FI-CO initial load from the Central Finance system. The system extracts the data from the source system into staging tables CFIN_ACC* in Central Finance via RFC. Upon successful extraction, you'll work on the data to resolve any mapping and posting errors prior to posting into Central Finance. Figure 3.6 illustrates the FI-CO initial load. The following describes how the data flows between the systems:

❶ **Beginning the initial load extraction**
You'll begin the initial load extraction process from the Central Finance system. Upon executing the extraction program in Central Finance, the system makes an RFC to the source system.

❷ **Extracting finance data from the source system**
In the next step, the system reads accounting tables in the source system. Depending on the source system, for SAP ERP, the initial load extraction program reads accounting tables BKPF and, BSEG, and dependent tables and starts grouping them into multiple packages. The data selection criteria used is based on what's maintained in the general replication settings (VCFIN_SOURCE_SET) in the source system. The package size specified in this setting is used for optimum performance. Ensure that the system extracted all packages before simulating the data.

❸ **Updating staging tables in Central Finance**
Packaged information transferred from the source system is updated in Central Finance staging tables CFIN_ACC*.

❹ **Business mapping**
SAP Master Data Governance foundation is used to maintain and perform business mapping. During posting, business mapping is applied prior to posting.

❺ **Posting to the Universal Journal**
After resolving errors, documents are posted to the Universal Journal in the Central Finance system with the initial load posting. SAP Application Interface Framework isn't used in FI-CO initial load simulations or posting.

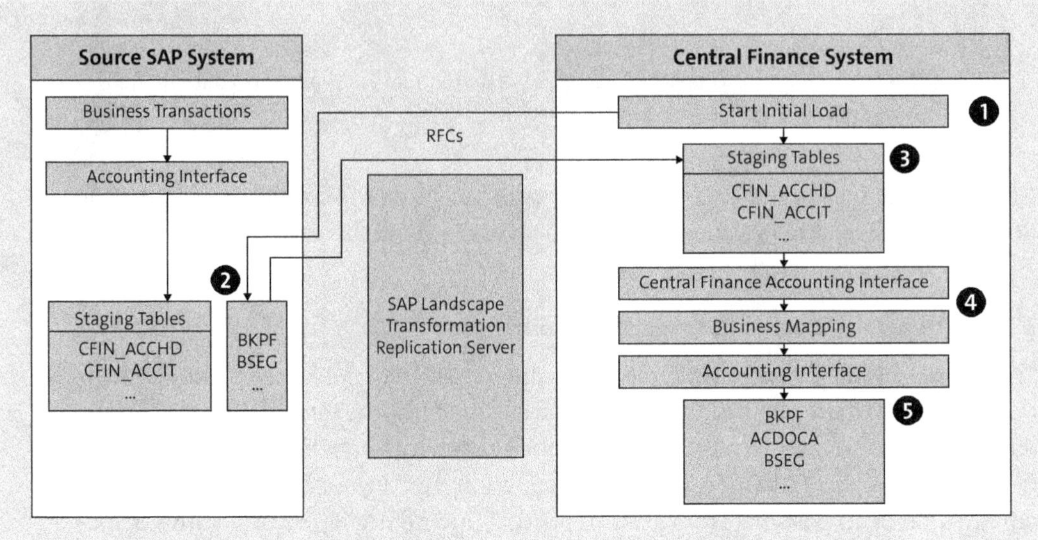

Figure 3.6 FI-CO Initial Load Data Flow: SAP-Based Source System

3.2.2 Replication through SAP Landscape Transformation Replication Server

As shown in Figure 3.6, the classic finance initial load of FI-CO documents uses a direct RFC connection to transfer data from the source system to Central Finance. On the other hand, online or real-time replication uses SAP Landscape Transformation Replication Server to transfer the source system data to Central Finance. Figure 3.7 shows the real-time replication scenario. Let's understand the data flow of online replication using SAP Landscape Transformation Replication Server:

❶ **Initial load and log of migration status**

Once the initial load is extracted for the initial load, a log of the migration status table (table CFIN_MIG_LOG) in the source system populated with the company code and timestamp. Any new financial postings in the source system are recorded in the source staging tables (table CFIN_ACC*) for online replication.

❷ **SAP Landscape Transformation Replication Server**

SAP Landscape Transformation Replication Server pulls the data from the source system depending upon the advanced filter setting in Transaction LTRS and transfers the data to the SAP Landscape Transformation Replication Server. In addition, SAP Landscape Transformation Replication Server creates the insert, update, and delete triggers on the application table at the source system. SAP Landscape Transformation Replication Server also generates logging table for each application table in the source system, which captures any insert and update done on an application table at the source system. The collected finance data from the source system is then transferred to Central Finance using SAP Landscape Transformation Replication Server.

❸ **Business mapping**
SAP Master Data Governance foundation is used to maintain and perform business mapping. During posting, business mapping is applied prior to posting source data into Central Finance.

❹ **SAP Application Interface Framework**
Central Finance integrated with SAP Application Interface Framework to offer business users error monitoring and resolution capabilities. SAP Application Interface Framework provides a structured way of displaying messages generated during the transfer of data from the source system to Central Finance. SAP Application Interface Framework provides robust selection criteria such as namespace, interface, creation date, and time.

❺ **Posting to the Universal Journal**
After resolving master data or configuration errors, documents are posted to the Universal Journal in the Central Finance system.

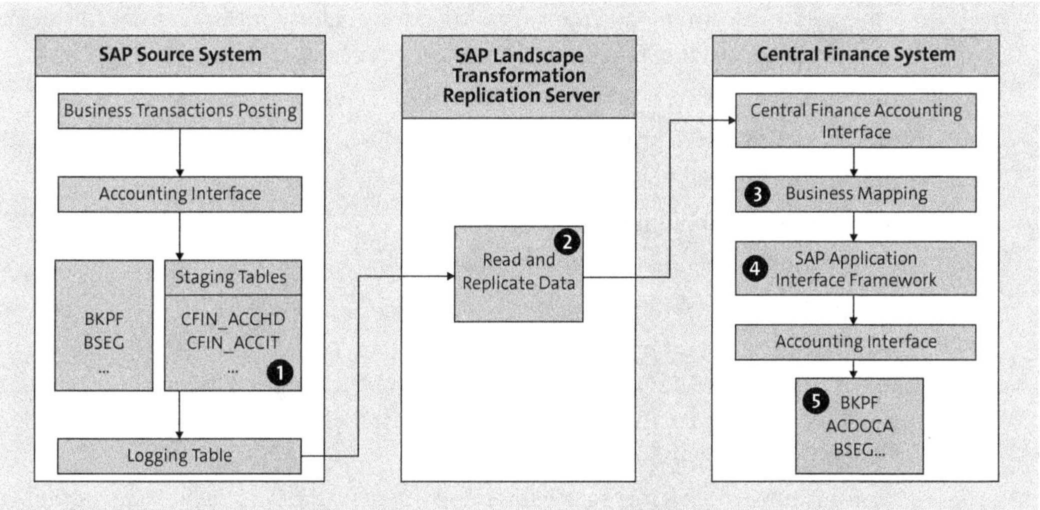

Figure 3.7 SAP Landscape Transformation Replication Server Architecture

3.3 Third-Party System Integration with the Central Finance System

Central Finance supports replicating journal entries and changes to journal entries from a third-party system to Central Finance using a third-party interface. In addition, AVL is supported from the SAP S/4HANA 2021 release. If you're on SAP S/4HANA 1909 or 2020, SAP recommends applying SAP Note 3063148. In this section, we'll discuss the following important aspects of third-party system integration with Central Finance:

- Third-party replication scenarios supported in Central Finance
- Limitations of third-party scenarios

- Requirements for setting up a third-party system
- Architecture of data flow in third-party system integration
- External bolt-on software for third-party integration scenarios

We'll discuss each scenario next.

3.3.1 Supported Third-Party Replication Scenarios

The third-party interface supports replicating three types of data in Central Finance:

- **Third party system interface for posting financial data**
  You can replicate both journal entries and reversals from a third-party system to the Central Finance system continuously:
  - General journal entries, including source references, segment, profit center, division, and trading partner
  - Customer or vendor open item, including references to source references, segments, profit centers, divisions, trading partners, discounts, payment information, indirect product taxes and withholding taxes
  - Reversal documents with reference to unique reference and reversal reason
  - Journal entries with profitability characteristics
  - Replicating invoice clearing information
  - Replicating withholding tax information
  - Replicating FI documents into multiple ledgers
  - Support of extensibility
- **Third-party system interface for changing FI data**
  With this interface, you can replicate journal entry changes from a third-party system to the Central Finance system continuously. The interface supports changes originating from the following:
  - Journal entries, accounts receivables documents, accounts payable documents
  - Reversal or cleared documents
  - Changes made to withholding information
- **Third-party system interface for replicating AVL**
  You can replicate selective logistics data to the Central Finance system.

3.3.2 Limitations of Third-Party Replication to Central Finance

Although third-party replication is supported in Central Finance, there are many limitations you should be aware of:

- Tax consistency checks aren't supported for third-party systems interface.
- Not all comparison reports are available for the third-party system interface. Only two transactions—Transaction FINS_CFIN_DFV_FI_NUM and Transaction

FINS_CFIN_DFV_FI_DOC—are available. These reports compare data from the Universal Journal and the accounting document cluster table BSEG to SAP Application Interface Framework tables:
  - CFIN_EX_AIFROOT
  - /AIF/STD_IDX_TBL
  - CFIN_EX_AIFACCTI
  - CFIN_EX_AIF_IDX
- SAP doesn't offer any standard reports or tools to clean up or reset SAP Landscape Transformation Replication Server staging table data.
- No standard tools are available to load SAP Landscape Transformation Replication Server staging tables. Customers have to build their own tools to bring third-party system data into the SAP Landscape Transformation Replication Server staging tables.
- No profitability analysis mapping tool is available similar to costing-based profitability analysis mapping tool for SAP-based source systems.
- Adding new fields to existing structures requires additional code via BADI_FINS_EX_INTF.
- No data profiling or downloading tools are available for third-party data types to SAP Landscape Transformation Replication Server structures.
- Classic initial load or IDR aren't available for the third-party system interface.
- Deferred tax replication isn't supported.
- The document splitting functionality isn't supported in the third-party system interface.
- The reverse and repost function (Transaction FINS_CFIN_CREV) isn't supported for the documents from the third-party system interface to Central Finance.

3.3.3 Setting Up Third-Party System Integration

In this section, we'll explain the important prerequisites you'll have to consider when handling third-party system integration with Central Finance.

Financial Data Structure and Supporting Fields

To replicate third-party system data, inbound staging tables are created first in the SAP Landscape Transformation Replication Server to load the data from the third-party system. SAP has provided the utility program to create the inbound staging tables that contain the accounting document data loaded from the third-party system to the Central Finance system. Table 3.3 shows SAP-delivered staging tables for third-party financial data replication.

Staging Table Name	Description	Purpose
/1LT/CF_E_HEADER	Document header	Stores fields related to header data of accounting document
/1LT/CF_E_ACCT	Accounting items	Item data of the accounting document
/1LT/CF_E_DEBIT	Debtor items	Item data of the customer document
/1LT/CF_E_CREDIT	Creditor items	Item data of the supplier document
/1LT/CF_E_PRDTAX	Product tax items	Product tax data
/1LT/CF_E_WHTAX	Withhold tax items	Withholding tax data fields replication
/1LT/CF_E_EXTENT	Customer extension	Additional customer-specific data for the journal entry
/1LT/CF_E_COPA	Costing-based profitability analysis characteristics	Characteristics in costing-based profitability analysis
/1LT/CF_E_CLRITM	Clearing items	Clearing items-relevant fields replication
/1LT/CF_E_CLRWHT	Withholding tax items within clearing	Replicating withholding tax information within clearing

Table 3.3 Staging Tables for Third-Party Financial Data Replication

Table 3.4 shows SAP-delivered staging tables for third-party financial data *changes* replication.

Staging Table Name	Description	Purpose
/1LT/CF_E_HEADER	Document header	Stores fields related to header data of accounting document
/1LT/CF_E_ACCT	Accounting items	Item data of the accounting document
/1LT/CF_E_DEBIT	Debtor items	Item data of the customer document
/1LT/CF_E_CREDIT	Creditor items	Item data of the supplier document
/1LT/CF_E_WHTAX	Withhold tax items	Withholding tax data fields replication
/1LT/CF_E_DOCCHG	Document changes	Changed relevant data replication
/1LT/CF_E_EXTCHG	Customer extension	Fields that are customer-specific fields

Table 3.4 Replicating Third-Party System Document or Journal Changes

[+]

Tip

SAP recommends implementing SAP Notes based on release, which add staging tables and fields required for FI data replication on the SAP Landscape Transformation Replication Server for both posting financial data and changes coming in for financial data. Table 3.5 lists SAP Notes that provide guidance on installing SAP Landscape Transformation Replication Server tables for third-party system interfaces for Central Finance.

Release	SAP Note
SAP S/4HANA 2023	3340008
SAP S/4HANA 2022	3169099
SAP S/4HANA 2021	3067156
SAP S/4HANA 2020	2922300
SAP S/4HANA 1909	2713300
SAP S/4HANA 1809 and 1709 FPS 2	2610660
SAP S/HANA 1709 FPS 1	2527831
SAP S/4HANA 1610 FPS2	2462424

Table 3.5 Third-Party Installation SAP Notes

Extraction of Source Data from a Third-Party System to SAP Landscape Transformation Staging Table

Extracting and uploading the data from a third-party source system to SAP Landscape Transformation staging tables isn't delivered as part of the Central Finance standard solution. You're responsible for extracting data from a third-party system and populating the staging tables at SAP Landscape Transformation using business object data services (BODS), partner product insightsoftware, or any other similar solution. There could be cases where a third-party accounting data field isn't aligned with the corresponding SAP data element. In such cases, data conversion should be done before loading the data into inbound staging table as per SAP data element standards.

[«]

Note

SAP Business One and SAP Business ByDesign are supported via the SAP Landscape Transformation staging tables approach similar to the third-party staging tables.

Now that you have a fair understanding of SAP Landscape Transformation staging tables and populating the tables for third-party financial data replication, let's discuss the architecture of third-party replication in Central Finance.

3.3.4 Architecture of Third-Party System Integrating with Central Finance

In a third-party interface integrating with Central Finance, you'll commonly have three systems. Figure 3.8 illustrates the data flow from a third-party system to SAP Landscape Transformation and then to Central Finance. Let's discuss each component next:

❶ **Third-party system**
Once the initial load is extracted, the log of migration status table (table CFIN_MIG_LOG) in the source system is populated with the company code and timestamp. Any new financial postings in the source system are recorded in the source staging tables (table CFIN_ACC*) for online replication.

❷ **SAP Landscape Transformation**
In SAP Landscape Transformation, database triggers are created on the header table /1LT/CF_E_HEADER, so accounting data header table /1LT/CF_E_HEADER should be inserted with accounting data at the end. Take care if a custom tool is built to load the third-party data into the inbound staging tables. After loading accounting data into inbound staging tables, the tables are transferred into Central Finance in two separate phases:

- SAP Landscape Transformation initial load to transfer the legacy data by adding table /1LT/CF_E_HEADER by choosing data provisioning option **Start Load** or **Start Replication**. Read the data from inbound staging tables and transfer to the Central Finance system by setting up the parallelization.
- SAP Landscape Transformation Replication Server is used to transfer the accounting documents daily to the Central Finance system. Here, a database trigger is being created. There is an option to disable the delete and update trigger as well if it's required. In addition to database triggers, a logging table is also created for table /1LT/CF_E_HEADER in the SAP Landscape Transformation system. Any new data loaded into inbound staging tables will transfer to the Central Finance system in replication mode. The read engine reads the data from the logging table and transfers the data into the Central Finance system. There are options to transform data at SAP Landscape Transformation Replication Server before loading to the Central Finance system or filtering the required data per the project scope.

❸ **SAP Application Interface Framework**
Central Finance integrated with SAP Application Interface Framework to offer business users error monitoring and resolution capabilities. SAP Application Interface Framework provides a structured way of displaying messages generated during transfer of data from the source system to Central Finance. The SAP Application Interface Framework tool provides robust selection criteria such as namespace, interface, creation date and time. With emergency correction mode, you can update underlying source data prior to posting.

❹ **Business mapping**
SAP Master Data Governance foundation is used to maintain and perform business

mapping. During posting, business mapping is applied prior to posting source data into Central Finance.

❺ **Posting to the Universal Journal**
After resolving master data or configuration errors, documents are posted to the Universal Journal in the Central Finance system.

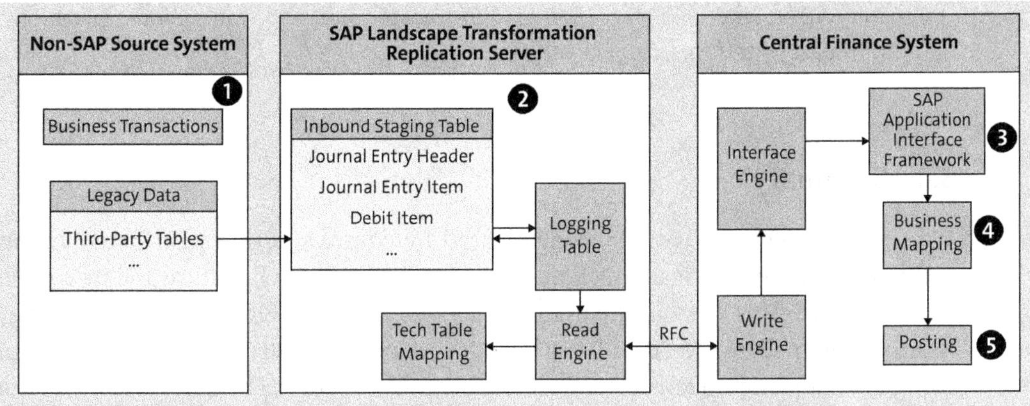

Figure 3.8 Non-SAP-Based System Architecture

[+]

Tip

SAP doesn't support third-party interfaces for SAP-based source systems for the following reasons:

- To avoid risks involved with data inconsistencies, compliance, and reconciliation issues
- To avoid any maintenance conflicts, the standard interface and third-party interface maintenance cycles
- To prepare for upcoming future functions of SAP S/4HANA functions readiness
- To prevent incompatibilities

3.3.5 External Bolt-On Software for Third-Party Source Systems

Third-party system integrator insightsoftware offers finance master data and transaction replication features for a third-party system connecting to Central Finance. The product offers prebuilt extractors that feature the following:

- Data load to SAP Landscape Transformation Replication Server staging tables
- Data profiling
- Data mapping
- Data transformation

- Data consolidation
- Initial load of master data replication and ongoing replication of master data
- Master data harmonization when multiple source systems are involved

Tip

Consult SAP Notes 3396547 and 3396547 for master data and transaction replication for third-party interfaces, respectively.

3.4 Summary

We began the chapter with challenges faced by companies in a distributed system architecture. Later, you learned about the Central Finance architecture and its components in detail. This lays out the foundation of your Central Finance learning journey. Next, you learned how the data flows for FI initial load using RFC and online replication using SAP Landscape Transformation Replication Server for FI-CO document replication, respectively. We also delved into the architectural components of third-party interface components. In addition, we discussed SAP limitations on connecting to a third-party interface in Central Finance. The building blocks of central architecture will help you understand the following chapters. In the next chapter, you'll learn the master data aspects in the Central Finance implementation scenario.

Chapter 4
Master Data Management

In general, master data management (MDM) activities in any organization are the most time-consuming. It involves creating, maintaining, and troubleshooting master data–related issues to facilitate accurate transactional postings. Central Finance relies heavily on accurate master data to replicate transactional data from the source system to Central Finance. As data is transferred from various source systems, companies face a wide variety of challenges to harmonize master data and streamline master data processes. We'll cover solutions in this chapter.

Although the complexity involved with creation and maintenance of master data exists in all enterprises, companies with heterogenous landscapes have additional challenges with synchronizing master data across systems. Central Finance most often is a heterogenous landscape connecting multiple source systems. To facilitate accurate reporting and smooth replication of data from the source systems to Central Finance, it's imperative that you streamlines master data processes for creating, maintaining, and replicating master data to Central Finance. Inefficient master data processes or nonharmonized master data can lead to problems such as incorrect transactional posting in Central Finance, inaccurate reporting or qualify of reporting, delays in period-end and year-end closing, incomplete or inconsistent data between connected systems, time-consuming manual master data activities and manual correction activities, and nonstandardized processes or nonstreamlined processes.

As a result of these anomalies, there is a strong appreciation for having standardized processes for creation and maintenance of master data in Central Finance. To ensure consistent master data across different systems, stakeholders want to harmonize master data between different source systems and the Central Finance system. A harmonized master data set is a common master data structure with a consistent view of key data for current and future systems connected to the Central Finance system. At the end, you need well-defined master data with the following:

- Master data setup that supports Central Finance replication with minimal interruption
- Security roles and relevant SAP Fiori apps to support the master data strategy
- Consistent master data maintained in the enterprise landscape

- Maintenance of master data with appropriate workflow approvals in place
- A golden record that represents the source of truth
- All fields available to support the initial load and replication aspects in Central Finance
- Master data tools and templates to support Central Finance operational activities
- Error-resolution content for SAP Application Interface Framework setup and available readily to resolve master data–related errors
- Well-defined process to update key mapping and value mapping in Central Finance

Now that you understand the desired master data goals in Central Finance, let's consider the master activities performed in an enterprise. At a broader level, multiple activities are involved in syncing master data across the system landscape. Let's explore typical activities required in master data management (MDM):

- **Profiling**
  Data quality is an important aspect of accurate reporting in any enterprise system. Data profiling is a process of examining existing data and its dependencies, thereby exploring opportunities to improve data quality. In addition, you'll understand data quality pain points for cleansing for duplicates, uniformity, and correction opportunities prior to extraction of data for Central Finance.
- **Governance**
  With governance, you'll have a streamlined process to create and manage master data across the enterprise. This includes obtaining necessary approvals from stakeholders to create and maintain master data via workflows to manage master data lifecycles. Moreover, the governance establishes an explicit ownership of master data and transparent processes, thereby eliminating confusion among stakeholders.
- **Process**
  You'll create and maintain master data centrally to maintain consistency using the appropriate tools. This includes automation jobs, workflow approvals, and ongoing maintenance tasks.
- **Replication**
  You'll distribute master data to connected systems either on batch processing or near real time. Manual efforts are reduced wherever possible to avoid errors or delays in processing or replicating master data
- **Monitor**
  You'll establish ongoing monitoring and generate necessary reports to review trends and improve inefficiencies in master data processes. Monitoring identifies where to focus on improvements, especially when considering high-volume master data.

- **Archive**
  When master data is no longer in use or is residing beyond a company's retention period, you'll establish an effective archiving of master data processes to improve system performance and maintain the system with only active master data.

In this chapter, first you'll learn about SAP Master Data Governance. Then, we'll transition into key considerations for creation, mapping, and maintenance of master data organizational elements from a Central Finance context. Finally, you'll learn Central Finance business mapping to support replication of transactional data from the source systems to the Central Finance system.

4.1 SAP Master Data Governance for Central Finance

As part of SAP's digital core initiative, data management is a critical component that governs master data in enterprise systems. SAP Master Data Governance is an MDM offering that governs master data centrally to consolidate, create, modify, and synchronize data among connected systems. In addition, the solution provides you with a framework of tools to tailor and extend standard solutions across your landscape.

SAP Master Data Governance offers three key features to support MDM, as shown in Figure 4.1:

❶ **Central data governance**
With this feature, you'll centrally govern and distribute master data to avoid duplicates and improve data quality. This includes on-premise systems, cloud applications, third-party systems, and across all line of business (LoB) processes. Governing includes creating, maintaining, and deleting master data and relevant business processes to support the master data processes.

❷ **Master data consolidation**
You'll consolidate master data from decentralized systems on a continued basis to sustain the accuracy of master data and data quality. Before you consolidate, the quality of data must be assessed and corrected. Consolidation of data happens enterprise-wide whenever mergers and acquisitions occur or when process reengineering is adopted.

❸ **Data quality evaluation**
You'll measure data quality using various SAP-provided tools such as SAP Smart business, along with data services to enrich data or remediate any issues. Once data quality is evaluated, you'll optimize business processes around master data to make better business decisions.

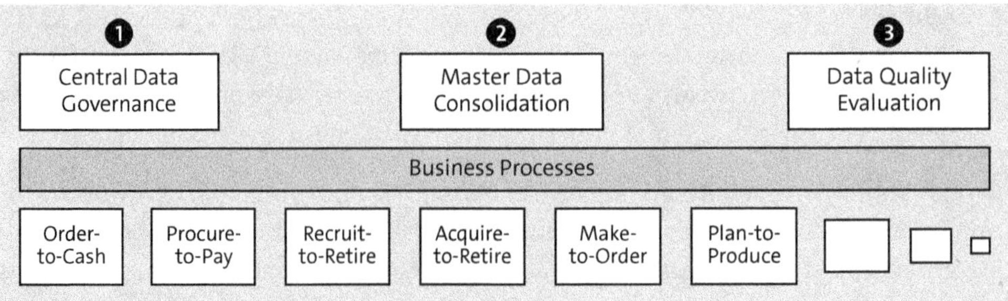

Figure 4.1 SAP Master Data Governance Functions

In the following sections, you'll learn key components of SAP Master Data Governance and how it integrates with Central Finance.

4.1.1 Functions

For a successful Central Finance implementation, enterprise master data quality plays a vital role as master data needs to be synchronized with connected source systems and the Central Finance system. SAP Master Data Governance comes with multiple domains to cater to various business use cases. In addition, you'll implement your own domains for specific requirements. Figure 4.2 shows SAP-delivered SAP Master Data Governance domains and extensible domains by the customer. These out-of-the-box domains ❶ provide the customer with the most important master data elements and their attributes, as you can see in Table 4.1. In addition to the standard delivered domains, you'll create your own custom domains to meet business requirements ❷.

Master Data	What Information Is Stored?
Customer	Telephone numbers, addresses, email addresses, contact names, and so on
Vendor/Supplier	Telephone numbers, addresses, email addresses, contact names, and so on
Material	Description, units of measure (UoM), pricing data, financial data, and so on
Financials	General ledger accounts, cost element hierarchies, profit centers, cost centers, and so on
Retail Article	Single and generic articles, listings, hierarchies, characteristics for variants, and so on
Technical Asset	Equipment; functional locations data; maintenance, repair, and operations (MRO) bills of materials (BOMs); and so on

Table 4.1 SAP-Delivered Domains

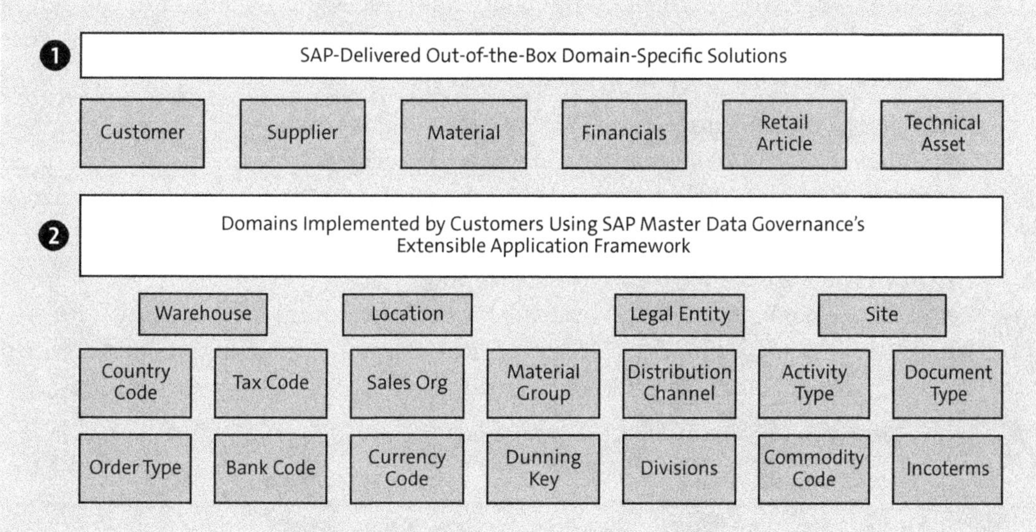

Figure 4.2 SAP Master Data Governance Domains

As discussed earlier, "light" SAP Master Data Governance supports key mappings and value mappings in Central Finance, which is included as part of the Central Finance license. A full suite of SAP Master Data Governance capabilities enables you to create, change, and distribute master data to all relevant systems to keep your master data in sync. In addition, the SAP Master Data Governance system is installed as a separate system or resides within the Central Finance system. A full SAP Master Data Governance solution requires a separate license and must be considered as part of an overall MDM strategy.

4.1.2 Deployment Options

Central Finance can leverage full features of SAP Master Data Governance to govern master data centrally to create, change, and distribute it for all systems involved in the enterprise. If SAP Master Data Governance is chosen as an MDM solution, SAP provides two common deployment options to implement SAP Master Data Governance:

- SAP Master Data Governance deployed in a separate system
- SAP Master Data Governance deployed within the Central Finance system

Let's look at each option's advantages and disadvantages.

SAP Master Data Governance Deployed in a Separate System

In this approach, you'll deploy SAP Master Data Governance in a separate system. All mapping information is stored in a separate system, and master data is distributed to the other systems in the landscape. In this case, the master data repository acts as a data hub where you'll run all master data processes. Figure 4.3 illustrates the Central Finance scenario where SAP Master Data Governance is installed in a separate instance. Let's understand data flow in this case:

❶ **Master data replication from SAP Master Data Governance**
Implement full SAP Master Data Governance processes in a separate instance to create, modify, and delete master data. Subsequently, you'll replicate master data to all connected systems in the landscape.

❷ **Read SAP Master Data Governance mapping**
Central Finance interface will read SAP Master Data Governance mapping from prior postings to post data. The SAP Master Data Governance mapping includes key mapping, value mapping, and cost object mapping.

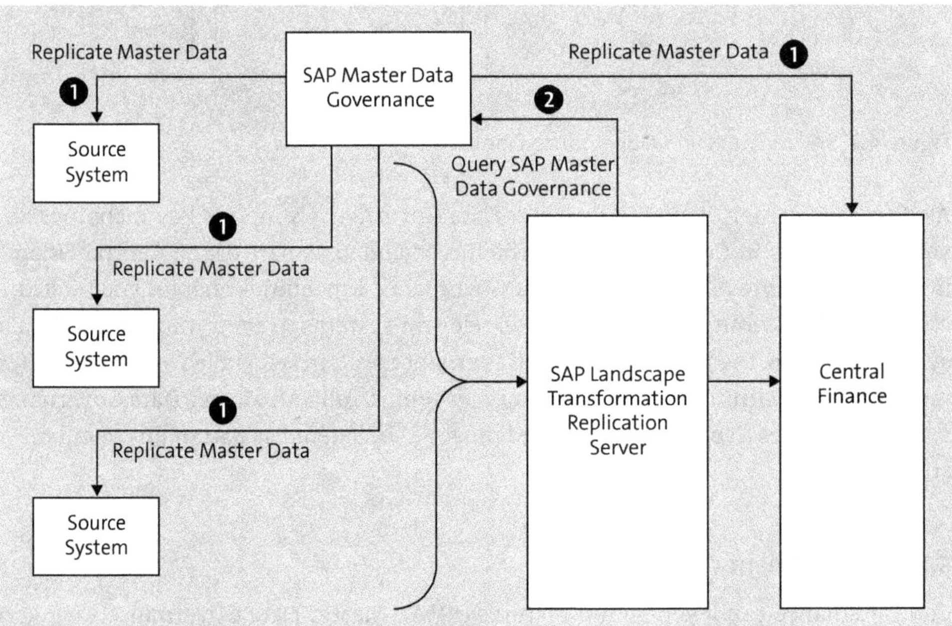

Figure 4.3 SAP Master Data Governance Deployed in a Separate Instance

[+] **Tip**

SAP recommends installing SAP Master Data Governance in a separate instance to take full advantage of its capabilities. This also helps during maintenance and upgrade cycles without interrupting Central Finance.

SAP Master Data Governance Deployed in the Central Finance System

In the second approach, you'll deploy SAP Master Data Governance in the Central Finance system as a master data hub. All mapping information is stored in the same

Central Finance system, and then master data is distributed to all other systems in the landscape. You'll run all master data processes within the Central Finance system. Figure 4.4 illustrates this scenario:

❶ **Master data replication from SAP Master Data Governance housed in Central Finance**
First, you'll implement full SAP Master Data Governance processes to create, modify, and delete master data centrally within the Central Finance instance. Subsequently, you'll replicate master data to all other connected systems in the landscape, including source systems.

❷ **Read SAP Master Data Governance mapping**
Central Finance will read SAP Master Data Governance mapping prior to posting data natively. Unlike in the earlier approach, no replication of master data is required in this case within Central Finance instance. This also minimizes activities related to remote connections within Central Finance. The system provides both key mapping and value mapping between the source system and the Central Finance system.

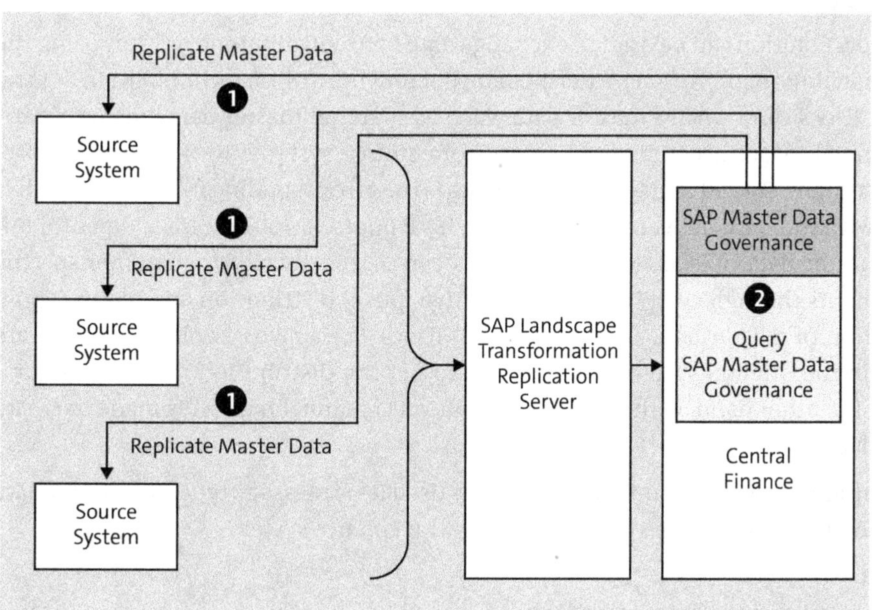

Figure 4.4 SAP Master Data Governance Deployed in the Central Finance System

[+]

Tip

This isn't an ideal option, especially considering long-term data volume and performance considerations. The system sizing must be considered prior to installing SAP Master Data Governance on a Central Finance box.

4.1.3 Master Data Creation and Maintenance Using Other Tools

Besides the full SAP Master Data Governance solution, there are other tools available to create and maintain master data. SAP Master Data Governance acts as a central master data hub encompassing complete processes, including governance and consolidation. Depending on the volume of master data, the solution can be either automated with scheduler background jobs or enabled with near real-time replication options. In either case, you'll need a close monitoring process to check the status to avoid master data replication errors. In addition, you'll define a governance process with workflow process steps that can be collaborated for a single master data object or multiple master data objects. In addition, the tools should have the capability to process mass data either interactively or in batch mode. This is helpful especially during month-end closing to resolve transactional replication errors related to master data in SAP Application Interface Framework. Choosing the right tool depends on various business requirements, the master data volume, and the long-term MDM strategy. In this section, you'll get an overview of other master data tools used to create and maintain master data.

Application Link Enabling Distribution Model

SAP's IDocs solution can be used to exchange data between different systems using the application link enabling (ALE)/intermediate documents (IDocs) technology. In Central Finance, IDocs can transfer master data, such as material master data, business partners, general ledger accounts, and so on from source systems to the target Central Finance system. One advantage of using IDocs is the error-handling capability, which is built in with standard delivered functionality. SAP offers various message types for different master data objects. Moreover, IDocs can be enhanced for customer-specific requirements through custom code. Both active (push) distribution and request (pull) distribution of master data are possible with IDocs. For active distribution, IDocs are created in the original system and distributed per the distribution model configuration. On the other hand, with request distribution, a manual request is made to create master data.

Following are a few important and commonly used IDoc message types for master data replication from the source system to the Central Finance system:

- `GLMAST`
  General ledger account master data.
- `MATMAS`
  Material master data.
- `CREMAS`
  Vendor master data.
- `DEBMAS`
  Customer master data.

- `COSMAS`
  Cost center master data.
- `PRMAS`
  Profit center master data.
- `BUPA_INBOUND_MAIN`
  Business partner master data.

SAP Data Services

SAP Data Services offers data transformation services for transferring data from source systems to a Central Finance system. It's an extract, transform, and load (ETL) tool to facilitate data transformation services. In addition, SAP Data Services requires a separate license. Most companies use SAP Data Services to migrate data from an old system to a new system in a greenfield or bluefield implementation. In addition, SAP Data Services can be used as an intermediate staging solution for data profiling. As mentioned earlier, Central Finance supports only replication of cost objects and project systems WBS master data. Other master data, such as general ledger accounts, customers, vendors, cost centers and so on, aren't supported out of the box. For large volumes of data, you must build interfaces using the SAP Data Services technology, which is composed of repositories and transformation rules to transfer master data to the Central Finance system or any other system. The relevancy rules vary by different requirements. Some key benefits of SAP Data Services are as follows:

- Can be used to integrate various source systems to consolidate master data from various sources
- Supports SAP Application Interface Framework error monitoring
- Can be used as an alternative to the SAP Master Data Governance solution
- Supports non-SAP ERP systems, including third-party applications
- Automates data transformation from different source systems
- Data profiling and data enriching capabilities for cleansing, deduping, standardizing, and validation activities to ensure consistent and accurate data
- Deduping of master data to eliminate duplicate data from various source systems
- Provides dashboards to view data from various sources
- Can be used as a migration tool from legacy or homegrown systems
- Robust security features to protect sensitive data during data transformation, thereby complying with local and regulatory requirements
- User-friendly interface to profile, manage, or transfer data
- Workflows to automate tasks

Migrate Your Data: Migration Cockpit

SAP offers the Migrate Your Data (app ID F3473), a migration cockpit, to migrate data from source systems or staging tables to the Central Finance system. The legacy transfer migration cockpit (Transaction LTMC) to create master data and transaction data in SAP S/4HANA is depreciated from SAP S/4HANA 20221 onward. Consult SAP Note 3355075.

SAP provides relevant templates and mapping information for master data fields. With this app, the system walks you through a sequence of steps to upload master data and transactional data, as shown in Figure 4.5. This process requires manual intervention to create master data. With this app, you can do the following:

- Create migration projects to migrate data and monitor status of the overall migration.
- Choose migration objects related to the Central Finance master data scope.
- Perform mapping tasks for the migration objects.
- Use a simulation option before loading data into SAP S/4HANA.
- Monitor the progress of master data migration.

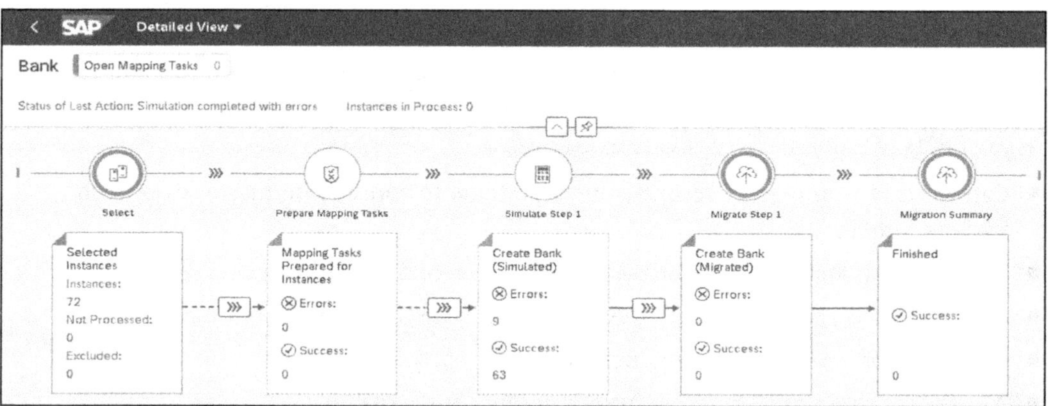

Figure 4.5 Migrate Your Data App

Third-Party Migration and Custom Programs

You can choose any third-party MDM tools such as IBM InfoSphere or insightsoftware SourceConnect (previously called Magnitude SourceConnect). These products offer solutions for governing master data, including harmonizing and replicating master data into Central Finance. The insightsoftware SourceConnect tool provides master data replication and transactional data replication solutions from both source SAP and non-SAP systems to a Central Finance system.

Custom programs generally aren't recommended from a leading practice standpoint. In addition, the custom programs require more maintenance requirements and add complexity to the MDM processes. Nevertheless, companies can build their own custom programs to create, maintain, and replicate master data into Central Finance systems.

4.2 Master Data Cleansing in Source Systems

As you integrate Central Finance with multiple source systems, an understanding of the quality of master data in source systems is essential for a successful implementation of the project. Master data quality and inconsistent processes pose a challenge to bringing harmonized master data into Central Finance. In addition, consistent master data and relevant processes ensure minimal errors during data replication from the source system to Central Finance. During the prepare and explore phases, you'll involve relevant business stakeholders in the master data cleansing activities as they bring in necessary knowledge to perform the data profiling activities. Data profiling is an essential activity to be performed before beginning the initial load and online replication. The list of master data cleansing varies with every customer and business requirements. Typical master data cleansing activities include the following:

- Identify master data design elements in scope for Central Finance. This needs to be determined based on the initial load scope and scope of online replication objects, so these actions limit the scope of data profiling.
- Assess master data quality to find duplicate, inconsistent, or incorrect data.
- Determine the level of cleansing required.
- Identify duplicate or redundant master data.
- Isolate unwanted attributes that aren't used or that aren't supposed to transfer to Central Finance.
- Perform data profiling and relevant assessment activities.
- Establish rules around data quality.
- Plan and execute master data cleaning activities.
- Enrich data for missing information.
- Harmonize master data (N:1 consolidation).
- Perform master data validation to ensure the data meets Central Finance standards and requirements.
- Leverage Master Change Delta (MCDelta), which helps track changes made in master data to understand the data consistency and integrity of data in an enterprise. The tool comes with predefined logic to help in data profiling.

In this section, we'll begin with important master data elements requiring cleansing or considerations in the Central Finance context. Then, we'll discuss the scope of master data, enhancement options, and other programs.

4.2.1 Master Data Elements

In this section, we'll discuss the two most important master data elements: general ledger master data and customer/vendor master data for cleansing. We'll also discuss processes for other master data elements at a high level.

General Ledger Master Data

The general ledger master is a critical master data element that often is one of the most time-consuming activities when harmonizing master data to Central Finance. Prior to beginning a Central Finance implementation journey, you'll ensure that the general ledger master is accurate and complies with Central Finance minimum requirements. As part of the cleansing activities, you'll perform the following:

- **Closing or blocking general ledger accounts**
  You'll perform an assessment of which general ledger accounts aren't used or are no longer needed. Often these general ledger accounts haven't been used recently. You can evaluate by looking at the general ledger balance transactions (Transaction FAGLB03 or Transaction FS10N) to identify account balances and recent transactions. If an account is created and used by mistake, you'll transfer balances to other correct accounts. In addition, you can block general ledger accounts at the chart of accounts (CoA) or at the company code level. Archiving general ledger accounts can also be considered based on company policy.
- **Open item–managed general ledger accounts**
  In legacy SAP ERP systems, often general ledger accounts are created with open item–managed general ledger flags incorrectly. In some cases, accounts are created with an open item–managed flag, but they aren't managed actively. This means open items aren't cleared consistently or are left open for a long time. From a master data standpoint, you can convert those open item–managed general ledger accounts to line-item managed general ledger accounts. SAP provides Transaction FAGL_ACTIVATE_OP (report FAGL_SWITCH_TO_OPEN_ITEM) to convert an open item–managed general ledger to a line-item general ledger for SAP ERP 6 EhP3 onward. This transaction isn't supported for lower releases. If you're using SAP S/4HANA as a source system, then you'll run program FINS_SWITCH_TO_OPEN_ITEM to convert open items. This program can either activate or deactivate an open item general ledger account. In addition, you can involve the business in closing open items that are no longer active.

[+]

Tip

Consult SAP Notes 1356457 and 1770786 for additional guidance and steps to be followed when using Transaction FAGL_ACTIVATE_OP.

Refer to SAP Note 2745769 for additional information on report FINS_SWITCH_TO_OPEN_ITEM in SAP S/4HANA.

[!]

Warning

Caution must be exercised as performing these activities will interrupt ongoing business operations. Therefore, the activities must be planned during weekends or off-business hours.

Customers and Vendors

Like general ledger accounts, you'll carry out customer and vendor master data cleansing activities in the source systems. This often requires help from business stakeholders as they possess the necessary knowledge to cleanse customer and vendor records. Following are the typical steps to follow when cleansing customer or vendor master data:

- Identify inconsistent customer and vendor data in the source system with business users' help.
- Analyze inconsistent fields when comparing to corresponding fields in Central Finance. Ensure all business partners' important fields are aligned with master data in the source system.
- Identify duplicate data. Based on volume, you may use SAP Data Services or any other similar tools to identify duplicates.
- Merge duplicate customers/vendors or block incorrect master data, so they aren't brought into Central Finance. This includes merging account groups. When merging multiple account groups, the number range must be set up similar to the source account group number range.
- Update missing or incomplete customer and vendor information.
- Evaluate address information to ensure it's accurate, especially for tax calls. In addition, if withholding tax is in scope for replication, then you'll evaluate that relevant data is set up correctly.
- Evaluate one-time accounts mapped to one-time accounts in Central Finance.
- As a leading practice, keep business partner numbering the same as customer and vendor numbers.
- SAP recommends setting up branch/head office relationships for the customer and vendor master consistently between the source system and the Central Finance system.
- Determine how localization requirements impact business partner master data.
- Perform audit or internal control reporting requirements.

[+]

Tip

Before starting the Central Finance project, consider customer/vendor integration (CVI) to convert customers and vendors to business partners. A business partner is a role-based master data object that comprises both customers and vendors with a single interface. The CVI conversion isn't mandatory but recommended to have consistent business partner data.

Other Master Data Elements

You can carry out a similar exercise for other master data elements such as cost center, profit center, and so on. For example, you can identify inactive or unused cost centers by running SAP-delivered reports to assess recent activities. Likewise, you can check cost center and profit center hierarchies that are no longer used. As part of cleanup, you carry out blocking cost centers or profit centers, making zero balances, or archiving master data objects per company policy.

4.2.2 Scope of Master Data

The first step in Central Finance begins with understanding the scope of its master data organizational elements because that drives the reporting requirements in Central Finance. The complexity may increase when Central Finance is connected to multiple source systems. Transactional data relies on master data to replicate documents from the source system to Central Finance. The scope for master data is determined based on the scope of process scenarios included as part of a Central Finance implementation.

Second, once the scope is defined, you'll assess master data quality in each source system, including an understanding of the master data definition and its usage. In addition, you need to understand any consolidation processes or simplifying processes and their relevant impacts on the quality of master data to be replicated to the Central Finance system. Typically, a Central Finance implementation will have several master data objects in scope, and you'll conduct workshops with relevant stakeholders for those master data objects accordingly. Table 4.2 provides a comprehensive list of master data objects you might consider in a typical Central Finance implementation. This isn't a complete list as every project varies with master data requirements.

Tip

The following master data objects aren't transferred to Central Finance and aren't supported:

- Sales orders
- Business process
- Real estate objects
- Networks
- Asset master data

Financial Accounting (FI)	Controlling (CO)	Bank Account Management	Business Partner (Implemented in FI View)	Other Master Data/Organizational Elements
▪ Chart of accounts (CoA) ▪ General ledger accounts ▪ Substitution/ migration accounts ▪ Business area ▪ Functional areas ▪ Business partners ▪ Projects/WBS elements ▪ Cost objects ▪ Tax codes	▪ Profit centers ▪ Profit center hierarchies ▪ Segments ▪ Cost centers ▪ Cost center hierarchies ▪ Cost elements and cost element groups ▪ Activity types ▪ Statistical Key figures ▪ Projects and WBS elements	▪ Bank keys ▪ Bank accounts ▪ House banks	▪ Customers ▪ Vendors ▪ Branch/head office business partners ▪ Joint venture accounting (JVA) ▪ Production Revenue Accounting (PRA) ▪ Authorization groups ▪ Industry keys ▪ Extended withholding tax ▪ Third-party vendors from employees created with the standard program ▪ Trading partners (Company ID)	▪ Material master ▪ Costing-based profitability analysis characteristics ▪ Plants ▪ JVA master data ▪ UoMs

Table 4.2 Master Data Objects in Central Finance

4.2.3 Enrichment and Validation of Master Data

Master data can be enriched to support Central Finance replication activities. Although minimal master data fields are replicated to Central Finance, you may enrich certain master data in certain fields during master data replication, as follows:

- Using business add-ins (BAdIs) or other code, you can adjust or enhance customer or vendor data.
- Validate rules to verity data for critical mapping fields, which includes identifying validation rules and cross-system validation.
- Trigger additional actions during master data replication.

- Build workflow for critical master data creation or changes to ensure the changes are approved by relevant business stakeholders.
- Adopt error handling if errors or warnings occur during replication of master data.

4.2.4 Other Programs

For unique situations, custom programs can be built for cleansing-related activities. This includes identifying duplicate records, finding inconsistent data in tables, and fixing inconsistent master data. In general, custom programs aren't recommended from a leading practice standpoint. However, based on master data quality checks and volume, sometimes it may be necessary to build custom programs as standard tools that aren't available extensively.

4.3 Harmonizing Master Data and Organizational Elements

When connecting multiple source systems to Central Finance, one of the challenges project teams face is harmonizing master data. If harmonization of master data isn't performed correctly, the process will lead to inaccurate mapping of master data that results in incorrect reporting. *Harmonization* refers to consolidating multiple objects through mapping to one object in Central Finance (N:1 mapping). On the other hand, if an object is no longer part of Central Finance design elements, in some instances, you'll clear such objects through the initial load and online either via mapping or using enhanced mapping via a BAdI.

In this section, you'll learn key considerations for master data elements and their parameters in Central Finance. In addition, we'll also explore important configurations for harmonization.

4.3.1 Chart of Accounts

Among many design elements, the chart of accounts (CoA) design is the most important design element for Central Finance implementations. When designing the CoA in Central Finance, it's important to keep both current and future requirements in mind. In addition, a CoA can be structured to support all requirements of an enterprise. During the design phase of the project, there are several key aspects to be considered when implementing Central Finance:

- Consider only the required CoA based on leading practices when gathering requirements.
- Consider accounts that support analysis and reporting standpoints.
- Evaluate current accounts used and their use cases in source systems to ensure you have all requirements gathered completely.

- Establish governance around CoA creation and maintenance.
- Maintain account grouping, which is critical for determining how to classify accounts.
- Take advantage of the flexibility to expand on future needs.
- Understand and fulfill cross-functional requirements.
- Ensure the CoA follows standard naming conventions and complies with industry standards and local statutory requirements.
- Fulfills organizational structural requirements such as segment reporting and account hierarchy.

In SAP, the CoA is structured in three types, as listed here and shown in Figure 4.6:

❶ **Operational CoA**

As the name implies, the operational CoA is used to capture day-to-day operational activities within an enterprise. An operational CoA is required. You'll assign the operational CoA to the company code. From the Central Finance standpoint, you'll design an operational CoA while keeping consolidated reporting requirements in mind. You'll ensure both operational and statutory requirements are met with enough details. In addition, you'll also design the CoA by considering the technical attributes to fulfill the Central Finance requirements. Furthermore, the CoA must fulfill both current and projected future needs.

From the Central Finance context, the operational CoA design is critical as it serve as a global CoA for multiple entities. The global CoA provides a unified or centralized CoA view for enterprises while existing posting logic in local systems stays the same. Central Finance supports both 1:1 or N:1 CoA scenarios. However, the effort involved in harmonizing the CoA simply can't be underestimated.

❷ **Country CoA**

If a company follows local accounting standards, then you'll handle those with an alternative CoA for the countries that require it. Assigning a country-specific CoA to a company code is optional.

❸ **Group CoA**

This comprises general ledger accounts to represent a consolidated group to facilitate group reporting. Optionally, you'll assign a group CoA to an operational CoA. This is helpful especially when you perform intercompany eliminations and group reporting tasks.

[+]

Tip

When creating a new CoA in Central Finance, you should consider new posting logic in SAP S/4HANA with the Universal Journal, especially considering the ledger approach, currencies, and so on. This is very important when you consider ledger group-specific postings.

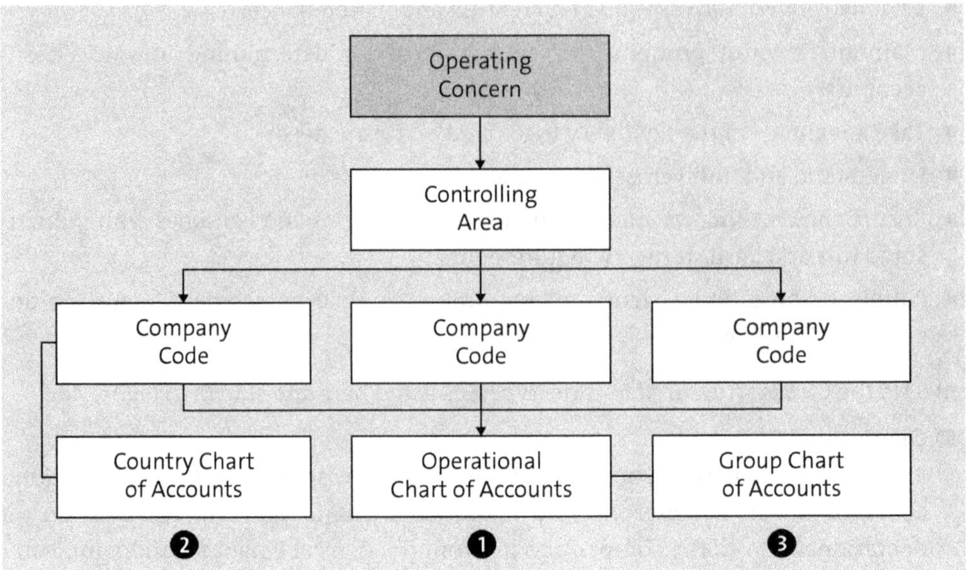

Figure 4.6 Chart of Accounts Design

4.3.2 General Ledger Accounts

In general, accurate master data maintenance and mapping of master data are key to any successful Central Finance implementation. General ledger account mapping is an important mapping object that requires careful analysis and understanding the attributes that will impact replication. The general ledger attributes must be reviewed and aligned prior to mapping in Central Finance. Furthermore, the future state of the mapping must be clearly defined, documented, and signed off from project stakeholders during the explore and design phase of the project. In this section, you'll learn about the general ledger account attributes to consider for the Central Finance implementation.

Field Status Groups

In SAP, the field status group configuration determines whether a field value is required or optional, and whether the field is hidden during a financial document posting. General ledger accounts are assigned with a field status group under the **Create/Bank/Interest** tab of the general ledger master data screen. From a Central Finance perspective, the field status groups must be harmonized if multiple source systems are involved. Otherwise, you'll see documents fail in SAP Application Interface Framework or cause inconsistencies. Table 4.3 provides mapping rules for field status groups from a Central Finance view.

Source System	Target Central Finance System	Central Finance Recommendation
Field status group	Identical field status group	Supported and the recommended approach
Field status defined as optional	Field status defined as required	Supported but can lead to errors if the source value isn't populated
Field status defined as required	Field status defined as optional	Supported and recommended when harmonizing with other source SAP ERP systems

Table 4.3 Mapping Rules for Field Status Groups

Figure 4.7 shows an example field status group showing **Additional account assignments**. The field status group assigned to the general ledger account must have the same fields **Suppress**, **Req. entry**, and **Opt. entry** columns. SAP recommends setting most fields as optional to avoid errors and to facilitate reporting in Central Finance. A harmonized field status can be set up the same as the source system, especially when the accounts are posted directly in the Central Finance system.

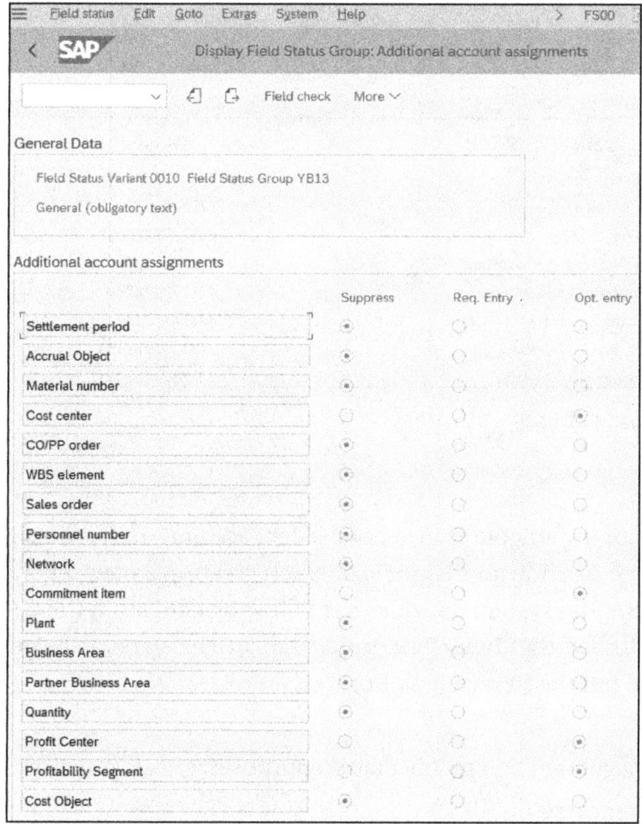

Figure 4.7 Field Status Groups

General Ledger Account Groups

Similarly, general ledger accounts are grouped by account groups in SAP. In addition, the account groups control field status and number range assignments. The field status determines which fields are required, optional, or suppressed during general ledger account creation. For example, you can group all asset-related accounts into one group and all customers and vendors into another group. When creating general ledger master records, the account group is a mandatory field. The account group controls the following features:

- Grouping accounts by the CoA
- Defining general ledger account number ranges
- Specifying which fields are required, optional, or suppressed

In Central Finance, account groups must be created prior to creating and mapping general ledger accounts. Figure 4.8 shows how example account groups are classified per nature or type of account. General ledger account groups are required prior to creating general ledger account master data in Central Finance.

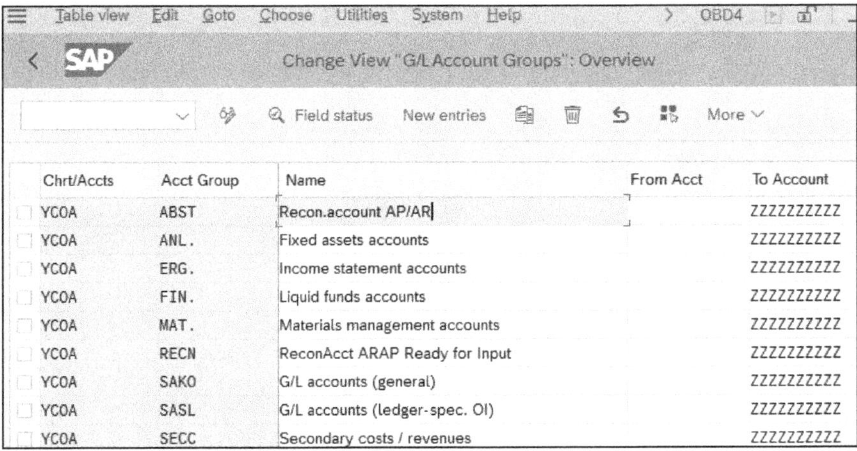

Chrt/Accts	Acct Group	Name	From Acct	To Account
YCOA	ABST	Recon.account AP/AR		ZZZZZZZZZZ
YCOA	ANL.	Fixed assets accounts		ZZZZZZZZZZ
YCOA	ERG.	Income statement accounts		ZZZZZZZZZZ
YCOA	FIN.	Liquid funds accounts		ZZZZZZZZZZ
YCOA	MAT.	Materials management accounts		ZZZZZZZZZZ
YCOA	RECN	ReconAcct ARAP Ready for Input		ZZZZZZZZZZ
YCOA	SAKO	G/L accounts (general)		ZZZZZZZZZZ
YCOA	SASL	G/L accounts (ledger-spec. OI)		ZZZZZZZZZZ
YCOA	SECC	Secondary costs / revenues		ZZZZZZZZZZ

Figure 4.8 General Ledger Account Groups

General Ledger Account Types

One of the most important mapping attributes in general ledger account master data is the general ledger account type, as shown in Figure 4.9. When creating a new general ledger account in SAP S/4HANA, the system provides the following choices in the **G/L Account Type** field, located under the **Type/Description** tab of the general ledger account master data. This field has the following options:

- **C Cash Account**
  Used for bank account management and reconciliation purposes.
- **N Nonoperating Expense or Income**
  Income statement account used on the FI side only (doesn't flow to CO).

- **P Primary Costs or Revenue**
  Profit and loss (P&L) account (e.g., salary expenses).
- **S Secondary Costs**
  P&L account used within CO.
- **X Balance Sheet Account**
  Balance sheet account (e.g., assets and liabilities).

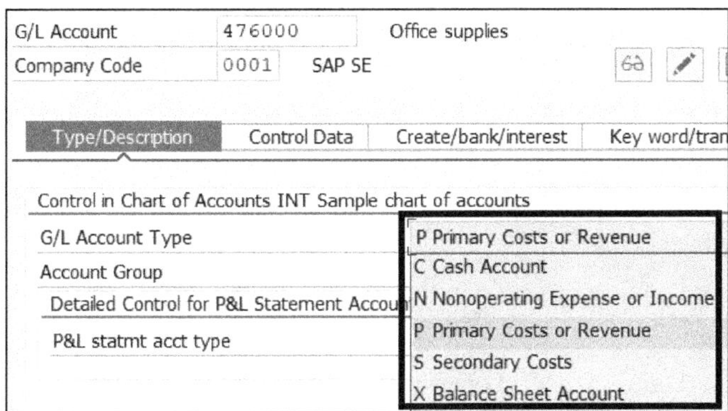

Figure 4.9 General Ledger Account Type

From the standpoint of Central Finance, when mapping general ledger accounts, the account type must match between the source system and the target systems according to the mapping recommendations listed in Table 4.4.

Source System	Target Central Finance System	Central Finance Recommendation
Balance sheet accounts	Balance sheet accounts	Supported
P&L account without cost element created	Nonoperating expense or income account	Supported
P&L account defined with cost element	Primary costs or revenue account	Supported
Secondary cost element	Secondary cost element	Supported

Table 4.4 Account Type Mapping

Tip

Secondary cost elements are now defined as general ledger accounts in SAP S/4HANA without any time intervals. However, the secondary cost elements in the source system are created with validity dates. As a result, attributes of cost elements changed during transfer of the initial load data are based on the initial load time period selected.

Cost Element Categories

For P&L accounts, SAP recommends you map accounts with the same account categories to avoid mapping errors. One slight deviation is that P&L accounts defined without a cost element are now mapped with the new account type nonoperating expense or income account in SAP S/4HANA. In general, a 1:1 mapping is supported in Central Finance. Table 4.5 provides several mapping recommendations for cost element categories.

Source System	Target Central Finance System	Central Finance Recommendation
Cost Element Category 1	Cost Element Category 1	Supported
Cost Element Category 11/12	Cost Element Category 11/12 or Cost Element Category 1	Supported
Cost Element Category 42	Cost Element Category 42	Supported (other cost element categories aren't supported)
Cost Element Category 43	Cost Element Category 43	Supported (other cost element categories aren't supported)
Cost Element Category 21	Cost Element Category 21	Supported (other cost element categories aren't supported)
Cost Element Category 41	Cost Element Category 41	Supported (other cost element categories aren't supported)
Cost Element Category 31	Cost Element Category 21	Supported (other cost element categories aren't supported)
Cost Element Category 90	No cost element category	Not supported
No cost element category	Cost Element Category 90	Not supported

Table 4.5 Cost Element Category Mapping

[+]

Tip

SAP recommends keeping cost elements mapping with the same cost element category. In case if this scenario isn't feasible, and source cost element (category 01) is mapped to Central Finance cost element (category 11, 12), SAP recommends clearing source cost objects during replication. In addition, SAP recommends that profitability segments must be defined as default cost objects for cost elements (category 11, 12) in Transaction OKB9 (Automatic Account Assignment). In some cases, the source cost object must be cleared during replication using a Central Finance BAdI.

Figure 4.10 displays both primary and secondary cost elements in SAP S/4HANA. The cost element categories **50**, **51**, **52**, and **61** must be set up the same way as in the source.

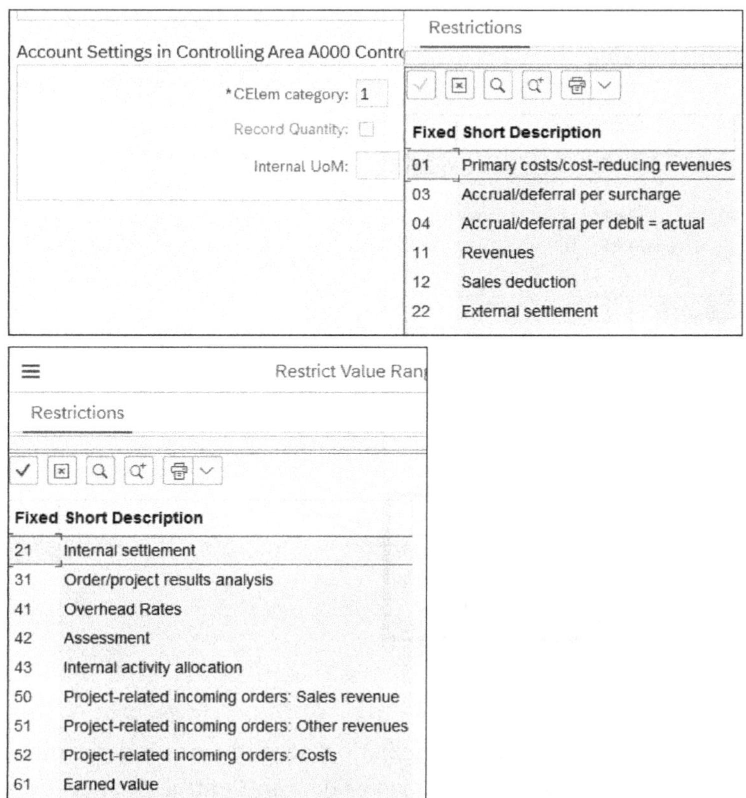

Figure 4.10 Cost Element Categories

In SAP S/4HANA, balance sheet accounts are denoted by the **Apply Acct Assignments Statistically in Fixed Asset Acct / Material Acct** indicator under the **Control Data** tab of the general ledger account. For the design changes in balance sheet accounts with cost element category **90**, consult SAP Note 2274763.

[+]

Tip

For various account assignments scenarios, refer to SAP Note 2310493.

General Ledger Reconciliation Accounts

The mapping rule for the **Recon. Account for Acct Type** indicator on a general ledger account must be identical between the source and target systems, with some exceptions. The following options are available on the **Recon. Account for Acct Type** dropdown list, as shown in Figure 4.11:

- **A Assets**
- **D Customers**
- **K Vendors**
- **V Contract Accounts Receivable**

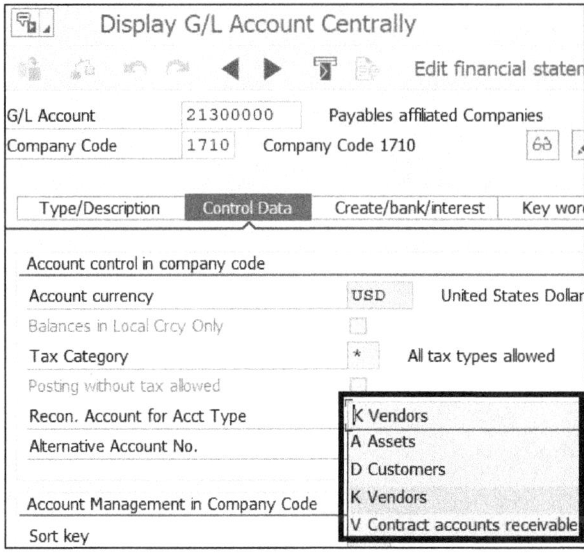

Figure 4.11 Reconciliation Account

In Central Finance, customer and vendor reconciliation accounts are redetermined from corresponding mapped business partner master data in Central Finance. Conversely, if there is a need to use a source system account, use Central Finance customizing activity **Define Handling of Reconciliation Accounts per Company Code** to prevent default redetermination in Central Finance. However, this may result in inconsistency if an open item is posted with the Redetermine if possible option, but the settings will be changed to 1 Keep for subsequent follow-on documents such as clearing or reversal. You must carefully evaluate the requirement prior to making the change. The customizing of this activity is covered in Chapter 7, Section 7.4.3. Central Finance doesn't support replicating the contract accounts receivable and payable (FI-CA) subledger component (reconciliation account type V Contract Accounts Receivable). Nonetheless, replication of financial postings resulting from closing and transferring of reconciliation keys are supported (reference procedure BKPF-AWTYP = FKKSU).

In a Central Finance context, asset accounting data replication isn't supported, so an asset reconciliation account must be mapped to a general ledger account. In addition, asset transactions from an asset reconciliation account in the source system are posted to the Universal Journal with posting keys 40 and 50 rather than asset-related posting keys. Moreover, the system doesn't transfer asset subledger information, such as asset numbers, to Central Finance.

[+]

Tip

SAP offers the central asset accounting functionality that can be activated at the company code level. With this functionality, you can manage asset accounting processes, including asset closing activities in Central Finance. Consult SAP Note 3084402 for more information on activating central asset accounting. This feature requires disabling asset accounting in the source system except for assets relevant to logistics processes.

Table 4.6 shows various Central Finance replication scenarios and mapping rules for reconciliation accounts values in the control data of the general ledger account master.

Source System	Target Central Finance System	Recommendation
Nonreconciliation account, for example, a basic general ledger account	Reconciliation account	Not allowed
Reconciliation account = A Assets	Nonreconciliation account (e.g., a basic general ledger account)	Allowed since asset accounting subledger information is stripped away (because asset accounting replication isn't supported in Central Finance yet)
Reconciliation account = C Customer	Reconciliation account = C Customer	Supported
Reconciliation account = K Vendor	Reconciliation account = K Vendor	Supported
Contract accounts receivable = V Contract Accounts Receivable	Contract Accounts Receivable = V Contract Accounts Receivable	Not supported (refer to SAP Note 2184567, question 40)

Table 4.6 Reconciliation Accounts Mapping Rules

Open Item–Managed General Ledger Account

The **Open Item Management** indicator under the **Control Data** tab of general ledger account master data, shown in Figure 4.12, is set on certain balance sheet accounts to manage accounts on an open item basis. For example, goods receipt/invoice receipt (GR/IR) clearing accounts or payroll clearing accounts are managed on an open item basis. From the Central Finance mapping point of view, general ledger accounts with the **Open Item Management** indicator activated in the source system must be mapped to similar account settings in Central Finance.

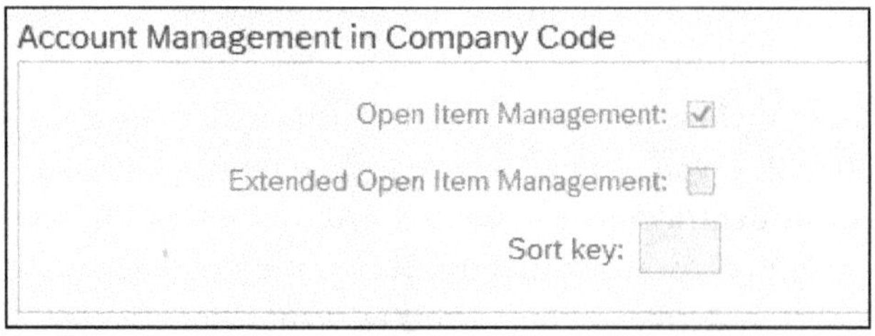

Account Management in Company Code

Open Item Management: ☑

Extended Open Item Management: ☐

Sort key:

Figure 4.12 Open Item Management Flag

Table 4.7 shows the mapping rules for the **Open Item Management** flag. If an open item account isn't properly managed, we recommend reviewing historical data to assess the volume and quality of this data. SAP recommends cleanup of all prior year open items that are no longer relevant. The cleanup task requires a lot of communication with business stakeholders to align on remediation goals.

Source System	Target Central Finance System	Central Finance Recommendation
Non-open item managed account	Non-open item managed account	Supported.
Non-open item managed account	Open item managed account	Allowed, but extensive testing is required to confirm the functionality. This mapping includes the transfer of account balances.
Open item managed account	Non-open item managed account	Not supported. This scenario will lead to inconsistencies. While replicating documents to Central Finance, the system will display an error. In addition, follow-on documents will fail.
Open item managed account	Open item managed account	Supported.

Table 4.7 Open Item Managed Accounts

Manage Balances in Local Currency Only

If the **Manage Balances in Local Currency Only** indicator has been enabled under the **Control Data** tab of the general ledger account master record, the system allows you to update only in local currency when documents are posted to the account. This indicator is generally set with accounts such as cash discount clearing accounts and GR/IR clearing accounts. In a Central Finance scenario, this indicator must be set up identically between the source and target systems. SAP particularly recommends the indicator be enabled for GR/IR accounts in Central Finance to avoid replication errors. Figure

4.13 shows **Balances in Local Crcy Only** indicator in the **Control Data** tab of the general ledger account maser record.

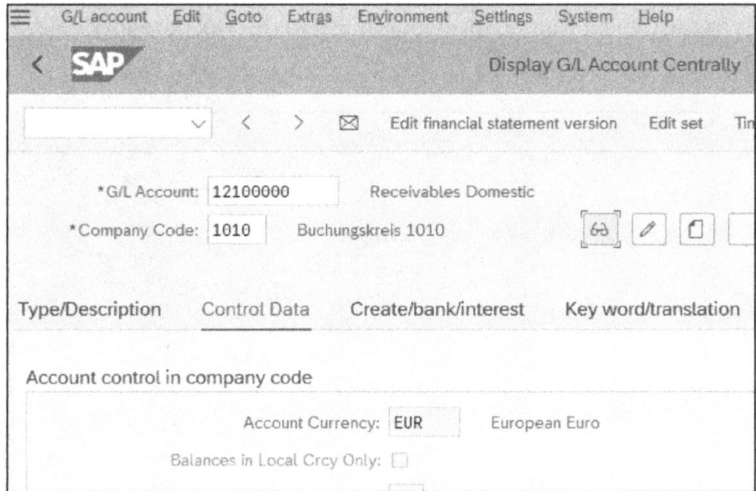

Figure 4.13 Balances in Local Currency Indicator

Table 4.8 describes the Central Finance recommended settings for the **Balances in Local Crcy Only** indicator.

Source System	Target Central Finance System	Recommendation
Manage Balances in Local Currency Only indicator enabled	**Manage Balances in Local Currency Only** indicator enabled	Supported and recommended
Manage Balances in Local Currency Only indicator *not* enabled	**Manage Balances in Local Currency Only** indicator *not* enabled	Supported and recommended
Manage Balances in Local Currency Only indicator enabled	**Manage Balances in Local Currency Only** indicator *not* enabled	Supported with limitations (SAP recommends testing all use cases)
Manage Balances in Local Currency Only indicator *not* enabled	**Manage Balances in Local Currency Only** indicator enabled	Supported with limitations (SAP recommends testing all use cases)

Table 4.8 Mapping Rules for Balances in the Local Currency Indicator

Tax Category in General Ledger Account Master Record

The **Tax Category** indicator under the **Control Data** tab of the general ledger account master record determines whether the general ledger account is tax relevant. SAP recommends keeping the indicator settings identical between the source and target systems. However, some exceptions are allowed. Table 4.9 provides the **Tax Category** values mapping rules when finance documents are replicated to the Central Finance system. The symbols in the table designate possible values in the **Tax Category** field, as shown in Figure 4.14.

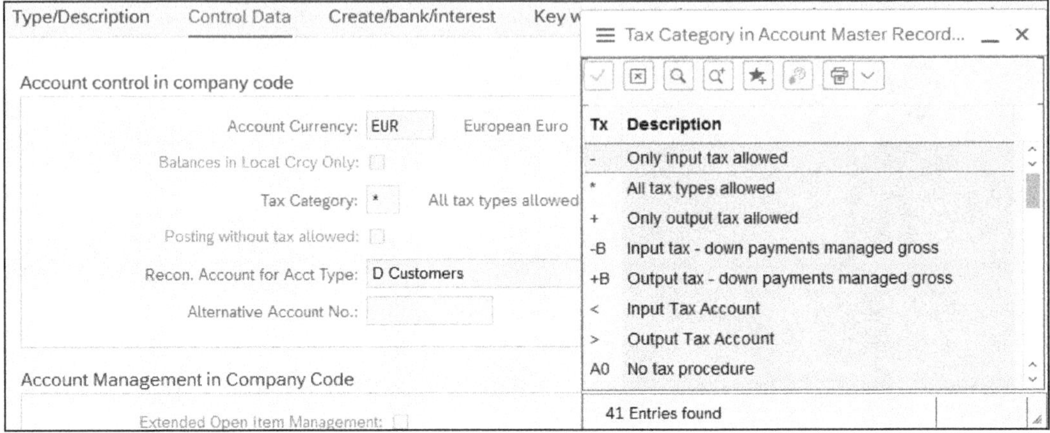

Figure 4.14 Tax Category

Source System	Target Central Finance System	Central Finance Recommendation
- Only input tax allowed	**- Only input tax allowed**	Supported.
- Only input tax allowed	*** All tax types allowed**	Supported.
- Only input tax allowed	**+ Only output tax allowed**	Not supported.
*** All tax types allowed**	**- Only input tax allowed**	Supported documents posted with an input tax are allowed. The source system sending a different **Tax Category** indicator isn't supported.
*** All tax types allowed**	**+ Only output tax allowed**	Supported documents posted with an output tax are allowed. The source system sending a different **Tax Category** indicator isn't supported.
*** All tax types allowed**	*** All tax types allowed**	Supported.

Table 4.9 Mapping Rules for Tax Category Settings

Source System	Target Central Finance System	Central Finance Recommendation
+ Only output tax allowed	* All tax types allowed	Supported.
+ Only output tax allowed	+ Only output tax allowed	Supported.
+ Only output tax allowed	- Only input tax allowed	Not supported.
> Output tax account	* All tax types allowed	Supported.
< Input tax account	* All tax types allowed	Supported.
* All tax types allowed	> Output Tax Account	Supported documents posted with an **Output Tax Account** indicator are allowed. The source system sending a different **Tax Category** indicator isn't supported.
* All tax types allowed	< Input Tax Account	Supported documents posted with an **Input Tax Account** indicator are allowed. The source system sending a different **Tax Category** indicator isn't supported.

Table 4.9 Mapping Rules for Tax Category Settings (Cont.)

Posting without the Tax Allowed Indicator

The **Posting without tax allowed** indicator under the **Control Data** tab of the general ledger account record (refer to Figure 4.14) allows postings to the Universal Journal without a tax code populated on the document line item. During posting, the system verifies the tax code values against this indicator. Sometimes, an account is used for both taxable and nontaxable situations. For example, foreign customers may not use a jurisdiction code, and thus the tax code may not be relevant. In such cases, the indicator allows not having a tax code populated on expense and revenue accounts.

In Central Finance, identical **Posting without tax allowed** settings are recommended. Table 4.10 shows the supported scenarios in Central Finance for this indicator.

Source System	Target Central Finance System	Central Finance Recommendation
Posting without tax allowed indicator enabled	**Posting without tax allowed** indicator enabled	Supported

Table 4.10 Mapping for Posting without Tax Allowed

Source System	Target Central Finance System	Central Finance Recommendation
Posting without tax allowed indicator *not* enabled	**Posting without tax allowed** indicator *not* enabled	Supported
Posting without tax allowed indicator enabled	**Posting without tax allowed** indicator *not* enabled	Caution (confirm all processes works with this setting)
Posting without tax allowed indicator *not* enabled	**Posting without tax allowed** indicator enabled.	Supported

Table 4.10 Mapping for Posting without Tax Allowed (Cont.)

Sort Key

The **Sort key** is assigned to all general ledger accounts under the **Control Data** tab in the company code segment of the general ledger master record. The **Sort key** field is helpful for sorting accounting documents using various sort sequences, for instance, cost accounts by cost center, customer or vendor accounts by assignment number, and asset accounts by asset number. In addition, it helps to assign a specific value, for example, a document number on assignment field during posting. The assignment field is filled in manually or automatically on a document line item using the definition of the **Sort key** via Transaction OB16. In addition, business partners also can contain their own sort keys. From a Central Finance replication standpoint, the **Sort key** logic from the source system takes priority. As a result, the assignment field (ZUONR) populated from the source system is kept upon replication. Note that Central Finance sort key logic takes precedence only if the assignment field is blank in the source system or if the field is cleared using Central Finance BAdI logic.

Note

SAP recommends following a consistent approach when defining and assigning sort keys to ensure consistency is followed across the company codes.

Blocking Indicator

SAP recommends keeping accounts without any chart level or company code blocks for the initial load to prevent any errors during the initial load simulations or posting. The same applies during the online replication. If there are valid reasons, consider the initial load and online replication activities prior to blocking, and come up with alternative mapping to prevent errors.

Post Automatically Only

In SAP, accounts such as GR/IR, tax accounts, discount accounts, and so on are set up with the **Post Automatically Only** indicator on the **Create/bank/interest** tab of general ledger account master, as shown in Figure 4.15. On these accounts, manual accounts aren't permitted by the system. Although this indicator doesn't impact the replication, SAP recommends identical settings between the source system and Central Finance to avoid data inconsistencies.

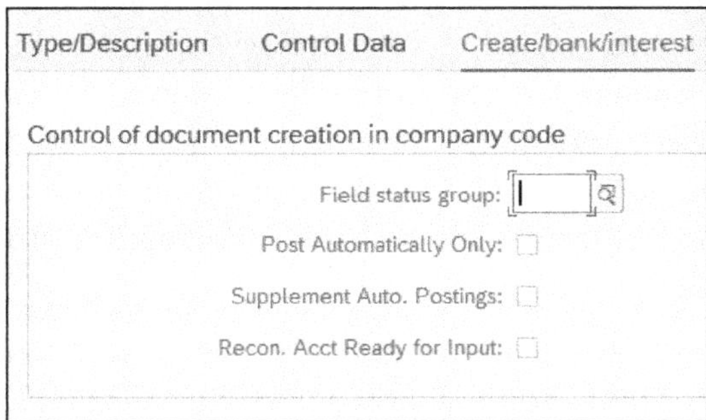

Figure 4.15 Post Automatically Only

Special General Ledger Indicator

When a special general ledger indicator is posted with a transaction, the system chooses a parallel subledger account rather than the normal reconciliation account at the time of posting. Typically, special general ledger accounts are used for down payment or bill of exchange types of transactions. From a Central Finance context, SAP recommends identical special general ledger indicators setup between the source system and the Central Finance system. In addition, you'll maintain special general ledger indicators mapping in Central Finance. Table 4.11 lists the Central Finance support for special general ledger indicator scenarios.

Source System	Target Central Finance System	Central Finance Recommendation
Account assignment for special general ledger is posted on a transaction.	The account assignment for the special general ledger indicator and account are the same, and the special general ledger indicator is mapped.	Supported
Account assignment for special general ledger is posted on a transaction that differs between the source system and the Central Finance system.	The account assignment for the special general ledger differs between the source system and the Central Finance system.	Not supported

Table 4.11 Central Finance Special General Ledger Posting Scenarios

Tip

Central Finance doesn't support the special general ledger indicator **W Bill of exchange** (UMSKZ). SAP Note 2288901 provides a possible workaround to overcome this scenario.

4.3.3 Cost Centers

In SAP, the *cost center* organizational unit refers to a responsibility center where costs are incurred and reported. Cost centers are created to cater for several functional requirements such as to allocate, collect, and report costs for services received or performed, and so on. A cost center is assigned to the standard hierarchy, which is a collection of cost centers grouped by its purpose. In addition, the cost center category identifies the type of category defined in an organization such as administration, production, and distribution.

From a Central Finance perspective, cost center master data must be created and mapped in Central Finance prior to performing the initial load or starting online replication. Both cost centers and profit centers were created with reference to a controlling area, which can be harmonized if multiple controlling areas are configured in the source system. With the Central Finance implementation, you can redesign cost centers and profit centers based on the future state of the organizational structure considering current and future reporting requirements. Figure 4.16 shows the cost center master data attributes. Ensure that cost centers are created in the correct **Company Code** ❶ and with the correct **Profit Center** ❷ assignment.

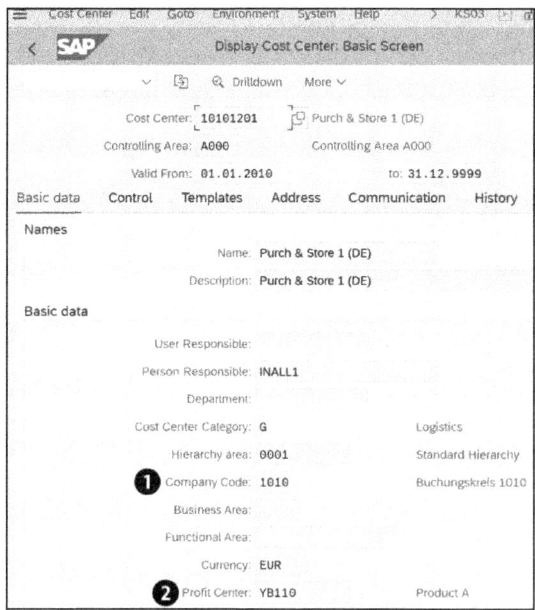

Figure 4.16 Cost Center Master Data

Cost centers are used in both FI-CO postings and CO postings in the source system. When business areas and functional areas aren't populated during the transactional postings, the system tries to drive from cost objects such as cost centers and orders or general ledger accounts. Therefore, ensure functional area and business area are maintained in the cost center and other cost objects per business requirements. In addition, if cost centers are no longer active or not used, lock the cost center for postings on the **Control** tab.

[«]

Note

During FI-CO initial load postings, if general replication settings are configured for monthly balances or document summarization is on in the source systems, the system doesn't transfer CO account assignments. In such cases, you'll maintain default account assignments in Transaction OKB9 to read the default assignment.

4.3.4 Profit Centers

In Central Finance, profit centers and segments must be derived correctly in both the initial data load and in continuing online replication. The system first checks the mapping action for profit centers and segments and then maps the profit centers and segments accordingly. Financial documents with income statement line items posted in the source system often contain account assignment objects, such as cost centers, WBSs, and internal orders. Profit centers and segments are derived from the account assignment master data. When documents are transferred to Central Finance, profit centers are derived from the mapped account assignment objects, such as cost centers or WBSs, from Central Finance. If there is a deviating profit center, the system overwrites it from the derivation rule sequence to maintain consistency. On the other hand, if profit centers have conflict in mapping between profit center and profit center assigned in cost center mapping, this will result in CO account assignments error having different profit centers (**Message no. BK 134**). Therefore, you must align the mapping of profit center and mapping of cost center with the same profit center assigned. In other words, you cannot mix different profit centers between profit center mapping and cost center mapping.

When the profit centers in document line items lack an account assignment object, the system looks for a corresponding mapped profit center from Central Finance. During the initial load, if the system can't find a profit center, a dummy or default profit center is derived from the maintained default values in Transaction FAGL3KEH. To understand how CO account assignments function in the Central Finance scenario, consult SAP Note 2310493. Figure 4.17 shows an overview of the profit center.

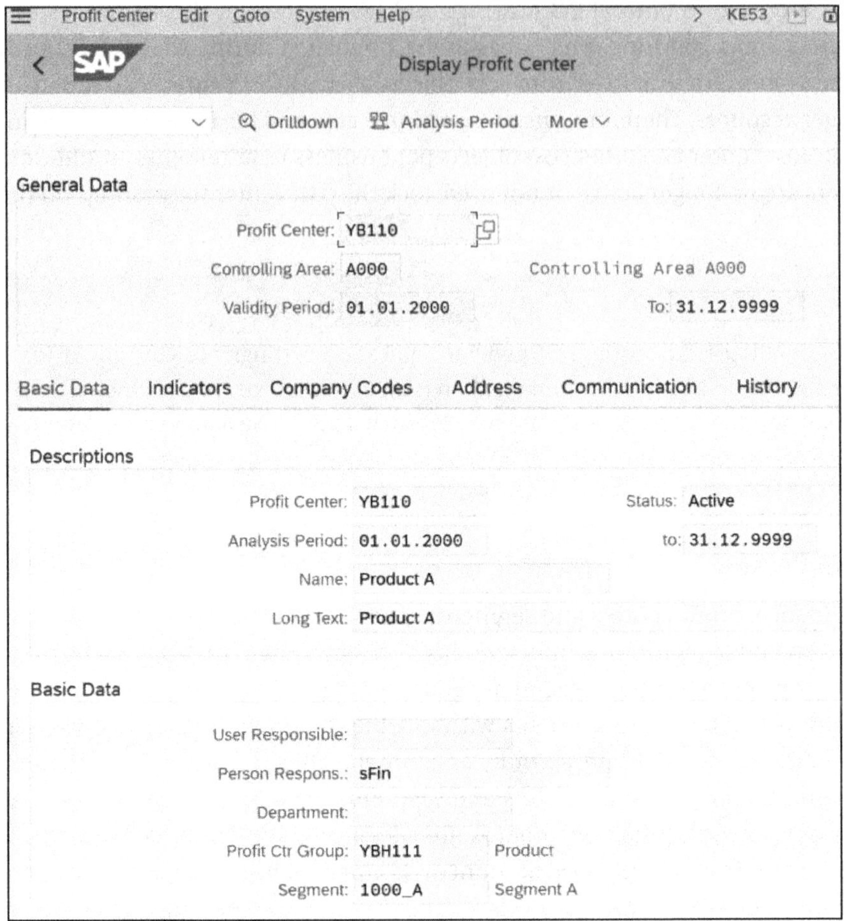

Figure 4.17 Profit Center Overview

[»]

Note

Like general ledger accounts, cost center and profit center attributes should be discussed in discovery workshops in coordination with business teams. These organization elements must be thoroughly reviewed with end reporting in mind. In addition, consider the technical requirements of mapping with the source system.

4.3.5 Functional Area

In SAP, functional areas (FKBER) are used by management to analyze the cost of sales accounting with functions in an organization. For example, a functional area could be production, engineering, marketing, and so on. With functional areas, you can generate a contribution margin report by functional area. To use functional areas for cost of sales accounting, you must first create functional areas and subsequently assign them in the master data. A functional area can generally be derived from the assigned master data objects. In addition to assigning functional areas in master data, for complex requirements, validation and substitution rules can be used to derive and populate functional areas during posting. From a sequence of derivation standpoint, the system derives a functional area in a specific sequence: first the controlling account assignment object, then the general ledger account or substitution rule, and finally—if entered—during document posting. Consult SAP Note 3003021, which summarizes the sequence of functional area derivation.

From the Central Finance context, during replication, functional area isn't redetermined automatically. Based on the mapping action entity and BAdI enhancement, you can derive functional areas.

4.3.6 Business Partners

In SAP S/4HANA, business partners provide a comprehensive view of master data for customers, vendors, and general business partner information. Figure 4.18 shows an example business partner. In SAP ERP, customer master and vendor master records are created and maintained separately. On the other hand, a business partner in SAP S/4HANA can be assigned with multiple roles such as a general role to maintain basic information, FI customer, FI vendor, credit management, and so on. For Central Finance, you'll maintain business partners mapping for the initial load and real-time replication. If your source systems have the same customer or vendor numbers across systems, but they require a unique mapping in Central Finance, the SAP Master Data Governance consolidation tool can be used to analyze and configure rules for mapping. Based on the rules configured, a golden record is maintained in the Central Finance system. In addition, SAP provides multiple options to create business partners in Central Finance:

- The Manage Business Partner app (app ID F3163) can be used to create and maintain business partners in SAP S/4HANA directly.
- Business partners can be created centrally with the SAP Master Data Governance standalone system and then be distributed to Central Finance. Mapping is maintained for the replication and initial load activities.
- SAP Master Data Governance can be used directly in the SAP S/4HANA system where Central Finance is implemented. In this case, Central Finance will use mapping of master data from SAP Master Data Governance in the same box.
- Other tools, such as SAP Data Services or IDocs can also be used to replicate.

From a Central Finance context, you must consider the following:

- Identifying customers and vendors requires harmonization in Central Finance. Sometimes, this is a daunting task if the volume is high. In such cases, you need to identify appropriate tools to achieve this task.
- Identify reporting requirements of customers and vendors whether central payment is active or not active. If no reporting requirements are identified, evaluate options of using a default value for mapping. However, avoid dummy master data wherever possible.
- Find reporting requirements from a profitability analysis reporting standpoint, especially considering the customer as a characteristic or a dimension.
- Update missing or incomplete customer and vendor information to align with business partners. The other option is to implement a BAdI to remove unnecessary data or to update incomplete data during replication.
- Evaluate one-time accounts mapped to one-time accounts in Central Finance.
- The leading practice suggests keeping business partner numbering the same as the customer and vendor numbers.
- SAP recommends having branch/head office relationships for the customer and vendor master set up consistently between the source system and the Central Finance system.

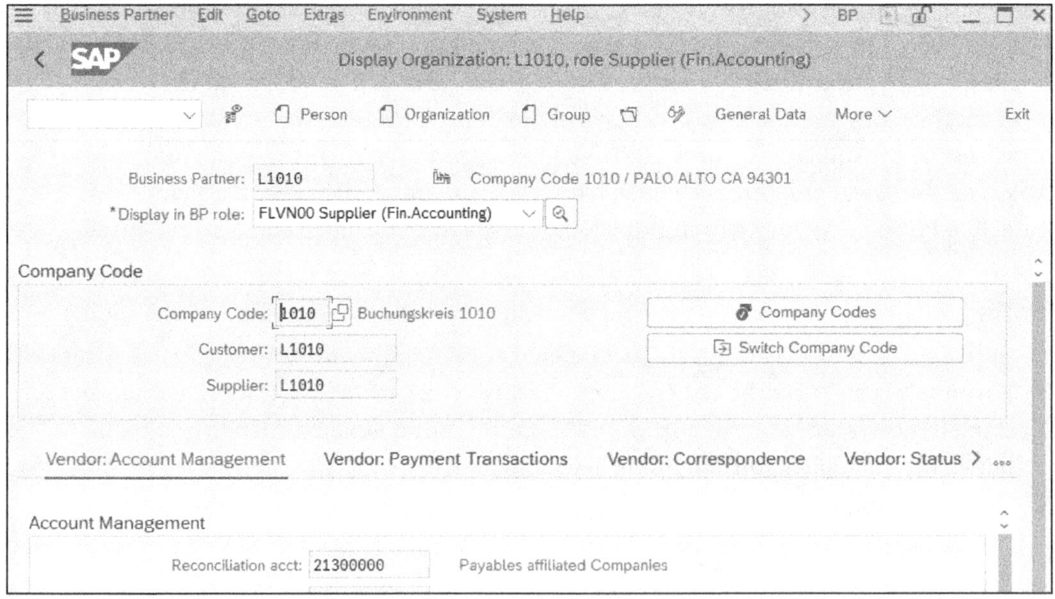

Figure 4.18 Business Partner

4.3.7 Material Master

When the material master is used in source system financial transactions, the transactional data contains material master-relevant information such as quantities, UoMs, and so on. In such cases, you'll need to create material masters and map them based on mapping action settings in Central Finance. Harmonizing material masters is the most consuming activity in a Central Finance environment. This will result in consistent data across the enterprise and provide additional reporting capabilities, especially in product profitability analysis. If material isn't included in the scope, you won't be able to report product profitability in Central Finance. While harmonizing material master, you'll come across challenges such as these:

- While replicating the material master data, validation checks in the source system not available in Central Finance
- Same material number used differently in different source systems
- Enabling consistent numbering schemes
- Creating consistent product hierarchies
- Grouping materials and aligning with the common reporting information model (CRIM)
- Identifying material types and material fields relevant for reporting
- Maintaining material master views (basic, plant, accounting views)

You need to maintain relevant mapping if SAP Master Data Governance is implemented for material master maintenance.

4.3.8 Master Data Hierarchies

In addition to master data, you'll create and maintain master data hierarchies to support business reporting requirements. SAP provides the Manage Flexible Hierarchies app (App ID F2759), as shown in Figure 4.19, which offers users the chance to create and maintain hierarchies quickly. This app does the following:

- Allows you to create and maintain master data hierarchies, especially for cost center, profit center, company code, and so on
- Supports a user-defined sequence of master data attributes for country, city, and region
- Allows mass creation of hierarchies through uploads
- Allows you to download master data and attributes used in hierarchies
- Allows mass updating of extended attributes of master data used in hierarchies

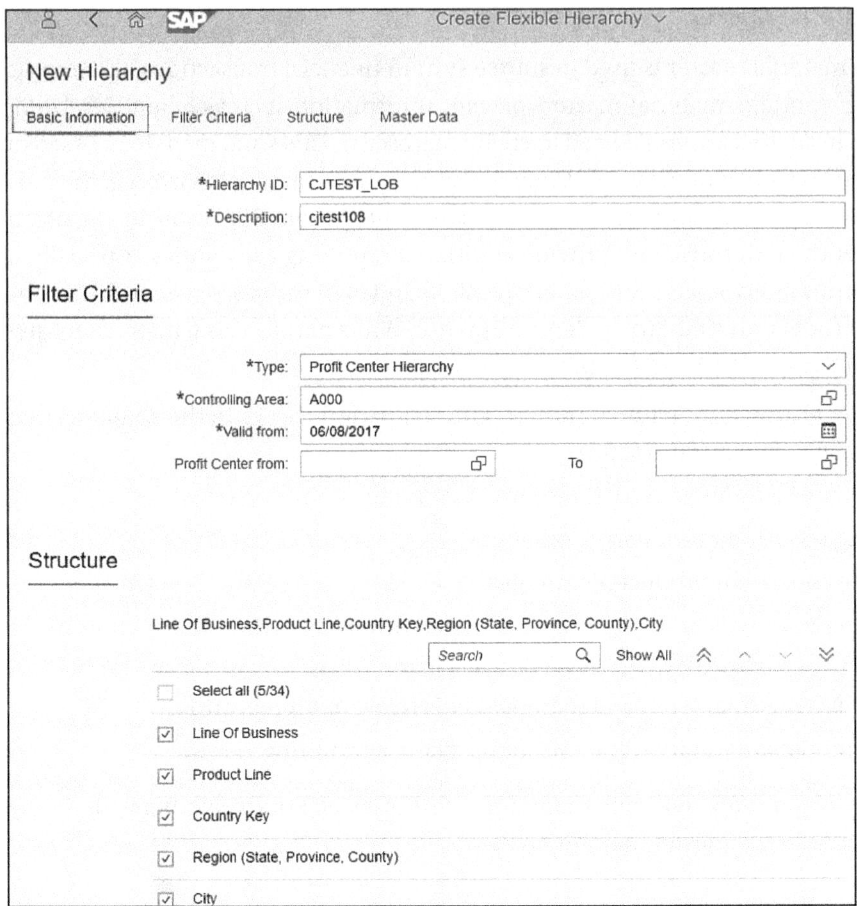

Figure 4.19 Manage Flexible Hierarchies

4.3.9 Company Code

One main advantage of Central Finance is its capability to design company codes with global reporting in mind. There has been some confusion among Central Finance user communities regarding whether you can harmonize company codes in Central Finance, that is, consolidating multiple companies and mapping to a single company code in Central Finance. When it comes to company code design, Central Finance supports 1:1 mapping between the source system and the Central Finance system. Although you can map company codes with the N:1 mapping scenario, Central Finance doesn't support this scenario out of the box. On the other hand, splitting a company code into multiple companies (mapping 1:N) is also not supported. In these scenarios, you can consider a customized solution. Extensive regression testing is required in almost all process areas, including reporting to confirm that the functionality is working as intended. Table 4.12 provides company code mapping scenarios in Central Finance.

4

Source System Company Code	Central Finance Company Code	Central Supported Scenario
Company code 1000	Company code 1000	1:1 is supported.
Company code 1000	Company code US01	1:1 is supported. Different naming conventions used in Central Finance are supported if the underlying configuration settings are consistent and identical.
Company codes 1000, 2000	Consolidated into US01	M:1 isn't supported; use a custom solution.
Company code US01	Split into company codes 1000 and 2000	1:M split isn't supported; use a custom solution.

Table 4.12 Company Code Mapping in Central Finance

[+]

Tip

To learn about restrictions on this topic, refer to SAP Note 2184567, question 39.

4.3.10 Controlling Area

With the Central Finance implementation, you can transition to a single controlling area to adopt simplified management accounting. Although you can define multiple controlling areas in Central Finance, the leading practice is to create a single controlling area. When you connect multiple source systems, every source system controlling area could be operating in a different global currency. You can harmonize these controlling areas that were set up with different global currencies with a single controlling area and a single global currency in Central Finance. Table 4.13 illustrates an example scenario of controlling area harmonization.

Source System		Central Finance	
Controlling Area	Controlling Area Global Currency	Controlling Area	Controlling Area Global Currency
1000	EUR	A000	USD
2000	USD		
3000	GBP		

Table 4.13 Controlling Area Harmonization

[+]

Tip

One prerequisite to harmonization of controlling areas is that all entities must be assigned to the same operating CoA. As a leading practice, SAP recommends activating cross-company code cost accounting to take advantage of one unified controlling area. With this approach, all CO master data is assigned to a company code. The intercompany transactions posted then can be used to generate assets and liability reporting automatically in real time in Central Finance.

4.3.11 Operating Concern

In SAP, you'll define an operating concern, which is a key organizational element used in margin analysis for analyzing and reporting the profitability analysis by various dimensions in Central Finance. The operating concern is the highest organizational unit in CO. SAP's leading practice is to assign all entities and countries to a single operating concern with N:1 mapping in Central Finance. By designing this way, profitability reporting dimensions such as product, customer, or geographic location can be aligned per CRIM. In Central Finance, margin analysis is the recommended approach over costing-based profitability analysis as margin analysis is integrated with the Universal Journal (table ACDOCA). Note that both margin analysis and costing-based profitability analysis can be enabled as there are no technical restrictions. However, costing-based profitability analysis isn't part of the Universal Journal. In addition, all features available in costing-based profitability analysis are now available with margin analysis. Therefore, you'll have challenges in justifying enabling costing-based profitability analysis in the Central Finance system.

4.4 Central Finance Mapping

During prepare and explore phases, you'll evaluate which master data elements need to be harmonized in the Central Finance system. Besides organizational elements such as ledgers and company codes, you'll determine master data mapping requirements for CoA, general ledger accounts, cost centers, profit centers, and so on. The harmonized data is created and mapped in SAP Master Data Governance in Central Finance. You'll create master data mapping only when it's required. Well-designed and harmonized master data enables consolidated financial reporting in Central Finance. SAP Master Data Governance mapping in Central Finance ensures that when data is replicated from different source systems, it's processed and transformed to Central Finance. In this section, we'll go through configuration for master data with step-by-step instructions.

4.4.1 Defining Technical Settings for All Involved Systems

The **Business System** links both the source system and Central Finance for key and value mapping in Central Finance. SAP recommends creating a business system for each source system and the Central Finance system for mapping purposes. Each source system is represented by the logical system created in the Central Finance system. Each logical system is then linked with the business system in Central Finance. If the Central Finance system is connected to a system landscape directory (SLD), then you can choose that business system. If no dedicated SLD is created, the Central Finance system finds the local business system. In this case, there is no SLD setup, yet you'll enable Central Finance business mapping with the local business system. This requires implementation of BAdI `MDG_IDM_GET_LCL_SYSTEM`. Therefore, it's mandatory to maintain the relationship between the business system and logical system in the SLD. Access to the SLD should be configured correctly in the Central Finance system. In this section, we'll cover the steps to access the SLD and get the business systems set up correctly. To set up the SLD, follow these high-level steps:

1. From the Central Finance system, use Transaction SLDAPICUST.
2. Maintain the SLD using Transaction SLDHTMLGUI.

After setting up the SLD, you can link the source logical system with the business system in this configuration. To configure this, navigate to Central Finance customizing activities menu path **SAP Customizing Implementation Guide • Financial Accounting • Central Finance: Target System Settings • Mapping • Define Technical Settings for All Involved Systems**. After reaching the configuration, you'll land at the **Change View "Define Business Systems": Overview** screen, as shown in Figure 4.20.

On this screen, click the **New Entries** ❶ button. You'll arrive at the **New Entries: Overview of Added Entries** screen. Maintain values for the **Business System**, **Logical System**, and **RFC Destination** fields ❷. Click the **Save** button to save the entries. Maintained values are reflected upon returning to the previous screen ❸.

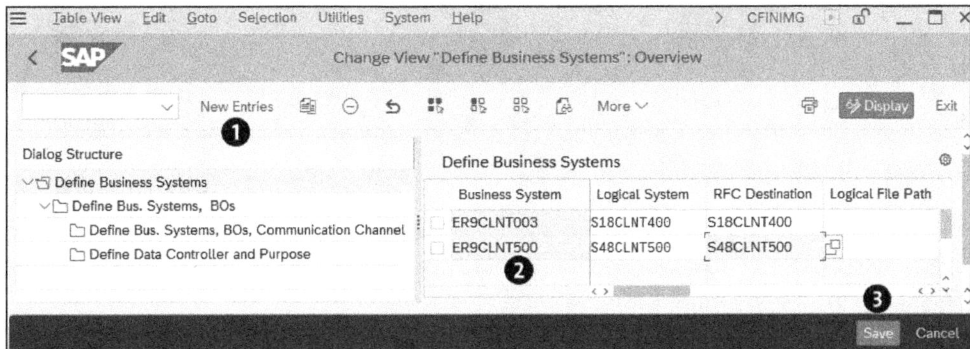

Figure 4.20 Define Technical Settings for All Involved Systems

[»]

Note
Consult SAP Note 2225086 on enabling Central Finance business mapping without the need to set up a dedicated SLD.

4.4.2 Define Mapping Actions

Central Finance offers mapping actions where you can set the mapping of a cost object or configuration element. For example, if you wanted the trading partner to be derived from business partner master data in Central Finance, you would set a *clear data* action for the company ID field in a mapping action. In this way, the system clears the incoming source value prior to posting documents in the Universal Journal. To define mapping actions, navigate to the Central Finance customizing activities menu path **SAP Customizing Implementation Guide • Financial Accounting • Central Finance: Target System Settings • Mapping • Define Mapping Actions**. After reaching the configuration, you'll arrive at the **Display View "Mapping Entity": Overview** screen, as shown in Figure 4.21.

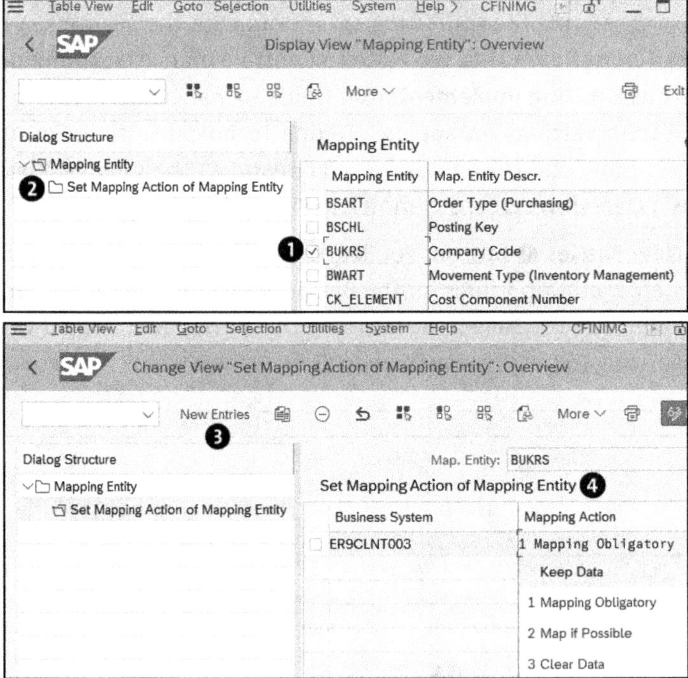

Figure 4.21 Mapping Action

On this screen, choose the mapping entity ❶, and double-click the **Set Mapping Action of Mapping Entity** ❷ folder icon located under **Dialog Structure** on the left subscreen. On the resulting screen, you'll maintain a new **Mapping Action** by clicking the **New**

Entries ❸ button. The **Mapping Action** ❹ dropdown list comes with four options, as listed in Table 4.14 along with their descriptions and recommended usages.

Mapping Action	Description	Recommendation
Keep Data	No mapping is done in this case as the incoming source value is kept.	The system defaults to this option if no mapping action has been assigned for the mapping entity. The underlying assumption is the same ID, and the same setup is maintained between the source system and the Central Finance system.
Mapping Obligatory	Mapping is required.	Maintain the required mapping; otherwise, the system raises an error message. The system supports the following: N:1 mapping: This is for multiple objects such as general ledger accounts, cost centers, and so on. 1:1 mapping: For example, you'll map company codes 1:1.
Clear Data	The incoming source value is cleared.	For example, the trading partner is derived from the target Central Finance system. These fields aren't considered in either the initial load or online replication.
Map If Possible	The system first checks whether a mapping is available. If no mapping has been maintained, the system keeps the original value from the source.	If the mapping is identical, you don't need to maintain entries in the mapping. The sender value is retained and transferred to the Central Finance system.

Table 4.14 Mapping Actions

4.4.3 Key Mapping

In Central Finance, master data mapping for general ledger accounts, customer accounts, and so on are maintained in a key mapping or an ID mapping SAP Master Data Governance table. These master data objects are generally created by users on day-to-day operations and used in transactional postings. SAP offers two types of tools to maintain, display, and upload key mapping data. The first tool is an SAP GUI transaction that can be accessed via the **SAP Easy Access** menu path **Accounting • Central Finance • Mapping • Manage Mappings** or by using Transaction FINS_CFIN_MAP_MANAGE, as shown in Figure 4.22 ❶. This transaction provides mass upload, download, and delete capabilities. You'll need the SAP_SFIN_CFIN_ADMIN role assigned to use the transaction. The second approach uses Web Dynpro–based Transaction MDG_KM_MAINTAIN ❷, which allows users to maintain one mapping entity at a time.

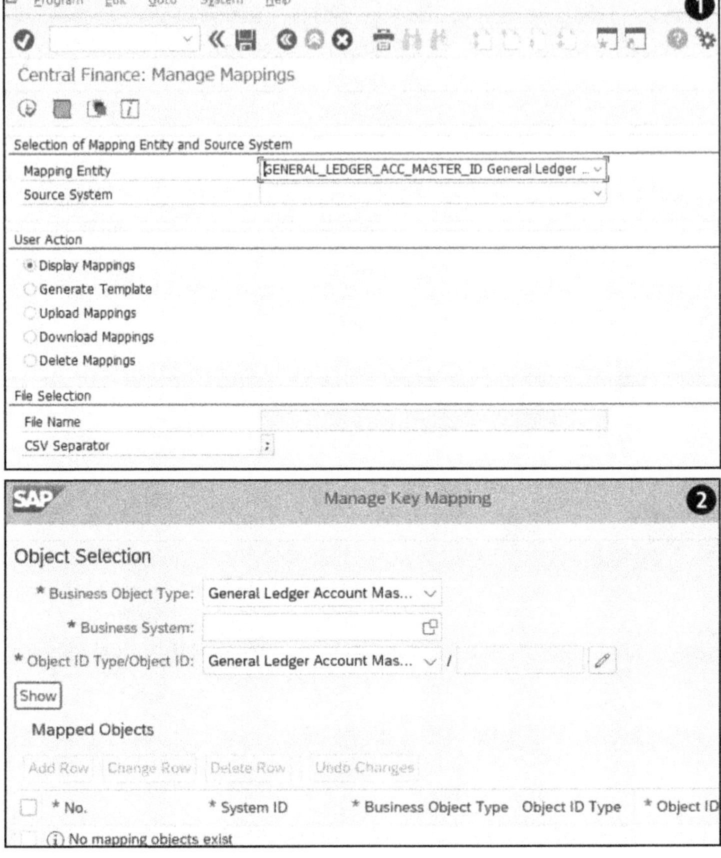

Figure 4.22 Key Mapping

With Transaction FINS_CFIN_MAP_MANAGE, you'll have the following options:

- **Display mappings**
  The mapping data is displayed in read-only mode.
- **Generate Template**
  You can download templates in CSV format for the selected mapping entity.
- **Upload Mappings**
  The downloaded template can be populated with mapping to upload for mass volume of mapping. When uploading mapping, the system displays logs of successful upload and failed errors.

- **Download Mappings**
  Downloads the selected mapping entity into a Microsoft Excel format or other formats.
- **Delete Mappings**
  Deletes mapping identified in the upload template.

The transaction provides the **Test Run** option to simulate the mapping prior to uploading data.

Transaction MDG_ANALYZE_IDM provides all maintained mapping key values for a business object. This is helpful to evaluate mapping and propose corrections if required.

4.4.4 Value Mapping

Value mapping in Central Finance refers to the code/type mapping defined in customizing activities. For example, customizing activities such as company codes, payment methods, document types, and so on are mapped in value mapping in Central Finance. In addition, most configuration settings must be identical between the source systems and target systems to prevent errors However, you can configure different code values as well. If the code values are different, then mapping is required for each field based on **Mapping Action** settings. For example, payment term 010 in the source system may be referred to as 100 in the target system. In this case, a value mapping is required, from 010 to 100 payment terms. Value mapping requires two customizing activities: assign code lists and maintain value mapping. Let's understand these customizing activities next.

Assign Code Lists

To maintain value mappings, one prerequisite configuration is to assign code lists. In assign code lists configuration, you assign data elements to each global data type that requires mapping. In addition, mappings must already exist in the configuration. For every global data type that requires mapping, you'll maintain an entry for each business system. To maintain entries in assign code lists, follow the IMG menu path **Financial Accounting • Central Finance • Central Finance – Target System Settings • Mapping • Define Value Mapping (Code Mapping) • Assign Code Lists to Elements and Systems.** You'll arrive at the **Change View "Assign Code Lists to Elements and Systems": Overview** screen, as shown in Figure 4.23. On this screen, click the **New Entries** ❶ button. On the resulting screen, maintain the **Type**, **Global Data Type**, **Business System**, **List ID**, **List Agency ID**, and **List Agency ID** fields. Note that the **Internal List ID** field is left empty for the global data type **BUKRS**. However, this field may be relevant for mapping when you have a context structure, such as a company- or country-specific mapping, linked to the mapping entity. For example, payment methods or payment terms are configured at the country level. For these fields, the **Internal List ID** field is populated with the country value. On the other hand, the **BLART**, **BSCHL**, **BUKRS**, **GSBER**, **KKBER**, and **KOKRS** global data types aren't set up with context structures. Once all fields are populated, click the **Save** button to save the entries. When you return to the initial screen, you can see the changes saved ❷, and additional changes can be made from this screen as well. Once any additional changes are made, click the **Save** ❸ button to save the entries.

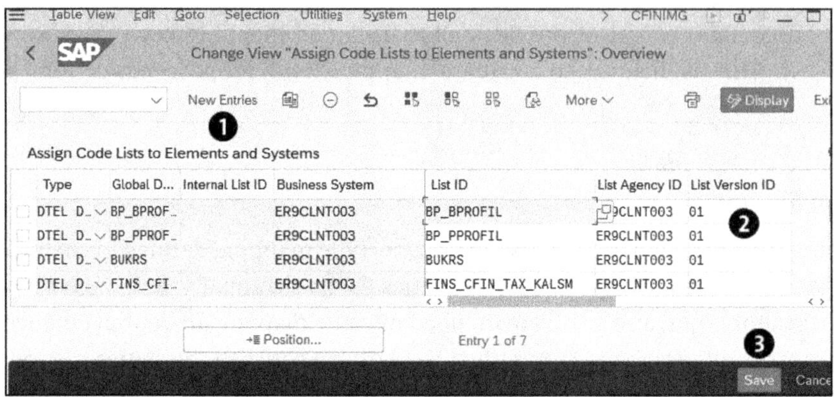

Figure 4.23 Assign Code Lists

Maintain Value Mapping

In this customizing activity, you maintain the value mapping for customizing activities. SAP provides several standard fields. For example, for each field, such as **BUKRS** or **BLART**, the source value is mapped to the target value. The prerequisite for this configuration is to assign code lists. To maintain a value mapping, follow the IMG menu path **Financial Accounting • Central Finance • Central Finance – Target System Settings • Mapping • Define Value Mapping (Code Mapping) • Maintain Value Mapping** or use Transaction CFINIMG. The system opens informational popup windows to caution you that the table is cross-client and that changes aren't permitted. Once you pass through the initial popup screens, the system opens the **Display View "Maintain Value Mapping Fields": Overview** screen. Select an object type and global data type, for example, **Data Element** and **BUKRS**, and click the **Display mapping relationships** icon in the middle of the screen, which opens the **Display View "Assign Code Lists": Overview** screen. Select **Edit**, and click **New Entries** to assign a code list, as shown in Figure 4.24 ❶. After maintaining code lists, double-click **Define Value Mapping** ❷ from the **Dialog Structure** on the left subscreen. On the resulting screen, maintain the mapping through the **Internal Code Value** field for the target system and through the **External Code Value** field ❸ for the source system. Click the **Save** ❹ button to save the entries.

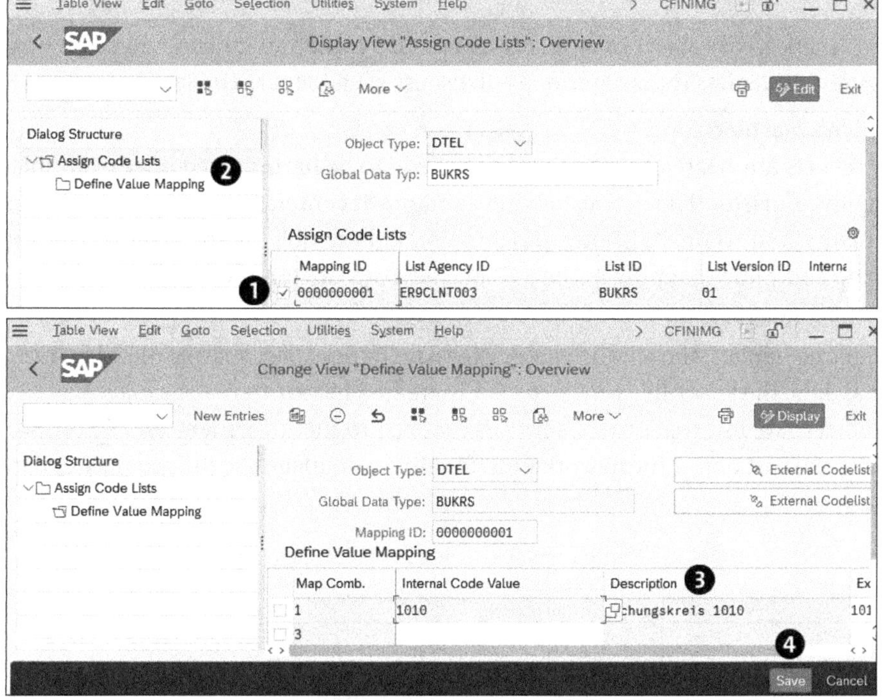

Figure 4.24 Define Value Mapping

[!]

Warning

Although key mapping or value mapping can be changed any time technically, SAP doesn't recommend changing either of these because it can result in data inconsistency and impacts subsequent follow-on documents failing in SAP Application Interface Framework. Moreover, this may lead to reconciliation issues between the source system and the Central Finance system.

4.4.5 Cost Object Framework

Cost objects are organizational elements in CO, and they help businesses to collect costs for detailed activities within an enterprise. In Central Finance, cost objects are created and mapped automatically after enabling the cost object framework. During financial postings replication, the system uses corresponding mapped cost objects to post the documents in Central Finance.

In this section, you'll learn about cost objects framework concepts and related configuration to support cost object replication. We'll begin with an overview of how SAP classifies long-lived objects versus short-lived objects. Then, we'll transition into cost object framework–related configuration: defining cost object mapping scenarios, defining mapping rules, and other miscellaneous topics in cost object replication.

Overview

In Central Finance, cost object mapping is separated by whether it's short-lived or long-lived. Figure 4.25 depicts this scenario. We'll discuss each scenario in detail next:

❶ **Long-lived master data**
These objects are relatively stable and only need to be harmonized once. Examples of long-lived master data are cost centers and profit centers, company codes, materials, plants, and so on. As shown in Figure 4.25, these objects are managed by key mapping and value mapping in the SAP Master Data Governance tool.

❷ **Short-lived master data**
As the name implies, the short-lived master data lives a short span of life. Like transactional data, the short-lived master data is created more often. Examples of short-lived objects are internal orders, services orders, production orders, and so on. SAP provides the cost object framework functionality to manage short-lived master data objects.

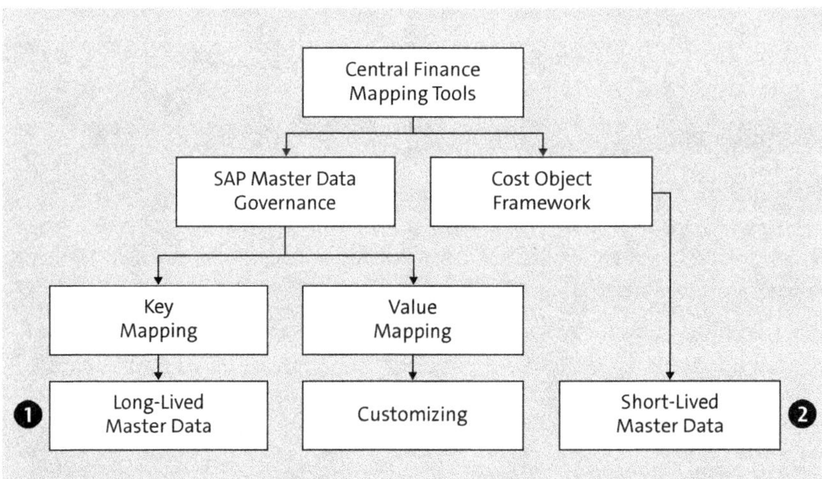

Figure 4.25 Long-Lived and Short-Lived Master Data

SAP provides the cost object framework functionality to create and change cost objects in Central Finance. With the cost object framework, you can do the following:

- Map a source system cost object to a Central Finance system cost object.
- Replicate cost objects from the source system to the Central Finance system automatically based on configuration rules.
- Replicate changes made to cost objects in the source system to corresponding cost objects in the Central Finance system.
- Simulate cost object mapping using smoke tests and via the initial load cost object mapping in SAP Landscape Transformation.

- Use flexible mapping options to either derive characteristics values from the local system or derive central characteristics from the target Central Finance system.
- Support customer fields replication.
- Support allowing subsequent changes to mapping rules once replication has started.
- Perform proof of concept via individual cost object replication.
- Correct cost objects after loading and replication.
- Delete cost objects and their assignments.
- Use BAdI options to enhance custom scenarios for above and beyond standard functionality.

Although the cost object framework supports cost object master data replication and changes, you should be aware of some limitations:

- Transfer of budget or plan information isn't supported.
- Replication of settlement rules assigned to cost objects isn't supported.
- Sales orders assigned with cost objects aren't replicated. Refer to Chapter 11 for information on the accounting view of logistics (AVL) for transfers of logistics data.

Tip

Consult SAP Note 2971596 for answers to frequently asked questions on the cost object framework functionality.

Now that you have the background of short-lived master data, in this section, you'll learn about the cost object framework and its underlying configurations such as the cost object mapping scenario and mapping rules. Figure 4.26 shows the cost object framework configuration overview:

❶ **Maintain cost object scenarios**
From configuration of the cost object framework, you'll complete the **Define Cost Object Mapping Scenario** customizing activity first. A scenario in this context contains default characteristics mapping for a specific type of cost object. SAP provides template scenarios.

❷ **Mapping rule**
Mapping rules apply to whether you want to clear or derive certain values from local characteristics. Rules are passed to SAP Master Data Governance to apply for mapping. *Derive from local* means the system gets the mapped value directly from the corresponding characteristics of the source object.

❸ **BAdI mapping**
If standard mapping doesn't fulfill the requirements, SAP provides `BADI_FINS_CFIN_CO_OBJECT_ASSIG` to map above and beyond the standard functionality.

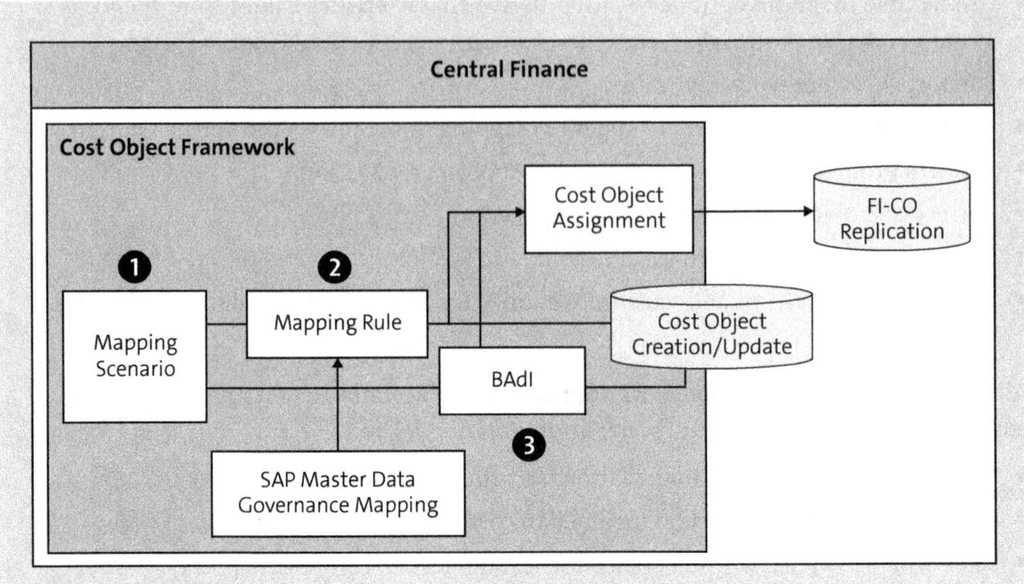

Figure 4.26 Cost Object Framework Configuration

The cost object framework transfers different type of cost objects master data from the source system to the Central Finance system. Table 4.15 displays Central Finance–supported cost object scenarios and supported cardinality.

Type of Cost Object in the Source System	Type of Cost Object in the Central Finance System	Supported Cardinality
PP Production order	Product cost collector	N:1
Product cost collector	Product cost collector	N:1 and 1:1
Internal order	Internal order	N:1/1:1
Service order (PM Order)	Service order (PM order)	N:1/1:1
QM order	QM order	N:1/1:1
PP production order	Internal order	N:1/1:1
Service order (PM order)	Internal order	N:1/1:1
Process order	Internal order	N:1/1:1
Process order	Product cost collector	N:1
CO production order	Internal order	N:1/1:1

Table 4.15 Cost Object Mapping Scenarios

[+]

Tip

Changes in cost objects in the source system replicated to the target system works only for the 1:1 scenario. Once the mapping scenario is activated, SAP doesn't support changing cardinality.

During replication of cost objects, the system reads cost object information from table AUFK and its dependent tables from the source system and create/update assignment table FINS_CFINT_ASGMT in Central Finance during transfer of master data. For the 1:1 scenario, table AUFK is updated from cost object replication automatically. However, for the N:1 scenario, the system creates a new order automatically. Subsequently, any new orders created get mapped in the standard cost object mapping table automatically. In general, for example, the N:1 scenario is used for production orders and maintenance orders. Later, during FI-CO document replication, if a cost object is required in a transaction line, the system uses the assignment table to find the corresponding target cost object in the Central Finance system.

Tip

For the latest supported cost object scenarios in Central Finance, refer to SAP Note 2180924. SAP Note 2308365 provides guidance on changes to cost object replication.

To create cost objects automatically, you must configure the cost objects mapping scenario and mapping rules. Figure 4.27 illustrates the process of how the system transfers cost objects from the source system to Central Finance. Let's explore the data flow of the cost object framework:

❶ **Maintain cost object scenarios**

First you'll complete the **Define Cost Object Mapping Scenario** customizing activity. A scenario in this context contains default characteristics mapping for a specific type of cost object. SAP provides template scenarios that can be copied to suit project requirements. Table 4.16 displays SAP-delivered template scenarios for the cost object framework.

Scenario Template	Cost Object in the Source System	Cost Object in the Central Finance System	Cardinality
SAP001	Production order	Product cost collector	N:1
SAP002	Product cost collector	Product cost collector	1:1
SAP003	Internal order	Internal order	1:1
SAP004	Service order (PM order)	Service order (PM order)	N:1
SAP005	QM order	QM order	N:1

Table 4.16 SAP-Delivered Scenario Templates

❷ **Maintain rules for scenario**
Once you've configured defining mapping scenarios, you'll configure **Define Mapping Rules for cost object scenarios**. This configuration establishes a mapping relationship between source cost objects and cost objects in the Central Finance system.

❸ **Cost object assignment**
After mapping rules have been configured, during cost object replication, the system creates target cost objects depending on the cost object mapping rule. Then, assignments between the source cost object and central cost object are created in assignment table FINS_CFIN_ASGMT in Central Finance.

❹ **FI-CO/CO document posting**
When FI-CO documents and CO documents are replicated over to Central Finance, the system derives cost objects from assignment table FINS_CFIN_ASGMT and posts the documents to the Universal Journal (table ACDOCA).

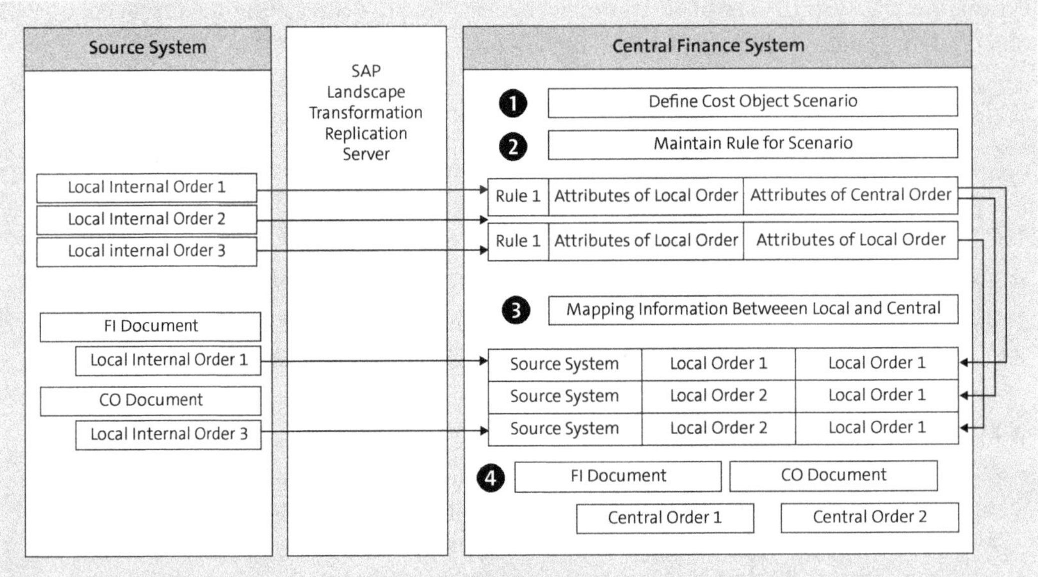

Figure 4.27 Cost Object Mapping Framework Overview

Defining the Cost Object Mapping Scenario

In this section, you'll learn about the customizing activity for **Defining Cost Object Mapping Scenario**, which maps the cost object category in the source system to a cost object category in the Central Finance system. After defining and activating the customizing activity, the system uses this information during cost object replication.

The leading practice is to copy from the template cost object scenarios delivered by SAP. Nevertheless, you can create a new cost object scenario from scratch. Prior to configuring a new cost object scenario, you'll require access to authorization object F_CFIN_GEN with actions, as shown in Table 4.17.

Type of Activity	Activity
CFIN_APPL	04
ACTVT	06 or 07

Table 4.17 Authorization Objects

To create a new cost object scenario, navigate to IMG customizing menu path **Financial Accounting • Central Finance • Central Finance: Target System Settings • Cost Object Replication • Cost Object Mapping • Define Cost Object Mapping • Define Scenarios for Cost Object Mapping** or navigate to it from Transaction CFINIMG. After entering the configuration activity, you'll arrive at the **Change View "Maintain Scenarios for Object Mapping": Overview** screen, as shown in Figure 4.28. On this screen, click the **New Entries** ❶ icon to create a new cost object scenario. You're taken to the **New Entries: Overview of Added Entries** screen. On this screen, enter values for these fields:

- **Scenario** ❷
  Enter a scenario name.
- **Scenario Description** ❸
  Enter the scenario description. The field has a maximum length of 100. .
- **Table Name** ❹
  Enter a table name to maintain the relationship between the source cost object and the central cost object. This table is created once the cost object scenario is activated. In addition, this table is used in the subsequent **Define Mapping Rules for Cost Object Mapping Scenarios** configuration (navigate from Transaction CFIN_MAPPING).
- **Source Object Category** ❺
  Choose the source object category from the dropdown list.
- **Central Object Category** ❻
  Choose the corresponding central object category from the dropdown list.
- **Cardinality** ❼
  Choose the cardinality between the source cost object category and central object category from the dropdown list, for example, 1:1 mapping for internal orders.
- **Status** ❽
  Activate the scenario once the source and central characteristics are selected. The system creates scenarios by default in **Inactive** status, which is ignored for replication. Scenarios with **Active** status replicate cost objects from the source system to the target system. The **Update** status is chosen for 1:1 cardinality to replicate cost objects only for updates.

Click the **Save** button to save the scenario. Once you've saved, double-click the **Source Characteristic** ❾ on the left pane by selecting the scenario you just saved.

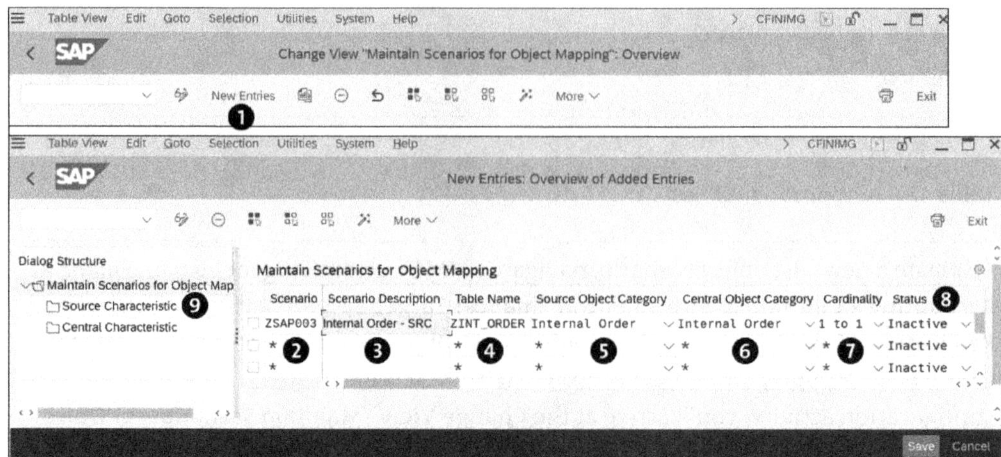

Figure 4.28 Maintain Scenarios for Object Mapping

Tip

The system doesn't take the **Only Update** scenario during the mapping rule conflict check. As a result, the system allows you to have overlapping mapping rules at the same time for active scenarios. For example, you've replicated a cost object, but you want to make changes to characteristics or mapping rules for any new cost objects created from that point onward. For this purpose, you'll create a new scenario. However, you'll change the status of the old scenario to **Only Update**. As a result, the old scenario is used when changes are made to the old cost objects replicated earlier.

Note that not all attribute changes are replicated to Central Finance. SAP limits the changes only to CO attributes. Consult SAP Note 2308365 for more on change replication.

On the resulting screen, as shown in Figure 4.29, the system displays frequently used fields ❶. However, you can also add other fields relevant for cost object replication per project requirements. To add a new characteristic, click the **New Entries** ❷ icon. Note that the source and cost object categories can be common among multiple scenarios, but the system allows one source cost object per scenario.

Next, you'll define central characteristics by double-clicking the **Central Characteristics** ❸ folder icon on the left pane. Like source characteristics, the system displays frequently used central fields ❹, but you can add other fields necessary per project requirements. To add a new central characteristic, click the **New Entries** ❺ icon, which displays a list of characteristics available. As you've noticed, certain fields have the **Derive from Local** checkbox selected, which means these fields keep values intact from the source system. On the other hand, for the fields without the indicator checked, the system derives values from the cost object mapping rule in the target system. Once editing is done on the scenario, click the **Save** ❼ icon to save the scenario.

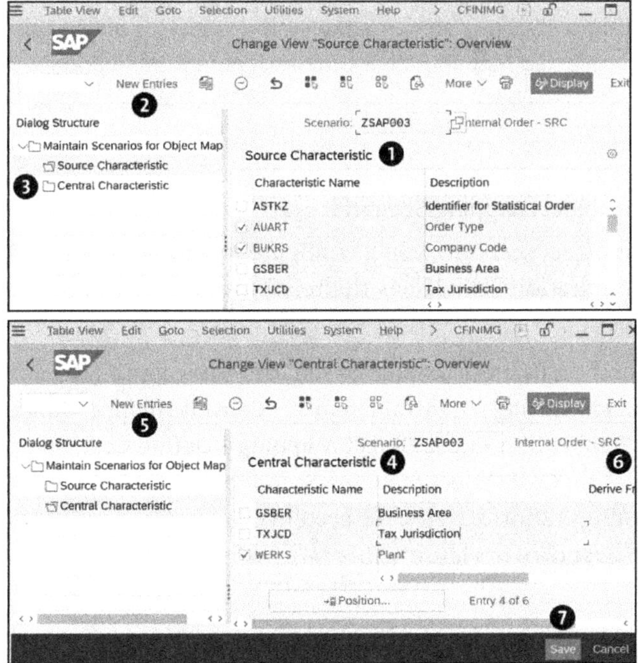

Figure 4.29 Maintain Scenarios for Object Mapping - Source and Central Characteristics

After saving the entry, double-click **Maintain Scenarios for Object Mapping** folder icon on the left pane. As shown in Figure 4.30, you'll activate the scenario by selecting **Active** from the dropdown list under the **Status** column ❶ located on the **Change View "Maintain Scenarios for Object Mapping: Overview** screen. Next, click the **Activate** ❷ icon. Click the **Save** button to save the entries.

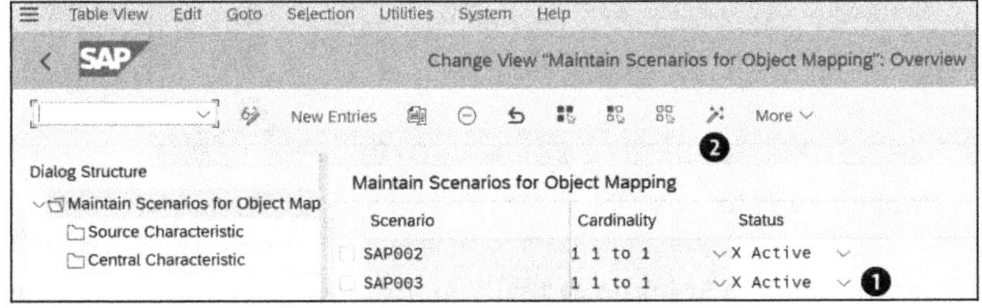

Figure 4.30 Define Scenarios for Cost Object Mapping - Activation

[+]

Tip

If you want to copy an existing scenario, click the **Copy** icon on the **Change View "Maintain Scenarios for Object Mapping": Overview** screen. Then, repeat the steps just described.

If you want to delete an existing cost object scenario, the system allows it only if no assignment data exists for the selected scenario. Check the assignment table prior to deleting cost object scenarios.

Define Mapping Rules for Cost Object Mapping Scenarios

Next, after configuring the cost object scenario, you'll configure mapping rules in this customizing activity. This configuration establishes the relationship between source cost objects and central cost objects.

To configure mapping rules for a cost object mapping scenario, navigate to IMG customizing menu path **Financial accounting • Central Finance • Central Finance: Target System Settings • Cost Object Replication • Cost Object Mapping • Define Cost Object Mapping • Define Mapping Rules for Cost Object Mapping Scenarios** or use Transaction CFINIMG. After entering the configuration activity, the system takes you to the **Central Finance Mapping Entry** screen, as shown in Figure 4.31.

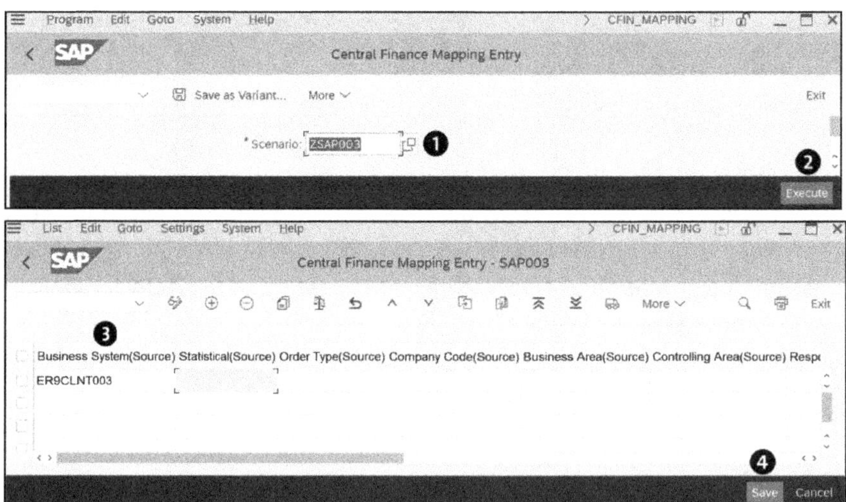

Figure 4.31 Define Mapping Rules for Cost Object Mapping Scenarios

On this screen, enter the value in **Scenario ❶** that you just created from the previous configuration, and then click the **Execute ❷** button. You'll arrive at the **Central Finance Mapping Entry - SAP003** screen. On this entry screen, you can see fields listed on the left for the source system and corresponding fields listed on the right for the Central Finance system. You'll populate matching characteristics to establish mapping between the source system and the Central Finance system ❸. Once mapping is done, click the **Save ❹** button to save the configuration.

[+]

Tip

The system allows you to make changes to mapping rules. However, note that if a rule is already used and cost objects have been created with that rule already, changes won't be reflected on the replicated cost object or in the assignment table. SAP provides Transaction CFIN_CO_MAPPING_CORR (Correct Cost Object Mapping) to remap the cost object per new rules defined. In addition, cost objects already created with old mapping rules aren't impacted. The system provides an option to create a new cost object and change the old assignment with newly created cost objects. Caution must be exercised when making such changes. You'll go through these changes in a test system prior to changing in the live system.

Correcting Cost Object Mapping

As discussed in the earlier section, in some cases, mapping rules changes may be required for cost object scenarios after replication. This happens after you've discovered some process changes or mistakes in the mapping rules. In such situations, SAP provides Transaction CFIN_CO_MAPPING_CORR (Correct Cost Object Mapping) to correct the assignment between the source cost object and target cost object. Once the mapping changes are made, with this transaction, the system creates a target cost object and then updates the cost object assignment table.

For demonstration purposes, let's assume you have a cost object mapping rule created with source order type **Y600** mapped to target system order type **0600** initially. Later, you'll find that the correct order type mapping is **0400** in the target system. Figure 4.32 shows the initial mapping rule ❶ and subsequent change in mapping rule ❷.

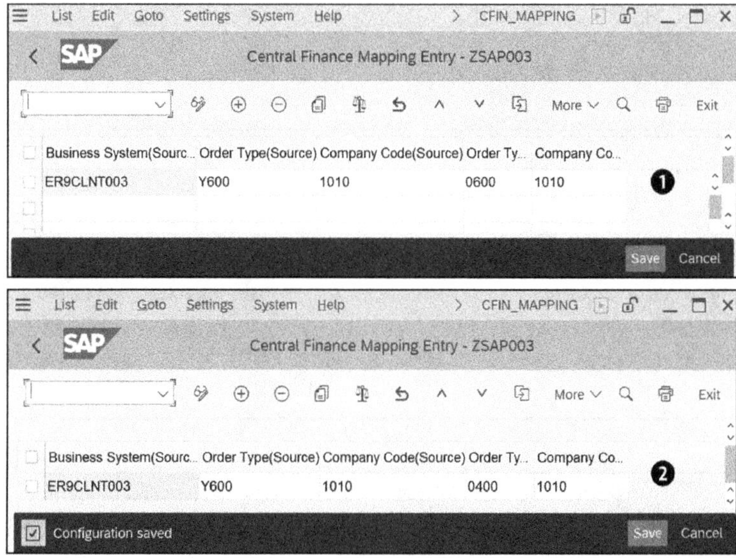

Figure 4.32 Cost Object Mapping Rule: Correction

After a successful replication of a cost object from the source system to the target system, to perform a subsequent correction on the cost object created earlier, navigate to IMG customizing menu path **Financial Accounting • Central Finance • Central Finance: Target System Settings • Cost Object Replication • Cost Object Mapping • Define Cost Object Mapping • Correct Cost Object Mapping** or use Transaction CFIN_CO_MAPPING_CORR. After entering the configuration activity, the system takes you to the **Cost Object Mapping Correction** screen, as shown in Figure 4.33. On this screen, enter values in the following fields:

❶ **Logical System ID**
Enter the source system where the correction is required.

❷ **Object Category**
Enter the cost object category in this field.

❸ **Object ID**
Enter the cost object ID, for example, internal order number or range of internal orders in this field. By leaving this field empty, the system takes all cost objects in the object category.

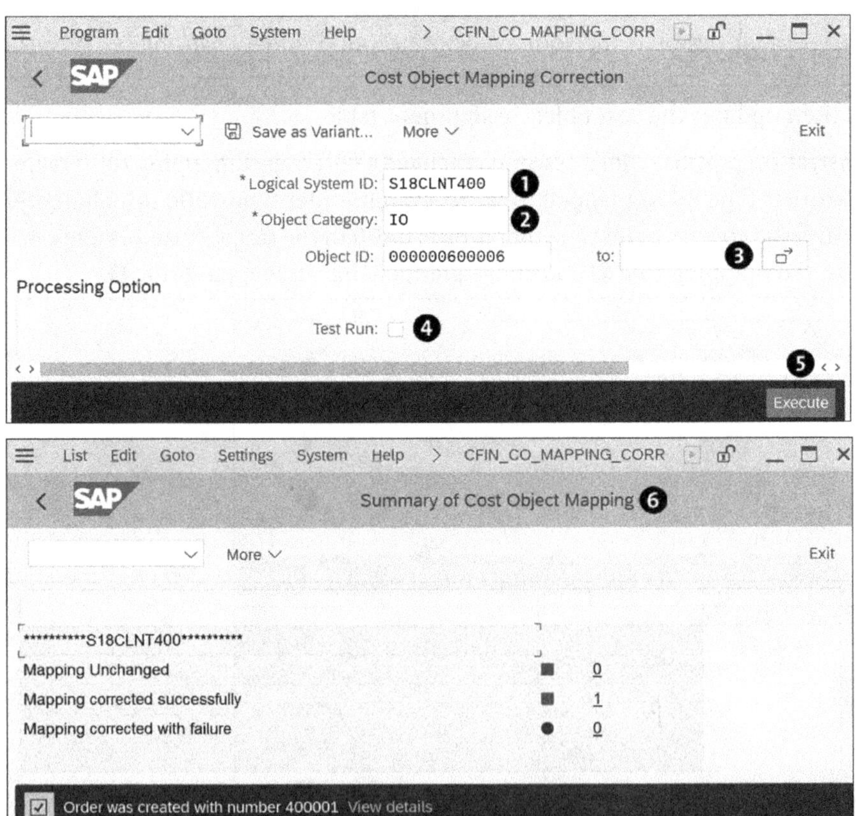

Figure 4.33 Correct Cost Object Mapping

You can run this in **Test Run** ❹ to see the correction results prior to running in productive mode. This step is always recommended to ensure no errors are uncovered during the test run. Click the **Execute** ❺ button. Next, the system takes you to the **Summary of Cost Object Mapping** ❻ screen. On this screen, the system displays results in a summarized form. You can navigate to the details by clicking the number for the cost object category.

[+]

Tip

After the new mapping rule is corrected using the transaction, the target cost objects created earlier under the old mapping rule won't be impacted or deleted. The change involves creating a new cost object in the target system and changing the assignment to the newly created cost object. In addition, there are limitations for cost object assignment scenarios with status **U** (**Update Only**), which isn't included in the correction report.

Once the mapping is corrected successfully, a new cost object is created in Central Finance assignment table FINS_CFINIT_ASGMNT, as shown in Figure 4.34. The new cost object **6000021** ❶ is generated in the mapping assignment table. On the new cost object, the order type **0600** ❷ has been assigned rather than the old order type **0400**. In addition, a new assignment has been generated for a new cost object.

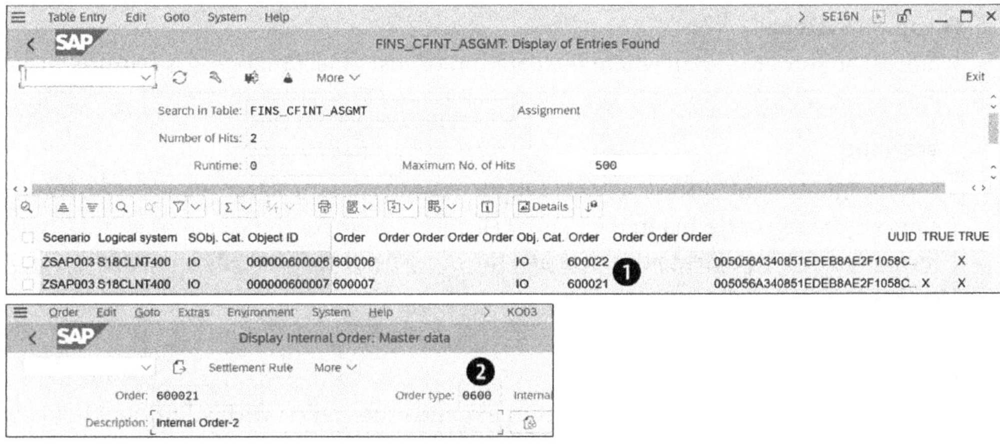

Figure 4.34 Correct Cost Object Mapping Table View

[!]

Warning

This transaction should be used with caution as this may cause inconsistency in the tables. You'll use the transaction only when you know the changes is absolutely required to support reporting.

For individual cost object testing, SAP provides Transaction CFIN_CO_OBJ_REPL, which can be used to test the scenario.

Deleting Cost Object Mapping and Cost Objects

SAP provides Transaction CFIN_CO_MAPPINR_DEL to clean up existing assignments and created cost objects. The transaction deletes entries in tables AUFK and FINS_CFINIT_ASGMNT only when no transaction data exists for the cost objects. This transaction is used when you want to clean up redundant cost objects that are no longer used, thereby improving performance. The system deletes cost object assignments synchronously, and cost objects are deleted asynchronously.

To delete cost object mapping and cost objects, navigate to IMG customizing menu path **Financial Accounting • Central Finance • Central Finance: Target System Settings • Cost Object Replication • Cost Object Mapping • Define Cost Object Mapping • Delete Cost Object Mapping and Cost Objects** or use Transaction CFIN_CO_MAPPING_DEL. After entering the transaction, the system takes you to the **Delete Mapping and Cost Objects in Central System** screen, as shown in Figure 4.35. On this screen, enter values in the following fields:

❶ **Scenario**
Enter the scenario name to clean up the cost object assignment.

❷ **Logical System ID**
Enter the source system where the deletion of cost object mapping is required.

[!]

Warning

Leaving the **Logical System ID** field empty deletes all cost object mapping from all source systems connected to the Central Finance system for the scenario selected. It's recommended to populate this field and the other fields to avoid any unintended execution of transaction.

❸ **Allow Cost Object Selection**
Choose the checkbox, and the system expands the screen with additional fields to enter.

❹ **Object ID**
Enter the cost object ID, for example, the internal order number or range of internal orders in this field. By leaving this field empty, the system takes all cost objects in the object category selected. The object category is also selected automatically based on the scenario chosen earlier.

❺ **Commit Package size**
Set the system default package size to **1000**. You can adjust the value per system landscape requirements depending on the volume.

Once all fields are populated, choose **Test Run** ❻ to see the simulated results prior to running in productive mode. This step is always recommended to ensure no errors are uncovered during the test run. Click the **Execute** ❼ button. The system prompts with a popup confirmation screen with the question **Deletes all Central Systems objects: Do you want to continue?** Click the **Yes** button to continue.

Next, the system takes you to the **Summary of Cost Object Mapping Deletion** screen. On this screen, the system displays results in a summarized form. You can navigate to the details by clicking **Deleted mappings** and **Deleted cost objects** ❽.

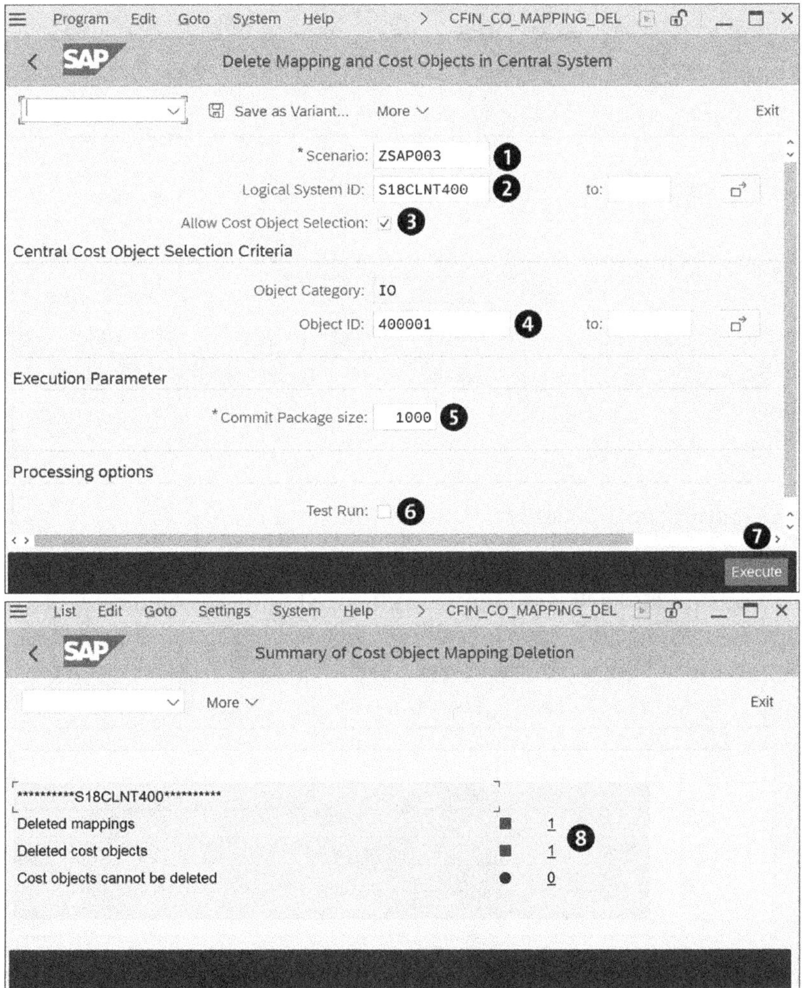

Figure 4.35 Delete Cost Object Mapping and Cost Objects

After clicking on the number for the deleted mapping and deleted cost objects, the system takes you to the screens shown in shown in Figure 4.36: **Deleted Mappings** ❶ and **Deleted Cost Objects** ❷.

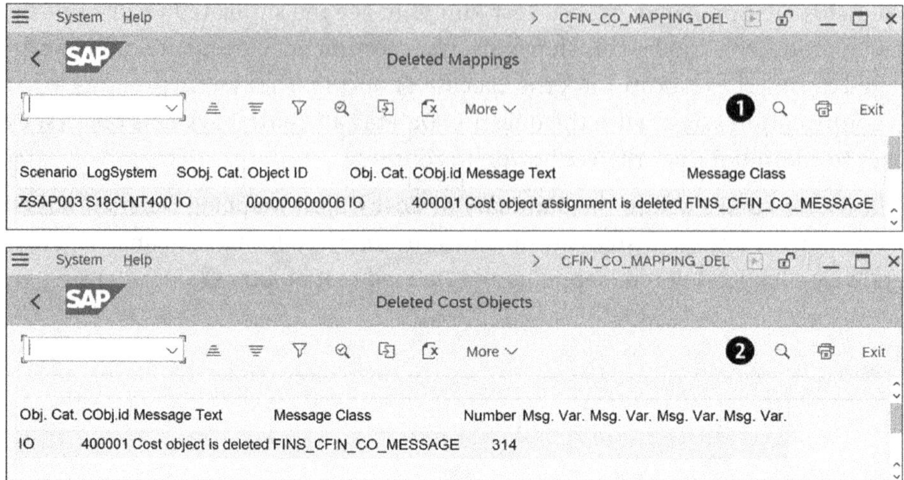

Figure 4.36 Delete Cost Object Mapping and Cost Objects: Table View

Tip

The transaction is supported only for cost objects replicated using the cost object framework in Central Finance. The system doesn't allow you to undo the deletion once cost objects are deleted.

4.4.6 Complex Mapping with a Central Finance BAdI

SAP provides a BAdI to perform above and beyond mapping or master data–related enhancements where standard delivered functionality can't fulfill the requirements. In Central Finance, several BAdIs are available related to master data enhancement, mapping, and validation. Table 4.18 lists SAP-delivered BAdIs along with methods that can be implemented in Central Finance for various master data enhancement or master data clearing scenarios.

BAdI	Description	Methods Available/ Enhancement Spot
BADI_FINS_CFIN_MAPPING_RULE	Determine mapping action.	DETERMINE_RULE
BADI_FINS_CFIN_MAPPING_PROCESS	Enhance standard mapping.	ES_FINS_CFIN_MAPPING

Table 4.18 Central Finance BAdIs for Master Data

BAdI	Description	Methods Available/ Enhancement Spot
BADI_FINS_CFIN_AC_ INTERFACE	Enhance standard processing of FI-CO posting data. Method MAP_VALUES can be used for validation of master data much earlier than posting.	PREPARE_INPUT_DATA MAP_VALUES ADJUST_POSTING_DATA CALL_AC_INTERFACE
BADI_FINS_CFIN_CO_ INTERFACE		GET_PROCESS_CONTROL MAP_REPLACE_CO_POSTING_ DATA/ ES_FINS_CFIN_CO_ INTERFACE
BADI_FINS_CFIN_CO_ OBJECT_ASSIG	SAP Master Data Governance mapping and WBS mapping to skip master data.	GET_PROCESS_CONTROL MAP_ORDER_MASTER_DATA

Table 4.18 Central Finance BAdIs for Master Data (Cont.)

4.4.7 Configuring Enhanced Business Mapping

SAP Central Finance provides customer-defined field mapping to support accounting interface replication. In Central Finance, it's a common business requirement to map or clear certain source fields data values. However, not all standard fields are available as mapping entities to enable mapping in the Central Finance system. SAP provides an enhanced business mapping entity functionality to add a new mapping entity without the technical team's help. To accomplish this, follow a three-step process:

1. Maintain the value mapping.
2. Define the mapping entity for the global data type.
3. Map the action.

Let's explore how to add a new mapping entity with the preceding steps in the following sections.

Maintain Value Mapping

The first step in this process is to create a value mapping definition. Without a value mapping, the system displays the error message **Type DTEL and Global Data Type are not enabled for MDG value mapping**. Therefore, you must enable global data types in the value mapping first. To maintain value mapping, navigate to Central Finance customizing activities menu path **SAP Customizing Implementation Guide • Financial Accounting • Central Finance: Target System Settings • Mapping • Define Value Mapping (Code Mapping) • Maintain Value Mapping**. After reaching the configuration, you'll be prompted with an information popup window cautioning you that the table is cross-client. Click the **Continue** icon. The system then takes you to the **Change View "Maintain Value Mapping Fields": Overview** screen. On this screen, click the **New Entries** button. You'll arrive at the **New Entries: Overview of Added Entries** screen, as shown in Figure 4.37. In our example, let's create a mapping entity for the planning level field (FDLEV). On this screen, maintain the following fields ❶:

- **Object Type**
  Select **DTEL Data Element** from the dropdown list.
- **Global Object Type**
  Enter "FDLEV" for the planning level field.
- **Client Dep**
  Select this indicator.
- **Input Help**
  Enter "CL_MDG_CODE_LIST_PROVIDER".

When finished, click the **Save** ❷ button to save the entry.

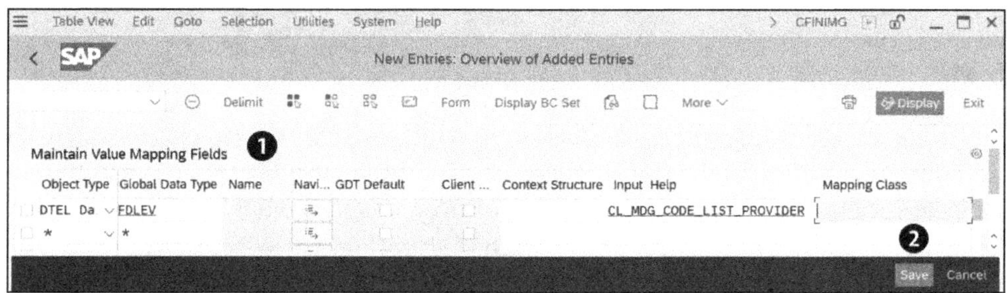

Figure 4.37 Value Mapping: Global Data Type

Define Mapping Entities (Enhanced Configuration)

In this customization, you'll create a mapping entity for the planning level field. SAP recommends that customer-defined mapping entities begin with Y or Z. To define a mapping entity, navigate to Central Finance customizing activities menu path **SAP Customizing Implementation Guide • Financial Accounting • Central Finance: Target System Settings • Mapping • Advanced Settings • Define Mapping Entities (Enhanced Configuration)**. After reaching the configuration, you'll arrive at the **Change View: "Define Mapping Entities": Overview** screen. On this screen, click the **New Entries** button. The system then takes you to the **New Entries: Overview of Added Entries** screen, as shown in Figure 4.38. Here, enter values for fields located under **Define Mapping Entities** ❶:

- **Mapping Entity**
  Enter mapping entity "Z_FDLEV" (starts with Y or Z for customer defined).

- **Type**
  Select **DTEL Data Element** from the dropdown list.
- **Global Data Type**
  Enter "FDLEV" as the global data type. Click the **Save** button to save the changes.

Now, choosing the row you just saved, double-click **Define Mapped Fields (Customer)** ❷ on the left pane located under **Dialog Structure**. This brings up the **New Entries: Overview of Added Entries** screen shown at the bottom of Figure 4.38. On this screen, enter the following values:

- **Structure**
  Enter "ACCIT" for the accounting interface item information, and then press Enter. This will make the **Type Name** field value automatically populated.
- **Field Name**
  Enter "FDLEV" in this field.

Click the **Save** ❸ button to save the entry.

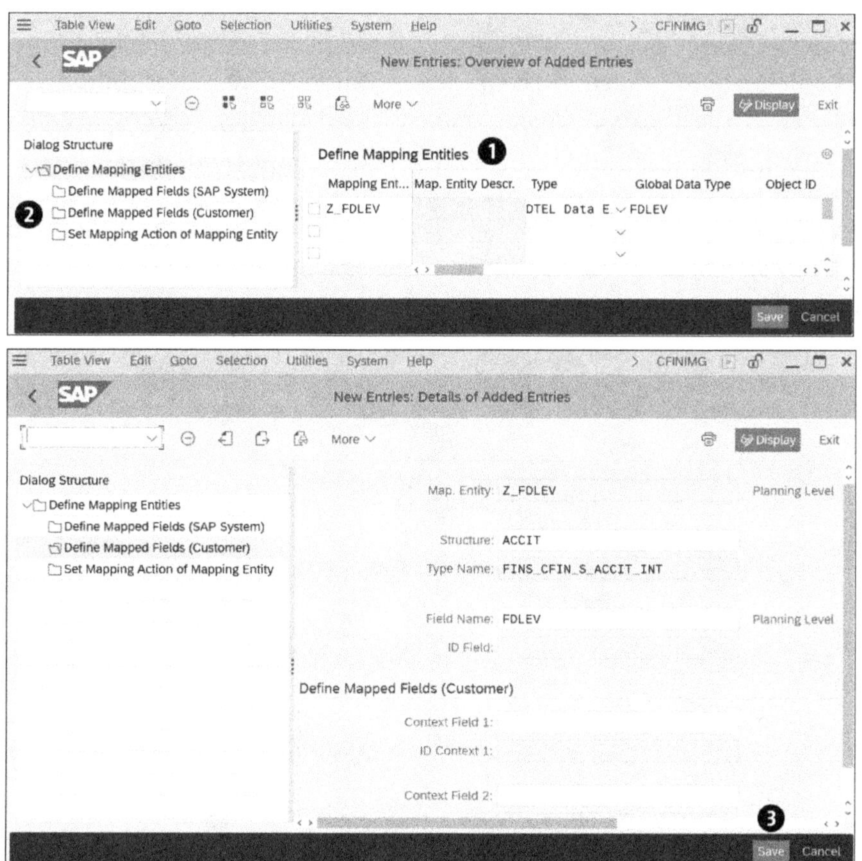

Figure 4.38 Define Mapping Entities (Enhanced Configuration)

Set the Mapping Action of the Mapping Entity

Next, the mapping action setting is required to complete the configuration. **Mapping Action** comes with four possible options, as listed in Table 4.19.

Mapping Action	Purpose
Mapping Obligatory	This will set the planning level field as mandatory. The value can come from the source, but there must be a corresponding mapping .
Keep Data	Use this option if you want to retain data coming from the source system.
Clear Data	This option clears the value coming from the source system.
Map if Possible	The system tries to map the field value. If the field value isn't mapped, the system takes the original data from the source system, and no error will be raised.

Table 4.19 Mapping Action

We'll continue the mapping action setup from the previous **Define Mapping Entities (Enhanced Configuration)** configuration step. As shown in Figure 4.39, from the **New Entries: Overview of Added Entries** screen, double-click **Set Mapping Action of Mapping Entity** ❶ in the left pane located under **Dialog Structure**. On the resulting screen, click the **New Entries** button. This brings up the **New Entries Overview of Added Entries** screen. Here, maintain values in the **Business System** field, and select the appropriate value from the **Mapping Action** ❷ dropdown list per your project requirements. Click the **Save** ❸ button to save the entry.

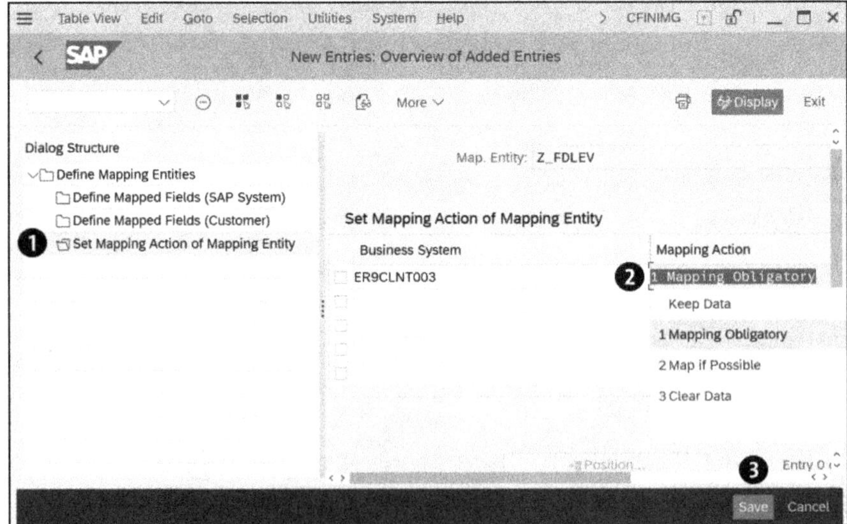

Figure 4.39 Set Mapping Action of Mapping Entity

When you post an FI document with the planning level at the line item in the source system, the document is replicated with the error message that the planning level isn't mapped (see Figure 4.40 ❶). Now that you've defined the planning level field in the mapping entity, this field is available for mapping in Transaction FINS_CFIIN_MAP_MANAGE under the **Mapping Entity** ❷ dropdown list.

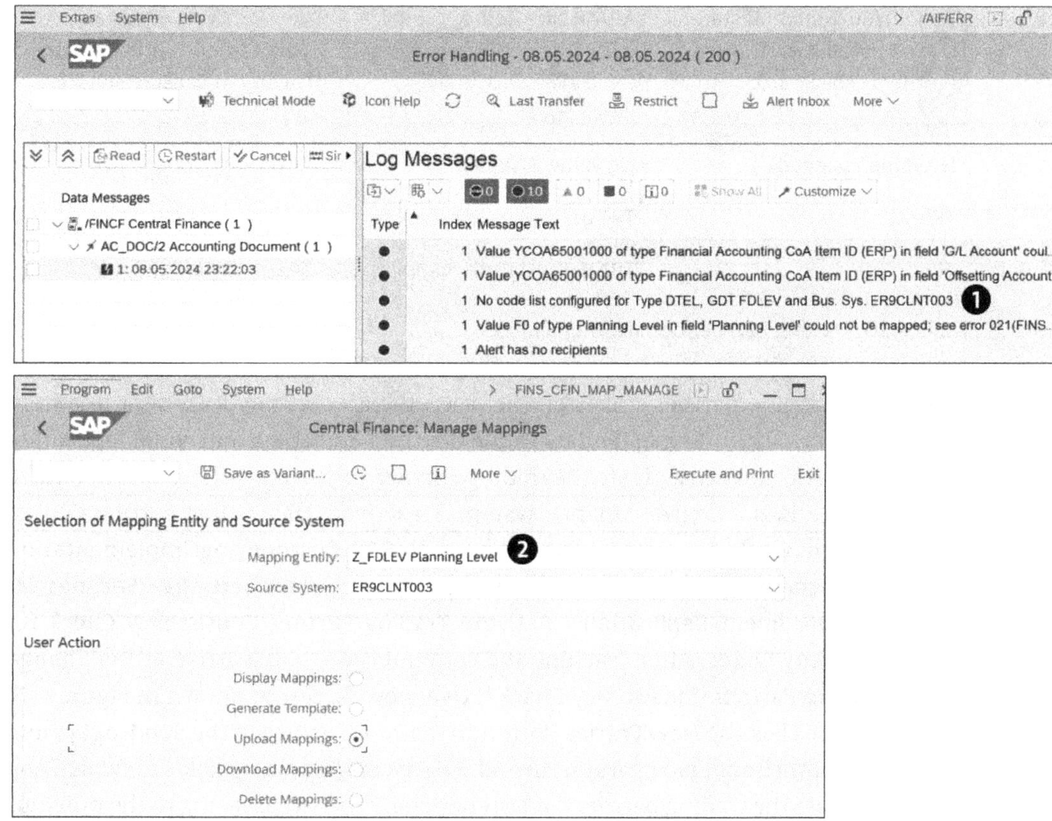

Figure 4.40 Mapping with Planning Level Mapping Entity

4.4.8 Trading Partner Consistency Check

The trading partner consistency check can be activated from the release SAP S/4HANA 2022 onward. With this new feature, you can activate or deactivate trading partner consistency checks for a company code. SAP supports the trading partner consistency check by displaying the error message **Trading partner &1 in AP/AR line item is inconsistent with master data** in SAP Application Interface Framework. Table 4.20 lists a scenario in which the trading partner consistency check is performed, and the system displays the error during replication of the document to Central Finance. Trading partner is assigned to the vendor/customer master data, and then it's populated to the corresponding AP/AR accounting document line items. On the other hand, the trading partner is mapped in the Central Finance system through key mapping but isn't derived again while posting the FI accounting document from the vendor/customer master data in the Central Finance system. It can lead to different trading partners assigned to the FI accounting document line item compared to the trading partner used in the vendor/customer master data in the Central Finance system.

Vendor/Customer Master Data Defined with Trading Partner in the Source System	AP/AR Line Items Posted with Trading Partner in Central Finance after Replication	Vendor/Customer Master Data Defined with Trading Partner in Central Finance System
No value assigned	No value assigned	TP1
TP0	TP1 derived from mapping	No value assigned
TP0	TP1 derived from mapping	TP2

Table 4.20 Trading Partner Consistency Check

To keep the trading partner consistent between the accounting document line item and the vendor/customer master data in the Central Finance system while replicating the FI document, you can activate the trading partner consistency check for the source company code in the Central Finance system. To activate the trading partner consistency check, navigate to customizing menu path **SAP Customizing Implementation Guide • Financial Accounting • Central Finance: Target System Settings • Settings for Accounting Document Replication • Activate Trading Partner Consistency Check for Source Company Codes**. After reaching the configuration, you'll arrive at the **Change View: "Trading Partner Consistency Check": Overview** screen, as shown in Figure 4.41. On this screen, click the **New Entries** button to maintain values in the **SendLogSys** and **Src CoCode** for the sender logical system and source company code fields for which you want to activate the trading partner consistency check. Upon returning to the previous screen, you'll notice the trading partner defined.

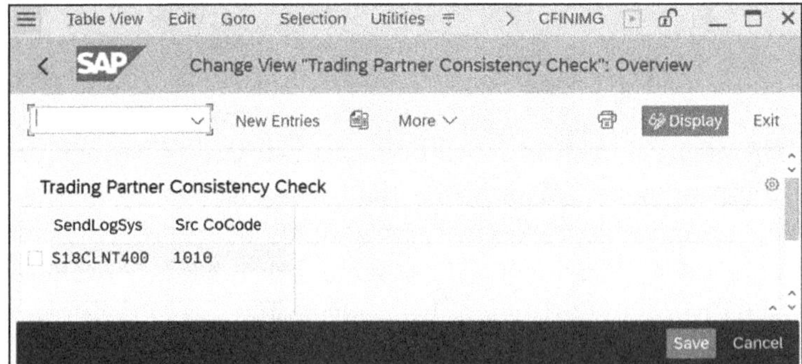

Figure 4.41 Activate the Trading Partner Consistency Check for the Source Company Codes

After activating the trading partner consistency, post an account document in the source system where the trading partner value on the line item is different from what's in the Central Finance mapping. As shown in Figure 4.42, the SAP Application Interface Framework screen displays the error message **Trading partner &1 in AP/AR line item is inconsistent with master data**.

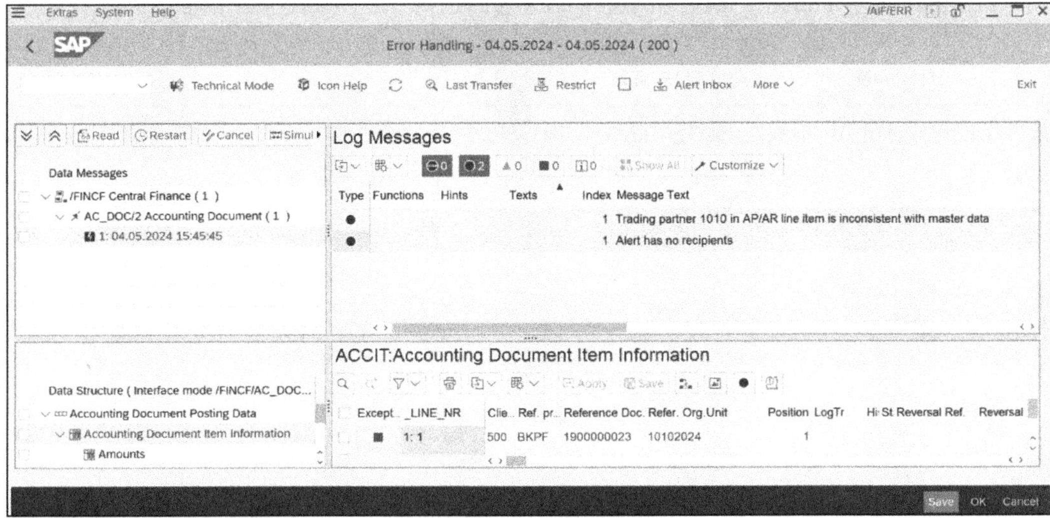

Figure 4.42 Trading Partner Replication Error

Note

If central payment is active for the source company code and intercompany AP/AR functionality is active, SAP doesn't support deactivating the trading partner consistency check once activated.

Tip

For additional guidance on trading partner derivation in FI, consult SAP Note 2672390.

4.5 Audit Log for Key Mapping

SAP provides audit capabilities to track changes to key mappings in Central Finance with Transaction FINS_CFIN_KM_AUDIT. With the transaction, you can review changes and deletion history for the key mapping objects. However, to view the audit log for key mapping, you must activate it first in configuration. In this section, we'll explore how to activate the configuration and view the changes to key mapping with the transaction.

4.5.1 Define Object Identifier

In this customizing activity, you'll identify the business object type for the mapping entity. To get to the business object type for the mapping entity, follow the Transaction SPRO menu path by navigating **Cross-Application Components • Processes and Tools for Enterprise Applications • Master Data Governance • Data Replication Settings • Key Mapping • Enhance Key Mapping Content • Define Object Identifiers**. As shown in Figure 4.43, the standard **BO Type** (business object type) **899** is set for the mapping entity **FIN_ACC_CHART_OF_ACC_ITEM_ID**.

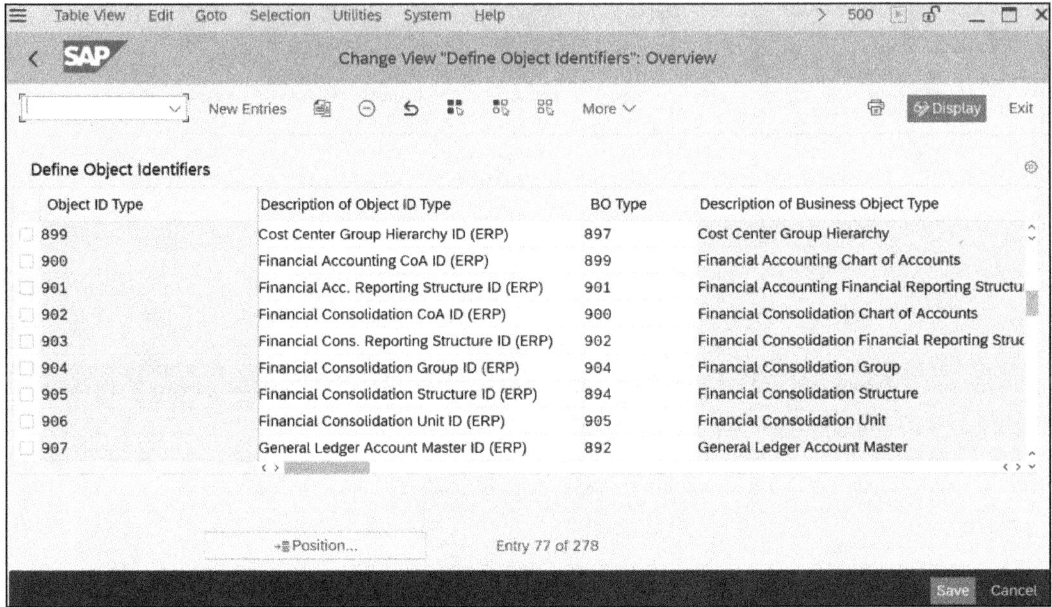

Figure 4.43 Define Object Identifier

4.5.2 Assign Business Objects to Main Contexts

In this step, you'll identify the main context to which business object types relevant to the mapping entity is assigned. To see the assignment, follow Transaction SPRO menu path by navigating **Cross-Application Components • Processes and Tools for Enterprise Applications • Master Data Governance • Data Replication Settings • Key Mapping • Enhance Key Mapping Content • Assign Business Objects to Main Contexts**. As shown in Figure 4.44, **ChartOfAccounts** ❶ is the **Main Context** for **BO Type 899** ❷ (for the mapping entity `FIN_ACC_CHART_OF_ACC_ITEM_ID`). SAP allows more than one **BO Type** or mapping entity to be associated with one **Main Context**. Once you activate the change log for the **Main Context**, it activates all the associated mapping entities. You can also see that **BO Type 900** is associated with **Main Context ChartOfAccounts**.

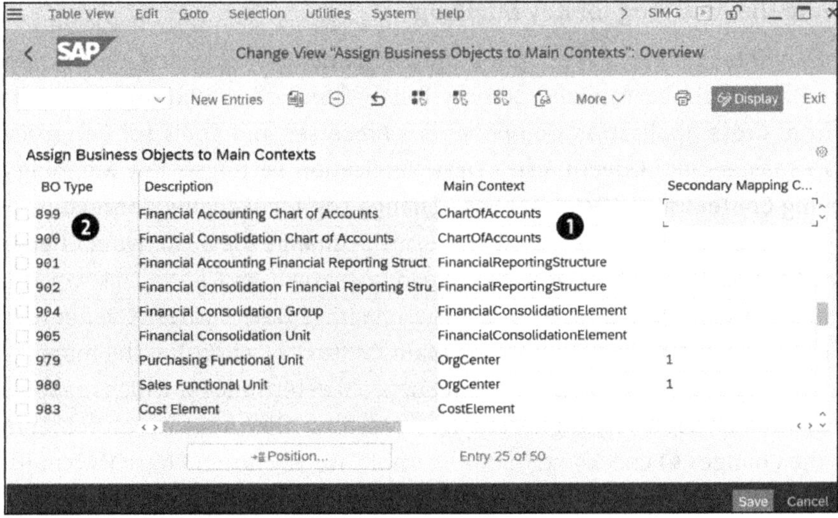

Figure 4.44 Assign Business Objects to Main Contexts

4.5.3 Define Mapping Contexts

Key mapping change logs are activated at the **Mapping Context** level, not at the level of mapping entity. Key mapping is performed using Transaction FINS_CFIN_MAP_MANAGE. This transaction allows you to display, download, or delete key mapping in Central Finance. In this configuration step, you'll ensure **Mapping Contexts** exist for the business object types. In addition, you can define a new mapping context for the Unified Key Mapping Service (UKMS). To see the mapping contexts that exist for the CoA mapping context, follow Transaction SPRO menu path by navigating **Cross-Application Components • Processes and Tools for Enterprise Applications • Master Data Governance • Data Replication Settings • Key Mapping • Define Mapping Contexts for UKMS • Define Mapping Contexts**. As shown in Figure 4.45, **ChartOfAccounts Main Context** exists in this configuration by the standard delivered.

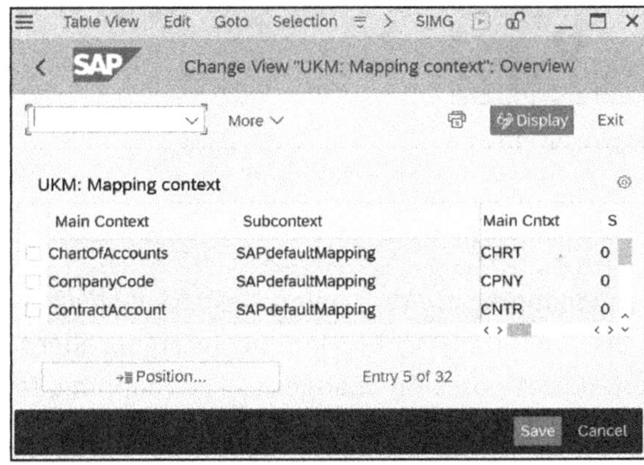

Figure 4.45 Define the Mapping Context

4.5.4 Activate the Audit Log for Key Mapping

To enable the audit log for key mapping, you must activate the change log for the mapping context. To reach the customizing activity, follow Transaction SPRO menu path by navigating from **Cross-Application Components • Processes and Tools for Enterprise Applications • Master Data Governance • Data Replication Settings • Key Mapping • Define Mapping Context for UKMS • Activate Change Log for Mapping Context**. After reaching the configuration activity, you'll arrive at the **Change View "Activate Change Log for Key Mapping Context": Overview** screen, as shown in Figure 4.46. On this screen, click the **New Entries** ❶ button. On the resulting **New Entries: Overview of Added Entries** screen, maintain values in the **Main Context** by choosing the mapping context from the dropdown menu. On this screen, you'll determine for which mapping contexts you want to allow logging of changes based on the project requirements. Next, select the **Changes** ❷ checkbox. In our example, we've chosen **ChartOfAccounts** key mapping objects to demonstrate the functionality. Click the **Save** ❸ button.

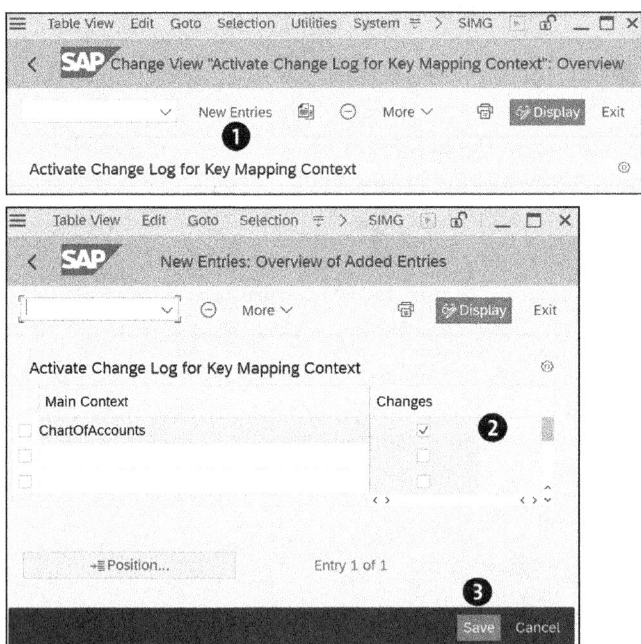

Figure 4.46 Activate Change Log for Mapping Context

4.5.5 Run Audit Log Report

In Central Finance, audit logs for the key mapping are very helpful, especially because they provide a time frame of change history for any replication inconsistency identified. The Central Finance target system doesn't prevent changes to key mapping, for example, to the general ledger account, in the middle of the month. SAP doesn't recommend changing mapping as it may introduce data inconsistencies in the system. In addition, this may lead to financial reconciliation issues. Nevertheless, if this happens, you need a mechanism to find the changes made to mapping and who made the changes. With the audit log report, you can track mapping changes.

After activating the change log for the mapping context, you can run the audit log report in Central Finance. As a prerequisite, the `SAP_SFIN_CFIN_AUDITOR` role must be assigned to the person who will be executing the transaction. To execute the audit log transaction, navigate to it from **SAP Easy Access** menu path **Accounting • Central Finance • Mapping • History of Key Mappings** or use Transaction FINS_CFIN_KM_AUDIT. Upon entering the transaction, you'll arrive at the **Central Finance: History of Key Mapping** screen, as shown in Figure 4.47. On this selection screen, enter values in the following:

❶ **Systems**
Enter the **Source System** value.

❷ **Value Options**
Choose the **Mapping Entity** CoA from the dropdown list.

❸ **Time Frame**
Enter the time frame in which you want to see the audit log.

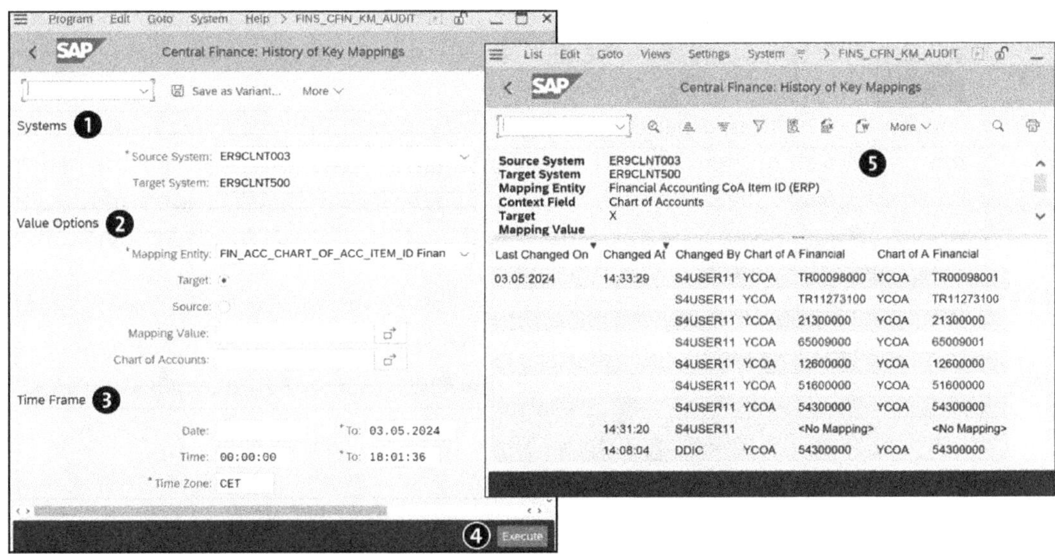

Figure 4.47 History of Key Mappings

Select the **Display List with Descriptions** indicator under **Output Options** as required. Click the **Execute** ❹ button. You'll arrive at the **Central Finance: History of Key Mappings** output screen ❺.

From Figure 4.47 choose the first row and click the **Display** button. The **Action** field shows whether the mapping is added, deleted, or changed, as shown in Figure 4.48.

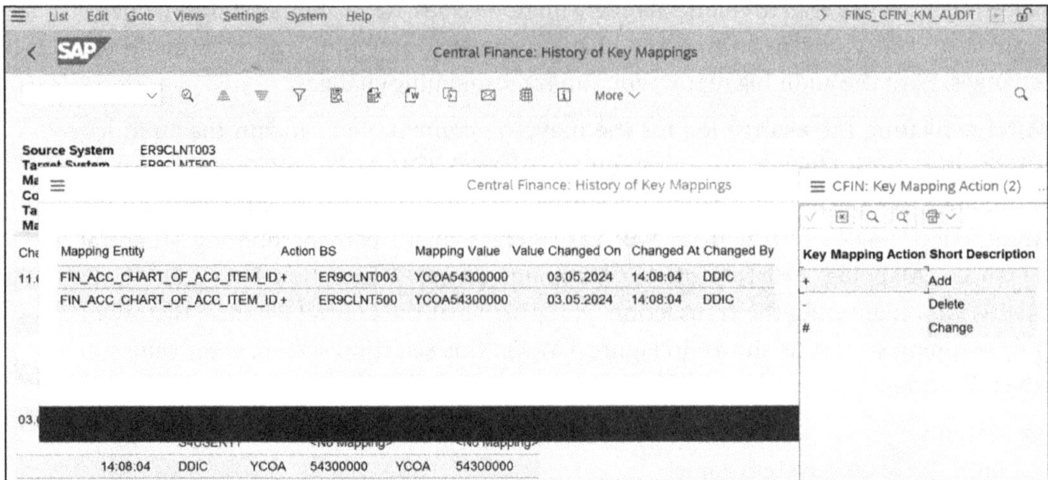

Figure 4.48 History of Key Mappings: Display Action

4.6 Summary

Master data maintenance and distributing master data in a heterogenous landscape is a challenging task. You'll need the right tools and technologies to achieve the desired goals. In addition, master data profiling, cleansing, and harmonizing are time-consuming activities, especially in larger enterprises. In this chapter, we started the introduction to master data and how you'll harmonize master data for the Central Finance scenario. Then, we walked you through the important cleansing activities you would carry out in the source system. After that, we discussed SAP Master Data Governance and relevant tools. Then, we transitioned into Central Finance mapping functionality. Finally, the chapter ended with running audit log reports for key mapping objects. In the next chapter, you'll learn configuring another important component in the SAP Central Finance landscape: SAP Landscape Transformation Replication Server.

Chapter 5

Configuring SAP Landscape Transformation Replication Server

5

SAP Landscape Transformation Replication Server is an important component of Central Finance that helps to replicate data from various source systems in real time. SAP Landscape Replication Server comes with powerful transformation and filter features to support replication of data from both SAP and non-SAP systems.

SAP Landscape Transformation Replication Server acts as a middleware platform in any landscape where data needs to be transferred and synchronized. Once the initial data is transferred, SAP Landscape Transformation Replication Server supports delta data replication as well. This makes reporting data much easier as the data can be replicated in real time or be scheduled for a specific time.

Why SAP Landscape Transformation for Central Finance? There are several replication tools available to transfer data, but SAP Landscape Transformation provides a central foundation for real-time finance for any SAP-based system or non-SAP-based system. Thus, SAP uses SAP Landscape Transformation Replication Server as the replication engine for the SAP S/4HANA for Central Finance foundation. SAP Landscape Transformation provides the following features:

- **Real-time data replication**
  SAP Landscape Transformation Replication Server enables real-time data replication from multiple source systems to the target Central Finance system. SAP Landscape Transformation Replication Server uses a trigger-based replication approach to pass data from the source system to the target system.
- **Heterogeneous system support**
  SAP Landscape Transformation Replication Server can replicate data from different kinds of source systems, including SAP and non-SAP systems, which enables a high degree of flexibility in data integration.
- **Data transformation and mapping**
  SAP Landscape Transformation Replication Server provides a range of data transformation and mapping capabilities, allowing users to customize the data per the requirements.

- **Monitoring**
  SAP Landscape Transformation Replication Server provides a range of monitoring capabilities that enable users to get the data replication status and resolve issues in real time.

From a technical perspective, SAP Landscape Transformation Replication Server is part of the Data Migration Server (DMIS) add-on for SAP NetWeaver. Installing the SAP Landscape Transformation Replication Server DMIS provides security authorization objects, functions, and features to support Central Finance activities. SAP released the first SAP Landscape Transformation Replication Server with version 1.0 (also called DMIS 2010) released in 2011. Since the first release, SAP Landscape Transformation Replication Server has evolved for more than a decade now with a lot of new features and greater product stability. The current version of SAP Landscape Transformation Replication Server is 3.0 (DMIS 2018) released in 2022. Support Package (SP) 12 is the current service pack for SAP Landscape Transformation Replication Server 3.0. In addition, SAP also released SAP Landscape Transformation Replication Server Replication Server for SAP S/4HANA 1.0 (DMIS 2020). Table 5.1 shows the different SAP Landscape Transformation Replication Server releases and corresponding SAP Note references.

As you can see in the table, mapping of service packs between DMIS 2020 and DMIS 2018 is presented. In addition, DMIS 2020 SP is identical to the corresponding DMIS 2018 SP released at the same time. Consequently, all information about features, compatibility, and corrections are identical to the corresponding DMIS 2018 SP.

DMIS 2020	DMIS 2018	DMIS 2018 Note
SP00	SP04	2883752
SP01	SP05	2955790
SP02	SP06	3010969
SP03	SP07	3130804
SP04	SP08	3151041
SP05	SP09	3254674
SP06	SP10	3314851
SP07	SP11	3383627
SP08	SP12	3438948

Table 5.1 Service Pack Level Mapping between DMIS 2020 and 2018

[«]

Note

For further guidance on mapping between DMIS 2018 and DMIS 2020 SP levels, consult SAP Note 2965075 - Release Information SAP Landscape Transformation Replication Server – DMIS 2020 (SAP Landscape Transformation Replication Server for SAP S/4HANA).

If the source system SAP ERP uses a third-party database with a runtime-database license, SAP recommends using the *business integration* scenario in SAP Landscape Transformation Replication Server Replication Server rather than using *data-based replication*. To facilitate this, SAP recommends implementing SAP Notes 2223621 and 2223801.

In this chapter, you'll learn SAP Landscape Transformation Replication Server's basic concepts, architecture, and step-by-step configuration for the important replication objects in Central Finance. The chapter then transitions to advanced features and expert functions. In addition, the chapter will clear up misconceptions about SAP Landscape Transformation Replication Server and leading practices to be followed when implementing Central Finance.

5.1 Getting Started with SAP Landscape Transformation Replication Server

In this section, you'll learn foundational topics in SAP Landscape Transformation Replication Server for Central Finance scenarios. We'll focus on the following areas:

- SAP Landscape Transformation Replication Server overview and concepts
- Prerequisites to be considered prior to SAP Landscape Transformation Replication Server configuration
- SAP Landscape Transformation Replication Server deployment options
- SAP Landscape Transformation Replication Server architecture
- Central Finance replication tables scope

5.1.1 SAP Landscape Transformation Replication Server Overview

Figure 5.1 illustrates a high-level overview of SAP Landscape Transformation Replication Server features. As you observe, SAP Landscape Transformation Replication Server acts as a middleware with three key functions:

❶ **Data transfer**
SAP Landscape Transformation Replication Server provides data transfer capabilities either in real time or on a scheduled basis. In Central Finance, real-time replication is a common use case scenario. In addition, besides the initial transfer of data,

the solution offers a delta data replication capability. Furthermore, SAP Landscape Transformation Replication Server is also capable of scheduled data replication based on certain time intervals according to business requirements.

❷ **Transformation**
When transferring data from source systems to target systems, SAP Landscape Transformation Replication Server supports a filtering capability to bring only the data necessary from source systems. In addition, you'll be able to enrich data, convert data, or skip data before transferring data to target systems.

❸ **Operation and monitoring**
Once you've configured and activated SAP Landscape Transformation Replication Server features, you'll be able to monitor data with the following functions:

- Start replication
- Stop replication
- Resume replication
- Monitor with SAP Solution Manager from release 7.1 SP5 or higher
- Administrative and maintenance tasks
- Email alerts capability

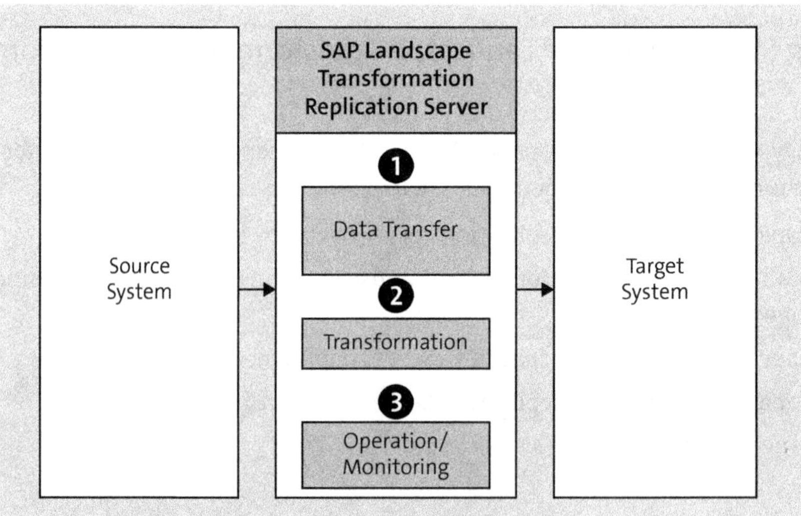

Figure 5.1 SAP Landscape Transformation Functionality

Moreover, SAP Landscape Transformation Replication Server can be applied in various situations where data needs to be transferred among different systems in a distributed network. Although SAP Landscape Transformation Replication Server can be catered for diverse business requirements, the most important use cases are as follows:

- Replicating data from SAP running on traditional relational databases (e.g., SQL server) into an SAP HANA database

- Replicating data from source systems to SAP Business Warehouse (SAP BW)
- Replicating data from the core data services (CDS) view and compatibility views, which is only applicable if SAP S/4HANA is a source system
- Replicating data into SAP HANA on Cloud Foundry, which is an SAP HANA database as a service (DBaaS)
- Replicating data from source system to cloud-based system
- SAP Landscape Transformation Replication server is now offered with SAP Business Technology Platform (SAP BTP) and supported on both on-premise and cloud platforms
- Replicating data from non-SAP source systems to SAP S/4HANA systems
- Replicating data from SAP S/4HANA as the source system to SAP S/4HANA systems
- Replicating data from the SAP cloud as the source system to SAP S/4HANA systems

The last three use cases are primarily used in Central Finance scenarios where SAP Landscape Transformation Replication Server enables replication of data from the source systems without disrupting existing processes. Figure 5.2 displays a high-level supported use case.

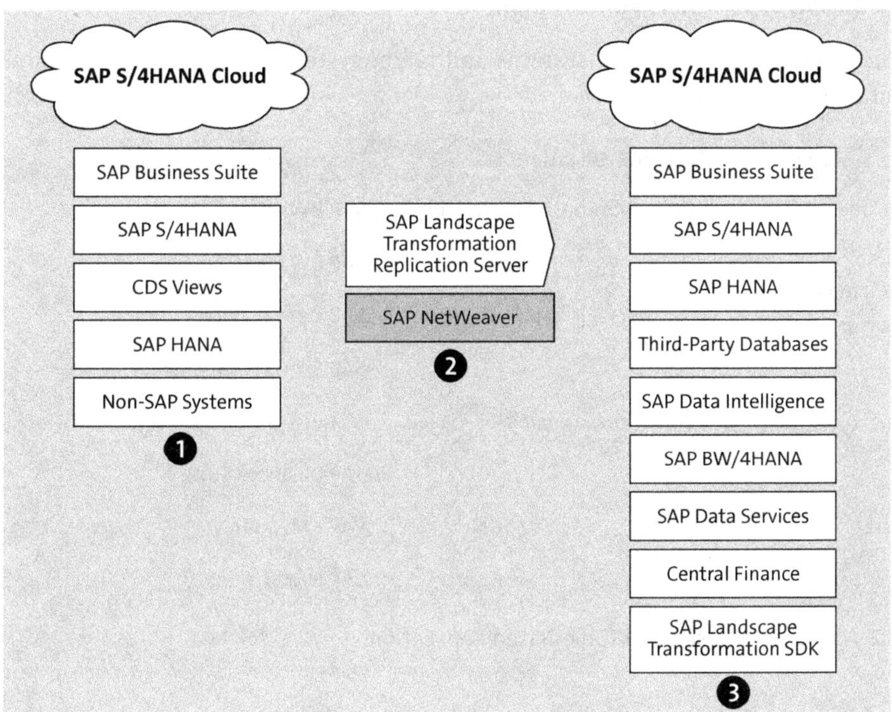

Figure 5.2 SAP Landscape Transformation Replication Server Overview

As you can see, SAP Landscape Transformation Replication Server supports multiple systems as source systems on the left ❶. If you want to load data to the target SAP

S/4HANA system 1809 higher release, the SAP NetWeaver version for the SAP Landscape Transformation Replication Server system must be 7.50 or higher ❷. A wide range of servers are supported as the target ❸.

Note

When integrating SAP Cloud as a source system, SAP Landscape Transformation Replication Server must be on SAP S/4HANA.

SAP Landscape Transformation Replication Server comes in various types of replications. From central finance context, let's look at two common use-case approaches:

- **Table, view or partition-based replication**
  With this approach, SAP Landscape Transformation Replication Server merely copies data from a source system table to a target system table. Data can be from a table, view, or partition based. You'll still be able to write custom transformation rules.
- **Application-based replication**
  Central Finance is an application-based replication scenario. This means data transferred from source systems is reposted using mapping and preconfigured rules from the target Central Finance system.

Table 5.2 display types of source systems and target systems using both application-based and table-based replication.

Source Systems	Type or Replication	Target Systems
SAP Business Suite SAP HANA SAP S/4HANA Native DB	Application-based replication	SAP BW
		SAP BW/4 HANA
		SAP Data Services
		SAP S/4HANA (Central Finance)
	Table, view, or partition-based replication	Native DB
		SAP Business Suite
		SAP S/4HANA
		SAP HANA

Table 5.2 Application-Based and Table-Based Replication

5.1.2 Deployment Options

SAP Landscape Transformation Replication Server can be deployed in multiple ways. SAP Landscape Transformation Replication Server can be deployed on any SAP system with an SAP NetWeaver ABAP stack of at least 7.02 or higher. From a Central Finance perspective, SAP Landscape Transformation Replication Server can be deployed in three different ways:

- In a separate system
- In a source system
- In the Central Finance system

In this section, you'll learn about each deployment option and its advantages and disadvantages.

SAP Landscape Transformation Replication Server Installed in a Separate System

SAP recommends installing SAP Landscape Transformation Replication Server in a separate system, especially when you have high data volume. In addition, having a separate system is ideal for many reasons. Figure 5.3 illustrates SAP Landscape Transformation Replication Server deployed as a separate system. When data from the source system ❶ is transferred to the Central Finance system ❸, SAP Landscape Transformation Replication Server ❷, which is set up as a separate system, replicates/loads the data.

Figure 5.3 Central Finance Deployed in a Separate System

Table 5.3 lists important pros and cons of installing SAP Landscape Transformation Replication Server in a separate server. You must carefully review prior to making the Central Finance landscape decisions.

Pros	Cons
A dedicated separate system can function independently without impacting the source system or Central Finance system, thereby improving performance.	High hardware/subscription costs are involved in maintaining a separate SAP Landscape Transformation Replication Server.
Flexibility in upgrading SAP Landscape Transformation Replication Server without impacting the source system or Central Finance system.	A separate license is required.

Table 5.3 SAP Landscape Transformation Deployed in a Separate Server

Pros	Cons
SAP Landscape Transformation Replication Server can be used for other scenarios besides Central Finance.	Additional maintenance effort is required.
Clear separation of security access and roles for segregation of duties is provided.	There is the potential for data access failures or network failures due to the Central Finance or source system not being in the same server.
Maintenance activities can be performed independent of the source system and Central Finance system.	
Flexibility in sizing is provided.	

Table 5.3 SAP Landscape Transformation Deployed in a Separate Server (Cont.)

Note

SAP strongly recommends installing SAP Landscape Transformation Replication Server in a separate instance for the various reasons we've highlighted in Table 5.3.

SAP Landscape Transformation Replication Server Installed in the Central Finance System

SAP Landscape Transformation Replication Server can be deployed in the Central Finance system, as shown in Figure 5.4. As you can see, data from source systems ❶ replicated to Central Finance with SAP Landscape Transformation Replication Server ❷ which is installed on target Central Finance system.

Figure 5.4 Central Finance Deployed on the Central Finance System

Table 5.4 lists important pros and cons of installing SAP Landscape Transformation Replication Server on a Central Finance system.

Pros	Cons
No separate license is required, if the license of the Central Finance system includes use of SAP Landscape Transformation Replication Server. Consult SAP Note 2707835 for additional information on SAP Landscape Transformation Replication Server licensing.	SAP Landscape Transformation Replication Server may result in performance bottlenecks if the landscape isn't sized up correctly.
SAP Landscape Transformation Replication Server accesses data directly within the Central Finance system, reducing the possibility of network failure.	There is a dependency on SAP Landscape Transformation Replication Server maintenance activities.
Security concerns are reduced.	Upgrading requires planning.
SAP Landscape Transformation Replication Server is delivered with SAP S/4HANA, so no DMIS add-on installation is required. Consult SAP Note 2906076 for further guidance on this topic.	SAP Landscape Transformation Replication Server can't be used for other replication scenarios such as SAP BW.
Chances for data write failure are decreased because SAP Landscape Transformation Replication Server resides within the same box.	
The existing setup can be used.	
Only local remote function call (RFC) connections are required, thereby reducing the probability of network failures during write.	

Table 5.4 SAP Landscape Transformation Deployed in a Separate Server

[«]

Note

Although installing SAP Landscape Transformation Replication Server on the Central Finance system provides some benefits, SAP strongly recommends installing SAP Landscape Transformation Replication Server in a separate instance on its own.

SAP Landscape Transformation Replication Server Deployed with a Source System

SAP Landscape Transformation Replication Server can be deployed in the source system, as shown in Figure 5.5. As you can see, data from source systems ❶ is replicated to Central Finance ❷ with SAP Landscape Transformation Replication Server being deployed on the source system.

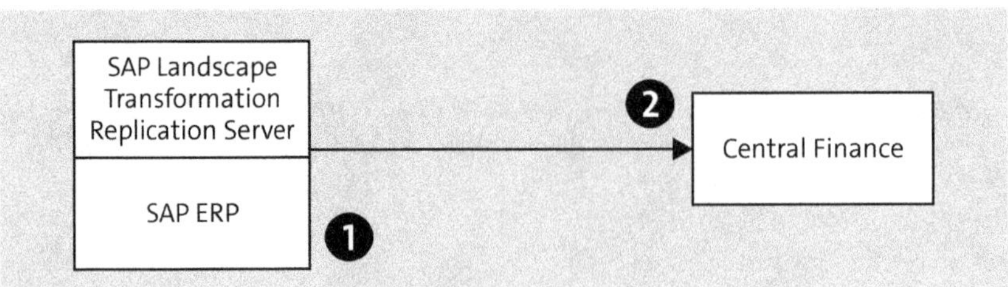

Figure 5.5 Central Finance Deployed with the Source System

Table 5.5 lists essential pros and cons of installing SAP Landscape Transformation Replication Server on the source system.

Pros	Cons
SAP Landscape Transformation Replication Server accesses data directly within the source system, thereby reducing network failures during read.	If multiple source systems are connected to Central Finance, then you'll be installing SAP Landscape Transformation Replication Server in all source systems, resulting in additional cost, maintenance, and complexity.
The existing SAP ABAP-based system can be used.	SAP Landscape Transformation Replication Server maintenance activities are dependent on the source system maintenance cycle.
Only local RFC connections are required, thereby reducing the probability of network failures during read.	Upgrading requires planning because the source system upgrade cycle must be assessed.
Security concerns are reduced as SAP Landscape Transformation Replication Server resides within same server.	SAP Landscape Transformation Replication Server may result in performance bottlenecks if the landscape isn't sized up correctly in the source system. This is often difficult to achieve because installing SAP Landscape Transformation Replication Server on an existing source system interrupts current systems and processes.
	There is no flexibility in sizing.

Table 5.5 SAP Landscape Transformation Deployed in a Separate Server

Note

This approach is the least recommended approach as the idea of Central Finance is non disruption of existing systems or processes. By installing SAP Landscape Transformation Replication Server on the source system, the existing landscape will be interrupted.

Although SAP Landscape Transformation Replication Server offers a wide range of use cases, we'll limit our discussions to Central Finance–supported objects from now on. In the next section, you'll learn what prerequisites to consider before configuring SAP Landscape Transformation Replication Server for Central Finance.

5.1.3 Prerequisites

Certain prerequisites are required for SAP Landscape Transformation Replication Server to replicate data from the source system to Central Finance. You'll begin with the installation of SAP Landscape Transformation Replication Server and then configure the SAP Landscape Transformation Replication Server for Central Finance. You'll also ensure you have the appropriate access in SAP Landscape Transformation Replication Server to carry out the configuration tasks.

Installing SAP Landscape Transformation Server and Data Migration Server Versions

First, you'll analyze the SAP Landscape Transformation Replication Server release and DMIS version based on the product availability matrix. SAP Landscape Transformation Replication Server must be installed on DMIS 2011 SP11 or higher with a recommended minimum release of SP13. However, for new installations of SAP Landscape Transformation Replication Server, SAP recommends using SAP Landscape Transformation Replication Server for SAP S/4HANA 1.0 (based on DMIS 2020) or SAP Landscape Transformation Replication Server 3.0 (based on DMIS 2018) on the highest available SP. Consult SAP Note 2154420 for recommendations and installation requirements.

[«]

Note

If you're using SAP Landscape Transformation Replication Server and you're interested in the latest information about corrections and changes introduced with the different SPs, consult SAP Note 2675613 – SAP Landscape Transformation Replication Server (DMIS 2010/DMIS 2011/DMIS 2018/DMIS 2020/SAP S/4HANA) installation, corrections and release information. This note provides information about all releases.

If the SAP Landscape Transformation Replication Server is installed on a separate system, SAP recommends installing the same SP level on all systems: the source system, target system, and the SAP Landscape Transformation Replication Server.

As it might not always be possible to install the new SP on all involved systems at the same time, several combinations are supported. If not, all systems are installed on the newly recommended SP, some new functionality may not be available. The project teams should evaluate appropriate versions while considering Central Finance requirements. Table 5.6 lists the compatibility matrix between the source system, target system, and replication server. In the table, you can replace DMIS 2018 SP with the respective DMIS 2020 SP. Refer to SAP Note 2965075 for additional guidance on this.

Source System	Replication Server	Target System
DMIS 2010 SP10 DMIS 2011 SP22 and higher DMIS 2018 SP07 and higher SAP S/4HANA 1709 any version SAP S/4HANA 1809 any version SAP S/4HANA 1909 any version SAP S/4HANA 2020 any version SAP S/4HANA 2021 any version SAP S/4HANA 2022 any version SAP S/4HANA 2023 any version	DMIS 2018 SP12	DMIS 2011 SP22 and higher DMIS 2018 SP07 and higher SAP S/4HANA 1709 any version SAP S/4HANA 1809 any version SAP S/4HANA 1909 any version SAP S/4HANA 2020 any version SAP S/4HANA 2021 any version SAP S/4HANA 2022 any version SAP S/4HANA 2023 any version

Table 5.6 Source and Target Systems Compatibility Matrix with SAP Landscape Transformation Releases

Note

For the latest information on the compatibility matrix, refer to the SAP Note released for your specific SAP Landscape Transformation Replication Server DMIS service pack. For example, you'll refer to SAP Note 3438948 for the release information of SAP Landscape Transformation Replication Server DMIS 2018 SP12.

After installing SAP Landscape Transformation Replication Server, you must ensure the following:

- Check whether all the DMIS-related SAP Notes are applied in the source and SAP Landscape Transformation Replication Server system. In addition, SAP strongly recommends running the CNV_NOTE_ANALYZER_SLT report, which is available in both the SAP Landscape Transformation Replication Server and source systems. This report will display any missing SAP Notes in both source system and SAP Landscape Transformation Replication Server. If the report isn't available in the source system or SAP Landscape Transformation Replication Server, refer to SAP Note 3016862 - DMIS Note Analyzers with Separated Scenarios for ABAP-Based Migration and Replication Technology (DMIS2011/DMIS2018/DMIS2020/SAP S/4HANA). The new report replaces the old report DMC_NOTE_ANALYZER. With this new report, SAP separated the notes download capability by scenarios. For example, now SAP Landscape Transformation Replication Server has its own transaction versus SAP S/4HANA Migration Cockpit or ABAP Integration for SAP Data Intelligence.
- Keep SAP Landscape Transformation Replication Server content uploaded with the latest content in the SAP Landscape Transformation Replication Server using report DMC_UPLOAD_OBJECT. Consult SAP Note 2154420 - SAP Landscape Transformation

Replication Server for SAP Central Finance for SAP Source System for additional guidance on this topic.

- Check that there is enough background work processes available in the source system, SAP Landscape Transformation Replication Server, and target Central Finance system for the high-volume data loads. You can check the available work process via transaction SM50.

Analyzing SAP Notes Requirements

After identifying compatible SAP Landscape Transformation Replication Server releases and DMIS versions, you'll review any missing SAP Notes with program CNV_NOTE_ANALYZER_SLT (earlier releases report DMC_NOTE_ANALYZER), which checks for any missing notes in the system for the installed SAP Landscape Transformation Replication Server version. In addition, the program also installs any identified missing SAP Notes. You must have an active OSS account connected to the system for the program to function. Figure 5.6 displays the **Analyze SAP Notes for scenario SLT** initial screen. On this screen, choose the value in the **RFC Destination** ❶ field. Next, choose either the **Source System** or **Central System** ❷ radio button, and then click **Execute** ❸. The system displays the results screen ❹ with SAP Notes to be implemented if applicable. In our example, the system is up-to-date and doesn't list any SAP Notes.

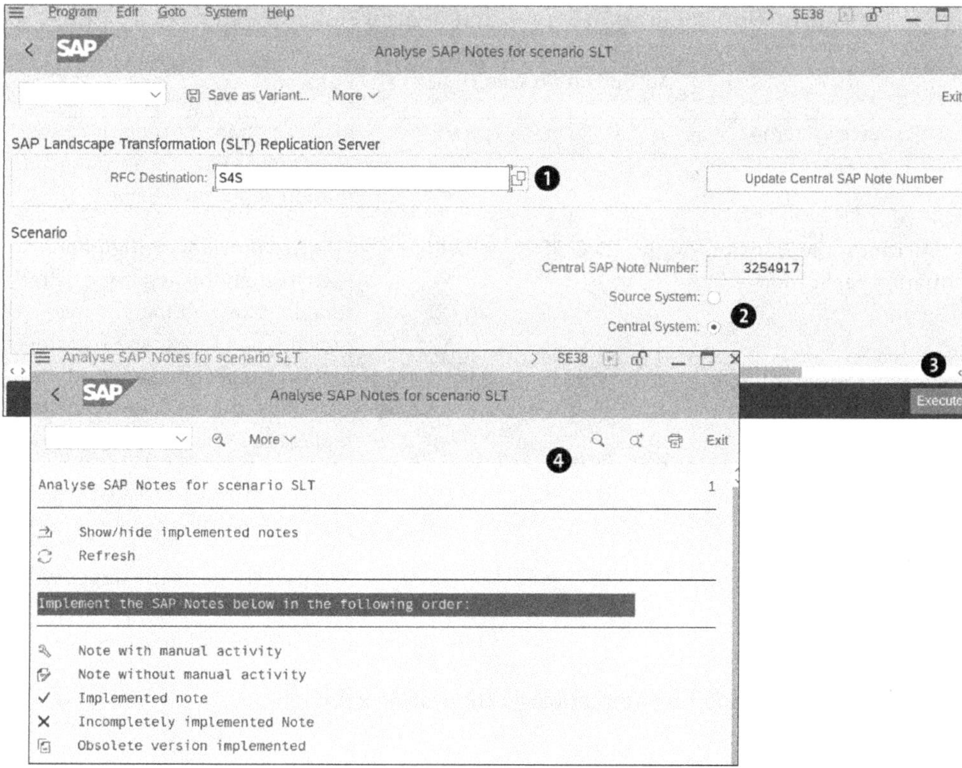

Figure 5.6 SAP Landscape Transformation Note Analyzer

Non-SAP Third-Party System Integration with Central Finance

To load and replicate data from third-party systems to Central Finance, SAP provides preconfigured content that loads the data from a staging area. SAP provides the staging area to connect multiple third-party systems that have data models that are different from the Central Finance system data model. Extracting and loading the data to the staging area depends on the third-party system data structure. SAP doesn't provide tools to bring data from third-party systems to the staging area. Before transferring data from the non-SAP system, certain prerequisites must be met:

- SAP Landscape Transformation Replication Server is installed and configured.
- SAP recommends installing SAP Landscape Transformation Replication Server 3.0 or SAP Landscape Transformation Replication Server for SAP S/4HANA 1.0.
- Necessary authorization is granted for the RFC user.

Authorization

To replicate data from the source system to Central Finance, you'll require appropriate authorization to configure SAP Landscape Transformation Replication Server, access remote systems, and so on. You'll need two types of users: an initial user (RFC user) to provide connectivity access and a replication user who performs the activity. Table 5.7 lists required authorization objects in the source system, SAP source system, SAP Landscape Transformation Replication Server, and Central Finance system.

	Authorization Role	Purpose
SAP source system	SAP_IUUC_REPL_REMOTE	RFC user connects the source system from SAP Landscape Transformation Replication Server.
SAP Landscape Transformation Replication Server	SAP_IUUC_REPL_ADMIN	This role provides configuration such as create, change, and delete functions. In addition, the role provides SAP Landscape Transformation Replication Server administrations to stop, start, or resume replication.
	SAP_IUUC_REPL_DISPLAY	This role provides users with display configuration data.
Central Finance	SAP_IUUC_REPL_REMOTE	RFC user connects the target system from SAP Landscape Transformation Replication Server.

Table 5.7 SAP Landscape Transformation: Authorization Role

[«]

Note

Appropriate authorization is required for the RFC user in both the source and target systems for SAP Landscape Transformation Replication Server. In addition, the SAP Landscape Transformation Replication Server configurator should have proper authorization to perform SAP Landscape Transformation Replication Server activities. You can access the security guide for additional information on security requirements at *http://s-prs.co/v590101* under the **Implement** section.

5.1.4 Concept and SAP Landscape Transformation Replication Server Activities

In Central Finance, real-time replication is enabled via SAP Landscape Transformation Replication Server. Prior to enabling real-time replication, you'll load initial data. Don't confuse the initial load of SAP Landscape Transformation Replication Server with financial accounting-Controlling (FI-CO) initial load postings. The FI-CO initial load is carried out from Central Finance with an RFC call over to the source system. On the other hand, you'll execute the initial load performed from SAP Landscape Transformation Replication Server as in the examples of the initial load of cost object replication and CO internal postings. Figure 5.7 illustrates the concept of load and replication activities:

❶ **Load data**

In this step, the system allows a one-time data transfer of a source system table. As a prerequisite, you'll create relevant predefined load objects. As a general practice, this step can be combined with the replicate data step.

❷ **Replicate data**

With this step, you'll be able to perform both load and replication of source system data to Central Finance.

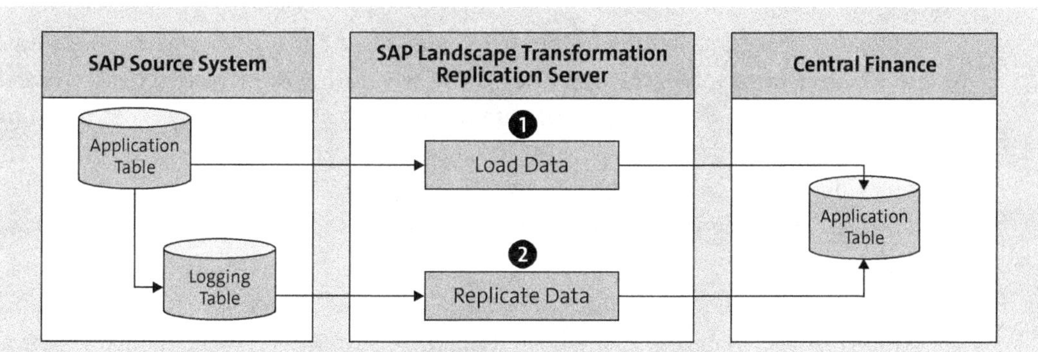

Figure 5.7 Load and Replication Concept

[»]

Note

Best practice is to use start load for simulation. For a real load, choose to start replicate data as a data provisioning option in SAP Landscape Transformation Replication Server.

[»]

Note

For SAP-based source systems, the FI-CO initial load is initiated from Central Finance without involving SAP Landscape Transformation Replication Server. In addition, the project systems (PS) initial load and replication happens via application link enabling (ALE) technology.

Now that you've learned about the initial load and replication sequence, let's take a deep dive into SAP Landscape Transformation Replication Server architecture next.

5.1.5 SAP Landscape Transformation Replication Server Architecture

In this section, we'll discuss the SAP Landscape Transformation Replication Server architecture and how SAP Landscape Transformation Replication Server transfers data from source systems to target systems. From a Central Finance perspective, you must understand data load and replication from two types of source systems: SAP-based source systems and non-SAP-based source systems. We'll take you through the architectural concept of both types of replications from the Central Finance context.

SAP-Based Source System

Figure 5.8 shows the architecture concept of SAP-based source systems replicating data to Central Finance:

❶ **Staging tables**

When business transactions are posted in SAP-based source systems, relevant business transactions tables are updated via the accounting interface for replication. In addition, relevant database triggers are invoked for inserts, updates, and deletions for each trigger operation on the tables associated with Central Finance objects. The tables list source system trigger tables for Central Finance objects. Each of these tables has dependent tables to extract and load data to SAP Landscape Transformation Replication Server.

❷ **DB triggers**

When business transactions are posted in source systems, relevant business transaction tables are updated via the accounting interface. In addition, relevant database triggers are invoked for inserts, updates, and deletions for each trigger operation and then updates the Central Finance staging tables.

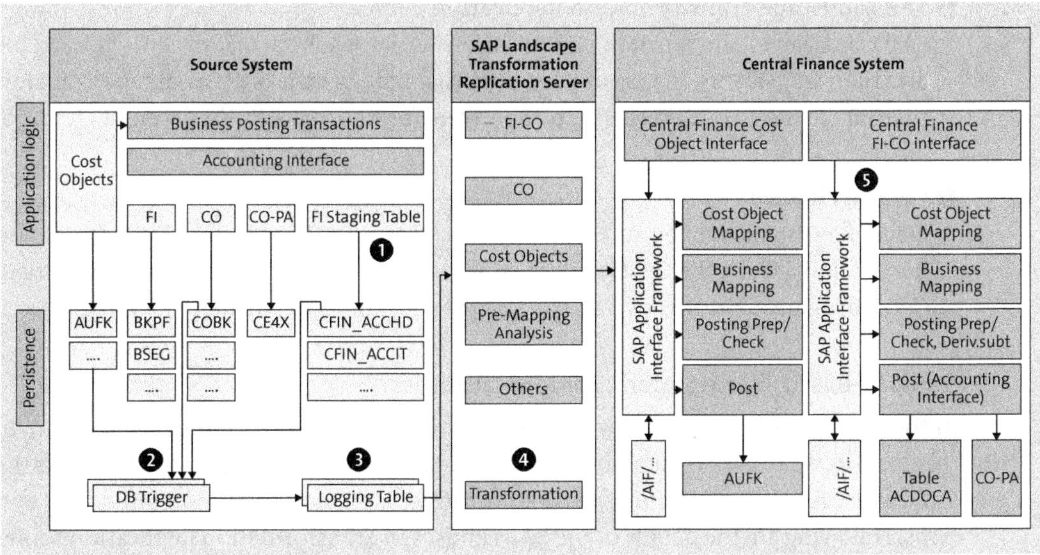

Figure 5.8 SAP Landscape Transformation Architecture: SAP-Based Source Systems

❸ Logging table

The application data is first replicated to the logging table before data is sent to the target system through SAP Landscape Transformation Replication Server. The logging table is generated when the mass transfer ID (MT ID) for the replication object is activated in SAP Landscape Transformation Replication Server. When this happens, database triggers for inserts, updates, and deletes are also created in parallel. Figure 5.9 shows Logging Table and Trigger for tax replication table T001. In addition, when SAP Landscape Transformation Replication Server is defined and activated, function modules are created in the source system to transfer source application data to the logging table. Then, the system transfers data from the logging table to SAP Landscape Transformation Replication Server. The logging table contains both application and technical information. The application data represents business data, whereas the technical data is used for delta processing.

Display Details: T001

Object Name: **T001** Object Type: **Transparent Table**

System	Entity	Name	Status
Source System	Logging Table	/1CADMC/00000206	■
Source System	Trigger	/1LT/00000206*	■
Replication Server	Migration Object	Z_T001_00B	■
Replication Server	Function Group	/1LT/S48500000000000031	■
Target System	Target Table	T001	■

Figure 5.9 Database Triggers and Logging Table for Tax Table T001

❹ **SAP Landscape Transformation Replication Server**
SAP Landscape Transformation Replication Server reads data from the logging table and then transforms, maps, and converts that data per the SAP Landscape Transformation Replication Server setup. The transformed data is then sent to Central Finance.

❺ **Central Finance**
Once data is transferred to Central Finance from SAP Landscape Transformation Replication Server, the Central Finance accounting interface is called to post data in Central Finance. Before posting, the system goes through business mapping.

Non-SAP-Based Source System: For Journal Entries

In the previous section, you saw how data flows from an SAP-based source system. In this section, we'll discuss data flow from a non-SAP-based system. Figure 5.10 shows the SAP Landscape Transformation Replication Server architecture for non-SAP-based systems. Following are the details of the SAP Landscape Transformation Replication Server architecture and its components:

❶ **Extract and load data**
As you can see, the concept of DB triggers and logging tables are the same as in the SAP-based system. However, due to the different structure and data models of non-SAP source systems, you must first extract and load data to staging tables.

❷ **Staging tables**
SAP provides stagging tables for non-SAP source systems to stage data for replication. Once business transactions are posted in source systems, data from relevant tables must be extracted and mapped to the inbound staging table format. This part is typically handled from the customer's side. SAP doesn't provide any tools to extract and map because the data models vary with every third-party ERP system.

❸ **DB triggers**
Once data is extracted, mapped, and updated to the staging table, database triggers are invoked to update the logging table. The nature of database triggers could be inserts, updates, and deletions for each trigger operation.

❹ **Logging table**
The application data is first replicated to the logging table before data is sent to the target system through SAP Landscape Transformation Replication Server. The logging table is generated when a MT ID is activated in SAP Landscape Transformation Replication Server. In addition, when SAP Landscape Transformation Replication Server is defined and activated, function modules are created in the source to transfer application data to the logging table. The logging table houses both application and technical information.

5 SAP Landscape Transformation Replication Server read engine
Data is read from the logging table by the SAP Landscape Transformation read engine and sent to the target system using an RFC connection.

6 SAP Landscape Transformation Replication Server write engine
Once data mapping and transformation takes place, the SAP Landscape Transformation Replication Server write engine calls to send information to Central Finance.

7 Central Finance
Once data is transferred to Central Finance from SAP Landscape Transformation Replication Server, the Central Finance accounting interface is called to post data in Central Finance to the Universal Journal. Before posting, the system goes through business mapping.

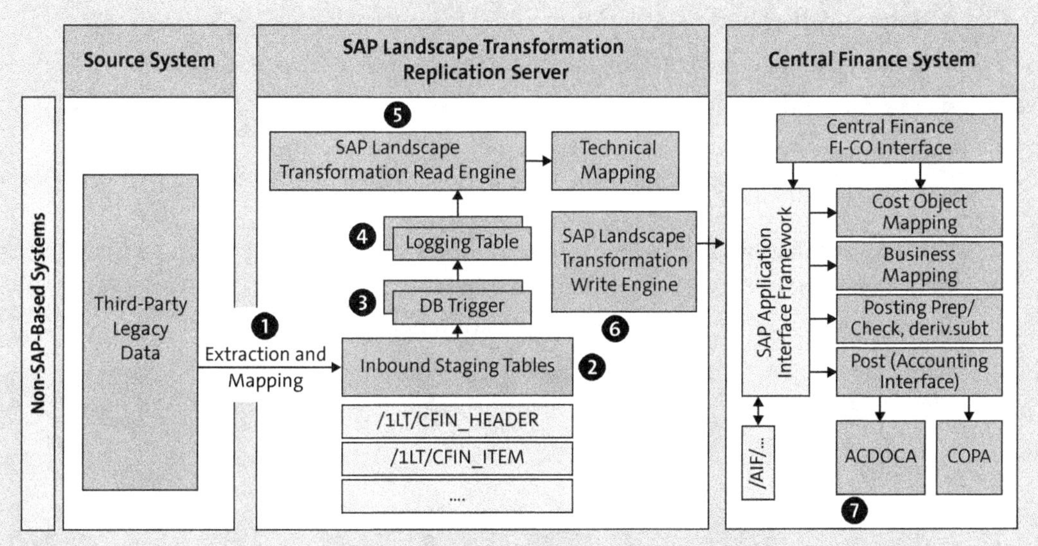

Figure 5.10 SAP Landscape Transformation Architecture: Non-SAP-Based Source Systems

[«]

Note

For non-SAP systems replication, although real-time or near-real-time replication is technically feasible, additional configurations and tools are necessary. In general, non-SAP system replications are scheduled in batches or intervals. How frequently data is extracted and loaded for third-party systems depends on the data volume and complexity involved.

5.1.6 Central Finance Tables Scope

In this section, we'll cover the SAP Landscape Transformation Replication Server tables delivered by SAP to support SAP-based and third-party replication scenarios.

SAP-Based Source System: SAP Landscape Transformation Replication Server Tables

SAP Landscape Transformation Replication Server is used in Central Finance to replicate data from the source system to Central Finance. The application varies for different application components such as the following:

- FI postings
- CO postings
- Commitments replication
- Sales and distribution (SD) documents (accounting view of logistics)
- Material management (MM) documents (accounting view of logistics)
- Tax data and relevant settings
- Product costing
- Down payments

Table 5.8 displays a comprehensive list of tables set up in SAP Landscape Transformation Replication Server using Transaction LTRC. Section 5.2.1 covers step-by-step instructions on setting up load and replication objects for the first table in the list.

Area	Type of Object	Table	Load Object	Replication Object
Tax and Withholding Tax	Company code configuration	T001	CFI_T001_L	CFI_T001_R
	Withholding tax types	T059P	CFI_T059P_L	CFI_T059P_R
	Withholding tax code (enhanced functions)	T059Z	CFI_T059Z_L	CFI_T059Z_R
	Minimum and maximum amounts for withholding tax	T059MINMAX	CFI_T059MINMAX_L	CFI_T059MINMAX_R
	Company code-specific information per withholding tax type	T001WT	CFI_T001WT_L	CFI_T001WT_R

Table 5.8 Central Finance Tables for SAP Landscape Transformation Setup

Area	Type of Object	Table	Load Object	Replication Object
Tax and Withholding Tax (Cont.)	Rounding rules for company code, withholding tax type and currency	T001RWT	CFI_T001RWT_L	CFI_T001RWT_R
	Standard accounts table	T030	CFI_T030_POOL_L (SAP ECC source) CFI_T030_TRANS_L (SAP S/4HANA source)	CFI_T030_POOL_R (SAP ECC source) CFI_T030_TRANS_R (SAP S/4HANA source)
	Cross-client FI configuration	T000F	CFI_T000F_L	CFI_T000F_R
	Document types	T003	CFI_T003_L	CFI_T003_R
	Country code	T005	CFI_T005_L	CFI_T005_R
	Tax key	T007A	CFI_T007A_L	CFI_T007A_R
	Tax processing in accounting	T007B	CFI_T007B_POOL_L (SAP ECC source) CFI_T007B_TRANS_L (SAP S/4HANA source)	CFI_T007B_POOL_R (SAP ECC source) CFI_T007B_TRANS_L (SAP S/4HANA source)
	Tax jurisdiction code	TTXD	CFI_TTXD_POOL_L (SAP ECC source) CFI_TTXD_TRANS_L (SAP S/4HANA source)	CFI_TTXD_POOL_R (SAP ECC source) CFI_TTXD_TRANS_R (SAP S/4HANA source)
Finance	FI-CO posting	CFIN_ACCHD	CFIN_ACCHD_L	CFI_ACCHD_R
	IDR	CFIN_ACCHD_IDR	CFI_ACCHD_IDR_L	CFI_ACCHD_IDR_R

Table 5.8 Central Finance Tables for SAP Landscape Transformation Setup (Cont.)

Area	Type of Object	Table	Load Object	Replication Object
Controlling	Cost object	AUFK	CFI_AUFK_L	CFI_AUFK_R
	CO secondary posting	COBK	CFI_COBK_L (SAP ECC Source) CFI_COBK_1511_L (SAP S/4HANA Source)	CFI_COBK_R (SAP ECC Source) CFI_COBK_1511_R (SAP S/4HANA Source)
	Activity rate (from SAP source systems to Central Finance)	COST	CFI_COST_L	CFI_COST_R
	Activity rate (from Central Finance to SAP source systems	COST	N/A	CFI_COST_R
	Commitment	CFIN_CMT_H	CFI_CFIN_CMT_H_L	CFI_CFIN_CMT_H_R
	Cost estimate	KEKO	CFI_KEKO_L	CFI_KEKO_R
	EC-PCA postings	CFIN_ACCHD	N/A	CFIN_ACCHD_R
	Simulation of table AUFK	AUFK	CFIN_SIM_AUFK_L	N/A
	Simulation of table COBK	COBK	CFIN_SIM_COBK_L	N/A
	Simulation of commitment	CFIN_CMT_H	CFI_CFIN_CMT_SIM	N/A

Table 5.8 Central Finance Tables for SAP Landscape Transformation Setup (Cont.)

[»]

Note

As you can see from Table 5.8, the Project System component isn't listed as it uses ALE technology to replicate project system's master data to Central Finance and not SAP Landscape Transformation Replication Server. In addition, joint venture accounting (JVA) uses the Data Replication Framework (DRF), so it's not listed here.

Non-SAP-Based SAP Landscape Transformation Replication Server Tables: Journal Entries

Table 5.9 displays SAP-delivered staging tables that are installed when you follow SAP-recommended SAP Note 3340008 for the third-party integration in SAP S/4HANA 2023.

Staging Table Name	Description
/1LT/CF_E_HEADER	Document header table
/1LT/CF_E_DEBIT	Debtor items
/1LT/CF_E_CREDIT	Creditor items
/1LT/CF_E_PRDTAX	Product tax items
/1LT/CF_E_WHTAX	Withholding tax items
/1LT/CF_E_COPA	Costing-based profitability analysis criteria
/1LT/CF_E_CLRITM	Clearing items
/1LT/CF_E_CLRWHT	Withholding tax within clearing
/1LT/CF_E_EXTENT	Customer extension on header level
/1LT/CF_E_EXT_IT	Customer extension on item level
/1LT/CF_E_DOCCHG	Changed fields
/1LT/CF_E_EXTCHG	Changed customer fields

Table 5.9 Non-SAP Staging Table

Non-SAP-Based SAP Landscape Transformation Replication Server Tables: Accounting View of Logistics Data

You'll create SAP Landscape Transformation Replication Server staging tables for third-party logistics data replication after executing the programs mentioned in SAP Note 3079411. You'll download program code and execute the programs to generate the staging tables. Table 5.10 lists the program names and corresponding file names you'll run to generate the SAP Landscape Transformation Replication Server staging tables listed in Table 5.11.

Replication Object	Program Name	File Name
Sales Document	ZFINS_CFIN_AV_GEN_TABLE_SO	ZFINS_CFIN_AV_GEN_TABLE_SO.txt
Customer Invoice	ZFINS_CFIN_AV_GEN_TABLE_CI	ZFINS_CFIN_AV_GEN_TABLE_CI.txt

Table 5.10 Programs Used to Generate SAP Landscape Transformation Replication Server Staging Tables for the Logistics Accounting View for Third-Parties

Replication Object	Program Name	File Name
Purchase Order	ZFINS_CFIN_AV_GEN_TABLE_PO	ZFINS_CFIN_AV_GEN_TABLE_PO.txt
Supplier Invoice	ZFINS_CFIN_AV_GEN_TABLE_SI	ZFINS_CFIN_AV_GEN_TABLE_SI.txt

Table 5.10 Programs Used to Generate SAP Landscape Transformation Replication Server Staging Tables for the Logistics Accounting View for Third-Parties (Cont.)

Table 5.11 lists the SAP Landscape Transformation Replication Server staging tables for the logistics accounting view for third-party systems.

Replication Object	SAP Landscape Transformation Replication Server Staging Tables	Staging Table Description
Sales order	/1LT/CF_E_SODOC	Sales Document: Header Data
	/1LT/CF_E_SOITEM	Sales Document: Item Data
	/1LT/CF_E_SOPART	Sales Document: Partner
	/1LT/CF_E_SOSLN	Sales Document: Schedule Line Data
Customer invoice	/1LT/CF_E_CIDOC	Customer Invoice: Header Data
	/1LT/CF_E_CIITEM	Customer Invoice: Item Data
Purchase document	/1LT/CF_E_PODOC	Purchase Order: Header Data
	/1LT/CF_E_POITEM	Purchase Order: Item Data
	/1LT/CF_E_POSCH	Purchase Order: Schedule Line Data
	/1LT/CF_E_POACC	Purchase Order: Account Assignment Data
	/1LT/CF_E_POMRD	Purchase Order: Material Document Data
	/1LT/CF_E_PORSI	Purchase Order: Supplier Invoice Data
Supplier invoice	/1LT/CF_E_SIDOC	Supplier Invoice: Header Data
	/1LT/CF_E_SIITEM	Supplier Invoice: Item Data
	/1LT/CF_E_SIACC	Supplier Invoice: Account Assignment Data

Table 5.11 Non-SAP Staging Table: Accounting View of Logistics

Replication Object	SAP Landscape Transformation Replication Server Staging Tables	Staging Table Description
Supplier invoice (Cont.)	/1LT/CF_E_SIGLAC	Supplier Invoice: General Ledger Account Data

Table 5.11 Non-SAP Staging Table: Accounting View of Logistics (Cont.)

Note

For additional guidance on non-SAP integration with Central Finance, refer to SAP Note 2932517.

Now that you've learned the basic concepts of the SAP Landscape Transformation Replication Server architecture, let's deep dive into SAP Landscape Transformation Replication Server configuration.

5.2 Setting Up Replication

At a high level, SAP Landscape Transformation Replication Server configuration involves four steps. In this section, you'll learn how to set up replication for Central Finance replication objects. Before you begin the SAP Landscape Transformation Replication Server configuration, you must have access to the SAP_ICCU_REPL_ADMIN role, which is used to create, display, and change configuration settings in SAP Landscape Transformation Replication Server. Figure 5.11 outlines the four-step SAP Landscape Transformation Replication Server configuration activities:

❶ **Create configuration**
Create an MT ID using Transaction LTRC for relevant Central Finance replication objects. MT ID is a unique identifier configured to identify the source system, which is used to transfer data between systems. In addition, the SAP Landscape Transformation Replication Server connector connects to SAP Landscape Transformation Replication Server using MT ID. SAP recommends specifying the application **CFIN** when SAP Landscape Transformation Replication Server configuration is defined using Transaction LTRC for Central Finance. As a result, the system checks whether predefined migration content exists for the table added in the MT ID for replication. On the other hand, if no predefined objects exist for any table, the system issues an error message.

❷ **Copy predefined load and replication objects**
Copy the predefined load and replication objects defined in the specific project and subproject for a specific Central Finance table. You'll carry out this step using report IUUC_REPL_PREDEF_OBJECTS. To copy predefined load replication objects, you can

also use Transaction LTRC and by navigating **MT ID Expert Functions • Processing Steps • Preparatory Functions • Create Predefined Load and Replication Objects.**

❸ **Filters and transformation**
In this optional step, add filters and transformation logic under predefined event-based and field-based rules using the Migration Workbench (Transaction MWB) for the copied predefined load and replication created under template projects and subprojects.

❹ **Start load and replication**
Once the SAP Landscape Transformation Replication Server configuration is completed, begin loading and replication of relevant objects. First, choose **Start Load/ Start Replication** located under the **Data Provisioning** tab in Transaction LTRC. SAP provides seven data provisioning options, as discussed in the "Data Provisioning in SAP Landscape Transformation Replication Server" section.

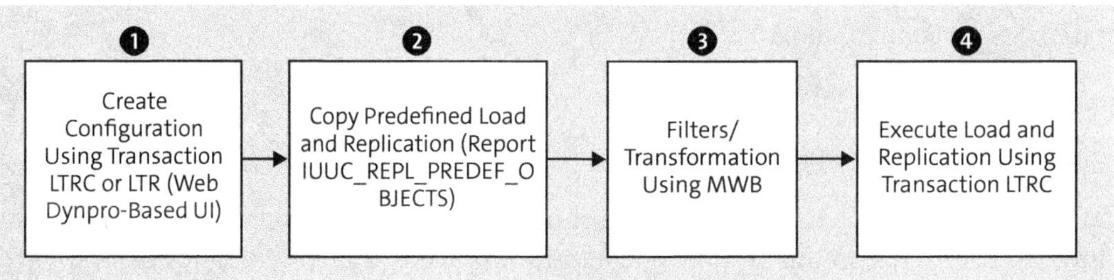

Figure 5.11 SAP Landscape Transformation Replication Setup

In Transaction MWB, SAP provides the template project and subproject to create, edit, and migrate SAP Landscape Transformation Replication Server objects. Leading practices suggest copying these template projects and subprojects to create your own Z* project. Table 5.12 displays the SAP-delivered project and subprojects for Central Finance.

Project	Subproject	SAP Landscape Transformation Replication Server Object Details
REPL_CFIN	REPL_CFIN	Central Finance objects for SAP ERP source
	REPL_CFIN_AS	Account view logistics
	REPL_CFIN_SRC	Required source-to-source step for replication of purchase orders
	REPL_CFIN_AE	Third-party system interface for the account view in logistics
	REPL_CFIN_ST	Third-party system integration for accounting documents

Table 5.12 Project and Subprojects in Transaction MWB

Next, let's get into the details of the SAP Landscape Transformation Replication Server configuration step-by-step.

5.2.1 Replicating Tax Tables

During the initial load and replication, you must activate tax consistency as a prerequisite for central payment to ensure the smooth transfer of financial data from the source system to the Central Finance system. In addition, uniform tax data helps in centralized tax reporting. One prerequisite for the tax consistency check is that you must configure and activate tax tables in SAP Landscape Transformation Replication Server. We'll take one tax table (i.e., table T001), as an example, to walk you through the detailed configuration steps in the following sections. Table 5.13 lists the required source system tax tables to be replicated to Central Finance for tax consistency checks. The third column shows the Central Finance equivalent tables upon replication.

Source Table	Description	Replicated Table in Central Finance
T001	Company code	FINS_CFIN_T001
T000F	Cross-client FI configuration	FINS_CFIN_T000F
T003	Document type	FINS_CFIN_T003 (supported from SAP S/4HANA 2022 onward)
T005	Country code	FINS_CFIN_T005
T007A	Tax code	FINS_CFIN_T007A
T007B	Tax processing key	FINS_CFIN_T007B
TTXD	Tax jurisdiction code	FINS_CFIN_TTXD

Table 5.13 Tables Replicated for Tax Consistency Check and Central Payments

[«]

Note

For the latest tax tables update, consult SAP Note 2494127.

Create Configuration

In this section, as a first step, you'll create an MT ID using the SAP Landscape Transformation Replication Server cockpit (Transaction LTRC). Upon opening the transaction, you'll arrive at the **SAP LT Replication Server - Cockpit Overview** screen, as shown in Figure 5.12. Click the **Create** ❶ icon to create a new configuration for tax table replication. The system takes you to the **Create Configuration** screen shown in the bottom of the figure. This screen is divided into the following configuration screens, which are configured sequentially:

1. **Specify General Data**
   Primarily captures the unique configuration ID and description.
2. **Specify Source System**
   Enter source system information such as RFC destination and other parameters.
3. **Specify Target System**
   Enter target Central Finance information such as RFC connection and other important parameters.
4. **Specify Transfer Settings**
   Maintain key parameters related to job settings and replication options.
5. **Review and Create**
   Review and create the configuration. The system creates an MT ID.

Let's look at each configuration screen activity next.

Specify General Data

On the **Specify General Data** configuration screen, after clicking the icon to create a new configuration (Figure 5.12 ❶), you'll enter values in the following fields:

- **Configuration Name** ❷
  Enter a configuration name specific to the Central Finance replication object.
- **Description** ❸
  Enter a description of the configuration. This is a free-form text field used for identification purposes.
- **Authorization Group** ❹
  Enter the specific authorization for the configuration. If you specify an authorization group, SAP checks and uses authorization object `S_DMIS_SLT` by default. Ensure you've added this authorization object to the role.
- **Application** ❺
  Choose an application from the dropdown list. Table 5.14 lists SAP-provided values in the dropdown list and their explanations. Choose **CFIN Central Finance (Data-based replication)**. The other application options from the dropdown aren't in scope for Central Finance replication objects.

Application	Explanation
CFIN	Central Finance (data-based replication) to replicate data from the SAP source system to the Central Finance system. It doesn't include the SAP source system using a third-party database with a runtime license.
CFIN_PI	Central Finance (business integration) to implement the Central Finance scenario for the SAP source system using a third-party database (i.e., MySQL with a runtime license).

Table 5.14 Application in SAP Landscape Transformation

Application	Explanation
TDF	Tax declaration framework especially used in localization (e.g., Brazil). For additional information on TDF, refer to SAP Notes 3436114 and 2190963.

Table 5.14 Application in SAP Landscape Transformation (Cont.)

Once all information is entered, click the **Next** button to continue to the next configuration screen.

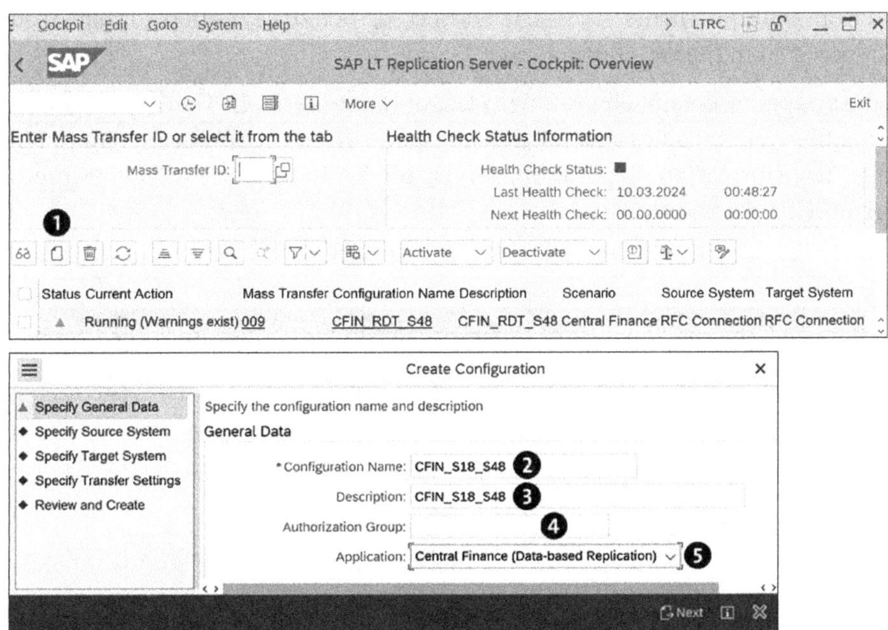

Figure 5.12 Create Configuration: Specify General Data

Specify Source System

On the **Specify Source System** configuration screen, as shown in Figure 5.13, choose the **RFC connection** ❶ radio button. Next, enter values in **RFC Destination** ❷. Then, choose the **Allow Multiple Usage** ❸ checkbox to enable multiple configurations to replicate from the same source system. As a result, the system allows multiple replication objects to be configured.

Tip

SAP recommends choosing **Allow Multiple Usage** always to allow a 1:N replication scenario. SAP refers to a 1:N scenario where you'll be able to create multiple MT IDs to replicate data from the same source system for different replication objects. In addition, the same table can be replicated with a maximum of four configurations (i.e., four MT IDs). The **Allow Multiple Usage** checkbox is enabled as a default setting from the

DMIS 2013 SP13 release onward. If the checkbox isn't selected, the system allows only one configuration (i.e., 1:1 configuration), which will be a performance issue, so SAP doesn't recommend the 1:1 configuration.

If the source system is SAP S/4HANA 2020 or higher, there is no limit on the number of SAP Landscape Transformation Replication Server configurations to replicate the same table to the target Central Finance system.

Ensure that the **Read from Single Client** ❹ checkbox is deselected if you want to replicate data from multiple clients. If this flag is selected, then the system reads data from the client specified in the **RFC Destination**.

Finally, the system by default chooses **Copy List of Namespaces** ❺. This means, during replication, the system copies a list of all namespaces from the source system to SAP Landscape Transformation Replication Server to facilitate creating the technical objects required for replication.

Note

SAP recommends leaving **Copy List of Namespaces** selected. Otherwise, if the system does copy namespaces, it may result in inconsistencies or configurations that don't work properly.

Click the **Next** icon to continue with the next configuration screen.

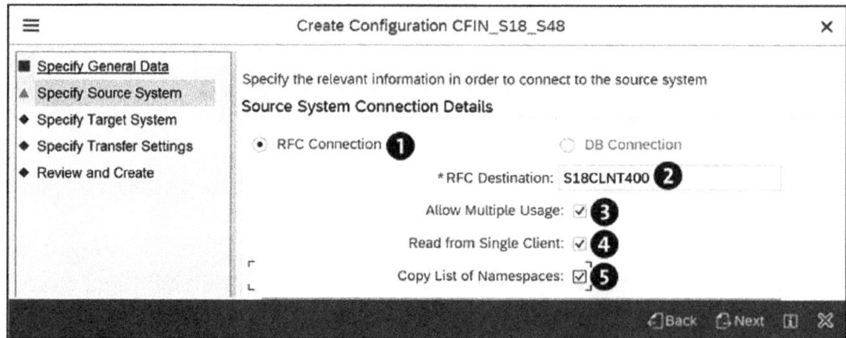

Figure 5.13 Create Configuration: Specify Source System

Specify Target Data

On the **Specify Target System** configuration screen, as shown in Figure 5.14, choose the **RFC Connection** ❶ radio button. Next, choose **Scenario** from the dropdown list. The system provides two possible scenarios to choose from, as shown in Table 5.15. Choose **ABAP-Based SAP System (Target Tables Not Managed by SLT)** ❷ from the dropdown list for the Central Finance scenario. This standard RFC scenario allows you to replicate data to the existing tables in the target system.

Scenario	Description
RFC ABAP-Based SAP System	Target tables are not managed by SAP Landscape Transformation Replication Server.
RFM ABAP-Based SAP System for replication flow management (RFM)	Target tables are managed by SAP Landscape Transformation Replication Server. This option is not used in a Central Finance replication scenario.

Table 5.15 Specify Target System

Next, select a destination from the **RFC destination** dropdown list ❸, and click the **Next** icon to continue.

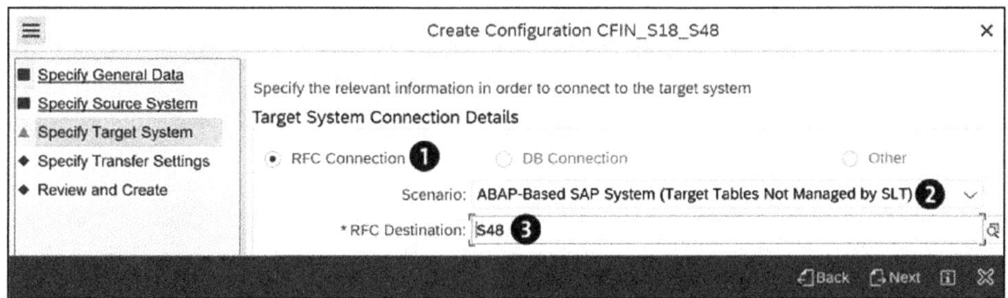

Figure 5.14 Create Configuration: Specify Target System

Specify Transfer Settings

In the **Specify Transfer Settings** configuration screen, you'll define several parameters relevant to the initial load and replication of data. This screen is further divided into three major subsections: **Data Transfer Settings**, **Job Options**, and **Replication Options**. We'll review each section next.

In the **Data Transfer Settings** subsection, as shown in Figure 5.15, you'll choose the value for the **Initial Load Mode** ❶ field from the dropdown list. SAP recommends choosing **Performance Optimized**, and this value is chosen by default for Central Finance. Table 5.16 lists the options provided under **Initial Load Mode**.

Initial Load Mode Options	Description
0	Resource optimized is used for reading type 3 for all tables. For SAP and SAP HANA–based systems, the system uses reading type 1.
1	Performance optimized is used for reading type 4 for cluster tables or 5 for all other tables.

Table 5.16 Initial Load Mode

Next, enter values in the following fields under the **Job Options** subsection, as shown in Figure 5.15:

- **Number of Data Transfer Jobs ❷**
  Enter a value in this field to run the number of jobs in SAP Landscape Transformation Replication Server during source system table replication.
- **Number of Initial Load Jobs ❸**
  Enter the maximum number of data transfer jobs for the initial load. Ensure you have enough jobs available to run replication jobs when selecting this option. This option also varies with landscape and data volume requirements.
- **Number of Calculation Jobs ❹**
  Enter the number of jobs used to perform calculations for the access plan. The calculation jobs are triggered only during the initial load and aren't relevant for replication.

  Finally, in the **Replication Options** section, you'll specify how you want to run the replication. The system provides the options listed in Table 5.17. Choose the **Real Time ❺** radio button, which is the recommended option.

Replication Options	Purpose
Real Time	As the name implies, data is replicated in real time from the source system to target system.
Schedule by Interval	This option provides a fixed interval to replicate data from the source system.
Scheduled by Time	You'll schedule a specific time to replicate data. The system checks the logging table for any new data to be replicated on the scheduled time interval.
On Demand	You'll have an option to replicate data using RFCs as required.

Table 5.17 Replication Options

The **Approval Required for All Data Provisioning Actions ❻** checkbox isn't required to be checked. However, if you select this, then any action scheduled for the object must be approved before the action starts.

Once you've completed all parameters in the **Specify Transfer Settings** configuration screen, choose the **Next ❼** icon to continue to the next the **Review and Create configuration** screen.

Review and Create

In this final step, navigate to the **Review and Create** screen, as shown in Figure 5.16 ❶. The system displays all configurations you've just completed in the previous sections. Once you've reviewed all the information, click the **Create ❷** icon to create the configuration. The bottom screen shows a successfully created MT ID ❸.

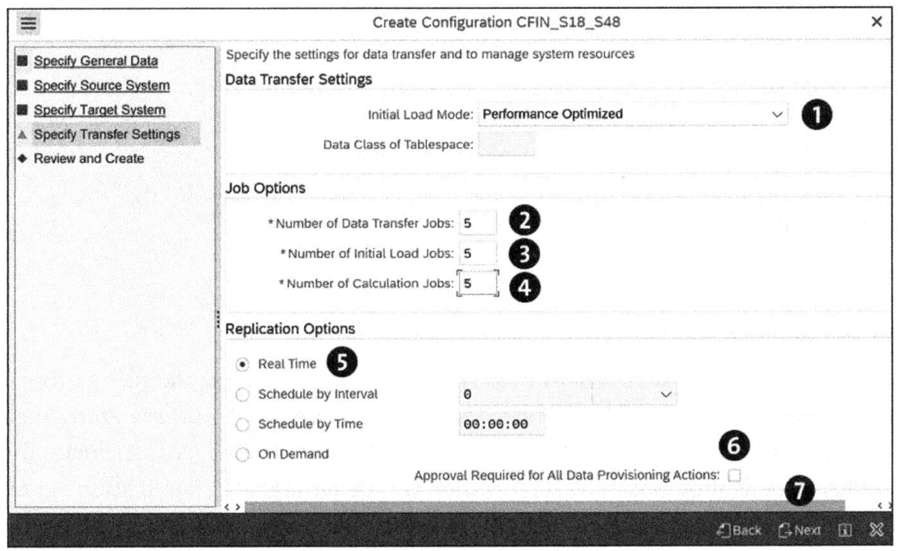

Figure 5.15 Create Configuration: Specify Transfer Settings

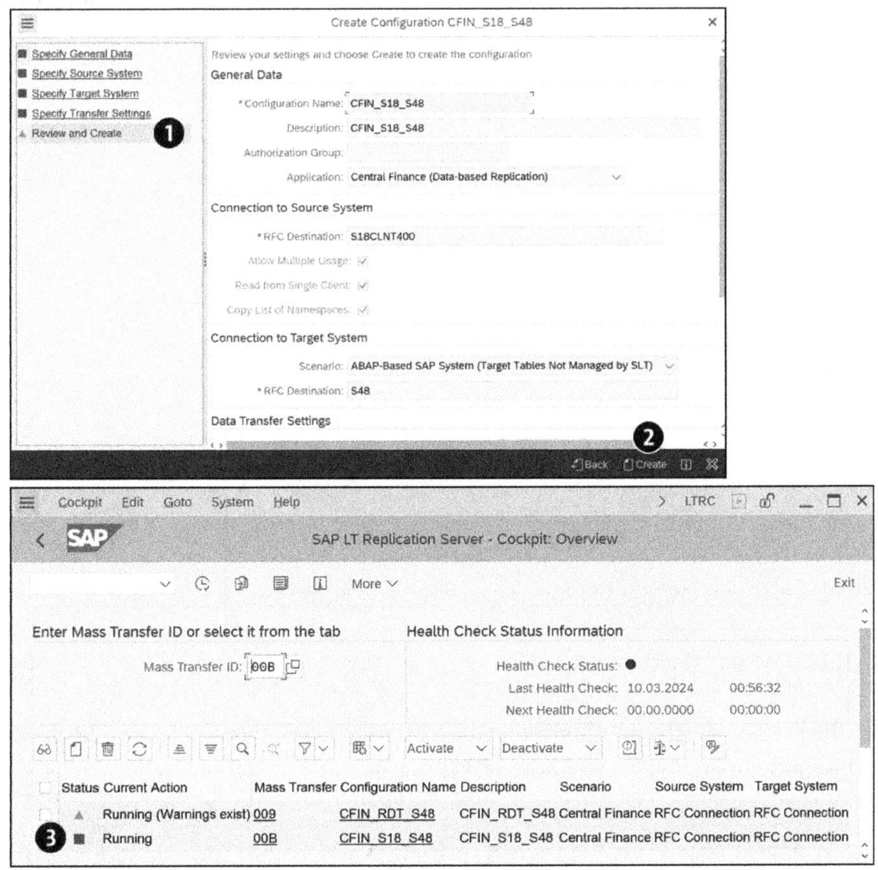

Figure 5.16 Create Configuration: Review and Create

Note

We've walked you through the steps of creating and configuring using the SAP Landscape Transformation Replication Server cockpit. The same steps can be followed with Transaction LTR, which is based on the Web Dynpro user interface. Consult SAP Note 1972009 for further guidance on creating an SAP Landscape Transformation Replication Server configuration.

Create a Predefined Load Object

This configuration can be done in two ways: using Transaction LTRC and navigating to **MT ID Expert Functions • Processing Steps • Preparatory Functions • Create Predefined Load and Replication Objects** or using program IUUC_REPL_PREDEF_OBJECTS to create the initial load and replication objects for a specific MT ID. On the **SAP LT Replication Server - Cockpit 00B - CFIN_S19_S48** screen, double-click **Create Predefined Load and Replication Objects** ❶ located under **Processing Steps** in the left pane, as shown in Figure 5.17. After navigating away from the transaction, the system takes you to the **Create Predefined Replication Objects** screen. On this screen, choose **Copy Predefined Object** ❷ button.

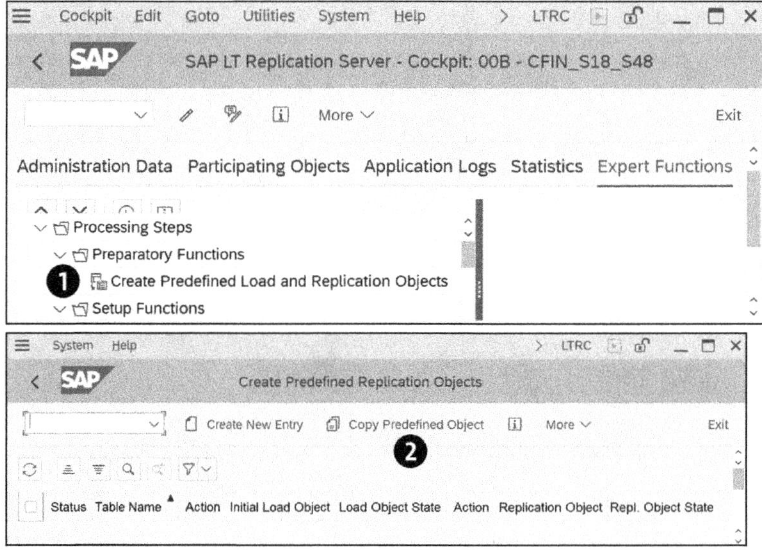

Figure 5.17 Create a Predefined Object

On the resulting **Copy existing Predefined Object** screen, as shown in Figure 5.18, enter values in the following fields in the **Source Object** area:

❶ **Project**
Enter value **REPL_CFIN** in this field, which is unique across the landscape for Central Finance replication.

❷ **Subproject**
Enter value **REPL_CFIN** in this field.

❸ **Predefined Object**
Enter the predefined object for the initial load. In this example, enter predefined object "CFI_T001_L" for the initial load of Table T001. The naming convention of predefined objects follows: CFI_<Tablename>_L, where CFI stands for Central Finance and L stands for load.

❹ **Table Name**
Enter value "T001" for the company code table replication corresponding to the predefined object chosen from the earlier step.

Next, choose the **Create Predefined Load Object** ❺ radio button and then click the **Continue** ❻ icon.

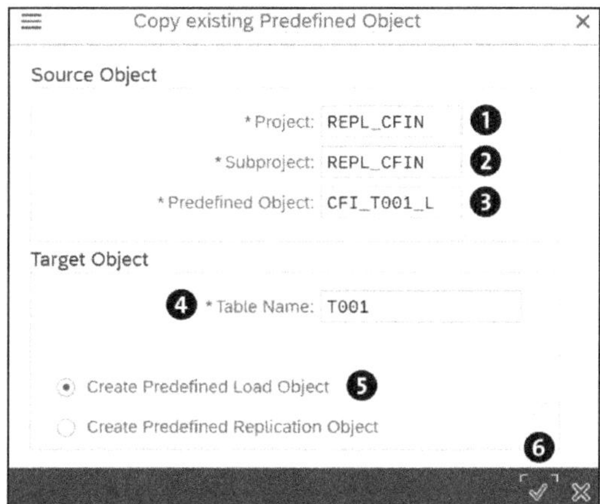

Figure 5.18 Create a Predefined Load Object

After completing this step successfully, you'll notice the predefined load object shows up in the list on the **Create Predefined Replication Objects** screen.

Create Predefined Replication Object

Next, to create the predefined replication object, choose **Copy Predefined Object**.

On the resulting **Copy existing Predefined Object** screen, as shown in Figure 5.19, enter values in the following fields in the **Source Object** section:

❶ **Project**
Enter value "REPL_CFIN" in this field, which is unique across the landscape for Central Finance replication.

❷ **Subproject**
Enter value "REPL_CFIN" in this field.

❸ **Predefined Object**
Enter the predefined object for the initial load. In this example, enter "CFI_T001_R" for the initial load of Table T001. The naming convention of the predefined object follows: CFI_<Tablename>_R where CFI stands for Central Finance and R stands for replication.

❹ **Target Object**
Enter **Table Name** of table for replication, in this example, table value of "T001" is entered.

Next, choose the **Create Predefined Replication Object** ❺ radio button, and then click the **Continue** ❻ icon to create the predefined replication object for table T001.

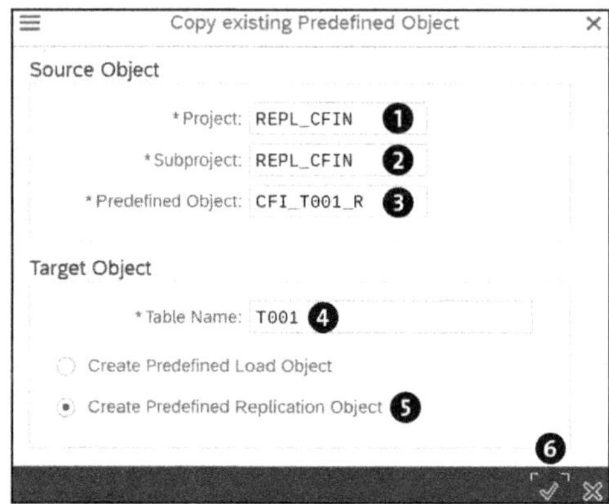

Figure 5.19 Create the Predefined Replication Object

After completing this step successfully, you'll notice the predefined load and replication objects show up on the **Create Predefined Replication Objects** screen ❶, as shown in Figure 5.20. **Load Object State** ❷ and **Repl. Object State** ❸ must be activated for load and replication objects.

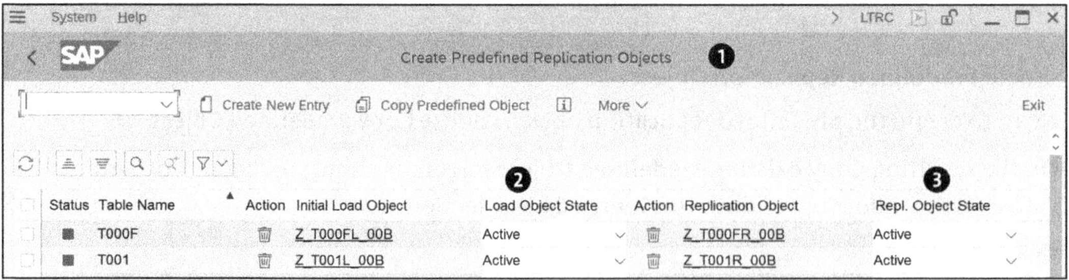

Figure 5.20 Load and Replication Object Status

Filters and Transformation Using the Migration Workbench

It's necessary to implement a runtime filter, which is defined in the Migration Workbench (Transaction MWB), a hierarchical migration object, which helps to filter the data in the initial load and real-time replication for the Central Finance target system. It controls the data transfer from the source system to the Central Finance system per business requirements.

Start Replication

In this step, you'll start loading and replication for the identified replication object. Figure 5.21 shows starting replication for table T001:

❶ **Start Load**
This option performs source data calculation, and data is loaded and sent to Central Finance.

❷ **Start Replication**
The system creates a logging table and database triggers in the source system. In addition, the initial load is carried out followed by delta replication. Real-time replication begins after the SAP Landscape Transformation Replication Server initial load is completed.

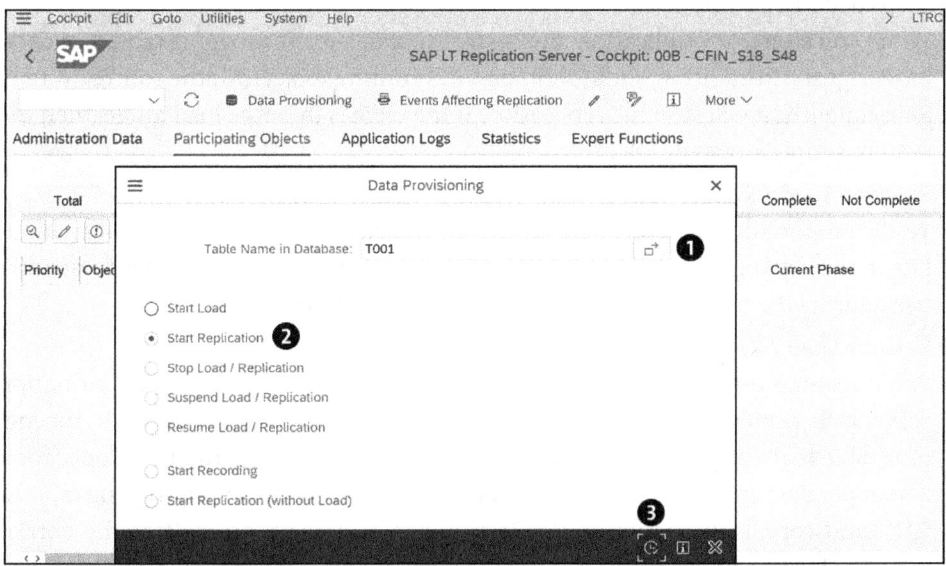

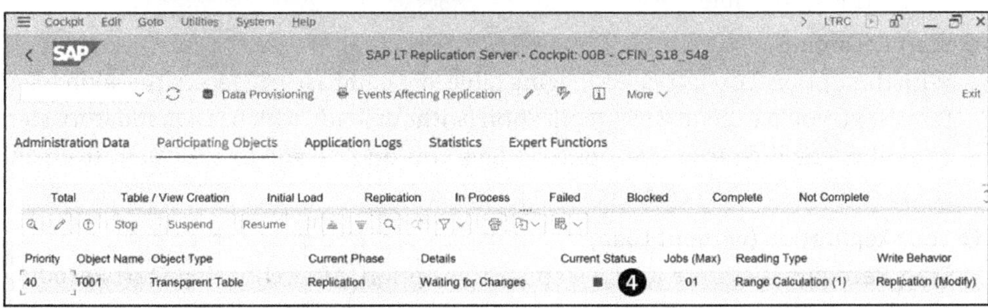

Figure 5.21 Start Replicating Tax Table T001

Once you've selected the **Start Replication** radio button, click the **Execute** icon ❸ to continue. The bottom of Figure 5.21 ❹ displays the **Current Status** of the replication.

Data Provisioning in SAP Landscape Transformation Replication Server

As discussed earlier, SAP provides seven data provisioning options to control table replication and initial load data transfer. You must also understand the differences between these data provisioning options as explained here and shown in Figure 5.22:

❶ **Start Load**
This option performs source data calculation, and data is loaded and sent to Central Finance.

❷ **Start Replication**
The system creates a logging table and database triggers in the source system. In addition, the initial load is carried out followed by delta replication. Real-time replication begins after the SAP Landscape Transformation Replication Server initial load is completed.

❸ **Stop Load / Replication**
When you stop load/replication, the system first deletes the logging table and database triggers. In addition, the system de-links runtime objects in the source system. Subsequently, it will stop the replication. If the table is in initial load mode, then the system simply stops the load.

❹ **Suspend Load / Replication**
You'll temporarily pause replication for various reasons. However, data will accumulate in the logging table in the source system. For example, this option is frequently used during the maintenance window of any connected systems.

❺ **Resume Load / Replication**
You'll resume replication from suspended status. SAP Landscape transformation server calls reader module. Any new unprocessed data that accumulated in the logging table read for processing is sent to SAP Landscape Transformation Replication Server per data creation sequence. After data is transferred from the logging table to SAP Landscape Transformation Replication Server, the system deletes the entries from the logging table.

❻ **Start Recording**
With this option, you'll create a logging table and database triggers in the source system. Next, you'll execute start replication (without load), which is a mandatory step. Typically, you'll choose this option when you don't want to load data from the source system, but only start real-time replication.

❼ **Start Replication (without Load)**
Start Recoding is a prerequisite step for this option. After choosing **Start Recoding**, you'll start real-time replication.

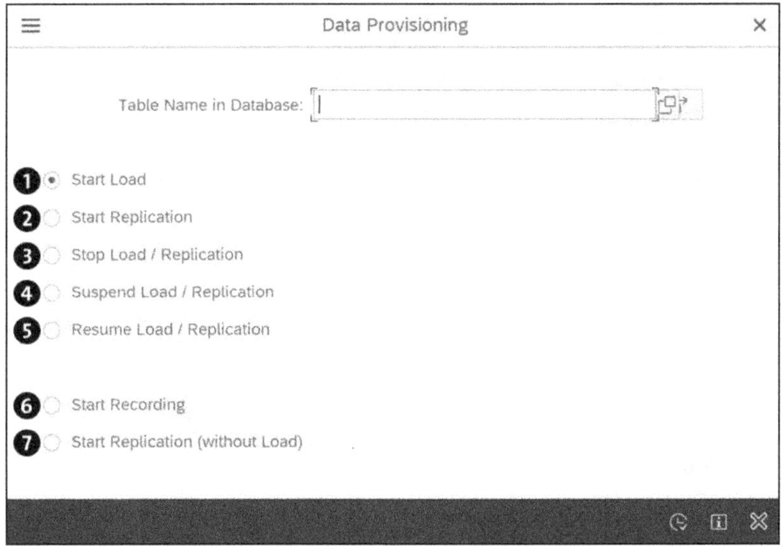

Figure 5.22 Data Provisioning

5.2.2 Replicating Withholding Tax Tables

Central Finance supports withholding taxes posted with invoices and at the time payments in the source system are replicated as part of financial postings to Central Finance during the initial load and online replication. In addition, when an invoice with withholding taxes is reversed, the relevant withholding data reversals are also transferred to Central Finance. Withholding tax calculation is performed only in the source system, and no recalculation occurs in Central Finance. However, if central payment is active, the system calculates withholding tax at the time of payment clearing in the Central Finance system. SAP recommends consistent withholding tax configuration, especially withholding tax types and withholding tax codes, to be identical properties between the source system and Central Finance system to avoid errors in SAP Application Interface Framework during replication. However, the system allows for a different tax types and tax code mapping. MDG business mapping must be carried out in central finance appropriately to support different tax types and tax codes.

Prior to replicating withholding tax details along with FI documents, for tax consistency and central payment functionality, you must set up SAP Landscape Transformation Replication Server tables for withholding tax replication. Structure ACCIT_WT holds specific withholding tax fields that are transferred to Central Finance.

Table 5.18 lists the required SAP Landscape Transformation Replication Server objects to be set up for withholding tax load and replication.

Source Table	Description	Replicated Table in Central Finance
T001	Company code	FINS_CFIN_T001
T001RWT	Rounding rules for company code and withholding tax type	FINS_CFIN_T001RW
T001WT	Company code-specific withholding tax data	FINS_CFIN_T001WT
T030	Standard accounts	FINS_CFIN_T030
T059MINMAX	Minimum and maximum amounts for withholding tax	FINS_CFIN_T059MM
T059P	Withholding tax codes	FINS_CFIN_T059P
T059Z	Withholding tax types	FINS_CFIN_T059Z

Table 5.18 Withholding Tax Tables

From an SAP Landscape Transformation Replication Server perspective, you'll set up these tables like you've configured the tax tables discussed in Section 5.2.1.

[»]

Note

Central Finance supports only the replication of extended withholding taxes from SAP S/4HANA 1809 FPS0 onward. Prior to this release, you'll need to apply withholding tax relevant SAP Notes to use this feature. Consult SAP Note 2668261 for the latest withholding tax tables list.

Classic withholding tax replication is not supported in Central Finance. Once you've set up withholding tax tables replication in SAP Landscape Transformation Replication Server, SAP recommends running the tax consistency report for any missing configuration or inconsistencies. If your source system is running on classic withholding tax, you must migrate to extended withholding taxes.

Refer to SAP Note 2623514 for further guidance on relevant SAP referenced notes that may be required. In addition, this SAP Note comes with a guide released to customers on withholding tax functionality from a central finance context.

5.2.3 Configuring for Table AUFK: Cost Object Replication

Recall we've covered concepts of cost object framework extensively in Chapter 4, Section 4.4.5. In this section, as part of cost object replication, you'll do cost object simulation first to ensure that there are no missing configurations, master data, and data mapping in the Central Finance system.

In terms of SAP Landscape Transformation Replication Server configuration for table AUFK cost object load and replication, you'll repeat the same steps we've explained in the tax table replication in Section 5.2.1. The key difference is that you'll select different predefined objects. For simulation, you'll choose only load objects. Table 5.19 lists the load and replication objects for cost object replication.

Type of Replication	Table	Load Object	Replication
Cost object	AUFK	CFI_AUFK_L	CFI_AUFK_R

Table 5.19 Predefined Load and Replication Objects for Table AUFK

As shown in Figure 5.23, for simulation of cost objects, on the **Create Predefined Replication Objects** screen, enter the **Predefined Object** as "CFIN_SIM_AUFK_L" ❶ and the **Table Name** as "AUFK" ❷. Then, choose the **Create Predefined Load Object** ❸ radio button. Click the **Execute** button to continue.

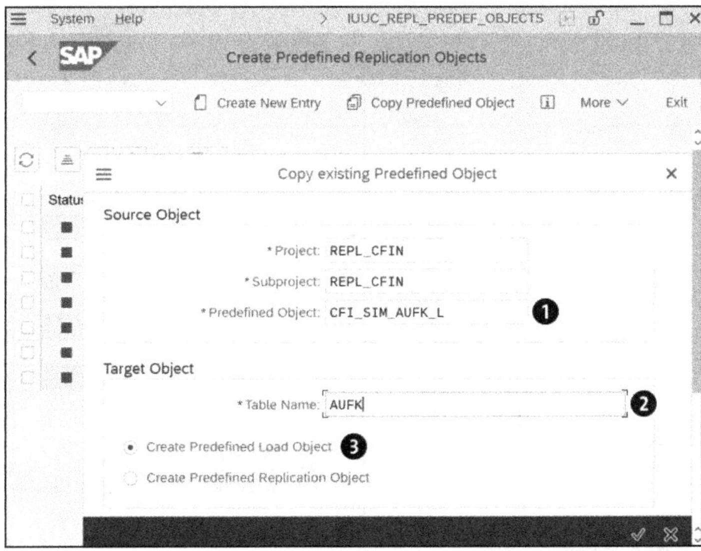

Figure 5.23 Create a Predefined Load Object for Cost Object Simulation

Note

For simulation of cost objects, you'll create only a load object. For a real run, both load and replication must be created with different predefined objects provided with Central Finance.

5.2.4 Configuring for Table CFIN_ACCHD: Financial Postings

Performing classic initial load is covered in Chapter 8, Section 8.3.1. After the classic FI-CO initial load is carried out, you'll set up SAP Landscape Transformation Replication Server to perform delta load (from logging table) and also enable real-time replication.

The first prerequisite is to create MT ID as we've explained in the steps for tax table replication. In this section, we'll carry out predefined object creation followed by MWB filter rules if required for object CFIN_ACCHD replication. Table 5.20 lists the load and replication objects for the replication object CFIN_ACCHD.

Type of Replication	Table	Load Object	Replication
Financial postings	CFIN_ACCHD	CFI_CFIN_ACCHD_L	CFI_CFIN_ACCHD_R

Table 5.20 Predefined Load and Replication Objects for CFIN_ACCHD

To carry out predefined object creation, as shown in Figure 5.24, enter the **Predefined Object** as "CFI_CFIN_ACCHD_L" ❶ and the **Table Name** as "CFIN_ACCHD" ❷ in the **Create Predefined Replication Objects** screen. Finally, choose the **Create Predefined Load Object** ❸ radio button, and then click the **Continue** button.

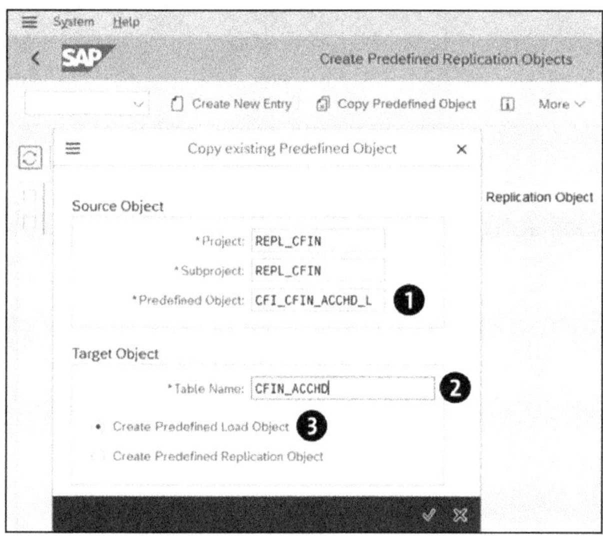

Figure 5.24 Create Predefined Load Object for Financial Replication

As a result, the system creates the template predefined load object for CFIN_ACCHD, as shown in Figure 5.25. The system generates the mapping rule SKIP_CFIN_ACCIT and transfer rule SKIP_CFIN_CLNT successfully. These rules will have the standard code copied from the standard object, which can be modified per project requirements.

Like load objects, you'll carry out predefined object creation for the replication object, as shown in Figure 5.26. On the **Create Predefined Replication Objects** screen, enter the **Predefined Object** as "CFI_CFIN_ACCHD_R" ❶ and the **Table Name** as "CFIN_ACCHD" ❷. Finally, choose the **Create Predefined Replication Object** ❸ radio button, and then click the **Continue** button.

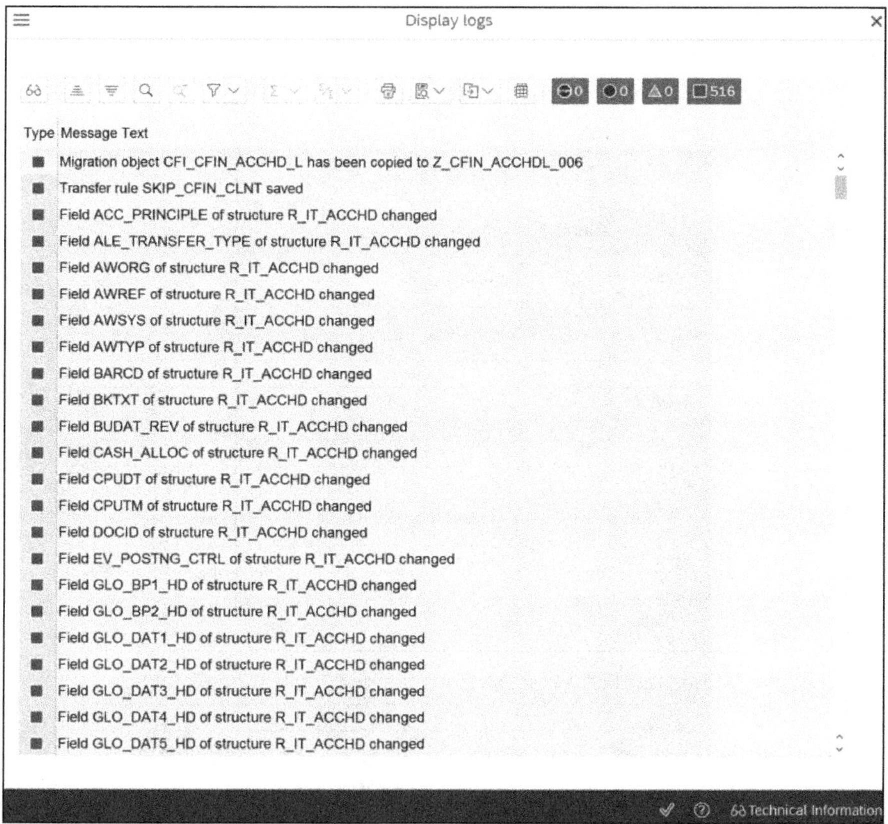

Figure 5.25 Copy Predefined Load Objects: Confirmation for Load Object

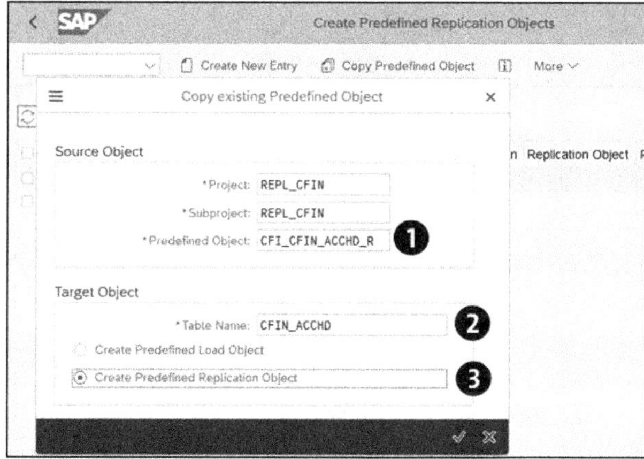

Figure 5.26 Create a Predefined Load Object for the Financial Replication Object

As a result of this action, the system creates a template predefined replication object for `CFIN_ACCHD`, as shown in Figure 5.27. As you've also observed from the figure, mapping rule `SKIP_CFIN_OPFLG` is `generated`. Likewise, the rule will have the standard code copied from the standard object, which can be modified per project requirements.

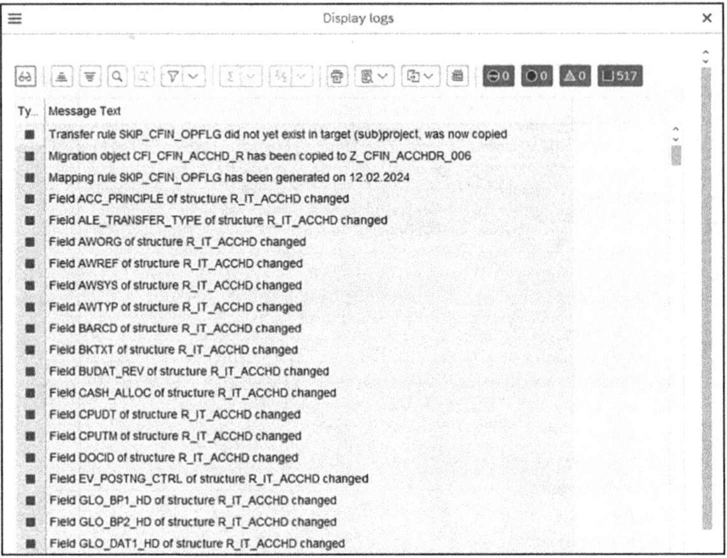

Figure 5.27 Copy Predefined Load Objects: Confirmation for Replication Objects

Once both predefined load and replication objects are generated, you must activate them. As shown in Figure 5.28, you'll activate **Load Object State** ❶ and **Repl. Object State** ❷.

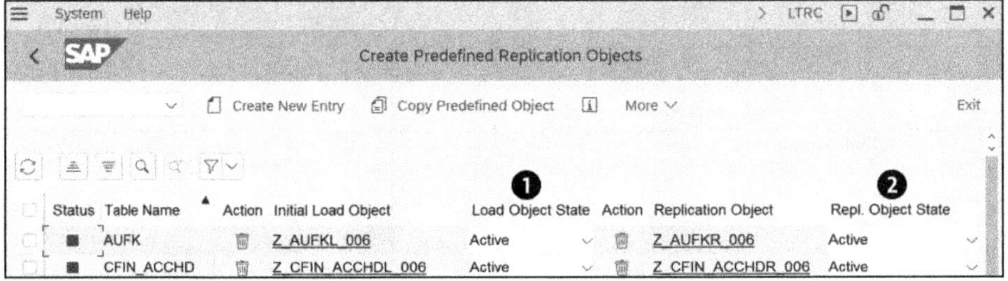

Figure 5.28 Copy Predefined Load and Replication Objects CFIN_ACCHD: Status

In the next step, you can modify the copied predefined event-based rule to load/replicate selected data per the filter criteria you've defined earlier. You'll perform this activity using Transaction MWB. On the **MWB: Create/Edit Migration Objects** screen, as shown in Figure 5.29, select the template **Project** and **Subproject** ❶, which are created during the predefined load and replication objects configuration for the **CFIN_ACCHD** MT ID. Then, click the **Change** button to continue. You'll arrive at the **DMC Tool: Change Subprojects - (Saved)** screen, as shown at the bottom of Figure 5.29. On this screen, click the **Conversion Logic** ❷ button to navigate to **DMC Tool: Maintain Transfer Rule (Change) – Saved** screen.

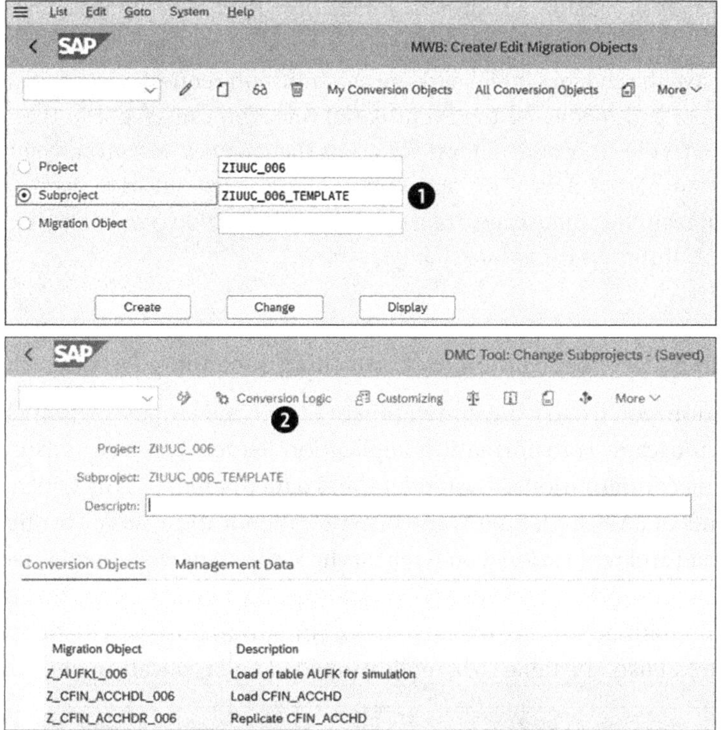

Figure 5.29 Editing the MWB Template Object for CFIN_ACCHD

Figure 5.30 shows the event-based rule to load and replicate subset financial data depending on custom filter requirements. In the **Rule Maintenance** window, select **SKIP_CFIN_ACCIT** ❶. You'll modify or add the custom code by selecting **Code Editor** ❷.

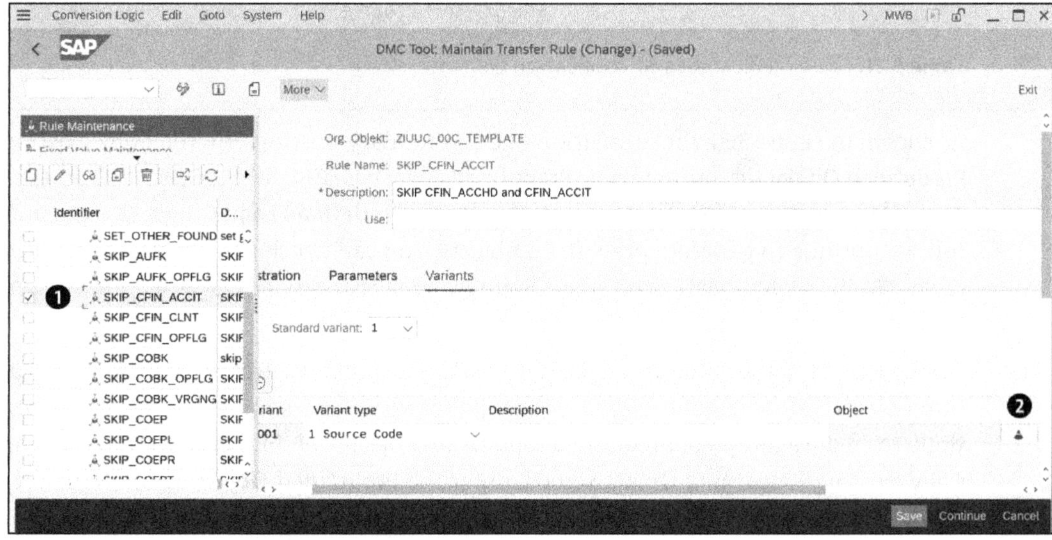

Figure 5.30 Skip Rule for CFIN_ACCHD

Tip

You'll get a popup to save the changes in the package or a transport request while editing the template project and modifying the event-based rule. You can save it in the local package or log in to the transport request to record the changes. However, you shouldn't move the transport request. This activity must be carried out as a client-dependent activity in each SAP Landscape Transformation Replication Server client. The idea to record the transport is to capture change history.

5.2.5 Configuring Table COBK Replication for Controlling Secondary Postings

With table COBK replication, you'll carry out simulation of CO secondary posting, which is performed in SAP Landscape Transformation Replication Server. This is to ensure that there are no missing configurations, master data, and data mapping in the Central Finance system. In terms of SAP Landscape Transformation Replication Server configuration for table COBK load and replication, you'll repeat the same steps we've explained in the tax table replication in Section 5.2.1 for MT ID creation. The key difference is that you'll select different predefined objects for table COBK. For simulation, you'll choose only load objects. Table 5.21 lists the table COBK replication load and replication objects.

Type of Replication	Table	Load Object	Replication
CO secondary posting simulation	COBK	▪ CFI_SIM_COBK_L	▪ N/A
CO secondary posting	COBK	▪ CFI_COBK_L (SAP ERP source) ▪ CFI_COBK_1511_L (SAP S/4HANA source)	▪ CFI_COBK_R (SAP ERP source) ▪ CFI_COBK_1511_R (SAP S/4HANA source)

Table 5.21 Table COBK Load and Replication Objects

As shown in Figure 5.31, for simulation of CO secondary postings, on the **Copy existing Predefined Object** screen, enter the **Predefined Object** as "CFI_SIM_COBK_L" ❶ and the **Table Name** as "COBK" ❷. Then, choose the **Create Predefined Load Object** ❸ radio button to continue to generate predefined objects. You can check the simulated result of table COBK in SAP Application Interface Framework with interface **CO_DOC_SIM** version 1.

For an actual load, as you can see in Figure 5.32, you'll use a different load and replication objects SAP S/4HANA release 1511 or higher setup as a source system. Click **Continue** to create a template predefined replication object for table COBK. On the resulting **Copy existing Predefined Object** screen, enter the **Predefined Object** as "CFIN_COBK_1511_L" ❶ and the **Table Name** as "COBK" ❷. Then, choose the **Create Predefined Load**

Object ❸ radio button to continue to generate predefined objects. You can check the simulated result of table COBK in SAP Application Interface Framework with interface **CO_DOC** version 1.

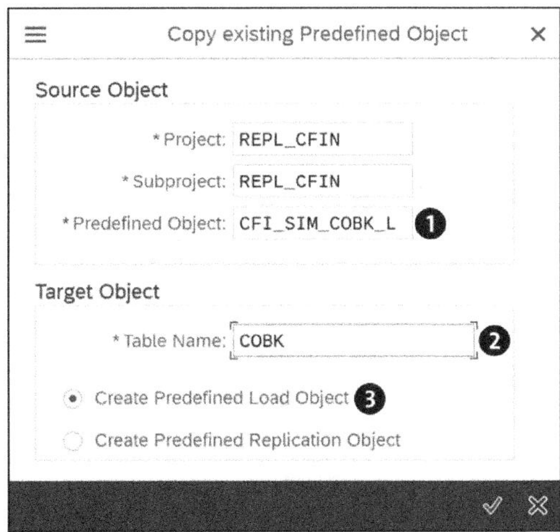

Figure 5.31 Copying Existing Predefined Objects for the Table COBK Simulation

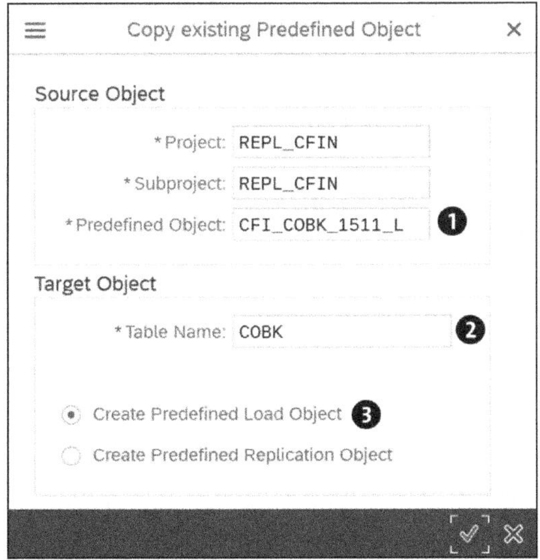

Figure 5.32 Copying the Existing Predefined Objects for the Table COBK Simulation Load

For replication of table COBK, as shown in Figure 5.33, on the **Copy existing Predefined Object** screen, enter the **Predefined Object** as "CFIN_COBK_1511_R" ❶ and the **Table Name** as "COBK" ❷. Then, choose the **Create Predefined Replication Object** ❸ radio button, and click the **Continue** button. This will create both predefined load and replication objects by copying from the standard.

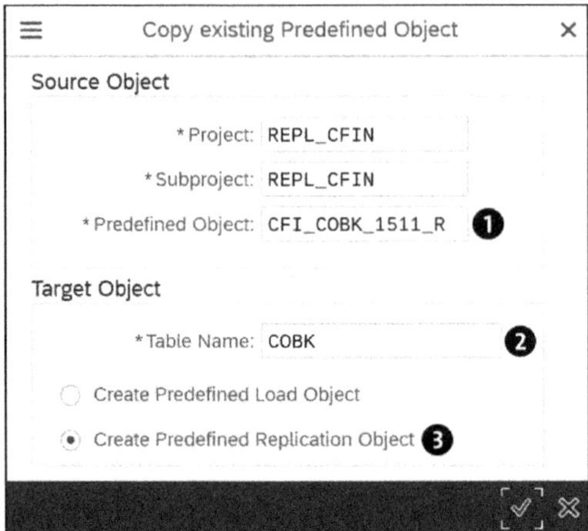

Figure 5.33 Copying Existing Predefined Objects for Table COBK Replication

Note

SAP has pre-delivered rules such as SKIP_COBK, SKIP_COBK_VRGNG, and SKIP_COEP where you can control what data will be loaded/replicated in the target Central Finance system. These rules can be modified under the template project and subproject created in the predefined load and replication objects. These activities can be carried out using Transaction MWB.

5.3 Advanced Replication Settings

The advanced settings allow you to apply transformation rules during the transfer of data from the source system to the target system. In this section, you'll learn about the advanced settings in SAP Landscape Transformation Replication Server. For Central Finance, the following features are used frequently:

- Filters/rules
  - Initial load filters
  - Field-based rules
  - Event-based rules
- Controlling triggers

Let's review each of the filters and rules next.

5.3.1 Understanding Filters

With filters, you can replicate a subset of the data from the source system to the target system or specify a specific condition to transfer data. Broadly, SAP Landscape

Transformation Replication Server provides three filter options for the target Central Finance system:

- **Initial load filter**
  A trigger-based filter created with Transaction LTRS. With the initial load filter, the system implements a trigger at the source system. As a result, only filtered data from the source system is transferred to SAP Landscape Transformation Replication Server.
- **Event-based rule**
  As the name implies, you'll be able to create an event-based rule to derive a field value to the receiving structure based on requirements. It's a runtime filter defined in the MWB. For example, `SKIP_AUFK` is an event-based rule for filtering cost object data.
- **Field-based data transformation rule**
  SAP provides a field-based data transformation rule to derive the field value of the receiver structure per required by business requirements.

Let's look at each of the filter options in detail next.

5

Initial Load Filters

With the initial load filter, you can extract a subset of data from the source system, thereby improving the performance of the initial load and replication. To apply the initial load filter, navigate to the settings using Transaction LTRS. The transaction takes you to the **SAP LT Replication Server: Configuration Overview** screen, as shown in Figure 5.34. On this screen, choose **MT ID** ❶, navigate to the **Advanced Replication Settings** ❷ area on the left pane, and choose **Performance Options** ❸. Next, choose the table, in our example, **COBK** ❹, and then go to the **Filter Options** ❺ tab, and select **Filter Initial Load Only** ❻.

As you can see, you can define multiple filters. Each *filter* uses an `OR` condition while each *filter condition* uses an `AND` condition. In addition, `DATS` and `TIMS` can be configured with the following format:

- `DATS`: YYYYMMDD
- `TIMS`: HHMMSS

Once you make the changes, click the **OK** button at bottom to save the entries.

[+]

Tip

If you're adding a filter in **Filter Options** while a table is in replication mode, regeneration of the migration option is required. For further guidance on this topic, consult SAP Note 2776058.

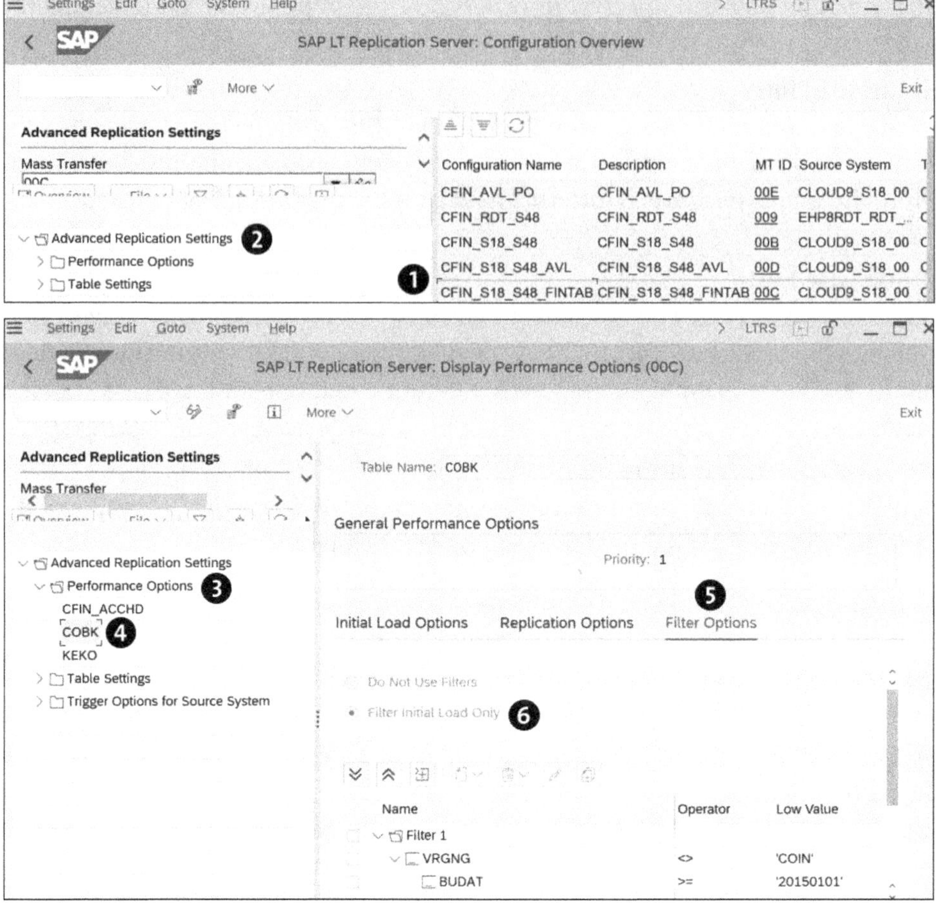

Figure 5.34 Initial Load Filter

Event-Based Rule

The event-based rule is a runtime filter defined in Transaction MWB that helps to filter the data in the initial load and real-time replication for the Central Finance target system. As a result of implementing the event-based rule, the system implements logic in the runtime objects on the SAP Landscape Transformation Replication Server to filter out irrelevant records. Furthermore, using this method, the data is transferred from the source system to SAP Landscape Transformation and is filtered out during the SAP Landscape Transformation runtime. SAP has provided the pre-delivered rule for each finance object associated with the Central Finance target system through SAP Landscape Transformation Replication Server.

To implement a runtime filter using Transaction MWB, navigate to the template subproject for the MT ID where the rules are to be defined, as shown in Figure 5.35. Once you select the **Project** ❶ and **Subproject** ❷ information by pressing [F4] on the **MWB: Cereate/Edit Migration Objects** screen, click **Change**, and you'll arrive at the **DMC Tool: Change Subprojects – (Saved)** screen, as shown at the bottom of Figure 5.35. On this screen, enter a value in **Descriptn** ❸, and select the **Conversion Logic** ❹ icon.

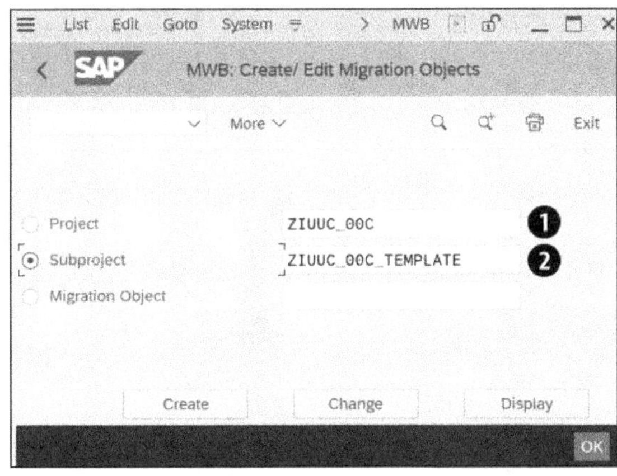

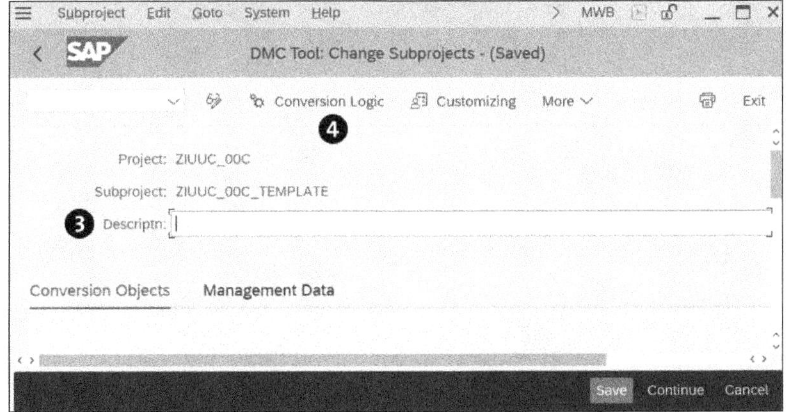

Figure 5.35 MWB Create Edit Migration Objects

As shown in Figure 5.36, the template **Event Based Rule** ❶ is adapted to filter the data from the source to the Central Finance system through SAP Landscape Transformation Replication Server. Next, to add the custom code, select **Code Editor** ❷. You'll add the necessary code in the **MWB: Maintain Rule Coding** ❸ screen.

As shown in Figure 5.37 ❶, event-related processing rules can be created for events BEGIN-OF-LOOP, BEGIN-OF-RECORD, END-OF-LOOP, and END-OF-RECORD. As an example, SKIP_AUFK is created for the event **Start of Record** ❷ for the cost object initial load and replication.

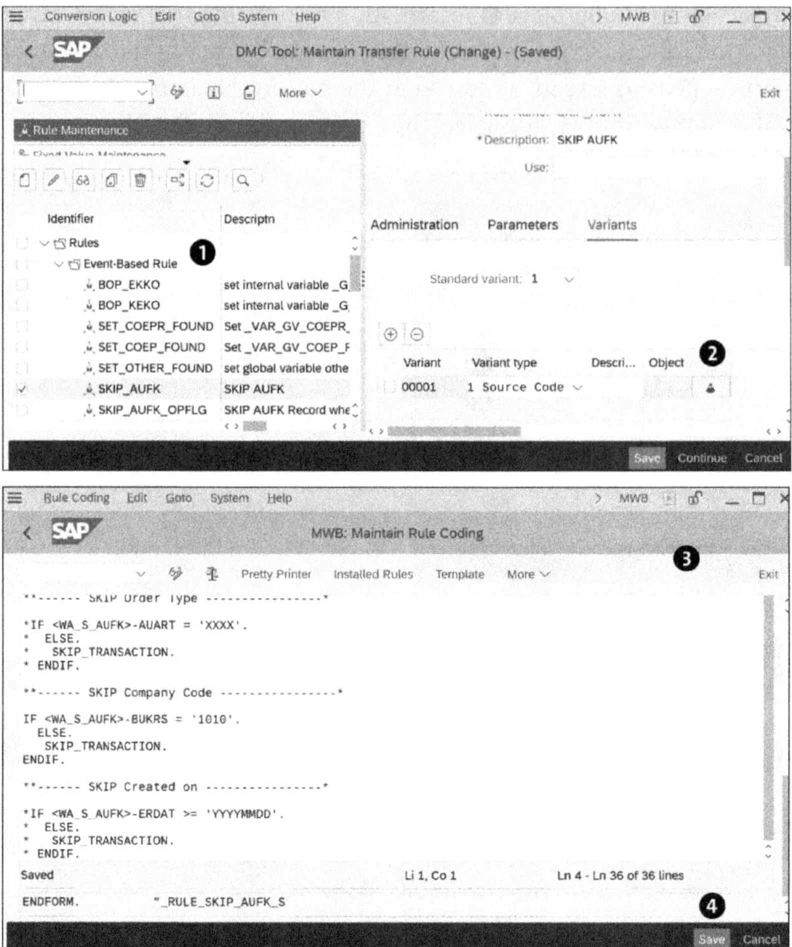

Figure 5.36 MWB Event-Based Rules

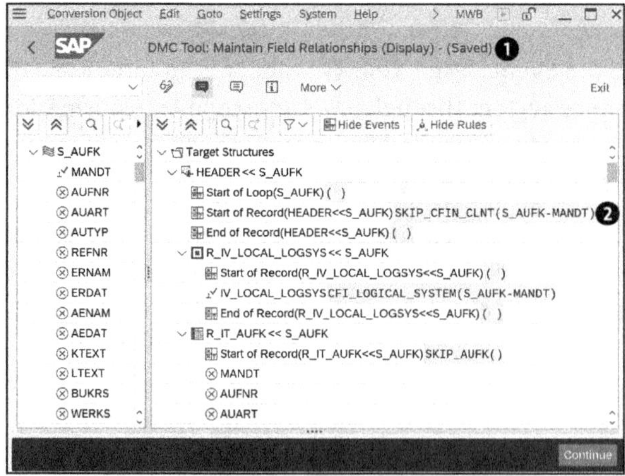

Figure 5.37 MWB Event-Based Rule Assignment

Field-Based Data Transformation Rule

In this section, you'll learn how a field-based rule is created, which derives the field value of the receiver structure as required by a standard delivered logic by SAP Landscape Transformation Replication Server. A preconfigured field-based rule is created for the Central Finance objects to get loaded/replicated through SAP Landscape Transformation Replication Server. As you can see from Figure 5.38 ❶, **CFIN_LOGICAL_SYSTEM** ❷ is the field-based rule to get the logical system from the sender. If the Central Finance target system is connected with multiple source systems, SAP Landscape Transformation Replication Server will have a logical system field value for each sender system.

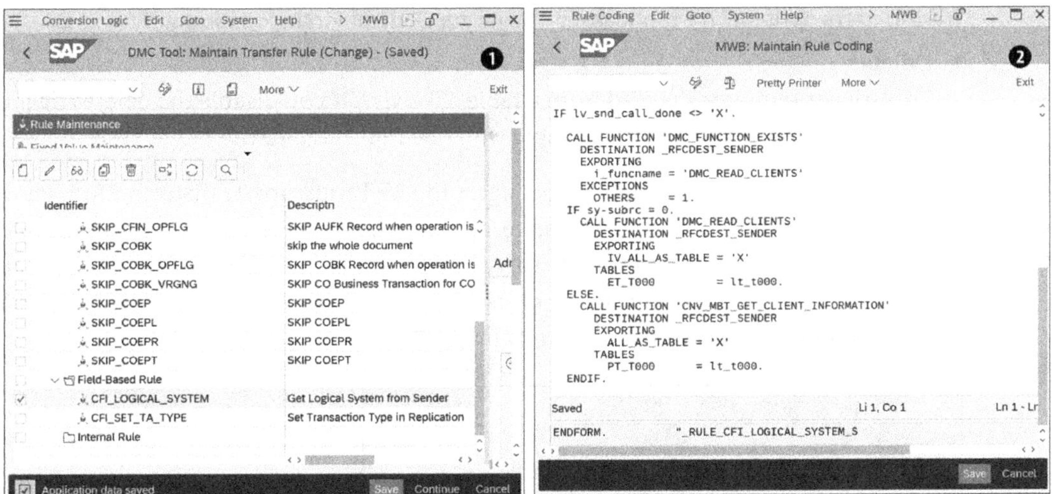

Figure 5.38 Field-Based Data Transformation Rule

Figure 5.39 shows field-based rule **CFI_LOGICAL_SYSTEM**, which is assigned to field LOGSYS_SENDER. This field derives the logical system value for each posting in the Central Finance target system.

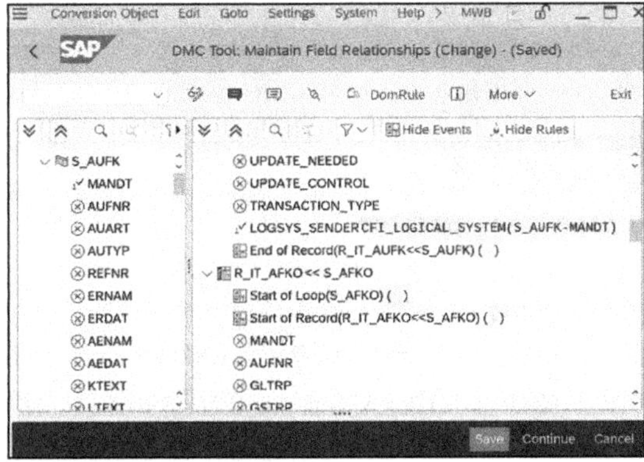

Figure 5.39 Field-Based Data Transformation Rule: Logical System Assignment

5.3.2 Controlling Database Triggers

There are three types of database triggers created when adding a table for replication in the SAP Landscape Transformation Replication Server configuration:

- **Insert**
  As the name suggests, insert triggers inserts of data into a table.
- **Update**
  This trigger changes a record on an existing table. There is a provision to disable the update and delete database triggers of the table in the SAP Landscape Transformation Replication Server. Note if you disable the update trigger then the UPDATE operation in the table won't replicate to the target Central Finance system.
- **Delete**
  This trigger removes a record from a table. Likewise, if you disable the delete trigger, then the DELETE operation in the table won't be replicated to the target system.

Let's explore how to disable database triggers in SAP Landscape Transformation Replication Server. Prior to making the change, as a prerequisite, the SLT_EXPERT user should have the appropriate authorization to carry out this task. To enable authorization, go to Transaction SU3. On the **Maintain User Profile** screen, add an entry for user **SLT_EXPERT** with **Parameter value** = **X**, as shown in Figure 5.40.

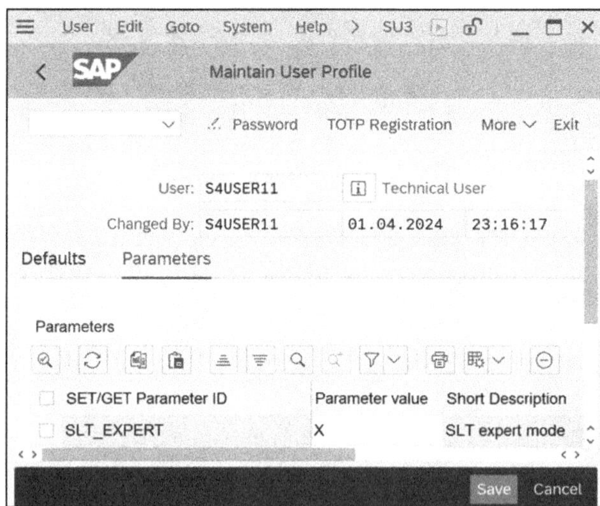

Figure 5.40 Maintain User Profile SAP Landscape Transformation Expert

Next, execute Transaction LTRS, and choose the **MT ID**, as shown in Figure 5.41 ❶. Under **Advanced Replication Settings**, right-click on **Table Settings** ❷ to add a table. This brings up the **Add Table for Table Settings** screen. On this screen, enter "AUFK" as the **DB Table Name** ❸, and click the **Continue** ❹ icon.

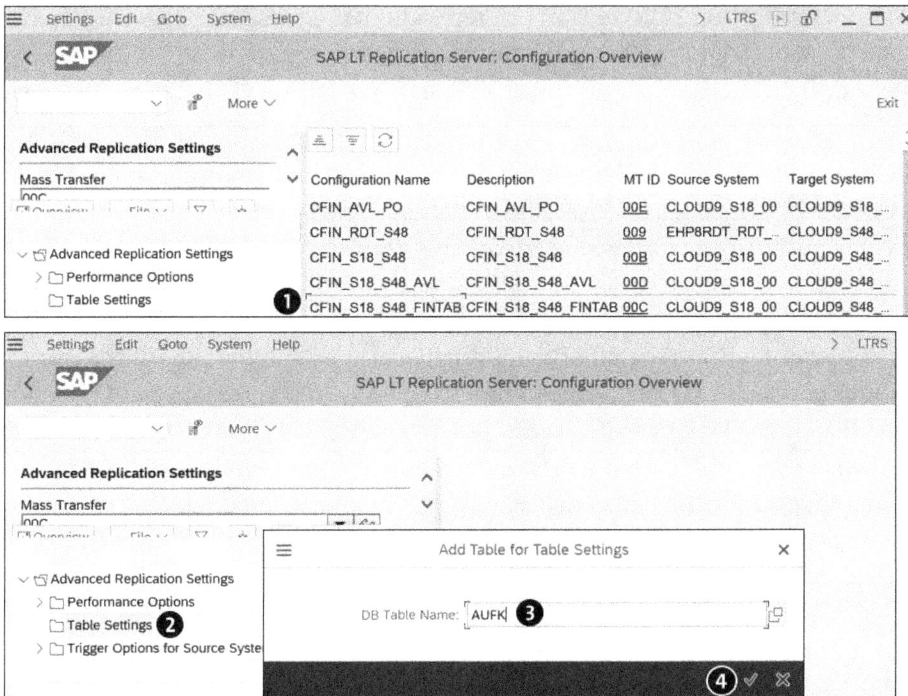

Figure 5.41 Database Trigger: Add Table AUFK

This action leads to the **SAP LT Replication Server: Change Table Settings (00C)** screen, as shown in Figure 5.42.

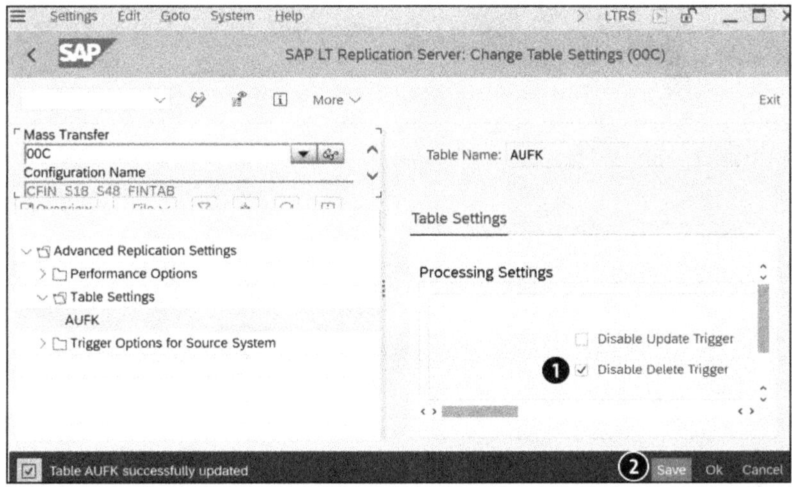

Figure 5.42 Database Trigger: Table Settings

On this screen, select the **Disable Update Trigger** checkbox or the **Disable Delete Trigger** checkbox per project requirements. If you select **Disable Update Trigger**, after adding the table in Transaction LTRC to start the replication, the database trigger for update won't be created. Likewise, **Disable Delete Trigger** works the same way as the **Disable Update trigger**. In our example, we chose **Disable Delete Trigger** ❶ checkbox. Click **Save** ❷.

If triggers for a table already exist in the source system, and you want to disable the update or delete trigger per the project requirements, then you'll follow the steps mentioned previously. However, if you need to adjust a table trigger per setting—for example, you disable the delete trigger for AUFK and replication of the table was already working—you'll be able to redefine the trigger using report IUUC_REDEFINE_DB_TRIGGERS. After reaching the report, you'll arrive at the **Redefine the DB triggers for one or multiple tables** screen, as shown in Figure 5.43. On this screen, enter selection criteria values in the **Mass Transfer ID** and **Table Name** ❶ fields, and click the **Execute** ❷ icon. When the system brings up the warning message popup window, click **OK** to continue. The triggers for the tables that you specify will be deleted and will be created again with a message. Click **Yes** to continue. Next, the system opens the **Display logs** ❸ screen with the confirmation that the triggers are redefined successfully for the table.

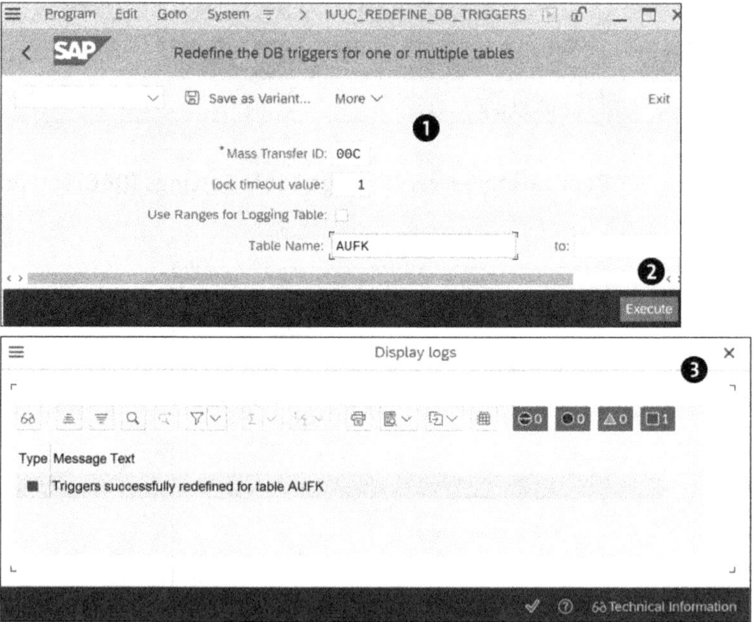

Figure 5.43 Redefine the Database Triggers for One or Multiple Tables

5.3.3 Monitoring

SAP Landscape Transformation Replication Server configuration MT ID is created to load/replicate the data from the source ERP (SAP or non-SAP) to the target Central Finance system. In Transaction LTRC, you have the option to enter configuration notes and activate/deactivate configurations on the toolbar. There are multiple ways to monitor the data transfer from the source system to the Central Finance system. Let's look at the monitoring options next.

Application Logs

The **Application Logs** tab allows you to view the initial load and replication logs for executed definition and replication tasks. Application logs show errors, warnings, and other information relevant for troubleshooting. In Transaction LTRC, choose the MT ID, and navigate to the **Application Logs** tab, as shown in Figure 5.44. As you can see from the figure, the application logs **Category** provides nine categories. With the categories, you'll be able to check logs captured for the data transfer. By default, this screen shows all logs. If you want to view application logs for a single replication, then you can also enter the table name in the **Object Name** field, which will then display the application logs for that table.

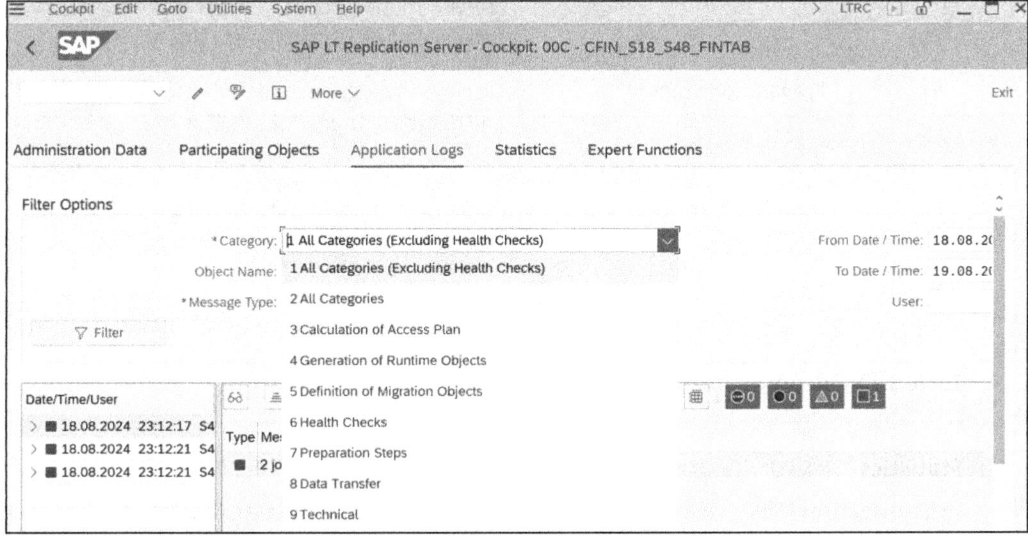

Figure 5.44 SAP Landscape Transformation Application Logs

Table 5.22 provides explanations of each **Category** listed from the dropdown menu in Figure 5.44.

Category	Category Description	Purpose
1	**All Categories (Excluding Health Checks)**	Displays all logs except the health checks.
2	**All Categories**	Displays all logs in all categories.

Table 5.22 Category of Logs

Category	Category Description	Purpose
3	**Calculation of Access Plan**	Displays the logs for calculation of the access plan executed during the calculation job which, runs during the initial load of any SAP Landscape Transformation Replication Server object.
4	**Generation of Runtime Objects**	Displays the runtime object generation logs performed during the initial load and real-time replication.
5	**Definition of Migration Objects**	Displays the logs created during definition of migration objects.
6	**Health Checks**	Displays the health check logs. Health check covers the connection between the source system and target Central Finance system, as well as whether all looks okay for each SAP Landscape Transformation Replication Server object added for replication/load in the specific MT ID.
7	**Preparation Steps**	Displays the logs created during logging tables and database triggers, as well as the number of jobs scheduled for the data transfer.
8	**Data Transfer**	Displays the logs created during the initial load and replication data transfer.
9	**Technical**	Displays the logs for the unexpected errors that occurred while retrieving the MT ID information.

Table 5.22 Category of Logs (Cont.)

Statistics

In Transaction LTRC, choose the MT ID, and navigate to the **Statistics** tab, as shown in Figure 5.45. Here, you can view the initial load and replication data transfer statistics. In addition, you can view unprocessed logging table records during real-time replication. Let's discuss each of the **Filter Options** next:

- **View Unprocessed Logging Table Records**
  Figure 5.45 shows the **View Unprocessed Logging Table Records** radio button selected. When you choose this button under **Filter Options**, the system allows you to check the unprocessed logging table records for all the tables in the MT ID if any exist. In addition, you'll be able view a specific table's statistics by entering the table name in the **Object Name** field. You'll enter a table here only if a logging table is generated for that object. By default, the system displays any unprocessed records with the oldest timestamp, along with a count of unprocessed records.

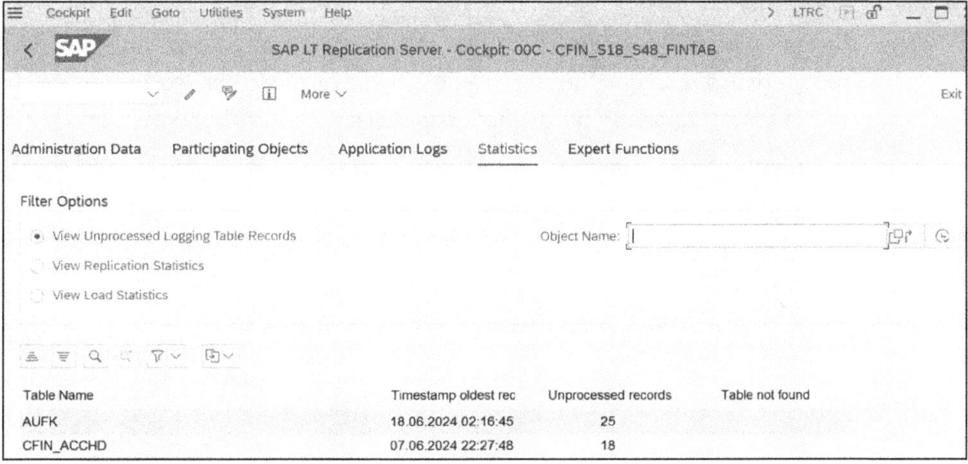

Figure 5.45 SAP Landscape Transformation Monitoring Statistics

- **View Replication Statistics**
  As shown in Figure 5.46, the **View Replication Statistics** button shows the replication of data transfer statistics, which includes the number of records inserted, updated, and deleted, along with latency information in seconds. To view replication statistics, you must enter a time interval in hours first. As a result, the system displays statistical information based on the time interval specified. In addition, you can view the statistical information by using the **Aggregate data by** field, which allows you to aggregate by days, hours, minutes, or weeks.

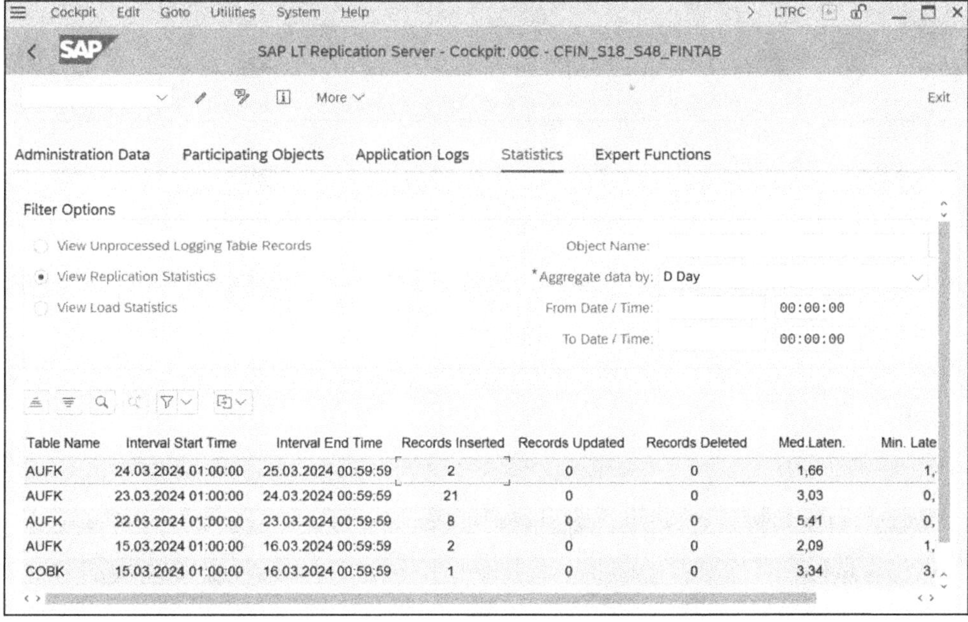

Figure 5.46 SAP Landscape Transformation Monitoring: View Replication Statistics

- **View Load Statistics**
  Here, you can view the initial load data transfer statistics, as shown in Figure 5.47. **View Load Statistics** includes the start and end time of the initial load for each SAP Landscape Transformation Replication Server object. In addition, you'll be able to view the statistics for the successfully finished initial load, initial load in process, or initial load in error status.

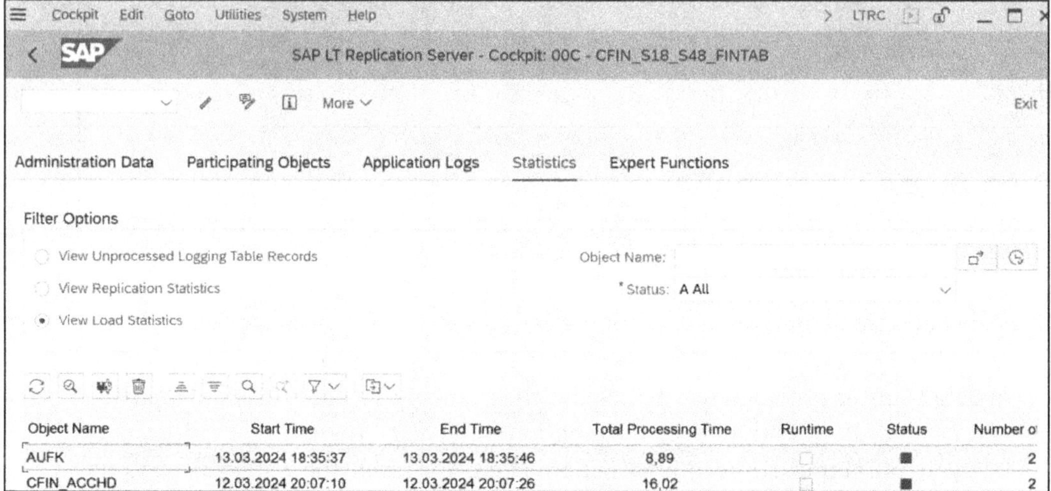

Figure 5.47 SAP Landscape Transformation Monitoring: View Load Statistics

Table 5.23 provides an explanation of each field for the View Load Statistics radio button under Filter Options.

Field	Description
Object Name	Table added in the MT ID
Start Time	Timestamp at which the initial load started
End Time	Timestamp at which the initial load finished
Processing Time	Difference between start time and end time
Runtime Information	Runtime information checkbox available for read and write data of the object
Status	Status of the object
Number of Portions	Number of portions of an object
Portions in Process	Number of portions for which the initial load is in process
Portions Finished	Number of portions for which the initial load is finished

Table 5.23 Load Statistics

Field	Description
Portions with Errors	Number of portions for which an error occurred
Records Inserted	Number of records inserted into the target table
Records Updated	Number of records updated into the target table
Records Deleted	Number of records deleted from the target table
Records Calculated	Number of rows read from the source table during access plan calculation
Records Read	Number of records read from the source table
Read Time (Seconds)	Time required to read data from the source system
Write Time (Seconds)	Time required to write data to the target system
Conversion Time (Seconds)	Time required to process the data in SAP Landscape Transformation Replication Server
Net Runtime (Seconds)	Total of read, write, and conversion time
Read Time (Percent)	Percentage of net runtime to read the data from the source system
Write Time (Percent)	Percentage of net runtime to write the data to the target system
Conversion Time (Percent)	Percentage of net runtime to process the data in SAP Landscape Transformation Replication Server

Table 5.23 Load Statistics (Cont.)

Enabling Notifications for the SAP Landscape Transformation Replication Server Configurations

When technical issues occur when replicating source data to the Central Finance system in real time, SAP Landscape Transformation Replication Server provides a notification functionality to send notifications on health checks to relevant people who can take appropriate action to fix the technical issues. You can enable the notification from Transaction LTRC by navigating to **Mass Transfer ID (Load/Replicate the Data) • Expert Functions • Functions.** On this screen, you'll specify settings for notifications for a specific MT ID created to replicate/load data from the source system to the Central Finance system. Alternatively, you can access the same settings using Transaction IUUC_REPL_EN. Upon reaching the transaction, you'll arrive at the **Notification Settings** screen, as shown in the Figure 5.48. On this screen, enter the **Mass Transfer ID** ❶, and then click **Execute** ❷.

On the resulting **Specify Settings for Notifications Mass Transfer ID 00C** screen, you can enter the frequency of the notification to be sent. In addition, you'll be able activate the notifications for different types of health checks. To send email notification to recipients, you must enable the email configuration as well with the help of Basis consultant.

Under the **Notification Status** ❸ section, you can check if the notification is enabled or not. If not, then the system displays the message **Notifications are not active for mass transfer MTID**. Furthermore, you can activate the notification setting by clicking on **Activate** and saving the settings. On the other hand, if notification is enabled, then the system displays the message **Notifications are active for mass transfer MTID**. You can also deactivate the notification settings by clicking on **Deactivate** and then saving the settings.

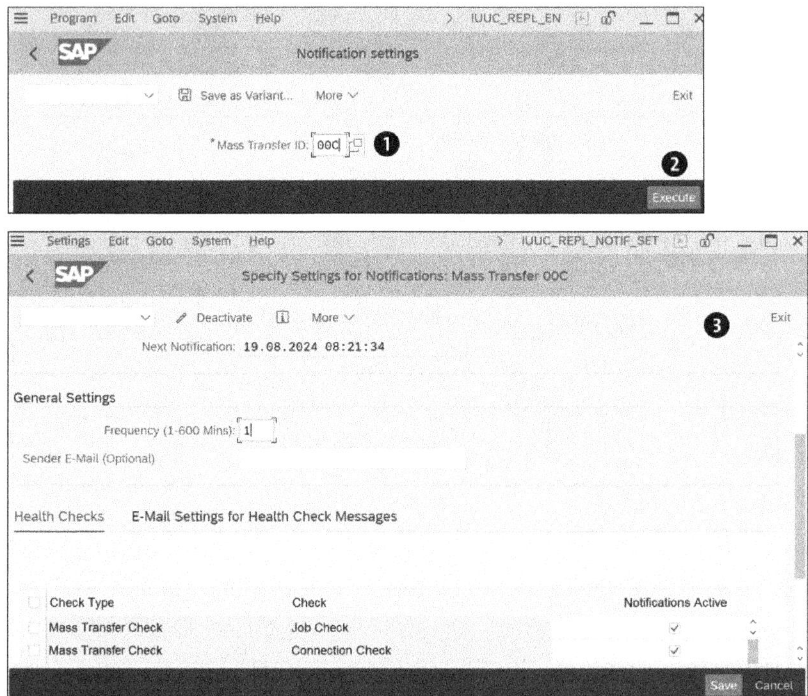

Figure 5.48 Notification Settings for SAP Landscape Transformation MT ID

Table 5.24 provides the health checks for which notifications can be activated in the notification settings.

Health Check Type	Checks Carried Out
Job Check	Checks if the master job is running for the MT ID
Connection Check	Checks if the connections to the source and target systems are working for the MT ID setup

Table 5.24 SAP Landscape Transformation Health Checks

Health Check Type	Checks Carried Out
Tigger Check	Checks if triggers for each table added in the MT ID are active and whether any issue exists
Table Status Check	Checks the status of the table added in the MT ID and whether triggers for each table exist
Maintenance Event Check	Checks if maintenance events exist
Logging Table Check	Checks if a logging table exists for the table added in the MT ID and if the number of entries in the logging table exceeds the defined threshold
Replication Check	Checks if replication is running
Latency Check	Checks if the length of time taken for the change in the source system is committed to the target system

Table 5.24 SAP Landscape Transformation Health Checks (Cont.)

In Transaction LTRC, navigate to the **Expert Functions** tab, as shown in Figure 5.49 ❶.

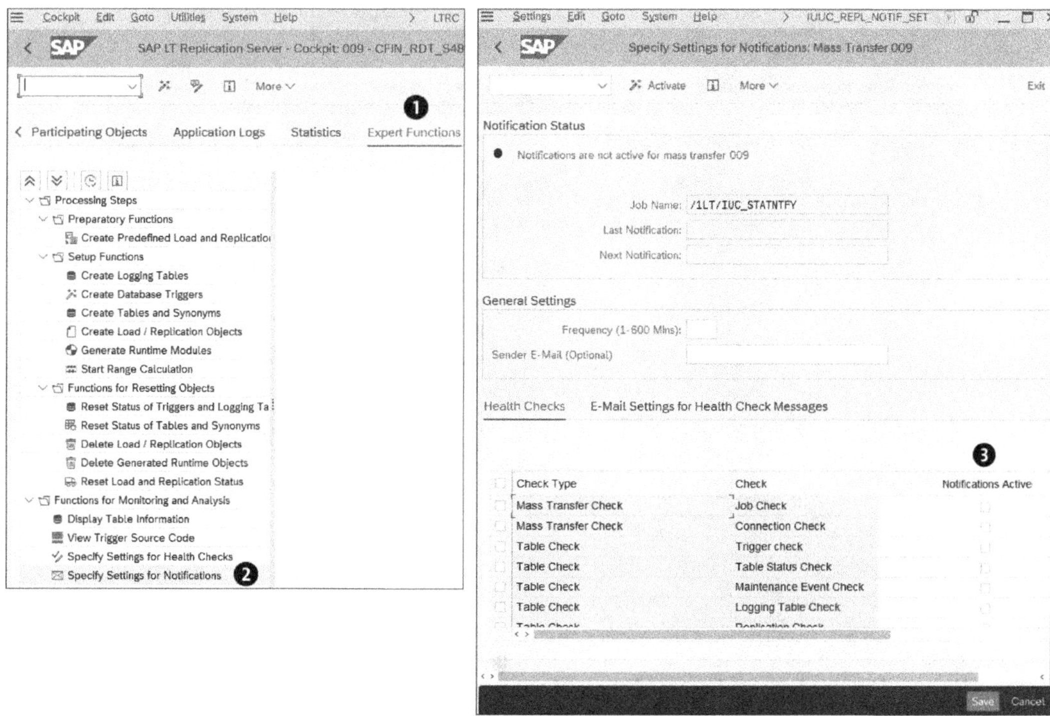

Figure 5.49 Specify Settings for Notification Settings

On this screen, go to **Specify Settings for Notifications** under the **Functions for Monitoring and Analysis** ❷ folder icon in the left pane. On the resulting screen, you'll be able to activate and configure notifications for health checks. The system provides you with a view of the **Check Type** and **Check** columns to see different types of checks. In addition, select an appropriate checkbox in the **Notifications Active** ❸ column to include the results of a check to be sent via email.

As shown in Figure 5.50, navigate to the **Settings for Latency Check** tab from Transaction LTRC.

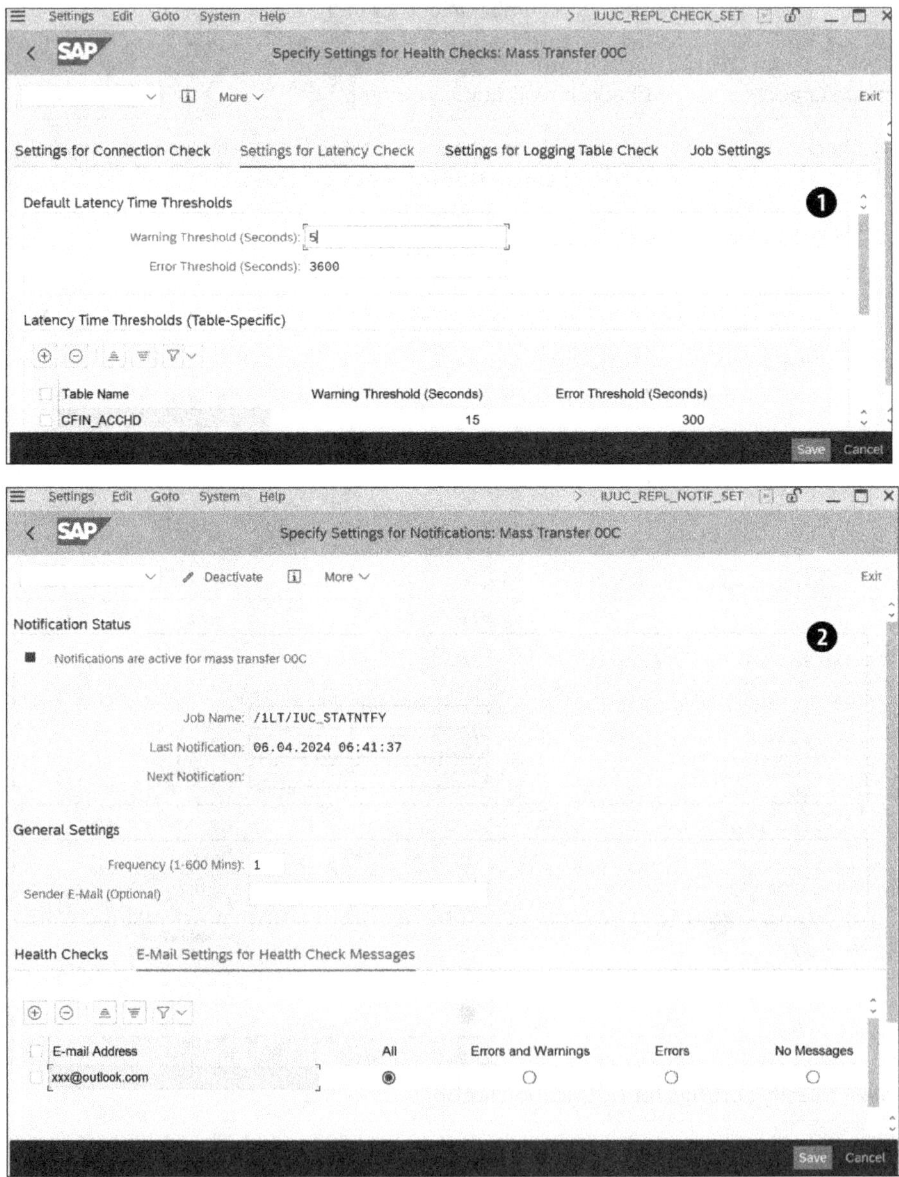

Figure 5.50 Health Checks for MT ID

Here, you can define threshold values for the replication latency to send via email. If the latency threshold is higher than the defined threshold, the system sends a notification to relevant people. As you can see from the screen ❶, the system provides latency times under **Latency Time Thresholds (Table-Specific)**:

- **Warnings Threshold (Seconds)**
  Specify warning thresholds in this column.
- **Error Threshold (Seconds)**
  Specify error thresholds in this column

You can maintain email notification options under the **E-Mail Settings for Health Check Messages** tab ❷.

The notification monitor batch job `/1LT/IUC_STATNTFY` can be run for the following:

- Connection check
- Table status Check
- Replication check and other checks

The last notification shows the time and date when the notification was sent to the recipients added under the **E-Mail Settings for Health Check Messages** tab option shown earlier in Figure 5.50.

Figure 5.51 shows the output of successfully run batch job `/1LT/IUC_STATNTFY` with its status showing **Finished**.

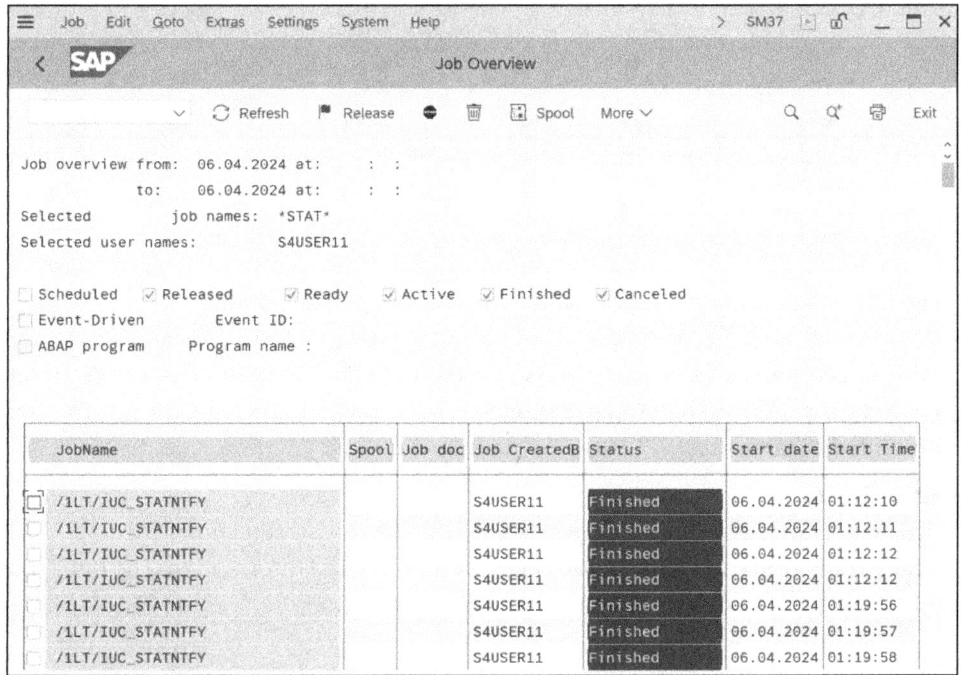

Figure 5.51 Notification Batch Job Output

SAP Landscape Transformation Replication Server Monitoring

In addition to Transaction LTRC, SAP provides Transaction LTRO, which gives you a single screen to monitor SAP Landscape Transformation Replication Server information. The main difference between Transaction LTRC and Transaction LTRO is that with Transaction LTRC, you'll have to switch one configuration to another configuration to see the relevant status. On the other hand, with Transaction LTRO, all relevant information is displayed in a single screen. After reaching Transaction LTRO, you'll arrive at the **SAP LT Replication Server - Monitoring** screen, as shown in Figure 5.52.

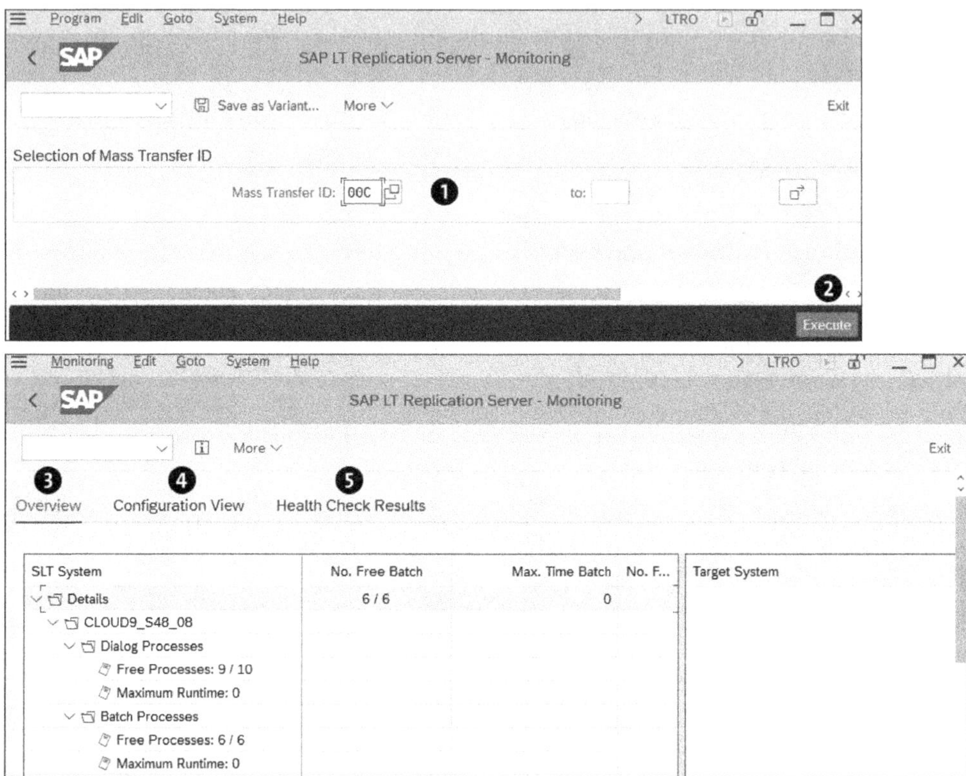

Figure 5.52 SAP Landscape Transformation Replication Server Monitoring

On this screen, enter a value in the **Mass Transfer ID** ❶ field, and click **Execute** ❷ to continue. On the resulting **SAP LT Replication Server – Monitoring** screen, the information is presented in three tabs:

- **Overview** ❸
  This tab is divided into multiple sub-screens. The left screen shows the SAP Landscape Transformation Replication Server system-related information. The right sub-screen shows target systems and related information.

- **Configuration View ❹**
  This tab displays the status of the master job and all configurations selected.
- **Health Check Results ❺**
  This tab displays the results of all SAP Landscape Transformation Replication Sever health checks and notifies the relevant people when any issues or unusual activity occurs in SAP Landscape Transformation Replication Server.

Monitoring with SAP Solution Manager

SAP Solution Manager can be integrated with SAP Landscape Transformation Replication Server to view replication-related notifications and alerts. The minimum prerequisite for this functionality is that SAP Solution Manager must be at a 7.1 SP5 or higher release level. Once SAP Landscape Transformation Replication Server is configured with SAP Solution Manager, it offers the following:

- Visibility of connections established to source systems
- Status of replication over a 24-hour period
- Status of master jobs and load jobs
- Database trigger status

[+]

Tip

Consult SAP Notes 1558756 and 2081759 for additional guidance on SAP Landscape Transformation Replication Server monitoring in SAP Solution Manager.

5.4 Expert Functions

From Transaction LTRC, choose the MT ID, and navigate to the **Expert Functions** tab, as shown in Figure 5.53 ❶. As shown in left pane, **Expert Functions** are delivered with several functions ❷. We've seen **Expert Functions** frequently used when setting up predefined load and replication objects for Central Finance replication objects. However, here several other functions can be leveraged for troubleshooting SAP Landscape Transformation Replication Server. For instance, you can use the **Functions for Resetting Objects** options for troubleshooting or to reset the status of a table, reset a trigger, reset a logging table, reset load and replication status, and so on. As you've seen in the previous section, the **Functions for Monitoring and Analysis** function can be used to send replicated related notifications to relevant users. Although **Repair Functions** are rarely used, they can be used to change the connection settings to target a system for a specific mass ID. We've limited our discussions on **Expert Functions** as a detailed discussion is beyond the scope of this book.

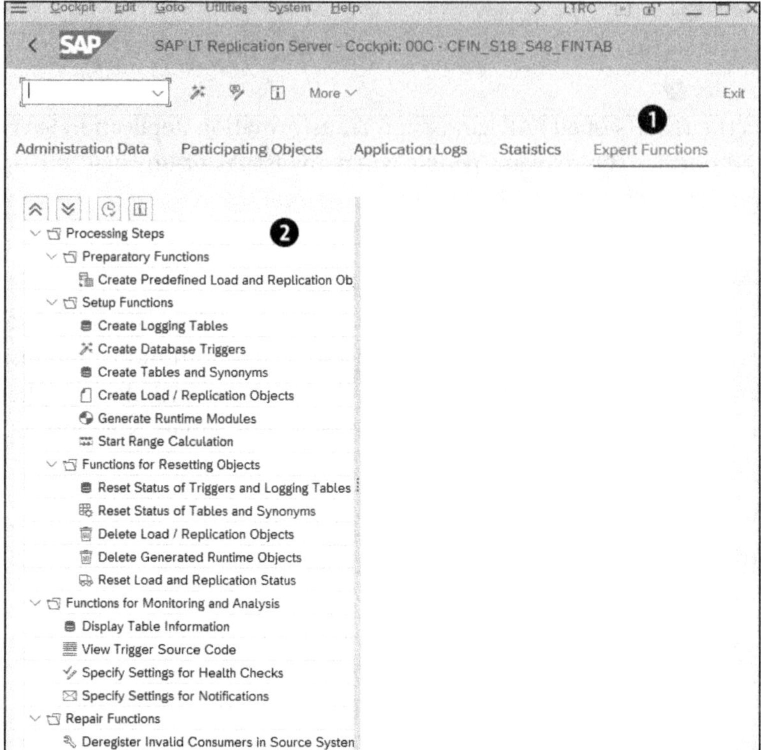

Figure 5.53 SAP Landscape Transformation: Expert Functions

5.5 Summary

The SAP Landscape Transformation Replication Server enables online replication from source systems to the Central Finance system. It's an essential integration component that connects SAP Landscape Transformation Replication Server and its architecture. The chapter provided learners with installation options recommended by SAP. Deployment options are important considerations when choosing SAP Landscape Transformation Replication Server options in Central Finance. This is covered in the second section of this chapter. Later, we walked you through the step-by-step configuration instructions for setting up SAP Landscape Transformation Replication Server objects in Central Finance. At the end of the chapter, you learned advanced settings such as setting up filters, controlling triggers, and monitoring SAP Landscape Transformation Replication Server. The monitoring section covered important resources needed for troubleshooting and monitoring initial load and replication objects. In the next chapter, we'll dive deeper into another important component of Central Finance: SAP Application Interface Framework.

Chapter 6
Configuring SAP Application Interface Framework

Central Finance receives replicated data from various source systems. SAP Application Interface Framework is used as a single centralized place to analyze, monitor, and fix errors for all Central Finance replication scenarios. SAP Application Interface Framework not only can manage Central Finance interface scenarios, but the tool can manage any incoming data from other interfaces too. The SAP Application Interface Framework tool is predelivered with Central Finance.

In SAP, interfaces enable the transfer of data either between SAP systems or between an SAP system and a third-party system. SAP integrates with other SAP systems or third-party systems with technologies such as IDocs, SAP Process Integration XML, application programming interfaces (APIs), and SAP Data Services. When data is processed between systems in an integrated environment, one common requirement of implementing interfaces is monitoring and error handling capability. As a feature-rich tool, SAP Application Interface Framework comes with error handling and monitoring capability included as a standard functionality in SAP S/4HANA. SAP Application Interface Framework offers the following key features for managing interfaces:

- **Error monitoring**
  SAP Application Interface Framework comes with a user-friendly interface to navigate to monitor and correct error messages. In addition, the tool offers mass processing of SAP Application Interface Framework messages. Messages can be displayed with role-based security authorization.
- **Alerts management**
  SAP Application Interface Framework also offers an alerts management functionality to send alerts or email messages.
- **Customization**
  SAP Application Interface Framework provides developers/technical teams with extensive customization options to develop reusable interface components, which can be leveraged across different interfaces. For example, messages can be assigned SAP transactions and field labels can be displayed with customized text.

- **Authorization**
  With SAP Application Interface Framework, you can assign different authorizations to users based on business requirements.

Central Finance uses the SAP Application Interface Framework tool to implement interfaces, monitor errors, and resolve errors. In Central Finance, while transferring data from the source systems, you'll come across errors related to master data, configuration, security, and technical. Table 6.1 lists a few examples of such common errors during transfer of data from the Central Finance interface.

Type of Error	Description of Error	Example
Master data	■ Project master data mapping not mapped	■ **FINS_CFIN_PS009 Key mapping for project $1 is missing.**
	■ Posting period not open	■ **F5201 Posting period 001 2024 not open**
	■ Cost center is blocked for posting	■ **KI170 Cost center $1/$2 locked for primary posting at this time**
Configuration	■ Doc splitting errors	■ **GLT0002 Document Splitting: Items for clearing &1/&2/&3/&4 not found**
		■ **GLT2201 Balancing profit center in line item 002 not filled**
Security error	■ Authorization errors	■ **F5083 You have no authorization for posting in company code $1**
		■ **F5086 You're not authorized to post to this account**
Technical error	■ Unexpected error	■ **AR182 Hyphens in incorrect position (only permitted in 4th and 12th)**
		■ **7Q630 A number could not be determined for numbering group XX**

Table 6.1 Interface Common Errors

These types of errors are monitored and resolved promptly for faster replication processing via SAP Application Interface Framework, which plays a vital role in the Central Finance integration scenario. Therefore, you'll need a solid understanding of SAP Application Interface Framework configuration settings for error handling. In this chapter, you'll learn the core concepts of SAP Application Interface Framework, runtime groups, error handling, and alerts management. First, we'll review configuration settings for Central Finance error handling. Then, the chapter dives deep into alerts management offered with SAP Application Interface Framework. Finally, you'll understand how to address SAP Application Interface Framework performance issues when processing large volumes of messages in Central Finance implementations. The book focuses only on SAP Application Interface Framework features from a Central Finance implementation perspective.

6.1 Getting Started with SAP Application Interface Framework

During interface posting, you'll come across errors when interface posting fails due to various reasons. When the process doesn't work, you'll intervene and reprocess the error after solving for the root cause. SAP Application Interface Framework is delivered with rich capabilities to monitor interface errors. SAP Application Interface Framework offers companies with several benefits:

- SAP Application Interface Framework isn't just limited to Central Finance; any interface can take advantage of SAP Application Interface Framework features.
- Namespace and version have separate interfaces for easy identification.
- A user-friendly interface shows critical information in a single screen.
- Mass processing of error messages is possible after resolving issues.
- Alerts can be sent to responsible persons for faster resolution.
- Error resolution support content can be maintained via SAP Application Interface Framework messages.

Next, let's discuss the SAP Application Interface Framework architecture. Figure 6.1 shows the architecture of SAP Application Interface Framework in the Central Finance replication scenario. Data is replicated from the source system to Central Finance via SAP Landscape Transformation Replication Server. Within SAP Application Interface Framework, the following actions are carried out:

❶ **SAP Application Interface Framework**
When data is transferred from the source system, SAP Application Interface Framework goes through checks, key mappings, value mapping, calls to business application programming interface (BAPI), and standard functions.

❷ **Customizing**
SAP Application Interface Framework offers customizing to enable features in SAP Application Interface Framework. Typically, technical teams or super users carry out customizing activities in SAP Application Interface Framework.

❸ **Application log**
Next, SAP Application Interface Framework writes the status of Processed to application log. These messages are stored in SAP Application Interface Framework tables and displayed when executing Transaction /AIF/ERR or Transaction /AIF/IFMON. The application logs enable users to troubleshoot errors.

❹ **Monitoring and error handling**
Business users or super users typically give access to Transaction /AIF/ERR and Transaction /AIF/IFMON to monitor messages.

❺ **Alert management**
The application interface comes with an alerts management functionality. As the errors occur, the error messages are grouped and sent via email or to the My Inbox app (app ID: F0862) to recipients to take further action.

❻ **Security and authorization**
SAP Application Interface Framework comes with robust authorization features. You'll assign interfaces to different users to access error monitoring tools.

Once the messages go through these steps, data is transferred to the target system.

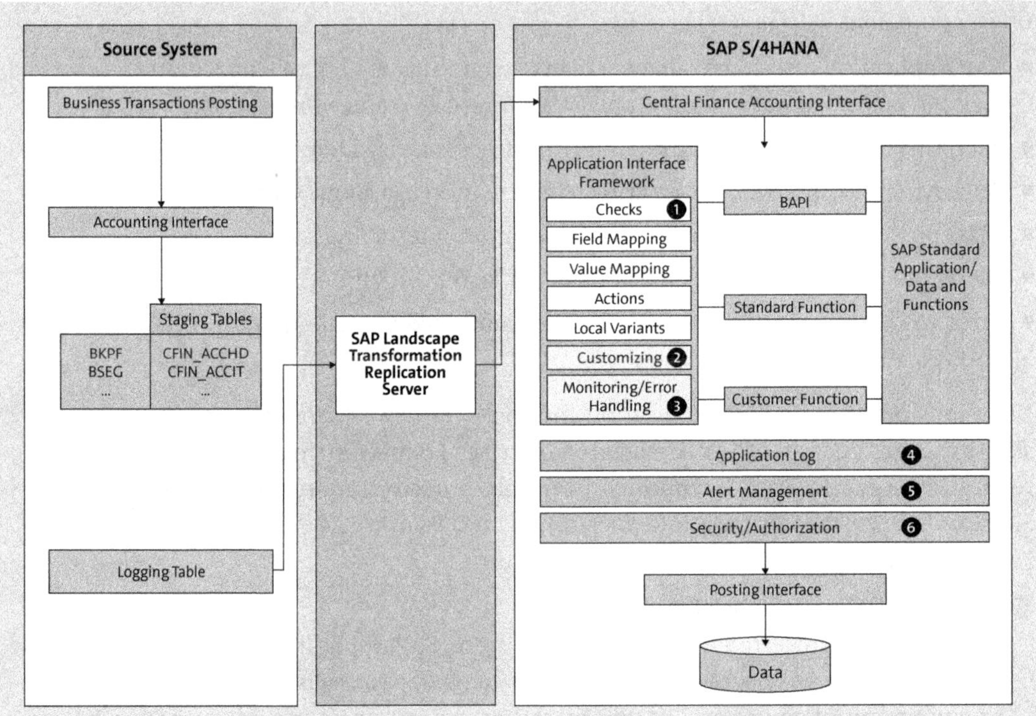

Figure 6.1 SAP Application Interface Framework Architecture

Now that you've learned the high-level architecture of SAP Application Interface Framework, let's discuss key terminology used in SAP Application Interface Framework next.

6.1.1 Terminology

It's important to understand key terminology used in SAP Application Interface Framework to effectively use the tool. Let's review the important terminology used with SAP Application Interface Framework from the Central Finance perspective:

- **Interface monitor and error handling**
  SAP provides a user-friendly interface to analyze, monitor, correct errors, restart, or cancel SAP Application Interface Framework messages. You'll access the interface with Transaction /AIF/ERR. Section 6.3 covers the interface in-depth.
- **Namespace**
  By SAP's definition, a *namespace* is used for logical structuring of objects and configuration within SAP Application Interface Framework. The namespace comes with interface definition, checks, mappings, and actions. In addition, you can provide access to users by namespace. For example, the SAP-delivered namespace for Central Finance is /FINCF. A custom namespace can be defined for a customer-specific requirement within SAP Application Interface Framework customizing activity.
- **Interfaces and versions**
  In SAP Application Interface Framework, recipients are responsible for interfaces to monitor errors occur in the interface. You'll assign a user or a user group responsible for the errors and then you'll assign the recipient to the interface. Each interface comes with versions. For example, accounting document interface **AC_DOC** has two versions—**2** and **10**—which uniquely identifies an interface. Table 6.2 lists Central Finance–delivered interfaces and versions by application area.

Application Area	Type of Replication	Interface	Version
Finance accounting (FI)	Accounting document	AC_DOC	2
		AC_DOC	10 – bgRFC
	Accounting document change	AC_DOC_CHG	2
		AC_DOC_CHG	10 – bgRFC
	Accounting document – external interface	AC_DOC_EX	2
	Accounting document change – external interface	AC_CHG_EX	2
	Temporary postings	AC_DOC_TMP	1
			10 – bgRFC
Controlling (CO)	Controlling document	CO_DOC	1
	Controlling document simulation	CO_DOC_SIM	1
	Cost object replication	CO_OBJ	1
	Cost object simulation	CO_OBJ_SIM	1

Table 6.2 SAP Delivered Interfaces

Application Area	Type of Replication	Interface	Version
Controlling (CO) (Cont.)	Commitment document	CMT_DOC	1
	Commitment document replication	CMT_SIM	1
	Cost center activate rate replication	CC_AR	1
	Material cost estimate replication	CE_MAT	1
	Profit center accounting (PCA)	PCA_DOC	1
	PCA simulation	PCA_DOC_SIM	1
	Project system master data	PS_OBJ	1
Logistics	Accounting view of sales order	AV_SO	1
	Accounting view of customer invoice	AV_CI	1
	Accounting view of purchase order	AV_PO	1
	Accounting view of supplier invoice	AV_SI	1
	Update purchase order history	UPDT_PO	1
SEPA Mandate	SEPA mandate creation	ISEPA_CR	1
	SEPA mandate change	ISEPA_CH	1
	SEPA mandate save replication	ISEPA_SR	1

Table 6.2 SAP Delivered Interfaces (Cont.)

- **Recipient**
  In SAP Application Interface Framework, recipients are responsible for interfaces that they own. Recipients monitor errors and take steps to resolve them as they occur in the interface. First, you'll define a user or a user group responsible for responsible for monitoring errors. Then, you'll assign the recipient to the interface.
- **Alerts**
  SAP provides a user-friendly interface to analyze, monitor, correct errors, restart, or cancel SAP Application Interface Framework messages. You'll see alerts in the interfaces monitor Transaction /AIF/IFMON. Section 6.3 covers interface monitoring in-depth.

- **Message status**
  SAP provides a user-friendly interface to analyze, monitor, correct errors, restart, or cancel SAP Application Interface Framework messages. You'll access the messages with Transaction /AIF/ERR or /AIF/IFMON.
- **Value mapping**
  Derive configuration-related mapping from a destination structure for the source structure values. Central Finance uses SAP Master Data Governance value mapping for configuration items.
- **Actions**
  In SAP Application Interface Framework, actions are steps to process a message. This includes business logic carried out during runtime and calls made to function modules to execute certain tasks.

6.1.2 Content Installation

To use SAP Application Interface Framework in Central Finance, you must install predelivered SAP Application Interface Framework content and relevant business configuration sets (BC sets) for relevant interfaces first. Based on the type of replication, in recent versions of SAP (e.g., SAP S/4HANA 2023), SAP predelivered relevant SAP Application Interface Framework contents for Central Finance scenarios, so there's no need to extract SAP Application Interface Framework content. Nevertheless, you'll apply relevant content based on replication scope and SAP's recommendation for replication objects. Table 6.3 displays the SAP Application Interface Framework deployment scenario by different application areas and types of replication. Before proceeding with SAP Application Interface Framework content extraction, ensure you've activated business function FINS_CFIN in Central Finance via Transaction SFW5.

Application Area	Type of Replication	Deployment Scenario ID	SAP Application Interface Framework Interface
Finance	Accounting documents (from SAP systems)	SAP_AIF_0017	AC_DOC – Accounting documents AC_DOC_CHG – Accounting document change AC_DOC_TMP – Temporary postings
	Accounting documents (from third-party systems)	SAP_AIF_0026	AC_DOC_EX – Accounting documents AC_CHG_EX – Accounting document change AC_DOC_TMP – Temporary postings

Table 6.3 Predelivered SAP Application Interface Framework Content

Application Area	Type of Replication	Deployment Scenario ID	SAP Application Interface Framework Interface
Accounting view of logistics (AVL)	Sales document	FINCF_AV_SO	AV_SO
	Customer invoice	FINCF_AV_CI	AV_CI
	Purchase document	FINCF_AV_PO	AV_PO
	Supplier invoice	FINCF_AV_SI	AV_SI
Controlling	Activity rate – source system to central system	SAP_AIF_0023	CC_AR
	Activity rate – Central Finance system to the source system	SAP_AIF_0030	CC_AR_C2S
	Activity type	SAP_AIF_0040	ACTY_TP
	Material cost estimate	SAP_AIF_0018	CE_MAT
Other	Update purchase order history	SAP_AIF_0014	UPDT_PO
	SD clearing status transfer	SAP_AIF_0051	SDBL_CLRST

Table 6.3 Predelivered SAP Application Interface Framework Content (Cont.)

To apply the predelivered SAP Application Interface Framework content, navigate to **SAP Easy Access** menu path **Cross Application Components • SAP Application Interface Framework • Administration • Interface Development • Content Delivery • Deploy Content Transport** or use Transaction /AIF/CONTENT_EXTRACT. After reaching the transaction, you'll arrive at the **AIF Content Transport-Deploy** screen, as shown in Figure 6.2. On this screen, select **Deployment Scenario ID** from the dropdown list under **Deployment Scenario** section ❶. As discussed earlier, Table 6.3 provides a list of deployment scenarios and replication types. In this example, we've selected deployment scenario ID **SAP_AIF_0017** for accounting documents replication. In the **Parameter** section ❷, you'll select **No automatic registration** checkbox if you want to register for automatic updates. The system updates table /AIF/ICD_L_ACT_D when a new version is updated in the system. The **Registration in AIM table** checkbox is used when you require simulation of zero downtime. SAP doesn't recommend using both checkboxes at the same time, which may cause data loss or inconsistencies. In the third section, SAP provides options to record customizing and workbench transports ❸ to transport SAP Application Interface Framework content to other clients.

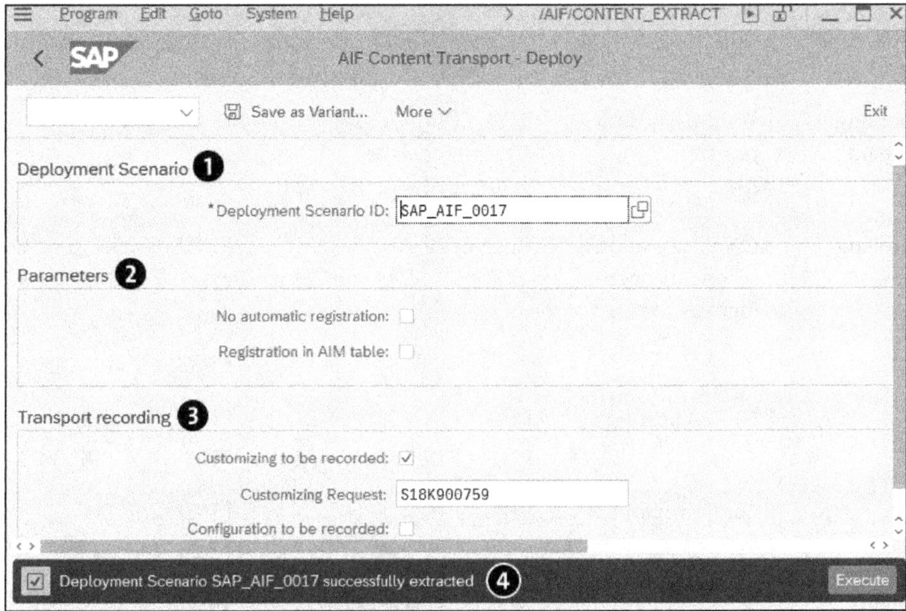

Figure 6.2 Activation of SAP Application Interface Framework Content

[+]

Tip

For predelivered SAP Application Interface Framework content, there is no need to lock the `/FINCF` namespace via Transaction FINS_CFIN_AIF_SETUP.

When you select the **Workbench Transport** checkbox, the **Recording in RFC Destination** indicator is displayed. Choosing this indicator brings up the **RFC Destination** field to be entered. Click the **Execute** icon. Upon activation of SAP Application Interface Framework content, the system provides a confirmation message **Deployment scenario SAP_AIF_0017 successfully extracted** ❹.

Next, you'll install BC sets for Central Finance replication scenarios. Table 6.4 provides SAP-delivered BC sets for relevant SAP Application Interface Framework interfaces. BC sets permit predefined configuration settings to be saved and reused.

Area	Type of Replication	BC Set Name	SAP Application Interface Framework Interface
SAP Application Interface Framework Configuration General	General content definitions for namespaces, applications, and so on	FINS_CFIN_AIF_GEN	
Accounting document	Third-party data transfer	FINS_CFIN_AIF_RAL_AC_DOC prerequisite: FINS_CFIN_AIF_DOC_SER	AC_DOC_EX
Controlling	CO objects and CO secondary documents	FINS_CFIN_AIF_CO	/FINCF CO_DOC 1 /FINCF CO_DOC_SIM 1 /FINCF CO_OBJ 1 /FINCF CO_OBJ_SIM 1
	EC-PCA postings	FINS_CFIN_AIF_PCA FINS_CFIN_AIF_PCA_SIM	/FINCF PCA_DOC 1 /FINCF PCA_DOC_SM 1
	Commitment documents	FINS_CFIN_AIF_CMT	/FINCF CMT_DOC 1 /FINCF CMT_SIM 1
Payments	SEPA mandate	/FINCF ISEPA_CH 01 /FINCF ISEPA_CR 01 /FINCF ISEPA_SR 01	FINS_CFIN_AIF_SEPA
JVA	Joint venture accounting	FINS_CFIN_AIF_JVA	/FINCF JVA_DOC /FINCF JVA_MD_JO /FINCF_MD_PA /FINCF_MD_VE

Table 6.4 BC Sets

To apply the SAP-delivered BC sets, navigate to **SAP Easy Access** menu path **SAP Menu • Tools • Business Configuration Sets • Display and Maintain BC Sets** or use Transaction SCPR3. Upon entering the transaction, you'll arrive at the **Business Configuration Sets: Display** screen, as shown in Figure 6.3. On this screen, navigate to **Goto**, and choose the **Activation Transaction** ❶ submenu item. On the resulting screen, you'll be prompted to enter **BC Set** name ❷ and press [Enter]. Click the **Activation** icon ❸. The system prompts with a window to record transport information.

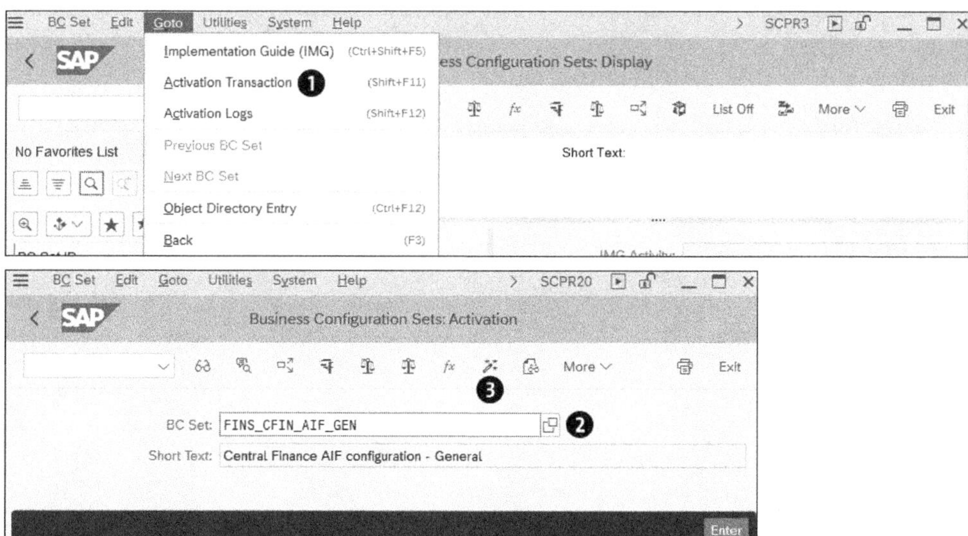

Figure 6.3 Activation of BC Sets

Upon clicking the **Activation** icon, the system prompts with the **Activation Options** screen, as shown in Figure 6.4 ❶. In this screen, you have options to select **Activation Options** ❷:

- **Overwrite Data**
  By default, the system selects the **Overwrite All Data** radio button.
- **Select Activation Mode**
  Leave the **Default Mode Recommended** option selected.
- **Deletion Functionality**
  Leave the default settings here.
- **Processing Options**
  Leave the default settings here

After reviewing all activation options, click the **Continue** ❸ icon. If activation is successful, the system returns a message and BC Set data updated in customizing tables.

Alternatively, BC set activation can be carried out using Transaction SCPR20 accessed from the same menu path as Transaction SCPR3. After activation of the BC sets, SAP recommends locking the /FINCF namespace by navigating from **SAP Easy Access** menu to **SAP Menu • Accounting • Central Finance • Error Handling • Setup for Central Finance** or using Transaction FINS_CFIN_AIF_SETUP. After reaching the transaction, as shown in Figure 6.5, choose the **Prepare configuration** ❶ radio button, and click the **Execute** ❷ button. Next, repeat the same steps for the **Complete configuration** radio button.

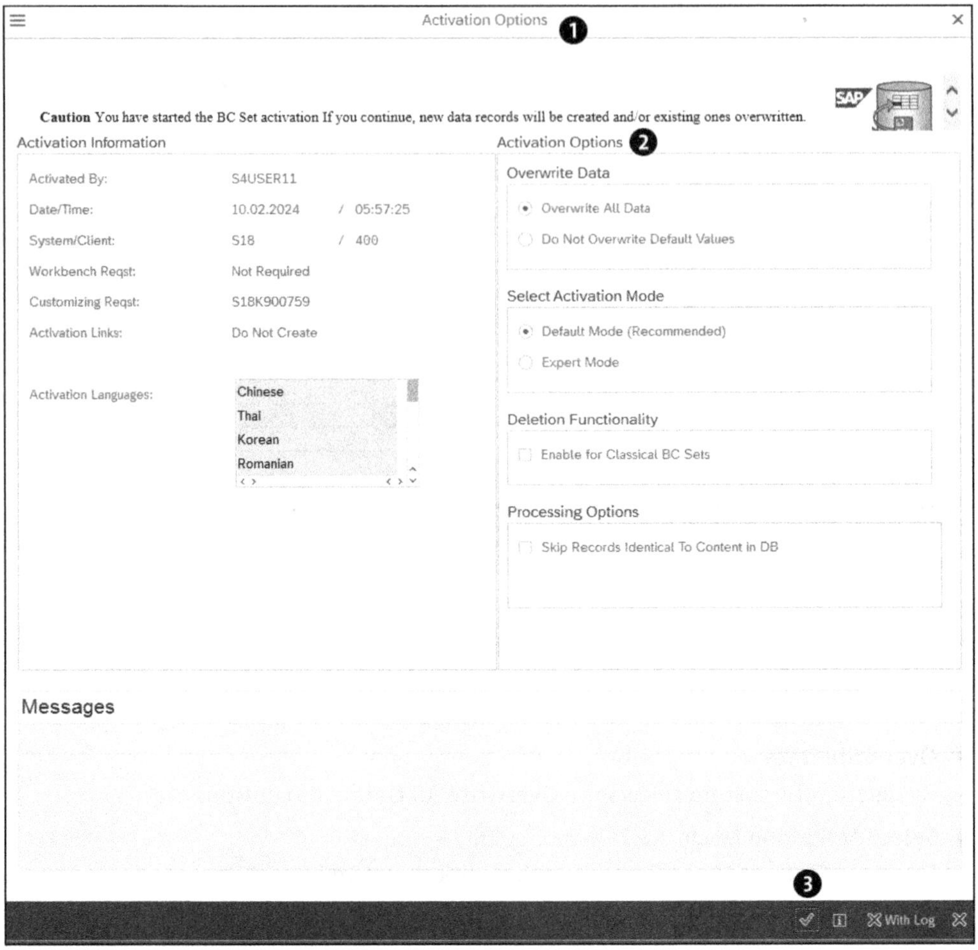

Activation Options ❶

Caution You have started the BC Set activation If you continue, new data records will be created and/or existing ones overwritten.

Activation Information

Activated By: S4USER11
Date/Time: 10.02.2024 / 05:57:25
System/Client: S18 / 400
Workbench Reqst: Not Required
Customizing Reqst: S18K900759
Activation Links: Do Not Create
Activation Languages: Chinese, Thai, Korean, Romanian

Activation Options ❷

Overwrite Data
- Overwrite All Data
- Do Not Overwrite Default Values

Select Activation Mode
- Default Mode (Recommended)
- Expert Mode

Deletion Functionality
- Enable for Classical BC Sets

Processing Options
- Skip Records Identical To Content in DB

Messages

❸ With Log

Figure 6.4 BC Sets: Activation Options

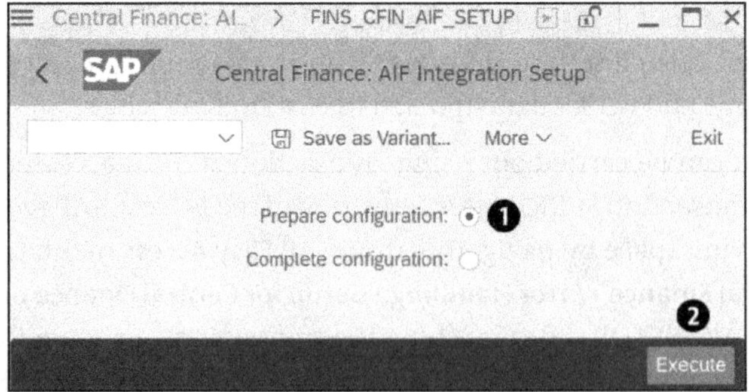

Figure 6.5 Setup for Central Finance FINS_CFIN_SETUP

6.1.3 Roles

Appropriate SAP Application Interface Framework roles are required for a specific function to be carried out in SAP Application Interface Framework. SAP delivers predefined template roles that can be copied and reused in Central Finance scenarios. Table 6.5 provides the list of the most important, yet SAP-delivered SAP Application Interface Framework roles relevant for Central Finance.

Role Type	Template Role	Usage/Tasks
All authorization	SAP_AIF_ALL	The role offers access to all transactions, especially transactions such as Transaction /AIF/SETUP, which provides access to all activities in SAP Application Interface Framework. SAP recommends this role is used only for testing purposes.
Architect/developer	SAP_AIF_ARCHITECT SAP_AIF_DEVELOPER	To develop interface activities.
Administrator	SAP_AIF_ADMIN	To perform advanced configuration tasks such as the following: ■ Custom functions, hints, message texts ■ Interface determination ■ Archiving, and so on
Auditor	SAP_AIF_AUDITOR	Auditor role displays SAP Application Interface Framework messages archived and blocked for data protection purposes.
Business user	SAP_AIF_POWER_USER	Responsible for monitoring interface and error handling. Typical tasks include the following: ■ Start/restart/cancel SAP Application Interface Framework message. ■ Edit/correct messages.
Message processor	SAP_AIF_PROCESSING	To process SAP Application Interface Framework messages.
Others	SAP_AIF_PROCESSING	Assigned SAP Application Interface Framework processor.
	SAP_AIF_USER	Assigned to an end user or business user.
	SAP_AIF_TEST_TEMPL	SAP Application Interface Framework test template (nonproductive).
	SAP_AIF_DATA_PROT_O	SAP Application Interface Framework data protection officer.

Table 6.5 Template SAP Application Interface Framework Roles Delivered by SAP

The SAP Application Interface Framework foundation is built with SAP NetWeaver system, so it inherits security settings from SAP NetWeaver. SAP Application Interface Framework is used by different stakeholders in a Central Finance project. To cater for different needs, SAP has predelivered authorization objects. Table 6.6 displays SAP-delivered authorization objects for SAP Application Interface Framework that can be leveraged for Central Finance replication scenarios.

Authorization Object	Authorization Object Description	Usage
/AIF/PROC	Process error message	Processed by specific user. For example, assign to RFC user that was used by SAP Landscape Transformation Replication Server
/AIF/CUST	Customizing	To perform SAP Application Interface Framework customizing activities
/AIF/ERR	Error handling	Limit authorization by read only or to a specific namespace and interface
/AIF/TECH	Technical error handling	Supports technical mode of error handling
/AIF/EMC	Emergency corrections	Used in emergency correction mode in Transaction /AIF/ERR
/AIF/CFUNC	Custom Functions	To maintain custom functions in the monitoring and error handling
/AIF/HINTS	Custom hints	To maintain custom hints
/AIF/IFDET	Interface determination	To maintain maintenance view of interface determination
/AIF/VMAP	Value mapping	To display and maintain value mapping
/AIF/LFA	File adopter	To access files and directories on the application server for file adapter Transactions /AIF/LFA_UPLOAD_FILE and /AIF/LFA_CHECK_SEND
/AIF/SER	Serialization	To maintain current external index of serialization
/AIF/CDLOG	Change log	To access change log using Transaction /AIF/CHANGE_LOG

Table 6.6 SAP-Delivered Authorization Objects

Authorization Object	Authorization Object Description	Usage
/AIF/CTEXT	Custom message texts	Access to custom message texts
/AIF/CLINK	Custom data links	Access to custom data links for error handling in SAP Application Interface Framework
/AIF/PERSD	Persistence message deletion	Access to delete SAP Application Interface Framework persistence message deletion via Transaction /AIF/PERS_DEL
/AIF/DPA	Processor assignment	Assign, unassign, or change the assignment of processors
/AIF/RAL	Read log	Access to read log Transaction /AIF/READ_LOG
/AIF/RAL_C	Read log customizing	Access to maintain fields related to read log Transaction /AIF/READ_LOG_CUST

Table 6.6 SAP-Delivered Authorization Objects (Cont.)

Note

Consult the SAP Application Interface Framework master guide (*http://s-prs.co/v590102*) for additional guidance on field names and authorization objects settings to be used for these SAP Application Interface Framework authorization objects.

SAP Help Portal also offers additional resources on SAP Application Interface Framework, which can be accessed at *http://help.sap.com/aif*.

In addition, consult SAP Note 1530212, which provides answers to frequently asked questions on SAP Application Interface Framework.

6.1.4 Recipient Handling

In SAP Application Interface Framework, interfaces are assigned to recipients for SAP Application Interface Framework message monitoring and error resolution. Recipients' assignment also allows users to view interfaces in the interface monitor transaction. Prior to assigning recipients to interfaces, you'll identify responsible owners who will monitor errors. Recipient handling is managed in three steps, as shown in Figure 6.6. Let's consider the steps next:

❶ **Define interface**
First you'll define interfaces in SAP Application Interface Framework. SAP predelivered recipients for Central Finance, as shown in Table 6.7.

Type of Replication	Recipient for Alert
Central Finance all interfaces	CFIN_RECIPIENT
Accounting documents	CFIN_AC_DOC
	CFIN_AC_DOC_EX
AVL	CFIN_AV_CI
	CFIN_AV_PO
	CFIN_AC_SI
	CFIN_AC_SO
Profit center replication	CFIN_PCA_DOC
	CFIN_PCA_DOC_SIM
Sales and distribution clearing transfer and down payment status	CFIN_SBDL_CLRST
	CFIN_SD_DPST

Table 6.7 SAP Application Interface Framework Recipients for Central Finance

❷ **Assign users to namespace**
In the next step, for Central Finance, you'll assign the /FINCF namespace/interface to a user.

❸ **Alert users**
Once SAP Landscape Transformation Replication Server is turned on, the system sends alerts to recipients either via My Inbox (app ID: F0862) or emails for error messages based on configured settings in Central Finance. The system sends alerts for the first error message of message category. Any subsequent alerts are only sent out if the old alert is either confirmed or expired.

Figure 6.6 Define and Assign Recipients

Now that you understand the concept of recipient handling, let's learn how to assign recipients in the Central Finance system. To assign recipients to Central Finance namespace, navigate to **SAP Easy Access** menu path **SAP Menu • Cross Application Components • SAP Application Interface Framework • Administration • Configuration • Recipients of a User** or use Transaction /AIF/RECIPIENTS. Alternatively, you can assign business users with the Assign Recipients to Users app (app ID: F2949). After reaching the transaction, you'll be prompted with the **Determine Work Area: Entry** screen, as shown in Figure 6.7.

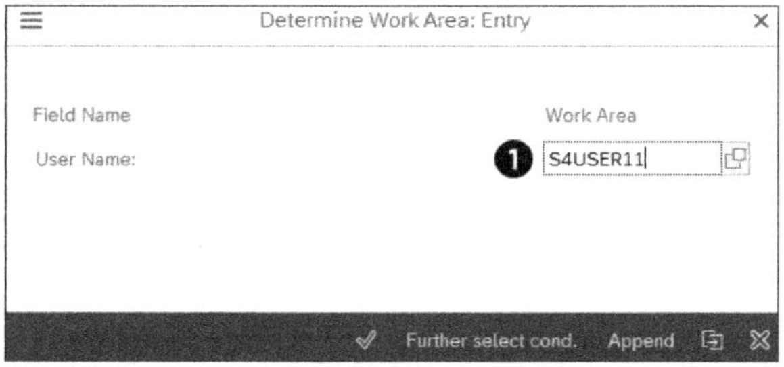

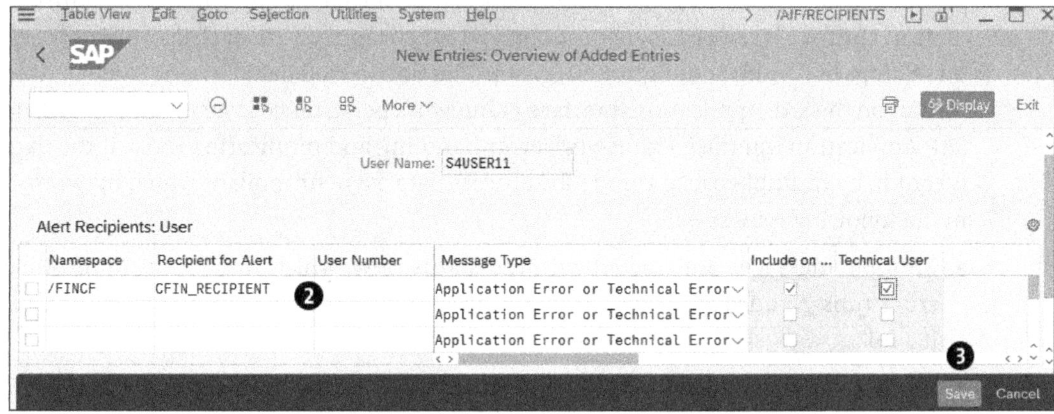

Figure 6.7 Assign Recipient to Namespace

On this screen, enter a value in the **User Name** ❶ field, and click the **OK** button. Click the **New Entries** button. On the resulting **New Entries: Overview of Added Entries** screen, enter the following information:

- **Namespace**
  Enter "/FINCF" for Central Finance.
- **Recipient for Alert**
  Enter "CFIN_RECIPIENT" ❷.
- **Message Type**
  Leave the default value **Application Error or Technical Error** chosen from the dropdown.

Then, choose the **Include on Overview Screen** checkbox. If you're assigning an IT or technical person, then choose the **Technical User** checkbox. Once all information is filled in correctly, click the **Save** ❸ button to save the entry.

After performing this exercise, you can see the activated SAP Application Interface Framework interface in Transaction /AIF/FIMON.

Note

In the example, we've seen assigning a specific user to a namespace. On the other hand, you can assign a role to a recipient, thereby eliminating the need for individual user maintenance.

6.2 Understanding Runtime Groups

SAP Application Interface Framework is delivered with its own runtime and persistence layer. When the source system data is transferred via SAP Landscape Transformation, the data is created asynchronously or synchronously for each document based on settings in runtime configuration group. The transferred data is then stored in XML format on the SAP Application Interface Framework persistence layer and displayed via SAP Application Interface Framework error handling and monitoring tools. If the data is sent in by multiple background jobs process data asynchronously, which may result in the following issues:

- Multiple messages are created with the status **New**, which may result in multiple stuck runs. A run is defined in the runtime configuration: the system executes once it's full.
- The maximum number of messages per run is defined in the runtime configuration.
- Locking errors may occur.
- Messages are processed synchronously, that is, at the same time and will result in serialization errors.

To avoid the preceding issues, SAP provides runtime configuration group configuration to process XML messages with the appropriate sequence. A runtime configuration group is identified by a combination of a namespace and a three-digit identifier. In general, a runtime group feature contains information such as the following:

- Number of messages executed for each packet received from the source system
- Number of messages executed per run
- Whether a run is executed either synchronously or asynchronously
- Whether to activate or deactivate the runtime group

Based on runtime configuration group parameters, you have better control over how SAP Application Interface Framework messages are queued and processed, thereby improving performance of SAP Application Interface Framework. In addition, when no runtime group is assigned for Central Finance namespace, the system uses default runtime group 400. The default runtime group processes 100 messages with each packet of 20 messages. Table 6.8 displays runtime groups defined for Central Finance. As you can see, multiple runtime configuration groups can be configured for various requirements based on Central Finance replication scenarios.

Runtime Configuration Group	Description	Usage
400	System runtime configuration is provided by default	Default runtime configuration. The system processes 100 messages with 20 messages in each package.
RES	Restart	You can define an automatic restart of SAP Application Interface Framework a message at frequent intervals. This isn't recommended if resolution of the error takes too long, or error counts are too high as it consumes system resources.
RPR	Reprocess	You can also set up automatic reprocessing. When the clearing document is posted before the initial load is reset/reversed in the source, then the reset clearing document replicates to the Central Finance system and pulls the missing cleared items from the source as open items. Consult SAP Note 2710447 for automatic reload of missing cleared items.
ZCF	Central Finance runtime configuration for finance	You can define a runtime configuration for Central Finance postings. There is an option to configure default messages to process.

Table 6.8 Central Finance Runtime Configuration Groups

Tip

Consult SAP Note 2578106 on usage restriction for runtime configuration groups 400 through 499.

Now that you've learned the basics of runtime configuration group concepts, let's discuss how it helps uses achieve serialization when processing SAP Application Interface Framework messages and the relevant setup required to enable runtime configuration groups in Central Finance.

6.2.1 SAP Application Interface Framework Serialization Concept

When SAP Application Interface Framework processes financial postings, it's essential that documents transferred from the source system are processed in the correct sequence in SAP Application Interface Framework. This means the SAP Application Interface Framework doesn't process a message until the processor document is processed successfully.

In Central Finance, bulk processing of SAP Application Interface Framework refers to a high volume of data processed via SAP Application Interface Framework. In general, bulk processing of SAP Application Interface Framework happens in the following scenarios:

- After a financial initial load is completed and documents accumulated since the initial load are started.
- If financial postings accumulated when SAP Landscape Transformation Replication Server is down for planned maintenance. During SAP Landscape Transformation Replication Server planned downtime, the system accumulates documents in the source system logging table.
- Starting or resuming SAP Landscape Transformation Replication Server.

Figure 6.8 illustrates the basic concept of serialization with an invoice and its follow-on documents posted in the right sequence, assuming central payment isn't active yet. When an invoice and follow-on documents are posted in the source system, from the functional order, SAP Application Interface Framework must process in the order of posting in the source system as follows:

❶ First, the customer invoice is posted in the source system.

❷ Next, the invoice is cleared in the source system which clears the customer open item.

❸ Reset the clearing document posted due to an error in clearing or to reverse an incorrect invoice.

❹ Reversal documents reverse invoice documents.

With serialization active, messages related to invoice transactions are processed according to the timestamp posted per the source system data. Timestamp information sent in the payload of the message to SAP Application Interface Framework (stored in table CFIN_AIF_SER). Serialization makes sure SAP Application Interface Framework doesn't start to process a message until the preceding message is fully processed. Otherwise, the follow-on message (e.g., reversal) will end up with an error in SAP Application Interface Framework.

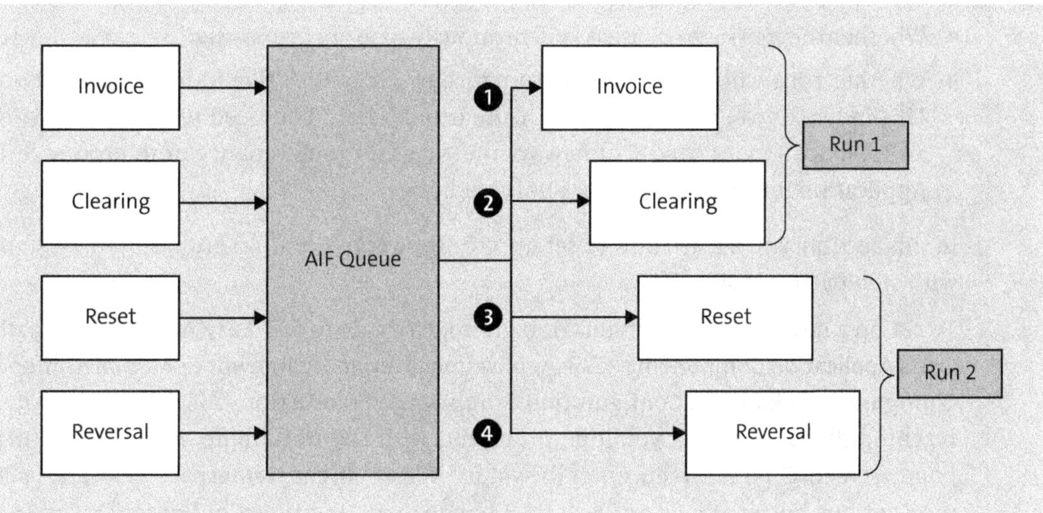

Figure 6.8 Serialization Concept

Note

Consult SAP Note 2679070 for the steps to follow when you start or resume SAP Landscape Transformation Replication Server to prevent serialization. This note now includes SAP Application Interface Framework application engine `XML bfRFC runtime`, which is supported from SAP S/4HANA 2023 onward.

Refer to SAP Note 2257865 regarding scheduling with multiple jobs when a mass volume is involved from the source system.

SAP S/4HANA now supports the new SAP Application Interface Framework application engine to prevent serialization errors.

6.2.2 Configuring Runtime Groups

As you recall, to process the source system data, SAP Application Interface Framework uses a runtime configuration group to process messages. SAP Application Interface Framework uses the XML runtime, which uses runs to process messages. A run contains multiple packages. In Central Finance, to transfer data, a runtime configuration is required that is identified by a namespace and a three-digit identifier. The runtime configuration serves primarily four purposes:

- Number of messages executed per each package.
- Number of messages executed per run. By default, the system processes 100 messages per package, and 20 messages are in each package. However, this value can be adjusted per the user-specific runtime configuration group.
- Whether the messages process synchronously or asynchronously.
- Whether runs will be executed automatically with the **Active** indicator turned on. This indicator can be deactivated in situations where you need to stop processing SAP Application Interface Framework messages for maintenance or to process SAP Application Interface Framework bulk loads.

In this section, you'll learn how to set up a runtime configuration groups in the system with step-by-step instructions.

To set up a new runtime configuration group, navigate to **SAP Easy Access** menu path **Cross Application Components • SAP Application Interface Framework • Administration • Configuration • Runtime Configuration Group** or use Transaction /AIF/PERS_CGR. After reaching the transaction, you'll be prompted with the **Determine Work Area: Entry** popup screen. On this screen, enter the value "/FINCF" in the **Namespace** field, and click the **Continue** button. You'll arrive at the **Change View "Runtime Configuration Group": Overview** screen, as shown in Figure 6.9. On this screen, click the **New Entries** ❶ button. On the resulting **New Entries: Details of Added Entries** screen, maintain the entries as follows:

- **Runtime Config. ID** ❷
  Enter a three-digit ID that identifies the runtime configuration group. The runtime configuration along with namespace makes a unique runtime configuration group within SAP Application Interface Framework.

- **Runtime Configuration Group** ❸
  Maintain the following fields under the **Runtime Configuration Group** section:
  - **Description**
    Enter the description of the runtime configuration group. The field has a maximum of 60 characters in length.
  - **Runtime Cfg Active**
    Choose this indicator to process SAP Application Interface Framework messages. If this indicator is off, then the system leaves the SAP Application Interface Framework messages with the **In Process** status.
  - **Background User**
    Enter a user who'll be processing the SAP Application Interface Framework message and is authorized asynchronously. This option is relevant only if you have a different person authorized to execute the job.

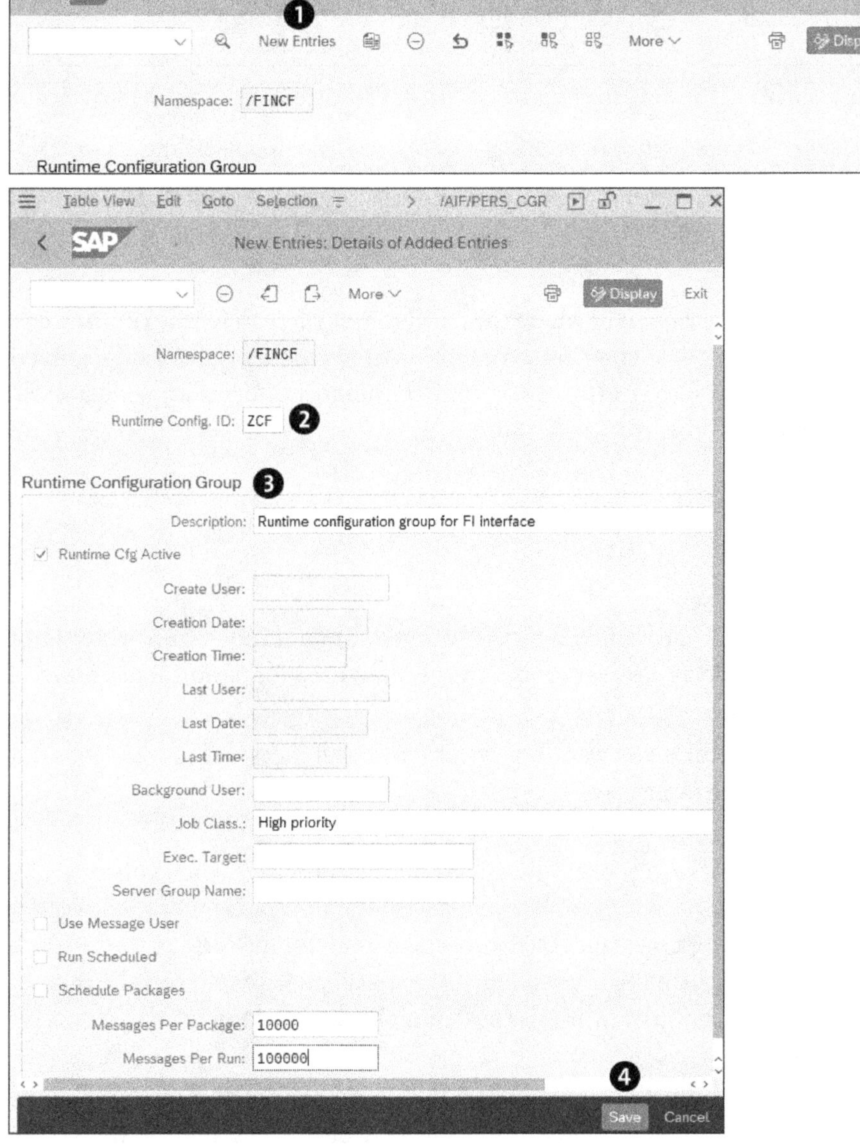

Figure 6.9 Runtime Group Configuration

- **Job Class**
  The value in this field sets the priority of the job while processing SAP Application Interface Framework messages. Table 6.9 list three choices on the dropdown table. Depending on the requirements of the runtime group, you'll choose the job class priority.

Job Class	Description
A – **High priority**	The runtime group with this option takes highest priority to process messages. For finance postings, this option is selected.
B – **Middle priority**	This option is suitable when jobs are scheduled at a frequent interval.
C – **Low priority**	The option is chosen when jobs don't require high priority.

Table 6.9 Job Classification

- **Exec. Target**
  Enter the name of the server where the background job runs for the runtime configuration group. The server information entered follows the naming conventions of *<Host Name>_<SAP client>_<System number>*, for example, CLOUD9_S18_00, which is broken down as follows:
  - CLOUND9 is the server name or host name.
  - S18 is the SAP client name.
  - 00 is the system number.
- **Server Group Name**
  Enter the server group name where the runtime configuration is run, which is different from the default server group. This is often used for performance reasons.
- **Use Message User**
  Select this indicator if you want the system to collect messages into the jobs that are created by the same user.
- **Run Scheduled**
  Select the checkbox to process messages asynchronously. When this option is selected, the system creates a job for each run. The run contains SAP Application Interface Framework messages to be processed in a bundled run and then in packages subsequently. On the other hand, if the checkbox isn't selected, the runs are processed synchronously, which is the default behavior of the system.
- **Schedule Packages**
  Like the **Run Scheduled** checkbox, select the checkbox to process packages asynchronously. A package contains 20 messages by default. As a result, the system creates one job per package. Conversely, if the checkbox isn't selected, the system processes the package synchronously. In addition, because of not selecting the checkbox, SAP Application Interface Framework processes packages in the same job as the run.
- **Messages Per Package**
  This option can be used for parallel processing. Enter the maximum number of messages to be processed per package by choosing the newly created runtime

group. Recall, the default runtime group contains 100 messages for each run and 20 messages in each packet within the run. With this setting, you can override default values. The system uses the value in this field and the value in the **Message Per Run** field to determine the maximum number of messages to process. Then, the system processes each package per definition. SAP recommends 20 messages for each package.

- **Messages Per Run**
  In addition to the **Message Per Package** indicator, this indicator provides parallel processing to improve performance of the system when transferring data. Enter the maximum number of messages per run. You can override the default message processing parameter of 100, which is recommended by SAP. The system uses the value in this field and the value in the **Message Per Run** field to determine the maximum number of messages to process.

To illustrate **Message Per Package** and **Message Per Run** fields further, see Table 6.10 with examples. Click **Save** ❹ to save your entries.

Message Per Run	Message Per Package	Incoming Messages	First Run	Second Run
50	10	90	5 packages (10 messages)	4 packages (10 messages) 1 package (5 messages)
100	20	190	5 packages (20 messages)	4 packages (20 messages) 1 package (10 messages)

Table 6.10 Message Processing

6.2.3 Assigning Runtime Groups

Once you've configured the runtime configuration, in this customizing activity, you'll assign the runtime configuration group to the replication object and namespace. To perform the customizing activity, navigate to customizing IMG menu path **SAP Customizing Implementation Guide • Financial Accounting • Central Finance: Target System Settings • Set Up Systems • AIF Runtime Configuration • Assign AIF Runtime Configuration Group to Replication Object** or choose Transaction CFINIMG. After reaching the configuration activity, you'll arrive at the **Change View "Assign Runtime Configuration to Replication..": Overview** screen, as shown in Figure 6.10. On this screen, maintain values in the **Namespace** and **Runtime Configuration group ID** ❶ fields for each active replication object. Figure 6.10 shows the **FI/CO Postings** replication object enabled with the runtime configuration object **ZCF** created in the previous step. Once the assignment of runtime configuration is assigned to a replication object, click the **Save** ❷ button to save the configuration.

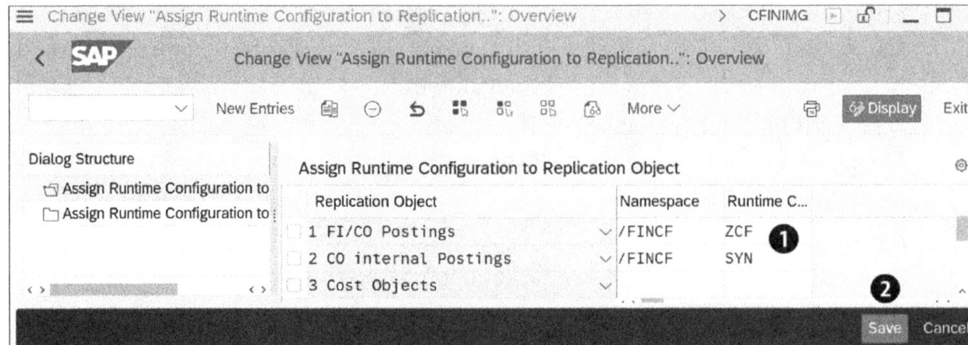

Figure 6.10 Assign the SAP Application Interface Framework Runtime Configuration Group to Replication Object

Tip

SAP's leading practice suggests creating a separate runtime configuration object for each replication object.

6.3 Error Handling

The basic idea of error handling is to display, analyze, and distribute error messages from the interface posting to relevant stakeholders. Central Finance error handling is required in both the initial load and during online replication scenarios. Table 6.11 shows the types of data transfer scenarios in Central Finance and what error handling tools are used.

Type of Posting	Error Handling Tool Used
Initial load of FI postings	Initial load comes with its own monitoring tool for simulation of mapping, simulation of posting, and final posting.
Initial load of cost objects, CO postings, commitments, projects/WBSs, EC-PCA	SAP Application Interface Framework
Online replication of FI postings, cost objects, CO postings, commitment replication, projects/WBSs, EC-PCA, cost estimates, cost rates, AVL, and JVA	SAP Application Interface Framework

Table 6.11 Error Handling for the Initial Load and Online Replication

In the next section, let's discuss the initial setup required to the SAP Application Interface Framework.

6.3.1 Setting Up the SAP Application Interface Framework Number Range

The first step in using SAP Application Interface Framework for Central Finance scenario is to generate an SAP Application Interface Framework number range. This is a required activity in Central Finance; otherwise, the system displays the error message **No Interval available for number range object /AIF/RUN**. To generate the SAP Application Interface Framework number range, navigate from **SAP Easy Access** menu path **SAP Menu • Cross-Application Components • SAP Applications Interface Framework • Administration • Tools • Setup** or use Transaction /AIF/SETUP. This transaction not only generates a number range but also creates predelivered customizing and other settings to use SAP Application Interface Framework in Central Finance. Upon entering the transaction, you'll arrive at the **AIF Setup** screen, as shown in Figure 6.11. On this screen, the system provides several checkboxes based on the SAP Application Interface Framework version. Select all checkboxes. The **Test Mode** ❶ checkbox can be used to check for any errors. Uncheck the **Test Mode** checkbox and click the **Execute** ❷ button. The system updates configuration tables and displays SAP Application Interface Framework objects along with a confirmation message at the bottom of the screen ❸.

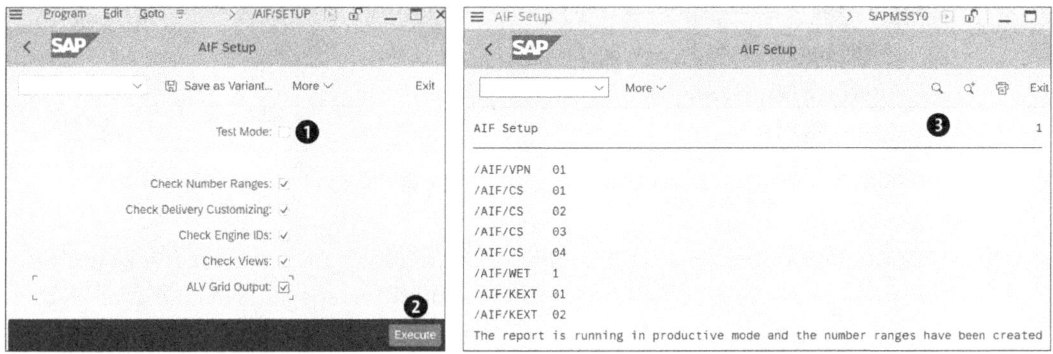

Figure 6.11 SAP Application Interface Framework Number Range

In Chapter 8, you'll learn about the initial load monitoring tool for financial posting. In the next section, you'll learn how to use error handling tools provided with SAP Application Interface Framework for monitoring and troubleshooting SAP Application Interface Framework messages.

6.3.2 Monitoring and Error Handling

When data is replicated from the source system, SAP Landscape Transformation Replication Server transfers the data in batches to Central Finance. The background jobs pick up data and processes in Central Finance. SAP Application Interface Framework is used to view data transferred from SAP Landscape Transformation. To display, analyze, and correct error messages, SAP provides monitoring and error handling Transaction /AIF/ERR. With this transaction, you can monitor and troubleshoot Central Finance replication errors. We'll focus on the most important features. To access this transaction, navigate from **SAP Easy Access** menu path **SAP Menu • Cross-Application Components • SAP Applications Interface Framework • Monitoring and Error Handling.** The transaction provides a plethora of selection parameters to limit data selection. Figure 6.12 displays the selection screen of Transaction /AIF/ERR. Alternatively, you can access a web-based user interface (UI) with Transaction /AIF/AIFX/ERR_WEB. The selection screen is divided into six subsections:

❶ **Application Selection**
The first field on the screen that uniquely identifies application ID in SAP Application Interface Framework. The value in this field is entered with SAP Application Interface Framework by default and delivered by SAP.

❷ **Application Specific Selection**
In this section, you'll select values in **Namespace**, **Interface Name**, and **Interface Version.** For Central Finance, SAP delivers the **/FINCF** namespace. Refer to Table 6.2 for a list of interface and versions delivered with Central Finance for different replication scenarios. For our demonstration purposes, you'll select the following values for finance replication from the **Select Interface** button, or enter it manually.

- **Namespace: /FINCF**
- **Interface Name: AC_DOC**
- **Interface Version: 2**

In addition, you have an option to select **Message Class** and **Message Number.**

❸ **More Specific Selection**
This section shows up when the finance interface **AC_DOC** and version **2** are chosen from the previous section. After entering application-specific selections, you'll press [Enter], which brings up this section. Table 6.12 provides a list of fields in this section and their usage.

Selection Field	Usage
Reference procedure (AWTYP)	Unique reference number from the sending application. For example, financial accounting documents will have BKPF, and initial load accounts receivable and accounts payable open items will have APAR, and so on.
Reference Document (AWREF)	Refers to the document number transferred from the source system.
Reference Org. Unit (AWORG)	Consists of the reference organizational unit.
Ref. Document Logical System (FIS_AWSYS)	Source logical system of reference document.

Table 6.12 More Specific Selections

Selection Field	Usage
Logical System of the Sender (LOGSYSTEM_SENDER)	Logical system where original documents originated from. This could be a source system or other external sources. Depending on the situation, this field may or may not be entered on transactional data.
Company Code in Sender System (BUKRS_SENDER)	As the name implies, company code of sender or source system.
Company Code (BUKRS)	Mapped Central Finance company code.
Transaction Code (TCODE)	Transaction code used in source data posting.

Table 6.12 More Specific Selections (Cont.)

Note

In general, the provided selection fields satisfy most customer requirements. If additional fields are required for selection, you can add additional fields with an interface-specific enhancement.

❹ **Generic Selection**

This section primarily offers two important selection criteria: period/date selection and/or **Message GUID**. The latter selection is useful to debug SAP Application Interface Framework messages. For easier period selection, SAP Application Interface Framework offers either a date and time stamp selection option or simply select **Today**, **Last 2 Weeks**, **Current Month**, **Current Year**, or **All** button options for a quicker selection.

❺ **Status Selection**

Messages processed from SAP Application Interface Framework contain a specific status. The selection screen offers the following statuses to choose from:

- **In Process**

  When you choose this indicator, the system displays SAP Application Interface Framework messages with the **In Process** status. When the **In Process** status checkbox is selected, the system displays additional substatuses to choose from:

  - **Processing**
  - **Waiting to be processed**
  - **Auto Reprocessing**
  - **Blocked by predecessor**
  - **Error in job processing**

- **Processed Successfully indicator**

  When selected, the system displays all posted transactions successfully.

- **Processed with Warnings indicator**
  When selected, the system displays all posted transactions with warnings.
- **Application Errors indicator**
  This displays application-specific errors.
- **Technical Errors indicator**
  This displays technical-specific errors.
- **Canceled Messages**
  This displays canceled messages.

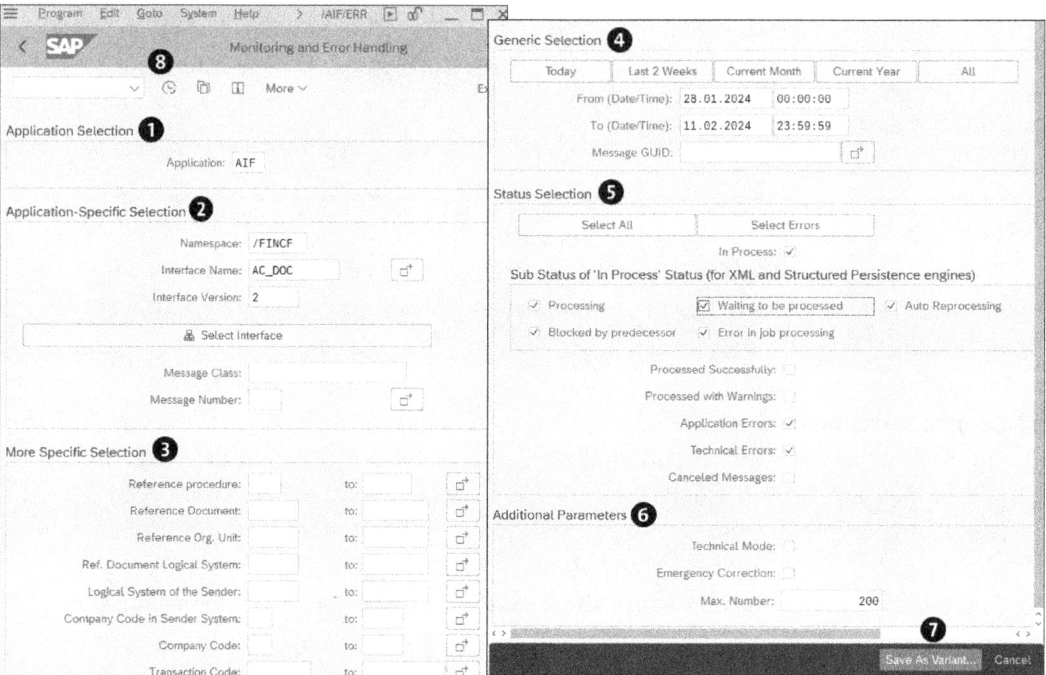

Figure 6.12 Monitoring and Error Handling

❻ **Additional Parameters**

In this section, you can choose the following:

- **Technical Mode indicator**
  This shows whether you'll execute the transaction in a normal mode or in a technical mode. If this indicator is selected, the system displays the output screen in a technical mode. This is especially helpful for technical users to troubleshoot SAP Application Interface Framework messages from a technical viewpoint. This requires the /AIF/TECH authorization object assigned to the role or user to select this indicator.
- **Emergency Correction indicator**
  This allows you to update source data prior to posting. The SAP recommendation

isn't to use this approach; rather, identify the root cause, and fix the data in the right way. Nevertheless, this option is available for users to use. This requires the /AIF/EMC authorization object assigned to the role or user to select this indicator.

- **Max. Number field**
  This allows you to enter a maximum number of messages to be displayed upon executing the transaction. By default, the system displays 200 messages. You can override this number to a higher number. However, keep performance considerations in mind when you choose a larger number to display SAP Application Interface Framework messages.

Note that you can save the layout with the **Save as Variant** ❼ button. Once you've entered the selection parameters, click the **Execute** ❽ icon. The system displays the **Error Handling** screen, as shown in Figure 6.13. This screen is divided into four subscreens:

❶ **Data Messages** view displays data messages for the selected criteria from the selection screen. As you can see in Figure 6.13, each SAP Application Interface Framework message is displayed with a status icon in a hierarchy that is grouped by namespace and interface. In addition, the submenu bar provides users with the chance to read, restart, and update user changes, or cancel SAP Application Interface Framework messages. Furthermore, the screen layout can be modified to include the most important fields. Data from this subscreen, including GUID information, can be copied to a Microsoft Excel document by pressing Ctrl+Y for further analysis. You can expand or collapse the selected node. In addition, by double-clicking a message or group of messages, the system brings up corresponding details in the right pane. Table 6.13 explains different icons of **Data Messages** and their meaning in Transaction /AIF/ERR (Monitoring and Error Handling).

Icon	Meaning
	Processing messages that resulted in an ABAP dump or failed while processing. You'll evaluate errors in Transaction ST22.
	Automatic reprocessing of the SAP Application Interface Framework message.
	Error in job processing is displayed when the SAP Application Interface Framework message fails due to a predecessor message failure, or the system is still in the process of processing the message (queued).
	Blocked by predecessor messages aren't processed or completed yet.
	Waiting to be processed messages are scheduled in a background job, but not kicked off yet.
	Restart is required due to changes made to SAP Application Interface Framework message.

Table 6.13 Meaning of Symbols in Transaction /AIF/ERR: Monitoring and Error Handling

Icon	Meaning
▲	*Warning message* an underlying message may have been posted yet.
⚡	*Error* means an SAP Application Interface Framework message failed due to application or technical errors.
🏁	*Success* refers to an SAP Application Interface Framework message being processed successfully.
✓	*Canceled* refers to an SAP Application Interface Framework message being canceled.
⇨	*In process* messages results in with or without any message in SAP Application Interface Framework. You can see application logs for any errors. If no logs are found, an ABAP dump may have occurred. You can view ABAP dumps via Transaction ST22. In addition, the **In Process** message waits for the predecessor message to be posted.

Table 6.13 Meaning of Symbols in Transaction /AIF/ERR: Monitoring and Error Handling

❷ **Log Messages** view displays log messages based on selected message from the **Data Message** view. If you select multiple messages or all messages, then the system displays all messages on this view. This section typically displays application errors, technical errors, warnings, success messages, information messages, along with message numbers and long text. In addition, this subscreen allows users to navigate to custom functions that provide users with additional functionality. **Log Messages** view also provides custom hits and functions to display icons based on defined actions

❸ **Data Structure** view shows a tree of structures and subtables of the selected messages. The system provides the tree structure to navigate different table views. In addition, you can view data in technical mode.

❹ **Data content** view displays transaction details for the selected message from the data structure in a table. If there are errors in a message, this view highlights the fields containing errors for easy identification. In addition, this is where you'll edit source data if you select the **Emergency Correction** checkbox in the selection screen.

[»]

Note

Table `CFIN_AIF_IDX_TBL` stores status and substatus fields for SAP Application Interface Framework messages (GUIDs). You can access this table from the Central Finance cockpit (Transaction FINS_CFIN_COCKPIT).

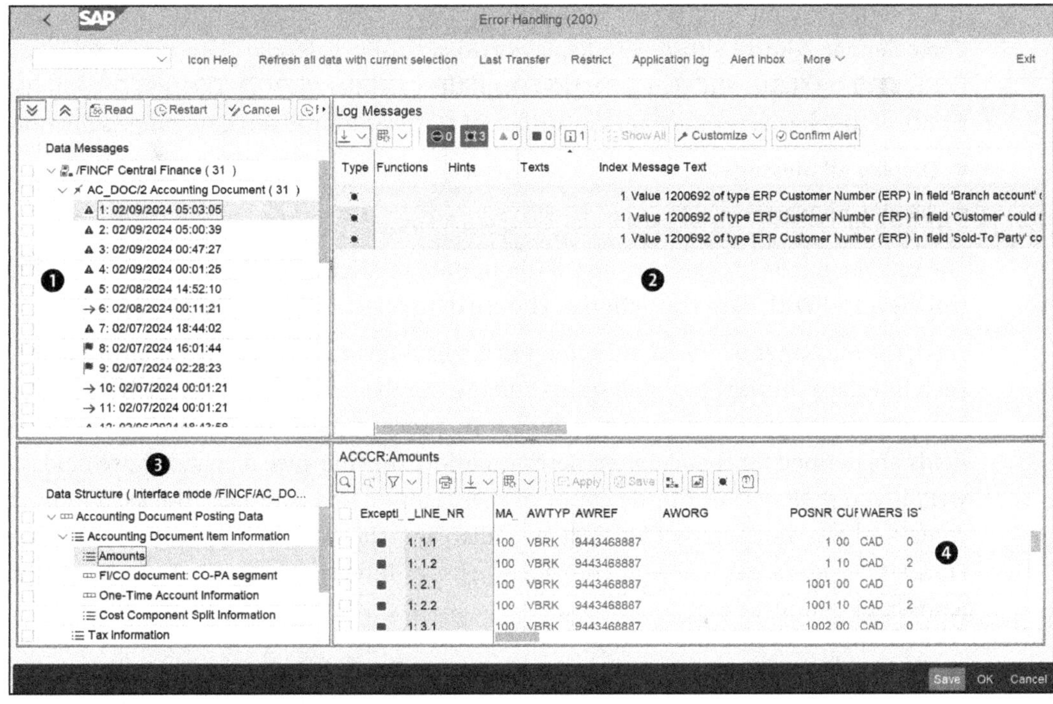

Figure 6.13 Error Handling

6.3.3 Interface Monitoring

Interface monitoring (Transaction /AIF/IFMON), is a user-friendly interface that provides both business and IT users with a single place to monitor all messages irrespective of interfaces. The transaction can be accessed either from SAP GUI or with the Overview Message Monitoring app (app ID: F4515). To access the SAP GUI transaction, navigate it to from **SAP Easy Access Menu** path **Cross Applications Components • SAP Application Interface Framework • Interface Monitor**. The interface monitoring offers the following features:

- Display overall status of messages by interface
- Display count of by status such as error
- Personalized view
- Alerts management
- Calendar to restrict by dates and date range
- Navigate to message summary, which provides message counts by message type and application log
- Navigate on select messages to monitoring and error handling (Transaction /AIF/ERR) from the message summary screen

The interface monitoring screen is divided into two subscreens, as shown in Figure 6.14. The calendar control subscreen displays two sections: calendar date range selection fields ❶ or by status and date selection from the calendar view ❷. The first section has a button to select error messages by different statuses for a given data range selection:

- **Display all messages**
- **Display only errors**

The second button toggles between **Without date restriction** to show the calendar control view and **With date restriction** to choose date selection range fields.

Next, the **Messages overview** subscreen ❸ displays all interfaces responsible for a user. Each interface shows the overall status and the number of messages for each status category, such as application errors, successfully processed, in process, and so on. If key fields are defined for the recipient determination, the interface displays those fields as well. If you double-click on the count of messages on any interface, the system navigates you to Transaction /AIF/ERR and displays all the SAP Application Interface Framework messages selected. Table 6.14 shows interface monitor **Data Messages** icons and their meaning:

Icon	Meaning
	Count of messages.
	Alert off, no glowing bulb means no alert exists.
	Alert on glowing bulb.
	Notification of email or alert default setting.
	Receive email for every error SAP Application Interface Framework message regardless of alert status.
	No email or alert sent for the interface.
	Displays count of successfully posted messages.
	Displays count of canceled messages.
	Shows count of **In Process** messages.
	Displays count of error messages.
	Error shown to technical user.
	Displays count of messages processed successfully with warning messages.
	Green status means all SAP Application Interface Framework messages were successfully processed in the interface or canceled manually.

Table 6.14 Meaning of Symbols in Transaction /AIF/IFMON (Interface Monitor)

Icon	Meaning
○▲○	Yellow status means the interface processed messages, but SAP Application Interface Framework messages contain warning messages.
●○○	Red status means the interface contains application errors or technical errors.
●●●	Gray status means the interface contains **In Process** messages.
○○○	Undefined status means the interface does contain any messages.

Table 6.14 Meaning of Symbols in Transaction /AIF/IFMON (Interface Monitor) (Cont.)

Finally, the **Message Summary** button ❹ navigates to the **Message Summary** screen. When you click on the count of messages or status icons, the system takes you to the **Monitoring and Error Handling** screen (Transaction /AIF/ERR).

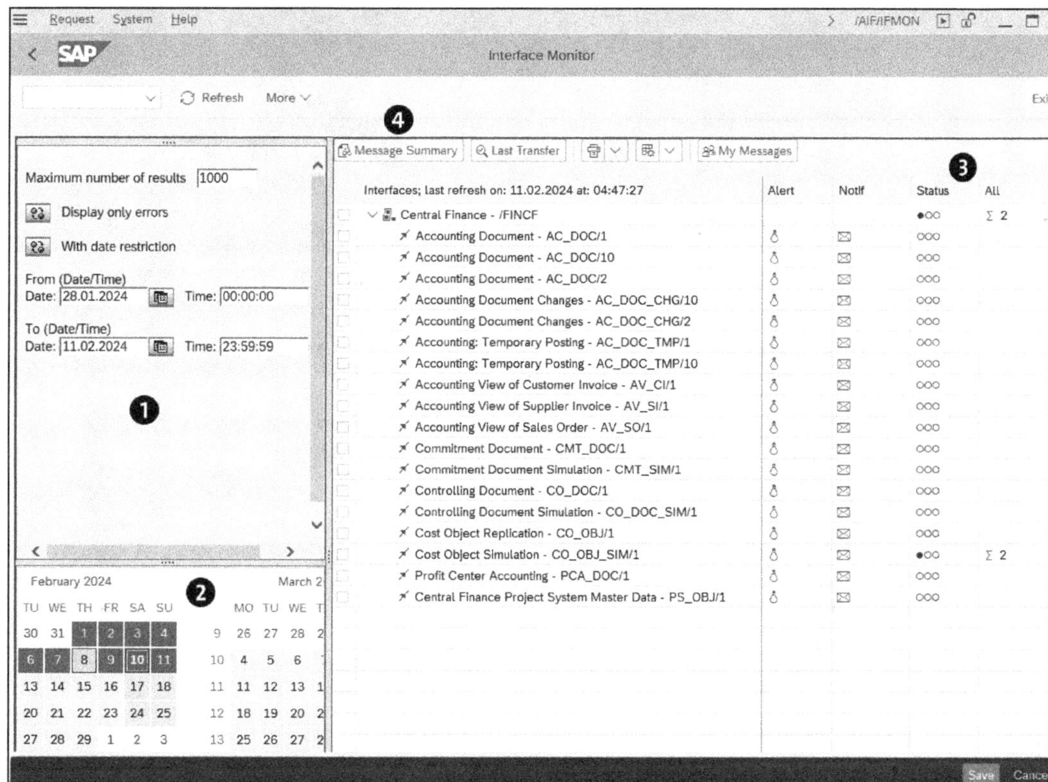

Figure 6.14 Interface Monitoring

Note

To view messages on the monitoring messages interface, you must assign SAP Application Interface Framework recipients. SAP Note 3358732 provides guidance on how to do this.

When you click the **Message Summary** button from the **Interface Monitor** screen, the system navigates you to the **Message Summary** screen, as shown in Figure 6.15.

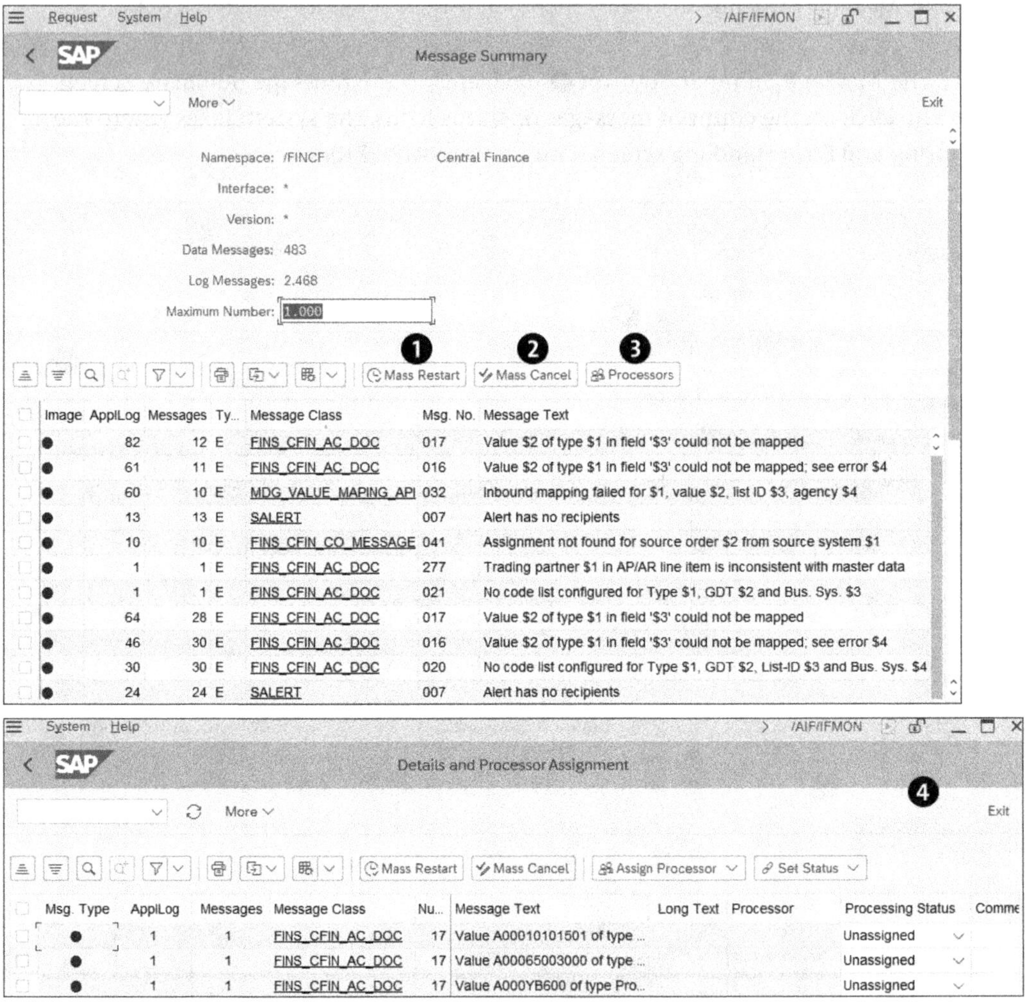

Image	ApplLog	Messages	Ty...	Message Class	Msg. No.	Message Text
●	82	12	E	FINS_CFIN_AC_DOC	017	Value $2 of type $1 in field '$3' could not be mapped
●	61	11	E	FINS_CFIN_AC_DOC	016	Value $2 of type $1 in field '$3' could not be mapped; see error $4
●	60	10	E	MDG_VALUE_MAPING_API	032	Inbound mapping failed for $1, value $2, list ID $3, agency $4
●	13	13	E	SALERT	007	Alert has no recipients
●	10	10	E	FINS_CFIN_CO_MESSAGE	041	Assignment not found for source order $2 from source system $1
●	1	1	E	FINS_CFIN_AC_DOC	277	Trading partner $1 in AP/AR line item is inconsistent with master data
●	1	1	E	FINS_CFIN_AC_DOC	021	No code list configured for Type $1, GDT $2 and Bus. Sys. $3
●	64	28	E	FINS_CFIN_AC_DOC	017	Value $2 of type $1 in field '$3' could not be mapped
●	46	30	E	FINS_CFIN_AC_DOC	016	Value $2 of type $1 in field '$3' could not be mapped; see error $4
●	30	30	E	FINS_CFIN_AC_DOC	020	No code list configured for Type $1, GDT $2, List-ID $3 and Bus. Sys. $4
●	24	24	E	SALERT	007	Alert has no recipients

Msg. Type	ApplLog	Messages	Message Class	Nu...	Message Text	Long Text	Processor	Processing Status	Comme
●	1	1	FINS_CFIN_AC_DOC	17	Value A00010101501 of type ...			Unassigned	
●	1	1	FINS_CFIN_AC_DOC	17	Value A00065003000 of type ...			Unassigned	
●	1	1	FINS_CFIN_AC_DOC	17	Value A000YB600 of type Pro...			Unassigned	

Figure 6.15 Interface Monitoring Message Summary

This screen provides an overview of all log messages with a count of occurrences and their statuses. You can navigate to Transaction /AIF/ERR (Monitoring and Error Handling) by double-clicking any line of the message summary screen. In addition, this screen offers important functionalities such as the following:

❶ **Mass Restart** button allows users to restart error messages in mass volume. This is possible only with an appropriate authorization object assigned to the user.

❷ **Mass Cancel** button allows users to cancel SAP Application Interface Framework messages in mass volume.

❸ **Processors** button shows details of SAP Application Interface Framework messages along with processor assignment, as shown at the bottom of Figure 6.15 ❹.

6.3.4 Restarting or Canceling SAP Application Interface Framework Messages from Monitoring and Error Handling

From Transaction /AIF/ERR (Monitoring and Error Handling), you can restart, cancel, or process with user changes a single or multiple SAP Application Interface Framework message(s). Choose the messages under **Data Messages** that you want to process, as shown in Figure 6.16 ❶. The following actions can be performed:

- The **Restart** icon ❷ allows you to restart a single message or multiple messages at a time. In general, you'll use this icon when a problem is solved, and you'll manually restart the message.
- In some situations, you'll cancel the error message due to duplicates or data issues. The **Cancel** icon ❸ allows you to do this.
- If a user change is made on data through emergency correction mode, then you'll use the **Repost with User Changes** icon ❹ to process such changes.

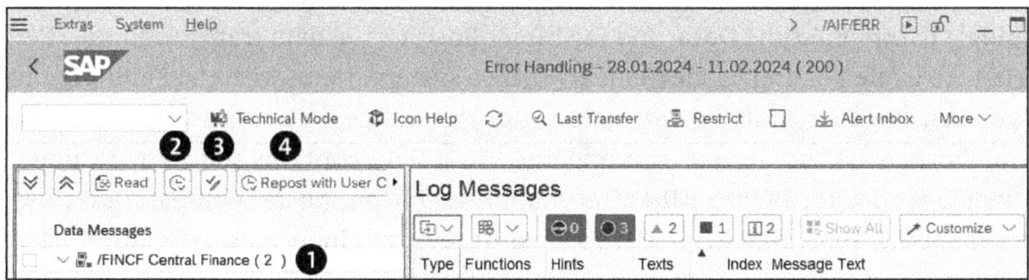

Figure 6.16 Restart or Cancel SAP Application Interface Framework

[+]

> **Tip**
> SAP provides Transaction /AIF/ADAPT_STATUS to change the **In Process** status of an application interface message. You need to review the scenario carefully before using this transaction, as the action may impact other dependent messages.

6.3.5 Analyze SAP Application Interface Framework Message

SAP also provides report FINS_CFIN_AIF_ANALYZER to analyze SAP Application Interface Framework errors in an alternative way. The report shows differences between the source system and the Central Finance system for the transferred transaction. After reaching the report via Transaction SE38 or Transaction SA38, the system displays the **Central Finance Analyze AIF Message** screen, as shown at the top of Figure 6.17. On this screen, enter the value in the **AIF Message Guid** ❶ field for any message in AC_DOC interface under namespace /FINCF. Next, click on the **Execute** ❷ button.

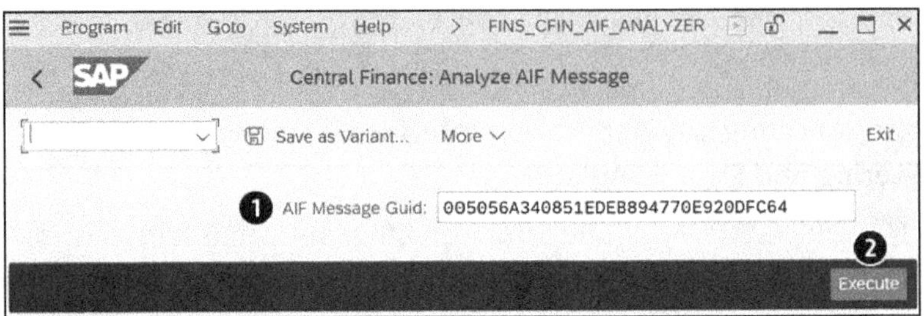

Figure 6.17 Analyze SAP Application Interface Framework Messages

On the resulting screen, you see the source system data and SAP Application Interface Framework data for each structure associated with the interface, as shown in Figure 6.18. For example, you can see **Transaction Type**, **Message Status**, **Customer BAdI Status**, and so on. **Transaction Type** displays **Accounting Document**, which refers to the AC_DOC interface holding the FI document replicated from the source system. **Message Status** displays the **Error** status of the replicated document that is in error. **Customer BAdI Status** displays **Inactive**, which means the BAdI isn't implemented in the Central Finance system. The system goes through a four-stage process: **Customer Prepare**, **Customer Mapping**, **Customer Adjust Data**, and **Customer Post** for the data transferred into Central Finance while posting in the Central Finance system. Next, with the **Simulate** button, you can see the simulated data value after the source data transformation in the Central Finance system. In our example, the VBUND field contains the source value of **1710**, which is transformed to 1010 after mapping. During the data preparation phase before mapping, the value isn't transformed in Central Finance as the system keeps **1710**. However, at the value mapping phase, trading partner **1710** is mapped with **1010** through SAP Master Data Governance mapping in Central Finance. This value is kept during the posting phase until it's committed to the database.

In addition, in Transaction /AIF/ERR, you can see the transformed value after mapping or any custom data transformation under each accounting document structure through this report. You can further analyze the data transformation by simulating the SAP Application Interface Framework GUID. Furthermore, you can also debug the GUID in simulation phase: First, you activate the breakpoint by clicking on the **inactive** button. After clicking this button, the system will disable the **active** button. Now you can click on the **Simulate** button. The system will open the debugger from where you can debug the SAP Application Interface Framework GUID.

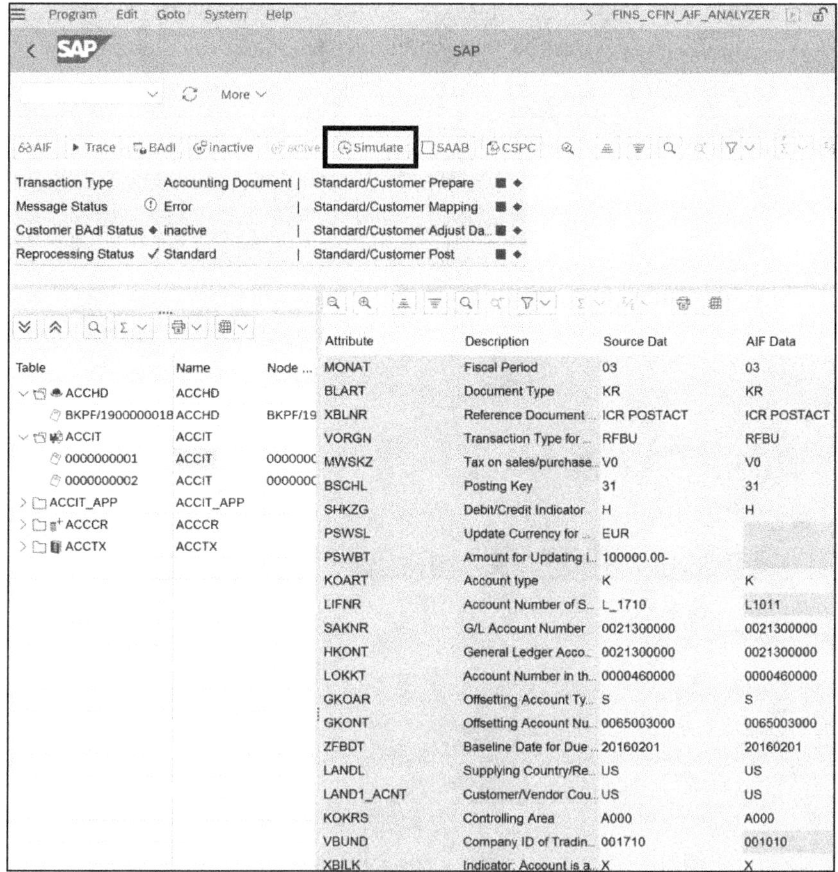

Figure 6.18 Analyze SAP Application Interface Framework Messages - result

6.3.6 Emergency Correction

Emergency correction mode allows you to make changes to data directly on the **Monitoring and Error Handling** (Transaction /AIF/ERR) screen. Authorization object /AIF/EMC must be assigned to the processor who is carrying out the emergency corrections. On the selection screen, after entering the namespace, interface, and other relevant selection criteria, press Enter to bring up the **Emergency Correction** indicator shown at the bottom of the screen (Figure 6.19 ❶). After choosing the indicator, click the **Execute** ❷ button.

On the **Error Handling** resulting screen, choose the message and data structure on the left pane. Then, go to the **Data Content** section on the bottom-right pane. The highlighted field **KUNNR** is showing in red color on line number **1.2**, as shown in Figure 6.20 ❶. When you double-click the highlighted field, the system brings up a popup screen where you can enter **Change Value** ❷. After entering a new value, click the **OK** ❸ button to continue. Next, click the **Save** ❹ icon to save the changes. Finally, click the **Repost**

with User Changes ❺ button to post the message. Note that if you use the **Restart** icon by mistake, you'll lose the changes made, and you'll have to repeat the steps again.

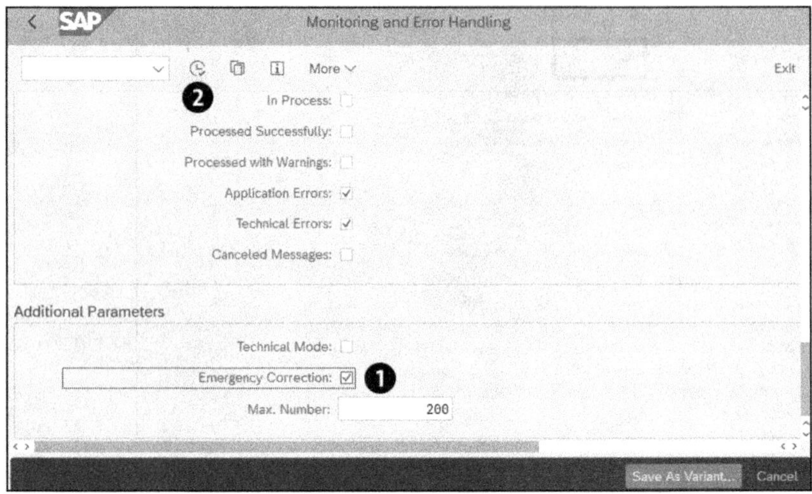

Figure 6.19 Emergency Correction

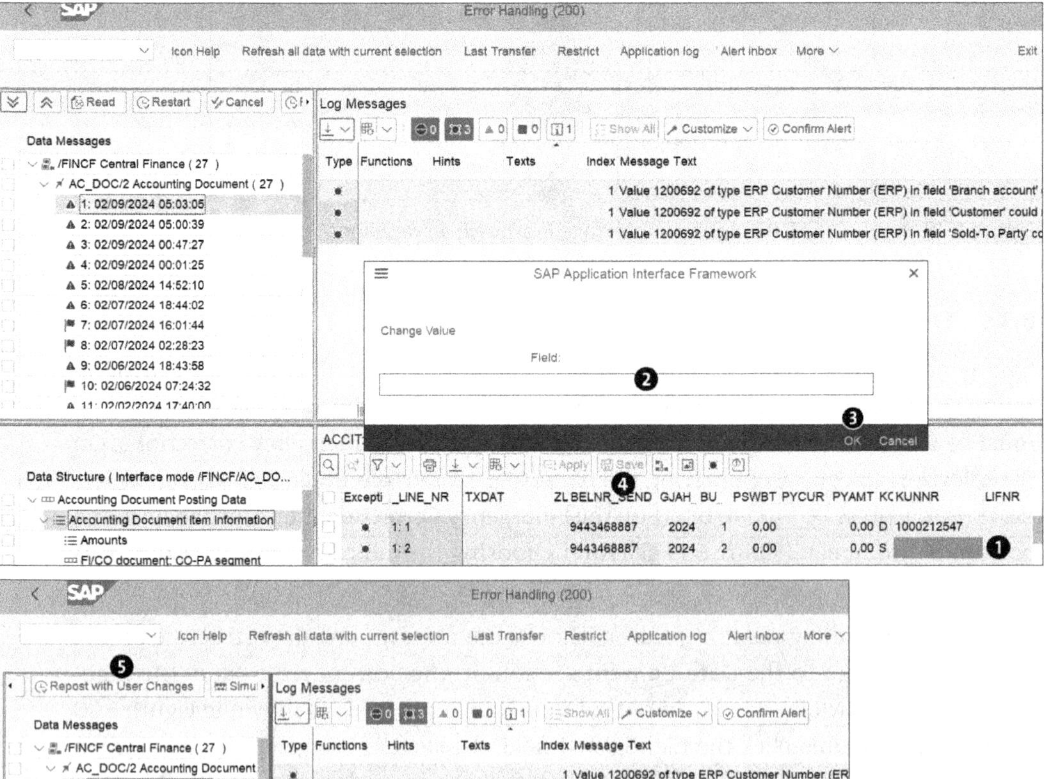

Figure 6.20 Emergency Correction Restart

[«]

Note

As mentioned earlier, emergency corrections must be avoided in the production system as they may lead to data inconsistencies between the source system and Central Finance system. It's strongly recommended to identify the real root cause of the error such as incomplete data mapping or configurations to fix and restart the messages rather than going through emergency correction mode. Generally, emergency correction mode is used in test systems for simulating results.

6

6.3.7 Debugging SAP Application Interface Framework Messages

While analyzing SAP Application Interface Framework errors, it's challenging to find the root cause in some cases. For standard unresolved errors, you'll open an OSS message with SAP support. On the other hand, if there is custom code involved, you can engage an SAP ABAPer to debug an SAP Application Interface Framework message. To debug an SAP Application Interface Framework message stuck with a technical or application error, there is an option using class /AIF/CL_APPL_ENGINE_XML to debug the error message. This is the easiest method, but the main drawback of this approach is to have **Goto statement** authorization in the system, which might be difficult to get in the production system. As a best practice, you can reproduce the error in a lower environment and try to debug in that system.

To begin debugging an SAP Application Interface Framework message, as shown in Figure 6.21, you need to set the breakpoint at line **55** in **Class /AIF/CL_APPL_ENGINE_XML** and **Method /AIF/IF_APPLICATION_ENGINE~RESTART**.

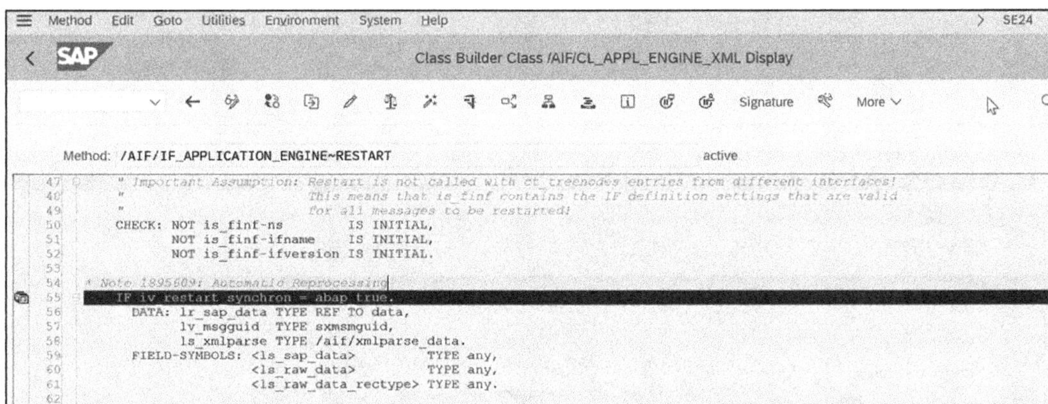

Figure 6.21 SAP Application Interface Framework Debugging: Setting the Breakpoint

Next, choose the SAP Application Interface Framework message that is in error from the Transaction /AIF/ERR (Monitoring and Error Handling) screen, as shown in Figure 6.22.

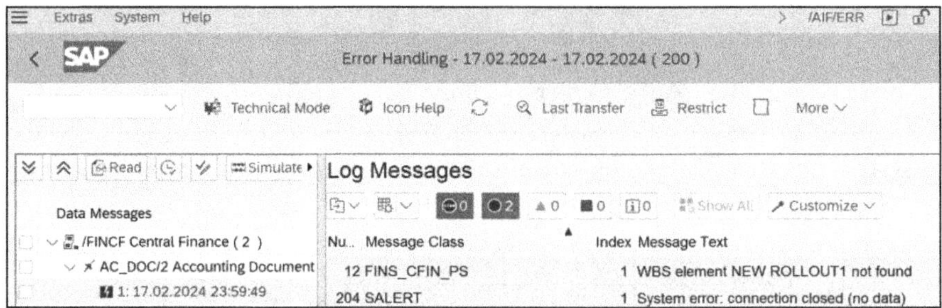

Figure 6.22 SAP Application Interface Framework: Debugging Restart

After restarting the SAP Application Interface Framework message, the debugger will stop in method /AIF/IF_APPLICATION_ENGINE~RESTART of class /AIF/CL_APPL_ENGINE_XML at line number 55, as shown in Figure 6.23 ❶. Next, in the debugger, jump to line number 63 by navigating to **Goto statement** ❷ or using Shift+F12.

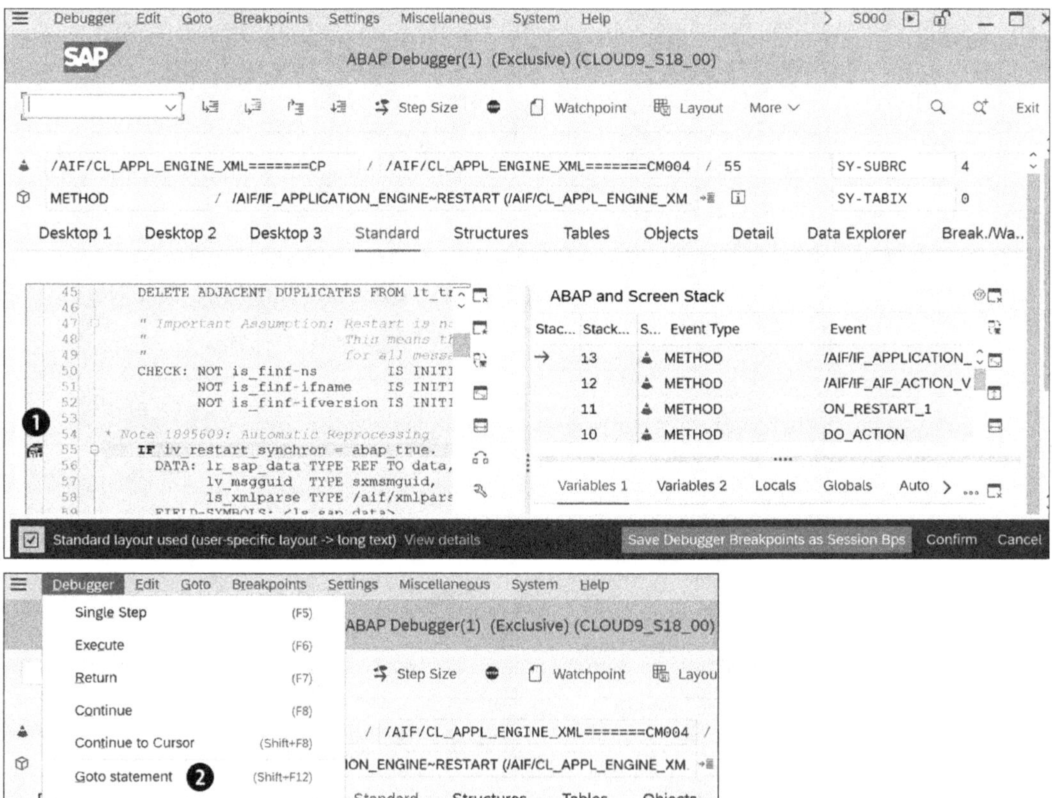

Figure 6.23 SAP Application Interface Framework Debugging: Goto Statement

Once you've reached this line, based on the error message in the SAP Application Interface Framework log, you can set the breakpoint on a particular message class and number. In our example, the SAP Application Interface Framework message failed with error message class **FINS_CFIN_PS** number **012**, so you need to set the breakpoint at this message class and number. Next, click on **Continue** by navigating from toolbar **Debugger • Continue** or using [F8]. The system will reach the point on which the error is displayed, as shown in Figure 6.24 ❶. This way, you can backtrack the call stack to identify the root cause of the problem.

In many cases, you'll debug an SAP Application Interface Framework message in a system where **Goto statement** authorization isn't granted. In such cases, the following function module approach may be useful to debug SAP Application Interface Framework messages:

1. Execute function module /AIF/RESTART_MSG. from Transaction SE37. This brings up the **Test Function Module: Initial Screen** ❷. On this screen, enter values in the **Import Parameters** fields.
   - **IV_MSGGUID**
     Enter SAP Application Interface Framework Message GUID information.
   - **IV_RTCFGR_NS**
     Enter Central Finance namespace "/FINCF".
   - **IV_NS**
     Enter Central Finance namespace "/FINCF".
   - **IV_IFNAME**
     Enter interface name "AC_DOC".
   - **IV_IFVERSION**
     Enter interface version "2".
2. After entering the required details on the function module, click on the **Debugging** button. The system takes you to SAP Application Interface Framework debugging mode inside the debugger.
3. Now, you can put the breakpoint at a particular message class and number, or you can put the debugger at the BAdI implementation if you want to debug your own custom code logic.
4. Press [F8] to reach the point where the breakpoint is added. Continue to analyze the code to find the root cause.

[+]

Tip

While debugging using the debug method, the SAP Application Interface Framework GUID error will go into **In Process** status occasionally if the debugger session is terminated suddenly due to technical errors or due to lost system connection. In such situations, go to Transaction /AIF/ERR, restart the GUID, and reset the error status back to error status before analyzing the GUID using the same functional module approach

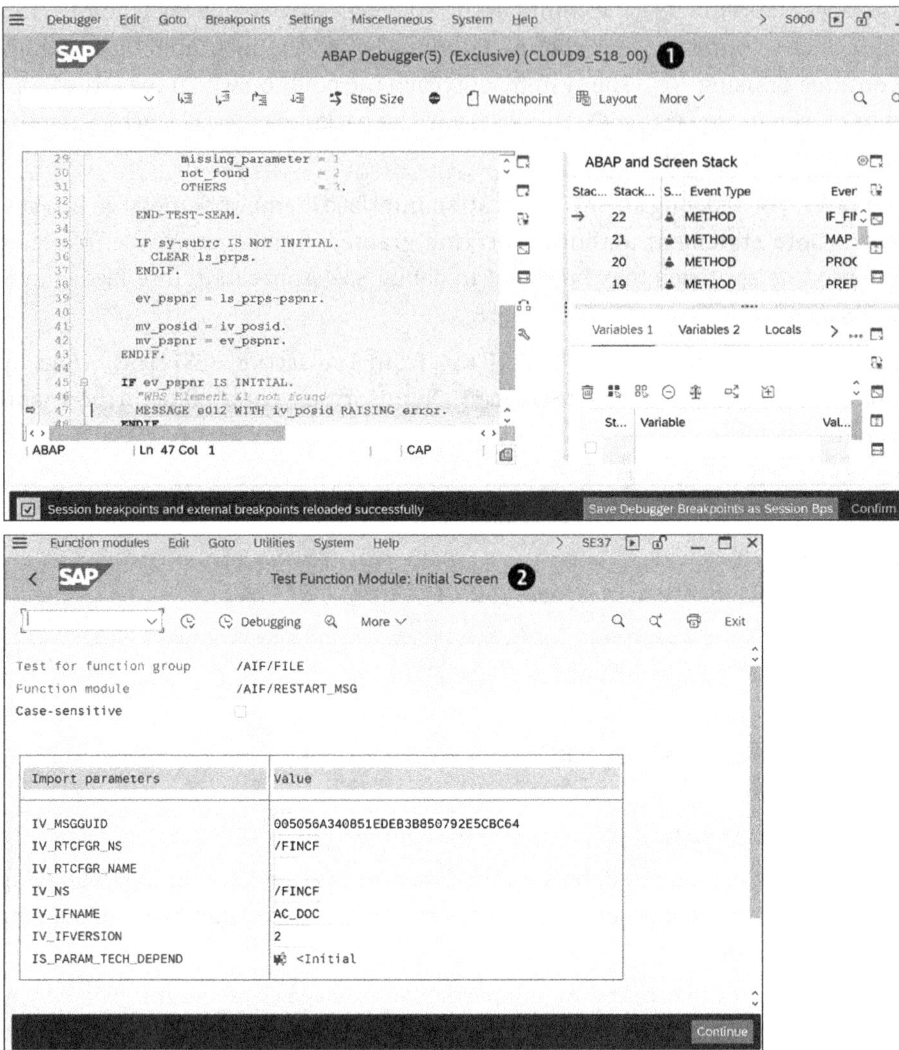

Figure 6.24 SAP Application Interface Framework Debugger: Setting Up Message Class and Number

6.4 Reprocessing SAP Application Interface Framework Messages

SAP Application Interface Framework provides a reprocessing functionality to restart SAP Application Interface Framework messages via a background job. First, you identify the list of error messages that you want to be included in reprocessing of SAP Application Interface Framework messages after discussing with relevant business stakeholders. For example, Table 6.15 provides a list of message classes and message numbers you'll include in reprocessing SAP Application Interface Framework messages. Whenever any of these errors occur, based on the configuration time interval, the background job will restart the SAP Application Interface Framework messages. During the time interval, you'll fix the problem, so the interface picks up the changes and processes the SAP Application Interface Framework messages.

Message Class	Message Number	Message Description
00	172	You're not authorized to use transaction $1
F5	83	You have no authorization for posting in company code $1
F5	351	Account $1 $2 is blocked for posting
F5	507	G/L account $1 is not defined in chart of accounts $2
F5	102	Customer & is not defined in company code &
F5	104	Vendor & is not defined in company code &
F5	28	Account & & is currently blocked by user &
F5	351	Account & & is blocked for post
KI	260	Cost center $1/$2 blocked against direct postings on $3
FINS_ACDOC_POST	050	Referenced invoice not yet posted (CC &1, FY &2, Doc. &3, Item &4)
FINS_CFIN_AC_DOC	016	Value $2 of type $1 in field '$3' could not be mapped; see error $4
FINS_CFIN_AC_DOC	017	Value $2 of type $1 in field '$3' could not be mapped

Table 6.15 SAP Application Interface Framework Messages for Reprocessing

To achieve this, you'll carry out the following configuration:

1. Create a runtime configuration group defined through **SAP Easy Access** menu by navigating to **Cross Applications Components • SAP Application Interface Framework • Administration • Configuration • Runtime Configuration Group** or using Transaction /AIF/PERS_CGR. Refer to Section 6.2.2 for detailed instructions on this step. For this exercise, you'll create a runtime configuration group, for example, **RES**.
2. Perform the **Define Reprocessing Action** configuration. To configure this, navigate to configuration from **SAP Easy Access Menu** by navigating to **Cross Applications Components • SAP Application Interface Framework • Administration • Configuration • Define Reprocessing Action** or using Transaction /AIF/REP_AC_DEF.

3. After entering the transaction, the system prompts you with a popup screen to enter **Namespace**: enter the Central Finance namespace "/FINCF", and then click the **Continue** button.
4. The system brings up another warning screen with the **All Fields which define the area have initial values. Check Input Values** message. Click **Continue** on this screen.
5. Next, you'll arrive at **the Change View "AIF Automatic Reprocessing: Define Reprocessing Action": Overview** screen, as shown in Figure 6.25. On this screen, click the **New Entries** ❶ icon.
6. On the resulting **New Entries: Details of Added Entries** screen, enter values in the following fields ❷:
   - **Repr. Action Name**: Enter the name of the automatic reprocessing action "RESTART".
   - **Function Module**: Enter the SAP-delivered function module "/AIF/RESTART_MSG".
   - **Namespace**: Enter the Central Finance namespace as "/FINCF".
   - **Runtime Config. ID**: Enter the runtime configuration group ID "RES" to define how messages are processed in SAP Application Interface Framework.
7. Once all fields are entered, click the **Save** ❸ button to save the entries.

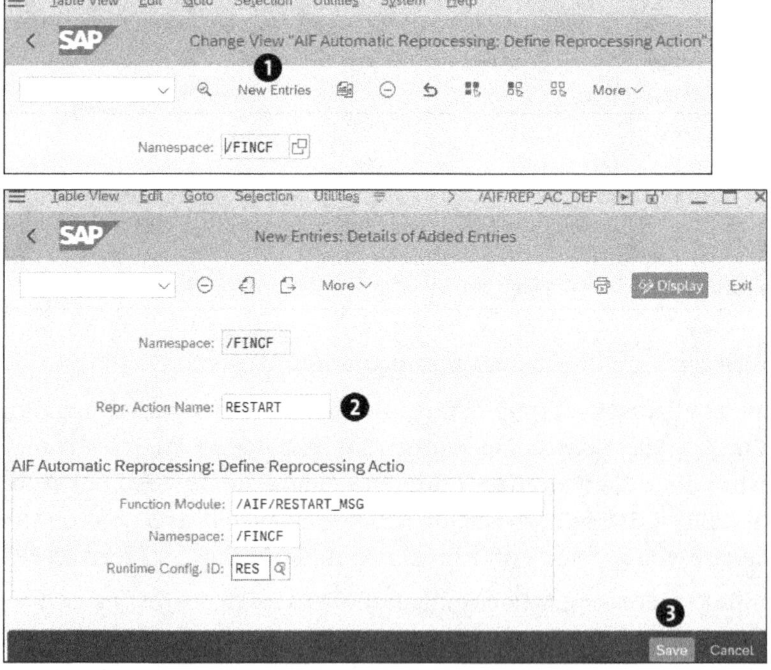

Figure 6.25 SAP Application Interface Framework: Automatic Reprocessing

8. Next, you'll assign the reprocessing action for the defined automatic reprocessing action. Navigate to the configuration from **SAP Easy Access** menu and choose **Cross Applications Components • SAP Application Interface Framework • Administration • Configuration • Assign Reprocessing Action** or use Transaction /AIF/REP_AC_ASGN.
9. After reaching the transaction, the system prompts with the **Determine Work Area: Entry** screen. On this screen, enter values in the **Namespace**, **Interface Name**, and **Interface Version** fields. Click the **Continue** button.
10. On the resulting **Change View "AIF Automatic Reprocessing Assign Reprocess: "Overview** screen, click the **New Entries** button, as shown in Figure 6.26 ❶.

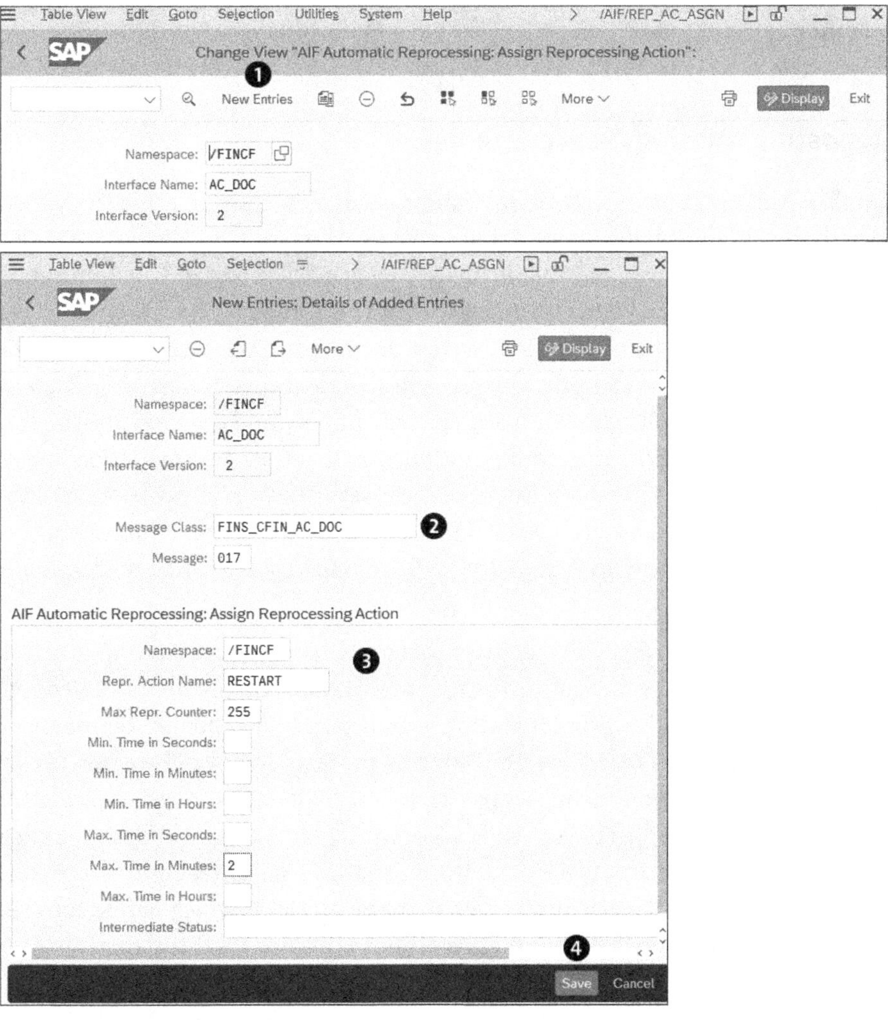

Figure 6.26 Assign Reprocessing Actions

11. On the resulting **New Entries: Details of Added Entries** screen, enter values in the following fields:
    - **Message Class**: Enter the message class, for example, "FINS_CFIN_AC_DOC" ❷.
    - **Message**: Enter the message number, for example, "017".
    - **Namespace**: Enter Central Finance namespace "/FINCF".
    - **Repr. Action Name**: Enter the reprocessing action name you've created in the previous step as "RSTART" ❸.
    - **Max. Repr. Counter**: Enter a maximum number of reprocessing trials as "255".
    - **Max. Time in Minutes**: Enter the maximum time as "2". You can maintain minimum and maximum waiting time with the fields here.
12. After entering values in all fields, click the **Save** ❹ button to save the entries.

6.5 Reposting Missing Cleared Items

During online replication, a follow-on document such as a reset clearing document ends up with the **Reference document not yet posted** (Message IDs **CFIN_AC_DOC 051**) or **Reset Clearing Failed** (Message ID **/FINS_CFIN_AC_DOC 015**) error in SAP Application Interface Framework. The most common reason for this error is the document posted before the initial load extraction period, so the document isn't available in Central Finance. Other reasons could be due to sequencing of documents posted because of serialization issues. However, a reset is being carried out now, and the system is looking for the original document. In this configuration, you'll set up automatic loading of missing cleared items so that the system posts the reset automatically without manual user intervention. Alternatively, to perform manual loading of cleared items, you can execute SAP-delivered report RFINS_CFIN_CORR_MISSING_CLRITM to load missing cleared items.

To achieve this, in the first step, you'll define the reprocessing action. To configure this, use **SAP Easy Access** menu by navigating to **Cross Applications Components • SAP Application Interface Framework • Administration • Configuration • Define Reprocessing Action** or using Transaction /AIF/REP_AC_DEF. As a prerequisite, you'll create the runtime configuration group **RPR** using Transaction /AIF/PERS_CGR. After entering the transaction, the system prompts you with a popup screen to enter the **Namespace**. Enter the Central Finance namespace of "/FINCF", and then click the **Continue** button. The system brings up another warning screen with the **All Fields which define the area have initial values. Check Input Values** message. Click **Continue** on this screen.

Next, you'll arrive at **the Change View "AIF Automatic Reprocessing: Define Reprocessing Action": Overview** screen, as shown in Figure 6.27. On this screen, click the **New Entries** ❶ icon. On the resulting **New Entries: Details of Added Entries** screen, enter values in the following fields ❷:

- **Namespace**
  Enter the Central Finance namespace of "/FINCF".
- **Repr. Action Name**
  Enter the name of the open item reprocessing action "LOAD_OI".
- **Function Module**
  Enter SAP-delivered function module "FINS_CFIN_AIF_REPROC_LOAD_ITEM" ❸.
- **Runtime Config. ID**
  Enter runtime configuration group ID "RPR" to define how messages are processed in SAP Application Interface Framework.

Once all fields are entered, click the **Save** ❹ button to save the entries.

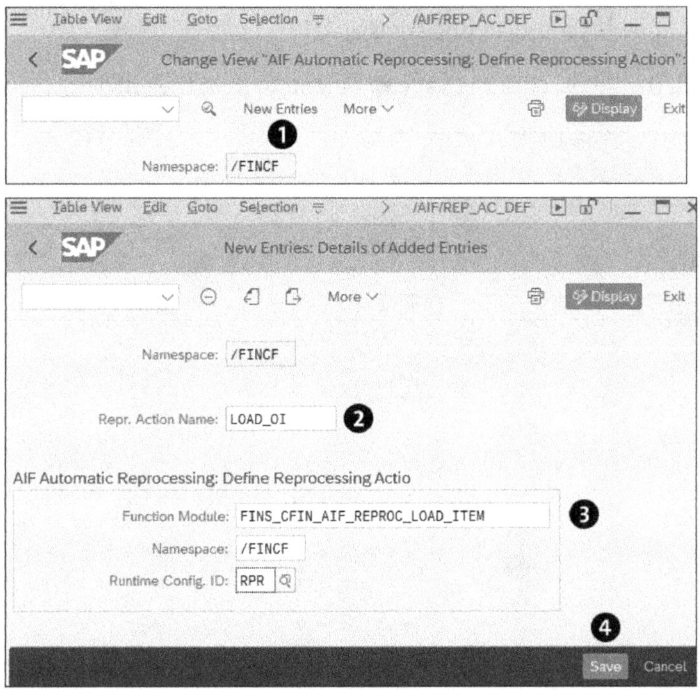

Figure 6.27 SAP Application Interface Framework: Load Open Item

Next, you'll assign the reprocessing action for the defined automatic reprocessing action for **LOAD_OI**. To assign the reprocessing action, navigate to configuration from the **SAP Easy Access** menu by navigating to **Cross Applications Components • SAP Application Interface Framework • Administration • Configuration • Assign Reprocessing Action** or using Transaction /AIF/REP_AC_ASGN. After reaching the transaction, the system prompts you with the **Determine Work Area: Entry** screen. On this screen, enter values in the **Namespace**, **Interface Name**, and **Interface Version** fields. Click the **Continue** button. On the resulting **Change View "AIF Automatic Reprocessing Assign Reprocess: "Overview** screen, click the **New Entries** button, as shown in Figure 6.28 ❶. Next, on the resulting **New Entries: Details of Added Entries** screen, enter values in the following fields:

- **Message Class**
  Enter the message class, for example, "FINS_CFIN_AIF" ❷.
- **Message**
  Enter the message number, for example, "015".
- **Namespace**
  Enter the Central Finance namespace of "/FINCF".
- **Repr. Action Name**
  Enter the reprocessing action name you created in the previous step: "LOAD_OI" ❸.
- **Max. Repr.** Counter
  Enter the maximum number of reprocessing trials as "1".

After entering values in all the fields, click the **Save** ❹ button to save the entries.

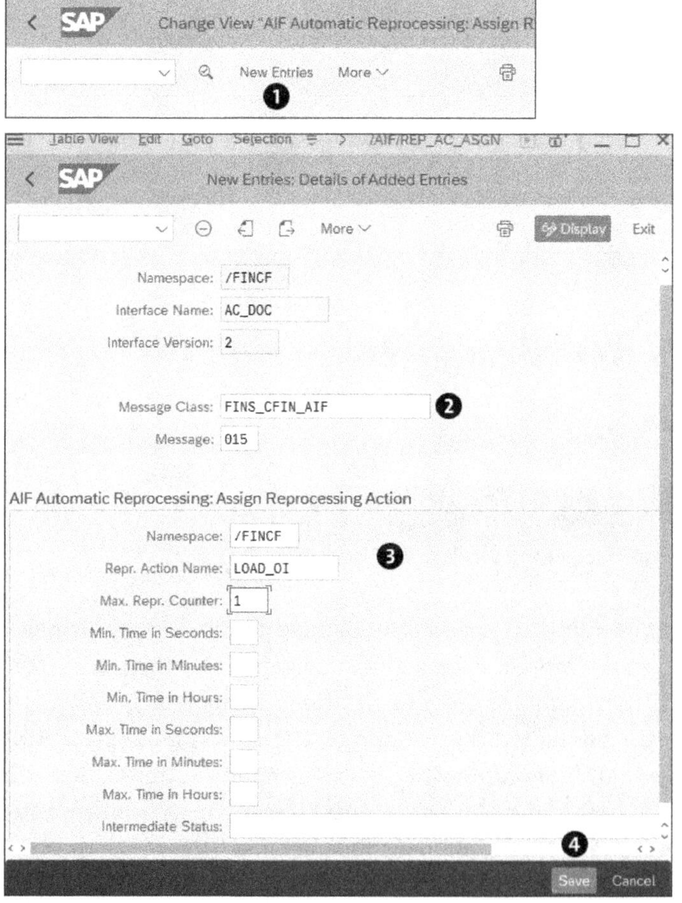

Figure 6.28 Assign Reprocessing Action: Load Open Items

[+]

Tip

Note that documents loaded and posted automatically by the report RFINS_CFIN_CORR_MISSING_CLRITM use the actual system date, which is different in the source system. This is standard behavior. If you want to change the default logic, you can implement custom logic via BAdI BADI_FINS_CFIN_AC_INTERFACE by replacing system date in posting date field BUDAT with the desired date.

Consult SAP Note 2710447 for additional guidance on automatic reload of missing cleared items.

6.6 Retriggering SAP Application Interface Framework Messages

During financial posting replication from the source system to Central Finance, there are situations where you have to resend SAP Application Interface Framework messages. SAP offers Transaction FINS_CFIN_SLT_TOUCH to resend SAP Application Interface Framework messages from the source system to the Central Finance system. The Transaction is executed from central finance system. In addition, it reads the source system staging table CFIN* and sends a new SAP Application Interface Framework message via SAP Landscape Transformation Replication Server. For lower releases, you'll execute the report FIN_CFIN_SLT_ADD_LOGTAB from the source system. The following are common reasons when you would use the report to retrigger SAP Application Interface Framework messages in the source system.

- An SAP Application Interface Framework message was canceled inadvertently.
- There are missing SAP Application Interface Framework messages although a record exists in the source staging table. This could happen when SAP Landscape Transformation Replication Server has an unplanned outage that caused documents to be missing in SAP Application Interface Framework.
- You've determined the SAP Application Interface Framework message contains corrupted data or some garbage characters that is leading to errors.

After running Transaction FINS_CFIN_SLT_TOUCH in the central finance system, you'll arrive at the **Retrigger SLT Replication** screen, as shown in Figure 6.29. On this screen, fill in the **Source Logical System**, **SLT Replication Object** and **Logging Table** fields under the **Generic selection** ❶ section. Next, On the **Additional parameters** ❷ section, enter values in the **Company Code**, **Fiscal Year**, and **Document Number** fields. Next, click the **Execute** ❸ button to continue. After executing the Transaction, the system shows the results of documents transferred from the staging table for replication. Highlight the row you want to retrigger and click the **Post** ❹ button. The system retriggers the selected document, and you can find the retriggered document in Transaction /AIF/ERR.

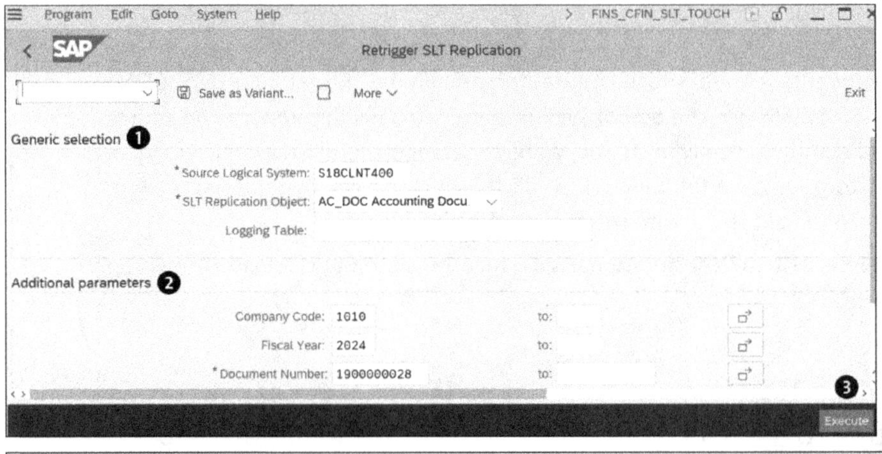

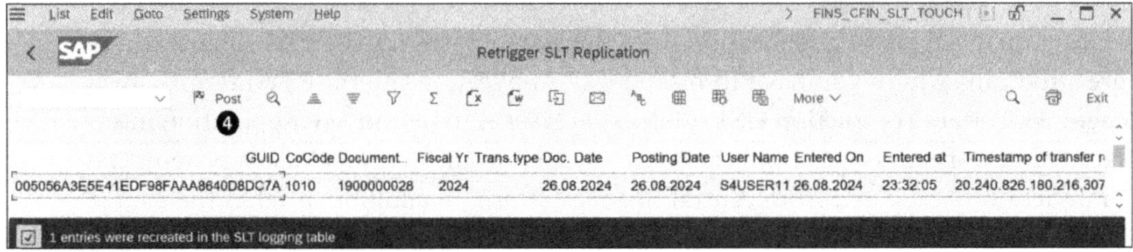

Figure 6.29 Retrigger SLT Replication

> **Note**
>
> To prerequisite to run this report is to ensure you've configured **Maintain RFC Assignments and Settings for Source Systems** for RFC Usage of **SLT Logging** with RFC destination. In addition, remote function call user assigned with the role SAP_IUUC_REPL_ADMIN and user executing transaction has access to the source systems company code.
>
> Note if staging table `CFIN*` is either archived or deleted, you won't be able to use this Transaction.

6.7 Alerts Management

SAP Application Interface Framework management is delivered with alert management, which can be used to send alerts and emails to relevant stakeholders in Central Finance. With alerts management, you can monitor the interface of Central Finance replication objects. By setting up recipients with appropriate authorization, responsible persons can receive emails or alerts in the My Inbox app (app ID: F0862). The interface sends the first error message of the interface. The next message is sent after the first alert is either confirmed or expires. When the source system data is replicated via SAP Landscape Transformation Replication Server to Central Finance, as shown in

Figure 6.30, sometimes posting of accounting documents fails for various reasons in Central Finance. Posting failures that appear in SAP Application Interface Framework can be grouped by message categories and can be alerted to responsible stakeholders. As you can see in Figure 6.30, you can group FI master data errors into one message category to alert the finance team. Likewise, you can group customer and vendor business partner errors into a different message category, which can be routed to the business partner master data team. Similarly, all configuration-related errors can be grouped into a different message category for IT teams. Message recipients are set up to receive alerts based on requirements. The responsible teams take appropriate action to resolve errors. This will expedite the error resolution process by eliminating manual communication through emails.

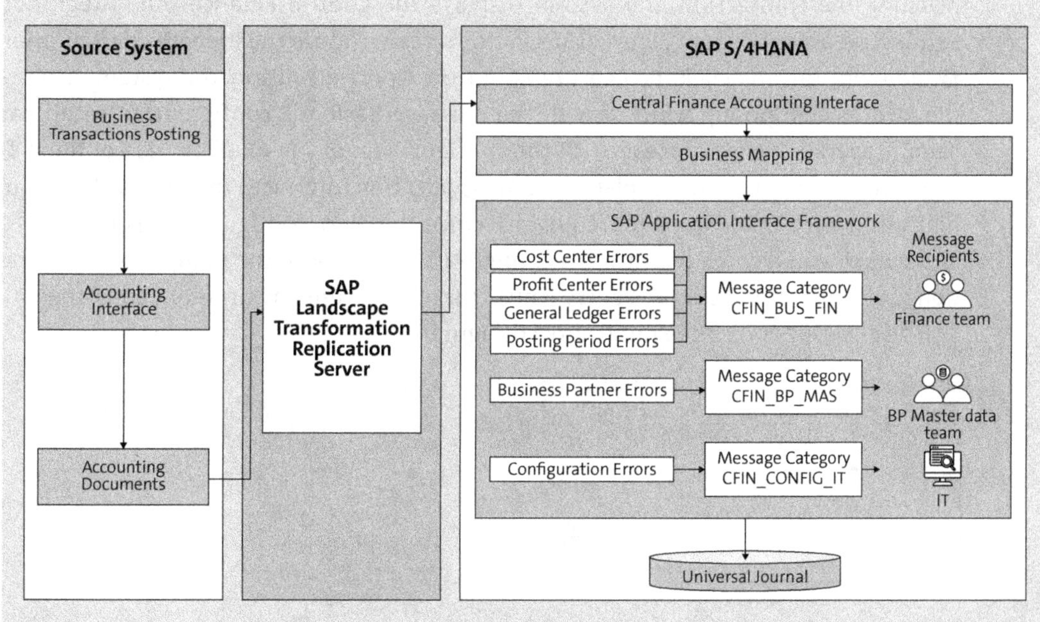

Figure 6.30 SAP Application Interface Framework: Alerts Management

To set up alerts in SAP Application Interface Framework, you'll require the following configuration in the sequence specified:

- Unlock namespace
- Define message category
- Assign message category
- Configure alerts
- Create customer assignment table
- Set up table `/AIF/ALRT_USRNOT`
- Define and assign recipients

- Email settings
- BAdI enhancement based on business requirements

Now, let's explore each configuration to enable SAP Application Interface Framework alert management.

6.7.1 Unlock Namespace /FINCF

Central Finance namespace **/FINCF** must be unlocked first to perform SAP Application Interface Framework alert/email-related configuration. To unlock the namespace, go to the configuration from **SAP Easy Access** menu by choosing **SAP Menu • Accounting • Central Finance • Error Handling** or using Transaction FINS_CFIN_AIF_SETUP. After reaching the transaction, the system displays the **Central Finance: AIF Integration Setup** screen, as shown in Figure 6.31. On this screen, choose the **Prepare configuration** ❶ radio button, and click the **Execute** ❷ button. Upon execution, the system displays the **Information** popup window with the message **Basic AIF configuration setup has been prepared. Please proceed with the configuration of AIF** ❸. Click the **Continue** ❹ icon. This enables you to complete the SAP Application Interface Framework alerts configuration with mandatory prerequisites. Once the alert configuration is completed and tested, you'll come back to this configuration to choose the **Complete configuration** radio button and save the entry. Thus, this will prevent any unintended changes to SAP Application Interface Framework configuration.

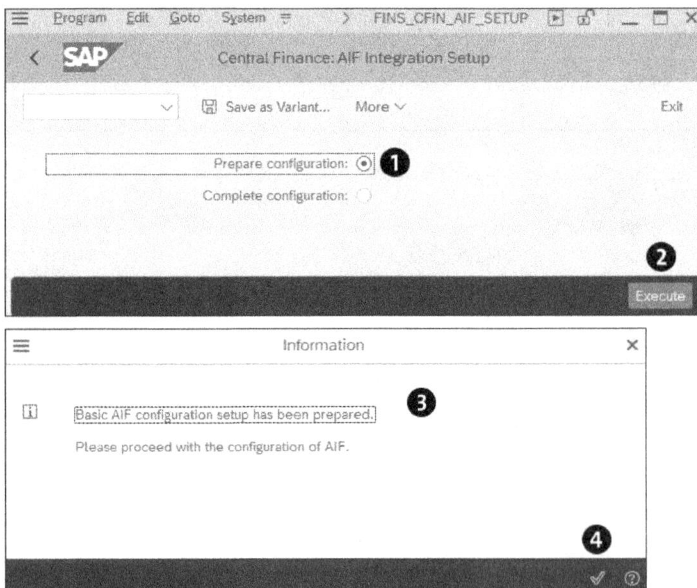

Figure 6.31 Unlocking SAP Application Interface Framework Configuration

6.7.2 Define Message Category

In this step, you'll create a message category that is used in alerts for users. A message category can group similar messages for the purpose of assigning the users for alerts. To create a message category, navigate to the configuration from the **SAP Easy Access** menu by navigating to **Cross Applications Components • SAP Application Interface Framework • Administration • Configuration • Define Message Category** or using Transaction /AIF/MSG_CAT_DEF. After entering the transaction, the system prompts you with a popup screen to enter **Namespace.** Enter the Central Finance namespace as "/FINCF", and then click the **Continue** button. Next, you'll arrive at the **Change View "AIF Message Category": Overview** screen. On this screen, click the **New Entries** icon. On the resulting **New Entries: Details of Added Entries** screen, enter values in the following fields, as shown in Figure 6.32:

- **Namespace**
  Enter the Central Finance namespace as "/FINCF".
- **Message Category**
  Enter a message category, for example, "CFIN_BP_MAS", ❶ to manage business partner errors.
- **Msg. Cate. Desc** ❷
  Enter the message category description, for example, "Business partner master data".

Once all values are entered, click the **Save** ❸ button to save the entry. Repeat the steps for other message categories you want to create for sending alerts.

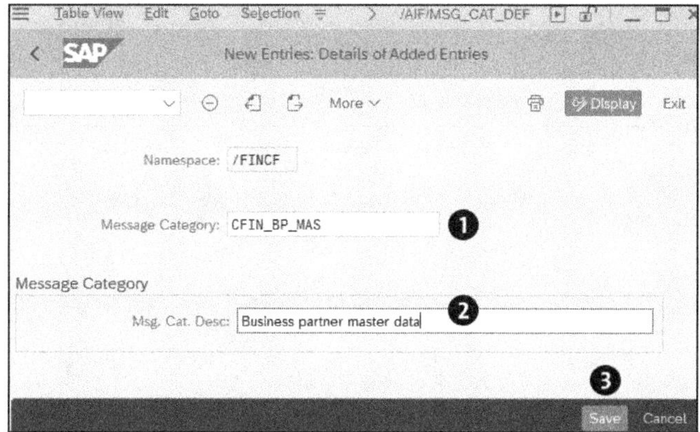

Figure 6.32 Message Category

6.7.3 Assign Message Category

Next, you'll assign the message category you've just defined with the message class and message number. To assign the message category to messages, navigate to configuration from **SAP Easy Access** menu by choosing **Cross Applications Components • SAP**

Application Interface Framework • Administration • Configuration • Assign Message Category or using Transaction /AIF/MSG_CAT_ASGN. After reaching the transaction, the system prompts you with the **Determine Work Area: Entry** screen. On this screen, enter values in the **Namespace** and **Message Category** fields, as shown in Figure 6.33 ❶. Click the **Continue** button. On the resulting **Change View "Message Category Assignment: "Overview** screen, click the **New Entries** button. Next, on the resulting **New Entries: Details of Added Entries** screen, enter values in the following fields:

- **Message Class**
  Enter the message class, for example, "KI" ❷.
- **Message**
  Enter the message number, for example, "260".

Once all message classes and messages are captured, click the **Save** ❸ button to save the entry.

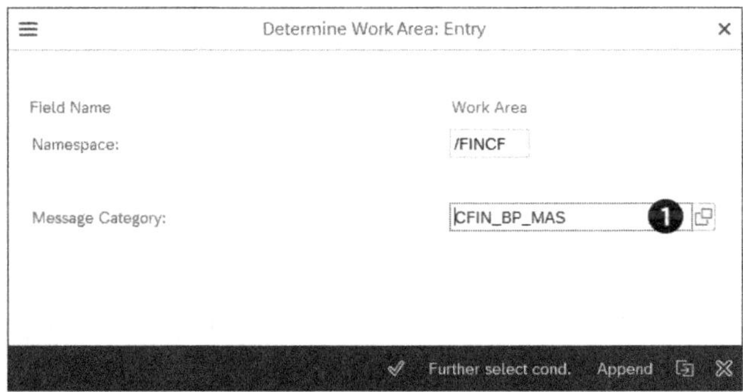

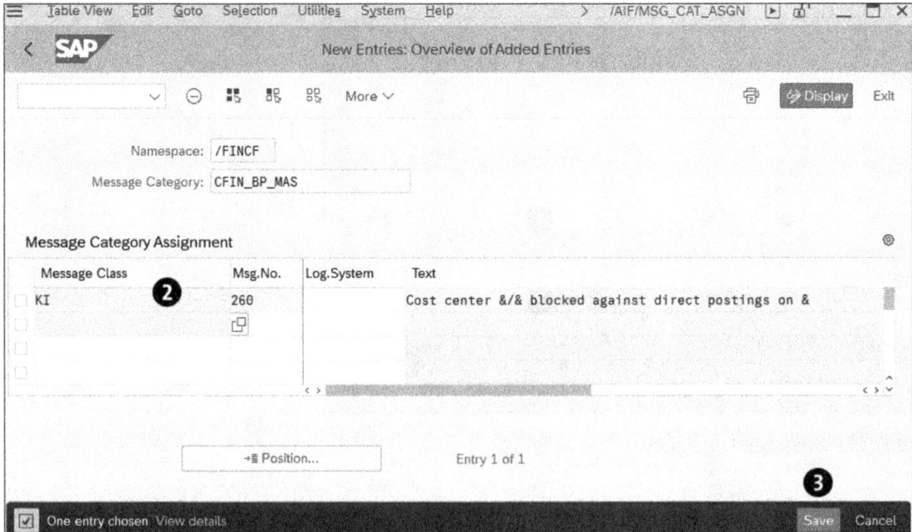

Figure 6.33 Assign Message Category

6.7.4 Define Recipients

In this step, you'll define recipients who are responsible for the interface to receive alert messages in case errors occur in SAP Application Interface Framework. Recipients need to be decided before configuring recipients in this step. To define recipients, navigate from the SAP Application Interface Framework customizing (Transaction /AIF/CUST) menu path **SAP Application Interface Framework • Error Handling • Define Namespace Specific Features**. After reaching the customizing activity, the system prompts you with the **Determine Work Area: Entry** screen. On this screen, enter value "/FINCF" for Central Finance in the **Namespace** field. Click the **Continue** button. You'll arrive at the **Change View "Define Interface Specific features: "Overview** screen, as shown in Figure 6.34. On this screen, click **Define Recipients** ❶ under **Dialog Structure** on the left pane and click the **New Entries** ❷ button. Next, on the resulting **New Entries: Details of Added Entries** screen, enter values in the following fields ❸:

- **Recipient for Alert**
  Enter the value "CFIN_BP_MAS" in this field.
- **Alert Recipient Description**
  Enter "Business partners errors" as the description.

Click the **Save** ❹ button to save the entry.

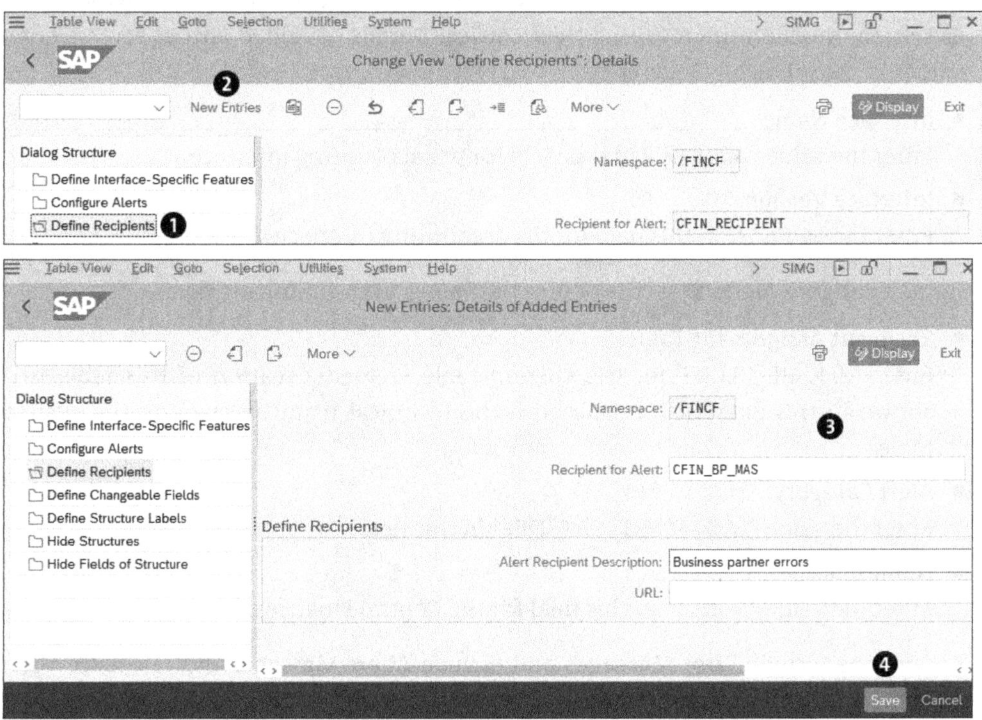

Figure 6.34 Define Recipients

6.7.5 Configure Alerts

After creating recipients, you'll assign each interface with an alert category. In addition, to link the interface, recipients, message category, and fallback recipient, SAP suggests you create a custom assignment table that is later assigned in this configuration. SAP provides two ways to create the custom table by copying from a template table:

- By copying from table /AIF/T_ALRT_DEF for key field dependent recipients
- By copying table /AIF/RECA_TMPL if there are message categories and a fallback recipient such as CFIN_RECIPIENT is used as a fallback recipient

In our example, we used template table /AIF/RECA_TEMPL to copy to a custom assignment table (ZFI_AIF_RECIP) as we've defined the message category earlier. Once you've created the assignment table, you'll assign interface and recipients to this step. To link the interface with the custom assignment table, navigate from SAP Application Interface Framework customizing (Transaction /AIF/CUST) menu path **SAP Application Interface Framework • Error Handling • Define Namespace Specific Features.** After reaching the customizing activity, the system prompts you with the **Determine Work Area: Entry** screen. On this screen, enter the value "/FINCF" for Central Finance in the **Namespace** field. Click the **Continue** button. You'll arrive at the **Change View "Define Interface Specific Features: "Overview** screen, as shown in Figure 6.35. On this screen, click **Configure Alerts** under **Dialog Structure** on the left pane, and click the **New Entries** ❶ button. Next, on the resulting **New Entries: Details of Added Entries** screen, enter values in the following fields ❷:

- **Interface Name**
  Enter the value "AC_DOC" in this field for the accounting interface.
- **Interface Version**
  Enter the value "2" in this field for the accounting interface version.

In the **Configure Alerts** ❸ section, enter the values in the following fields:

- **Recipient Assignment Table**
  Enter "ZFI_AIF_RECIP" for the custom table created. Creation of this table isn't shown in this exercise. Consult with the technical team on creating the custom table.
- **Alert Category**
  Enter the value "/AIF/ALERT_CAT_DEF" in this field.
- **Namespace**
  Enter the value "/FINCF" in this field for the Central Finance namespace.

Choose the **Include Error Messages** and **Include Abort Messages** indicators ❹. Then, click the **Save** ❺ button to save the entry.

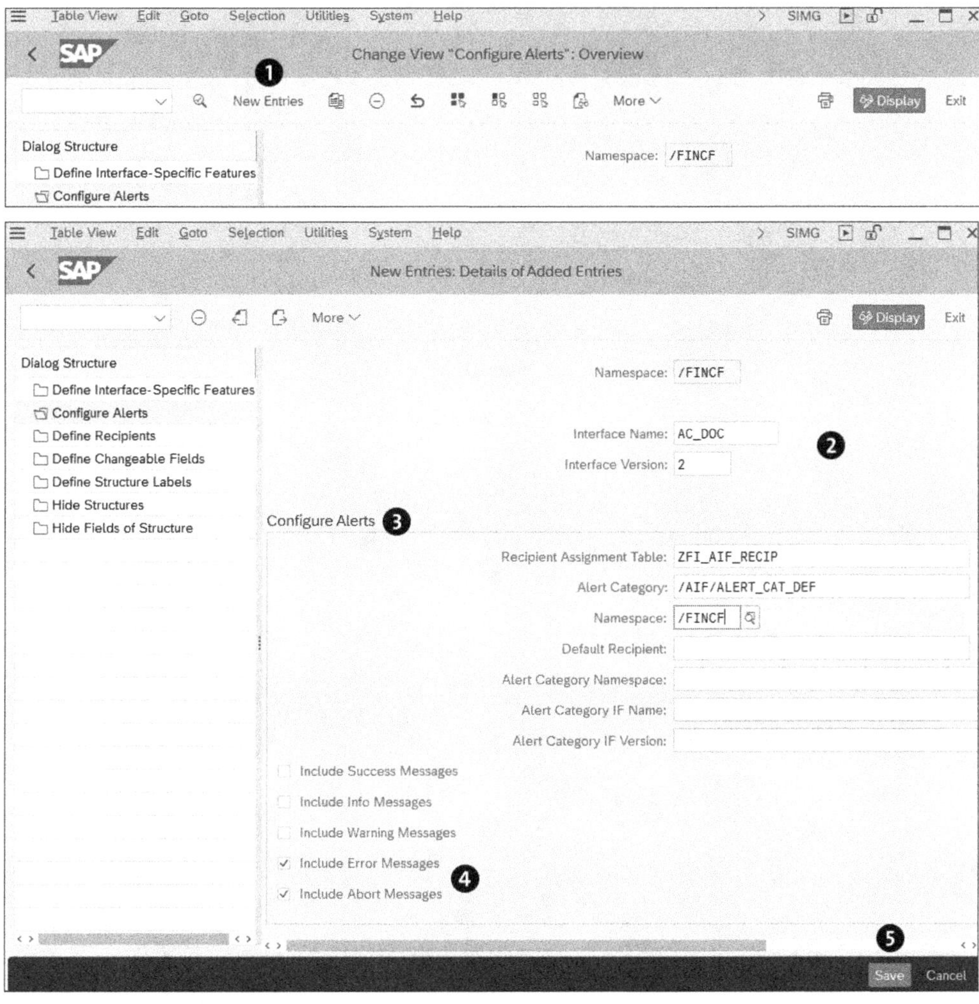

Figure 6.35 Configure Alerts

6.7.6 Custom Table Assignment

In this step, you'll assign custom table ZFI_AIF_RECIP maintained with assigned message categories to recipients. The custom assignment table links message categories and recipients. You'll maintain the entries with the help of the technical team as the entries maintained through Transaction SE11. After reaching the transaction, enter the custom table name in the **Table View** field, as shown in Figure 6.36 ❶. Navigate to the table maintenance view, and then add the key fields you want to assign with **Recipients** and **Message Category**. Notice that we've assigned **CFIN_RECIPIENT** ❷ as the **Fallback** recipient to receive the alerts. Click the **Save** ❸ button to save the entry.

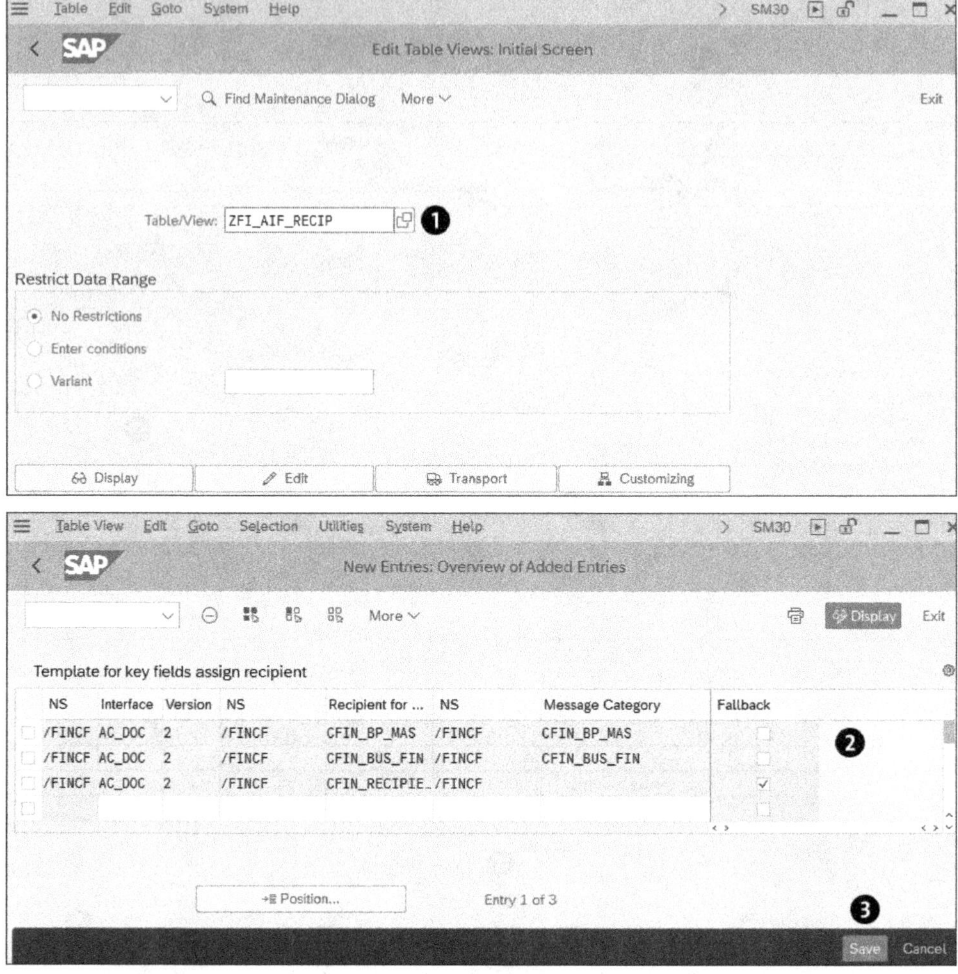

Figure 6.36 Custom Assignment Table

6.7.7 Table ALRT_USRNOT

In this step, you'll maintain entries in table /AIF/ALRT_USRNOT (maintenance view /AIF/V_ALRT_UN). This table contains email settings necessary for sending alerts by namespace, interface, version, recipient, and so on. You'll maintain the table with the technical team's help. Figure 6.37 displays the table /AIF/ALRT_USRNOT ❶ entry screen, and the second screen shows entries maintained ❷ on this table. **Send Email** and **Send Always** columns show which interface and recipients are set up to receive alerts.

To check the status and recipient of an alert raised due to an error, the interface monitor will glow with a light bulb icon in Transaction /AIF/IFMON, as shown in Figure 6.38. You can also view the alert in Transaction /AIF/ERR. In addition, if email settings are configured, an email is sent for the next alert. You can also access SAP Application Interface Framework logs through Transaction SLG1 (Analyze Application Logs) with object **Alert** and subobject **Processing**. Moreover, Transaction SOST (Send Requests Overview) can be used to view the status of the SAP Application Interface Framework email sent to recipients.

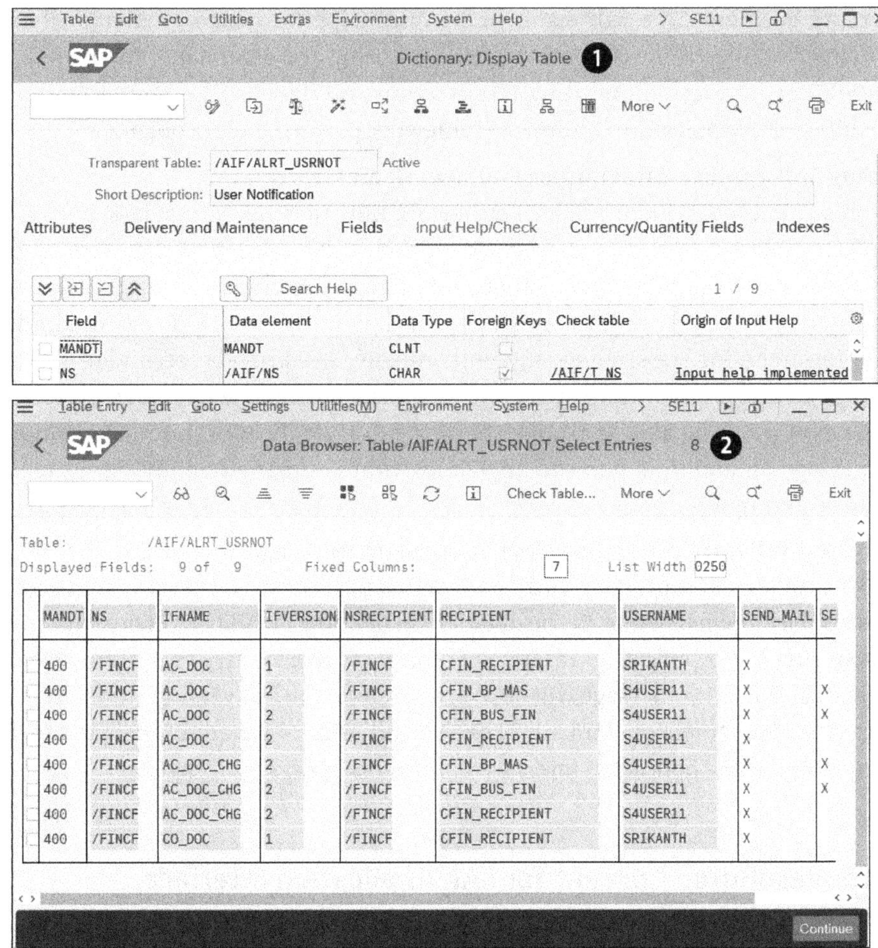

Field	Data element	Data Type	Foreign Keys	Check table	Origin of Input Help
MANDT	MANDT	CLNT			
NS	/AIF/NS	CHAR	✓	/AIF/T_NS	Input help implemented

MANDT	NS	IFNAME	IFVERSION	NSRECIPIENT	RECIPIENT	USERNAME	SEND_MAIL	SE
400	/FINCF	AC_DOC	1	/FINCF	CFIN_RECIPIENT	SRIKANTH	X	
400	/FINCF	AC_DOC	2	/FINCF	CFIN_BP_MAS	S4USER11	X	X
400	/FINCF	AC_DOC	2	/FINCF	CFIN_BUS_FIN	S4USER11	X	X
400	/FINCF	AC_DOC	2	/FINCF	CFIN_RECIPIENT	S4USER11	X	
400	/FINCF	AC_DOC_CHG	2	/FINCF	CFIN_BP_MAS	S4USER11	X	X
400	/FINCF	AC_DOC_CHG	2	/FINCF	CFIN_BUS_FIN	S4USER11	X	X
400	/FINCF	AC_DOC_CHG	2	/FINCF	CFIN_RECIPIENT	S4USER11	X	
400	/FINCF	CO_DOC	1	/FINCF	CFIN_RECIPIENT	SRIKANTH	X	

Figure 6.37 SAP Application Interface Framework: Table ALRT_USRNOT

Message Summary | Last Transfer | My Messages

Interfaces; last refresh on: 09/12/2022 at: 10:38:24	In Process	Alert	Mail	Status...	All Messages	Warnings	Errors
Central Finance - /FINCF	35			●OO	Σ 20952157	240	8
Accounting Document - AC_DOC/1				OOO			
Accounting Document - AC_DOC/2	18			●OO	Σ 20812701	240	8
Accounting Document Changes - AC_DOC_CHG/1				OOO			
Accounting Document Changes - AC_DOC_CHG/2	17			OO■	Σ 139456		

Figure 6.38 Alerts in the Interface Monitor

If the standard solution doesn't meet your requirements, SAP provides BAdIs /AIF/ALERT_DET_RECIPIENTS and /AIF/ALERT_DET_RECIPIENTSFILT to enhance the logic of determining recipients and filter-dependent custom logic to trigger alerts.

6.8 Improving SAP Application Interface Framework Performance

SAP Application Interface Framework XML messages grow significantly based on business activities in the source system. The volume could be in the millions per day, especially in retail business. This will result in significant performance problems when accessing large volumes of messages through monitoring and error handling (Transaction /AIF/ERR) or interface monitoring (Transaction /AIF/IFMON). Let's explore how you can address SAP Application Interface Framework performance problems:

- **Archiving SAP Application Interface Framework messages**
  SAP stores SAP Application Interface Framework XML messages in table /AIF/PERS_XML, and this table may occupy disk space quickly if the volume of transactions is high. To prevent an out of memory situation, SAP recommends archiving SAP Application Interface Framework XML messages on a periodic basis. SAP recommends considering archiving the strategy of your company. SAP provides archiving object /AIF/PERSX to archive table /AIF/PERS_XML entries. Once you've archived an error message, you won't be able to troubleshoot the SAP Application Interface Framework message. In addition, you can align with your company's archiving strategy and plan accordingly.
- **Compressed SAP Application Interface Framework message storage**
  SAP also provides compressing SAP Application Interface Framework messages before they become persistent in the database. This comes with the versioning concept of storing SAP Application Interface Framework messages. The system stores two versions. The latest message will be overwritten on the last version stored in the table. Consult SAP Notes on 2274361, 2279909, and 2274361 for additional guidance on compressed SAP Application Interface Framework message storage.

6.9 Error Resolution Content for SAP Application Interface Framework

SAP Central Finance supports SAP Application Interface Framework error resolution support content to be available in SAP Application Interface Framework error messages. With this feature, users can reach additional error resolution content quickly. The tool helps users expedite the error resolution process in SAP Application Interface Framework messages.

6.9.1 Maintain Error Resolution Support Content

In this configuration, you'll assign URLs that provide additional error resolution content specific to an error message. By assigning a URL in Transaction /AIF/ERR, you'll see the URLs configured with a specific message number. By clicking on the URL link, the system takes you directly to that URL to view the error resolution content. As a prerequisite, you must activate the error resolution support content. In this configuration, you'll maintain the error message class, message number, and a URL to be displayed for the error message, as shown in Figure 6.39. The configuration also allows you to enter a sequence number to display multiple URLs for a single message.

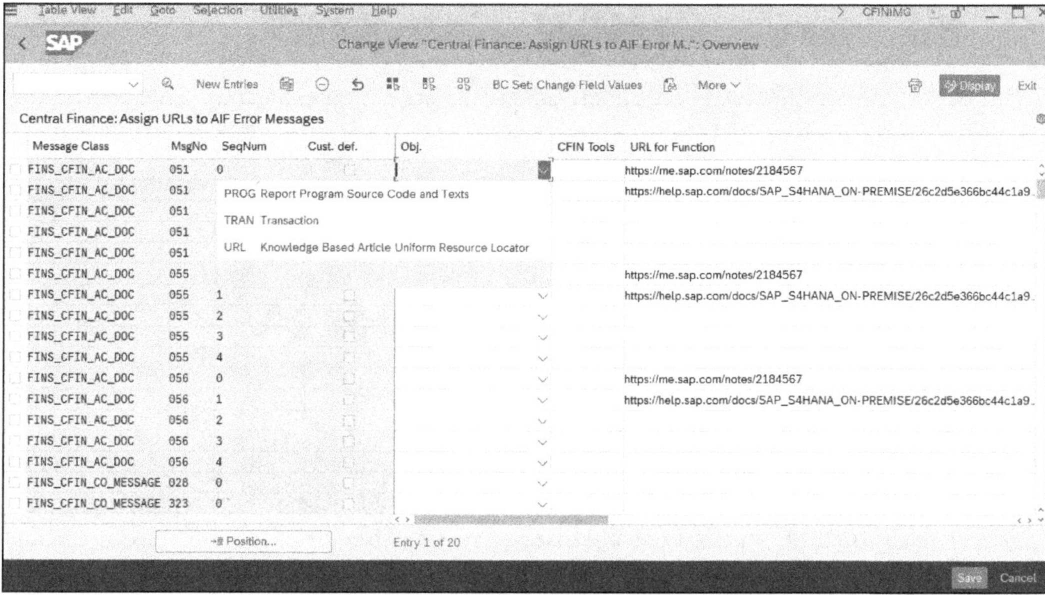

Figure 6.39 Maintain Error Resolution Support Content

6.9.2 Upload Error Resolution Content

In this program, you can view, upload, delete, or work with existing resolution support content and upload URL assignments in bulk. This tool is useful when you want to upload a lot of messages to save time. Figure 6.40 shows the selection screen ❶ and output screen ❷ based on the user selection criteria. The user action has an option to download, which can be used to export to a Microsoft Excel sheet to review the details further.

[+]

Tip

Consult SAP Note 3389718 for additional guidance on the Central Finance error support and solutions.

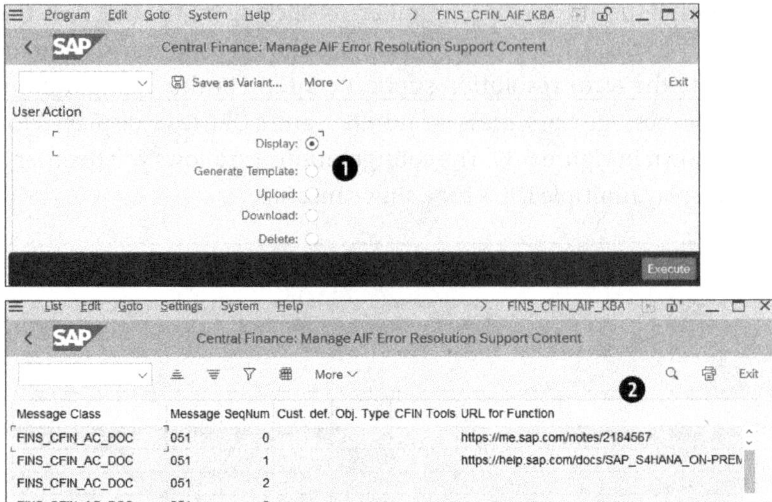

Figure 6.40 Upload Error Resolution Support Content

6.10 Central Finance Cockpit

Central Finance is now delivered with a Central Finance cockpit that comprises all the important support tools in a single place. The cockpit is accessed via report FINS_CFIN_TOOLS_COCKPIT in the Central Finance system. After reaching the cockpit through Transaction SA38 or SE38, you'll see the **Central Finance: Operation and Support Tools(Target System)** screen, as shown in Figure 6.41.

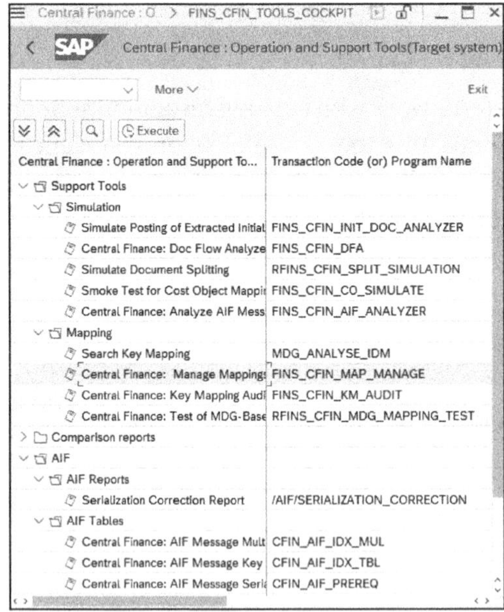

Figure 6.41 Central Finance Cockpit

To see this tool, SAP recommends implementing SAP Note 3358225. Upon implementing the SAP Note, you'll see the availability of the cockpit supporting various transactions and reports, as shown in Table 6.16.

Category	Description of Transaction or Report	Transaction/Report
Simulation	Simulate Posting of Extracted Initial Load Data for Individual Document	Report FINS_CFIN_INIT_DOC_ANALYZER
	Central Finance: Doc Flow Analyzer	Transaction FINS_CFIN_DFA
	Simulation Document Splitting	Report RFINS_CFIN_SPLIT_SIMULATION
	Smoke Test for Cost Object Mapping and CO Document Replication	Report FINS_CFIN_CO_SIMULATE
	Central Finance: Analyze SAP Application Interface Framework Message	Report FINS_CFIN_AIF_ANALYZER
Mapping	Search Key Mapping	Transaction MDG_ANALYSE_IDM
	Central Finance: Manage Mappings	Transaction FINS_CFIN_MAP_MANAGE
	Central Finance: Key Mapping Audit	Transaction FINS_CFIN_KM_AUDIT
	Central Finance: Test of MDG-Based Mapping	Report RFINS_CFIN_MDG_MAPPING_TEST
Comparison FI Report	Comparison of FI Document Header	Transaction FINS_CFIN_DFV_FI_NUM
	Comparison of FI Document Lines	Transaction FINS_CFIN_DFV_FI_DOC
	Comparison of FI Balances	Transaction FINS_CFIN_DFV_FI_BAL
	Comparison of FI Clearing Status	Transaction FINS_CFIN_DFV_CLR

Table 6.16 Central Finance Cockpit

Category	Description of Transaction or Report	Transaction/Report
Comparison CO Report	Comparison of CO Document Header	Transaction FINS_CFIN_DFV_CO_NUM
	Comparison of CO Line Items	Transaction FINS_CFIN_DFV_CO_DOC
	Comparison of CO Balances	Transaction FINC_CFIN_DFV_CO_BAL
SAP Application Interface Framework report	Serialization Correction Report	/AIF/SERIALIZATION_CORRECTION
SAP Application Interface Framework tables	Central Finance: SAP Application Interface Framework Message Multi-Index Table (Sender Doc)	CFIN_AIF_IDX_MUL
	Central Finance: SAP Application Interface Framework Message Key Fields	CFIN_AIF_IDX_TBL
	Central Finance: SAP Application Interface Framework Message Serialization (Prereq. Doc)	CFIN_AIF_PREREQ
SAP Application Interface Framework transactions	SAP Application Interface Framework Reprocessing Action Definition	/AIF/REP_AC_DEF
	SAP Application Interface Framework Reprocessing Action Assignment	/AIF/REP_AC_ASGN
	SAP Application Interface Framework Runtime Configuration	/AIF/PERS_CGR
	SAP Application Interface Framework Interface Monitor	/AIF/IFMON
	SAP Application Interface Framework Monitor and Error Handling	/AIF/ERR
	SAP Application Interface Framework Customizing	/AIF/CUST

Table 6.16 Central Finance Cockpit (Cont.)

Category	Description of Transaction or Report	Transaction/Report
Commitment Reports	Delete Replicated Commitment Documents for Controlling Area and Company Code	FINS_CFIN_CMT_DOC_CLEANUP
	Delete Replicated Commitment Documents for Purchase Order or Purchase Requisition	FINS_CFIN_CMT_DOC_DEL
	Replicate Individual Commitment	FINS_CFIN_CMT_REPL
Commitment tables	Central Finance: SAP Application Interface Framework Message Key Fields for Commitment Doc (Simulation)	CFIN_AIF_CMT_SIM
	Central Finance: SAP Application Interface Framework Message Key Fields for Commitment Doc	CFIN_AIF_IDX_CMT
	Commitment Posting Reference	FINS_CFINT_CMTRL
Individual replication	Execute Single Step in Mass Data Framework	FINS_MASS_DATA_TEST
	Replicate Individual Cost Object from the Source System	FINS_CFIN_REPL_CO_OBJ
	Replicate Individual Controlling Document from the Source System	FINS_CFIN_REPL_CO_DOC
Operational tools	Correction Program to Repost Missing Cleared Items in Central Finance	RFINS_CFIN_CORR_MISSING_CLRITM
	Reopen Technically Cleared Items Referenced by a Clearing SAP Application Interface Framework Message	RFINS_CFIN_REOPN_TECH_CLRD_DOC

Table 6.16 Central Finance Cockpit (Cont.)

Tip

Besides the Central Finance cockpit, SAP also provides a list of most common errors (100+) and relevant SAP Notes collected from various past projects. The errors list can be downloaded from SAP Note 3194497, **Attachments** tab.

6.11 Summary

In this chapter, we introduced SAP Application Interface Framework. We began with covering the SAP Application Interface Framework terminology and architecture. Then, we discussed the importance of the serialization concept, especially to process the messages in the correct order in SAP Application Interface Framework. This led to learning about runtime group configuration and assignment. We delved into understanding the components in monitoring and error handling and interface monitoring transactions. Next, the chapter transitioned into miscellaneous topics of reprocessing of SAP Application Interface Framework messages when errors occur, reprocessing missing cleared items, and retriggering SAP Application Interface Framework messages. You've also learned how to configure alerts management to send alerts and email messages to recipients for the interfaces they're responsible for. Finally, we touched on archiving and error resolution content subtopics. In the next chapter, we'll discuss key design elements and configuration of the source and Central Finance systems.

Chapter 7
Configuring Central Finance and Source Systems

Central Finance offers a nondisruptive way to replicate financial documents from source ERP systems. From a Central Finance perspective, important design elements for Central Finance must be understood first. Next, for Central Finance to work seamlessly with the source ERP systems, we require configuration of source systems, SAP Landscape Transformation Replication Server, and the Central Finance system based on business needs.

In the Central Finance scenario, transactional finance postings are sent from source ERP systems to the Central Finance system. During reposting of financial documents, the Central Finance system redetermines configurations and mappings prior to posting those documents in Central Finance. In addition to predefined configurations and mapping rules, Central Finance offers customer-specific enhancements prior to posting documents into the Central Finance system. Therefore, an understanding of source and central systems configuration is critical for establishing a Central Finance environment. Prior to configuring the source system and the Central Finance system, you must have the following:

- An understanding of licensing requirements for any software to be installed, including Central Finance and additional add-on tools such as SAP Analytics Cloud, SAP Analysis for Microsoft Office, group reporting, or external tax calculation engine
- An understanding of the system landscape, including interfaces, third-party software in scope and relevant configuration requirements, as well as the type of source systems to be connected, which can be either SAP-based or non-SAP based systems
- An understanding of key design topics, including ledgers, currencies, and document splitting
- An understanding of master data strategy and relevant configuration requirements
- An understanding of any data conversion or selective data transmission requirements that may drive additional configuration requirements
- Necessary security authorization objects and role requirements for connecting Central Finance with other systems

Note that the chapter covers general settings required to enable replication scenarios and a few financial accounting (FI)-related configurations. Configurations related to specific components (e.g., configuration related to Controlling [CO] covered in Chapter 10) are covered in their respective chapters. We've configured the SAP S/4HANA system as a source system. You'll notice throughout that the chapter screenshots reflect the SAP Fiori theme in SAP S/4HANA. In the previous chapters, you've learned SAP Landscape Transformation Replication Server and SAP Application Interface Framework configurations. In this chapter, we begin with prerequisites and important design element considerations for Central Finance. Then, we'll dive deeper into both the source system and the Central Finance system configurations. Finally, the chapter transitions into understanding configuration consistency checks and how to run them after performing configurations and business mappings.

7.1 Getting Started with Configuration

Although you can connect a single source to Central Finance, in general, Central Finance is a distributed landscape system where multiple systems are integrated together to distribute both financial and nonfinancial data from source enterprise resource planning (ERP) systems to the Central Finance system. Whether you're using Central Finance as a sidecar deployment option or as an operating solution approach to leverage SAP S/4HANA features for business processes, Central Finance prerequisites and required configurations must be taken into consideration.

From the Central Finance configuration requirement standpoint, the chapter groups technical and functional implementation activities into six broad groups, as shown in Figure 7.1. Besides Central Finance components, other applications that integrate with Central Finance must be included in the scope for configurations. Nevertheless, the most important component configurations are listed here:

❶ **Source system configuration**
This is where the main business processes run. Central Finance supports both SAP-based source systems and non-SAP-based systems to transfer data. To send data from source systems, you'll need to understand relevant source system release, SAP Notes requirements, and configurations requirements in the source system.

❷ **SAP Landscape Transformation Replication Server configuration**
This acts as the integration platform used for sending data from the source system to the Central Finance system. To facilitate replication, certain configurations and network connections must be completed prior to replicating data.

❸ **Central Finance system configuration**
An understanding of both technical and functional configuration is required prior to integrating Central Finance with other systems.

❹ **Security and Basis activities**
Different types of users and roles are required to configure and connect SAP Landscape Transformation Replication Server, the source system, and the Central Finance system. In addition, SAP Basis administrators are involved in system installations, connections, support applications, and database-related activities in the Central Finance environment involving all systems.

❺ **SAP Master Data Governance**
A full feature SAP Master Data Governance solution is available to manage master data in a Central Finance scenario. However, Central Finance implementations have the option to use the SAP Master Data Governance solution or customer chosen master data management tool to manage and replicate master data. If SAP Master Data Governance is in scope, additional configurations and testing are required.

❻ **Other interfaces and components**
Besides the core components, based on your requirements, other components or interfaces needs to be configured. This isn't limited to analytic reporting platforms, third-party bolt-on software, and so on because custom solutions must be considered as well.

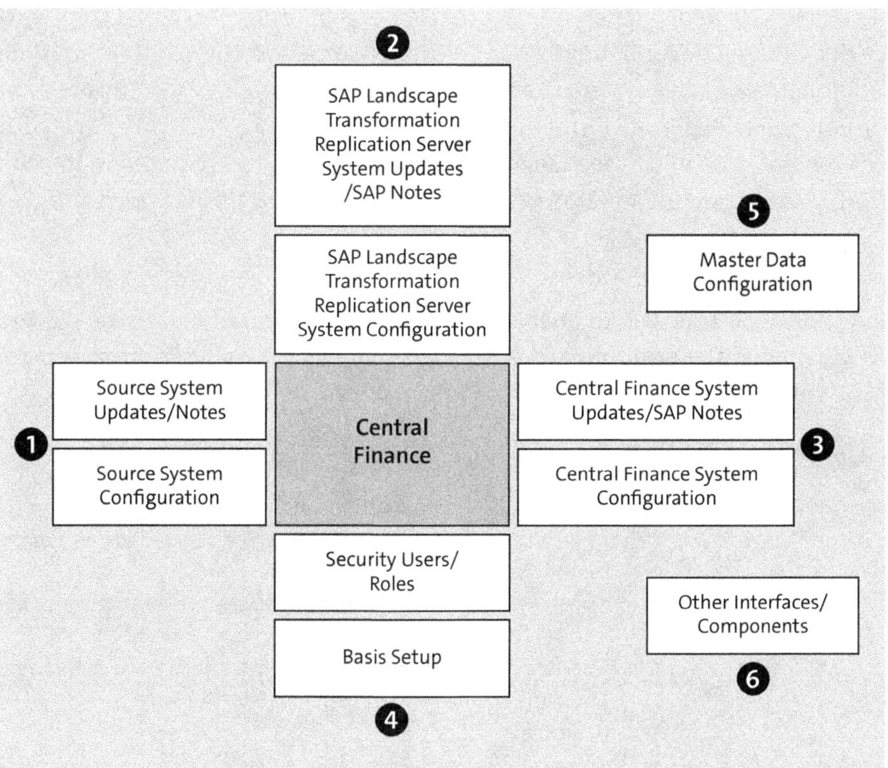

Figure 7.1 Central Finance Configuration Overview

The section provides an overview of prerequisites and step-by-step configuration knowledge necessary to carry out configuration in the source system and Central Finance system. In addition, you'll learn security requirements and additional critical finance design elements. SAP Landscape Transformation Replication Server configuration is covered in Chapter 5. SAP Application Interface Framework configurations are covered in Chapter 6.

7.1.1 Prerequisites

To implement Central Finance, both functional and technical prerequisite requirements must be evaluated and considered for the systems involved in the Central Finance integration scenario. This is required to replicate data from the source systems to the Central Finance system. Often, customer source systems are installed with various releases and support levels. SAP recommends implementing the latest support packages or relevant SAP Notes, which contain development packages and objects relevant to support the Central Finance integration scenario. By implementing the required support pack levels and relevant SAP Notes, implementation teams can avoid time spent on errors on already known or resolved issues. Therefore, the recommendation is to follow SAP guidelines on minimum release/support level as a first step. Next, you'll perform baseline configuration in Central Finance and the required minimal configuration in the source system to support Central Finance scenarios. The prerequisites for SAP Landscape Transformation Replication Server were already covered in Chapter 5, Section 5.1.1. In this section, we'll explain prerequisites required in the context of SAP source systems, non-SAP source systems, and Central Finance systems.

Prerequisites for SAP Source Systems

From a technical perspective, to enable Central Finance integration, source SAP ERP systems must be at the minimum SAP release/support level. Table 7.1 provides supported SAP source system and their release information.

SAP-Based Source System	Support
SAP ERP release 600 and above	Supported. To support Central Finance, the minimum support pack required is SP08. SAP recommends SP09 as a minimum release support pack for the Central Finance business integration scenario. However, SAP strongly recommends upgrading to the latest service pack to avoid implementing hundreds of SAP Notes and relevant manual activities, which may take several days or even weeks to complete.

Table 7.1 Prerequisites for SAP-Based Source Systems

SAP-Based Source System	Support
SAP S/4HANA on-premise	Supported. For the SAP S/4HANA 1511 release, SAP recommends that SAP Landscape Replication be installed separately. SAP Simple Finance 1.0 (SAP_FIN700) isn't supported as a source system.
SAP S/4HANA as a public cloud	Supported. Refer to Chapter 1, Section 1.3.3, for restrictions.
SAP R/3 release 4.6C and 4.7, and SAP ERP 500	Additional efforts are required with a customer-specific project.

Table 7.1 Prerequisites for SAP-Based Source Systems (Cont.)

Source system enablement is key to the success of any Central Finance implementation. During a project's preparation or exploration phases, careful evaluation and assessment is required to avoid underestimation on source system preparation and efforts involved and their impact on the effort and project timeline. Often, customers with lower SAP ERP versions face the dilemma of whether to upgrade to the latest support pack level or to implement only relevant SAP Notes with the minimum recommended service pack level. Sometimes, this isn't an easy choice to make as both options require additional effort and time and may also disrupt existing operations in the source system. The choice depends on which SAP release the customer is in, the supported pack level, and the complexity of upgrading to a service pack due to either resource constraints or business process disruptions that the business isn't ready to accept yet. We'll explain these approaches here:

- In the first approach, if the SAP source system is on an older support pack level, SAP recommends upgrading to the recent support pack as this approach reduces the number of SAP Notes to be implemented in the source system significantly. However, on the other side of this approach, upgrading to a recent support pack might also impact all SAP components, thereby increasing the regression testing and user acceptance testing efforts. In addition, evaluation of additional SAP Notes is required since the release of support packs to reduce errors or unexpected behavior during the initial load or replication activities. To see the current service pack level on any SAP source system, as shown in Figure 7.2, click the **System** ❶ submenu in the **SAP Easy Access** menu. In the **System** menu, choose **Status**. This brings up the **System: Status** subscreen. On this screen, click the **Product Version Details** ❷ icon. Next, the system displays the **Installed Software** popup window. In this screen, you'll observe the service pack and release information for components **SAP_APPL** ❸, **SAP_BASIS** ❹, and **SAP_FIN** ❺ components.

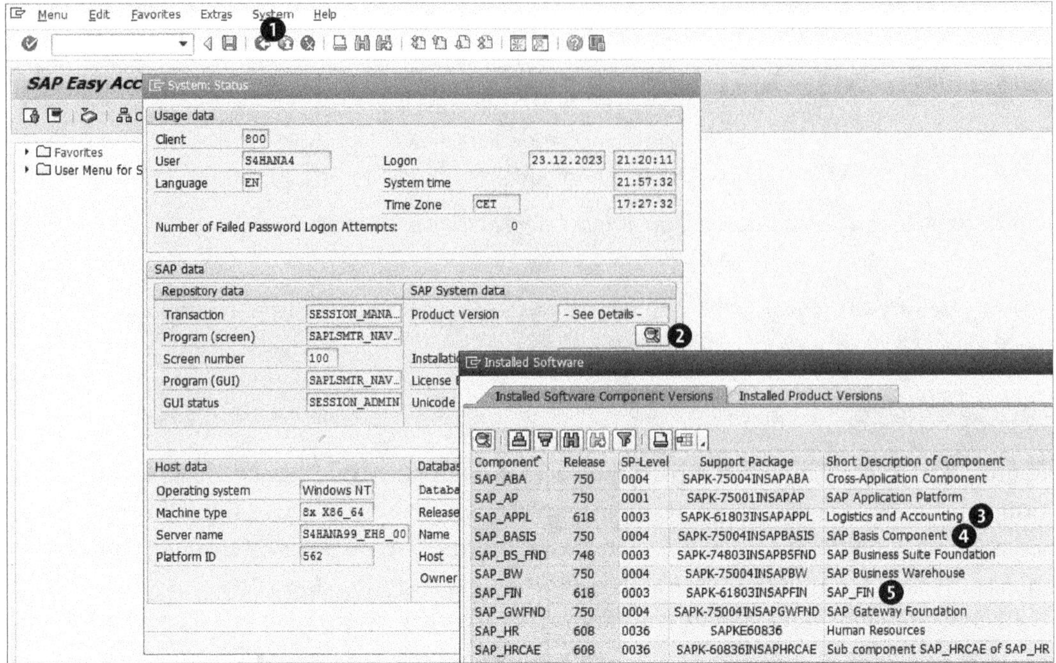

Figure 7.2 Release and Service Pack Level Check

- In the second approach, if you're on a higher service pack level, identify missing SAP Notes. Although the effort involved in implementing the missing SAP Notes may be minimal in this case, don't underestimate the effort required to implement dependent SAP Notes and to perform manual activities. The SAP Support Expert Search page (*https://support.sap.com/en/index.html*) can be used to identify the relevant notes for a given source system release and Central Finance component Figure 7.3 shows how to access the **Expert Search** page on SAP Support Portal. From the support page, click the **Access Expert Search** ❶ link to navigate to the **Expert Search** page ❷. Enter the following information to search for SAP Notes relevant to the source system under the Central Finance component:
  - **Component (Start with)**: Enter value "FI-CF" ❸ to search in the Central Finance component.
  - **Soft.Comp.**: Enter "SAP_APPL" ❹.
  - **Soft.Comp.Version**: Enter the software component version, for example, "617" ❺.
  - **Support Package (Greater Than)** ❻: Enter a support package that is greater than the current version of your source system.
  - **Release Status**: This defaults to **Released for Customer** ❼.

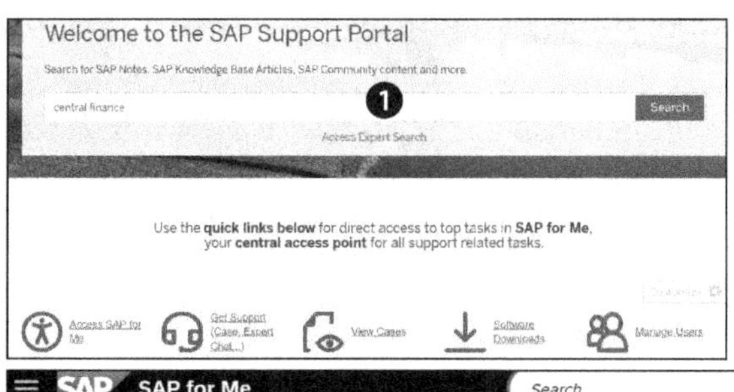

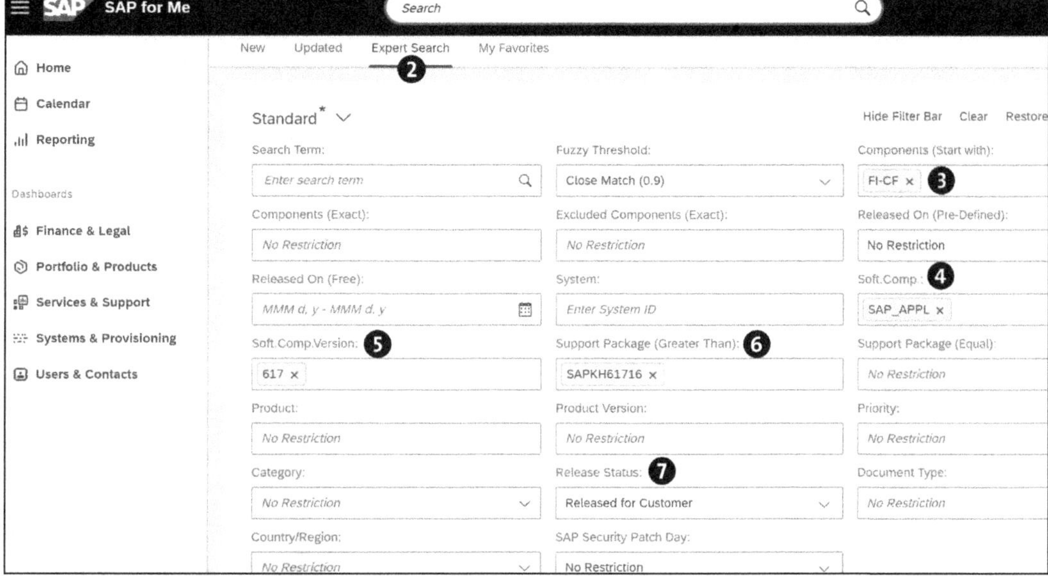

Figure 7.3 SAP Notes Advanced Search

Note

Consult SAP Note 2323494 for an overview of relevant SAP Notes applicable for the source system. This SAP Note references the other 260+ dependent SAP Notes based on your release level. However, this isn't a complete list. This SAP Note provides a starting point for identifying relevant notes and their applicability in the relevant source systems. A careful evaluation is required as the installation of these notes is time-consuming and complex due to sequencing and manual activities.

SAP offers new ways to upgrade or apply service packs using the SAP Note transport-based correction instruction (TCI) tool. Any corrections previously delivered with SAP Notes are done with either manual or automatic corrections and support packages upgrade (SP). With the TCI tool, now customers can implement SAP Notes or support packages quickly without many development skills due to an automatic installation process. To learn more about TCI-based application of SAP Notes and service packages and its restrictions, refer to SAP Note 2187425. The SAP Note comes with a detailed document attached for futher guidance.

Now that you've learned the technical aspects of the SAP-based source system requirements, let's look at a few prerequisites from a Central Finance perspective for the source system. The following list isn't comprehensive as it varies with every implementation, but it captures important prerequisites from a functional view:

- Classic withholding tax isn't supported with Central Finance. If the source system is running classic withholding tax, it must be updated to extended withholding tax. Consult SAP Note 2623514 for additional guidance on withholding tax requirements on Central Finance.
- Contract Accounts Receivable and Payable (FI-CA) replication isn't supported.
- Extended document splitting is not supported.
- Company code merger (M:1) or split (1:M) scenario is not supported.
- Replication of special purpose ledgers.

From a technical perspective, the unsupported scenarios are:

- Central Finance does not support the simple finance 1.0 software component version SAP _FIN_700
- Any highly customized, industry-specific solution
- Custom code such as user exists or enhancements in the source system

Prerequisites for Non-SAP Systems as a Source System

Central Finance supports integration with non-SAP systems (third-party systems). The following scenarios are supported with the standard interface for the third-party system. The third-party interface uses the standard error handing through SAP Application Interface Framework.

- Replicate initial FI documents.
- Replicate changes made to FI documents.
- Replicate accounting view of logistics (AVL).
- Execute comparison reports in Central Finance.

To enable replication for a non-SAP system such as Oracle e-Business Suite, PeopleSoft, and JD Edwards, SAP delivers preconfigured content that facilitates loading data from a non-SAP system to a staging area in SAP Landscape Transformation Replication Server. The staging area connects to non-SAP source systems. The following prerequisites and functional requirements must be considered for integrating non-SAP system to the Central Finance system:

- First, SAP Landscape Transformation Replication Server must be installed with DMIS_2011 version or higher to support a non-SAP system. Refer to Chapter 5, which covers prerequisites for SAP Landscape Transformation Replication Server requirements.

- For application error handling, certain business configuration sets (BC sets) must be installed.
- SAP doesn't support or have a solution to download from a third-party source system. This requires additional interfaces or other integration software. Transferring data from a non-SAP system to SAP Landscape Transformation Replication Server intermediate tables requires custom interfaces, SAP BusinessObjects data services, otherdata services, or third-party tools such as Magnitude SourceConnect.
- SAP doesn't offer any master data profiling for third parties. Customers build their own interfaces or programs to perform this task.
- One-time account postings aren't available because they aren't part of standard replication structures.
- The Central Finance initial load function isn't available for third-party source systems. This includes general ledger balances and open items.
- Asset accounting documents replication isn't supported for the third-party source system.
- Deferred tax solution isn't supported with the standard interface that connects the third-party system to Central Finance.
- Classic withholding taxes aren't supported.
- Document splitting isn't supported for the third-party source system.

Besides these, you'll also need resources with source system knowledge in multiple areas such as master data, integration, and functional areas to support the Central Finance implementation. Some of the challenges could be mitigated with SAP Central Finance Transaction Replication 2.0 by Insightsoftware for SAP S/4HANA. However, this may involve additional licensing and effort.

[«]

Source System Notes

Consult SAP Note 2932517 for frequently asked questions on Central Finance third-party system integration. This SAP Note gets updated frequently, so you may want to add it to your favorite's list. In addition, the following SAP Notes provide more help:

- SAP Note 3340008 offers guidance on SAP Landscape Transformation Replication Server release information for third-party interface for SAP S/4HANA 2023.
- SAP Notes 3396547 and 2977332 offer guidance on Insight software master data replication and finance transaction replication respectively.
- SAP Central Finance adoption content provides a template to organize SAP OSS Notes to be implemented in the correct order with dependent or prerequisite notes. This saves time to avoid errors such as implementing hundreds of SAP Notes, their dependencies, and manual activities.

Prerequisites for the Central Finance System

To facilitate Central Finance initial load and online replication, several prerequisites must be met from the Central Finance system. This includes both technical and functional prerequisites.

From a *technical perspective*, the following prerequisites must be evaluated prior to implementing Central Finance:

- You have licensing for Central Finance and other external software that integrates with Central Finance. Identify which finance release is required based on business requirements.
- Identify and apply relevant SAP Notes in the Central Finance system. Include any recent notes on Central Finance components on the SAP Support page. A few important application areas with SAP Notes include the following:
  - Central payments and cross-system process control (CSPC)
  - A set of new notes to enable the joint venture accounting (JVA) functionality if in scope
  - New features introduced with pilot SAP Notes
- For performance optimization, review any new performance-relevant SAP Notes released.
- For interface connectivity, review any new SAP Notes released to support the interface integration.

Likewise, from a *functional perspective*, the following prerequisites must be evaluated prior to implementing Central Finance:

- Identify company codes that are in scope for Central Finance to help with relevant configuration and then plan accordingly.
- Identify processes that are in scope for Central Finance. This may lead to additional application of SAP Notes or configurations:
  - Centralized processes such as central payment or central credit and dispute management
  - A set of new notes to enable the joint venture accounting (JVA) functionality if in scope
  - New features introduced with SAP either with pilot SAP Notes or with a major release
- Identify when to activate central payment and other centralized processes.

[«]

> **Note**
>
> SAP Note 2148893 provides comprehensive documentation and a list of SAP Notes required in both the source system and the Central Finance system. On the other hand, SAP Note 3361761 is a collection of SAP Notes segregated by topic.

7.1.2 Security Requirements

Central Finance is a distributed system landscape that communicates among connected systems using remote function calls (RFCs). To transfer data between servers, RFC users must be given necessary access to the source system and the Central Finance system. Table 7.2 summarizes the SAP roles required.

Security Role	Systems	Purpose
SAP_IUUC_REPL_REMOTE	■ Source system ■ Central Finance system	Assigned to RFC user in the source system and Central Finance system
SAP_AIF_PROCESSING	Central Finance	Access to SAP Application Interface Framework
SAP_IUUC_REPL_ADMIN SAP_MWB_PROJECT_MANAGER	SAP Landscape Transformation Replication Server	Assigned to configuration user in SAP Landscape Transformation Replication Server

Table 7.2 Security Roles in Central Finance

Next, let's learn each role and its purpose in detail. With the template role SAP_IUUC_REPL_REMOTE assigned to an RFC; users can do the following:

- Initiate SAP Landscape Transformation Replication Server initial load activities.
- Replicate data from the source system to Central Finance.
- Trigger SAP Landscape Transformation Replication Server–relevant activities in the source system.

Figure 7.4 shows the assignment of the SAP_IUUC_REPL_REMOTE ❶ role to user role maintenance Transaction PFCG. As you've noticed, in the first part of Figure 7.4, you can navigate to view the RFC access generated using the **Change Authorization Data** ❷ icon. The second part of the figure shows authorization check RFC access for the S_RFC object ❸. The third screen shows the **User** tab ❹ and assignment of the RFC user to the authorization object ❺.

Figure 7.4 SAP Role SAP_IUUC_REPL_REMOTE

> **Note**
>
> SAP Note 2154420 provides instructions on manual extension of the RFC authorization in the SAP_IUUC_REPL_REMOTE role.

Likewise, SAP delivered a similar template role, SAP_AIF_PROCESSING, to access SAP Application Interface Framework from SAP Landscape Transformation Replication Server, as shown in Figure 7.5. When SAP Landscape Transformation Replication Server transfers data to Central Finance, the SAP Landscape Transformation Replication Server user must have access to SAP Application Interface Framework with appropriate authorization. This template role provides such authorization, which can be copied to a Z role and can be assigned to the SAP Landscape Transformation Replication Server RFC user.

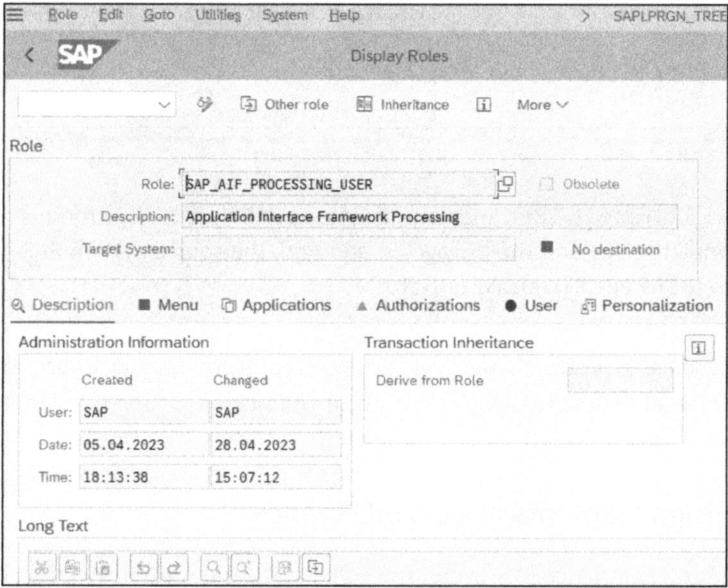

Figure 7.5 SAP_AIF_PROCESSING template role

In addition, SAP offers predefined template roles catered for technical and functional activities. Table 7.3 lists the template roles available, which can be copied into the customer namespace for adoption and assigned to appropriate users. Table 7.3 provides template roles for the source system and Central Finance.

System	Template Role Name	Usage
Source system	SAP_CFIN_RFC_USAGE_01	To extract data for FI
	SAP_CFIN_RFC_USAGE_02	To manage mapping
	SAP_CFIN_RFC_USAGE_04	To access comparison reports
	SAP_CFIN_RFC_USAGE_06	For central payment communication
	SAP_CFIN_RFC_USAGE_08	For side panels
	SAP_CFIN_RFC_USAGE_09	To extract data for management accounting
	SAP_CFIN_RFC_USAGE_10	To extract data for commitment
Central Finance	SAP_BR_GL_ACCOUNTANT_CFIN	SAP Fiori template role for the Central Finance general ledger accountant

Table 7.3 SAP-Delivered Template Roles for Central Finance

[»]

Note

Refer to source system SAP Note 2677866 and Central Finance SAP Note 3331552 for further guidance on template roles and their usage. In addition, the source system SAP Note 2677866 applies to SAP ERP 6.0 release onward.

Now that you understand the prerequisites and security requirements, we'll discuss the important design element considerations for Central Finance.

7.2 Important Design Elements in Central Finance

While implementing Central Finance, it's imperative that critical design decisions be made on foundational configuration elements. These baseline configuration requirements must be thoroughly evaluated and configured while considering legal and statutory requirements. Let's explore important design elements that impact Central Finance implementations.

7.2.1 Document Splitting

In SAP, document splitting is required for two primary reasons. First, the document splitting functionality enables companies to split the financial document line items, so complete financial statements can be drawn by specific characteristics such as profit center, segment, and so on. This leads to segment reporting to fulfill objectives of US

GAAP ASC 280 segment reporting (IFRS 8) requirements. Second, the document splitting functionality benefits managers to analyze revenue, expense, working capital, and so on to measure performance of a profit center, segment, or any other characteristic. In SAP, document splitting facilitates meeting segment reporting and management reporting requirements. Businesses post transactions through the entry view. When document splitting is active, the system posts both the entry view and general ledger view.

In the Central Finance context, document splitting is a critical design element that must be carefully evaluated in both the source system and Central Finance during the explore phase of the project. As a first step, you'll assess the quality of data by looking at the business transactions posted in the source system. Next, you'll determine what business reporting requirements are and how profit center and segments can be derived for balance sheet and P&L accounts in the Central Finance system. In addition, you'll check whether any coding block and extensibility fields are used and required as a document splitting characteristic. If a source system is activated with document splitting, then you'll run a configuration consistency check to find any inconsistencies.

When you add a company code in the **General Replication Settings** view of the `VCFIN_SOURCE_SET` view in the source system, the system also adds an entry in the `VCFIN_SOURCE_SPL` view in the background with the default value set as **No action**, as shown in Figure 7.6. This setting determines the behavior of document splitting for the maintained company code in the source system. This view can be accessed from Transaction SM30 or from **SAP Customizing Implementation Guide** menu path **Financial Accounting • Central Finance • Central Finance: Source System Settings • Handling of Document Splitting**.

In Figure 7.6, the field **Splt actn** has four options to choose from in the dropdown list as follows:

- **No action**
  This is the default value updated in the `VCFIN_SOURCE_SPL` view when a company code is added in `VCFIN_SOURCE_SET`. As the name implies, no document splitting check is performed in the source system during financial postings. In addition, no split information is transferred. If this option is chosen, upon replication of data, the Central Finance system tries to post the document with the configured rules in Central Finance. If data doesn't comply with the document splitting rules, the system raises a document splitting error in SAP Application Interface Framework.
- **E Run splitting process checks: raise error message**
  This option raises a hard error message if a financial transaction posted doesn't comply with the document splitting rules.
- **W Run splitting process checks: raise warning message**
  A warning message issues, but the system doesn't prevent posting financial postings in the source system; however, the document may fail in Central Finance.

- **X Transfer splitting information from source to target**
  No check is performed, but the system transfers splitting information from the source system to the Central Finance system. This step isn't applicable for the classic general ledger.

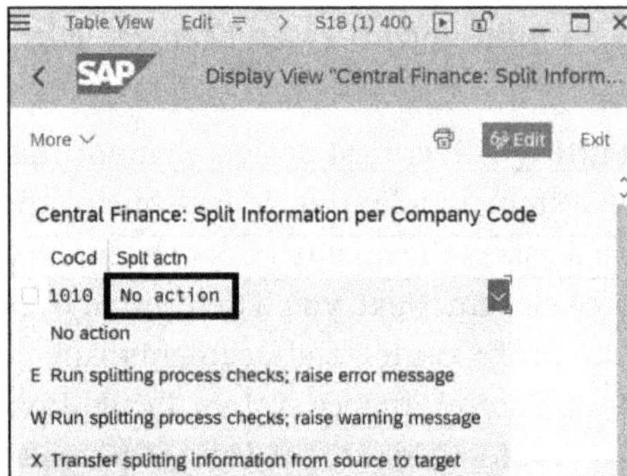

Figure 7.6 VCFIN_SOURCE_SPL view

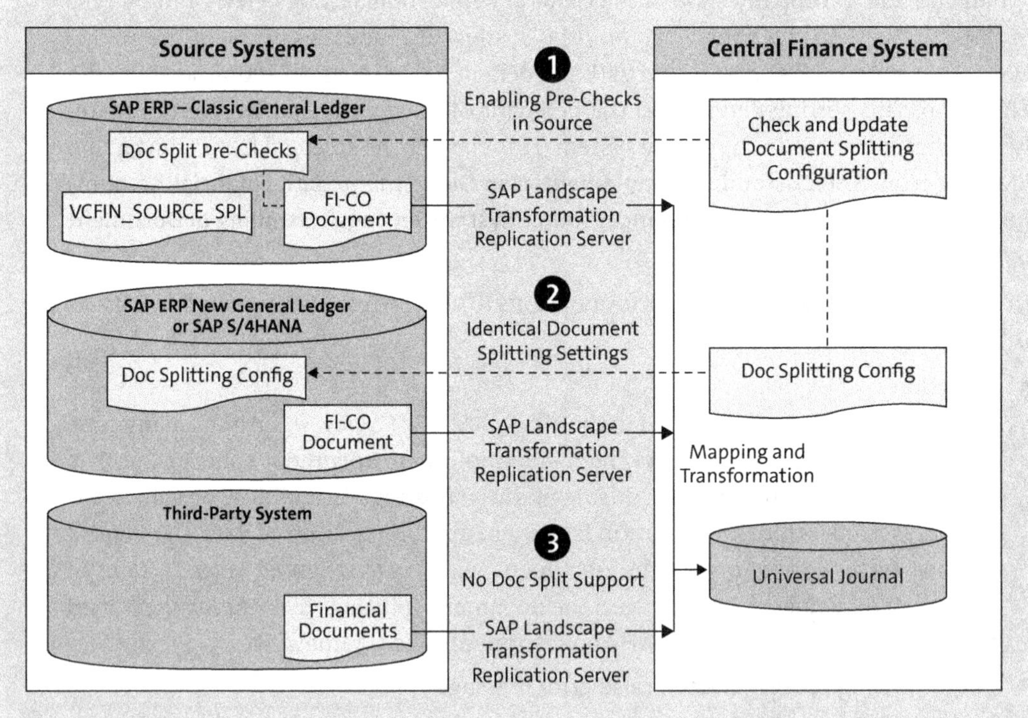

Figure 7.7 Document Splitting in Central Finance

Based on the source system, you'll come across three types of document splitting scenarios if document splitting is active in the Central Finance system, as shown in Figure 7.7:

❶ **SAP ERP with classic general ledger as the source system with no document splitting**
In this case, the document splitting functionality isn't available in the source system as the source runs on the classic general ledger. Although this scenario isn't supported for document splitting out of the box in Central Finance, SAP introduced document splitting prechecks to be carried out prior to posting financial documents in the source system. To enable these prechecks, SAP provides Transaction FINS_CFIN_SPLIT_CUST to transfer the document splitting Customizing activity from the Central Finance system to the source system. This transaction can be accessed from IMG menu path **SAP Customizing Implementation Guide • Financial Accounting • Central Finance: Target System Settings • Document Splitting • Check and Update Document Splitting Configuration**. During finance transactional postings, the system performs checks on postings against the transferred document splitting configuration settings based on settings you maintained in view `VCFIN_SOURCE_SPL` as discussed earlier. These prechecks (if values **E** or **W** were maintained in **Splt acttn** in view `VCFIN_SOURCE_SPL`) prevent most document splitting errors in the source system itself, but it may not prevent all errors. The idea is to prevent most document splitting errors proactively in the source system before the data is transferred to SAP Application Interface Framework in the Central Finance system.

Upon replicating a document from the source system to the Central Finance system, the system posts in the entry view and general ledger view in the Universal Journal. The general ledger view contains document splitting information based on configured rules derived from the Central Finance system.

[!]

Warning

If the **E Run Splitting process checks: raise error message** option is chosen in the `VCFIN_SOURCE_SPL` view, this will prevent posting in the source system, which may result in delayed posting to additional steps to correct and repost. This will go against the Central Finance theme of nondisruption to existing processes. You must evaluate overall requirements, including the impact of period-closing activities, before activating the document splitting prechecks.

❷ **SAP ERP with the new general ledger or SAP S/4HANA general ledger as the source system with document splitting active**
Both the source system and Central Finance system are activated with document splitting. In this case, for document splitting to work seamlessly, document splitting configurations must be identical between the source system and Central Finance system to avoid document splitting errors in the initial load or SAP Application Interface Framework. During replication, the system transfers split information from the source system (from tables `FAGL_SPLINFO` and `FAGL_SPLINFO_VAL`) to the Central Finance system if the **X Transfer splitting from source to target** option of `VCFIN_SOURCE_SPL` is set. This includes SAP S/4HANA Cloud Public Edition as a source system. However, in the case of SAP S/4HANA Cloud Public Edition, the system always transfers split information. For SAP S/4HANA general ledger and SAP ERP general ledger as source systems, the leading practice is to set **X Transfer splitting from source to**

target option of view VCFIN_SOURCE_SPL. You can confirm by viewing source staging accounting header table ACCHD-NOSPLIT with value C (Mode C) for the selected GUID. One exception is documents with the **Posting to General Ledger Only** status (BKPF-BSTAT = U) not transferred. For normal documents, upon transferring the document splitting information from the source system to Central Finance, the system splits items based on the corresponding transferred item information as base items. However, for clearing, items transferred will be reprocessed in Central Finance based on items found in table FAGL_SPLINFO, but not based on the data transferred from the source system. The system does this to ensure clearing balances to zero while considering currency settings and exchange rate applications. Thus, clearing items processed in Central Finance don't bear any influence on splitting based on the **X Transfer splitting from source to target** setting in view VCFIN_SOURCE_SPL.

SAP recommends running the Central Finance configuration consistency check (Transaction FINS_CFIN_CC) to resolve errors proactively. As you see at the top of Figure 7.8, the system displays the **Central Finance: Configuration Consistency Check Report** initial screen where you select the **Document Splitting** checkbox, as highlighted.

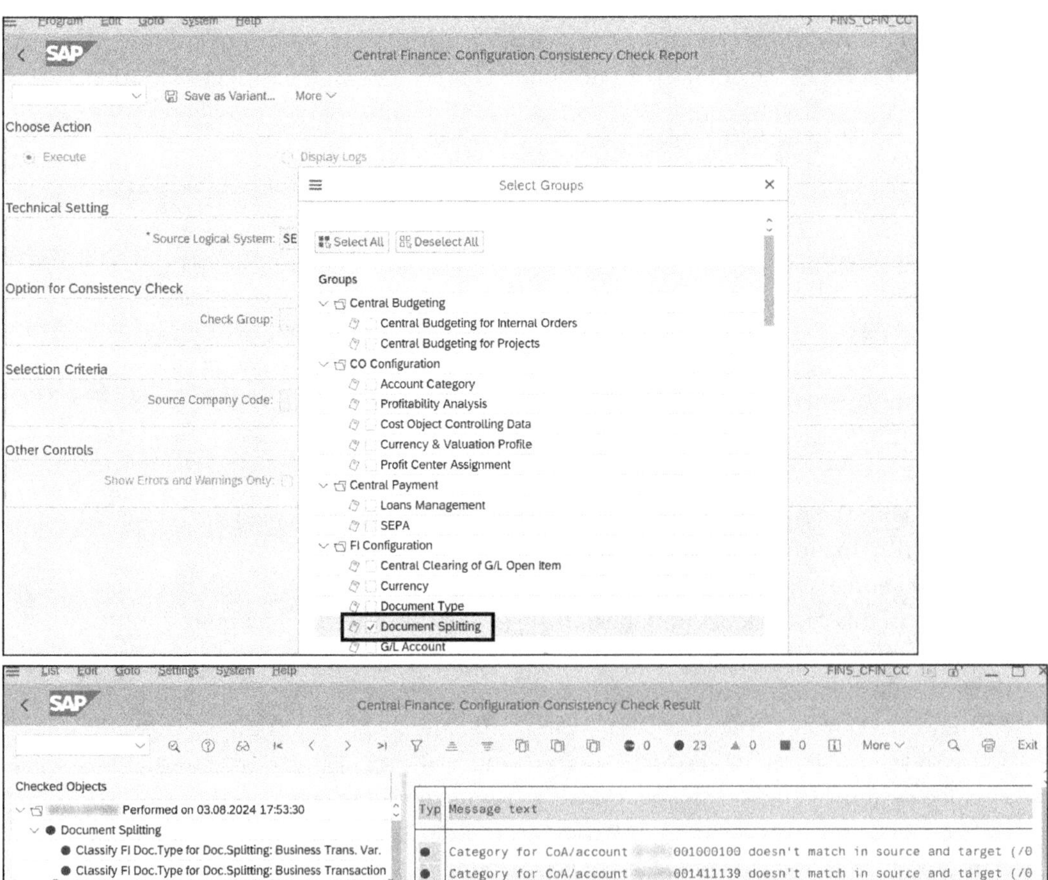

Figure 7.8 Document Splitting Configuration Consistency Check

Upon execution of the report with this checkbox, the system displays discrepancies found on the document splitting configuration between the source system and the Central Finance system for the following configuration items:

- **Classification of G/L Accounts for Document Splitting** to ensure general ledger accounts are configured correctly
- **Classification for Document Types for Document Splitting** to check whether business transaction variants are assigned to document types
- **Document Splitting Characteristics for General Ledger Accounting** to assign account objects for splitting

Besides the configuration consistency check, SAP also provides a simulation option in SAP Application Interface Framework to simulate document splitting prior to posting transaction to Universal Journal.

[+]

Tip

Central Finance doesn't support extended document splitting.

❸ **Third-party system as the source system**

The Central Finance third-party interface doesn't support document splitting of line items based on characteristics. In this case, the Central Finance business add-in (BAdI) `BADI_FINS_CFIN_EX_INTF` can be used to fill in document splitting information during the data transfer. Based on business transactions and scenarios, you can also consider additional document types in Central Finance if required to support third-party data.

[!]

Warning

For the source system with classic general ledger with no document splitting or varying document splitting configurations between the source system and Central Finance, you'll need custom enhancements based on requirements to support the initial load or online replication. Central Finance BAdI `BADI_FINS_CFIN_AC_INTERFACE` can be leveraged to enhance document splitting scenarios to prevent errors. BAdI logic can be implemented to derive document splitting characteristics, map to a different document type, or perform any custom solution required based on requirements. Custom enhancement requires extensive testing, especially for transactions originating from logistics components and follow-on postings. An example list of typical scenarios that may require further enhancement to support these scenarios is given here:

- When transferring the initial load open items and balances, the system doesn't bring in profit and loss (P&L) account line items, which hold account assignment objects to derive characteristics. In these cases, based on the business requirement, automatic account assignment (Transaction OKB9) or default profit centers maintained through Transaction FAGL3KEH can be used. A BAdI enhancement also can be used if the standard solution doesn't fit the business requirements.

- Mixed transactions are posted in the source system successfully, but the data transferred to Central Finance doesn't comply with document splitting rules. For example, a transaction with document type carrying vendor line item allowed posting in the source system, but it will fail in SAP Application Interface Framework in Central Finance due to the document splitting configuration.
- Missing or blank account assignment lines (e.g., missing profit center) are allowed in the source system, but they may result in a balancing profit center error in the Central Finance system.
- Payment documents are posted without any reference to an invoice document. During replication, the system is unable to inherit the profit center for document splitting.
- Transactions involving cross-company codes where one company code is activated for document splitting but the other company code isn't activated will fail in the Central Finance system.
- Follow-on documents are posted in the source system but fail in Central Finance due to differing document splitting configurations. Examples are cash discount, penalty interest, revaluation of open items, and so on.

7.2.2 Currencies

In this section, first we establish the basics regarding currencies. Then, you'll learn important currency considerations from a Central Finance implementation perspective.

Generally, currency design is driven based on legal and statutory requirements in SAP. When Central Finance is integrated with source systems, often the source system runs in SAP ERP or older versions. SAP ERP supports three currencies in FI ❶, two currencies in CO ❷, and three currencies in the material ledger ❸, as shown in Figure 7.9.

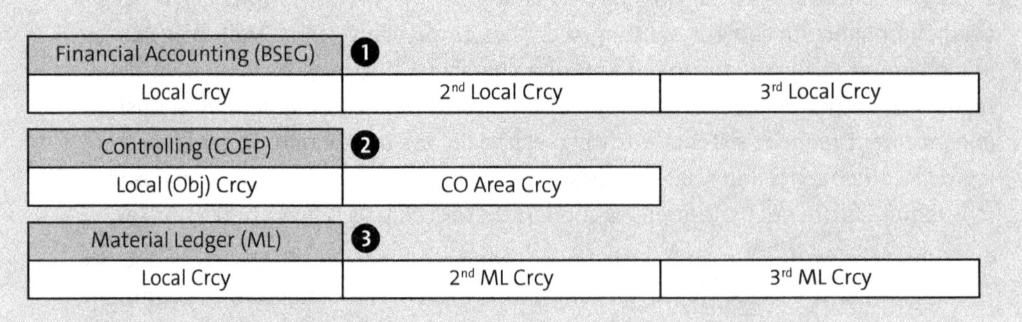

Figure 7.9 SAP ERP Currencies

In SAP S/4HANA, the supported currencies can be classified into five broad groups. Figure 7.10 shows currencies supported in the Universal Journal:

❶ **Document currency**
A currency (also called *transaction currency*) entered in the entry view of a document posting. Upon posting, the document currency amount stored in tables ACDOCA-WSL and BSEG-WRBTR with reference to the currency key are updated in document header table BKPF-WAERS. In the entry view, you can enter any currency per the business process.

❷ **Amount in balance transaction currency**
The amount in balance transaction currency amount converted currency of the general ledger account to support transaction key figures. This is done at general ledger account level. It's stored in tables ACDOCA-TSL and BSEG-PSWBT.

❸ **Local currency**
The local currency refers to the currency of the company code stored in table ACDOCA-HSL and relates to table BSEG-DMBTR in FI and object currency in CO (the **Diff. CCode Currency** indicator is not selected in the Controlling area configuration).

❹ **Global currency**
The currency defined in the controlling area configuration (TKA01-CTYP) and corresponding posting amounts stored in table ACDOCA-KSL used to generate consolidated financial statements. As a result of the company code assignment in the controlling area via Transaction OX19, the global currency amount in table ACDOCA-KSL is populated during posting automatically.

❺ **Freely definable currencies**
SAP delivered an additional eight freely definable currencies that have been available since the SAP S/4HANA 1610 release. However, only from release 1809, SAP has supported introducing a new freely definable currency. All freely definable currencies are dependent on the ledger and company code. Customers can define their own additional definable currencies based on the local and statutory requirements.

As shown in Figure 7.10, the first four currency fields—WSL, TSL, HSL, and KSL—apply to all ledgers and are required during posting.

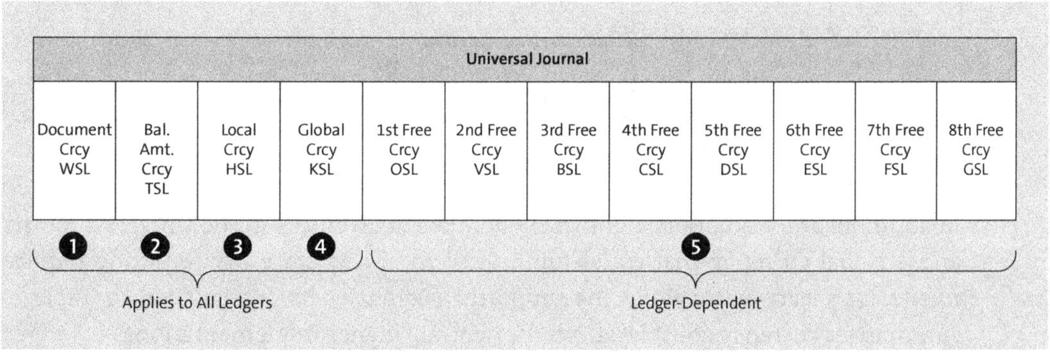

Figure 7.10 Currencies in the Universal Journal

Now that you've understood the basics of supported currencies, let's deep dive into supported currency scenarios from the Central Finance perspective. Although FI supports three currencies in SAP ERP, most source systems may be configured with a single or limited number of currencies. From a Central Finance design perspective, one key technical requirement up until SAP S/4HANA release 2022 is that FI integrated currencies (also called table BSEG currencies) configuration must be identical between the source system and the Central Finance system. In other words, you can't configure an additional integrated FI currency in the Central Finance system when that currency isn't configured in the source system. For example, if the source system isn't configured with group currency, then the system doesn't support configuring FI group currency in the Central Finance system. However, from SAP S/4HANA 2023 release onwards, SAP now supports deviating local currencies in Central Finance. As a result, you'll have an opportunity to define a second or third integrated FI currency in Central Finance that's different than source system. In addition, the system now supports having different currency keys between the source system and Central Finance. For follow-on documents, the system calculates amounts based on the exchange rates maintained in Central Finance. To support deviating FI integrated currencies, one prerequisite is that replicated clearing items are not summarized and the source system **Settings for the Transfer of Clearings** view VCFIN_CLR_SET is set up with the **With Line Items** indicator, as shown in Figure 7.11. This allows clearings in the target system when currency keys are different. Any new clearing line items result in not summarized. Conversely, if this flag is not set and postings are summarized, the system displays the **FINS_CFIN_AC_DOC279 - Activate clearing line-item creation for source company code** error. In addition, to support deviating currencies, SAP recommends several source system notes to be implemented in the source system. Consult SAP Note 2863836 to get the latest list of SAP Notes to be applied in source system to enable the deviating FI currencies functionality.

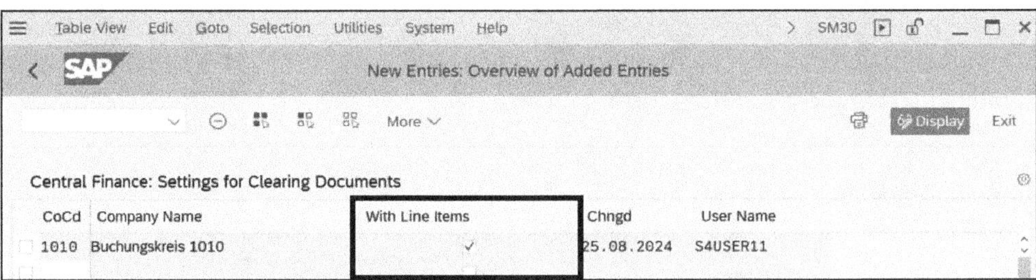

Figure 7.11 Settings for the Transfer of Clearings

In addition, SAP also offers eight freely definable currencies in the Universal Journal and a global currency that could fulfill most local and statutory reporting requirements. Let's learn more about the supported currencies in Central Finance. Table 7.4 summarizes currency considerations for Central Finance implementations.

Currency settings	Central Finance Recommendations
Currency and exchange rate configuration	Currency and exchange rate configuration must be configured as part of baseline FI-CO configuration prior to Central Finance activities. Evaluate configuration needs for the following: ▪ Currency types ▪ Exchange rate types ▪ Translation ratio ▪ Decimal places ▪ Currency conversion settings ▪ Valuation views
Company code currency settings	Must be identical for company code/ledger combinations between the source system and the Central Finance system. In other words, the first local currency must be identical between source system and Central Finance.
FI integrated currencies (BSEG-DMBE2, BSEG-DMBE3)	For clearing transfer activation and follow-on documents to post without errors, FI integrated currencies are now supported with SAP S/4HANA release 2023 onwards. In addition, you have the option to define the freely-defined currencies in the Universal Journal to fulfill Central Finance reporting requirements.
Controlling area currency (COEP)	Must be identical between the source and Central Finance systems. SAP recommends 30 for controlling area currency.
Currency transformation for controlling area currency	Although not supported by SAP, this is technically feasible. Thorough testing is required for any process failures.

Table 7.4 Currency Considerations

Currency requirements vary with the requirements of reporting. Figure 7.12 displays the following typical example scenarios when considering currency design in Central Finance implementations:

❶ **Using freely defined currencies in Universal Journal**
SAP S/4HANA provides up to eight freely defined currencies that can be used to fulfill local and statutory reporting requirements. Therefore, you're allowed to introduce a new freely defined currency during Central Finance implementation.

❷ **Deviating FI integrated currencies**
From the SAP S/4HANA 2023 release onward, Central Finance now supports deviating FI integrated currencies. Prior to this release, this functionality wasn't available, and the recommendation is to use freely defined currencies for any reporting requirements. Consult SAP Note 3433278 for additional information and restrictions on the support of this new feature.

❸ Fewer currencies in Central Finance
If the source system is assigned with more currencies and Central Finance is defined with fewer currencies, the system removes the additional currency when transferring the documents to Central Finance, as you can see in the figure. Consider the implications of merging company codes before considering this design approach.

❹ Mapping to a global currency and freely defined currencies
If the source system is defined with second and third integrated FI currencies, you can map them to a global currency and freely defined currency in Central Finance for the relevant company code.

		FI-Integrated Currencies (Table BSEG)			Universal Journal Currencies (Table ACDOCA)					
		DMBTR 1st Local Curr.	DMBE2 2nd Local Curr.	DMBE3 3rd Local Curr.	HSL Local Curr.	KSL Global Curr.	OSL Free Curr. 1	VSL Free Curr. 2	BSL Free Curr. 3	GSL Free Curr. 8
1	Source	10	30	40	N/A	N/A	N/A	N/A	N/A	N/A
	Central Finance	10	30	40	10	30	Z1	Z2	40	
2	Source	10			N/A	N/A	N/A	N/A	N/A	N/A
	Central Finance	10	30	40	10	30	Z1			
3	Source	10	30	40	N/A	N/A	N/A	N/A	N/A	N/A
	Central Finance	10	30		10	30	Z1			
4	Source	10	30	40	N/A	N/A	N/A	N/A	N/A	N/A
	Central Finance	10			10	30	40	Z1		

Figure 7.12 Central Finance Currency Considerations

Note

SAP introduced universal parallel accounting (UPA) to support all Universal Journal currencies across all SAP components, such as material ledger and asset accounting. This is a big leap forward in the currency and parallel accounting design considerations. However, UPA isn't supported fully in Central Finance as of writing this book. If UPA is activated in the source system, central finance transfers values of legal valuation ledger to central finance with several restrictions. Consult SAP Notes 3191636, 3305495, and 3265275 for information on this topic.

7.2.3 Ledgers

Ledger design is important to support local and statutory reporting as well as management reporting in Central Finance. SAP provides several approaches to account for parallel accounting needs in different releases of the general ledger as the product evolved. Table 7.5 shows the various options SAP offers for parallel accounting functionality. Although parallel accounts are equal to and as powerful as the ledger approach, SAP recommends using the ledger approach for new SAP implementations, including Central Finance, because it improves the integration of FI and managerial accounting in SAP. The account approach is still available and can be used where parallel ledgers can't be introduced without going through a full-scale project. The parallel ledgers released in new general ledger accounting is one of the foundational functionalities introduced with the new general ledger.

Options for Parallel Accounting	Classic General Ledger	New General Ledger (ERP)	S4/HANA General ledger
Parallel company codes	Available – Not recommended	Available – Not recommended	Available – Not recommended
Parallel special purpose ledger	Available	Available	Available
Parallel accounts	Available	Available	Available
Parallel ledgers	Not available	Recommended approach	Strongly recommended approach

Table 7.5 Parallel Accounting with Different Releases of the General Ledger

Parallel Ledgers

Now that you have a basic understanding of parallel accounting, let's review key design aspects from the Central Finance perspective. From Table 7.5, you can see that SAP strongly recommends the ledger approach, so we'll limit our discussions to parallel accounting with the parallel ledger approach. This recommendation applies to Central Finance implementations as well. With this approach, additional general ledgers or "independent sets of books" are created to meet parallel valuations for different accounting principles or varying fiscal year variant requirements. When transactions are posted to an account with a blank value in the **Ledger Group** field, the system posts values in all ledgers in the Universal Journal. For valuation differences, ledger-specific transactions are posted to a specific ledger. Figure 7.13 shows the concept behind the parallel ledger approach in SAP S/4HANA general ledger accounting.

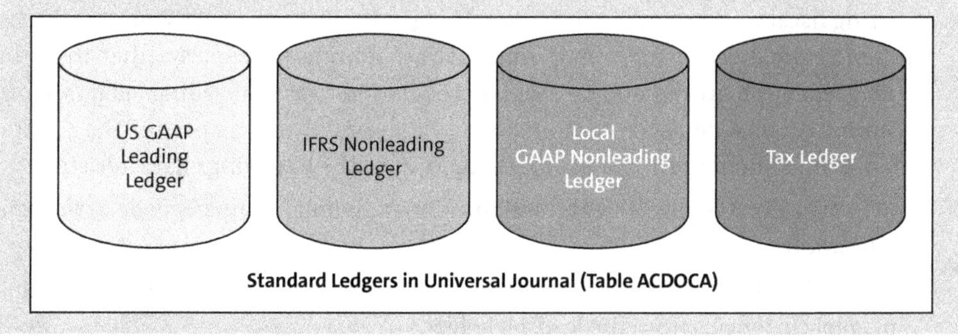

Figure 7.13 Standard Ledgers

While implementing Central Finance, a design decision must be made when considering ledger requirements. There are two types of ledgers in the SAP S/4HANA general ledger: standard ledger and extension ledger. As you've noticed from Figure 7.13, both leading and nonleading ledgers are called standard ledgers. In addition to standard ledgers, SAP includes extension ledgers that provide additional reporting capabilities without occupying additional data storage. From a Central Finance perspective, you identify ledger requirements, including extension ledgers, during the explore phase. In this section, let's discuss the three most important ledger design considerations:

- **Leading ledger**
  A leading ledger is a standard ledger with a full set of configurations and posting information. As the name implies, among the multiple standard ledgers defined, one ledger plays a leading ledger role. The leading ledger is the basis for deriving vital information such as the fiscal year variant and parallel currencies from company code. SAP delivers leading ledger 0L. Leading ledger is unique in FI and has the following characteristics:
  - The system assigns all company codes to the leading ledger and can't deactivate any company codes.
  - Usually, the leading ledger reflects the accounting principle of a consolidated company book of records.
  - The leading ledger is integrated with all subsidiary ledgers.
  - The leading ledger is always assigned to CO version 0. When values from the leading ledger are posted to CO, only version 0 is updated.
  - When migrating from the classic general ledger to the SAP S/4HANA general ledger, balances are migrated to the leading ledger by default.
  - In the account approach, only one ledger is defined, and the underlying ledger is leading ledger 0L.
  - Asset accounting book depreciation area 01 is assigned to leading ledger 0L
  - The default legal view for material ledger currencies is the leading ledger.
- **Nonleading ledger**
  A nonledger is a parallel ledger created to support a local accounting principle or a differing fiscal year variant. In the general ledger, the system allows you to maintain multiple standard ledgers to map different accounting principles. Other than the leading ledger, all other standard ledgers are referred to as nonleading ledgers. From a Central Finance perspective, you should be aware of this as part of the Central Finance implementation. Nonleading ledgers have the following characteristics:
  - Activate a nonleading ledger by assigning an individual company code in the configuration.
  - Nonleading ledgers derive posting relevant information, fiscal year variant, and parallel currencies from the leading ledger.

- Nonleading ledgers are used to fulfill varying fiscal year requirements per local and statutory requirements*. For example, the leading ledger uses a calendar year, whereas a nonleading ledger uses fiscal year starts from April to March.
- A unique document type and number range can be assigned to nonleading ledgers.

- **Extension ledger**
  Figure 7.14 illustrates the concept of the extension ledger in SAP S/4HANA. An extension ledger was introduced with SAP S/4HANA to support reporting and to generate financial statements for specific use cases. It persists delta values on top of an underlying ledger in the Universal Journal to provide users with the ability to post adjustment entries and report per management reporting requirements. An underlying ledger can be either a standard ledger or an extension ledger. Extension ledgers can stack multiple ledgers on top of each other to provide different reporting views. An important use case of an extension ledger is to create management account views that differ from external/legal views of accounting data. The extension ledger inherits currency settings and the fiscal year variant from the underlying base ledger. In addition, it can have separate open period variants and company code assignments. Figure 7.14 shows a standard ledger ❶ and an extension ledger ❷ that read values from another underlying extension ledger and a standard ledger.

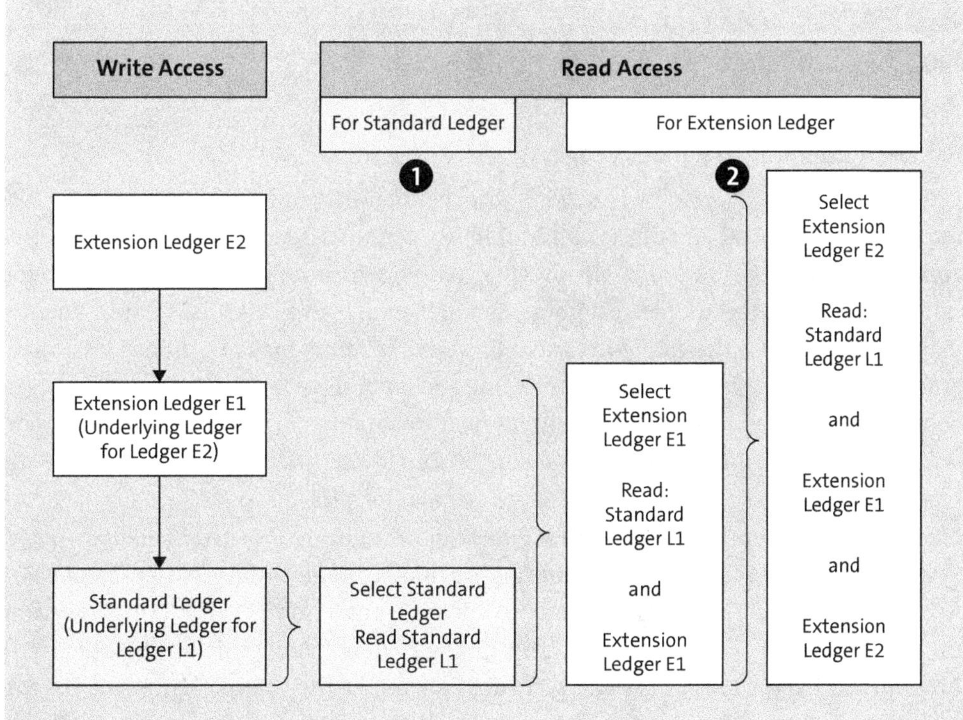

Figure 7.14 Extension Ledger Concept

Now that you understand the ledger definitions and concepts, let's review parallel ledger recommendations from the Central Finance viewpoint, as shown in Table 7.6.

Parallel Ledger	Central Finance Recommendations
Source system with parallel ledgers	SAP recommends the ledger approach in SAP S/4HANA.
Source system with account approach	SAP provides a parallel accounting account to the ledger approach configuration where you'll map the source system chart of accounts and general ledger accounts to a ledger group in Central Finance.
Deviating fiscal year variants	The initial load doesn't support deviating the fiscal year variant but instead uses the fiscal year variant of the leading ledger. Consult SAP Note 2764851 to understand certain limitations of deviating fiscal year variants.
Ledger group–specific posting	General ledger accounts defined with ledger group–specific clearing must match between the source system and target system.
Special purpose ledger	Special purpose ledger replication isn't supported.

Table 7.6 Ledger Key Considerations for Central Finance

Parallel Accounting to Parallel Ledger

If a source SAP ERP system is set up with parallel accounting using the accounts approach and Central Finance is configured with parallel approach, SAP provides this configuration (program SAPLCFIN_ACCT_LDGRP), which helps to establish the relationship between the source system general ledger accounts and the Central Finance ledger group. This is the case with SAP ERP source systems that run in the classic general ledger. In addition, parallel accounting includes separate valuations to support country-specific requirements. To use this relationship mapping, the fiscal year variant and currencies must be identical between the source and target systems. To define the relationship between parallel accounts to parallel ledgers, go to IMG menu path **SAP Customizing Implementation Guide • Financial Accounting • Central Finance: Target System Settings • Settings for Accounting Document Replication • Assign Source G/L Accounts to Central Finance** or navigate to it from Transaction CFINIMG. Upon opening the configuration, you'll arrive at the **Change View "Assignment of Source G/L Accounts to Target Ledger Group"** screen, as shown at the top of Figure 7.15. On this screen, click the **New Entries** ❶ button to assign validations groups to validation areas.

On the resulting **New Entries: Overview of Added Entries** screen, enter the values in the relevant fields as follows ❷:

- **Logical System**
  Enter the source logical system here.
- **ChtAccts**
  Enter the chart of accounts of the source system.
- **From G/L Account**
  Enter the general ledger account starting range.
- **To G/L Account**
  Enter the general ledger account ending range.

  The range of accounts maintained can be mapped to one target ledger group. The system doesn't consider accounts outside of this range. In addition, if you maintain multiple account ranges, ensure the accounts don't overlap.
- **Ledger Grp**
  Enter the ledger group of the target system where the ledger group is assigned with parallel ledgers. Leave it blank if a ledger-specific posting isn't transferred. If one posting with different accounts is mapped to multiple ledger groups, you'll notice multiple postings in Central Finance. If source postings contain a document type with an external number, SAP recommends you maintain the **Substitute Original Document Type in Central Finance** Customizing activity. Step-by-step instructions are covered in Section 7.4.3.

Click the **Save** ❸ button to save the configuration entry.

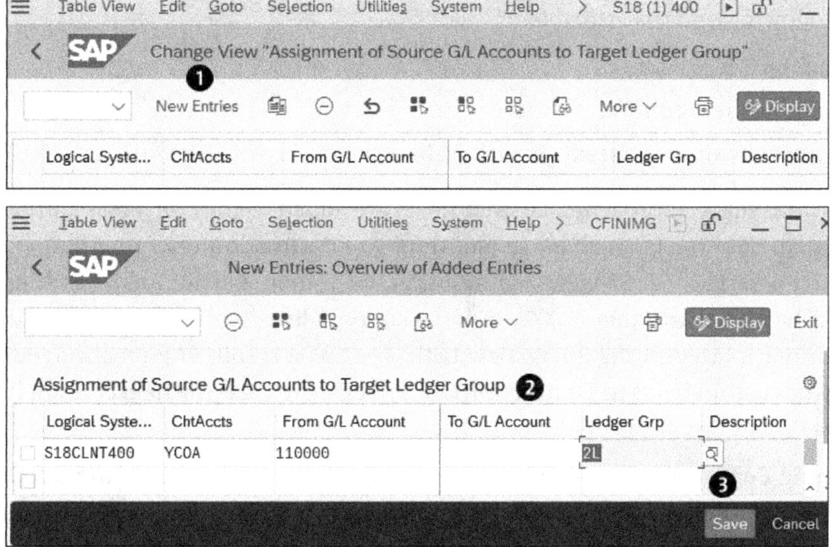

Figure 7.15 Assign Source General Ledger Accounts to Central Finance

[!]

Limitations

Although SAP supports parallel accounting to the parallel ledger approach in the Central Finance scenario, SAP recommends reviewing the following restrictions prior to configuration:

- Balances must tie to zero if line items of different ledger groups are posted in the same document.
- The cross-ledger clearing postings scenario isn't supported.
- The system posts ledger-specific postings to all ledgers in Central Finance.

SAP recommends avoiding postings of such nature in the source system.

7.2.4 Extensibility

SAP provides multiple extensibility options to meet business-specific needs by extending standard functionalities without disrupting core system while ensuring smooth transition during upgrades and maintenance cycles. Traditionally, SAP provided two ways to add custom fields:

- Add custom field using classic coding block (structure CI_COBL) extension approach via Transaction OXK3.
- Add used defined characteristics in COPA using Transaction KEA5.

In SAP S/4HANA, additional extensibility options are available using the Custom Fields and Logic SAP Fiori app (app ID: F1481). The app provides flexibility in creating custom fields on user interface, reports, email templates and so on. From the Universal Journal standpoint, following options available:

- *Coding block* business context.
- *Journal entry* business context
- *Market segment* business context, available since the release of SAP S/4HANA 2020

In all three extensibility approaches, custom fields are added to the Universal Journal (ACDOCA) and plan data (ACDOCP) tables. In addition, you'll run additional programs to update core data servies (CDS) views. In addition, customer-specific fields (coding-block) should be added to table ACDOCA only to support line-item reporting in SAP S/4HANA, but not to accounting entry view table BSEG. As a result, SAP recommends maintaining the custom fields in Transaction OBCY/OBCYX to clear the field content in BSEG during document replication.

When designing Central Finance, you must evaluate your requirements and adopt extensibility. From a Central Finance context, Table 7.7 lists support extensibility scenarios. In Central Finance, you need to adopt extensibility options like the source system to support smooth replication. In addition, when adding customs fields through

Custom Fields and Logic app, the system doesn't allow you to add a check table for user-defined characteristics. You can make a specific code list for the type and assign it to custom field. Consult SAP Note 3090299 for further guidance on this topic.

Extensibility Option in Source System	Central Finance System	Central Finance Support
Custom field created using classic coding bock approach via Transaction OXK3	Custom field transferred to Central Finance for FI and CO documents	Supported; source system equivalent coding block field must be maintained in Central Finance
Custom field with a technical name	Differencing technical name to central finance due to extensibility app creating prefix and suffix on custom field	Supported; technical field mapping required in SAP Landscape Transformation Replication Server or using a Central Finance BAdI
Key user extensibility via Custom Fields and Logic app using Coding Block business context	Custom field transferred to Central Finance for FI and CO documents	Supported; however, when the equivalent custom field is created using the Custom Field app, the technical field may change and require technical field mapping using SAP Landscape Transformation Replication Server
Key user extensibility via the Custom Fields and Logic app using the journal entry item business context	Custom field transferred to Central Finance for FI document only; CO document extension is not supported	Partially supported; technical SAP Landscape Transformation Replication Server or BAdI mapping is required for differing technical names
Key user extensibility via the Custom Fields and Logic app using market segment business context	Custom field transferred to Central Finance for FI and CO documents	Supported; SAP Landscape Transformation Replication Server technical mapping is required

Table 7.7 Extensibility in Central Finance

[!]

Restrictions

Custom field types of amount, quantity, or number are treated as characteristics in some applications. As a result, these applications do not treat these custom fields as key figures, which may lead to inconsistent results. If a key figure is included as custom field, you'll need to test it for applications such as balance carryforward, document splitting, and summarization. Consult SAP Note 2453614 for additional guidance.

7.2.5 Tax Configuration

Central Finance supports tax reporting using the replicated centralized data from Universal Journal. To perform tax reporting in Central Finance, when financial documents are replicated with tax data, especially sales/use tax, VAT, withholding tax, and deferred tax, tax-relevant data must be replicated consistently. To achieve uniform data, tax configuration settings must be identical between the source system and the Central Finance system. In addition, central payment activation requires consistent tax data to carry out central clearing activities and central tax reporting in Central Finance. On the other hand, if central payment isn't activated in Central Finance, all payment activities remain in the source system, and tax reporting is carried out in the source system. Before you configure tax configuration, ensure you get answers to these questions to scope the configuration activities accordingly:

- Is Central Finance used as merely a reporting system? Or are centralized operational activities such as central payment carried out in Central Finance?
- What tax scenarios are in scope: sales/use tax, VAT, GST, withholding, deferred tax, and so on?
- What's the future state of tax reporting? Will it be performed in Central Finance or the source system? Tax reporting is mandatory in Central Finance when central payment is active and with certain country-specific situations.
- Is external third-party bolt-on software used in managing taxes such as Vertex or Thomson Reuters ONESOURCE indirect tax solutions?

Based on the responses, you'll harmonize tax data and configure tax Customizing activities in Central Finance. Table 7.8 provides general guidelines on tax data harmonization prior to begin configurating tax customizing in Central Finance.

Tax object	Source System	Central Finance	Central Finance Recommendation
Tax procedure	▪ Country specific tax procedure configured ▪ Evaluate tax procedures used in other source systems and use which one	▪ Leading practice is to leverage SAP delivered tax procedure for countries maintained	▪ Tax calculation procedure (T005-KALSM) must be identical between source system and central finance system ▪ 1:1 tax procedure supported between source system and central finance system ▪ N:1 supported when multiple source systems connected ▪ 1:N mapping scenario supported if source system tax procedures are combined requires custom mapping

Table 7.8 Tax Objects Harmonization Recommendations

Tax object	Source System	Central Finance	Central Finance Recommendation
Account Key and condition type	▪ Condition type included FI tax calculation procedure and accounting type mapped to equivalent settings in central finance system		
Tax Jurisdiction codes	▪ Harmonization required to support tax jurisdiction codes if multiple source system involved using different naming		
Tac Code	▪ Tax codes properties (e.g., **Tax Type**, **CheckID**, etc.) that must be identical between the source system and Central Finance system ▪ Use country-delivered tax codes ▪ Tax codes from multiple source systems with N:1 and/or different properties may need custom mapping ▪ Align on accounts to which tax postings to be made are mapped in key mapping		
Plants Abroad	Active	Not active	Not supported
	Not active	Active	Supported
Registration for Indirect Taxation Abroad (RITA)	Active	Plants abroad is active	Supported; **Activate Conversion of RITA Postings to Plants Abroad Postings** configuration must be set up
	Active	Active	Supported
Withholding tax	Classic withholding tax	Enhanced withholding tax	Not supported; you must migrate source system extended withholding tax
	Enhanced withholding tax	Enhanced withholding tax	Supported
Deferred Tax	▪ Identify scope of countries require deferred tax legally ▪ Understand deferred tax replication limitations especially on initial load and replication ▪ Tax reporting requirements		
Localization	▪ Understand harmonization requirements if multiple source systems involved with varying local and statutory requirements ▪ Central Finance does not support localization in certain countries. For example, Nota Fiscal replication is not supported for Brazil Tax Authority (SEFAZ). However, this functionality request has been added to the Improvement Request List in Customer Influence		

Table 7.8 Tax Objects Harmonization Recommendations (Cont.)

Once tax customizing is completed, you'll run the tax consistency checks report early on to prevent potential inconsistencies in tax configuration and data. During replication of financial documents with tax data from the source system to the Central Finance system, the system calculates tax values based on the Central Finance system configuration. Recalculation happens in Central Finance merely for comparing calculated values in the Central Finance system to replicated values from the source system. Nevertheless, to accomplish consistent tax data in Central Finance, you'll enable the **Activate Tax Consistency for Company Codes** configuration in Central Finance. As a prerequisite, you'll set up trigger-based replication for tax-relevant tables in SAP Landscape Transformation Replication Server. SAP Landscape Transformation Replication Server configuration of tax tables is covered in Chapter 5.

Once SAP Landscape Transformation Replication Server is set up for tax tables, you'll enable the **Activate Tax Consistency for Company Codes** configuration in Central Finance. To activate tax consistency, go to IMG menu path **SAP Customizing Implementation Guide • Financial Accounting • Central Finance: Target System Settings • Settings for Accounting Document Replication • Activate Tax Consistency Check for Company Codes**, or navigate to it from Transaction CFINIMG. After entering the configuration, you'll arrive at the **Change View "Activate Tax Consistency Check for Company..": Overview** screen, as shown in Figure 7.16. On this screen, click the **New Entries** ❶ button. On the resulting **New Entries: Overview of Added Entries** screen, maintain values for the **Logical system** and **Src CoCode** ❷ fields. Click the **Save** ❸ button to save the configuration. If SAP Landscape Transformation Replication Server tables aren't configured for tax tables, you'll receive an error message on this screen.

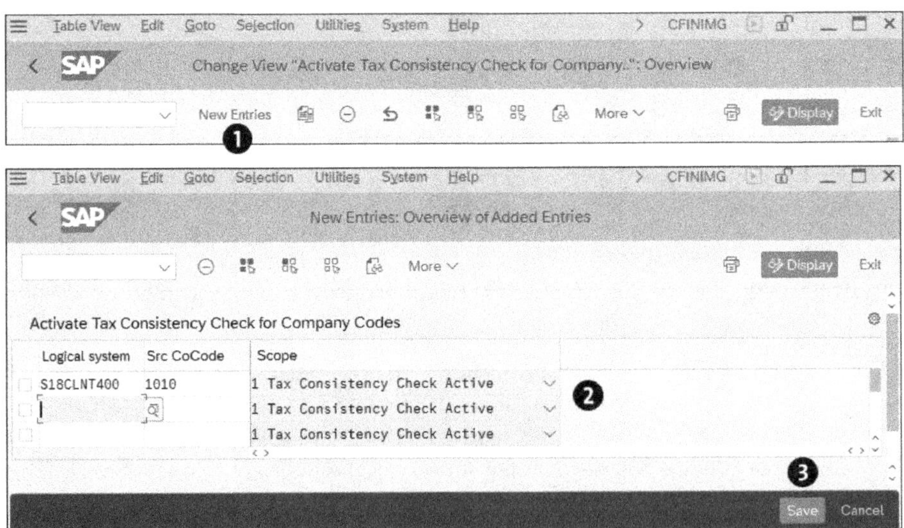

Figure 7.16 Tax Consistency Check

[+]

Tip

SAP Notes 2494127 and 2668261 provide the latest list of tables to be set up in SAP Landscape Transformation Replication Server for tax consistency checks and withholding tax consistency checks in Central Finance, respectively. If your source system is running classic withholding tax, you must update to extended withholding tax as the classic withholding tax isn't supported in Central Finance. For withholding taxes, refer to SAP Note 2623514 for required SAP Notes referred to in the **Reference** section of the OSS support portal.

7.2.6 Intercompany Transactions

Intercompany transactions or cross-company code transactions are business activities carried out between two or more affiliated companies in a company. You need to evaluate requirements for intercompany transactions involving both buying and selling and stock transfer orders (STO) scenarios involving logistics components. In the case of intercompany transactions with buying and selling, the system posts relevant financial transactions to accounts receivable (AR) and accounts payable (AP) subledgers in the source systems. During the financial postings, the system generates offsetting entries automatically based on configurations to balance the debits and credits and then assigns a cross-company number (BVORG field). Intercompany transactions are uniquely identified with a separate document type, document number, fiscal year, and company code. Likewise, STOs are posted to inventory accounts that are replicated to Central Finance. These intercompany postings in the source system are then replicated to Central Finance.

From the Central Finance perspective, to replicate intercompany transactions from source systems to the Central Finance system, you need to configure Central Finance to support the initial load and replication activities. Consider important design and configurations items related to intercompany transactions as follows:

- Chart of accounts created to support intercompany replication scenarios
- Intercompany automatic posting and clearing account configuration (Transaction OBYA)
- Initial load groups combined for company codes involving intercompany transactions to prevent initial load errors and reconciliation issues
- Harmonization of business partners master data
- Designing intercompany customer and vendor groups as required by business requirements
- Business partner design involving appropriate naming conventions (e.g., prefix with "IC") to differentiate the business partners from other customers
- Trading partner derivation rules
- Configuration of intercompany matching and reconciliation, and enabling relevant SAP Fiori apps

7.2.7 Document Summarization

In general, source systems with a high volume of transactions coming in from sales and distribution and materials management modules may lead to a higher data footprint in the entry view of the accounting line-item table (table BSEG). As a result, systems may experience performance problems with slow responses to critical processes. The SAP document summarization in FI feature allows you to store aggregated transactions in the accounting line-item table. From a Central Finance perspective, the document summarization configuration must be identical between the source system and the Central Finance system. This is to ensure that upon transferring line items to the Central Finance system, the system doesn't produce more accounting lines in the entry view (table BSEG) than what's posted in the entry of the document in the source system. In other words, if document summarization doesn't summarize some line items in Central Finance, the corresponding postings have been summarized in the source system. In such situations, a subsequent clearing document will only contain clearing data for document items known in the source system and not for additional items that may exist in Central Finance. This is to prevent documents from failing, especially when you have follow-on documents that don't result in line-item differences. To prevent summarization-related errors, it's important that document summarization settings are identical between source systems and the Central Finance system. Otherwise, the system displays the **FINS_CFIN_AC_DOC025** error in SAP Application Interface Framework with the **Source document &1 &2 &3 not posted due to inconsistent summarization** message. To troubleshoot this, you can check in BSEG-BUZEI_SENDER if multiple line items are posted with the same value in Central Finance.

Table 7.9 describes supported configuration scenarios for summarization settings between the source system and the Central Finance system.

Source System	Central Finance	Support
FI Summarization Dependent on the Object Type (Transaction OBCY)	FI Summarization Dependent on the Object Type (Transaction OBCY)	Supported
Enhanced Document Summarization in FI Dependent on the Object Type (Transaction OBCYX)	Enhanced Document Summarization in FI (Transaction OBCYX)	Supported. Summarization can be defined per company code to help harmonize configuration with various source systems. On the other hand, multiple logical systems with different summarization postings to the same company code isn't supported
Customer-specific fields (coding block CI_COBL)	Similar settings must be setup in Transaction OBCY or Transaction OBCYX settings	Makes sure the summarization of coding block fields doesn't result in line splitting

Table 7.9 Document Summarization Settings

To view the independent summarization of the entry view in your source system, navigate to IMG menu path **SAP Customizing Implementation Guide • Financial Accounting • Financial Accounting Global Settings • Document • Summarization • Independent Summarization of Entry View** or use Transaction OBCY. Upon executing the transaction, you're prompted with the **Determine Work Area: Entry** popup, as shown in Figure 7.17. On this screen, enter the **Reference procedure** under **Work Area** ❶ and continue. This leads to the screen that shows already defined summarizations if any ❷. Ensure you maintain identical settings between Central Finance and the source systems. Alternatively, you can check the summarization configuration in enhanced settings in Transaction OBCYX.

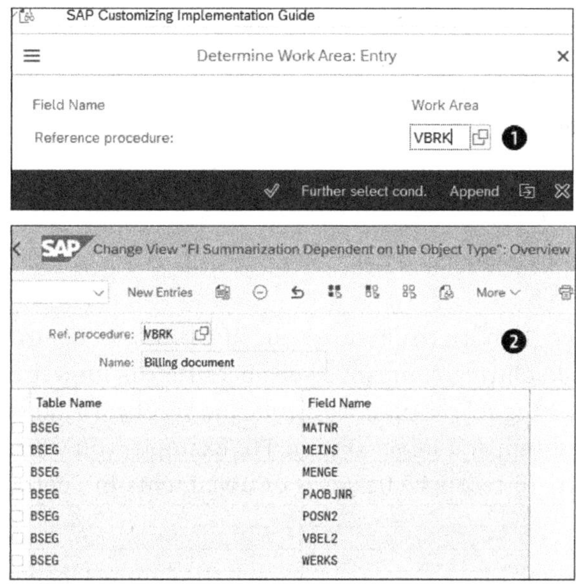

Figure 7.17 Independent Summarization of the Entry View

Note

SAP offers summarization at the Universal Journal (table ACDOCA) and a deferred summarization functionality. The functionality can be set up differently from the source system settings.

7.2.8 Merging or Splitting Company Codes

Central Finance doesn't support either merging (M:1) or splitting company codes (1:M) during transfer of documents from the source system to the Central Finance system:

- **Merging company codes**
  Technically, Central Finance allows merging company codes (M:1) in value mapping. If documents from two independent company codes are posted in the source system, this will result in two documents in the same company code in Central Finance.

However, if the source system company code is posted with cross-company code postings, documents will fail during transfer of follow-on documents during the initial load and real-time replication. If such a requirement arises, this must be taken as a custom solution and requires further evaluation and feasibility. The custom solution will include removing or deleting intercompany relationships (by clearing BVORG) to post the transactions successfully in Central Finance. In addition, trading partners (VBUND) cleared and rederived Central Finance derivation rules. One prerequisite to this is that the trading partners must be the same for the merged company codes. If a custom solution is considered, extensive testing is required in several functional areas:

  - Intercompany transactions
  - Central payment with payment processes
  - Central payment with down payment integration to the logistics processes
  - Consistency check reports
  - Comparison reports
  - Company specific FI-CO processes included as necessary

- **Splitting company codes**
  Likewise, splitting company codes (1:M) isn't supported out of the box in Central Finance. SAP doesn't recommend splitting company codes because the impact is much wider when it comes to posting follow-on documents and reconciliation of documents between the source system and target system. For example, you won't be able to use comparison reports to reconcile balances or documents in Central Finance.

Next, let's dive deeper into configuration of the source system and the Central Finance system. Most activities can be done independently, but there are a few configurations that depend on the other configurations. For example, the tax consistency check requires SAP Landscape Transformation Replication Server tax tables first.

[!]

Warning

From a leading practice standpoint, merging (M:1) or splitting (1:M) of company codes is *strongly* not recommended as these scenarios aren't supported in Central Finance out of the box. As Central Finance continues to evolve with new features and additional code constantly being added, the custom solution requires reevaluation of each major release or service pack upgrade.

7.3 Configuring Source Systems

This section covers configuration related to the SAP source system in Central Finance integration scenario. After upgrading to the latest service pack and/or applying relevant SAP Notes, the source system enables the Central Finance–relevant configuration menu to become visible to you. The source system configuration must be repeated for each source system connected to Central Finance. Both the source system configuration and Central Finance configuration can be carried out in parallel. However, there are some technical dependencies, such as RFC connections, required prior to configuration.

7.3.1 General Replication Settings

Before you begin Central Finance initial load and replication activities, the general replication settings must be carried out for each company code in scope in the source system. The underlying table for this configuration is table VCFIN_SOURCE_SET. This configuration is a prerequisite to transfer the initial load data and then to enable real-time replication of documents from the source system to target the Central Finance system. In addition, the same table is used for both the standard initial load and intermediate data retention (IDR).

To perform this activity, follow the IMG Customizing menu path **SAP Customizing Implementation Guide • Financial Accounting • Central Finance: Source System Settings • General Replication Settings** or navigate to it from Transaction CFINIMG in Central Finance. Alternatively, the settings can also be accessed using Transaction SM30 with appropriate access in the source system. Once the transaction is called up, the system presents the **Change "Customizing Central Finance Source Systems": Overview** screen, as shown in Figure 7.18. Take note of the following options:

❶ **New Entries**

On the initial screen, click **New Entries** to maintain entries for each company code in scope for Central Finance.

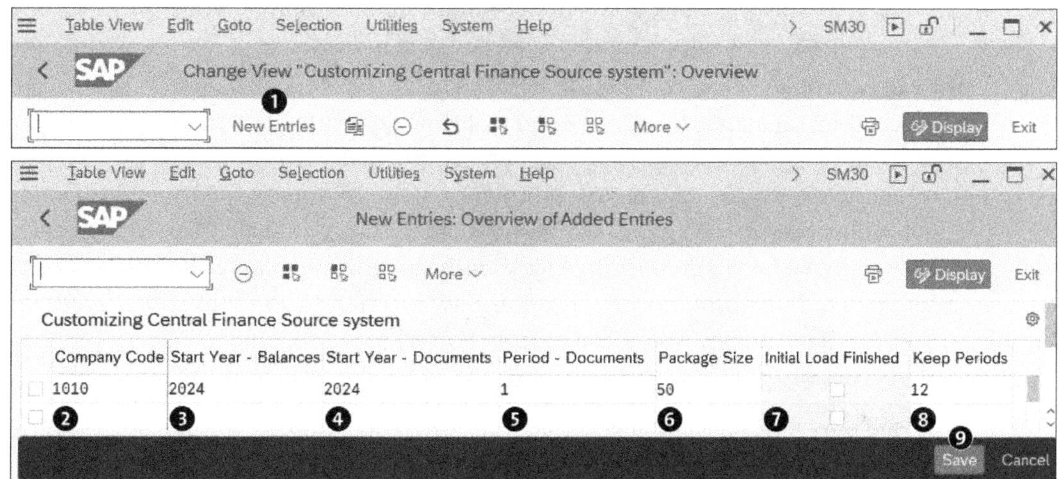

Figure 7.18 General Replication Settings

❷ **Company Code**
The Central Finance system extracts data by company code. In this field, enter each company code in scope for Central Finance.

❸ **Start Year - Balances**
In this field, capture the fiscal year to transfer balances from the source system during the initial load. As you've noticed, there is no option to enter monthly information for balances, so the system takes the balances starting from the first fiscal period of the year maintained in this field. The system transfers balances and open items with package keys starting with S and O.

❹ **Start Year - Documents**
This field indicates the fiscal year from which the detail transactions are to be transferred during the initial load. This field, along with the **Period - Documents** field, determines the period from which transactional details are to be transferred during the initial load.

❺ **Period – Documents**
This field, in conjunction with the **Start Year - Documents field**, determines the detail transactions to be transferred in the initial load. The system transfers transactions with package keys starting with D.

❻ **Package Size**
This field is used primarily for the extraction of initial load data. When the initial load extraction begins, the system uses this field value to determine which documents are packaged. The default value is 50 due to performance factors; however, this number can be changed based on the number of line items in a document. SAP recommends leaving the default value as 50.

❼ **Initial Load Finished**
This indicator is set after a successful initial load extraction and delta extraction is completed. Once this flag is set, a subsequent initial load extract ignores the company code checked with the Initial Load Finished indicator.

❽ **Keep Periods**
When Central Finance is activated and real-time replication is turned on, the staging tables are populated in the source system. This field defines how many periods data is retained in source staging tables. Program RFIN_CFIN_CLEANUP uses this field to determine how many periods will be retained in the Central Finance staging tables. One prerequisite to run this report is that the logging table must be defined in SAP Landscape Transformation Replication Server. Prior to executing this program, check if there are any unreplicated transactions in SLT logging table with program FIN_CFIN_SLT_LOGTAB_CHECK.

Note that the **GL Reconciliation Postings Transferred** field is set only when replication of CO postings isn't in scope, so the system transfers general ledger reconciliation postings to Central Finance via financial postings. This flag isn't available in the SAP S/4HANA setup as the source system and the flag are available only in the SAP ERP setup as a source system. In the case of SAP S/4HANA as a source system, by design, FI and CO are reconciled. On the other hand, for SAP ERP source systems, if CO secondary postings replication isn'taren't in scope, then general ledger reconciliations postings triggered in CO flow through FI-CO postings to Central Finance. In this case, SAP recommends selecting the **G/L Reconciliation Postings Transferred** indicator on VCFIN_SOURCE_SET view. On the other hand, if CO secondary postings are in scope, then general ledger reconciliations that are triggered in CO flow through the CO secondary postings. Thus, SAP recommends not activating the **G/L Reconciliation Postings Transferred** indicator when CO secondary postings are in scope. Otherwise, this will lead to duplicate postings.

Once all fields are configured per company code requirements, click the **Save** ❾ icon to save the entries. The information is updated to table VCFIN_SOURCE_SET.

[«]

Note

From a best practice standpoint, maintaining general replication settings (view VCFIN_SOURCE_SET) in the source system is considered a client-dependent activity. Therefore, the activity must be done in each client prior to the initial load extraction. In addition, the manual activity should be captured in the cutover plan.

7.3.2 Initial Load for Accounting Documents Extraction Settings

Central Finance introduced this new feature for how initial load data can be extracted from the source systems. The configuration is relevant for SAP S/4HANA or SAP ERP with the new general ledger as a source system. During the extraction step, the system offers to extract data in two ways:

- Extract data from the entry view (this is the default system behavior)
- Extract data from the general ledger view

The configuration offers the benefit of replicating freely defined currencies. If summarization is enabled in the source system, SAP recommends not activating these settings. By default, the system extracts accounting documents from the entry view. However, in this configuration, you can define how the system extracts accounting documents.

If there is ever a need to switch to extracting accounting documents from the general ledger view, navigate to IMG menu path **SAP Customizing Implementation Guide • Financial Accounting • Central Finance: Source System Settings • Initial Load for Accounting Documents: Extraction Settings** or use Transaction CFINIMG. After executing the transaction, the system opens the **Change View "Initial Load View Determination": Overview** screen, as shown in Figure 7.19. By clicking **New Entries** ❶ on this screen, you'll be taken to the **New Entries: Details of Added Entries** screen. In this screen, enable the **IL from G/L View** ❷ flag. Click the **Save** ❸ icon to save the configuration entry. The system saves the configuration into table VCFIN_IL_VIEW_DT.

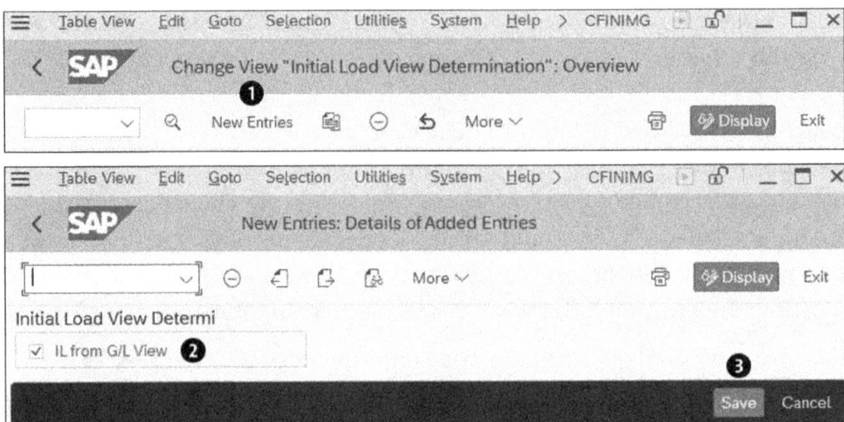

Figure 7.19 Initial Load for Accounting Documents Extraction Settings

Note

For non-SAP systems, only the entry view is replicated to Central Finance. However, there are some restrictions that come by enabling the **IL from G/L View** flag. The functionality isn't supported when the following functionalities are active in the source system:

- Document splitting
- Summarization

Refer to source system SAP Note 2907310, which provides additional guidance and some manual activities to enable this functionality.

7.3.3 Handling of Document Splitting

In Section 7.2.1, we discussed document splitting design concepts from a central finance context. In this configuration, you can define how SAP ERP source systems should handle document splitting when the source system isn't enabled with the document splitting functionality. To prevent document splitting errors, SAP provides precheck configuration, which helps to display a warning or an error message prior to posting financial documents in the source system. The configuration is enabled at the company code level in Central Finance. Once the configuration is enabled, every time a financial posting is made in the source system, the system checks whether the document complies with document splitting rules or simply sends the information to the Central Finance system. For example, if the system finds any noncompliant document splitting data, an appropriate error or warning message is raised based on settings maintained in view VCFIN_SOURCE_SPL.

Following are the options the configuration provides for document splitting:

- **Transfer splitting information from source to target**
  The system transfers document splitting information to the target system with the

posted main document. This scenario applies when document splitting configuration is identical between the source system and Central Finance system. The option is to avoid the splitting again in the Central Finance system.

- **Run splitting process checks; raise error message**
  In the second scenario, the system applies to the source system without the document splitting functionality, but the document splitting is activated in the Central Finance system. During financial posting, the system performs document splitting checks to ensure the entry data complies with the document splitting rules. Noncompliant data is prevented from being posted and raises appropriate error messages. For example, if the profit center is mandatory per document splitting configuration, but it's not populated on certain line items, the system will raise the balancing profit center error message: **GLT2201 Balancing field Profit center is not filled.**

[!]

Warning

Note the option may disrupt the source system process by preventing posting. Thus, it interrupts source system processes by preventing to achieve the goal of nondistruption to source systems. We suggest consulting with business teams on the impact caused by this approach during month-end closing and delays in posting failed documents.

- **Run splitting process checks; raise warning message**
  Likewise, this scenario also applies to the source system without the document splitting functionality, but the document splitting is activated in the Central Finance system. Instead of raising an error and preventing users from posting, the system raises a warning message. SAP recommends using this option to validate the feasibility of a functionality.

To enable document splitting validation, follow IMG menu path **SAP Customizing Implementation Guide • Financial Accounting • Central Finance: Source System Settings • Handling of Document Splitting** or navigate to it from Transaction CFINIMG in Central Finance. Figure 7.20 shows the entry view of **Handling of Document Splitting Configuration.**

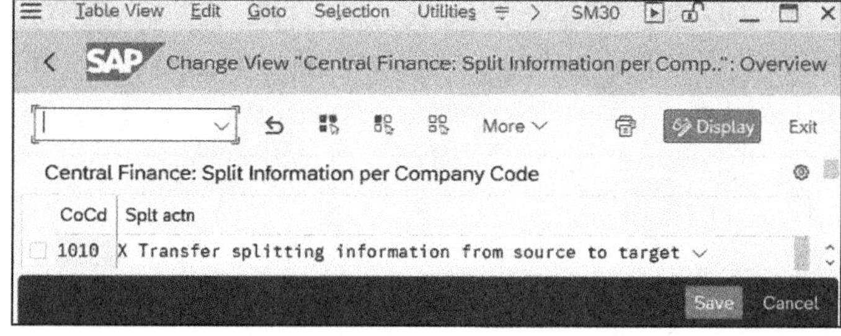

Figure 7.20 Handling of Document Splitting

On this configuration screen, maintain values in the **CoCd** field, and choose **X Transfer splitting information from source to target** from the **Split actn** dropdown field. Upon saving the information, the system stores it in the `VCFIN_SOURCE_SPL` view.

> **Note**
> If you're on an earlier version of Central Finance, consult the following SAP Notes on handling of document splitting configuration:
> - SAP Note 2793140 for the Central Finance system
> - SAP Note 2764175 for the source system

7.3.4 Set Up Connection to the Central Finance System

In this section, we'll cover connection prerequisites and related configurations for special scenarios such as the central payment and central budgeting functionality. If these special scenarios aren't in scope for a project, the configuration in this section isn't necessary. However, we recommend reviewing the long-term requirements and the possible configuration impact on data in the future.

Define a Logical System for Central Finance System

In a distributed network system such as a Central Finance landscape, you identify each system with a unique ID to transfer data between systems. In SAP, a logical system is required to transfer data between two or more systems, which generally uses one common database. In SAP terminology, a *logical system* refers to a client and must be unique across the company. This means they are defined as a cross-client across the system.

In this configuration, you'll define a logical system for the target Central Finance system. A similar configuration is done for the source system, as explained later in Section 7.4.2, from the target Central Finance system. To define a logical system for the target Central Finance system, navigate to Customizing IMG menu path **SAP Customizing Implementation Guide • Financial Accounting • Central Finance: Source System Settings • Set Up Connection to Central Finance System • Define Logical System for Central Finance System** or navigate to it from Transaction CFINIMG. Figure 7.21 shows how to add a logical system definition in the source system. Upon entering the configuration, on the **Change View "Logical Systems": Overview** screen, click **New Entries** ❶. You'll be taken to the **New Entries: Overview of Added Entries** screen. Here, maintain values for the **Log.System** and **Name of Logical System** ❷ fields. Click the **Save** ❸ icon to save the entries.

SAP recommends a logical system to keep the nomenclature consistent with the following guidelines. The logical system is made up of the following:

- Three-digit system ID beginning with S: <System-ID>
- Four digits to denote the client: CLNT
- Three-digit system represents client: <Client>

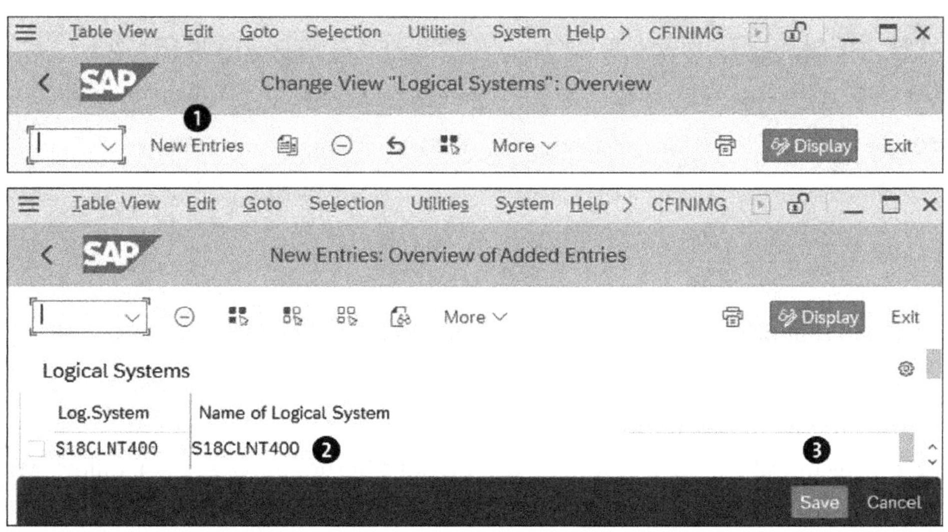

Figure 7.21 Define a Logical System for the Central Finance System

In our example, **S18** represents the system ID, and **CLNT400** identifies the client number.

> **Note**
> In SAP, for refresh or copying purposes, you can keep the logical system (LOGSYS) as the production logical system, name which simplifies copying. However, SAP doesn't recommend assigning the same logical system name to the preproduction or test environment. You'll use Transaction BDLC to update logical system names.

Set Up the Remote Function Call Destination for the Central Finance System

In this configuration, you'll create and maintain RFC destinations for Central Finance. In addition, you'll define parameters that enable RFCs to the Central Finance system. Generally, this activity is performed by the Basis team. RFC users are required, which are created by the security team.

As you recall, RFC connections require to and from communication such as reading and writing data between the source system and the Central Finance system. Therefore, RFC destinations are required in both source and Central Finance systems with connection type **3 RFC connection to ABAP system**. In addition, the RFC destination is required for Web Dynpro applications or OData service communications for relevant web or SAP Fiori apps. Table 7.10 provides SAP-recommended settings for RFC destinations for each source system for Central Finance integration scenarios.

Type of Parameter	Central Payment	Central Budgeting	Central Projects
User type	A (Dialog)	C (Communication)	C (Communication)
RFC type	**FUGR** (Function Group)	**FUGR** (Function Group)	**FUGR** (Function Group)
CFIN_APPL	10	18	09
ACTVT	16	16	16
Authorization object	F_CFIN_TRG	F_CFIN_TRG	F_CFIN_TRG

Table 7.10 RFC Settings Use Cases for the Central Finance System

To define technical parameters for RFC destinations in the source system, follow IMG menu path **SAP Customizing Implementation Guide • Financial Accounting • Central Finance: Source System Settings • Setup Connection to Central Finance System • Set Up RFC Destination for Central Finance System Central Finance System** or navigate to it from Transaction CFINIMG in the Central Finance system. Alternatively, you can also use Transaction SM59 to access the configuration with the appropriate security access. Upon executing the transaction, you'll arrive at the **Configuration of RFC Connections** screen, as shown in Figure 7.22. On the resulting screen, to create a new RFC destination, click **New Entries** ❶. You'll be prompted with the **Create Destination** popup window. In this screen, enter a value in the **Destination** ❷ field, and then choose **Connection Type 3 RFC connection to ABAP system** ❸. Click the **Save** icon to save entries.

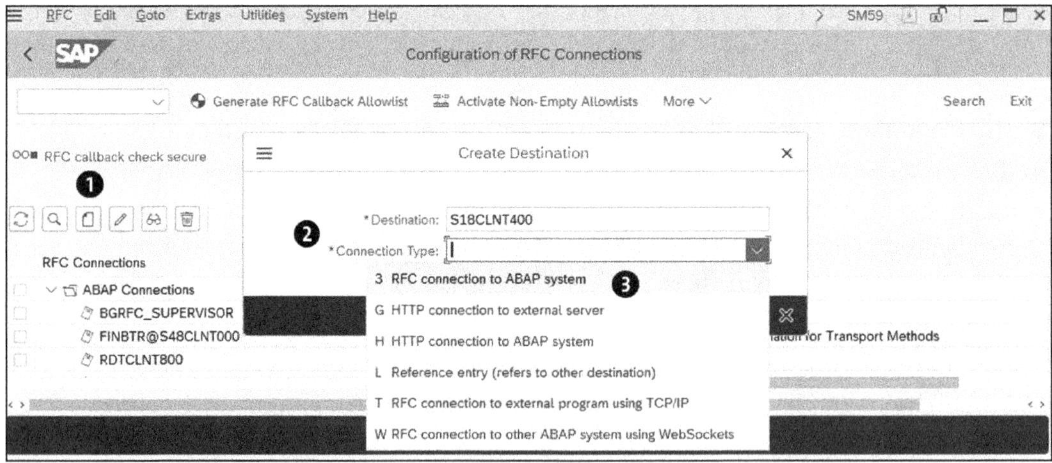

Figure 7.22 Setting Up the RFC Destination for the Central Finance System

[«]

Note

For Central Finance remote connections, SAP recommends using **RFC connection to ABAP system**, which natively integrates with SAP components. Other types of connections, such as HTTP and HTTPS, are used in situations where web browser–based applications are involved.

7

After adding a connection type successfully, to view the added entry, double-click the logical system or highlight it, and click **Display** icon. You'll be presented with the **RFC Destination S18CLNT400** screen, as shown in Figure 7.23. The most relevant settings in each tab are listed here:

❶ **Administration**
This tab shows administrative information such as who created the RFC destination and when it was created.

❷ **Technical Settings**
The following sections in this tab contain important parameters for connectivity:

- **Load Balancing Status**
  If the RFC destination is using load balancing, then you either activate or deactivate load balancing here. To load balance systems, additional appropriate settings must be maintained, and a relevant application server is specified here.
- **Save to Database as**
  The system allows you to enter a specific host name or an IP address. When the **Host** radio button is clicked, the system derives host information from ABAP instances defined in Transaction SM51.
- **Gateway Options**
  Enter the **Gateway Host** and **Gateway service** if relevant.

❸ **Logon & Security**
In this tab, you'll maintain security-related settings such as where secure network communications (SNC) settings derived from.

❹ **Unicode**
In this tab, you'll maintain whether the RFC destination switched to Unicode.

❺ **Special Options**
In this tab, the system provides advanced functions such as troubleshooting RFC connections with trace.

After configuration of the RFC destination, to perform the connection test, you'll check the RFC connection manually by either using **Connection Test** ❻ or choosing the **Utilities** ❼ menu and selecting **Connection Test** ❽. A successful connection test is displayed on the right side of Figure 7.23 ❾.

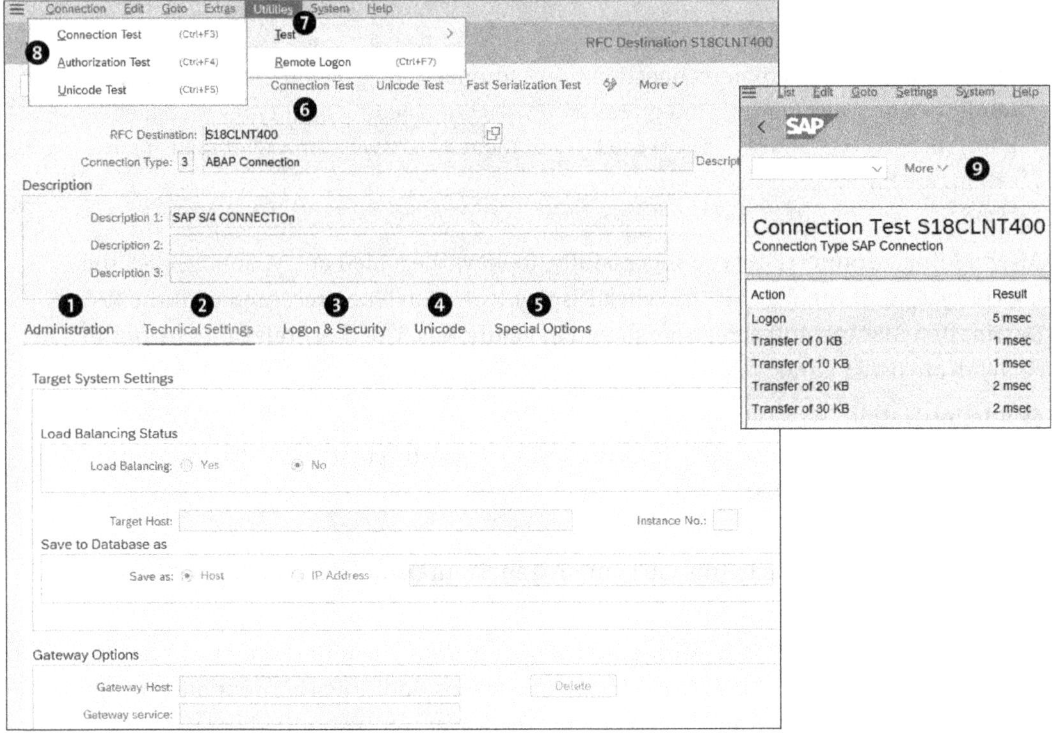

Figure 7.23 RFC Destination

Assign the Remote Function Call Destination for Displaying Objects from the Central Finance System

To display objects or read data from the Central Finance system, you must assign an RFC destination to the logical system we created earlier. The prerequisite for this step is to define RFC destinations and local systems first. SAP provides two ways to assign RFC destinations created earlier to logical systems to communicate between systems. In this configuration, you'll assign RFC destinations, created in the previous section, to the logical system. Your two options are as follows:

- Create or use a standard dialog destination with the following standard destinations provided with the transaction:
  - **Standard BAPI destination**
    Normal access with the RFC background user, usually via SAP all access using a Business Application Programming Interface (BAPI) program call.
  - **Standard dialog destination**
    Normal access with the RFC dialog user, usually via SAP all access-to-access transactions.

- An object type, method, and RFC destination are required for calling function modules or displaying objects with trusted access, as shown in Figure 7.24.

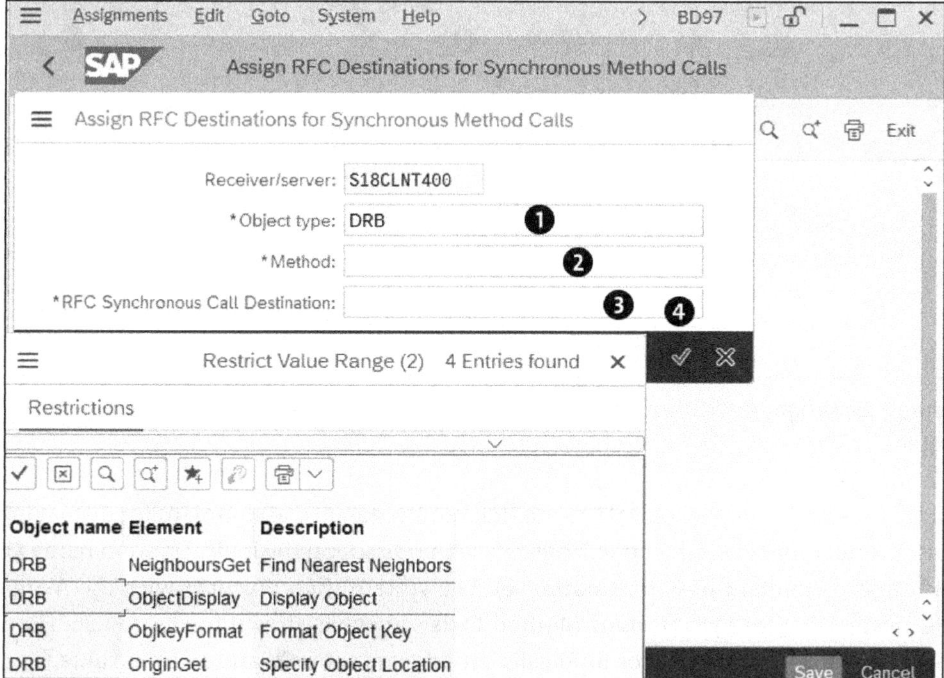

Figure 7.24 Assign the RFC Destination for a Special Method Destination

For example, for document relationship browser to display objects, you'll assign the following settings:

- **Object type** = **DRB** ❶
- **Method** = **ObjectDisplay** ❷
- **RFC Synchronous Call Destination** = **S18CLNT400** ❸

Click **Continue** ❹ to save the entry.

To assign RFC destinations to the logical systems for BAPIs, dialogs, and special methods, navigate to IMG menu path **SAP Customizing Implementation Guide • Financial Accounting • Central Finance: Source System Settings • Set Up Connection to Central Finance System • Assign RFC Destinations for Displaying Objects from Central Finance System** or navigate to it from Transaction CFINIMG. You can also reach the settings using Transaction BD97. Upon opening the configuration, you'll arrive at the **Assign RFC Destinations for Synchronous Method Calls** screen, as shown in Figure 7.25.

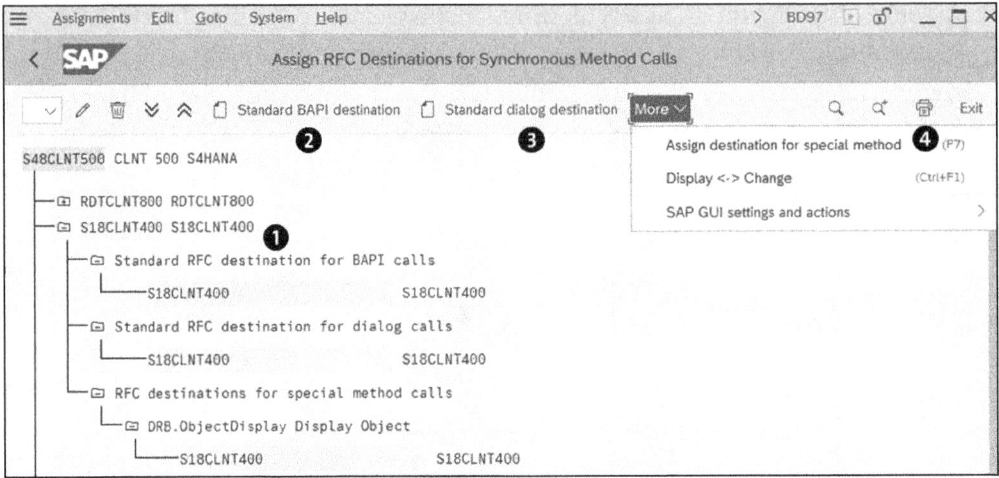

Figure 7.25 Assigning RFC Destinations for Displaying Objects from the Central Finance System

On this screen, locate the logical system that requires an RFC assignment for displaying objects under the hierarchy node. Then, place the cursor on the logical system name ❶, and choose **Standard BAPI destination** ❷. The system then prompts with the **Assign RFC Destination for Synchronous Method Calls** screen, as shown in Figure 7.26. Here, choose the **RFC destination for BAPI calls** dropdown menu ❶, and select a value from the **RFC Destination** ❷ column. Select **OK** ❸ to continue. The system returns to the previous screen with an added entry, as shown in Figure 7.25. Repeat the steps to assign RFC destinations for **Standard dialog destination** ❸ and **Assign destination for special method** ❹ buttons. After assigning all three RFC destinations to the logical systems, click the **Save** icon to save the entries.

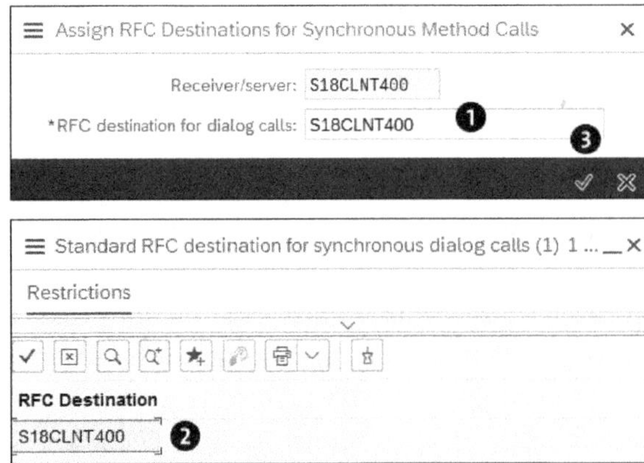

Figure 7.26 Assigning RFC Destinations for Synchronous Method Calls: Prompt

[«]

Note

Assigning RFC destinations to a logical system is considered a client-dependent activity, so the configuration needs to be repeated in each system.

Maintain Remote Function Call Assignments and Settings for the Central Finance System

For each logical system, RFC usage and RFC destinations must be maintained for certain Central Finance scenarios. Therefore, defining the logical system and RFC destination is a prerequisite for this step. In this configuration, you'll maintain RFC assignments and related settings for Central Finance to enable RFCs from the source system to the Central Finance system.

To maintain RFC assignments and settings for the Central Finance system, navigate to IMG menu path **SAP Customizing Implementation Guide • Financial Accounting • Central Finance: Source System Settings • Set Up Connection to Central Finance System • Maintain RFC Assignments and Settings for the Central Finance System** or navigate to it from Transaction CFINIMG. Once you reach the settings, click the **New Entries** button on the **Change View "Central Finance System": Overview** screen. You'll arrive at the **Change View "Central Finance System": Overview** screen, as shown in Figure 7.27. On this screen, click **New Entries** ❶. On the resulting **New Entries: Overview of Added Entries** screen, enter the logical system name under the **Leading System** column, and then press Enter. Then, highlight the logical system row you've just added ❷, and click on **RFC Destinations** under **Dialog Structure** located in the left pane ❸. In the subsequent screen, click **New Entries** to maintain values in **RFC Usage** and **RFC Destination** ❹ for the Central Finance system. The system offers the following **RFC Usage** options:

- **Central Payment**
- **Central Budgeting**
- **View Reference via GOS**
- **Central Projects – Asset Settlement**
- **Central Finance Navigation Router**
- **Central Asset Accounting**
- **Cost Object Status Check**

Which option to choose depends on the project scope. Once the **RFC Usage** and **RFC Destination** ❹ are maintained per project requirements, click the **Save** icon to save the entries.

[«]

Note

The **Set Leading Central Payment System** configuration is deprecated and now replaced with **Maintain RFC Assignments for the Central Finance System**. If the new configuration is missing in your system, apply any missing SAP Notes by consulting SAP Note 2792114 – Central Payment: RFC Usage on Source.

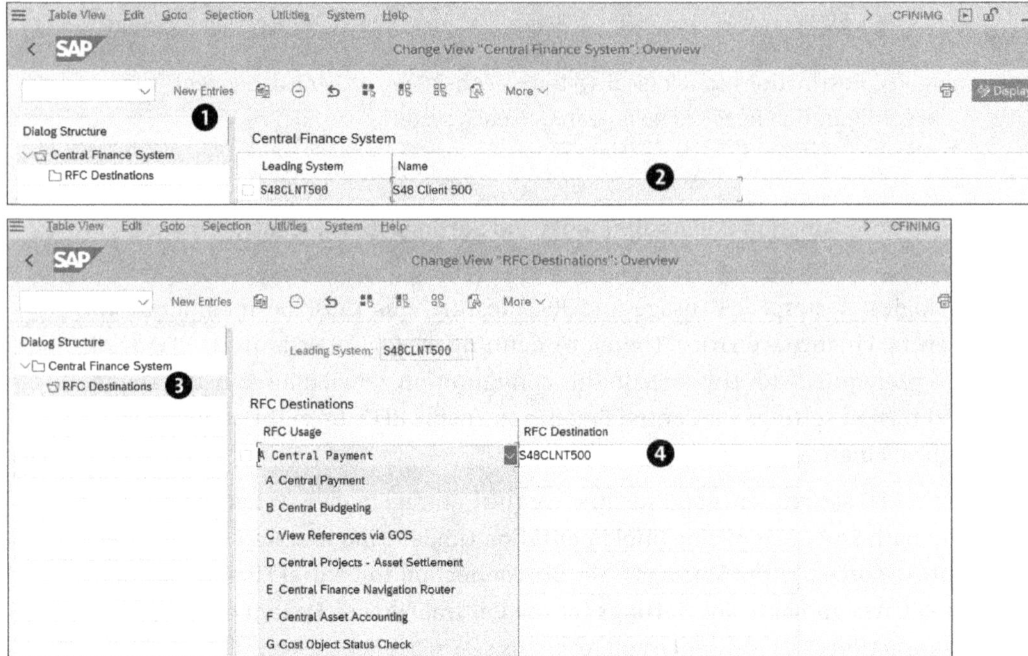

Figure 7.27 Maintaining RFC Assignments and Settings for the Central Finance System

Maintain an External Alias for Displaying Central Finance References Using Generic Object Services

In this configuration, you'll maintain an alias which is defined externally. External alias is defined in Transaction SICF. The system will replace the standard service path with the new path generated with the alias you've maintained during service calls. SAP uses generic object services (GOS) to reference the external alias. For example, for side panels, this helps users maintain any nontechnical name in an alias.

To maintain an external alias, navigate to Customizing IMG menu path **SAP Customizing Implementation Guide • Financial Accounting • Central Finance: Source System Settings • Set Up Connection to Central Finance System • Maintain External Alias for Displaying Central Finance References Using GOS** or navigate to it from Transaction CFINIMG. After reaching the settings, on the **Change View "Central Finance Key Value Customizing": Overview** screen, click the **New Entries** button, as shown in Figure 7.28 ❶. Next, on the **New Entries: Overview of Added Entries** screen, maintain the following relevant settings ❷:

- **Category**: Enter "SP_GOS_URL_PATH".
- **Key**: Enter the Web Dynpro Application ID here. SAP supports "FINS_CFIN_SP_ACDOC" and "FINS_CFIN_SP_CM".
- **Value**: Maintain the alias defined in Transaction SICF for service `WDR_CHIP_PAGE`. Click the **Save** ❸ icon to save the entries.

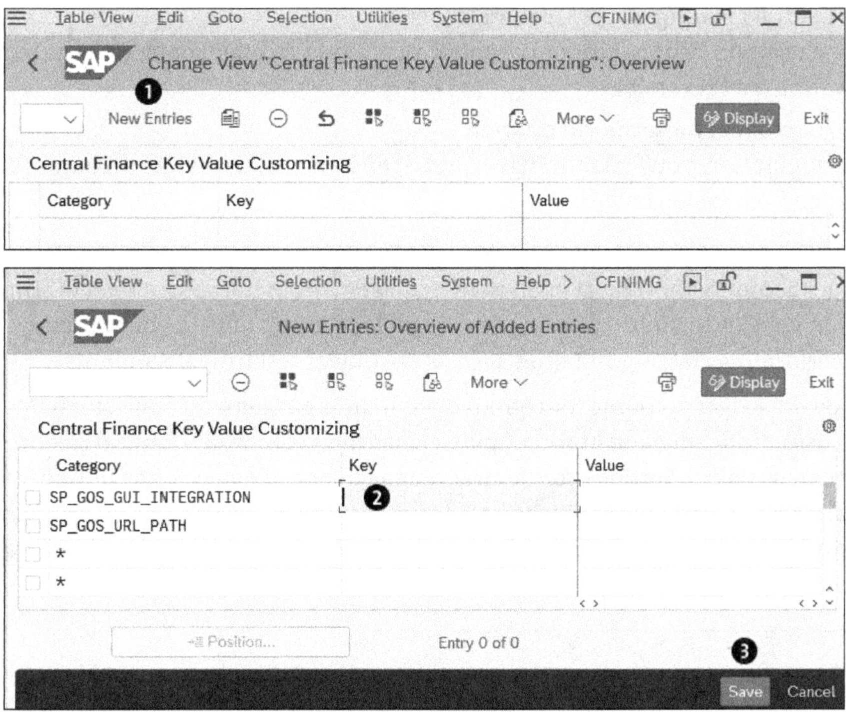

Figure 7.28 Maintain an External Alias for Displaying Central Finance References Using GOS

Note

External aliases start with a slash "/" whereas internal aliases are stored without a slash. For example, `/SAP/mySAPdotcom/ping` is an external alias to ping an ICF service.

7.3.5 Source System Settings for Central Management of Open Items in the General Ledger

SAP supports managing general ledger open items centrally in the Central Finance system. A prerequiste to this configuration is that you've first completed activation of central clearing of general ledger open items, as discussed in Section 7.4.6. With this functionality, the G/L open items are processed in the Central Finance system, and the corresponding open item posting in the source system is technically cleared with *ALE-Extern* automatically. For example, clearing related open item general ledger accounts such as goods receipt/invoice receipt (GR/IR) account activities can be performed in the Central Finance system using Transaction F.13 or FB05, and the corresponding clearing document in the source system is technically cleared.

To enable the central open item management functionality, Central Finance provides two configuration activities:

- **Activate Central Clearing of G/L Open Items**
  With the execution of this program, central open item management is activated. The activation isn't reversible.
- **Define G/L accounts for technical clearing**
  In this configuration, you'll maintain relevant open item general ledger accounts for central open item management.

After activating central open item management in Central Finance, you'll maintain general ledger accounts for central open item management in the source system. This configuration enables technical clearing of general ledger accounts in the source system when performing clearing activities such as posting or resetting documents.

To define general ledger accounts for central open item management, navigate to IMG menu **SAP Customizing Implementation Guide • Financial Accounting • Central Finance: Source System Settings • Settings for Central Management of Open Items in General Ledger • Maintain G/L Accounts for Technical Clearing** or navigate to it from Transaction CFINIMG. After reaching the configuration, maintain the company code and general ledger accounts for central open item management, as shown in Figure 7.29.

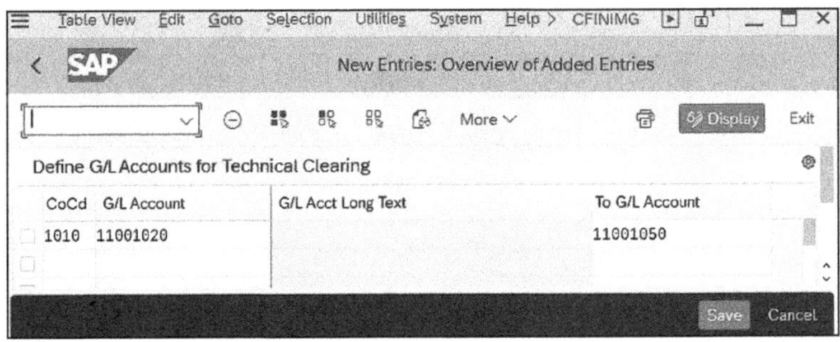

Figure 7.29 Define General Ledger Accounts for Technical Clearing

For the company code in scope, the configuration screen provides the option to maintain the general ledger account or range of general ledger accounts for central open item management. Alternatively, if the Central Finance customizing menu path is not available in your source system, you can maintain general ledger accounts for technical clearing through maintenance view `V_CFIN_CGL_ACC` via Transaction SM30.

[!]

Warning

SAP supports central open items once all prerequisite configurations are met. However, any historical open items before maintenance of general ledger accounts aren't technically cleared. Therefore, all historical items must be cleared in the source system prior to maintaining general ledger accounts in this configuration or running Transaction CFIN_CGL_TECH_CLR to set them as technically cleared (`ALE-extern`). In addition, AP/AR reconciliation and corresponding net cash discount clearing accounts are technically cleared as part of the central payment scenario, so they are excluded from this configuration.

7.3.6 Enhancements Using Source System Business Add-Ins

Central Finance offers enhancement opportunities where standard functionality doesn't fulfill project-specific requirements. As of writing this book, Central Finance provides three source system BAdIs that can be used for customer-specific enhancement opportunities. The source system BAdIs offer the following functionality during the Central Finance interfacing process:

- Profitability analysis information is updated in the CFIN_ACCPA_CHAR staging table during the transfer process. The BAdI allows customer-specific profitability analysis characteristics to be filled with values above and beyond those contained in standard characteristics. On the other hand, you can disable the transfer of profitability information.
- Customer-specific fields such as coding block fields aren't included as part of standard FI replication data. In such instances, the source system BAdI offers to transfer customer-specific fields of FI document line items to Central Finance. The customer-specific fields are enhanced with ACCIT, FINS_CFIN_S_ACCIT, and FIN_CFIN_ACCIT_APP structures and are updated in the CFIN_ACCIT_APP staging table.
- Add customer fields to the ACCIT structure. Consult SAP Note 2543899 on adding customer fields to the Central Finance structure.
- Add customer-specific CO fields during CO replication posting.
- Enhance customer-specific fields in commitment items stored in the CFIN_CMT staging table.
- Add customer-specific data for profit center accounting tables stored in the CFIN_PCA staging table.
- Add customer-specific data for cost component splits stored in table CFIN_ACCIT_CCS.

The BAdIs have methods that can be called at different places during the Central Finance interface process. Table 7.11 explains the source system–specific BAdIs and methods available for enhancements and their usage.

BAdI Name	Methods Available	Usage
BADI_FIN_CFIN_DOCUMENT: Add Information from Source System to Central Finance Documents	FI_ODC_ENHANCE_ONGONG FI_DOC_ADAPT_CCS_ONGOING	Enhance ongoing replication.
	FI_DOC_ENHANCE_INITIAL_LOAD	Enhance open items and balances during the FI initial load.

Table 7.11 SAP Source System BAdIs

BAdI Name	Methods Available	Usage
	CO_DOC_ENHANCE	Enhance CO documents during the initial load or online replication.
	CMT_ENHANCE	Enhance the initial load and online replication of commitment items.
	PCA_ENHANCE	Enhance the initial load and ongoing replication of profit center accounting.
BADI_FIN_CFIN_DP_MAP Enhance Inbound Processing of Reverse Mapping for SD Down Payment	ADJUST_MAP_SD_DOWN_PAYMENT	Adjust customer-specific logic for the reverse mapping of down payments.
BADI_FIN_CFIN_DFV	GET_IMPLCLNAME GET_ACCAS_IMPLCLNAME FILTER_BKPF FILTER_BSEG FILTER_COBK FILTER_COEP	Enhance selection from the source system for various reconciliation reports.

Table 7.11 SAP Source System BAdIs (Cont.)

7.4 Configuring the Central Finance System

In this section, you'll learn both technical and functional configurations relevant to the Central Finance system. Before configuring the Central Finance system, you must import all Central Finance–related SAP OSS Notes and baseline FI-CO configuration transports. Next, security users must be created for RFC communication in both the source and destination systems. The section is divided into subsections on the following topics:

- **Basic settings**
  This section discusses the basic settings that enable Central Finance functionality availability.
- **Setup systems**
  This section covers the technical configuration requirements for Central Finance such as RFC destinations and logical systems.

- **Settings for automatic document replication**
  This section walks through the configurations required to support replication scenarios.
- **Document splitting**
  This section shows you how to transfer documents splitting configuration to the source system for prechecks.
- **Initial load**
  This configuration section walks thorough the settings required to perform an initial load.
- **Central open item management**
  This section discusses how to activate central open item management.
- **Enhancements using Central Finance BAdIs**
  This section covers the delivered central finance system BAdIs.

Let's dive deeper into each configuration section next.

7.4.1 Basic Settings

In this section, you'll learn about the Central Finance business function and Web Dynpro services application link activation first. This forms the foundational basis to enable Central Finance configurations, so these settings are mandatory prior to carrying out Central Finance technical or functional configuration.

Activate Central Finance Business Function

Before you get started with Central Finance, you must activate the Central Finance business function in the target Central Finance system. Before switching the business function, ensure you have the Central Finance license obtained by your company. Although you can setup Central Finance for a proof of concept purpose, a valid Central Finance license is required to use the product. To activate the Central Finance business function, navigate to Customizing path **SAP Customizing Implementation Guide • Activate Business Functions** or use Transaction SFW5. After executing the transaction, you'll arrive at the screen shown in Figure 7.30. As you've observed, the system displays a warning message to alert you that only reversible business functions can be deactivated. Click the **Continue** button ❶. Then, navigate to business function **FINS_CFIN** ❷ with description **Central Finance** located under the **ENTERPRISE_BUSINESS_FUNCTIONS** node. Select **Activate Changes** ❸ to activate the business function. The program submits the activation in the background job. Once the activation is successful, the lightbulb icon turns yellow.

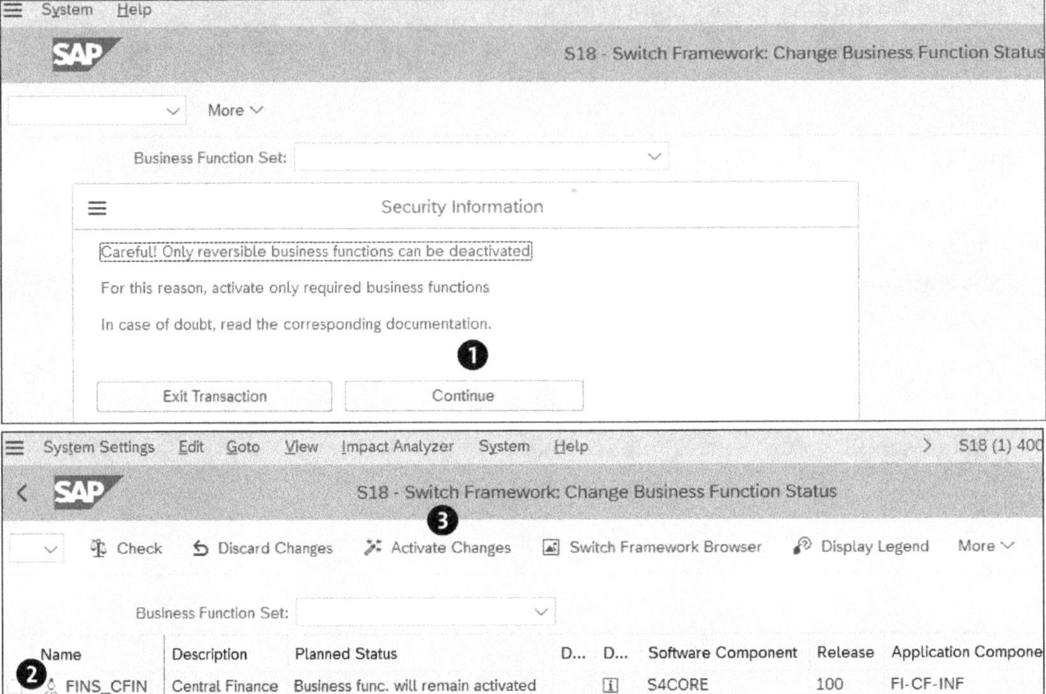

Figure 7.30 Activate Business Function

Activate the Web Dynpro Application

SAP provides an SAP Master Data Governance Web Dynpro application to create and edit key mapping. The service that enable this functionality is `MDG_BS_WD_ID_SERVICE`, which is initially delivered with **Inactive** status in SAP Master Data Governance central governance services due to security purposes. To activate the SAP Master Data Governance Web Dynpro services, go to Transaction SICF. On the **Define Services** entry screen, as shown in Figure 7.31, enter "MDG_BS_WD_ID_MATCH_SERVICE" in the **Service Name** ❶ field, and then choose the **Execute** ❷ icon. Make sure **SERVICE** is selected in **Hierarchy Type** field. Next, on the resulting screen, enter a description of the service in the **Description** ❸ field, which is used to apply additional filters to search the Web Dynpro service. Once the system filters **MDG_BS_WD_ID_MATCH_SERVICE** under the **Virtual Hosts/Services** column, right-click to select **Activate Service** ❹ from the context menu. As shown in Figure 7.31, the service is already activated in the demo system.

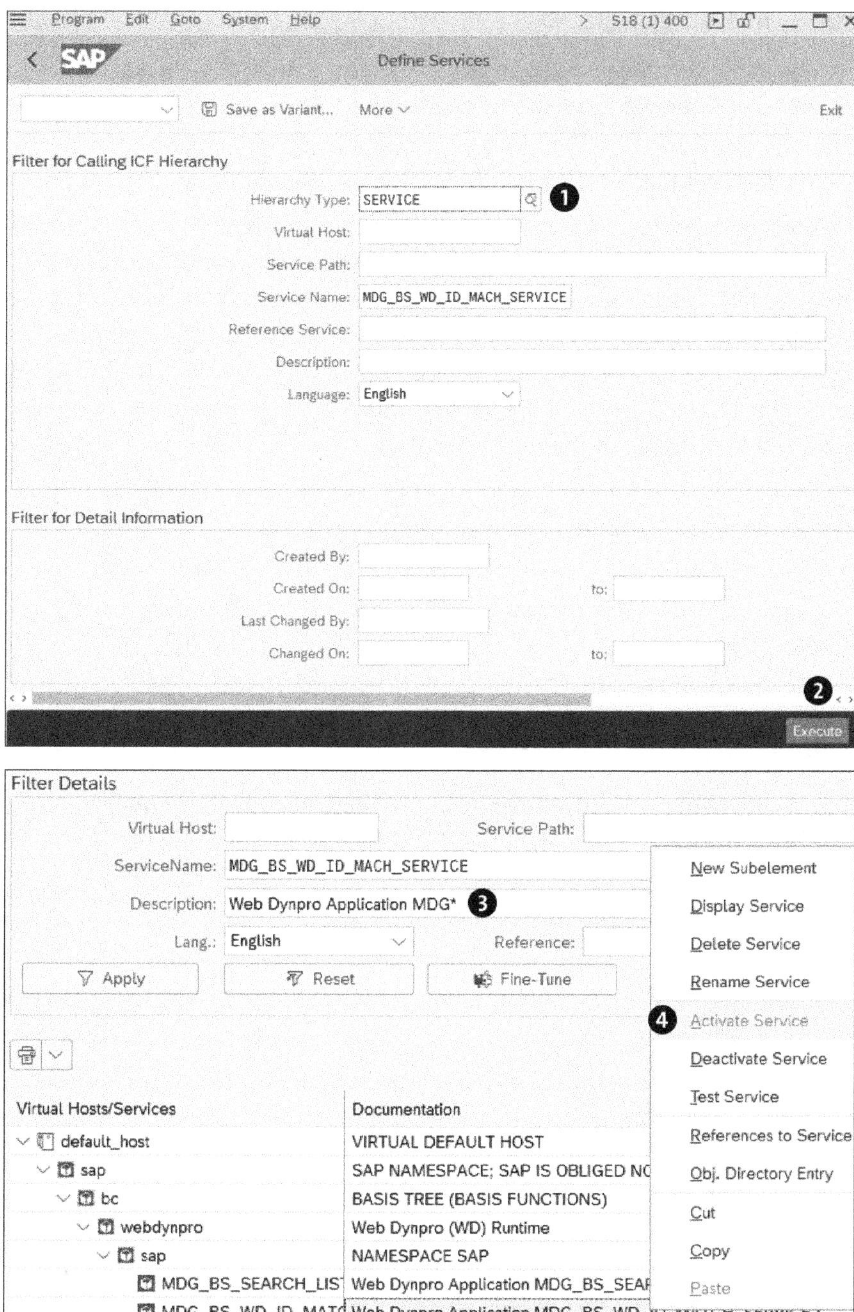

Figure 7.31 Web Dynpro Application Service Activation

7.4.2 Set Up Systems

This section covers the technical configuration for Central Finance systems. Let's walk you through each configuration in sequential order.

Set Up the Remote Function Call Destination for Source Systems

RFC connections are important for reading data from the source system and transferring that data to the Central Finance system. In this configuration, you'll define technical parameters for RFC destinations for different functional areas. This configuration is generally carried out by a Basis team. We've already discussed the configuration for the source system in Section 7.3 Likewise, you'll set up the RFC destination for source systems in this configuration. The following list and Table 7.12 describe the most relevant RFC destination settings as recommended by SAP:

- USER_TYPE
  User type tells whether the type is communication or dialog.
- RFC_TYPE
  This is the type of RFC authorization object. FUGR indicates function group, and FUNC indicates function module.
- RFC_NAME
  This indicates the name of an RFC object or function group.
- ACTVT
  This is the activity being performed. For example, 16 means execute, and 03 refers to display.
- CF_PURPOSE
  This is the activity usage in the source system.
- AUTHORIZATION OBJECT
  In SAP, an authorization object enables users to perform certain functions or actions.

	USER_TYPE	RFC Type	RFC_NAME	ACTVT	CF_Purpose	Authorization Object
Extracting Data from FI	C - Communication	**FUGR** - Function Group	FIN_CFIN_INITIAL_LOAD FIN_CFIN_CLEARING FIN_CFIN_RESET_CLEARING_FACI FIN_CFIN_PCA FIN_CFIN_DEFTAX	16 – Execute	01	S_RFC

Table 7.12 SAP-Recommended RFC Parameters

	USER_TYPE	RFC Type	RFC_NAME	ACTVT	CF_Purpose	Authorization Object
Extracting Data for Management Accounting	C - Communication	**FUGR** - Function Group	FIN_CFIN_CO FIN_CFIN_CO_SIMULATE	16 – Execute	09	S_RFC; User requires authorization object S_BTCH_ADM=Y.
Extracting Data for Commitments	C - Communication	**FUGR** - Function Group	FIN_CFIN_CMT	16 – Execute	10	S_RFC
Comparison Reports	A - Dialog	**FUGR** - Function Group	FIN_CFIN_DFV FIN_CFIN_CONFIG_CONSIS_CHK SDIFRUNTIME, FIN_CFIN_DFV_CMT FIN_CFIN_PCA	16 – Execute	04, 07	S_RFC; S_TCODE access for Transactions FB03, KSB5 User requires authorization objects F_BKPF, F_BKPF_BEK, K_VRGNG, K_TP_VALU, F_BKPF_KOA, and F_FAGL_LDR.
Central Payments	A - Dialog	**FUGR** - Function Group	FIN_CFIN_APAR_CPAY FIN_CFIN_APAR_SEPA FIN_CFIN_APAR_CCARD FIN_CFIN_APAR_DOWNPAYMENT FIN_CFIN_MM_DOWNPAYMENT SDIFRUNTIME	16 – Execute	06, 07	S_SFC; User requires authorization object F_BKPF_BUK.
Manage Mapping	C - Communication	**FUGR** - Function Group	FIN_CFIN_MAPPING_RFC	16 – Execute	02, 07	S_RFC

Table 7.12 SAP-Recommended RFC Parameters (Cont.)

	USER_TYPE	RFC Type	RFC_NAME	ACTVT	CF_Purpose	Authorization Object
Side Panel	A - Dialog	**FUGR** - Function Group	0012 3006_HRO FACI LC01 SDIFRUNTIME SDTX SG00 SWOR	16 – Execute	08	S_RFC; User requires authorization object S_TABU_NAM with **ACTVT**= 3, table AUFK. Side panel may also need additional authorizations.
Document Splitting	C - Communication	**FUGR** - Function Group	ABAP4_LEAVE_TO_TRANSACTION DDIF_FIELDINFO_GET FIN_CFIN_CONFIG_CONSIS_CHK, FIN_CFIN_DFV FIN_CFIN_DFV_VH_COMPANY_CODE FIN_CFIN_SPLITTING SDIFRUNTIME	16 – Execute	11	
Document Relationship Browser	C - Communication	**FUGR** - Function Group		16 – Execute		S_RFC

Table 7.12 SAP-Recommended RFC Parameters (Cont.)

To define technical parameters for RFC destinations for the source system, follow IMG menu path **SAP Customizing Implementation Guide • Financial Accounting • Central Finance: Target System Settings • Set Up System • Set Up RFC destination for Source Systems** or navigate to it from Transaction CFINIMG. You can also use Transaction SM59 to access the configuration. Upon opening the transaction, you'll arrive at the **Configuration of RFC Connections** screen, as shown in Figure 7.32.

After reaching the configuration, to create a new RFC connection, click **New Entries** ❶. You'll be prompted with the **Create Destination** popup window. In this screen, enter the client logical name in the **Destination** field, and then choose **Connection Type 3 RFC connection to ABAP system** ❷. Click **Save** ❸ to save entries.

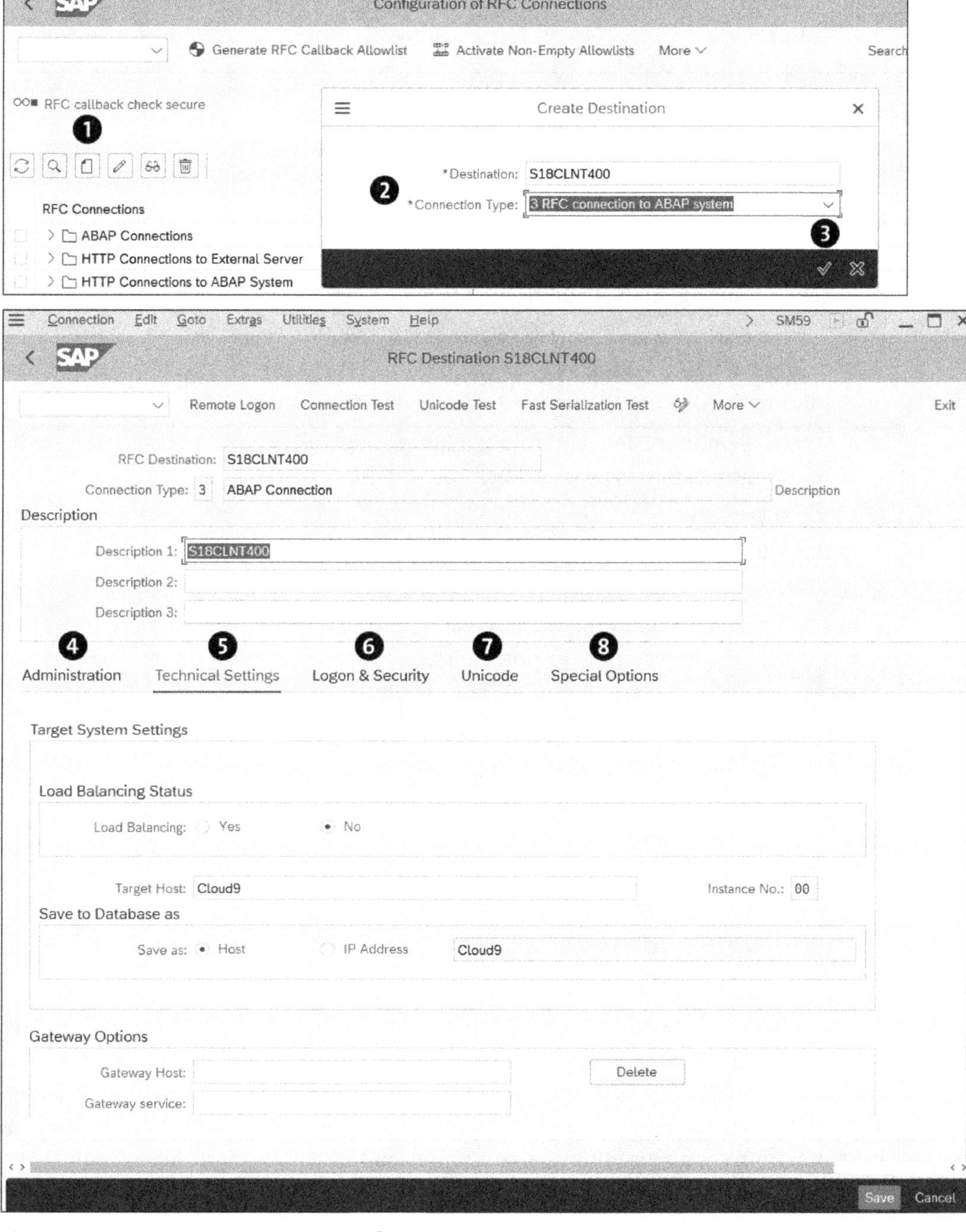

Figure 7.32 Set Up an RFC Destination for Source Systems

After adding a connection type successfully, to view the added entry, double-click the logical system or highlight it, and then click the **Display** icon. You'll be presented with the **RFC Destination S18CLNT400** screen, as shown in Figure 7.32. The most relevant settings in each tab are as follows:

❹ **Administration**
The tab shows administrative information such as who created it and when it was created.

❺ **Technical Settings**
This tab contains several sections:

- **Load balancing**: Activate or deactivate load balancing.
- **Save to Database as**: The system allows you to enter a specific host name or an IP address. When the **Save as Host** radio button is chosen, the system derives host information from ABAP instances defined in Transaction SM51.
- **Gateway options**: Here you can configure an RFC destination especially for TCP/IP communication via SAP Gateway.

❻ **Logon & Security**
In this tab, you'll maintain security-related settings.

❼ **Unicode**
In this tab, you'll maintain Unicode-related settings. Note when the source system is not Unicode, the system may not transfer data to the central finance system correctly.

❽ **Special Options**
In this tab, you'll perform advanced troubleshooting with trace.

Define a Logical System for the Source System and Central Finance Systems

We defined a logical system for the Central Finance system in Section 7.3.4. Similar to source system configuration, you'll also define a logical system for each connected source system client. As explained earlier, a logical system configuration is required to transfer accounting documents to the Central Finance system.

To define a logical system for each connected source system, navigate to Customizing menu path **SAP Customizing Implementation Guide • Financial Accounting • Central Finance: Target System Settings • Set Up Systems • Define Logical System for Source and Central Finance Systems**. Figure 7.33 displays logical system definition steps for the source system. After entering the configuration, on the **Change View "Logical Systems": Overview** screen, click the **New Entries** ❶ button. You'll be taken to the **New Entries: Overview of Added Entries** screen. On this screen, maintain values for **Log.System** and **Name of Logical System** ❷. Click the **Save** ❸ icon.

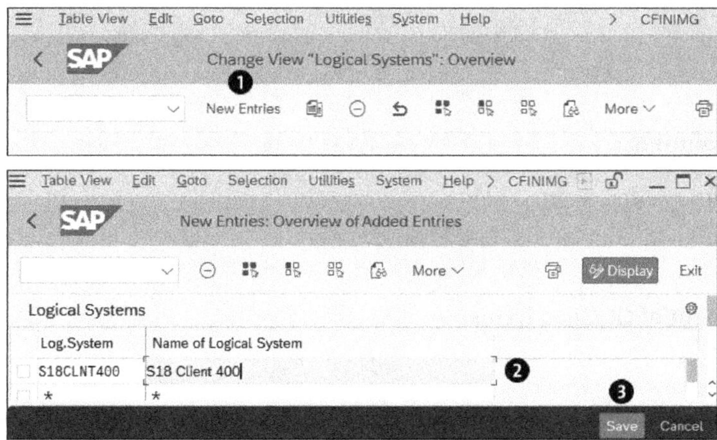

Figure 7.33 Define a Logical System for the Source and Central Finance Systems

Maintain Remote Function Call Assignments and Settings for Source Systems

For each logical system, RFC usage and RFC destinations must be maintained for certain Central Finance scenarios. Therefore, logical system and RFC destination definitions are prerequisite for this step. In this configuration, you'll maintain RFC assignments and related settings for Central Finance to enable RFCs from the source system into the Central Finance system. This is a prerequisite to start an initial load in Central Finance.

To maintain RFC assignments and settings for the Central Finance system, navigate to IMG menu path **SAP Customizing Implementation Guide • Financial Accounting • Central Finance: Source System Settings • Set Up Connection to Central Finance System • Maintain RFC Assignments and Settings for the Central Finance System** or navigate to it from Transaction CFINIMG. Once you reach the settings, as shown in Figure 7.34, click the **New Entries** ❶ button on the **Change View "Central Finance System": Overview** screen. On this screen, enter the logical system name under the **Leading System** column, and then press Enter. Then, highlight the row, an click on **RFC Destinations** in the **Dialog Structure** located in the left pane. In the subsequent screen, click **New Entries** to maintain **RFC Usage** ❷ and **RFC Destination** ❸. The system offers the following **RFC Usage** options as of writing this book based on SAP S/4HANA 2023:

- **1 Extracting Data for Financial Accounting**
- **2 Manage Mappings**
- **4 Comparison Reports**
- **6 Central Payment**
- **8 Side Panel**
- **9 Extracting Data for Management Accounting.**
- **10 Extracting Data for Commitments**

- 11 Document Splitting
- 12 Accounting Views
- 13 Activity Rates
- 14 Central Asset Accounting
- 15 Central Budgeting
- 21 Support Tools
- 22 SLT Logging
- 23 Central Management of GL Open Items
- 24 Extracting Plan Data

Select **RFC Usage** values that are in scope and click the **Save** button to save the entries.

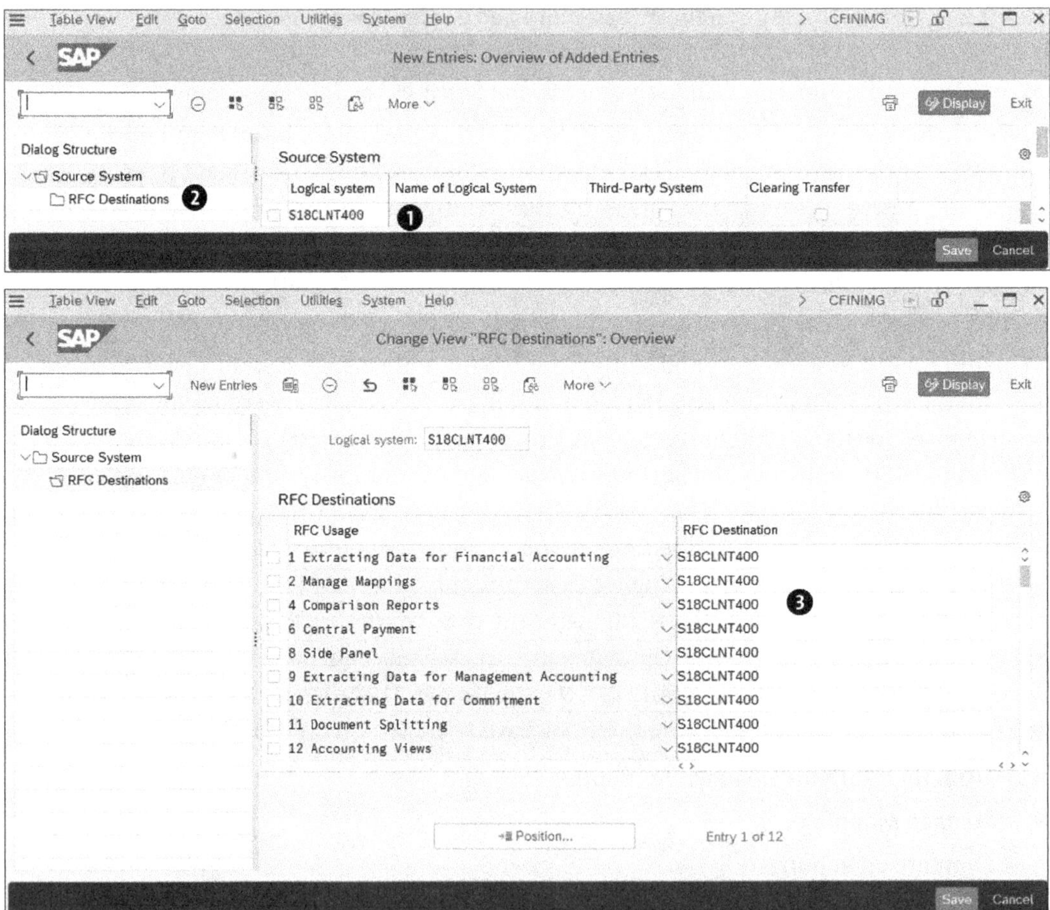

Figure 7.34 Maintain RFC Assignments and Settings for Source Systems: RFC Destination

Assign the Remote Function Call Destination to the Logical System for the Source System

To display objects or read data from the Central Finance system, you must assign an RFC destination to the logical system. The prerequisite for this step is RFC destinations and local systems defined earlier. SAP provides two ways to assign RFC destinations created earlier to logical systems:

- To create or use a standard dialog destination, use the following standard destinations provided with the transaction:
  - Standard BAPI destination: With this option, you'll assign the RFC destination to the logical system for the connected Central Finance system. You need normal access with an RFC background user. However, for BAPI program call, you need SAP All Access.
  - Standard dialog destination: Likewise, you'll assign the RFC destination to the logical system for the connected Central Finance system. You need normal access with an RFC background user.
- An object type, method, and an RFC destination are required for calling function modules or displaying objects with trusted access. For example, for the document relationship browser to display objects, you'll assign the following settings:
  - **Object type** = **DRB**
  - **Method** = **ObjectDisplay**

To assign RFC destinations, navigate to IMG menu path **SAP Customizing Implementation Guide • Financial Accounting • Central Finance: Source System Settings • Setup Connection to Central Finance System • Assign RFC Destinations for Displaying Objects from Central Finance System** or navigate to it from Transaction CFINIMG. Upon opening the configuration, you'll arrive at the **Assign RFC Destinations for Synchronous Method Calls** screen, as shown in Figure 7.35.

Next, on this screen, locate the logical system that requires RFC assignment for displaying objects under the hierarchy node. Place the cursor on the logical system name ❶, and click the **Standard BAPI destination** icon ❷. The system then opens the **Assign RFC Destination for Synchronous Method Calls** screen. On this screen, maintain the **RFC destination for BAPI calls** dropdown menu, and select the logical system value from the **RFC Destination** column. Select **OK** to continue. The system returns to the previous screen with the added entry. Likewise, repeat the steps to assign RFC destinations for **Standard dialog destination** ❸ and **Assign destination for special method** ❹ items. After assigning all three RFC destinations to the logical systems, click the **Save** icon to save the entries.

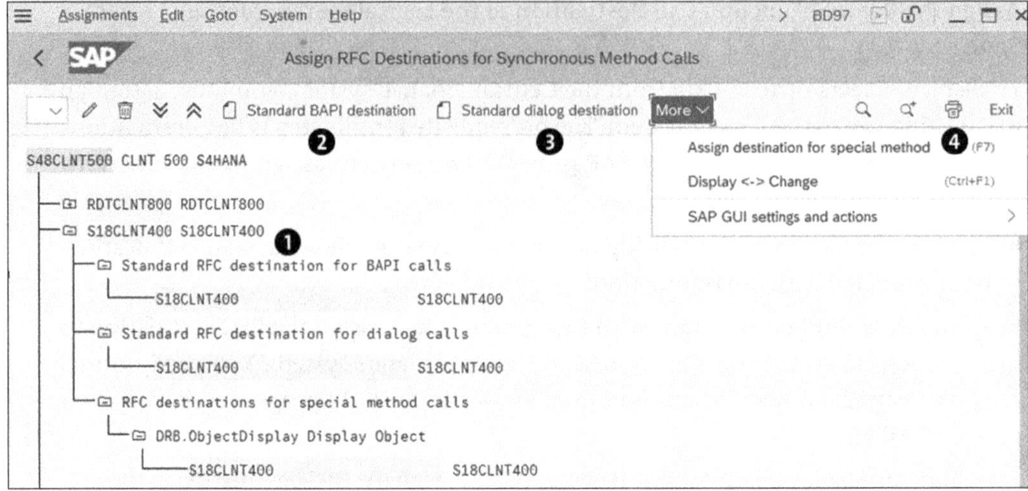

Figure 7.35 Assign RFC Destinations for Displaying Objects from Source Systems

Check the Logical System Assignment for the Central Finance Client

In this configuration, you'll check the logical system assignment for the Central Finance client. This is a client-dependent activity, so it's carried out in each client. Therefore, this task becomes part of the cutover activity and typically is carried out by Basis team. To check the logical assignment for the Central Finance client, navigate to the Customizing IMG menu path **SAP Customizing Implementation Guide • Financial Accounting • Central Finance: Target System Settings • Set Up Systems • Check Logical System Assignment for Central Finance Client** or navigate to it from Transaction CFIN-IMG. Upon opening the configuration, you'll arrive at the **Change View "Clients" Overview** screen, as shown in Figure 7.36. If no entry exists, maintain a new entry here. Next, on the **New Entries: Details of Added Entries** screen ❶, maintain the logical system for the client chosen ❷.

Following are the relevant settings made in this configuration:

- **Changes and Transports for Client-Specific Objects**
  Choose no changes allowed option here.
- **Cross-Client Object changes**
  Choose the **Cross-Client Settings Allowed** option based on your requirements.

Once changes are made, click **Save** ❸ to save the entries.

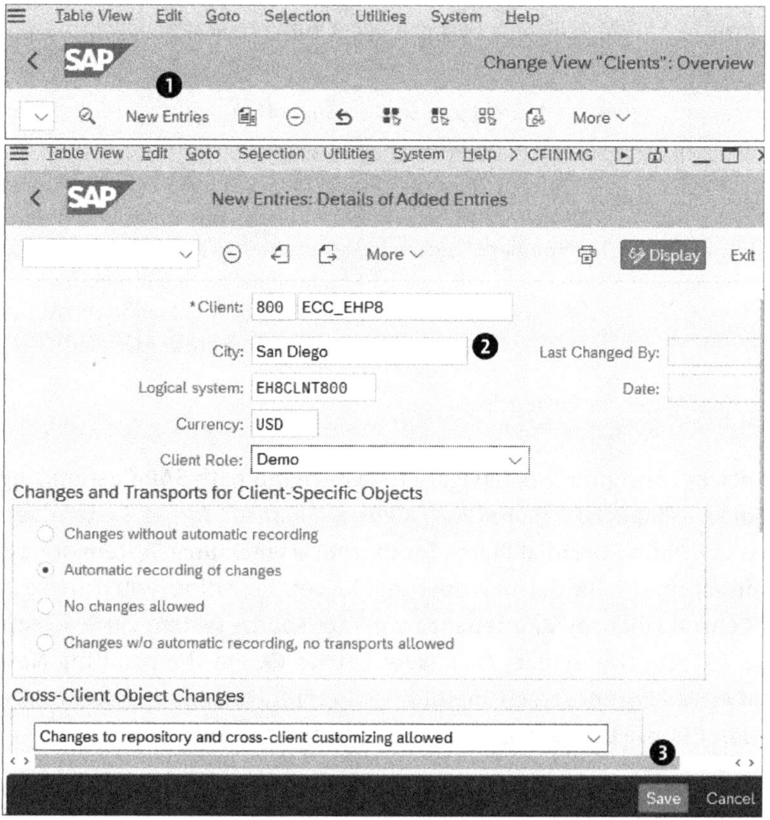

Figure 7.36 Check the Logical System Assignment for the Central Finance Client

[!]

Warning

The table is a cross-client activity, so caution is required regarding other systems in the landscape. Consider risks involved in any change or deleted activity as they may have an impact on other dependent systems. In addition, the user has the appropriate authorization to maintain the entries in this configuration.

Define Decimal Places for Currencies in the Source System

With this configuration, you'll maintain the number of decimal places for currencies in Central Finance. Otherwise, Central Finance displays errors during the initial load and online replication of documents, especially when lower decimal places are maintained in Central Finance than the source system for processes such as clearing transfer, central tax, and central payments. In such cases, with this configuration, you'll maintain differing decimal places in Central Finance. The Central Finance system then converts currency amounts correctly during transfer of financial data.

Table 7.13 displays possible scenarios for decimals in Central Finance.

Source System Decimals	Central Finance Decimal Settings	Central Finance Support
Identical decimal places as Central Finance setup	Identical decimal places as source system setup	Supported
Higher number of decimals	Lower number of decimals	Not supported
Lower number of decimals	Higher number of decimals	System may allow, however, additional testing is required to confirm that follow-on clearing scenarios work as intended

Table 7.13 Define Decimal Places for Currencies

To define decimal places configuration, navigate to IMG menu path **SAP Customizing Implementation Guide • Financial Accounting • Central Finance: Target System Settings • Set Up Systems • Define Decimal Places for Currencies in Source System** or navigate to it from Transaction CFINIMG. Upon opening the configuration, you'll arrive at the **Change View "Central Finance: Maintenance view for source system curre** screen, as shown in Figure 7.37. On this screen, click **New Entries** ❶. On the resulting **New Entries: Overview of Added Entries** screen, maintain values for the **Logical System**, **Crcy**, and **Decimals** ❷ fields. Choose the **Save** ❸ button to save the entries.

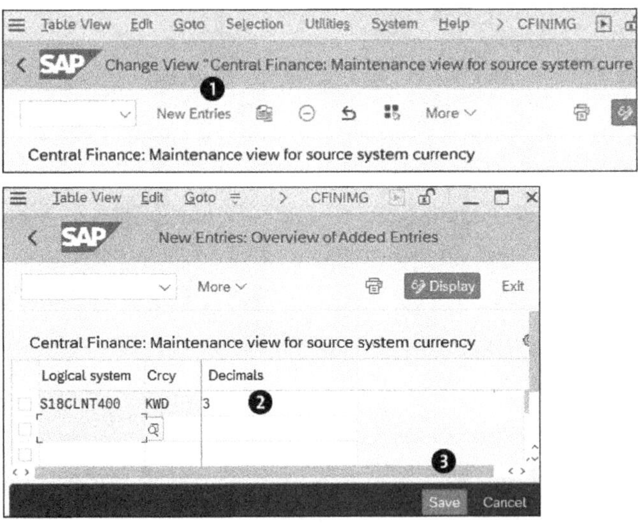

Figure 7.37 Define Decimal Places for Currencies in Source Systems

[»]

Note

The configuration is relevant only if decimals are different in the source system from what's configured in the target Central Finance system. If the currency between systems is the same, no entry is required in this configuration. During the design phase, we recommend reviewing all currencies in scope for Central Finance and identifying any currencies with differing numbers of decimals to configure in the system accordingly.

Review SAP Notes 2318183, 2325587, and 2217711 for further guidance on currency decimals.

7.4.3 Settings for Accounting Document Replication

In this section, we'll cover Customizing activities related to online accounting document replication for certain scenarios.

Define the Handling of Reconciliation Accounts per Company Code

During replication of accounting documents, by default, the system redetermines customer and vendor accounts from the business partner. This configuration provides an option to keep the source account to prevent redetermination. To carry out this Customizing activity, navigate to IMG menu path **SAP Customizing Implementation Guide • Financial Accounting • Central Finance: Target System Settings • Settings for Accounting Document Replication • Define Handling of Reconciliation Accounts per Company Code** or navigate to it from Transaction CFINIMG. Upon reaching the Customizing activity, maintain the value in the **Rec. Acct Handling** dropdown list, as shown in Figure 7.38. The system provides **Redetermine if possible** as a default option. Change it to **1 Keep** to override the default behavior.

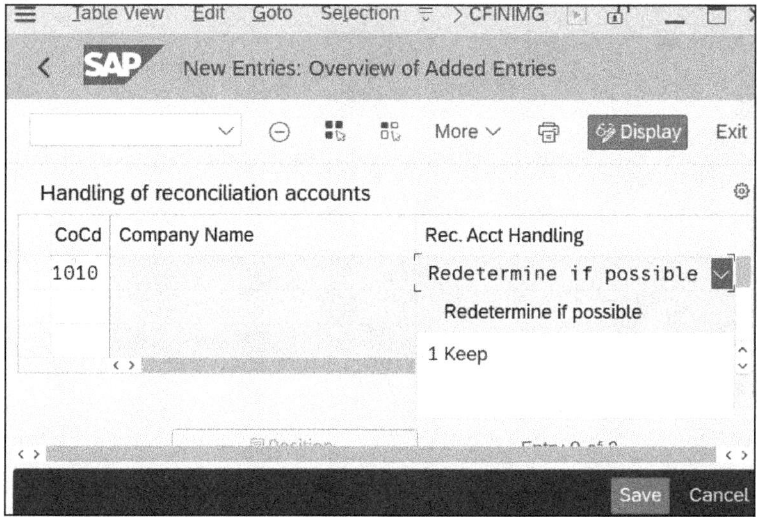

Figure 7.38 Define the Handling of Reconciliation Accounts per Company Code

Substitute the Original Document Type in the Central Finance System

With this Customizing activity, the system substitutes the original document posted in the source system to the document type maintained in this configuration. Specifically, this is required when those documents are posted with the document type setup with external numbering in the source system. The document types of setup in this Customizing use an internal number range. When source postings are transferred to Central Finance, the system substitutes with the document type maintained in this configuration. In particular, this Customizing activity applies to the following scenarios:

- Reversal and reposting and reposting
- Initial load
- Repost of missing cleared items
- Parallel accounting with mapping to ledger groups

To carry out this Customizing activity, navigate to IMG menu path **SAP Customizing Implementation Guide • Financial Accounting • Central Finance: Target System Settings • Settings for Accounting Document Replication • Substitute Original Document Type in Central Finance System** or navigate to it from Transaction CFINIMG. Upon reaching the configuration, as shown in Figure 7.39, click **New Entries** ❶. On the resulting **New Entries: Overview of Added Entries** screen, enter values in **Orig. Type** and **Subst Type** fields for original document types and substitute document types, respectively, and then press [Enter] ❷. Click the **Save** ❸ button to save the entry. The entries maintained in this configuration override the document type value mapping maintained in Central Finance.

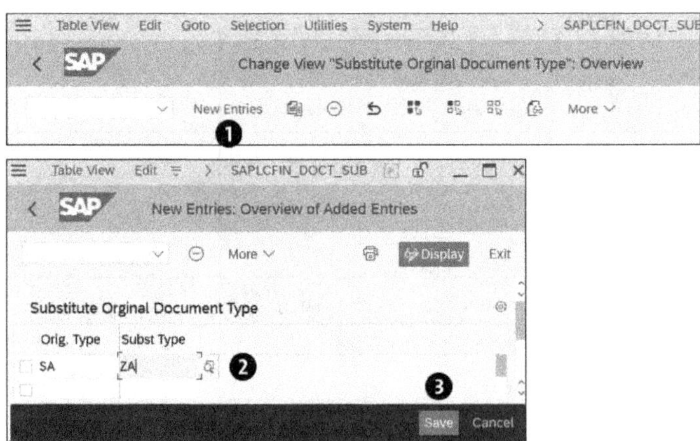

Figure 7.39 Substitute Original Document Type in Central Finance System

7.4.4 Check and Update Document Splitting Configuration

In this configuration, you'll transfer document splitting configuration information from Central Finance to the source system for enabling document split checks during posting in the source system. Recall from Section 7.2.1 that we discussed how the document splitting configuration scenario functions when the source system isn't active with document splitting. This Customizing activity addresses that scenario by transferring document splitting configuration to the source system for prechecks. During financial posting, the system checks these transferred settings to validate whether the document posted complies with document splitting configuration tables. Any new updates to the document splitting configuration in Central Finance can be transferred as well.

To transfer the document splitting settings, navigate to IMG menu path **SAP Customizing Implementation Guide • Financial Accounting • Central Finance: Target System Settings • Document Splitting • Check and Update Document Splitting Configuration** or navigate to it from Transaction CFINIMG. Upon opening the configuration, you'll arrive at the **Check and Update Document Splitting Configuration** screen, as shown in Figure 7.40.

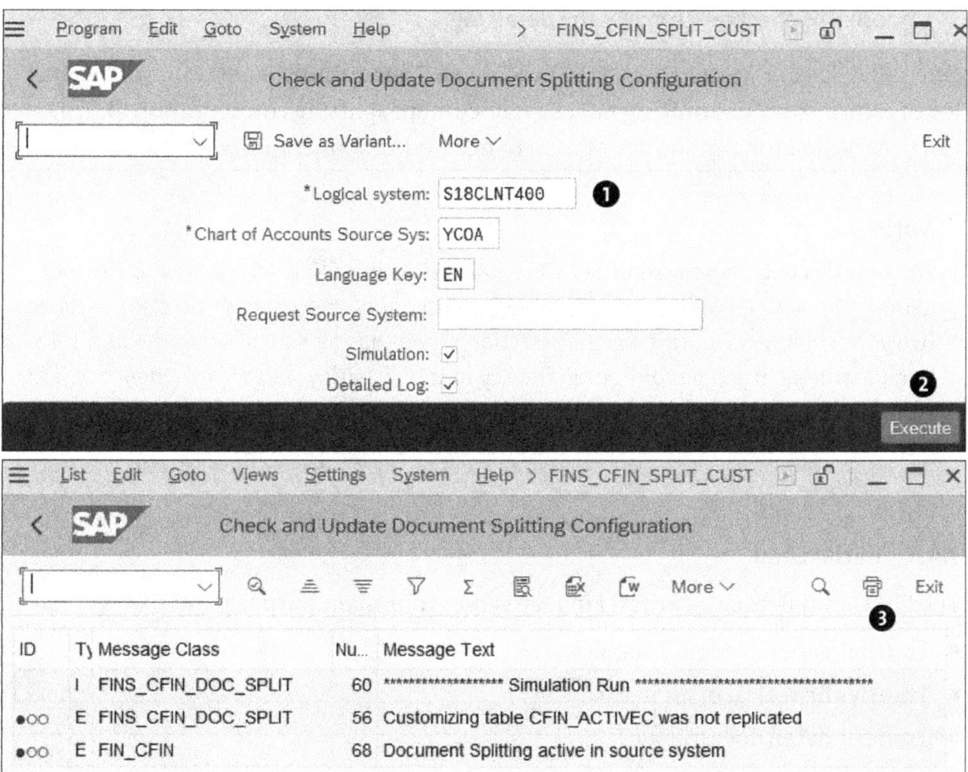

Figure 7.40 Check and Update Document Splitting Configuration

On this screen, enter values in the following fields and select the relevant checkboxes ❶:

- **Logical System**
  Enter a unique source logical system ID in this field.
- **Chart of Accounts Source Sys**
  Enter the source system chart of accounts that is in scope.
- **Language Key**
  By default, the system populates English as a language key to transfer text tables based on the user's logon language. However, this can be overwritten.
- **Request Source System**
  If you have an existing source system customizing transport, you'll enter that here, so the changes are recorded.
- **Simulation**
  This checkbox provides an option to run the report in a simulation mode, which doesn't update tables.
- **Detail Log**
  Choose this checkbox to view the detail log.

Once all fields are complete, choose **Execute** ❷. The resulting screen shows the output log of transferred Customizing tables for document splitting configuration ❸. This step can be repeated for any configuration change or delete operations.

Note

SAP provides two mapping entities, GENERAL_LEDGER_ACC_MASTER_ID and FIN_ACC_CHART_OF_ACC_ITEM_ID, for general ledger accounts and chart of accounts, respectively. As shown in Figure 7.40, the selection screen has a chart of accounts as a selection parameter. If you've configured the key mapping with your general ledger account, you'll need to change it to chart of accounts mapping entity FIN_ACC_CHART_OF_ACC_ITEM_ID. Otherwise, you'll run into the error **FINS_CFIN_MAPPING092**.

7.4.5 Initial Load

The FI-CO initial load in Central Finance serves four main purposes:

- Transfer general ledger balances.
- Transfer historical open items.
- Transfer detail documents.
- Aid in enabling real-time replication.

To transfer FI balances, open items, and detail documents from the source systems, the following prerequisite activities must be completed at a high level:

1. Establish baseline FI-CO organization data such as company code, controlling area, and relevant hierarchies (not covered in this book).
2. Configure Central Finance via Transaction CFINIMG (covered in different chapters and sections of this book).
3. Create and maintain master data objects (discussed in Chapter 4).
4. Configure and map master data (discussed in Chapter 4).
5. Complete general replication settings view `VCFIN_SOURCE_SET` (covered in Section 7.3.1).

In this section, you'll learn the initial load configurations in the Central Finance system.

Choose the Logical System

The initial load extract program of financial posting chooses the logical system defined in this configuration to transfer data from the source system to the Central Finance system. There are two ways to extract the initial load data if multiple source systems are involved:

- **Sequential extraction**
  With this approach, the following steps are followed:
  - Maintain the relevant logical system initial load extraction.
  - Run the initial load extraction.
  - Execute the delta run.
  - Set the initial load finished flag in `VCFIN_SOURCE_SET` located under **General Replication Settings** in the source system Central Finance menu.

  After executing these steps successfully, repeat them for the next logical system. This scenario is helpful in situations where you won't perform the initial load for a logical system at the same time.
- **Multiple separate initial load groups**
  With this approach, you'll configure separate initial load groups for each logical system. You'll run initial load activities in parallel.

To set up an active logical system for the initial load program to choose, navigate to IMG menu path **SAP Customizing Implementation Guide • Financial Accounting • Central Finance: Target System Settings • Initial Load • Initial Load Settings • Choose Logical System** or navigate to it from Transaction CFINIMG. Figure 7.41, shows the **Change View "Initial Load Logical Systems": Overview** screen upon executing the configuration. On this screen, click **New Entries** ❶. On the resulting **New Entries: Overview of Added Entries** screen, enter values in **Logical system** (for classic FI initial load) or **Logical System IDR** (for IDR initial load only) and **Size** ❷ fields. Choose **Save** ❸ to save the entry.

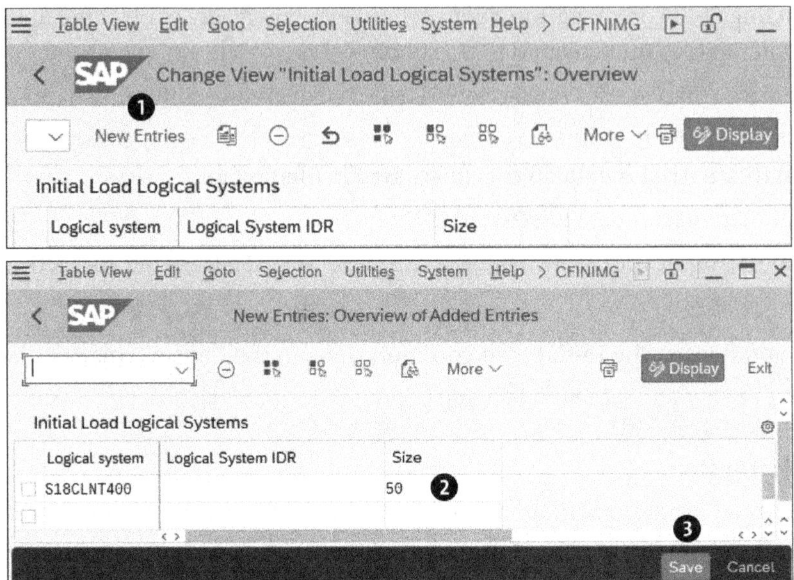

Figure 7.41 Choose the Logical System

The **Logical system** column is used for the standard initial load (covered extensively in Chapter 8, Section 8.3), and **Logical System IDR** applies to the intermediate data replication scenario, which is covered in Chapter 8, Section 8.5. The **Size** field refers to the package size used during the initial load steps of simulation of mapping, simulation of posting, and posting the initial load data. By default, 50 is the standard package size. In addition, refer to the package size for the initial load extraction step is defined in Section 7.3.1.

Define the Initial Load Group

In this Customizing activity, you'll create an initial load group that combines multiple company codes for initial load purposes. You'll group company codes with intercompany transactions in one group. In addition, you'll consider performance aspects to reduce the number of company codes in a group to the optimum level, so it doesn't impact performance during the initial load. The initial load groups are also used for initial load resetting purposes.

To set up an initial load group, navigate to IMG menu path **SAP Customizing Implementation Guide • Financial Accounting • Central Finance: Target System Settings • Initial Load • Initial Load Settings for FI • Define Initial Load Groups** or navigate to it from Transaction CFINIMG. Figure 7.42 shows the **Change View "Define Initial Load Groups": Overview** screen upon executing the configuration. On this screen, click **New Entries** ❶. On the resulting **New Entries: Overview of Added Entries** screen, enter values in the **Group ID** and **Group Description** ❷ fields. Next, choose **Assign Company Codes** ❸ in the dialog structure. On the resulting screen, click **New Entries**. On the resulting **Change View "Assign Company Codes": Overview** screen, maintain values in the **Logical System** and **Source Co Code** fields ❹, and then click the **Save** ❺ button to save the configuration.

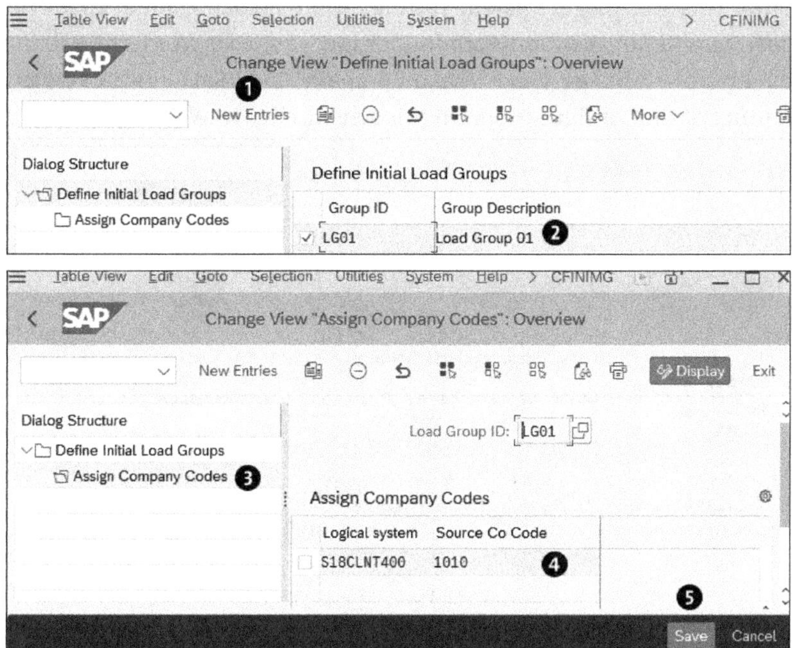

Figure 7.42 Define Initial Load Groups

Note

Transactions with cross-company codes must be in the same initial load group. You must evaluate transactions in the source system to identify intercompany transactions prior to adding relevant company codes in the initial load group.

Define Clearing and Substitution Accounts

In this configuration, you'll define migration clearing and substitution accounts for the transfer of initial load general ledger balances and balances on the receivables and payables. First, you'll define migration account at each company code level. During the initial load posting of balances, the account defined in this configuration is used as the offsetting account. After the initial load is completed, the balance on the migration clearing account should be zero.

In the second step of this configuration, you'll configure AP, AR, and open item–managed general ledger accounts as offsetting accounts. During the initial load posting, open items are posted into AP/AR accounts and offset to substation accounts. After the initial load is completed, the balance on the substation account should be zero.

To configure the migration clearing and substitution accounts two-step process, go to Customizing IMG menu path **SAP Customizing Implementation Guide • Financial Accounting • Central Finance: Target System Settings • Initial Load • Initial Load Settings • Define Clearing and Substitution Accounts** or navigate to it from Transaction CFINIMG. Figure 7.43 shows **Change View "Make Company Code Settings": Overview** screen upon executing of the configuration. On this screen, click **New Entries** ❶.

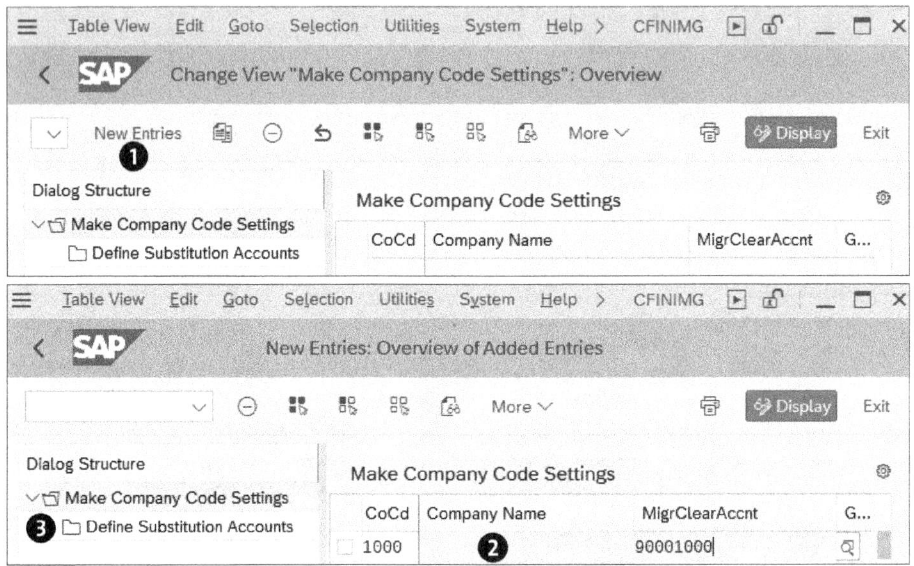

Figure 7.43 Define Clearing and Substitution Accounts: Migration Account

On the resulting **New Entries: Overview of Added Entries** screen, enter values in the **CoCd** field and press [Enter], then enter values in the **MigrClearAccnt** field, and press [Enter] again. Next, choose the checkbox or highlighting the line (for SAP GUI) ❷, and then double-click on the **Define Substitution Accounts** subnode ❸ located in the **Dialog Structure**. You'll arrive at the second section of Figure 7.44.

Next, on this **Change View "Define Substitution Accounts": Overview** screen, click **New Entries**, as shown in Figure 7.44 ❶. On the resulting **New Entries: Overview of Added Entries** screen, enter values for each reconciliation account for AP/AR, as well as open item–managed general ledger accounts ❷. You'll notice that the configuration offers users the chance to maintain reconciliation accounts or open item–managed accounts in ranges to avoid repetition of individual accounts. All accounts in the range are considered for the mapping. If the **To G/L Acct** field is left blank, the range is ignored by the system. Maintain corresponding substitution accounts. Choose **Save** ❸ to save the entries.

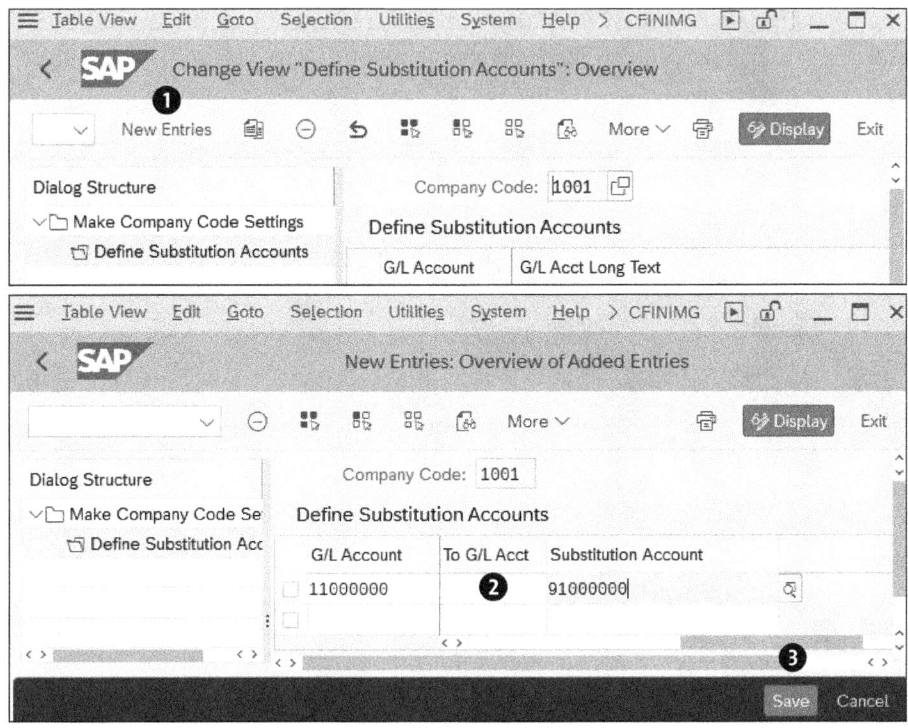

Figure 7.44 Define Clearing and Substitution Accounts

[«]

Note

Depending on the business requirements, the number of substitution accounts required varies. Besides AP, AR, and open item–managed general ledger accounts, no other general ledger accounts are mapped in this configuration. Otherwise, this will lead to migration and substitution accounts not tied to zero during reconciliation.

Reduce Package Size for Balance Load

To improve the performance of initial load balances, this optional configuration step provides settings that enable you to create small packages. In this configuration, you'll add additional fields that are appended to the package key. By adding the fields to the package key, the system creates small packages during the initial load of balances.

To reach the Customizing setting, navigate to Customizing IMG menu path **SAP Customizing Implementation Guide • Financial Accounting • Central Finance: Target System Settings • Initial Load • Initial Load Settings • Reduce Package Size for Balance Load** or navigate to it from Transaction CFINIMG. After reaching the settings, you can add the additional fields, as shown in Figure 7.45. On the **Change View "Balance Load: Additional Customizing for Packaging": Over** screen, click the **New Entries** ❶ button. On the resulting **New Entries: Overview of Added Entries** screen, enter values in **Logical System**, and then choose fields from the **Field Name in Totals Table** dropdown ❷. Choose **Save** ❸ to save the entries.

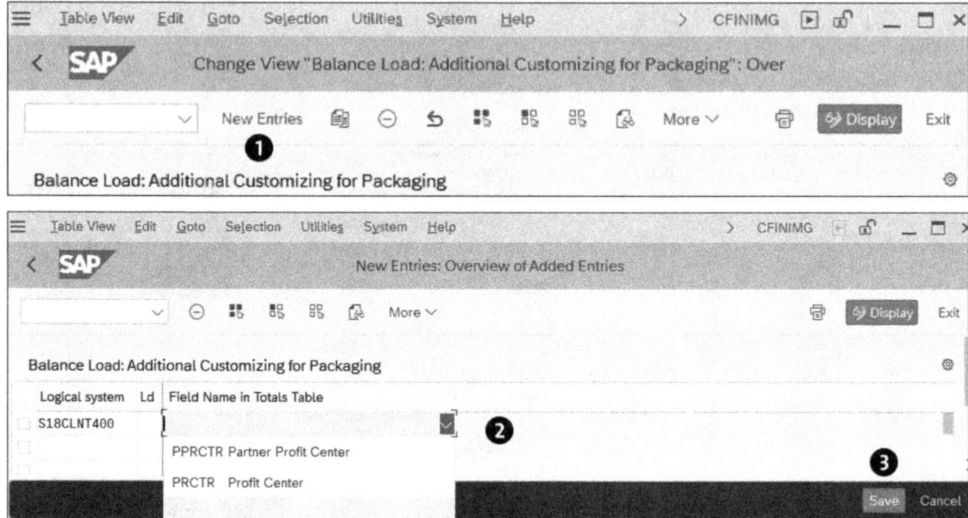

Figure 7.45 Reduce Package Size for Balance Load

[!]

Warning

SAP recommends this setting only when the initial load extraction, simulation, or posting steps are faced with out of memory dumps or you've observed extremely slow responses.

7.4.6 Activate Central Clearing of General Ledger Open Items

In this configuration activity, you'll first activate central open item management for each source system. Prior to activation, ensure you've all prerequisite SAP Notes applied by consulting SAP Note 3484064. Once prerequisites are met, then you'll maintain Customizing activity **Maintain G/L Accounts for Technical Clearing** in the source system, as we walked you through in Section 7.3.5. To activate the central open item management configuration, navigate to IMG menu **SAP Customizing Implementation Guide • Financial Accounting • Central Finance: Target System Settings • Central Management of Open Items in General Ledger • Activate Central Clearing of G/L Open Items** or navigate to it from Transaction CFINIMG. Figure 7.46 shows the **Change View "Central Management of G/L Open Items Customi.": Overview** screen. On this screen, maintain value in **SendLogSys** ❶ field to activate central clearing of general ledger open items and click the **Save** ❷ button to save the configuration.

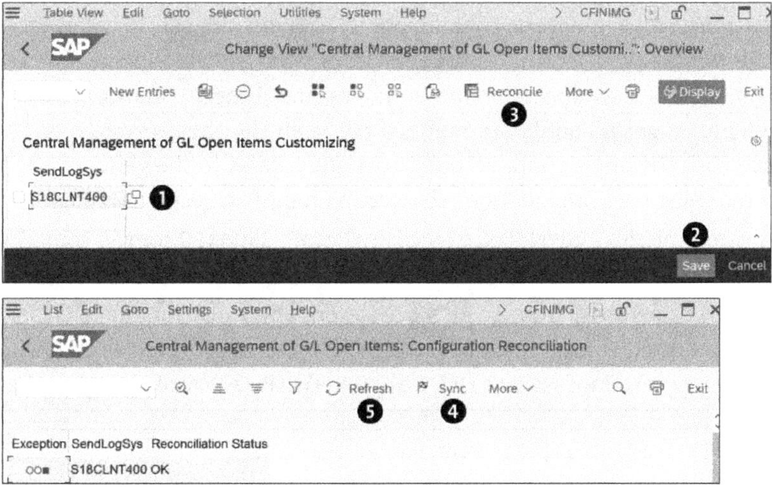

Figure 7.46 Activate Central Clearing of General Ledger Open Items

When you choose the **Reconcile** button ❸ after saving, the system compares mapping relationships in the source system and Central Finance system for the company code in scope for central clearing of general ledger open items and then displays any inconsistencies. In our example, the system displays no exceptions with a successful activation status. If any inconsistencies are found during activation, the system displays those errors with a red status. After fixing inconsistencies, you'll use the **Sync** ❹ button to check the status. The **Refresh** ❺ button can be used to display the status.

Alternatively, you can also activate central clearing of general ledger open items through maintenance view V_CFIN_CGL_CTRL via Transaction SM30 in Central Finance. For lower releases up to software component 105, Transaction CFIN_CGL_SWITCH can be used to activate central clearing to the general ledger.

Once activation is completed, you'll maintain source open items general ledger accounts for central clearing, as configured in Section 7.3.5.

Warning

Although central general ledger open item management is independent from central payment activation, the activation of central open item management is nonreversible. Therefore, extra caution should be taken into account prior to activating after evaluating all relevant process considerations.

Prior to activation of central open item management, the following prerequisites must be configured:

1. Define general ledger accounts for technical clearing.
2. Complete the initial load.
3. Clearing transfer is activated prior to the initial load. Refer to Chapter 9, Section 9.2.3 for clearing transfer activation.

4. Maintain consistency with how your general ledger accounts are set up, whether they are open item-managed accounts, non-open item managed accounts, or open items by ledger groups between the source system and Central Finance. You'll ensure the general ledger accounts are mapped between the source system and Central Finance with the appropriate account attributes. Refer to the general ledger open item mapping recommendations covered in Chapter 4, Section 4.3.2.
5. Apply the necessary missing SAP Notes. Consult SAP Note 3304480 for the source system notes and manual activities.
6. For historical open items exist prior to this activation, you must manually clear them or use SAP-delivered Transaction CFIN_CGL_TECH_CLR to set them as technically cleared (`ALE-extern`). Transaction CFIN_CGL_TECH_CLRLG is available to monitor the status of log.

7.4.7 Central Asset Accounting

With SAP S/4HANA release 2021 onward, Central Finance offers the central asset accounting functionality. The idea of this configuration is moving toward centralized processes in SAP S/4HANA, so you'll manage all asset accounting processes, including fixed assets closing tasks in Central Finance. This is part of SAP's future vision of bringing finance operations and shared services into Central Finance. By activating this feature, asset accounting processes are no longer supported in the source system except for processes that are integrated with the logistics component. Activation of central asset accounting involves the following steps:

1. **Configure the Central Finance system.**
   The first step involves configuring asset accounting in the Central Finance system. This is no different from configuring asset accounting in a typical greenfield environment. All scoped asset accounting processes and relevant configuration are completed in this step.
2. **Activate central asset accounting.**
   The second step involves activating central asset accounting by company codes that are in scope of the Central Finance system.
3. **Disable asset accounting in the source system.**
   Once central asset accounting is activated, disable asset accounting operations in the source system. The best way to prevent asset accounting operations is by removing authorizations.
4. **Complete the legacy data transfer.**
   Convert legacy master data, asset balances, and asset transactions to Central Finance. There are multiple ways to transfer legacy data from the source system to Central Finance.
5. **Repost the replicated asset postings.**
   This step is relevant only if Central Finance is active.

Now that you've know the process steps involved with central asset accounting, let's discuss the configuration required to support this functionality.

Activate Central Asset Accounting for Company Codes

As a first step, in this configuration, you activate central asset accounting in the Central Finance system. The activation is done at the company code level. Prior to activating central asset accounting, you must ensure the following:

- **14 Central Asset Accounting** is maintained under **RFC Usages** in Customizing activity **Maintain RFC Assignments and Settings for Source System.**
- Asset accounting is configured in the Central Finance system.
- If assets use cost objects and project work breakdown structures (WBSs) as account assignment objects, cost object replication and projects replication must be considered. If the central projects reporting scenario isn't active, then you'll map projects using Transaction FINS_CFIN_MAP_MANAGE.
- If the activation rate is used from the source system, the cost rate replication must be used.

To reach the Customizing setting, navigate to the Customizing IMG menu path **SAP Customizing Implementation Guide • Financial Accounting • Central Finance: Target System Settings • Central Asset Accounting • Activate Central Asset Accounting for Company Codes** or navigate to it from Transaction CFINIMG. After reaching the activation settings, you'll see the **Change View "Central Asset Accounting Activation": Overview** screen, as shown in Figure 7.47.

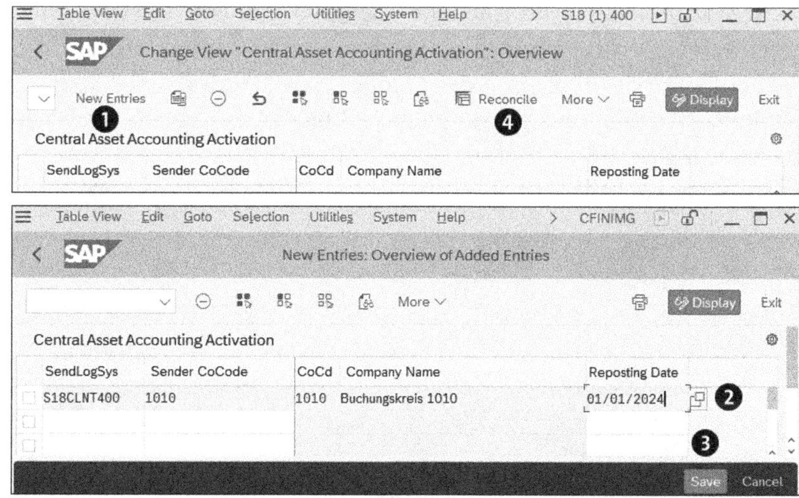

Figure 7.47 Activate Central Asset Accounting for Company Codes

On this screen, click **New Entries** ❶. On the resulting **New Entries: Overview of Added Entries** screen, enter the values as follows ❷:

- **SendLogSys**
  Enter the logical system of the source system that is in scope for central asset accounting.
- **Sender CoCode**
  Enter the company code in scope for central asset accounting.
- **Reposting Date**
  Pick a date to run reposting asset transactions from the source system to the central asset accounting system. In general, the posting date or invoice date is used for reposting asset transactions.

Choose **Save** ❸ to save the entries.

Upon a successful activation of central asset accounting, the **Reconcile** ❹ button can be used to generate a reconciliation report to view the activation status or any inconsistencies that may be present during activation.

[!]

Warning

Once central asset accounting is activated, it's nonreversible.

Define the Document Type for Reposting Replicated Asset-Related Transactions

In this section, you'll define the document type for reposting replicated asset-related transactions. SAP offers a repost replicated asset tool. To reach the Customizing setting, navigate to Customizing IMG menu path **SAP Customizing Implementation Guide • Financial Accounting • Central Finance: Target System Settings • Central Asset Accounting • Repost Replicated Asset Related Transactions • Define Document Type for Reposting Replicated Asset-Related Transactions** or navigate to it from Transaction CFINIMG. After reaching the activation settings, you'll see **Change View "Central Asset Accounting Activation": Overview** screen, as shown in Figure 7.48. On this screen, click **New Entries** ❶. On the resulting **New Entries: Overview of Added Entries** screen, enter the value in the **Document Type** field ❷, and press `Enter`. Click the **Save** button ❸.

[!]

Warning

Although activating central asset accounting seem straightforward, let's look at a few restrictions. SAP recommends reviewing these restrictions carefully and considering all process areas and systems prior to activating central asset accounting:

1. Central asset accounting isn't reversible once activated.
2. Combining company codes isn't supported.
3. Integration with the plant maintenance component isn't supported as the processes around this area are tightly integrated with logistics, which stays in the source system.
4. Asset under construction (AuC) legacy data with investment measure conversion isn't supported. A new AuC can be setup to transfer data.

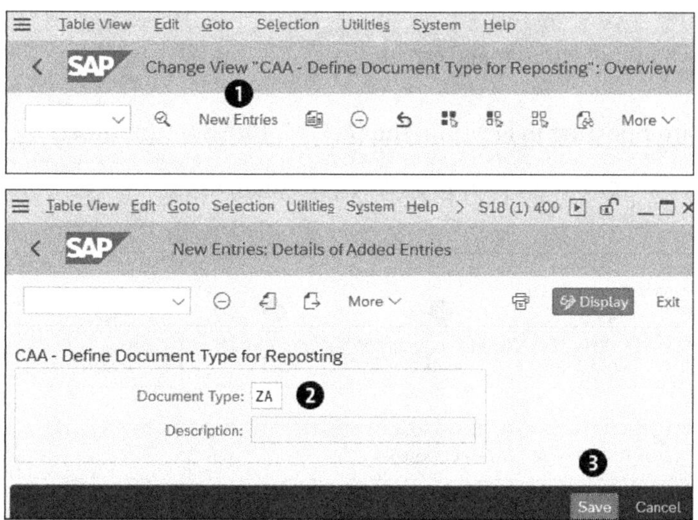

Figure 7.48 Define tool Document Type for Reposting Replicated Asset-Related Transactions

7.4.8 Enhancements Using Business Add-Ins in Central Finance

Central Finance delivers several BAdIs to transfer or enhance customer-specific data from the source system to Central Finance when the requirement isn't met with the Central Finance standard delivered functionality. These BAdIs come with methods and enhancement spots to implement custom logic at various stages prior to posting data into Central Finance, for example, Figure 7.49 shows methods that come with `BAdI_FINS_CFIN_AC_INTERFACE` and are most used in Central Finance implementations. This BAdI comes with six methods that can be called before posting financial data to the Universal Journal:

❶ `GET_PROCESS_CONTROL`
This method allows the standard logic or influence behavior of business processes to be skipped during runtime.

❷ `PREPARE_INPUT_DATA`
As you've noticed from Figure 7.49, this method is called before master data mapping. For example, you can influence values of master data prior to SAP Master Data Governance mapping.

❸ `MAP_VALUES`
This method can be implemented to map above and beyond Central Finance standard mapping (SAP Master Data Governance) for specific customer requirements. This happens after SAP Master Data Governance mapping.

❹ `ADJUST_POSTING_DATA`
This method is the last method used prior to the accounting interface. If there is a need to adjust any posting data, you'll implement this method.

❺ CALL_AC_INTERFACE
This method is used to add logic to manage errors.

❻ ADDITIONAL_POSTING
This method is called after posting in FI within the CALL_AC_INTERFACE method.

You'll call appropriate method based on your project requirements.

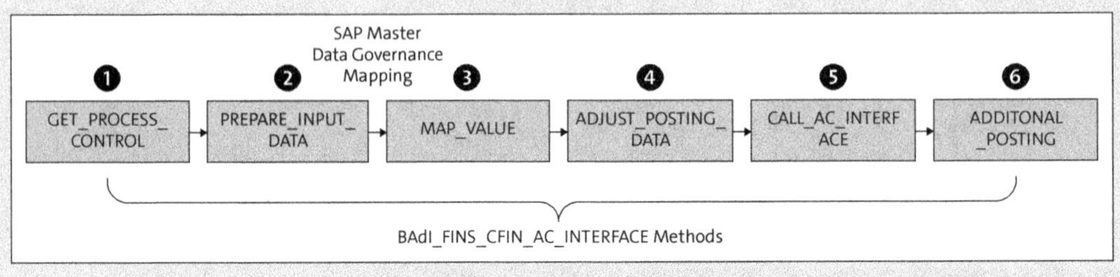

Figure 7.49 Central Finance BAdI_FINS_CFIN_AC_INTERFACE

Table 7.14 lists Central Finance BAdIs for customer-specific scenarios for various replication scenarios. We recommend reviewing the delivered BAdI documentation to understand the methods and restrictions, if any, prior to considering implementation.

BAdI Name	Purpose
BADI_FINS_CFIN_MAPPING_RULE	Used during the mapping step of posting
BADI_FINS_CFIN_MAPPING_PROCESS	Used for customer-specific mapping logic for mapping entities
BADI_FINS_CFIN_CMT_INTERFACE	Influences behavior of commitment document prior to posting
BADI_FINS_CFIN_AC_INTERFACE	Prepares or adjusts FI data prior to posting
BADI_FINS_CFIN_CO_INTERFACE	Prepares or adjusts CO document prior to posting
BADI_FINS_CFIN_CO_OBJECT_ASSIG	Influences cost object master data creation
BADI_FINS_CFIN_DFV	Enhances Central Finance comparison reports to add customer-specific fields or enhance reporting
BADI_FINS_CFIN_MAPPING	Used in key mapping above and beyond standard mapping
BADI_FINS_CFIN_EX_INTF	Used for influencing third-party data
BADI_FINS_CFIN_EX_INTF_CHG	Used for influencing third-party change data
BADI_FINS_CFIN_CURR_ADJ	Adjusts decimals for differing decimals between the source system and Central Finance system

Table 7.14 Central Finance BAdIs

BAdI Name	Purpose
BADI_FINS_CFIN_COPA_INTERFACE	Enhances profitability analysis data or logic
BADI_FINS_CFIN_COPA_MAPPING	Used for additional mapping for profitability analysis
BADI_FINS_CFIN_PS	Enhances project systems (PS) data
BADI_FINS_CFIN_DP_MAP	Performs outbound processing of reverse mapping for sales and distribution (SD) down payment
BADI_FINS_CFIN_CO_OBJ_EXT_NUM	Used to generate an external number for an internal order
BADI_FINS_CFIN_BC_OBJ_PROVIDER	Used in configuration for consistency checks
BADI_FINS_CFIN_BC_VALIDATION	Used in customer-defined validation for consistency checks
BADI_FINS_CFIN_ACTY_TP_REPL	Enhances activity type replication
BADI_FINS_CFIN_ACCTASGT_MAP	Enhances mapping process for account assignment object
BADI_FINS_CFIN_CAA_RT_EXTR	Used to support enhancements for central asset reposting

Table 7.14 Central Finance BAdIs (Cont.)

7.5 Running Consistency Checks

Now that you've completed configuration in the source and Central Finance systems, you'll run the configuration consistency check program to validate for any inconsistencies in configuration. SAP strongly recommends performing configuration consistency checks prior to beginning the initial load activities. The configuration consistency check offers the following benefits:

- Find inconsistencies in configurations between the source system and the Central Finance system.
- Confirm source postings transferred to the Central Finance system.
- Provide configurations to support additional customer-specific validation scenarios.
- Provide detailed messages for troubleshooting errors.
- Generate and download configuration consistency reports to share with teams for follow up.

In the following sections, you'll run consistency checks for the different functional areas.

7.5.1 Configuration Consistency Check

You'll run the central finance configuration consistency check report to detect inconsistencies in different functional areas between the source system and the Central Finance system. Prior to running this configuration consistency check report, the following activities are completed, as shown in Figure 7.50:

1. Complete the baseline configuration for financial and controlling areas.
2. Create the master data in Central Finance.
3. Complete the configuration object mapping activity (value mapping completed) in additional Central Finance system.
4. Complete master data mapping (key mapping).

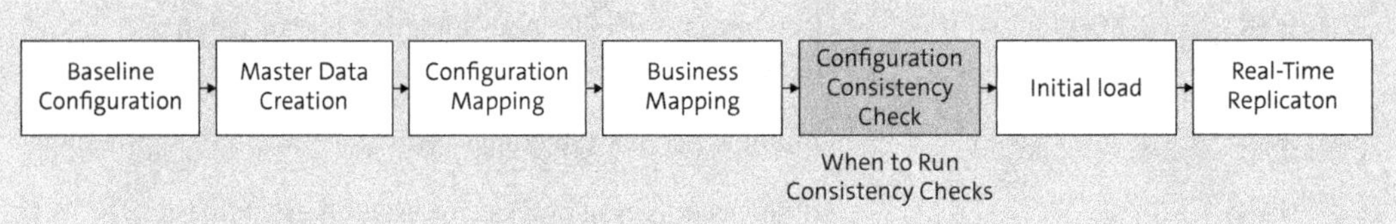

Figure 7.50 Configuration Consistency Check

Now that you've understood in what sequence you execute configuration consistency checks, let's execute the configuration consistency check report to understand it further.

In this step, first you'll execute a configuration consistency check to validate configuration settings between the source system and the Central Finance system. The report has options to run it either in the interactive mode or submitted as a background job if multiple company codes are involved. To execute the report, navigate to the Customizing menu path **SAP Customizing Implementation Guide • Financial Accounting • Central Finance: Target System Settings • Configuration Consistency Check • Execute Configuration Consistency Check** or navigate to it from Transaction CFINIMG. After executing the report, you'll arrive at the **Central Finance: Configuration Consistency** screen, as shown in Figure 7.51. On this screen, Choose **Execute** under the **Choose Action** ❶ area. Next, enter the value in **Source Logical System** ❷ under **Technical Setting**. Click the **All Categories** ❸ selection icon to choose the appropriate group to run the consistency check, as shown in Figure 7.52. In addition, enter values in **Source Company Code** ❹ or enter a range of company codes to run configuration for consistency checks. As you can see in Figure 7.51, the **Other Controls** section provides the **Show Errors and Warnings Only** ❺ checkbox, so the report displays only errors and warnings messages rather than showing everything. Finally, choose **Execute** ❻.

Next, when you choose the **Option for Consistency Check** icon ❸ from Figure 7.51, you're presented with the selection options shown in Figure 7.52. On this screen, choose appropriate checkboxes for functional groups based on your project scope.

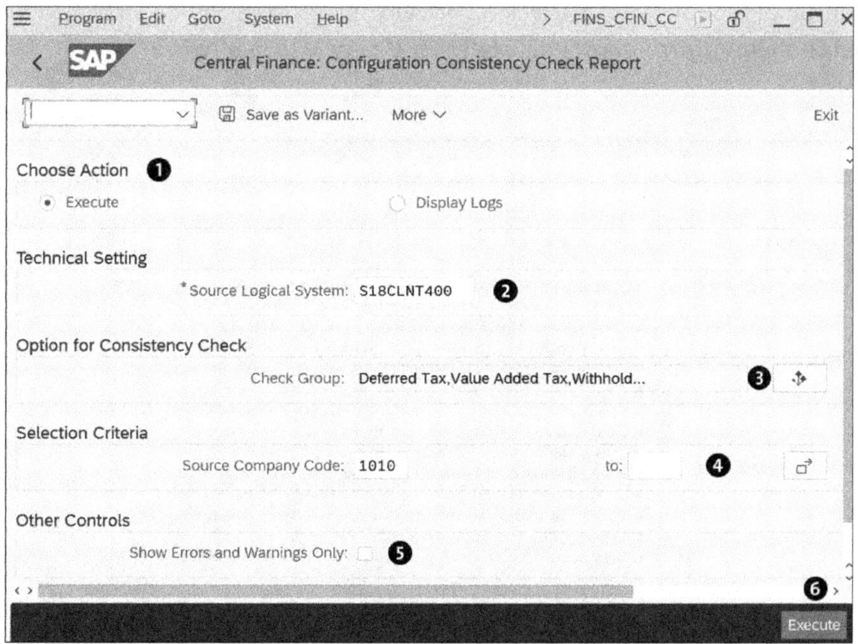

Figure 7.51 Execute Configuration Consistency Check Report

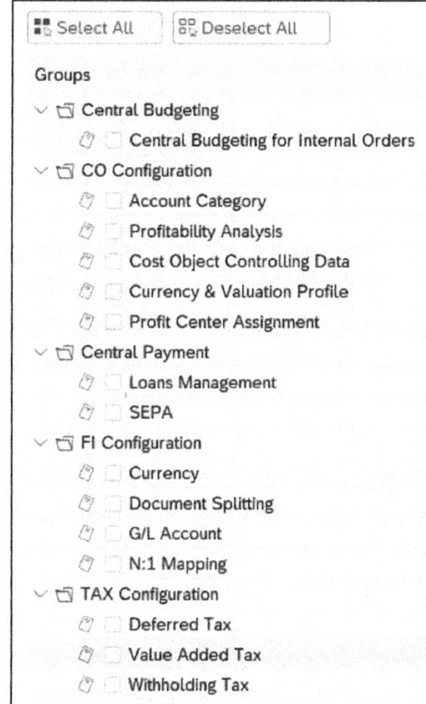

Figure 7.52 Configuration Consistency Groups

Upon execution of the consistency check report, any inconsistencies are displayed for the selected configuration or master data objects, as shown in Figure 7.53.

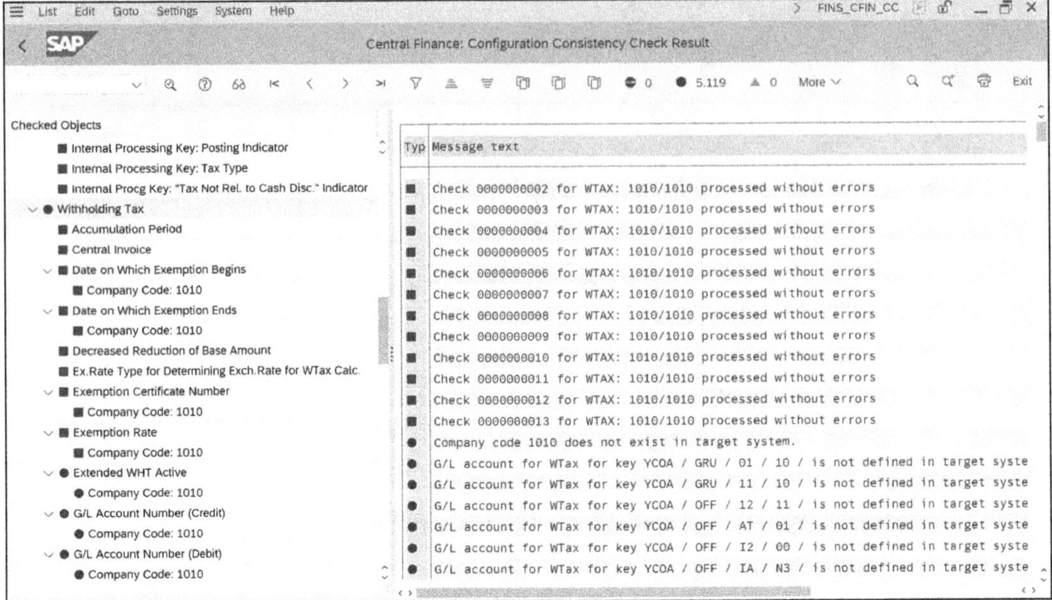

Figure 7.53 Configuration Consistency Results

Note

To execute the report, you'll require the relevant configuration access authorization role both in the source system and the Central Finance system.

In the next few sections, we'll discuss how configuration consistency checks help to identify and resolve inconsistency checks for different functional groups, as identified in Figure 7.52. Each group is organized with the type of check performed and what check is carried out in a tabular column.

Central Budgeting

After completing central budgeting scenario configurations, it's necessary to run the central budgeting consistency check to solve any inconsistencies between the source system and the Central Finance system. Table 7.15 lists key consistency check functions for central budgeting replicating scenario when the consistency check report is run.

Group	Type of Check Performed	Consistency Check Performed
Internal Orders	Order type (AUART) is used to differentiate based on business purpose.	Internal order type configuration (Transaction KOT2_OPA)
	Budget Profile (BPROF) provides control parameters for budget checking.	Checks whether a budget profile is assigned to order types and whether the budget profile is active for central budgeting for the internal order scenario
Projects/WBS	Project Profile (PROFIDPROJ) identifies certain types of projects.	Checks whether Customizing settings are identical between the source system and the Central Finance system
	Ensures Budget Profile settings are configured.	Checks whether a budget profile is assigned to the project profile and also whether the budget profile is active for functions for central budgeting projects scenario

Table 7.15 Central Budgeting Consistency Check

Configuring Financial Accounting

FI includes important configurations such as currency and document splitting. Table 7.16 lists FI configuration consistency checks for FI-CO-related objects.

Group	Type of Check Performed	Consistency Check Performed
Currency	General ledger account currency	Checks the general ledger master data currency between the source system and Central Finance system
	Company code currency	Checks that the local currency and additional currencies defined for the company code functions are the same between the source system and Central Finance
	Currency decimals	Checks whether decimal settings defined in the Central Finance system are equal or higher than the source system decimal settings

Table 7.16 FI Configuration Consistency Checks

Group	Type of Check Performed	Consistency Check Performed
Document Splitting	Activation check	Checks whether document splitting is consistently activated between the source system and Central Finance system
General Ledger Account	Open-item managed flag (XOPVW)	Checks that the field setting of the general ledger account is identical between the source system and Central Finance system
	Tax category (MWSKZ)	
	Clearing specific to ledger (X)	
	Balances in Local Currency Only (XSALH)	
Company Code	Mapping	Checks that the company code is mapped as 1:1 between the source system and Central Finance system (no support for M:1 or 1:M scenarios for company code mapping in Central Finance)

Table 7.16 FI Configuration Consistency Checks (Cont.)

Central Payment

Table 7.17 lists consistency checks performed for Central Finance primarily to ensure loans management isn't active. In addition, a similar check is done for Single Euro Payments Area (SEPA) mandate activation.

Group	Type of Check Performed	Consistency Check Performed
Loans Management (FS-CML)	Activation check	Checks that the FS-CML component isn't active since central payment doesn't use the FS-CML component
SEPA	Activation check	Checks relevant configuration settings for SEPA between the source system and Central Finance system

Table 7.17 Consistency Check for Central Finance

Configuring Tax

Often, replication of accounting data leads to errors due to missing tax configuration or incorrect mapping of tax codes. It's imperative that consistency checks should be run prior to Central Finance activities for the following tax areas:

- Value-added tax
- Withholding tax
- Deferred tax

Table 7.18 lists tax consistency checks performed by the system when the configuration consistency is run.

Functions	Type of Check Performed	Consistency Check Performed
Value-Added Tax	Checks company code settings	Country, VAT registration number, tax jurisdiction code, discount base amount is net value indicator, tax base amount is net of discount indicator, country ISO code, country is European Union member, country tax calculation procedure, country exchange rate type, length of first/second/third/fourth part of the tax jurisdiction code, external tax system indicator, determine taxes by line-by-line indicator, redetermine wrong jurisdiction code indicator, plants abroad, tax base amount changeable indicator, tax conversion local to document currency indicator
	Tax code settings	Error message for invalid tax amount indicator, tax type, EU code, tax code for sales taxes indicator, tax category when using external tax systems, relevant when using external tax system indicator, tax reporting country, MOSS tax reporting country for electronic service, discount base amount is net value indicator, tax base amount is net of discount indicator
	Internal processing key	Tax type, tax not deductible indicator, posting indicator, tax not relevant to cash discount indicator
Withholding Tax	Configurations check	Extended withholding tax active and relevant with holding configuration settings: tax type, tax codes, withholding tax base amounts

Table 7.18 Tax Consistency Checks

Functions	Type of Check Performed	Consistency Check Performed
Deferred tax	Configuration check	Tax transfer document, checks must clear, accept vendor invoices on full payment only, accept customer invoices on full payment only, create tax transfer document for 0% tax rate, tax not deductible, process taxes with higher tax rate first, process only cleared down payment clearing, date in tax transfer amount

Table 7.18 Tax Consistency Checks (Cont.)

Configuring Controlling

Likewise, the CO consistency check reveals any inconsistencies in the configuration. Table 7.19 provides a list of consistency checks performed on different functions.

Functions	Type of Check Performed	Consistency Check Performed On
Account Category	Account type: central	Checks for identical setting
	Account type: secondary cost element	
	Cost element category	
Cost Object Controlling	Statistical order	
	Plan integrated order	
Profitability Analysis	Definition of fields	Profitability analysis characteristics defined consistently
Currency and Valuation Profile	Configuration check	Checks whether currency and valuation profiles are assigned to the controlling area consistently when this setting is active in the source system
Profit Center Assignment	Cost center	Checks whether the profit center is assigned to the cost center or cost object consistently
	Cost object	

Table 7.19 Controlling Consistency Check

7.5.2 Customer-Defined Validations

SAP provides customer-defined validations to enhance configuration consistency check reports in addition to standard delivered validation groups. Once you configure customer-defined validation, the configuration consistency check report displays the SAP-delivered validation groups as well as customer-defined validation groups. This feature is useful when you need to check custom fields or fields not covered with a standard consistency check, for example, customer groups or brands (coding block field ZZ_BRAND). To enable this feature, SAP offers three objects to use in defining the business configuration consistency check: **Business Configuration Objects**, **Validation**, and **Validation Groups**. The tasks involved in creating customer-defined validation include the following:

1. Create a configuration object via **Manage Business Configuration Object**, which helps in the following:
   - Set of attributes is useful for various validation purposes.
   - Data provider reads data from the attributes used in the source system and Central Finance system.
   - Every attribute, for example, company code, is considered as one single configuration object that can be used with any validation groups.
2. Create a validation via **Manage Validation**. A validation helps with the following:
   - To define the check method for whether to use **Equality**, **Conditional indicator Equality**, or **Customer defined**.
   - To validate more complex scenarios with multiple attributes and complex conditions.
3. Create a validation group via **Manage Validation Groups** (also called check group), which is used for the following:
   - You'll create a definition for a set of validations from a business requirements standpoint. As a result, a validation group contains validations that are made up of configuration objects.
   - Validation groups are combined within a validation area
4. Assign Validations to Validation groups:
   - You'll assign validations to validations groups.
5. Assign Validation groups to Validation Areas:
   - You'll assign validation groups to validation areas.

Table 7.20 shows the configuration steps you'll follow to create a new customer-defined validation. We limit the discussion on this topic as this involves BAdI activity, which is beyond the scope of this book.

Configuration	Menu Path	Activity
Manage Business Configuration Objects	**SAP Customizing Implementation Guide • Financial Accounting • Central Finance: Target System Settings • Configuration Consistency Check • Manage Business Configuration Objects**	▪ In the first step, you'll define customer-defined business objects in customer namespaces starting with YY* or ZZ*. ▪ This activity requires creating a Data Dictionary (DDIC) structure or a class/interface type with relevant fields. For configuration objects, you need `BADI_FINS_CFIN_BC_OBJ_PROVIDER` implemented to enter the value in the **Configuration Object** and **BC Object Data Type** fields. Consult with the technical team on creating the appropriate DDIC structure for this activity.
Manage Validations	**SAP Customizing Implementation Guide • Financial Accounting • Central Finance: Target System Settings • Configuration Consistency Check • Manage Validations**	▪ In the second step, you'll define customer-defined validations. ▪ Implement BAdIs for validations with the correct check method. Consult the ABAP team on creating attributes.
Manage Validation Group	**Customizing Implementation Guide • Financial Accounting • Central Finance: Target System Settings • Configuration Consistency Check • Manage Validation Group**	In this configuration step, you'll define customer-defined validation groups to customer namespace YY* or ZZ*.
Assign Validations to Validation Groups	**SAP Customizing Implementation Guide • Financial Accounting • Central Finance: Target System Settings • Configuration Consistency Check • Assign Validations to Validation Groups**	In this step, the validation you've created earlier is assigned to a business validation group.

Table 7.20 User-Defined Validation Steps

Configuration	Menu Path	Activity
Assign Validations Groups to Validation Areas	**SAP Customizing Implementation Guide • Financial Accounting • Central Finance: Target System Settings • Configuration Consistency Check • Assign Validations Groups to Validations Areas**	In the final step, you'll assign validations groups created in an earlier step to a validation area.

Table 7.20 User-Defined Validation Steps (Cont.)

7.5.3 Permanent Consistency Checks

With this optional permanent consistency checks functionality, you'll be able to automate consistency checks on a predefined scheduled time and notify relevant stakeholders of configuration inconsistencies to receive a timely resolution of the issues. You can navigate to enable a permanent consistency check by following Customizing menu path **SAP Customizing Implementation Guide • Financial Accounting • Central Finance: Target System Settings • Configuration Consistency Check • Execute Configuration Consistency Check** or navigate to it from Transaction CFINIMG. As you can see in Figure 7.54, the program provides options to provision and schedule permanent consistency checks ❶ for selected objects ❷.

To achieve this, SAP recommends implementing BAdI BADI_FINS_CFIN_BC_MESSAGE_INT (**Enable Notification Channel for Permanent Consistency Check**) to trigger notifications to recipients by predefined configuration settings. With permanent consistency checks, Central Finance supports checking inconsistencies in the following objects and then sending an email or SMS notification:

- Company code
- General ledger account
- Cost center
- Cost element
- Cost object
- Document type
- Tax code
- Tax processing key
- Company code N:1 mapping

We limit our discussion on this topic as it requires enabling BAdIs to send notifications, so it's beyond the scope of this book.

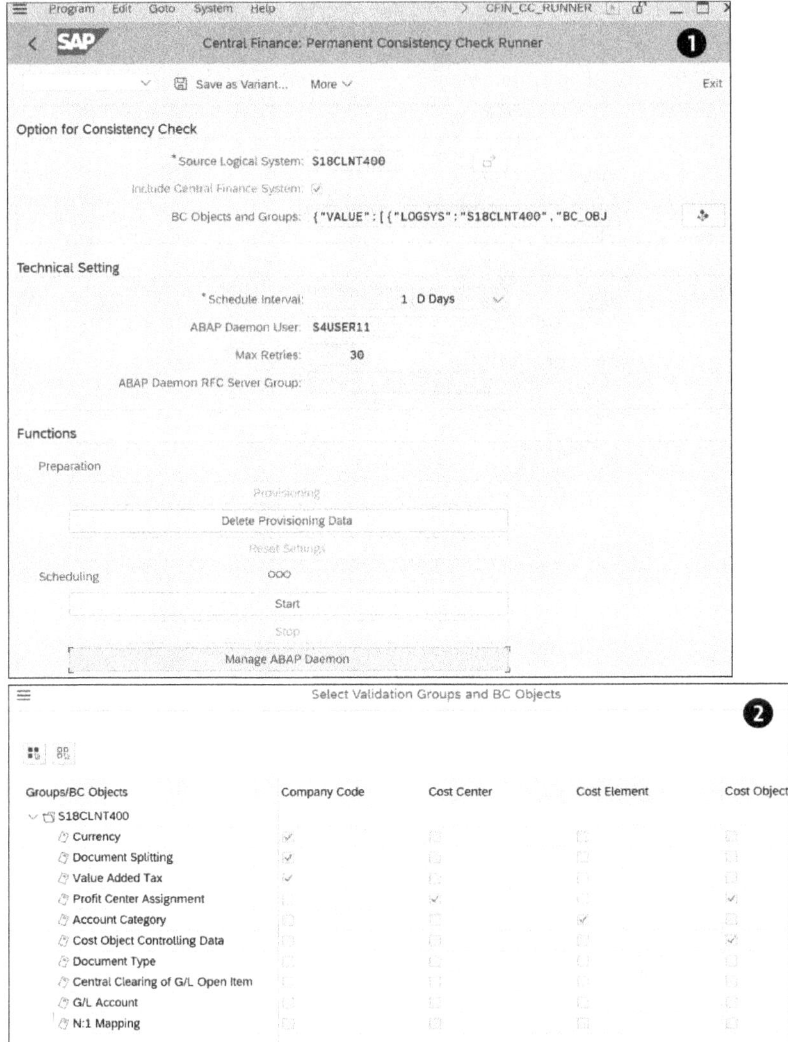

Figure 7.54 Permanent Consistency Checks

7.6 Summary

In this chapter, we discussed important design elements and their configurations in a Central Finance context. We highlighted prerequisites for both the source system and Central Finance system. Then, we went deeper into step-by-step configuration details on how to configure the source system and the Central Finance system. Next, the chapter walked you through configuration consistency checks for different functions that are essential before starting Central Finance activities. In addition, you've learned the customer-defined validations steps involved and related configurations. In the next chapter, you'll learn about the initial load, IDR, and real-time replication.

8

Chapter 8
Initial Load, Intermediate Data Retention, and Real-Time Replication

For any financial system transformation, you'll establish beginning balances for general ledger accounts on the day you go live with the new system. Likewise, in Central Finance, you'll establish general ledger account balances and bring open items and initial transactions prior to enabling real-time replication. Central Finance calls this process an initial load. In this chapter, we'll dive deeper into Central Finance initial load types for different application areas, processes, and execution steps.

In any enterprise resource planning (ERP) transformation project, you establish initial general ledger balances, convert open items, and bring transactional history to begin with. In SAP, balance conversion tools such as the Migrate Your Data - Migration Cockpit SAP Fiori app, data services, SAP Landscape Transformation Replication Server, SAP Rapid Data Migration, or other extract, transform, load (ETL) tools available for conversion of data. For Central Finance financial postings, SAP provides initial load to transfer initial general ledger account balances, open items, and transactional postings from certain periods based on selection settings from the source system to the Central Finance system. The initial balances can also be established with tools outside of Central Finance using third-party tools; however, this approach isn't recommended for the following reasons:

- Initial load comes with its own error handling and monitoring tools that are standard functionalities out of the box. The tools are constantly enhanced with new features almost every SAP release cycle. Otherwise. error handling and monitoring tools need to be built outside of Central Finance, which is a huge effort.
- Initial load uses SAP Master Data Governance mapping rules during initial load of accounts balance, open items, and transactional posting. If not, mapping needs to be built between source systems and the Central Finance system separately.
- Standard initial load brings in open items and open items clearing functionality for seamless central payment process and subsequent processes.
- SAP offers enhancements options using SAP-delivered business add-ins (BAdIs) both in the source system and Central Finance system, which can be leveraged.

- Data consistency across the source and Central Finance systems helps with reconciliation and reporting in Central Finance.
- Central Finance provides adoption content with best practices, preconfigured templates, and methodologies to accelerate the initial load process.

In this chapter, you'll learn the core concepts of initial loads, intermediate data retention (IDR), and real-time replication. Central Finance delivers the initial load for various objects and processes. You'll learn the initial load process for different applications in the relevant sections as follows:

- The initial load of cost objects (Section 8.2)
- The classic initial load of financial postings (financial accounting [FI]/Controlling [CO]) (Section 8.3)
- Resetting the classic FI-CO postings (Section 8.4)
- The initial load of FI postings using the IDR approach (Section 8.5)
- The initial load of CO postings (Section 8.6)
- The initial load of commitments (Section 8.7)
- Real-time replication key considerations (Section 8.8)

With that introduction, let's discuss the initial load and real-time replication steps next.

8.1 Initial Load and Real-Time Replication

Central Finance comes with the necessary tools to enable real-time replication. However, you must consider certain prerequisites prior to activating real-time replication, including performing the financial initial load. The following process steps of enabling real-time replication are organized into six major topics:

1. **Source system preparation**
   Besides upgrading to the minimum recommended service pack (SP) or applying necessary SAP Notes and cleansing activities in the source system, you must configure general replication settings view VCFIN_SOURCE_SET.
2. **SAP Landscape Transformation readiness**
   Before beginning Central Finance initial load activities, you must configure SAP Landscape Transformation for financial posting replication. In addition, prerequisite replication tables for cost object replication and tax consistency checks will also be carried out. After the SAP Landscape Transformation configuration, the key tasks include creating a database trigger and starting SAP Landscape Transformation replication for CFIN_ACCHD. The reason for the pause is to minimize the impact on the initial load reset.

3. **Central Finance system preparation**
   You'll perform several activities as part of the Central Finance system preparation as follows:
   - Applying missing SAP Notes for the required functionalities
   - Configuring baseline FI-CO activities
   - Creating and maintaining master data
   - Configuring relevant Central Finance activities located under Customizing Transaction CFINIMG
   - Enabling SAP Fiori apps and other reporting tools per requirements
4. **Initial load activities**
   You'll perform initial load activities include the following:
   - Extracting and monitoring initial load data from the source system
   - Simulating and monitoring master data
   - Simulating and monitoring posting
   - Executing final posting and monitor posting
   - Creating and maintaining master data
5. **Begin real-time replication**
   You'll start the replication activity.
6. **Monitor**
   You'll use SAP Application Interface Framework to monitor data posting.

Now that you've learned the high-level process of enabling real-time replication, let's get into the details regarding the prerequisites for the initial load in the source system, SAP Landscape Transformation system, and Central Finance system.

8.1.1 Initial Load Goals

First, let's discuss the initial load goals for Central Finance. When setting up Central Finance connections with source systems, the Central Finance initial load is necessary to transfer initial financial data, such as balances, open items, and transactions. From a technical perspective, the primary goal of Central Finance is to replicate financial data from an SAP source system or third-party systems to Central Finance in real time. Although the end goal of the initial load leads to enabling real-time replication in Central Finance, perhaps the initial load is a required step that updates the migration log table, which is necessary for online replication. By executing the initial load, you achieve three primary goals, as depicted in Figure 8.1:

❶ **Transfer general ledger balances**
Extract balances based on the year specified in the **Start – Balances** field in **General Replication Settings** (VCFIN_SOURCE_SET). See Chapter 7, Section 7.3.1, for more details.

❷ **Transfer open items**
Transfer all open items regardless of when they were posted.

❸ **Transfer detail documents**
Extract transactional details based on **Start Year-Documents** and **Period - Documents** fields in **General Replication Settings** (VCFIN_SOURCE_SET). See Chapter 7, Section 7.3.1, for more details.

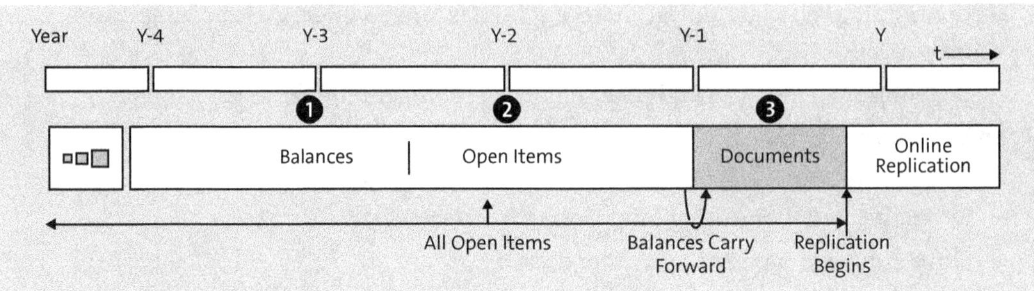

Figure 8.1 Enabling Real-Time Replication

[»]

Note

The initial load transfers all open items before the dates specified in **General Replication Settings**: **Start Year-Documents** and **Period – Documents** regardless of any prior years' open items being present in your source system. As a leading practice, any historical open items in prior years that are no longer relevant must be cleared prior to beginning the initial load activities.

8.1.2 Preparation and Prerequisites

Besides applying necessary system updates and SAP Notes, the first step for Central Finance begins with the initial load data extraction. Before you begin the FI initial load data extraction, ensure that the preparatory tasks and prerequisites are met in all connected systems. You'll learn about initial load readiness from each system in the following sections.

Source System

From a functional system-readiness perspective, prior to beginning the initial load, source system preparatory tasks must be completed. The preparatory tasks divided into the following categories:

- **Data cleansing**
  During the preparatory phase, you'll carry out cleansing activities in the source system for the following open items:

- Open item general ledger accounts
- Accounts receivable (AR) open items
- Accounts payable (AP) open items

- **Source system Customizing activities related to Central Finance**
Complete Customizing activities are required to prepare for the initial load. This includes setting up the **General Replication Settings** view VCFIN_SOURCE_SET in the source system as discussed in Chapter 7, Section 7.3.1. In addition, you'll complete the Customizing activities relevant to your replication scenarios that are in scope.
- **Consistency checks**
Run consistency checks to repair any inconsistencies in document indexes, transactions figures, and so on.
- **Run reconciliation reports**
Before the initial load extraction, all subledgers are reconciled to avoid any errors detected during the initial load process. You'll run reconciliation reports to capture a snapshot of the system before the initial load.
- **Timeout parameter**
During the initial load, the system packages open items, balances, and detail documents into a package. Depending on the volume of data extraction, you'll specify the timeout parameter for the initial load program to prevent job failure.

Table 8.1 summarizes key activities you'll perform before you begin the initial load extraction activities. The steps mentioned here are general guidance as this list could vary for each project, and specific tasks must be adopted based on your specific requirements. In addition, these process steps must be documented in the cutover plan.

Open Item Clearing/ Cleansing	Consistency Checks	Taking Snapshots before the Initial Load Extraction
■ Clear customer open items (Transaction F.32) ■ Clear vendor open items (Transaction F.44) ■ Clear general ledger open items (Transaction F-03) ■ Clear general ledger account for ledger groups (Transaction FB1SL) ■ Automatic clearing (Transaction F.13)	■ Run consistency check report RFINDEX to verify documents against indexes, documents against transaction figures, and indexes against transaction figures ■ Reconcile the general ledger with AP and AR subledgers using Transaction FAGLF03 or report TFC_COMPARE_VZ ■ Reconcile ledgers with report RGUCOMP4 or using Transaction GCAC ■ Run reports RM07MBST and RM07MMFI to reconcile material management (MM) with the general ledger	■ Financial statements using Transaction F.01 or report RFBILA00 ■ Totals report for cost centers Transaction S_ALR_87013611 ■ General ledger account balance using program RFSSLD00 ■ Compact document journal ■ Other specific reports per project requirements

Table 8.1 Source System Initial Load Preparation Activities

[»]

Note

Often, you'll come across open item-managed general ledger accounts either incorrectly setup or not managed consistently in the source system. This will lead to additional volume and errors during the initial load transfer of data from the source system to the Central Finance system. SAP provides program FINS_SWITCH_TO_OPEN_ITEM to either deactivate or activate the **Open Item Management** flag. You'll carry out the conversion activities as a separate project prior to starting the initial load activities. The program is available only after applying the latest SP and relevant manual activities in the source system. Consult SAP Note 2745769 on prerequisites, manual activities, and guidance for converting general ledger accounts. In addition, SAP Note 1770786 provides guidance on open item general ledger account conversion for lower SAP ERP releases.

System Landscape Transformation

Before you start the Central Finance initial load activities, you need to ensure all necessary SAP Landscape Transformation SAP Notes are applied. Next, SAP recommends setting up load and replication tables, as well as necessary filters, in SAP Landscape Transformation. Based on your requirements, you need multiple tables set up in SAP Landscape Transformation, as shown previously in Table 5.10 in Chapter 5. For the FI initial load, depending on your requirements, you need a minimum of the following tax tables set up in SAP Landscape Transformation (refer to Chapter 5, Section 5.1.6, for the tax tables):

- `AUFK`
  Use to load cost objects and enable replication of cost objects if in scope. Work breakdown Structure (WBS) replication is completed if in scope, which uses application link enabling (ALE) intermediate document (IDoc) technology instead of SAP Landscape Transformation.
- `CFIN_ACCHD`
  Use for the classic initial load and FI-CO replication.
- `CFIN_ACCHD_IDR`
  If the initial load is performed with IDR, then you'll activate the load and replication object for this table.
- `COBK`
  Use to support the CO secondary postings initial load and replication if in scope.

Central Finance System

From a Central Finance system standpoint, to support the FI initial load and real-time replication, perform the following preparatory activities:

- Ensure any missing SAP Notes are applied for the scoped item functionality.
- Complete baseline configuration for FI for the defined future state.
- Complete Central Finance Customizing activities and run configuration consistency checks.
- Installing necessary SAP Application Interface Framework content and Business Configuration Sets (BC Sets) based on your replication object's scope.
- Set recipients to manage errors in SAP Application Interface Framework.
- Enable the central project replication scenario if it's in scope.
- Complete the cost object load, and enable replication.
- Create and map master data in Central Finance.
- Deactivate the SAP Application Interface Framework runtime configuration for `CFIN_ACCHD` replication using Transaction /AIF/PERS_CGR before the initial load extraction.

8.1.3 Initial Load Sequence and Real-Time Replication Activation

The initial load consists of many steps that must be followed in a specific sequence to prevent initial load errors. After extracting the initial load data, you'll go through simulations steps prior to posting. During initial load posting, the system checks configurations and master data and then calls the accounting interface of the Central Finance system to repost. Thus, the postings deviate from the source system postings due to master data mapping, Customizing mapping, and other derivations that happen in Central Finance. In addition, follow-on documents such as reversals should follow posting sequentially with reference to predecessor documents to avoid missing errors. Therefore, it's critical that Central Finance initial load activities follow the recommended sequence. In this section, you'll learn the initial load sequence and list of steps to be followed during the initial load for SAP-based source systems.

Initial Load Sequence: High-Level Overview

At a high level, after completing FI and CO configurations and necessary prerequisites, you'll create and map master data. Then, you'll perform the initial load in a specific sequencing order as follows, for SAP-based source systems:

1. **Activating IDR recording**
   This step is relevant only if the initial load is carried out using the IDR approach. If you've chosen the classic initial load approach, this step isn't relevant and should be skipped.
2. **Initial load of WBS**
   If WBS replication is in scope, you'll complete the initial load of WBS. The initial load of WBS is carried out separately outside of Central Finance initial load activities.

3. **Cost object load**
   If transferring financial data references cost objects, the cost objects require both FI-CO and CO documents posting during the initial load. Therefore, you must load cost objects first. Recall what you've learned about cost object configuration in Chapters 4 and 8. The system reads cost objects from table AUFK and dependent tables. The initial load of cost objects is covered extensively in Section 8.2.
4. **Initial load of FI-CO**
   The initial load transfers general ledger balances, open items, and transactional data. During the initial load extraction, the system transfers financial data to table CFIN_ACCHD, table CFIN_ACCIT, and dependent staging tables in the Central Finance system. Then, you'll simulate and post to the Universal Journal in Central Finance. The system reads from these staging tables in Central Finance and calls the accounting interface of the central system to post the initial load documents.
5. **Initial load of CO documents**
   After FI-CO documents are posted, if CO document replication is in scope, the initial load transfers CO documents by reading CO data from table COBK, table COEP, and other dependent tables.
6. **Initial load of commitments posting**
   If commitments replication is in scope, you'll transfer commitment updates related to purchase requisitions and purchase orders from table source system to Central Finance.
7. **Reconciling initial load**
   After successful postings of the initial load, you'll reconcile the initial load with SAP-delivered comparison reports
8. **Activating online replication**
   For the subsequent FI initial load, you'll activate online replication. If central payment is in scope, you'll follow cutover steps to activate central payment.

Now that you've learned the initial load sequence at a high level, let's discuss the initial load steps in further detail for FI in SAP's recommended order.

Classic Financial Accounting Initial Load Sequence Steps

The initial load steps vary based on each project's requirements. The required steps are evaluated with project stakeholders and documented in the cutover plan during the explore phase of the project accordingly. Table 8.2 summarizes SAP-recommended initial load steps in chronological order. Following these sequence steps is important to avoid errors in the rest of the initial load. The table is organized based on each system's activity and corresponding reference sections in this book.

Sequence No.	Type of Activity	Source System	SAP Landscape Transformation	Central Finance	Recommendation/ Reference
1	Define replication objects and filters for table AUFK.		x		■ Chapter 5, Section 5.2.3
2	Perform the smoke test of cost objects.			x	■ Configuration of the cost object mapping scenario and mapping rules are prerequisites. ■ Error handling performed in SAP Application Interface Framework with namespace /FINCF and interfaces CO_OBJ_SIM and CO_OBJ. ■ Configuration is covered in Chapter 4, Section 4.4.5. ■ Initial load is covered in Section 8.2.
3	Simulate the initial load of cost object mapping.		x		
4	Process error messages for cost objects (table AUFK) simulation.				
5	Start the load and replication of cost objects (table AUFK).		x		
6	Process error messages for table AUFK.			x	
7	Make general replication settings.	x			■ Maintain values in table VCFIN_SOURCE_SET. ■ Manual activity in each client. ■ Chapter 7, Chapter 7.3.1
8	Define replication objects for table CFIN_ACCHD.		x		■ Chapter 5, Section 5.2.4
9	Deactivate SAP Application Interface Framework runtime configuration for table CFIN_ACCHD replication.			x	■ Deactivate runtime configuration /AIF/PERS_CGR (view V_CFIN_AIF_RT_CF). ■ Any documents that comes to SAP Application Interface Framework will be in **Pending** status. ■ Chapter 6, Section 6.2.2

Table 8.2 Classic FI Initial Load Sequence

Sequence No.	Type of Activity	Source System	SAP Landscape Transformation	Central Finance	Recommendation/ Reference
10	Start load and replication for table CFIN_ACCHD.	x			■ Chapter 5, Section 5.2.4
11	Extend the timeout parameter in the source system.	x			■ This is an optional step determined based on the volume of data being transferred. ■ Extend timeout parameter for the initial load extraction.
12	Extract data for the initial load.				■ Started from Transaction FINS_CFIN_LOAD_GRP_E. ■ Section 8.3.1
13	Monitor data extraction.			x	■ Monitored from Transaction FINS_CFIN_LOAD_GRP_M. ■ Section 8.3.1
14	Start new data extraction run to include delta records.			x	■ Started from Transaction FINS_CFIN_LOAD_GRP_E ■ Section 8.3.1
15	Monitor data extraction for the delta run.			x	■ Repeat to ensure the output of the run shows zero records identified. ■ Monitored from Transaction FINS_CFIN_LOAD_GRP_M.
16	Set the initial load **Finished** indicator.	x			■ This step prevents any subsequent delta runs for the company code. ■ Initial load **Finished** indicator is set in **General Replication Settings** (view VCFIN_SOURCE_SET), Transaction SM30.

Table 8.2 Classic FI Initial Load Sequence (Cont.)

Sequence No.	Type of Activity	Source System	SAP Landscape Transformation	Central Finance	Recommendation/ Reference
17	Simulate mapping.			x	▪ Depending on volume, this step may run longer the first time as the error messages are written to the database. ▪ Started from Transaction FINS_CFIN_LOAD_GRP_E. ▪ Section 8.3.1
18	Monitor simulation of mapping.			x	▪ Monitored from Transaction FINS_CFIN_LOAD_GRP_M. ▪ Section 8.3.1
19	Simulate posting.			x	▪ Started from Transaction FINS_CFIN_LOAD_GRP_E. ▪ Section 8.3.1
20	Monitor simulation posting.			x	▪ Monitored from Transaction FINS_CFIN_LOAD_GRP_M.
21	Post the initial load financial data.			x	▪ Started from Transaction FINS_CFIN_LOAD_GRP_E. ▪ Section 8.3.1
22	Monitor posting.			x	▪ Monitored from Transaction FINS_CFIN_LOAD_GRP_M. ▪ Section 8.3.1
23	Compare the initial load postings and expected CO postings in Central Finance.			x	

Table 8.2 Classic FI Initial Load Sequence (Cont.)

Sequence No.	Type of Activity	Source System	SAP Landscape Transformation	Central Finance	Recommendation/ Reference
24	Post process-deferred tax data after the initial load.			x	■ Optional step only if deferred tax is in scope for the project. ■ Run the report in Central Finance to update and adjust deferred tax data in table DEFTAX_ITEM as not all entries are created during the Central Finance initial load.
25	Activate the runtime configuration for table CFIN_ACCHD.			x	■ Helps in serialization of documents to go in **the** right order. ■ Execute report /AIF/PERS_RUN_EXECUTE in Central Finance, including runs in **New** status.
26	Process error messages for table CFIN_ACCHD postings.			x	■ Transaction AIF/ERR namespace /FINCF, and interface AC_DOC version 2.
27	Perform CO secondary postings.			x	■ Transaction CFIN_CO_INIT_PREP. ■ Section 8.6
28	Define replication objects for table COBK.		x		■ Transaction LTRC. ■ Chapter 5, Section 5.2.5
29	Perform smoke tests for CO documents.			x	■ Transaction CFIN_CO_SIMU. ■ Section 8.6.4
30	Perform simulation of the initial load of CO documents.			x	
31	Start load and replication of table COBK.		x		

Table 8.2 Classic FI Initial Load Sequence (Cont.)

Sequence No.	Type of Activity	Source System	SAP Landscape Transformation	Central Finance	Recommendation/ Reference
32	Process error messages for the table COBK transfer.			x	■ Transaction /AIF/ERR, namespace /FINCF, interface CO_DOC version 1.
33	Define replication objects for commitment replication (table CFIN_CMT_H).		x		■ Section 8.7
34	Simulate the initial load of commitment.			x	■ Section 8.7
35	Start load and replication for commitment.		x		■ Section 8.7
36	Process error messages for commitment replication.			x	■ Transaction AIF/ERR, namespace /FINCF, and interface CMT_DOC version 1. ■ Section 8.7
37	Run reconciliation reports.			x	■ Chapter 12, Section 12.3

Table 8.2 Classic FI Initial Load Sequence (Cont.)

Note

There are two ways to pause replication before posting to the Universal Journal. Step 9 in the preceding table follows deactivating the runtime configuration approach. This approach transfers the documents to SAP Application Interface Framework in **Pending** status. Upon executing persistence report /AIF/PERS_RUN_EXECUTE and starting SAP Application Interface Framework, documents post to the Universal Journal. Alternatively, table CFIN_ACCHD can be paused at SAP Landscape Transformation before posting to SAP Application Interface Framework and the Universal Journal. The latter approach reduces the effort involved if an initial load reset is required as there is less work in handling SAP Application Interface Framework. Refer to Section 8.1.6 for more options.

8.1.4 Empty Initial Load Concept

An initial load is a prerequisite to enable real-time replication. As you recall from the previous section, an initial load extraction triggers preparation for the online transfer of documents from the source system to the Central Finance system. Upon starting the Central Finance extraction program, the system populates an entry in table CFIN_MIG_LOG for each company code, which prepares any new postings to populate table CFIN_ACCHD and dependent staging tables (table CFIN_*) in the source system for online replication. The initial load is required to fill this source system's staging tables for the online replication route.

A full initial load is a very involved process considering error resolution, monitoring, and reconciliation activities. An empty initial load is useful when you don't want to go through a full initial load or you're not ready with all the master data, but you want to see the Central Finance replication functionality. Often, with an empty initial load, you'll perform a proof of concept on replication functionality to validate a specific functionality as part of the initial sprints. An empty initial load allows customers to carry out initial load postings with a future date to activate online replication instantly.

To perform an empty initial load, follow the IMG Customizing menu path **SAP Customizing Implementation Guide • Financial Accounting • Central Finance: Source System Settings • General Replication Settings** or navigate to it from Transaction CFINIMG. The settings can also be accessed using Transaction SM30 with appropriate access. Once the transaction is called up, the system presents the **New Entries: Overview of Added Entries** screen, as shown in Figure 8.2. In this screen, maintain entries as follows:

❶ **Company Code**
Enter each company code for which you want to perform an empty initial load.

❷ **Start Year - Balances**
Leave the field blank.

❸ **Start Year - Documents**
Choose a future year where no posting entries exist in that fiscal year.

❹ **Period - Documents**
Choose a future period where no posting entries exist in that period.

❺ **Keep Periods**
Enter value "12".

Leave the default values for other fields, and save the configuration values.

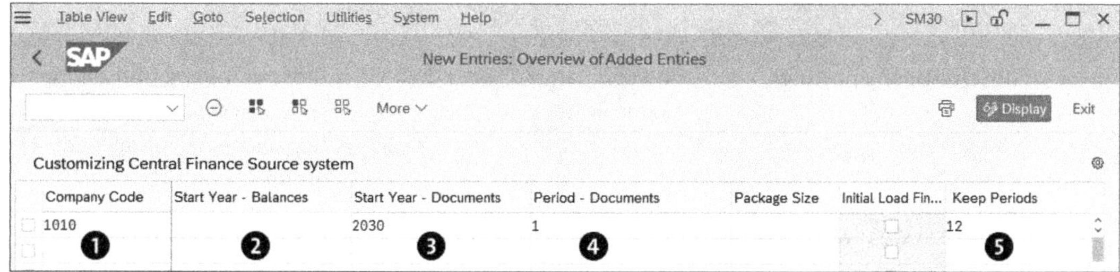

Figure 8.2 Empty Initial Load

[«]

Note

An empty initial load isn't recommended in a live production environment. In addition, an empty initial load is generally used for proof-of-concept purposes in test environments. However, SAP recommends copying live production data to a test environment and carrying out initial load activities to uncover potential errors and resolutions. Depending on the scope of companies and data volume, you'll plan multiple mock loads and test cycles with various functional scenarios until you've resolved errors and posted all initial load documents.

8.1.5 Standard Initial Load versus Intermediate Data Retention

SAP now offers two types of initial loads to extract balances and open items: classic initial load and intermediate data retention (IDR). Although extraction of balances and open items are the same between the two approaches, there are differences in how detail documents are handled. Table 8.3 summarizes the key differences between the two approaches.

Scenario	Classic Initial Load	IDR
Extracting balances and open items	No difference extracting balances and open items between the approaches	
Initial load detail documents	Uses target staging tables	Uses dedicated IDR tables
Reconciling	Possible lower data quality due to summarization, substitution rules, and so on	Improved data quality
Empty initial load	Supported	Not supported
Delta load	Delta load required	No delta load

Table 8.3 Classic Initial Load versus Intermediate Data Retention

Scenario	Classic Initial Load	IDR
Intercompany postings	All company codes in the same initial load group	Cross-company transactions split into separate transactions per each initial load group
Source system with UPA	Not supported with source system with universal parallel accounting (UPA)	IDR recommended if the source system is implemented with UPA
Source system as public cloud version	Not supported	Only available in IDR

Table 8.3 Classic Initial Load versus Intermediate Data Retention (Cont.)

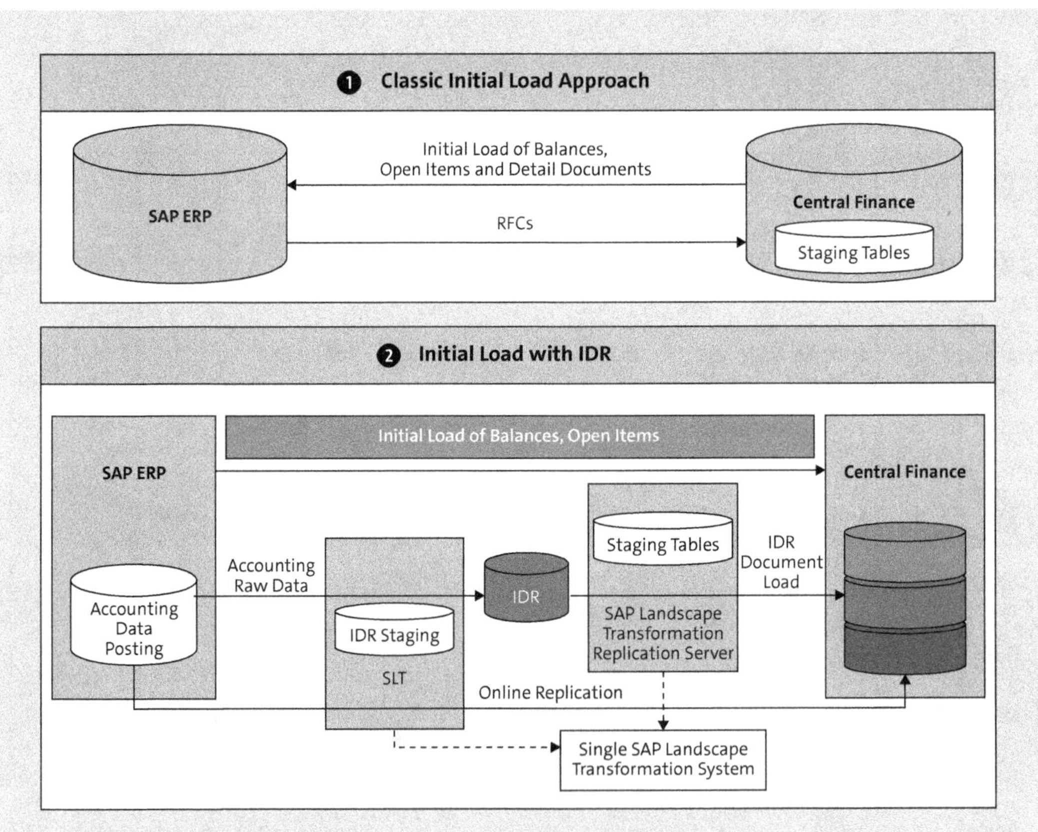

Figure 8.3 Classic Initial Load versus Intermediate Data Retention

Figure 8.3 illustrates how data flows between the classic initial load approach and the IDR approach:

❶ **Classic initial load**
This functionality has been available from the inception of Central Finance. As shown in the figure, the system transfers balances, open items, and documents through remote function calls (RFCs). If the source system is enabled with summarization or substitution rules, the original information is lost, resulting in a lower quality of data because FI to CO matching isn't always possible for reasons cited earlier.

❷ **IDR**
This functionality provides better quality data for documents because the system records raw data from the account interface. As you see in Figure 8.3, IDR activities of recording, extracting the initial load, and enabling ongoing replication are separated. You can start IDR recording much earlier, even before completing the Central Finance configuration. In addition, because IDR starts much earlier, you can troubleshoot and fix the errors much sooner.

8.1.6 Bulk Processing of Table CFIN_ACCHD/SAP Application Interface Framework Messages

During the initial load process, between the initial load extraction and initial load reconciliation, financial documents are accumulated and waiting to be posted in Central Finance via online replication through SAP Landscape Transformation Replication Server. Different approaches are available for handling accumulated postings, which is also called SAP Application Interface Framework bulk load. This volume, depending on the scope of company codes and data volume, could be in the millions. In general, there are three commonly used approaches for how to handle bulk volumes of data in table CFIN_ACCHD and SAP Application Interface Framework messages:

- In the first approach, you won't define and start table CFIN_ACCHD replication in SAP Landscape Transformation until the initial load is posted and reconciliated in Central Finance. Until that point, all accumulated documents are updated in the source system's table CFIN_ACCHD and dependent tables. These records haven't been transferred to the Central Finance system yet. Cutover steps typically have the following sequence:
  - Complete the initial load and reconciliation activities in the Central Finance system.
  - Deactivate runtime configuration in the Central Finance system.
  - Define and start table CFIN_ACCHD in SAP Landscape Transformation for the initial load. All documents transfer to SAP Application Interface Framework with **In Process** status in Central Finance.

- Activate logging table to SAP Application Interface Framework in **In Process** status runtime configuration in Central Finance.
- Run the AIF/PERS_RUN_EXECUTE report to start posting all **In Process** status SAP Application Interface Framework messages to the Universal Journal.

[»] **Note**

For the go-live of subsequent company codes, you'll add a filter for already-live company codes, so the system only brings in new company code data at go-live. This approach has the following benefits.

- Performance bottlenecks are minimized, making this approach strongly recommended for any volume of records.
- For the initial load reset, no deletion activity is involved in SAP Application Interface Framework.

- In the second approach, you'll start table `CFIN_ACCHD` in SAP Landscape Transformation and perform the FI initial load activities. You'll follow this sequence:
  - Deactivate runtime configuration in the Central Finance system.
  - Define and start table `CFIN_ACCHD` in SAP Landscape Transformation for the initial load before the initial load extraction.
  - Perform the initial load extraction, simulation, posting, and reconciliation activities. When the initial load extraction begins, all subsequent documents posted in the source system are transferred to SAP Application Interface Framework in **In Process** status.
  - Activate runtime configuration in Central Finance.
  - Run the AIF/PERS_RUN_EXECUTE report to start posting all **In Process** SAP Application Interface Framework messages to the Universal Journal.

[»] **Note**

This approach is found in standard SAP documentation. The major problem with this approach is customers may face huge performance bottlenecks due to the lower number of records transferred per second, so this approach isn't recommended. In addition, in case of an initial load reset, you'll have additional effort in deleting SAP Application Interface Framework messages and accounting tables.

- In the third approach, you'll start table `CFIN_ACCHD` in SAP Landscape Transformation and *pause* replication prior to the initial load extract. You'll follow this sequence:

- Deactivate runtime configuration in the Central Finance system.
- Define and start table CFIN_ACCHD in SAP Landscape Transformation Replication Server, and *pause* replication before the initial load extraction.
- Perform the initial load extraction, simulation, posting, and reconciliation activities. When the initial load extraction begins, all subsequent documents posted in the source system are transferred to the logging table in the source system.
- Resume table CFIN_ACCHD replication; this step transfers all documents sitting in the logging table to SAP Application Interface Framework in **In Process** status.
- Activate the runtime configuration in Central Finance.
- Run the AIF/PERS_RUN_EXECUTE report to start posting all **In Process** SAP Application Interface Framework messages to Central Finance. In addition, the SAP Application Interface Framework message status changes to **Success** if no configuration or mapping errors are found during posting.

8

[«]

Note

Although there is no deletion of SAP Application Interface Framework activity in an initial load reset, you'll still face performance bottlenecks with this approach, so it's generally not recommended.

8.2 Initial Load of Cost Objects

During the initial load data transfer, FI-CO documents and CO documents posted with cost objects in the source system require equivalent cost objects in Central Finance. Central Finance offers a cost object framework functionality to replicate short-lived cost objects from the source system to the Central Finance system. The prerequisites to carry out the initial load of cost objects are as follows:

1. The SAP Application Interface Framework interface configuration for table /FINCF/CO_OBJ is in Central Finance. You learned about cost object framework configuration in Chapter 4, Section 4.4.5.
2. Define mapping scenarios and mapping rules for cost objects, as covered in Chapter 4, Section 4.4.5.
3. Define predefined load objects and replication objects for table AUFK in SAP Landscape Transformation Replication Server, as explained in Chapter 5, Section 5.2.3. As a prerequisite, you copied the predefined load and replication objects from standard SAP Landscape Transformation Replication Server content. Table 8.4 shows the types of objects created in SAP Landscape Transformation Replication Server as a quick reference.

Predefined Object	Type of Object	Description
CFI_SIM_AUFK_L	Simulation load	Cost object simulation
CFI_AUFK_L	Initial load	Cost object initial load
CFI_AUFK_R	Real-time replication	Enabling real-time replication

Table 8.4 Cost Objects Created in SAP Landscape Transformation Replication Server

In this section, you'll focus on the initial load of cost objects and enabling real-time replication. Once all prerequisite configurations are in place, you'll be able to start the initial load of cost objects from SAP Landscape Transformation Replication Server. Cost object replication takes place via SAP Landscape Transformation Replication Server and is monitored in SAP Application Interface Framework.

8.2.1 Cost Object Replication Process

The overall cost object replication process is illustrated in Figure 8.4:

❶ **Cost object replication**
Based on Customizing settings for the mapping scenario and mapping rules, the system replicates cost objects from the source system to the Central Finance system. If the system doesn't find a matching rule for the given condition, the replication of the cost object will fail and display in SAP Application Interface Framework.

❷ **Cost object assignment**
Based on mapping rules, the system updates cost object mapping in cost assignment table FINS_CFINT_ASGMT in the Central Finance system. After the successful load/replication of cost objects, the assignments are created in this table.

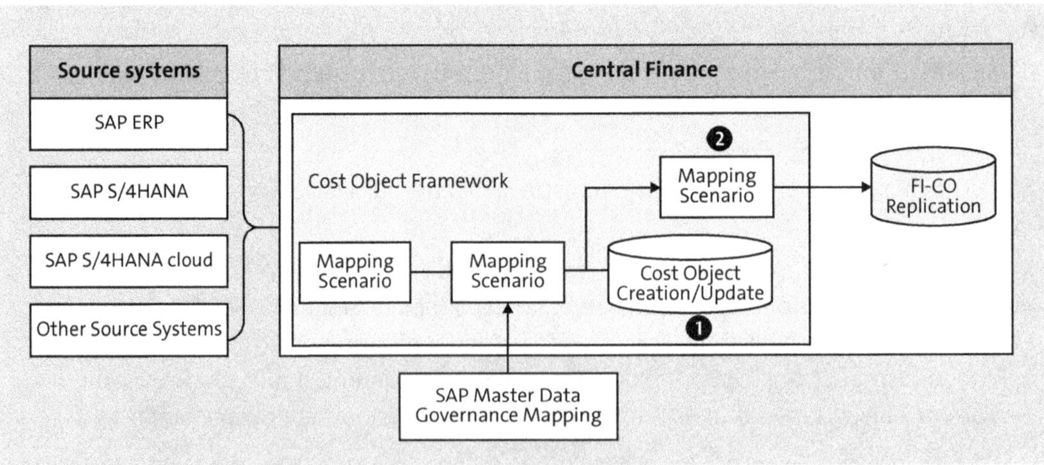

Figure 8.4 Cost Object Framework during Replication

Finally, prior to financial document posting, the system reads cost assignment table FINS_CFINT_ASGMT and fills the mapped cost object information into the posting document in Central Finance.

The initial load of cost objects is carried out with three steps:

1. **Smoke test of cost object**
   You'll execute the smoke test of the cost object from the Central Finance system. By executing the cost object smoke test, you have an opportunity to validate smoke test of mapping scenario, mapping rules, and other configuration. This step doesn't update tables. The smoke test call is made through an RFC that is often tested for a few records and has a maximum limit of 999 records. In addition, the smoke test program has its own in-built application logs for troubleshooting.
2. **Simulation of cost**
   You'll simulate the cost object executed from SAP Landscape Transformation Replication Server. Like the smoke test, when you execute the cost object simulation, you'll be able to validate the mapping scenario, mapping rule, and other relevant configurations. The cost object simulation can be carried out for higher volumes of data. In addition, as the name implies, the simulation step doesn't update tables.
3. **Initial load of cost object**
   Once the smoke test and simulation tests are successful, you'll create cost objects and cost object mapping with this step. You'll use SAP Application Interface Framework to monitor for any errors or restart after fixing errors.

[!]

Warning

The SAP Landscape Transformation Replication Server team is responsible for cost object simulation, initial load, and real-time replication of cost objects. If the same mass transfer ID (MT ID) is used for cost object simulation and cost object initial load/replication, you'll reset the transfer status of simulated cost objects table DMC_FM_RESTART in SAP Landscape Transformation before the actual initial load/replication.

Now that you've learned the basic concepts behind the initial load of cost objects replication, let's see the same steps in the system with step-by-step instructions.

8.2.2 Smoke Test of Cost Objects

In Central Finance, the smoke test program is available for both cost objects and CO secondary postings replication. The program is executed from the Central Finance system, which makes an RFC to the source system. As the name implies, the smoke test is used primarily to check configuration and mapping prior to performing the actual initial load. During the smoke test, the system doesn't create any cost objects; rather, it performs a sanity check to ensure all Customizing and master data exists. Although the smoke test is an optional step, SAP recommends this step to identify any missing configuration and master data. Prior to running the smoke test, ensure you've completed cost object mapping and relevant configuration, as explained in Chapter 5, Section 5.2.3.

To perform smoke tests for cost objects (table AUFK), navigate to Customizing menu path **Financial Accounting • Central Finance • Central Finance: Target System Settings • Initial Load • Initial Load Preparation for Management Accounting • Smoke Test for Cost Object Mapping and CO Document Replication** or use Transaction CFIN_CO_SIMU. After entering the transaction, you'll arrive at the **Smoke Test for Cost Object Mapping and CO Document Replication** screen, as shown in Figure 8.5.

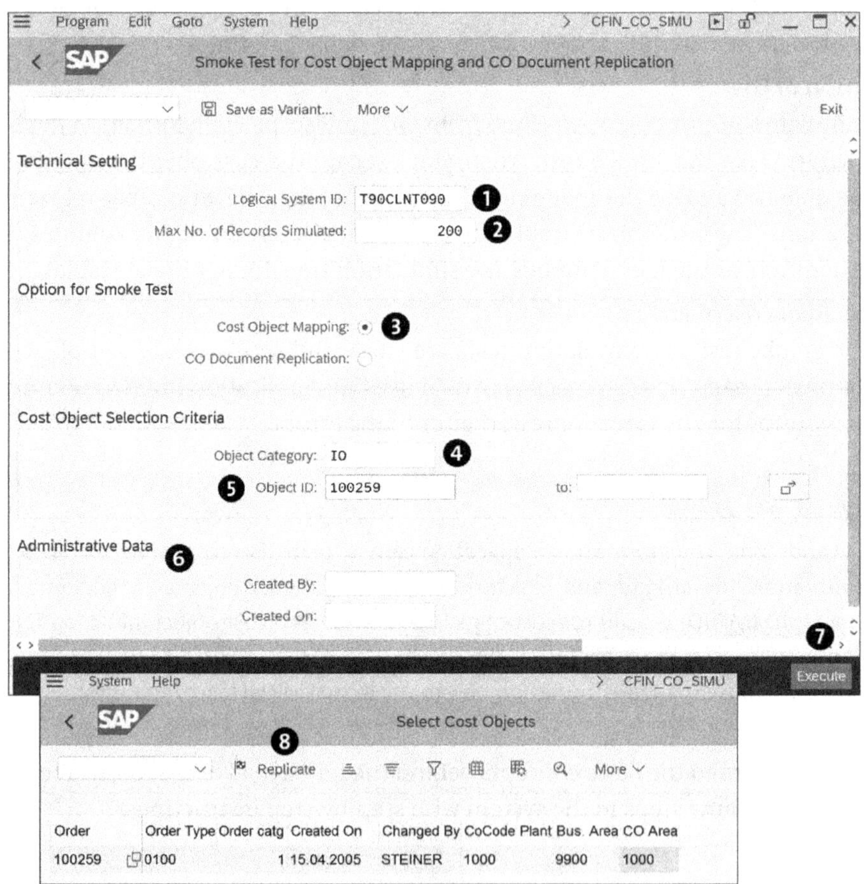

Figure 8.5 Smoke Test for Cost Object Mapping

On this screen, enter values in the following fields (in our specific example, we've used an internal order type **IO** to demonstrate the functionality):

❶ **Logical System ID**
Enter the source system from which the cost objects will be replicated to Central Finance.

❷ **Max No. of Records Simulated**
Enter a numeric value to represent the number of records to be simulated. As you recall, the program allows you to enter a maximum of 999 records.

❸ **Cost Object Mapping**
As you've noticed, the same program is used for both cost object mapping and CO document replication under **Options for Smoke Test**. Choose **Cost Object Mapping** for cost object smoke tests that display cost object–specific fields to enter next.

❹ **Object Category**
In the dropdown list, select **IO** for internal order replication. This is a required field.

❺ **Object ID**
If you know a specific internal order or a list of orders to simulate, enter internal orders in this field.

❻ **Administrative Data**
Under this section, choose the user ID of the person who created the internal order master data or choose the date when it was created.

Upon entering the necessary data in the selection screen, click the **Execute** ❼ button. This action takes you to the **Select Cost Objects** screen based on the chosen criteria from the previous screen. On this resulting screen, select the row, and click the **Replicate** ❽ button.

The **Smoke Test Result** screen, as shown in Figure 8.6, displays the status of orders ❶, which will be replicated over from the source system to the Central Finance system.

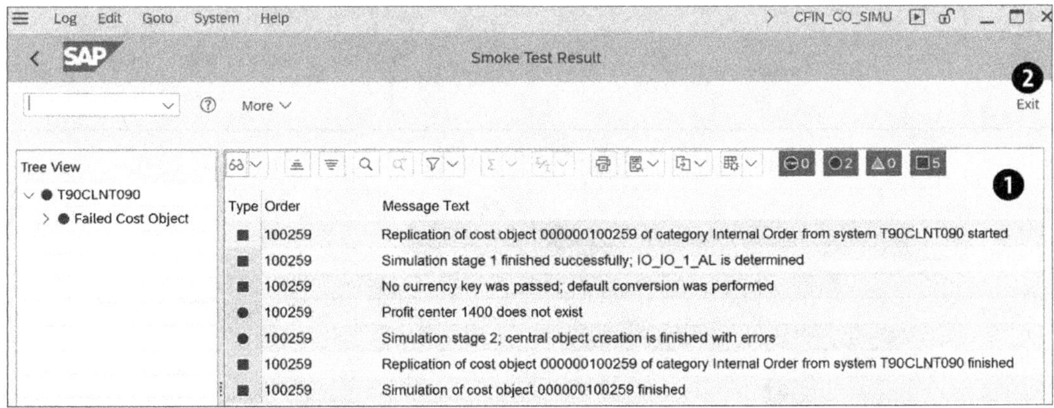

Figure 8.6 Cost Object Smoke Test Results

Any hard errors must be evaluated and addressed prior to running the actual initial load. As the smoke test doesn't create actual data in tables, you'll be able to repeat smoke tests multiple times to confirm all errors are resolved prior to running the actual initial load. Click the **Exit** ❷ button to return to the previous screen.

[»]

Note

The logical system will show up on the smoke test screen only when **Define Technical Settings for All Involved Systems** configuration is completed under the Central Finance Customizing menu (Transaction CFINIMG).

8.2.3 Simulation of Cost Objects

Like the cost object smoke test, you'll carry out the simulation of cost objects to check configuration and master data, but with a larger data volume. The cost object simulation doesn't update tables, and it's used primarily to check inconsistent configuration and load time for cutover purposes. The cost object simulation is carried out in SAP Landscape Transformation Replication Server and monitored from SAP Application Interface Framework. Although cost object simulation is an optional step, SAP recommends carrying out cost object simulation prior to performing the actual initial load.

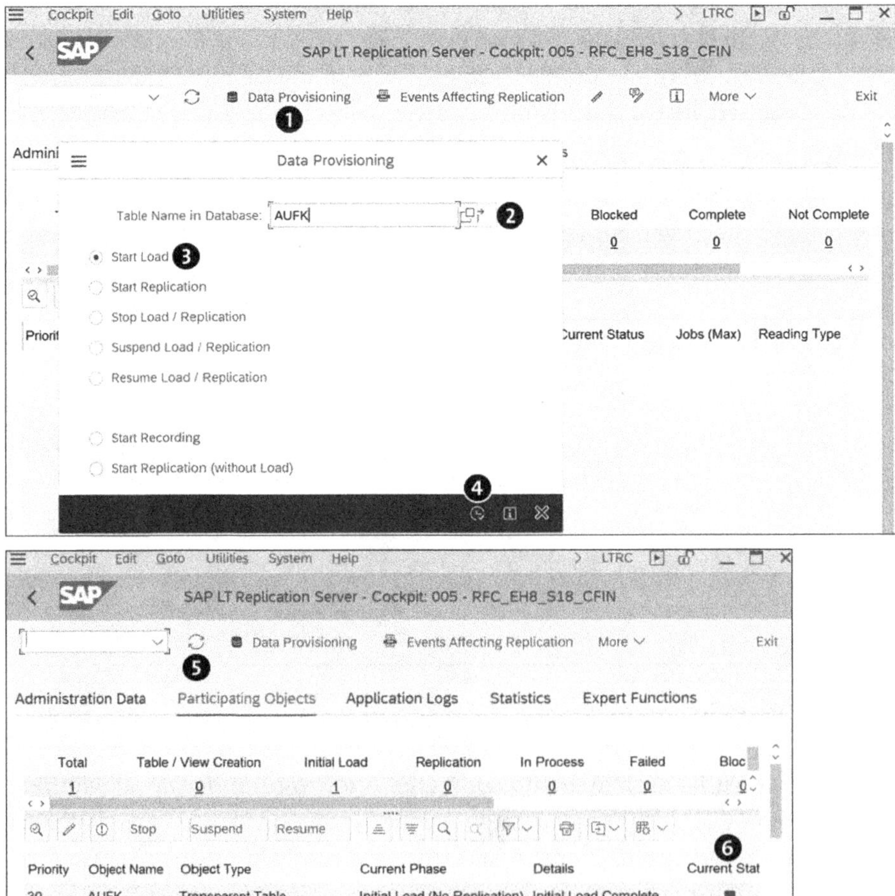

Figure 8.7 Simulation of Cost Objects

Prerequisites for configuring the MT ID for cost object replication (table AUFK) with predefined initial load object CFI_SIM_AUFK_L are covered in Chapter 5, Section 5.2.3. In this section, you'll carry out the simulation of the initial load of cost objects. To perform cost object simulation, access the SAP Landscape Transformation Replication Server cockpit using Transaction LTRC in the SAP Landscape Transformation Replication Server client. Choose MT ID **006**, and click on the **Participating Objects** tab, as shown in Figure 8.7 ❶. This brings up the **Data Provisioning dialog** box. To perform cost object simulation, enter "AUFK" in the **Table Name in Database** ❷ field, choose **Start Load** ❸, and click the **Execute** ❹ icon. This action starts the cost object simulation load. To check the status of the cost object simulation load, from the SAP Landscape Transformation Replication Server cockpit, click the **Refresh** ❺ icon. In addition, the **Current Status** ❻ shows **Initial Load (No Replication)**.

Successful cost object simulation will show up in SAP Application Interface Framework under interface CO_OBJ_SIM in Central Finance.

8.2.4 Initial Load of Cost Objects

Like cost object simulation, you'll execute the initial load of cost objects from SAP System Landscape Transformation Replication Server and transferred data and then monitor from SAP Application Interface Framework. Upon transfer of the initial load of cost objects, the system creates cost objects and cost object mapping in Central Finance.

The prerequisite requirements for configuring the MT ID for cost object replication (table AUFK) with the predefined initial load object CFI_AUFK_L and predefined replication object CFI_AUFK_R are covered in Chapter 5, Section 5.2.3. In this section, you'll carry out the real initial load of cost objects in SAP Landscape Transformation Replication Server. To perform the initial load/replication of cost objects, access the SAP landscape Transformation Replication Server cockpit using Transaction LTRC in SAP Landscape Transformation Replication Server. Upon entering the transaction, choose MT ID **006**, and click on the **Participating Objects** tab, as shown in Figure 8.8 ❶. This brings up the **Data Provisioning dialog** box. To perform the initial load of cost objects, enter "AUFK" in the **Table Name in Database** ❷ field, choose **Start Replication** ❸, and click the **Execute** ❹ icon. This action completes the initial load first and then enables real-time replication as well. In addition, the system creates a logging table and database triggers (insert, update, and delete) in the source system for real-time replication.

Note

As you've seen, the **Data Provisioning** dialog box, **Start Load** can be used to perform cost object simulation, but the system will only simulate and won't enable online replication. With the **Start Replication** option, however, the system both performs the initial load of cost objects and enables online replication without any manual intervention.

To check the status of the replication, from SAP Landscape Transformation Replication Server cockpit, click the **Refresh** ❺ icon. In addition, the **Current Status** ❻ column shows **Status**, as shown at the bottom of Figure 8.8.

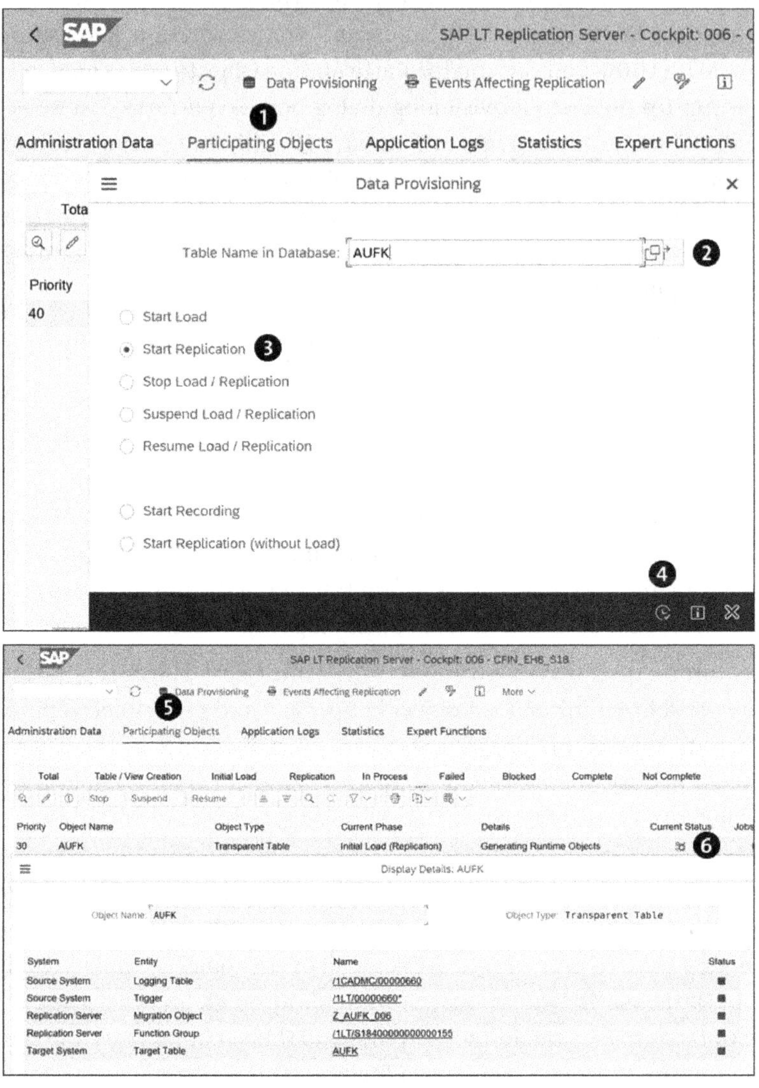

Figure 8.8 Initial Load of Cost Objects

In addition, a successful load will create entries in cost object table AUFK and assignment table FINS_CFINT_ASGMT in Central Finance.

[»]

Note

For additional information on cost objects replication, consult SAP Note 2971596, which provides answers to frequently asked questions.

8.3 Initial Load of Financial Accounting-Controlling Postings

In this section, you'll learn how the FI-CO initial load postings function in Central Finance. As you'll recall, Central Finance initial load transfers general ledger balances, open items, and accounting documents from the source system to Central Finance. The initial load is a required step in enabling real-time replication. The initial load has its own error resolution and monitoring tools to resolve errors and rerun the initial loads. Figure 8.9 illustrates the data flow for classic initial loads:

❶ **Beginning the initial load extraction**
You'll begin the initial load extraction process from the Central Finance system via the initial load group. Upon executing the extraction program in Central Finance, the system makes an RFC to the source system to pull the initial load data.

❷ **Extracting finance data from source system**
Depending on the source system, the system reads the relevant accounting tables and then groups the data in packages. How the system packages the initial load data is covered in Section 8.3.2.

❸ **Updating staging tables in Central Finance**
Packaged information then is sent to Central Finance staging tables (table CFIN*) for carrying out simulation and final posting activities.

❹ **Simulation of mapping and simulation of posting**
Central Finance comes with its own simulation and monitoring tools for the FI initial load. Once the initial load data is extracted and transferred to Central Finance

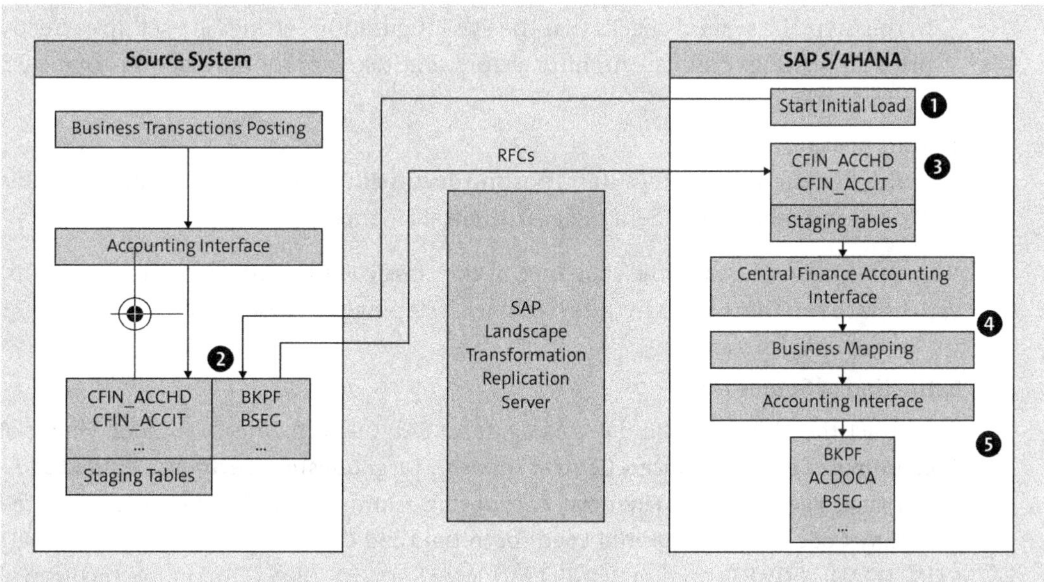

Figure 8.9 FI-CO Initial Load Data Flow

staging tables (table CFIN*), you'll perform mapping simulation and posting simulation to resolve any master data or configuration-related errors.

❺ **Posting to the Universal Journal**
Once you clear all errors from the posting simulation, you start posting the initial load data to the Universal Journal. The system calls the Central Finance accounting interface in the background to post the initial load data to the Universal Journal.

8.3.1 Performing the Classic Initial Load

From a high-level technical perspective, executing the FI classic initial load involves a four-step process. SAP recommends following this sequence to avoid errors:

1. **Start data extraction for the initial load.**
   In this mandatory step, the system extracts balances, open items, and documents based on the criteria set in **General Replication Settings** (VCFIN_SOURCE_SET) in the source system. You'll begin the extraction step from the Central Finance system. The system makes an RFC to the source system to transfer data to Central Finance staging tables. Extraction of the initial load job can be monitored with monitor data extraction. In addition, you'll repeat the extraction process in delta mode to transfer missing documents that weren't selected in the first run.
2. **Start the mapping simulation.**
   In this step, the system finds missing master data and Customizing settings required for posting. This step is recommended to identify and resolve master data errors.
3. **Start the posting simulation.**
   In this step, the system checks that the FI configuration settings are set up correctly prior to posting. You can monitor errors and take appropriate action to correct errors in this step.
4. **Start posting.**
   In this final step, the system starts posting documents to the Universal Journal (table ACDOCA) and other dependent tables in Central Finance.

Now that you understand the sequence of the classic FI-CO initial load steps, let's discuss how to run these steps in the system with step-by-step instructions.

Extraction of Source Data

To extract the FI initial load data, navigate to SAP Customizing IMG guide **Financial Accounting • Central Finance • Central Finance: Target System Settings • Initial Load • Initial Load Execution for Financial Accounting • Initial Load Execution for Selected Company Codes • Execute Initial Load for Initial Load Group** or use Transaction FINS_CFIN_LOAD_GRP_E (program RFINS_CFIN_MASS_DATA_MASTER). Upon execution of the transaction, you'll arrive at the **Execute Initial Load** screen, as shown in Figure 8.10. On this screen, choose the **Start Data Extraction** ❶ radio button. Next, enter values in the **Number of Batch Jobs** and **Server Group Name** fields under **Batch Job Control** ❷. Finally, enter a value in the **Initial Load Group** ❸ field, which you've created earlier. Refer to Chapter 7, Section 7.4, for step-by-step instructions on creating the initial load group. It's always recommended to execute the extraction job through background mode. Nevertheless, you can click **Execute** ❹ to start the extraction job interactively for smaller volume or empty initial loads. Depending on the volume to be extracted, the system may take a long time to run the extraction job. Therefore, it's not recommended to run the job interactively. When the job is submitted, you'll get a confirmation message on the resulting screen ❺, as shown at the bottom of Figure 8.10. You can return to the previous menu by pressing F3.

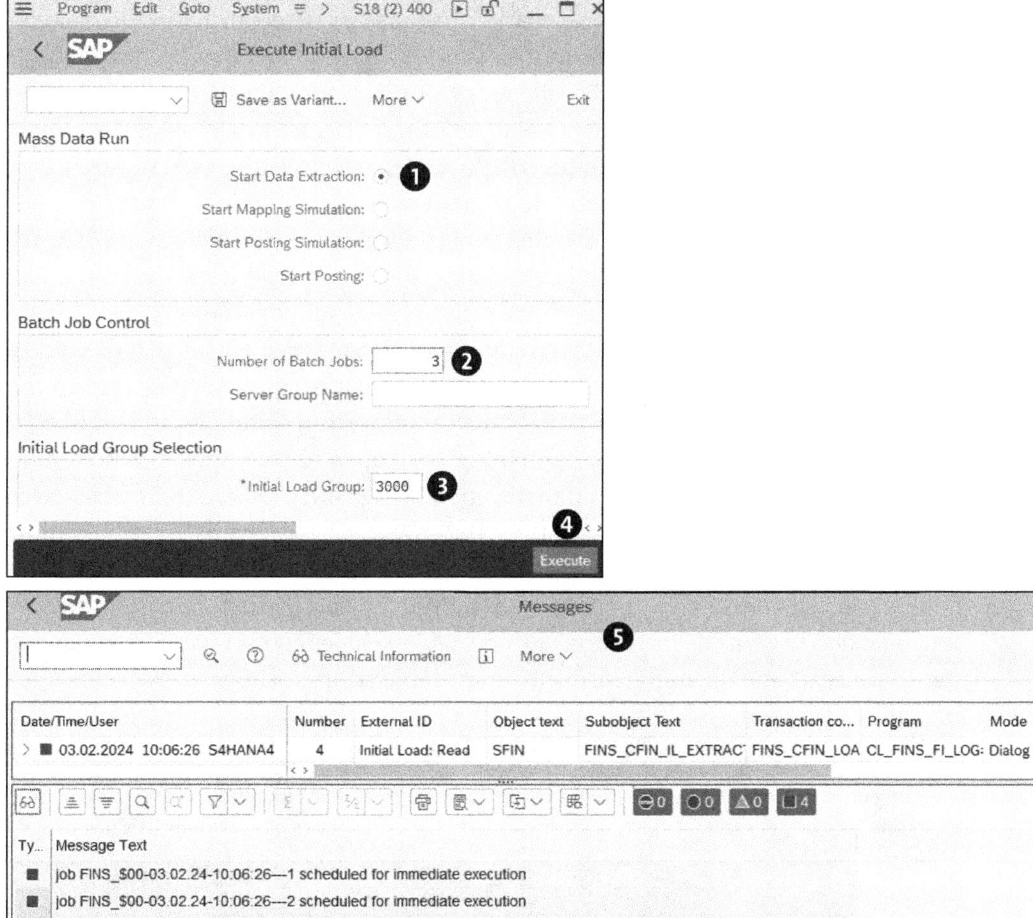

Figure 8.10 Execute the Initial Load

[+]

Tip

When executing a final initial load, the initial load group must be grouped with all involved company codes with cross-company code transactions. If you see error messages **RFC destination for system and usage not found** (message number FINS_CFIN_GEN101) or **Preparation for step CJ1_FI_DOC failed** (message number FINS_FI_MASS_DATA105), you must ensure you've configured **Choose Logical System** and **Maintain RFC Assignments and Settings for Source Systems** (covered in Chapter 7, Section 7.4).

Once the extraction job starts from Central Finance, the system initiates a call through an RFC to the source system. Note that SAP Landscape Transformation Replication Server isn't used for the classic initial load process. In the source system, for example, if the source system is in SAP ERP, the extraction program starts reading data from table BKPF, table BSEG, and other dependent tables and then starts packaging documents into different groups. In the case of general ledger balances, the system reads from table GLT0, table FAGLFLEXT, and other dependent tables depending on your source system. Once the packaging and sizing phases are completed, the system starts transferring data to Central Finance. While extracting the initial load data, the initial load program tries to match entries between FI (table BSEG) and CO (table COEP) wherever matching entries are found. If incorrect or incomplete entries are posted, the system may not pull all that information correctly. This also happens when you have summarization turned on, extract original raw data, or a substitution rule is implemented to remove or replace certain values. If the initial load open items or documents contain withholding tax data, the system transfers that along with the line items transferred. For the initial load deferred taxes, from SAP S/4HANA 2020 onward, the system transfers deferred taxes included with invoices, down payments, reversals, clearing documents, and tax transfer documents. The transferred data is then persisted in CFIN* staging tables in Central Finance. In addition, SAP excludes certain types of documents transferred through the initial load and online replication, as shown in Table 8.5.

Nonsupported Postings	Central Finance Support
Postings to CO-FI reconciliation ledger (general ledger reconciliation postings)	For SAP ERP systems, CO documents are replicated if the general ledger reconciliation flag isn't set in **General Replication Settings** view (VCFIN_SOURCE_SET). In this case, CO-FI postings are transferred through CO replication. For SAP S/4HANA source systems, by design, FI and CO are reconciled, so CO-FI reconciliation postings aren't needed any longer. As a result, SAP removed the general ledger reconciliation flag from the **General Replication Settings** view (VCFIN_SOURCE_SET).

Table 8.5 Nonsupported Postings in the Initial Load and Online Replication

Nonsupported Postings	Central Finance Support
Year-end closing postings	Reference transaction type BKPF-AWTYP = GLYEC and corresponding reversals BKPF-AWTYP = BKPF are acceted.
Clearings posting and clearing resets	Clearing postings and clearing resets aren't transferred for the initial load; ongoing replication is supported.
Recurring, parked, and sample documents	These postings aren't transferred in the initial load and online replication.
Noted items	These postings are supported, except for down payment requests and payment requests.
Balance carryforward items	Posting period (POPER) = 000, document status = C, and BTTYPE = RFBC are supported.
Closing operations	Activities carried out during the end of a fiscal year in some countries are supported.
Long texts (both header and item level) and attachments	These aren't replicated during creation nor if changes are made.
Supplementary tables	Information from supplementary tables, especially for special purpose ledgers, or country-specific information, for example Nota Fiscal in Brazil, is supported.
Noncompatible ALE postings	If ALE postings consist of incorrect or inconsistent postings, then the system may not properly transfer those documents.

Table 8.5 Nonsupported Postings in the Initial Load and Online Replication (Cont.)

The progress of the initial load extraction job can be monitored in two ways:

- **Using Transaction FINS_CFIN_LOAD_GRP_M (Data Extraction Monitor)**
  This is the preferred approach as it provides an intuitive screen to monitor the overall progress. Refer to the next section, "Monitoring Data Extraction," for further details.
- **Monitoring the background job using Transaction SM37**
  On the monitoring screen shown at the top of Figure 8.11, you can review the **Status** and **Duration(sec.)** ❶ fields by clicking the **Refresh** ❷ button. In addition, **Job log** provides additional information in case there is a job failure.

The system also updates table CFIN_MIG_LOG in the source system with the logical system, company code, and timestamp ❸. Any financial postings after the timestamp from this table are transferred to table CFIN* staging in the source system for online replication as the route later goes through SAP Landscape Transformation Replication Server.

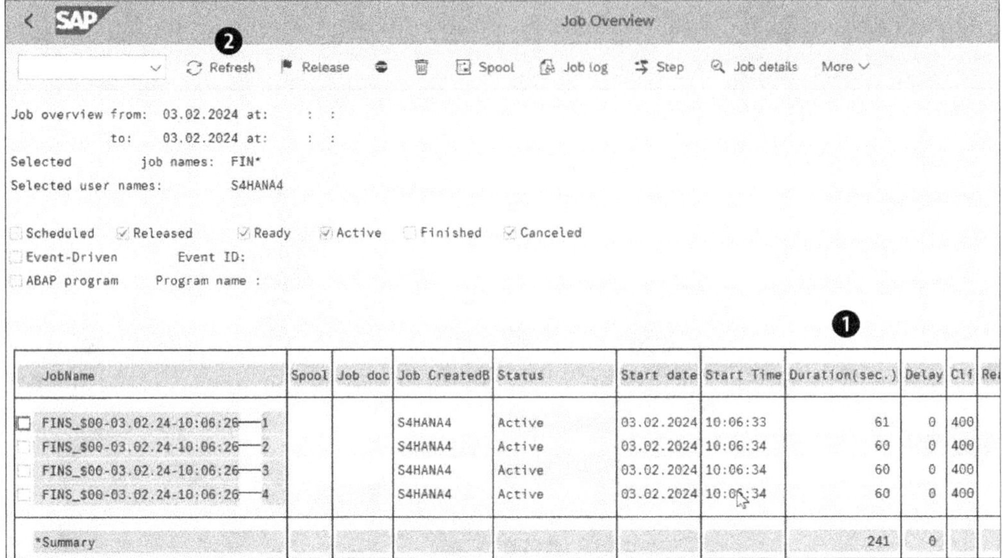

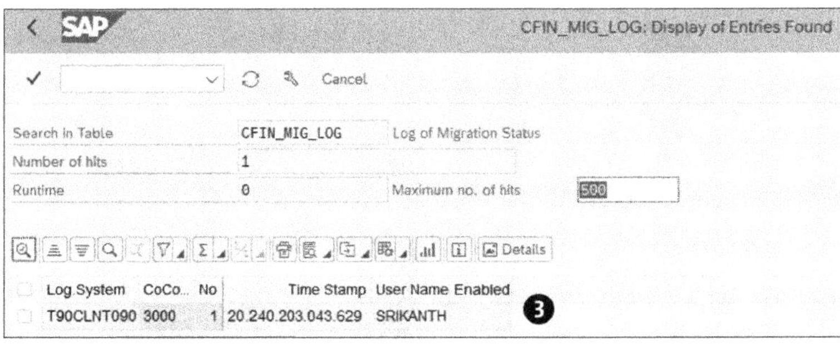

Figure 8.11 Execute the Initial Load Job Overview

[+]

Tip

When the initial load data is transferred to CFIN* staging tables in Central Finance, the system identifies the type of transaction with the reference transaction type field (CFIN_ACCIT-AWTYP) for the source system initial load data, as shown in Table 8.6. These identifiers are helpful especially when troubleshooting initial load errors.

Transaction Type	Identification in Table CFIN_ACCIT-AWTYP
Open items	APAR
Balances	SALDO
FI documents	FIDO
CO documents	CODO

Table 8.6 Transaction Types in Staging Table CFIN_ACCIT

Monitoring Data Extraction

To monitor data extraction, navigate to SAP Customizing IMG guide **Financial Accounting • Central Finance • Central Finance: Target System Settings • Initial Load • Initial Load Execution for Financial Accounting • Initial Load Execution for Selected Company Codes • Monitor Initial Load for Initial Load Group** or use Transaction FINS_CFIN_LOAD_GRP_M (program RFINS_CFIN_MASS_DATA_MASTER). Upon execution of the transaction, you'll arrive at the **Monitor Initial Load Data** screen, as shown in Figure 8.12.

On this screen, choose the **Monitor Data Extraction** ❶ radio button. Next, enter values in the **Initial Load Group** ❷ field, and click the **Execute** ❸ button. The system displays the status of the extraction job ❹, as shown at the bottom of Figure 8.12. Use the **Refresh** icon to see the progress of the extraction job. Once the job is finished, click the **Exit** ❺ button to go back to the menu.

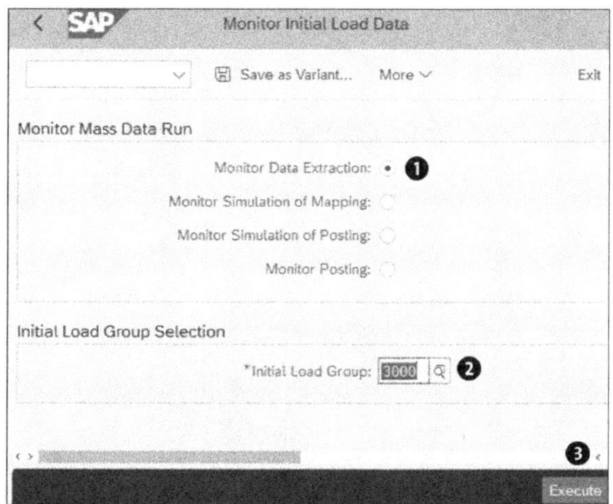

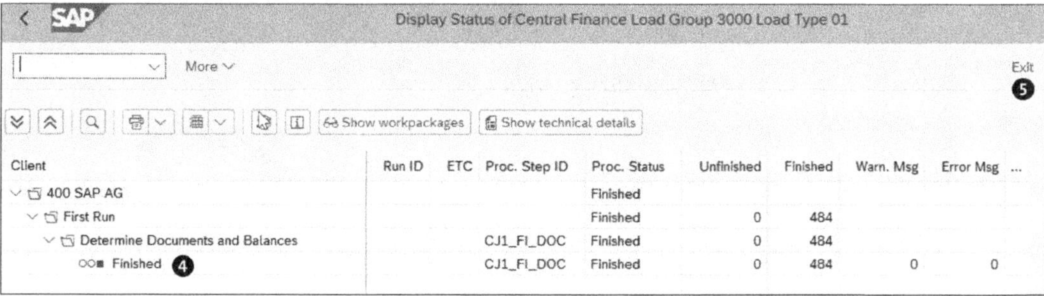

Figure 8.12 Monitoring the Initial Load Extraction

After completing the first round of the initial load extraction step, SAP recommends restarting extraction again in delta mode multiple times. Between the time of the initial load extracting job and writing the record to migration log table CFIN_MIG_LOG, there may be new documents posted in the source system that weren't captured either in the first-time initial load extraction run or the source staging table CFIN* (SAP Landscape Transformation route). SAP minimized the impact by moving the ongoing transfer update to the beginning phase of the program. Therefore, the missing documents will be few or none as we're looking at a fraction of seconds here. Consult SAP Note 2678658 for information on this.

These missing documents are captured by running a delta run. To start a delta run on the **Execute Initial Load** screen, as shown earlier in Figure 8.10 ❶, select the **Start Data Extraction** radio button. Click the **Execute** button to start the delta extraction for the initial load group. As shown in Figure 8.13, the **Monitor Data Extraction** step shows **First Run** ❶ and **Repeated Run 1** ❷ (delta run). Double-click on the **Finished** status under **Repeated Run** to view the output result of the delta run. Repeat the delta run multiple times until this screen shows **0** in the **Succ Msg** column ❸.

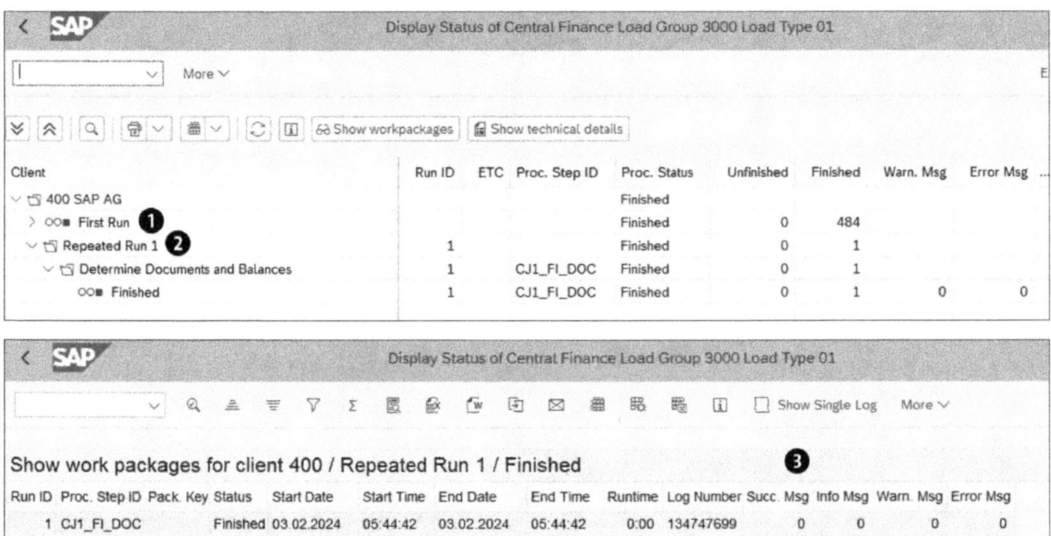

Figure 8.13 Initial Load Extraction: Delta Run

Once the delta run is successful, you need to select the **Initial Load Finished** checkbox in the **General Replication Settings** view (VCFIN_SOURCE_SET) in the source system to prevent subsequent inadvertent extraction runs by the system.

Mapping Simulation

From the same Transaction FINS_CFIN_LOAD_GRP_E (program RFINS_CFIN_MASS_DATA_MASTER), you start **Simulation of Mapping**. In this step, the system finds any mapping-related errors prior to posting documents to Central Finance. From the **Execute Initial Load Using Initial Load Groups** screen, select **Start Mapping Simulation**, as shown in Figure 8.14 ❶. Next, enter values in the **Number of Batch Jobs** ❷ and **Server Group Name** fields under the **Batch Job Control** section. Then, enter a value in the **Initial Load Group** field located under **Initial Load Group Selection** ❸. Click the **Execute** icon. The system displays the **Messages** window ❹ showing the job is scheduled. Return to the menu by pressing F3.

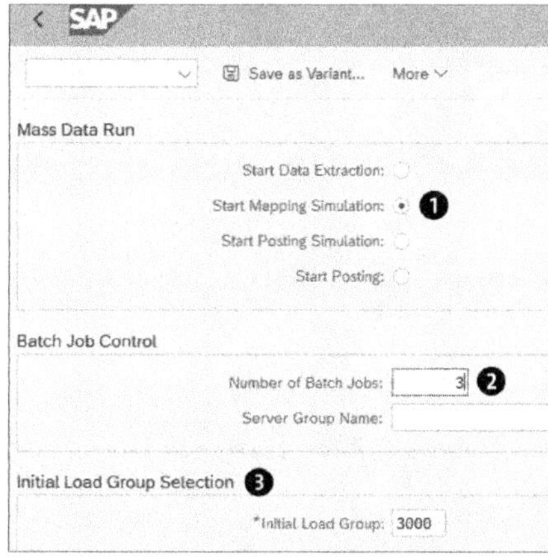

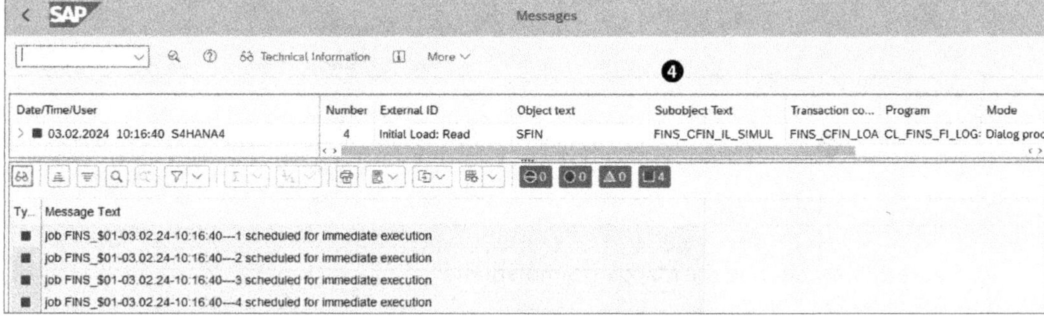

Figure 8.14 Start Mapping Simulation

Monitoring Mapping Simulation

Once mapping simulation is submitted, as shown in Figure 8.15, from Transaction FINS_CFIN_LOAD_GRP_M (program RFINS_CFIN_MASS_DATA_MASTER), select the **Monitor Simulation of Mapping** ❶ radio button on the **Monitor Initial Load Data** screen. Next, enter values in the **Initial Load Group** ❷ field, and click the **Execute** ❸ button. The system displays the **Simulate Mapping** status ❹, as shown at the bottom of Figure 8.15. Use the **Refresh** icon to see the progress of the extraction job. While the job is running, you can drill down to the mapping-related error messages by double-clicking the count ❺ shown in the **Error Msg** column. Once the job is finished, click the **Exit** icon to go back to the menu.

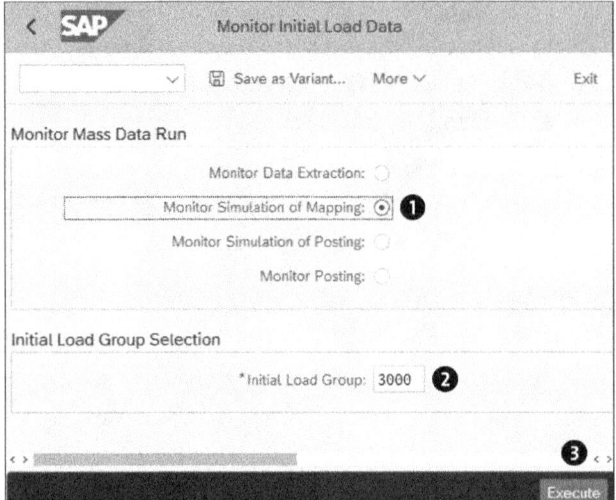

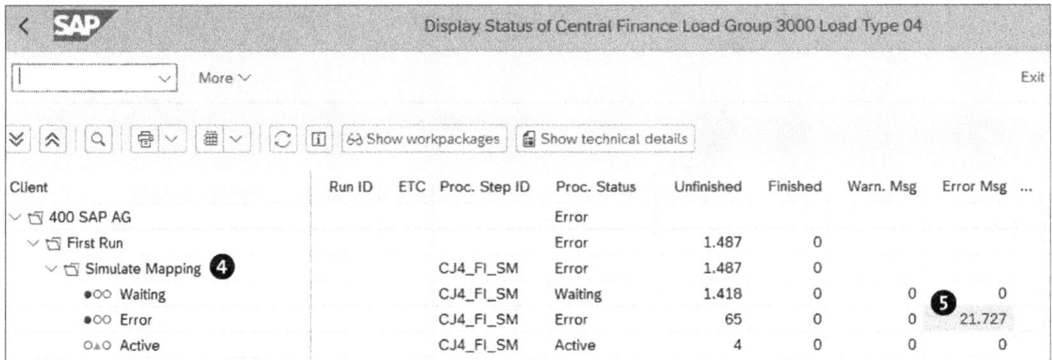

Client	Run ID	ETC	Proc. Step ID	Proc. Status	Unfinished	Finished	Warn. Msg	Error Msg
400 SAP AG				Error				
First Run				Error	1.487	0		
Simulate Mapping			CJ4_FI_SM	Error	1.487	0		
Waiting			CJ4_FI_SM	Waiting	1.418	0	0	0
Error			CJ4_FI_SM	Error	65	0	0	21.727
Active			CJ4_FI_SM	Active	4	0	0	0

Figure 8.15 Monitoring the Mapping Simulation

Posting Simulation

From the same Transaction FINS_CFIN_LOAD_GRP_E (program RFINS_CFIN_MASS_DATA_MASTER), you'll start the posting simulation. In this step, the system finds any FI configuration-related errors prior to the final posting of documents to Central Finance. From the **Execute Initial Load** screen, select **Start Posting Simulation**, as shown in Figure 8.16 ❶. Next, enter values in the **Number of Batch Jobs** and **Server Group Name** fields under **Batch Job Control** ❷. Enter a value in the **Initial Load Group** field, and click the **Execute** ❸ icon. The system displays messages ❹ of when the job is scheduled. Return to the menu by pressing F3 or clicking **Exit** ❺.

Monitor Posting Simulation

From Transaction FINS_CFIN_LOAD_GRP_M (program RFINS_CFIN_MASS_DATA_MASTER), select the **Monitor Simulation of Posting** radio button on the **Monitor Initial Load Data** screen, as shown in Figure 8.17 ❶.

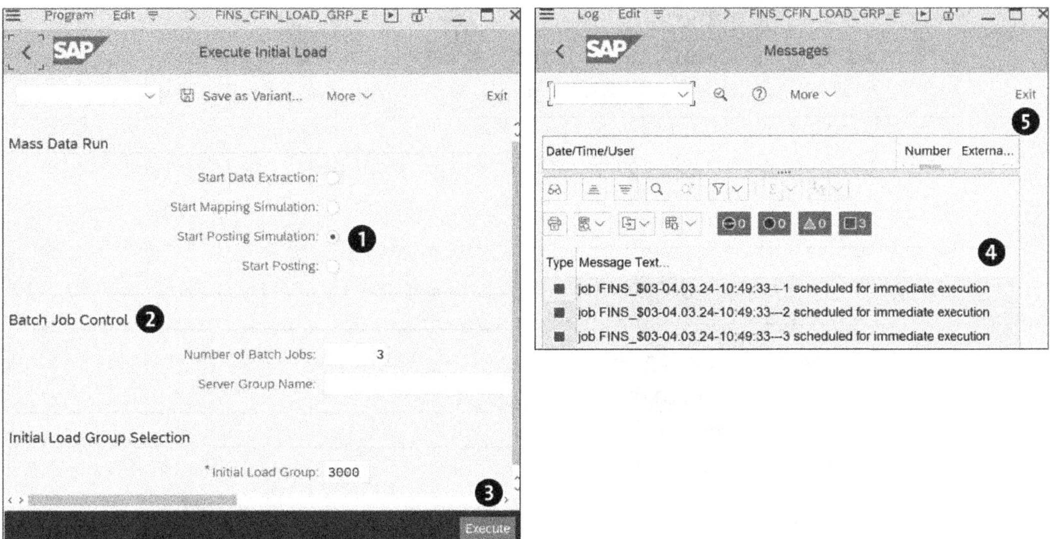

Figure 8.16 Execute the Initial Load for the Initial Load Group: Posting Simulation

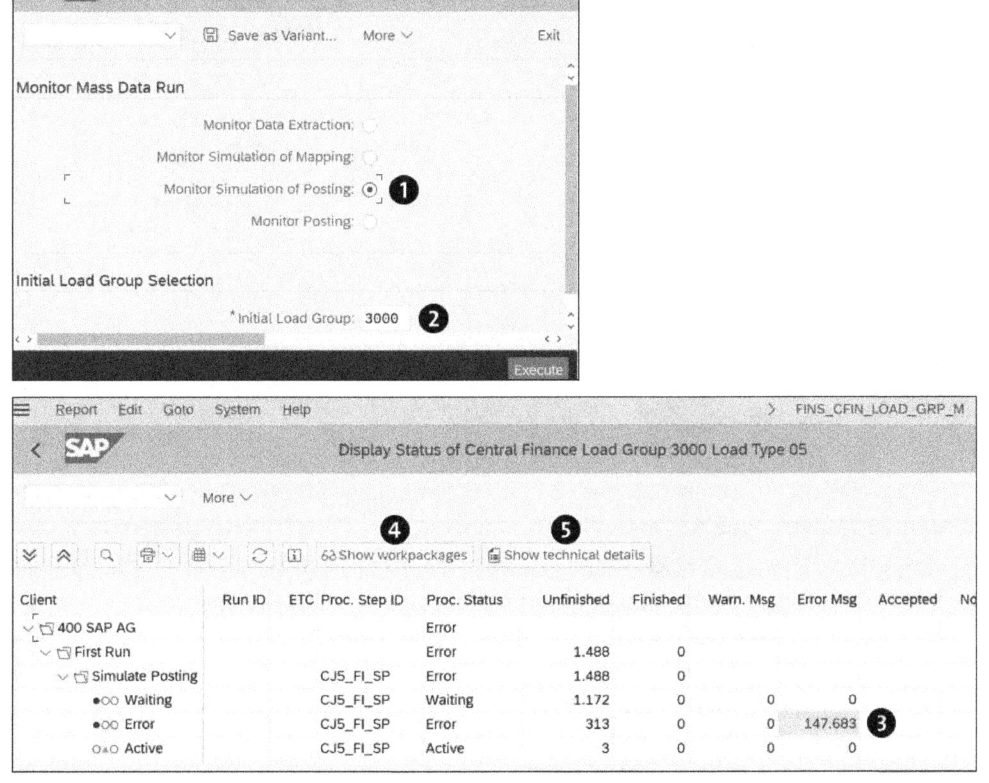

Figure 8.17 Monitor the Initial Load Group: Posting Simulation

Next, enter a value in the **Initial Load Group** ❷ field, and click the **Execute** button. The system displays the status of the posting simulation job. Use the **Refresh** icon to see the progress of the extraction job. While the job is running, you can drill down to the mapping-related error messages by double-clicking the count ❸ shown in the **Error Msg** column. The **Show workpackages** ❹ and **Show technical details** ❺ buttons also provide additional details.

When the **Show workpackages** button is chosen from Figure 8.17, the system takes you to the **Show work packages for client 400 / First Run / Error** screen, as shown in Figure 8.18. On this screen, the system displays all packages ❶. You can either select a row by clicking the **Show Single Log** ❷ button to display a single package or choose the **Show Error Overview** ❸ button to display an overview of all messages. When you choose the **Show Error Overview** button, the system displays all messages, as shown in Figure 8.18. You can select all messages or a single row and then click the **Show Detail** ❹ button.

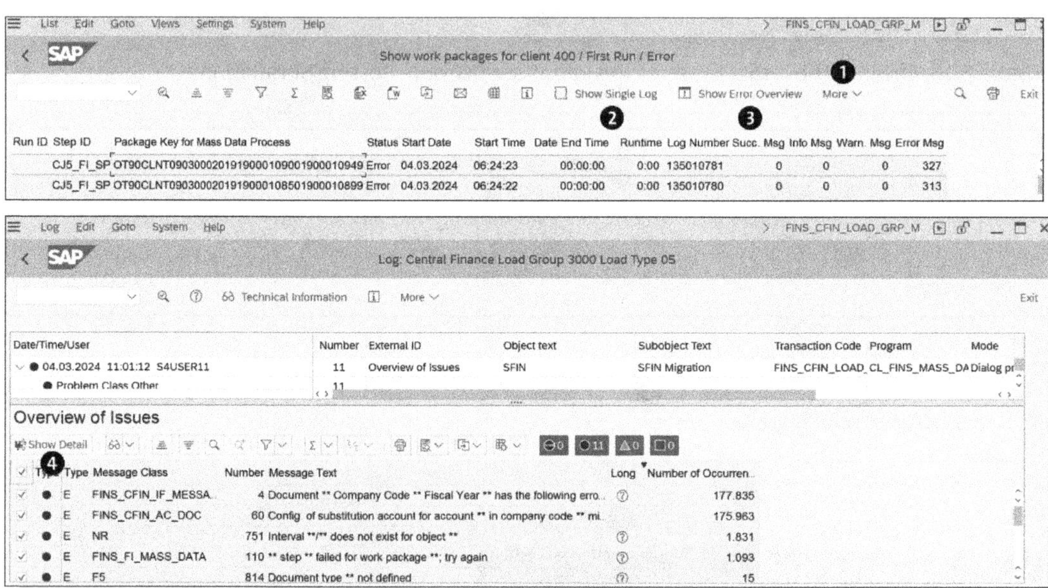

Figure 8.18 Show Work Package: Posting Simulation

The **Show Detail** button from Figure 8.18 leads to the **Logs for $03: Central Finance Load Group 3000 Load Type 05** screen, as shown in Figure 8.19 ❶. On this screen, you can review error **Message Text** ❷, **Long**, and package key showing under **Example Package with this Message** column ❸. Long text can be viewed with the **Question** icon in the **Long** column. You can further navigate to details from here by selecting the **Display** icon. Package keys are helpful to troubleshoot certain errors that can't be traced easily. Return to the menu by pressing F3 or clicking the **Exit** ❹ icon.

From a technical perspective, a package shows an error when `FINS_MASSPACKAGE-END_PROCESSING` isn't filled with the package. The same can viewed in the application log.

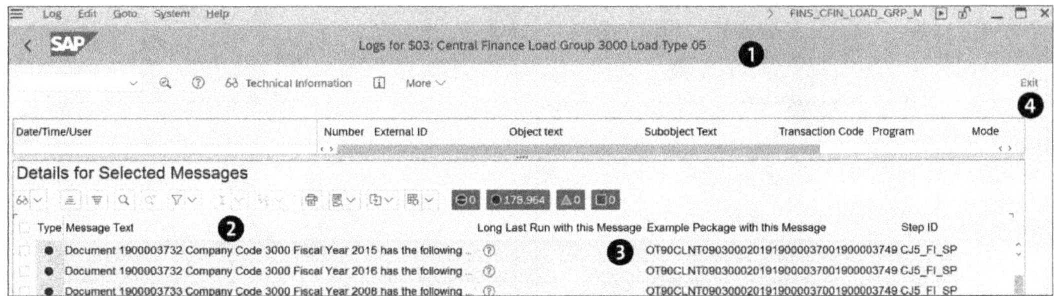

Figure 8.19 Show Details: Posting Simulation

Posting the Initial Load

From the same Transaction FINS_CFIN_LOAD_GRP_E (program RFINS_CFIN_MASS_DATA_MASTER), you'll start final posting. In this step, the system posts documents to the Universal Journal in Central Finance with the following behavior:

- Balances extracted by ledgers are posted to corresponding accounts in Central Finance, and the offset is posted to the migration clearing account defined for each company code in the **Define Clearing and Substation Accounts** Customizing activity.
- For balances of AR/AP reconciliation accounts, the system posts a redetermined AP/AR reconciliation account in Central Finance. The offset is posted to substitution and clearing accounts configured for each reconciliation account in **Define Clearing and Substation Accounts**.
- For individual documents extracted for the time frame specified in **Start Year - Documents** and **Period - Documents** fields, which are maintained in the source system view VCFIN_SOURCE_SET, the system posts all line items, including profit and loss (P&L) lines (as full documents) in Central Finance. Note that clearing document (BKPF-BSTAT ="A") and reset clearing document (BKPF-BSTAT="B") aren't posted as part of the initial load. For subsequent follow-on documents during online replication, the system posts as a plain inverse posting without a reference to the original document in Central Finance.
- Tax information is not transferred for initial load open items as both income and expense lines are not transferred. This applies to deferred tax as well. Therefore, you cannot perform tax reporting for initial load postings in Central Finance.

To perform final posting, from the **Execute Initial Load** screen, select **Start Posting**, as shown in Figure 8.20 ❶. Next, enter values in the **Number of Batch Jobs** ❷ and **Server Group Name** fields in the **Batch Job Control** section. Enter a value in the **Initial Load Group** ❸ field, and click the **Execute** ❹ icon. The system displays **Messages** ❺ that show the job schedule. Return to the menu by pressing F3 or clicking the **Exit** ❻ icon.

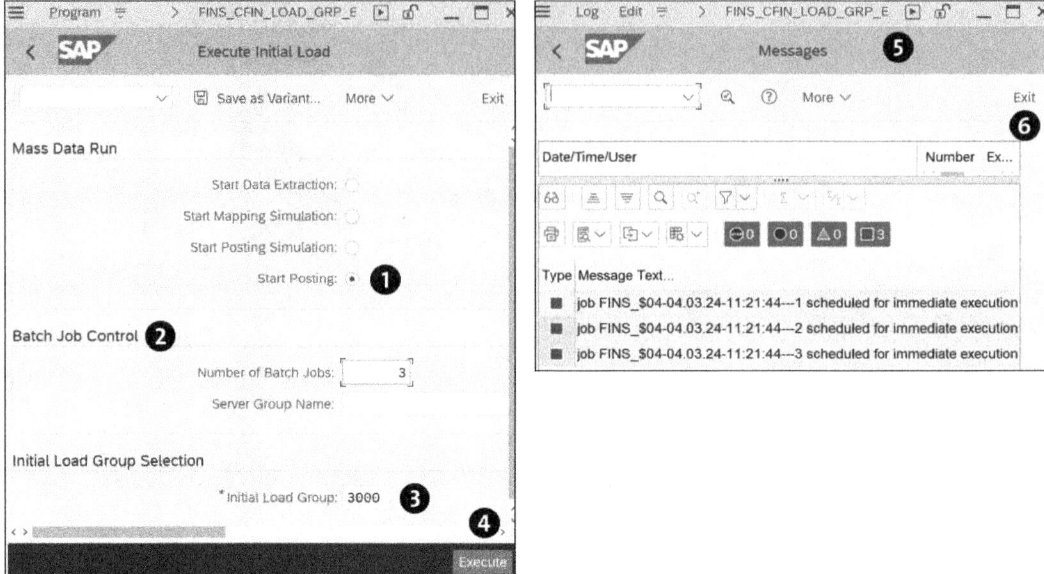

Figure 8.20 Execute the Initial Load: Start Posting

Upon posting the initial load data to the Universal Journal, the system deletes the data from staging table CFIN_ACC* for open items (CFIN_ACCIT-AWTYP=APAR) and balances (CFIN_ACCIT-AWTYP=SALDO). However, the system doesn't delete FI or CO documents (CFIN_ACCIT-AWTYP=FIDO and CODO) upon posting. You can use report RFINS_CFIN_CLEAR_INIT_LOAD to delete the initial load FI or CO documents after successful posting.

[+]

Tip

As the initial load transfers open items only, this applies to cross-company code postings as well. For example, if a cross-company code document is involved with two company codes, and one document is cleared, but the other company code document is still open, the system transfers open items with one company code, not as a cross-company code posting.

If there is a mismatch of document type configuration (intercompany postings indicator T003-XGSUB), the initial load fails with error message **Consolidated companies & and & are different**. To prevent intercompany-related errors in the initial load, review intercompany configuration (Transaction OBYA), information transferred in cross-company code numbers (CFIN_ACCIT-BVORG), document type configuration (intercompany postings indicator T003-XGSUB) in Transaction OBA7, and trading partner field (VBUND) derivation logic prior to posting the initial load.

Monitor Posting

From Transaction FINS_CFIN_LOAD_GRP_M (program RFINS_CFIN_MASS_DATA_MASTER), as shown in Figure 8.21, select the **Monitor Simulation of Posting** ❶ radio button on the **Monitor Initial Load Data** screen. Next, enter values in the **Initial Load Group** ❷ field and click the **Execute** button ❸. The system displays the posting simulation job status. Use the **Refresh** icon to see the progress of the extraction job. While the job is running, you can drill down to the mapping-related error messages by double-clicking the count ❹ shown in the **Error Msg** column. **Show workpackages** ❺ and **Show technical details** ❻ buttons also provide details.

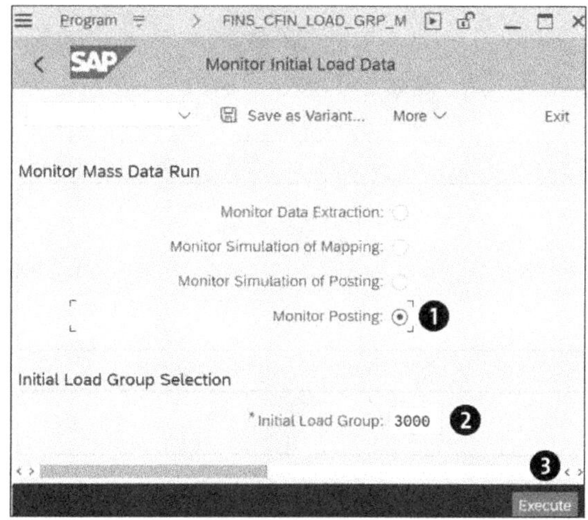

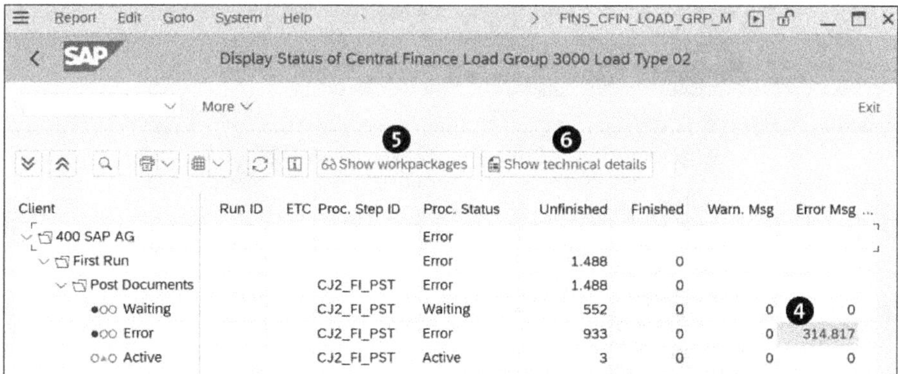

Client	Run ID	ETC	Proc. Step ID	Proc. Status	Unfinished	Finished	Warn. Msg	Error Msg ...
400 SAP AG				Error				
First Run				Error	1.488	0		
Post Documents			CJ2_FI_PST	Error	1.488	0		
Waiting			CJ2_FI_PST	Waiting	552	0	0	0
Error			CJ2_FI_PST	Error	933	0	0	314.817
Active			CJ2_FI_PST	Active	3	0	0	0

Figure 8.21 Monitor the Initial Load Group: Final Posting

Navigating to packages from the **Show workpackages** ❺ and **Show technical details** ❻ buttons for troubleshooting is the same as you've learned with the **Simulation of Posting** step, so we won't repeat the steps here.

Display Aggregated Initial Load Messages

In addition to the initial load simulation and monitoring tools, SAP also provides report RFINS_CFIN_DISPLAY_LOG to display aggregated messages from the FI initial load. Upon entering the report, the system takes you to the **Display Aggregated Initial Load Messages** screen, as shown in Figure 8.22. On this screen, choose **For Posting** ❶ in the **Display Aggregated Initial Load Messages** section. Next, enter values in the **ID of a Mass Data Run** ❷ and **Load Group** ❸ fields. Click the **Execute** ❹ button. On the resulting screen, you notice packages are displayed with the relevant error message class ❺, as shown at the bottom of the screen. You can navigate to individual messages by clicking the details icon. Click **Exit** ❻ to go to previous screen.

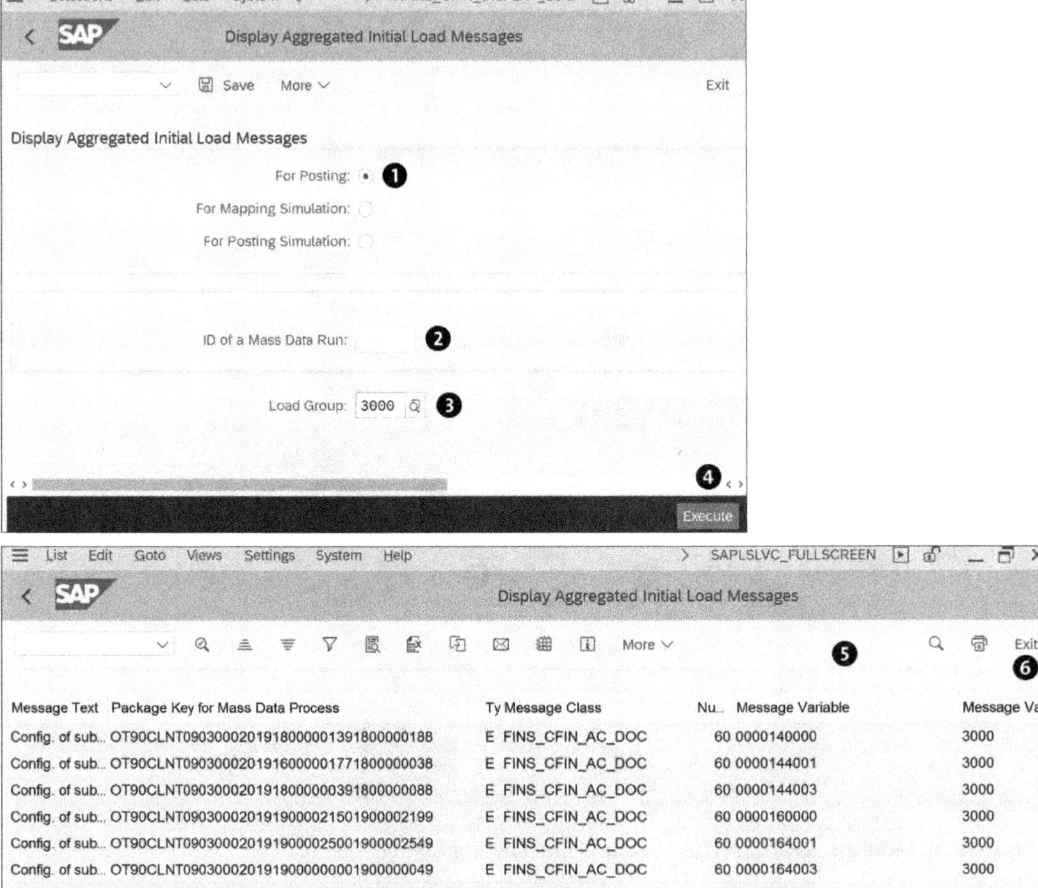

Figure 8.22 Display Aggregated Initial Load Messages

[«]

Note

To execute the report, you'll require the relevant configuration access authorization role both in the source system and Central Finance system.

8

Initial Load Analyzer

In addition to the initial load simulation and monitoring tools, SAP also provides the initial load analyzer tool (Transaction FINS_CFIN_INIT_DOC_ANALYZER) to simulate the initial load extraction for FI initial load documents. One prerequisite for this transaction is that you've completed the initial load extraction step in the Central Finance system. After reaching the transaction, you'll arrive at the **Simulate Posting of Extracted Initial Load Data for Individual Doc.** screen, as shown in Figure 8.23. On this screen, enter a value in the **Package Key** field under the **Package Key** radio button ❶. Alternatively, you can choose by source system fields. When you choose the radio button for type of package—**Document Number**, **Open Item**, or **Balance Load**—the system enables fields under the **Source system information** section ❷. Click the **Execute** ❸ button to continue. This leads to the **Documents for simulation and data tracing** screen, as shown in Figure 8.23. On the resulting screen, select a row you want to simulate, and then click the **Simulate Posting** button ❹ to analyze the errors for the selected individual document in simulation mode. This simulation works only for data before the initial load posting is completed. **Simulate Posting** doesn't post any documents in the Central Finance system.

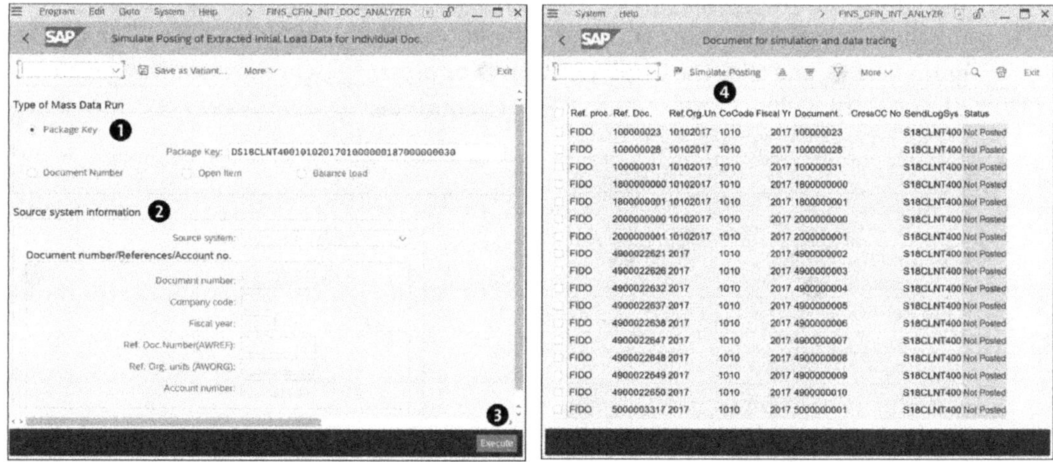

Figure 8.23 Initial Load Analyzer

Figure 8.24 displays the posting data simulation to analyze the errors further.

Document for simulation and data tracing > FINS_CFIN_INT_ANLYZR

Document for simulation and data tracing

More Exit

Transaction Type	Initial Load - FI Documents	Standard/Customer Prepare	■ ◆
Message Status	Error	Standard/Customer Mapping	■ ◆
Customer BAdI Status	◆ inactive	Standard/Customer Adjust Data	■ ◆
Reprocessing Status	✓ Standard	Standard/Customer Post	■ ◆
Splitting active	✓ active	Splitting method-->0000000012	■ ◆
Central payment active	✓ active	ENABLE_CENTRAL_PAYMENT	■ ◆
Disable technical clearing	✓ active	DISABLE_TECH_CLEARING_OF_OPEN_ITEMS	■ ◆

Table	Name
ACCHD	ACCHD
FIDO/0100000023/10102017/S	ACCHD
ACCIT	ACCIT
0000000001	ACCIT
0000000002	ACCIT
ACCIT_APP	ACCIT_APP

Attribute	Description	Staging Da	Source Dat	Prepare Data	Prepare Da	Map Values	Map Val
XBILK	Indicator: Account is a ba...	X	X	X	X	X	X
WWERT	Translation Date	20170104	20170104	20170104	20170104	20170104	2017010
VORGN	Transaction Type for Gen...	RFBU	RFBU	RFBU	RFBU	RFBU	RFBU
TAX_KALSM	Procedure (Pricing, Outp...	0TXD	0TXD	0TXD	0TXD	TAXD	TAXD
TAX_COUNTRY	Country/Region Key			DE	DE	DE	DE
SHKZG	Debit/Credit Indicator	S	S	S	S	S	S
RACCT_SEND...	General Ledger Account i...			TR10010000	TR10010000	TR10010000	TR1001

Figure 8.24 Initial Load Analyzer: Simulation and Data Tracing

Debugging a Financial Accounting Initial Load

Besides the initial load error monitoring and initial load analyzer tools, there are situations where you need to debug an initial load package to trace the root cause of an error in the code. This is often used when custom code is implemented, and this task is done by the technical team supporting the Central Finance project. You can debug the code with program FINS_MASS_DATA_TEST by calling it from Transaction SE38. After reaching the transaction, as shown in Figure 8.25, enter "FINS_MASS_DATA_TEST" in the **Program** field ❶, and click the **Display** button ❷ or press F7. On the resulting screen, put a breakpoint on line 88 ❸, and press F8 to continue.

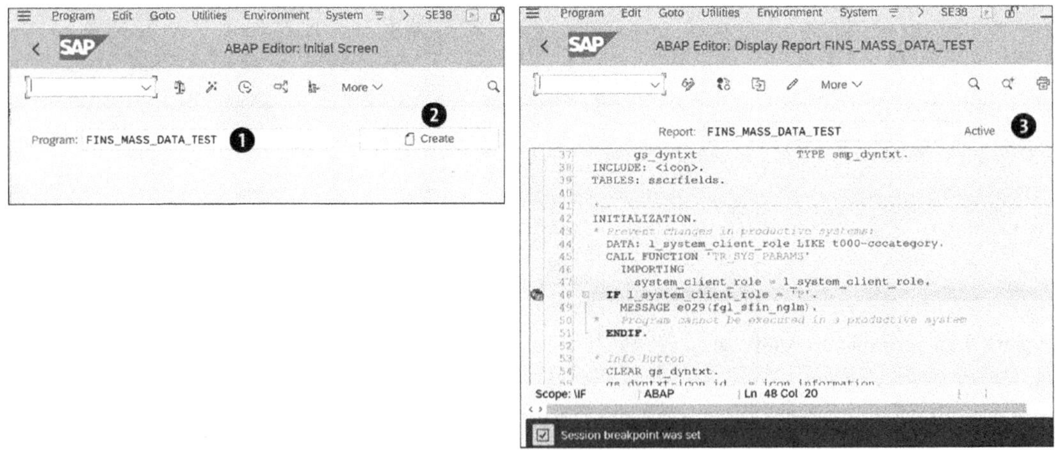

Figure 8.25 Debug the Initial Load: Set the Breakpoint

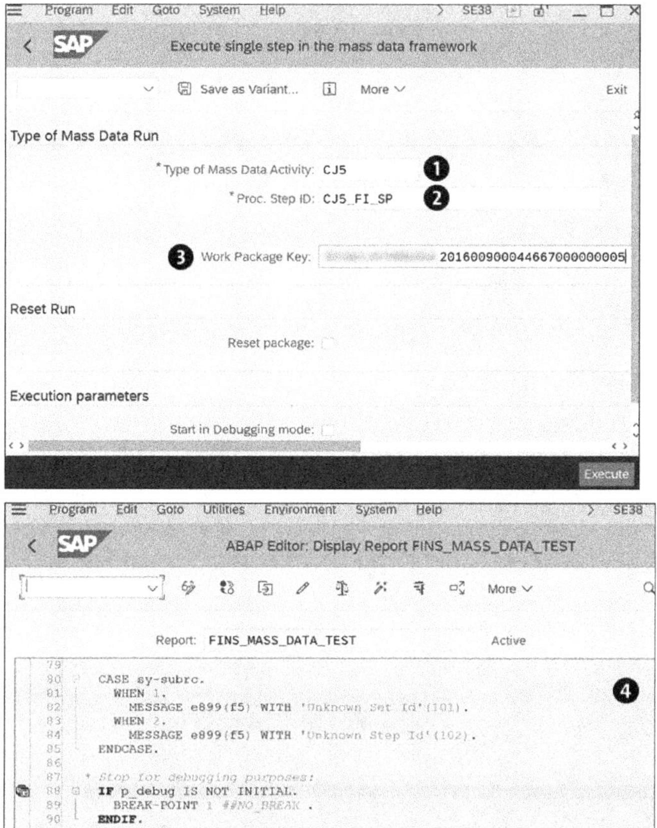

Figure 8.26 Debug the Initial Load: Mass Data

This brings up the **Execute single step in the mass data framework** screen, as shown in Figure 8.26. On this screen, enter values in the following fields:

❶ **Type of Mass Data Activity**
Enter "CJ5" to debug a package during posting simulation. You can debug packages of extraction documents (package D), open items (package key O or P), or balances (package S). You can debug these by choosing the appropriate **Type of Mass Data Activity**, as shown in Table 8.7, for the FI-CO initial load.

Type of Mass Data Activity	Process Step ID	Purpose
CJ1	CJ1_FI_DOC	Data extraction packages
CJ2	CJ2_FI_PST	Packages failing during final posting
CJ4	CJ4_FI_SM	Errors resulting in mapping simulation
CJ5	CJ5_FI_SP	To debug errors from posting simulation

Table 8.7 Type of Mass Data Activity

❷ **Proc. Step ID**
Enter "CJ5_FI_SP", as seen in our example.

❸ **Work Package Key**
Enter a package you want to debug. In our example, we've chosen a D package.

Click the **Execute** icon to continue. The system takes you back to the code to debug the code ❹. Put curser on line 92, and press Shift+F12 to continue to debug the error. Depending on the nature of the error and complexity, you'll repeat the steps multiple times until you find the root cause.

Once the code is debugged, press F8 to execute the code to see if the error disappears on the **Messages** screen, as shown in Figure 8.27.

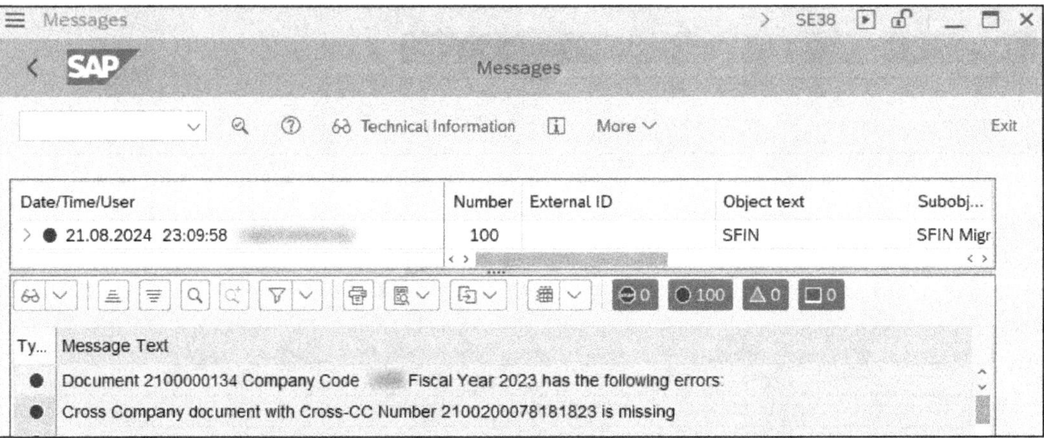

Figure 8.27 Debug the Initial Load: Error Screen

8.3.2 Additional Considerations for the Financial Accounting Initial Load

In this section, we'll walk through a few miscellaneous topics that are relevant for FI initial load activities.

Transferring Differing Fiscal Year Variants

Depending on your source system, you'll come across a scenario where the source system using the new general ledger is integrated with Central Finance with a deviating fiscal year variant associated for a nonleading ledger. For example, if the source system is implemented with leading ledger (0L) with a fiscal year variant of calendar year K4, then two ledgers are implemented for Central Finance: a leading ledger with fiscal year variant K4 and a nonleading ledger with fiscal year variant V3 (April to March). In such a scenario, the system loads initial balances that are extracted by the leading ledger and re-derives the period from the leading header fiscal year variant. The FI initial load is extracted based on the **General Replication Settings** (VCFIN_SOURCE_SET), not from the period derived from the target system fiscal year variant. As a result, nonleading ledger postings in Central Finance are shifted by a certain number of periods because the nonleading ledger has the same periods when compared to the leading ledger.

Note

Consult SAP Notes 3103722, 2916900, 2764851, and 3134105 for additional guidance on this topic

Performance Considerations

When extracting data from source systems and posting the initial load data in Central Finance, often you experience performance problems due to various factors ranging from source system landscape, limited work processes, volume of data, complex custom programs, sizing of servers, and so on. Besides these factors, during extraction of the initial load or processing the initial load postings, often you see memory overflow issues that lead to runtime errors which can be noticed in Transaction ST22 with messages TSV_TNEW_PAGE_ALLOC_FAILED, MEMORY_NO_MORE_PAGING, and so on. In such cases, you need to work with the Basis team to increase parallel work processes or increase timeout parameters.

SAP also recommends keeping a shorter period in **Period - Documents** of the **General Replication Setting** (VCFIN_SOURCE_SET) view. Better yet, keep this field value starting from the beginning of the current fiscal year. This reduces extraction time in the source system, processing time for the initial load process in Central Finance, and the effort involved in reconciliation and reporting. In addition, less data transferred means less master data maintenance and greatly improved mapping efforts are greatly improved.

Packaging and Packaging Type

During the initial load extraction in the source system and initial posting, the system uses packaging keys to group and process the FI-CO initial load data. Table 8.8 displays how packaging is composed of different elements for different types of FI postings during the initial load extraction and posting.

Balance	SS18CLNT4001010000001600000L						
Sections of package key	S	S18CLNT400	1010	0000160000	0L		
Position of package key and meaning	1	2-11	12-15	16-25	26-27		
	Balance	Logical system	Company code	General ledger account	Ledger		
Open Item	**PS18CLNT4001010202016000311151600033273**						
Sections of package key	P	S18CLNT400	1010	2020	1600031115	1600033273	
Position of package key and meaning	1	2-11	12-15	16-19	20-39		
	Open item	Logical system	Company code	Fiscal year	Document range (from – to)		
Open Item	**OS18CLNT40010102020000019600012 – at the initial load data extraction**						
Sections of package key	O	S18CLNT400	1010	2020	0000196000		Period 12
Position of package key and meaning	1	2-11	12-15	16-19	20-29		12
	Open item	Logical system	Company code	Fiscal year	Account		Period
Open Item	**OS18CLNT4001010202061000004256100000454 – at the initial load posting**						
Sections of package key	O	S18CLNT400	1010	2020	6100000425 6100000454		
Position of package key and meaning	1	2-11	12-15	16-19	20-39		
	Open item	Logical system	Company code	Fiscal year	Reference document number range (AWREF)		
Document	**DS18CLNT4001010202061000004256100000454 – at the initial load extraction**						
Sections of package key	D	S18CLNT400	1010	2020	6100000425 6100000454		
Position of package key and meaning	1	2-11	12-15	16-19	20-39		
	Document	Logical system	Company code	Fiscal year	Document number range		
Document	**DS18CLNT4001010202061000004256100000454 - at the initial load posting**						
Sections of package key	D	S18CLNT400	1010	2020	6100000425 6100000454		
Position of package key and meaning	1	2-11	12-15	16-19	20-39		
	Document	Logical system	Company code	Fiscal year	Reference document number range (AWREF)		

Table 8.8 Packaging Keys for the Initial Load Data

To improve performance or memory overflow, SAP introduced the new package type for open items (begins with P rather than O) from the SAP S/4HANA 1809 FPS1 release onward. The system now has new packaging logic that groups by document number rather than using the old logic of grouping by general ledger account. To extract with package type P for lower releases, before beginning the initial load, ensure you've implemented SAP Note 2715002. In addition, to improve performance or memory issues, SAP provides further smaller packages with package type T for postings involved with partner profit centers. In addition, SAP provides Customizing settings **Reduce Package Size for Balance Load** with maintenance view VCFIN_BALANCEFLD, which provides a field name to be added to the package key. Configuration of this step is covered in Chapter 7, Section 7.4.6. Consult SAP Note 2745324 for further guidance on this.

Financial Accounting-Controlling: Initial Load Enhancements

Central Finance provides BAdIs to enhance data quality during FI-CO:

- Central Finance comes with source system BAdI BADI_FIN_CFIN_DOCUMENT, which can be used to add additional logic during the initial load data extraction. For example, if source financial postings contain custom fields during extraction, BAdI method FI_DOC_ENHANCE_INI-TIAL_LOAD is used to transfer custom fields to Central Finance. If your source system isn't transferring custom fields, implement SAP Note 2576335 and prerequisites notes.
- SAP provides multiple BAdIs to enhance FI-CO initial data prior to the final posting in Central Finance (based on your requirements, refer to SAP Central Finance adoption content or BAdI documentation for further guidance):
  - BADI_FINS_CFIN_AC_INTERFACE
  - BADI_FINS_CFIN_INITLOAD_INTF
  - BADI_FINS_CFIN_MAPPING_RULE
  - BADI_FINS_CFIN_MAPPING_PROCESS
  - BADI_FINS_CFIN_COPA_MAPPING
  - ADI_FINS_CFIN_CURR_ADJ

8.4 Resetting the Financial Accounting-Controlling Initial Load

In this section, let's explore steps involved when a FI-CO initial load requires a reset.

8.4.1 When to Reset the Initial Load

In general, you'll carry out a FI-CO initial load reset for multiple reasons:

- When running multiple mock cycles, you need to validate the initial load data to meet business requirements.

- If an error occurs in the initial load either due to incorrect data selection or the system not bringing all documents due to missing SAP Notes in the source system.
- Expanded scope of company codes or functionality changes occur.

8.4.2 Resetting an Initial Load

An initial load reset requires activities performed in the source system, SAP Landscape Transformation, and Central Finance system. Table 8.9 explains the reset activities in the source system and sequence of activities you'll typically perform during the FI initial load reset.

Sequence No.	Activity	Transaction/IMG Menu Path
1	Run report RCFIN_DEL_MIG, which deletes the initial load data in the source system. SAP recommends checking whether SAP Note 2224892 is implemented to avoid any runtime errors when running the report. The SAP Note is relevant for the source system with SAP ERP 600 through 618 releases. Alternatively, you can reset the source system staging table data using report RFINS_CFIN_DEL_MIG_SRC from the Central Finance system, which has been available since SAP S/4HANA 1809 FPS01.	Transaction SE38 or SA38
2	Uncheck the **Initial Load Finished** indicator in the **General Replication Settings** (table VCFIN_SOURCE_SET).	Transaction SM30 or navigate to **SAP Customizing Implementation Guide • Financial Accounting • Central Finance: Source System Settings • General Replication Settings**

Table 8.9 FI Initial Load Reset : Source System Activities

During the FI initial load data transfer process, the system stores data in Central Finance staging tables and migration tables. In addition, you'll delete posted documents in the Universal Journal and dependent tables. Table 8.10 provides the sequence of steps you'll follow during a FI-CO initial load reset in the Central Finance system.

The initial load reset also requires activities to be performed in the SAP Landscape Transformation system, as explained in Table 8.11. These activities are required only when online replication is turned on.

Central Finance System		
Sequence No.	**Activity**	**Transaction**
1	Run report RFINS_CFIN_CLEAR_INIT_LOAD.	Transaction SA38 or SE8
2	Delete SAP Application Interface Framework messages if real-time replication is turned on.	Transaction /AIF/PERS_DEL_AUTH Transaction /AIF/PERS_DEL
3	Delete posted documents in the Universal Journal if the initial load postings and online replication are active.	Transaction FINS_CFIN_DOC_DELETE (program RFINS_CFIN_DOCUMENTS_DELETE)
4	Reset the company code/controlling area. This is a full system reset only if step 3 doesn't work for some reason.	Transaction OBR1 to reset data with the related company code

Table 8.10 FI Initial Load Reset: Central Finance System Activities

SAP Landscape Transformation		
Sequence No.	**Activity**	**Transaction**
1	■ Stop replication of table CFIN_ACCHD (CFIN_ACCHD_IDR if IDR is active). ■ Stop replication of tables AUFK and COBK if they are required to be reset.	Transaction LTRC
2	■ SAP recommends resetting the transfer status for data transferred through SAP Landscape Transformation. ■ Run report C_FM_RESTART_COPY_DELETE for table CFIN_ACCHD (and for tables AUFK and COBK if required to reset the application replication data).	Program DMC_FM_RESTART_COPY_DELETE via Transaction SE38

Table 8.11 FI INITIAL LOAD RESET: SAP Landscape Transformation System Activities

8.5 Initial Load Using the Intermediate Data Retention Tool

As we discussed in Section 8.1.5, the main difference between the classic initial load and IDR is how you extract documents. How you extract balances and open items remains the same between both approaches. With the IDR approach, documents are captured with the Start Recording to IDR System step, which you'll initiate from Transaction FINS_CFIN_MAN_REPL (program RFINS_CFIN_MANAGE_REPLICATION) in Central Finance. After starting the recording, financial transactions that are posted in the source system are captured in IDR staging tables (table CFIN_ACCHD_IDR and other dependent tables) in the source system and replicated to a separate IDR system. Later, you'll replicate data from the dedicated IDR system to Central Finance once you complete the initial load activities.

In this section, let's explore detailed steps carried out with the IDR approach.

8.5.1 Preparation

Before you begin IDR recording, you need to understand the systems and preparation steps involved with the IDR approach. As you seen in Figure 8.3, IDR involves four systems, as also described in Table 8.12.

Source Systems	IDR System	SAP Landscape Transformation	Central Finance
SAP ERP 6.0, SAP S/4HANA on-premise, and SAP S/4HANA Cloud releases.	A dedicated IDR system is recommended, which can be a minimum of SAP ERP 6.0 or higher release level than the source system.	The SAP Landscape Transformation system enables trigger-based replication for Central Finance as recommend in SAP Note 2154420.	SAP S/4HANA 2020 FPS1 or higher release.

Table 8.12 IDR System Components

Tip

SAP recommends consulting SAP Note 3017968 before you begin preparation for IDR. You need to request SAP support to release this SAP Note through an OSS incident. In addition, check whether you need to implement SAP Note 2987586 in your source system and Central Finance system, which enables the Data Dictionary (DDIC) objects required for the IDR persistency table `CFIN_ACCHD_IDR` based on your release levels.

There are several prerequisites you'll carry out for IDR preparation. Table 8.13 provides the preparation steps necessary for IDR.

Preparation Steps	Purpose
RFC destinations	Set up the RFC destination to connect the source system and IDR system.
Logical system	Create a logical system. Ensure the logical system assigned to RFC destinations is set up for the source system and IDR system with Customizing activity **Choose Logical System of the Initial Load**.

Table 8.13 Preparation Customizing Activities for IDR

Preparation Steps	Purpose
General Replication Settings	Maintain **General Replication Settings** view VCFFIN_SOURCE_SET using Transaction SM30 or navigating from Central Finance Customizing Transaction CFINIMG. Once IDR recording begins, SAP doesn't support changing this view unless you perform an IDR initial load reset.
Initial load groups	Maintain the initial load groups in the Central Finance Customizing activity. IDR supports cross-company postings. SAP recommends creating the initial load groups by keeping all involved company codes to support cross-company postings in the same group.
SAP Landscape Transformation setup	Configure and activate the SAP Landscape Transformation replication for the following: ▪ Source system to dedicated IDR system ▪ IDR system to Central Finance system For IDR, you need a load replication object in SAP Landscape Transformation for different types of postings: ▪ **FI-CO replication from source to IDR table CFIN_ACCHD_IDR** Load object CFI_ACCHD_IDR_L and replication object CFI_ACCHD_IDR_R. ▪ **Source system CFIN_MIG_LOG replication to IDR** Load object CFI_MIG_LOG_L and replication object CFI_MIG_LOG_R. ▪ **FI-CO posting from IDR to Central Finance (table CFIN_ACCHD)** Load object CFI_CFIN_ACCHD_L and replication object CFI_CFIN_ACCHD_R. ▪ **FI-CO postings from source to Central Finance (table CFIN_ACCHD)** Load object CFI_CFIN_ACCHD_L and replication object CFI_CFIN_ACCHD_R. The predefined load and replication objects are part of project REPL_CFIN and subproject REPL_CFIN.
Choose the logical system	To define the logical IDR system for the source system, navigate to the Customizing activity (Transaction CFINIMG) **Central Finance: Target System Settings • Initial Load • Initial Load Settings • Choose Logical System**. Perform Customizing activity **Choose Logical System** in addition to the source logical system.
Maintain RFC assignment and settings for the Central Finance system	To maintain the settings, navigate to the Central Finance Customizing activity (Transaction CFINIMG) **Central Finance: Target System Settings • Set Up Systems • Maintain RFC Assignment and Settings in the Central Finance System**. For this Customizing activity, you need to maintain the entry for both the IDR system and source system. In addition, RFC usage **Extracting Data for Financial Accounting** is required for the IDR logical system and IDR RFC destination as well.

Table 8.13 Preparation Customizing Activities for IDR (Cont.)

8

Now that you understand the preparation steps for IDR, let's dive deeper into starting recording and the subsequent steps you'll follow for IDR.

8.5.2 Performing the Intermediate Data Recording Initial Load

This section is divided into several sequential steps that are required to complete the IDR as follows:

1. Start recording to the IDR system.
2. Block previous posting periods.
3. Extract balances/open items.
4. Unblock previous posting periods.
5. Start replication,
6. Prepare the transfer from IDR to Central Finance.
7. Stop recording to the IDR system.
8. Reset the source staging tables for IDR.
9. Delete IDR persistent tables.
10. Delete IDR initial load data in Central Finance.

Let's explore each replication of these with step-by-step instructions.

Start Recording to the IDR System

IDR begins with the **Start Recording to IDR System** step from Transaction FINS_CFIN_MAN_REPL (program RFINS_CFIN_MANAGE_REPLICATION), as shown at the top of Figure 8.28. Before you begin, ensure you've configured the preparation steps identified in Table 8.13. To start recording, you can reach this transaction by navigating from the Central Finance Customizing activity (Transaction CFINIMG) **Central Finance: Target System Settings • Initial Load • Initial Load Execution for Selected Company Codes • Manage Intermediate Data Retention**. After reaching the transaction, on the **Manage Intermediate Data Retention** screen, enter a value in the **Initial Load Group** ❶ field. Ensure you've the initial load group created has all company codes involved with cross-company code transactions. Next, choose the **Start Recording to IDR System** ❷ radio button, and click the **Execute** ❸ button. The system prompts with a caution window with the message **Do you want to activate Start Recoding to IDR System for Initial Load Group LG02? If you activate this step, all postings in the source.** On this screen, click the **Yes** button to continue. The resulting screen shows **Display Logs** ❹ confirming that the recording of the IDR was successful.

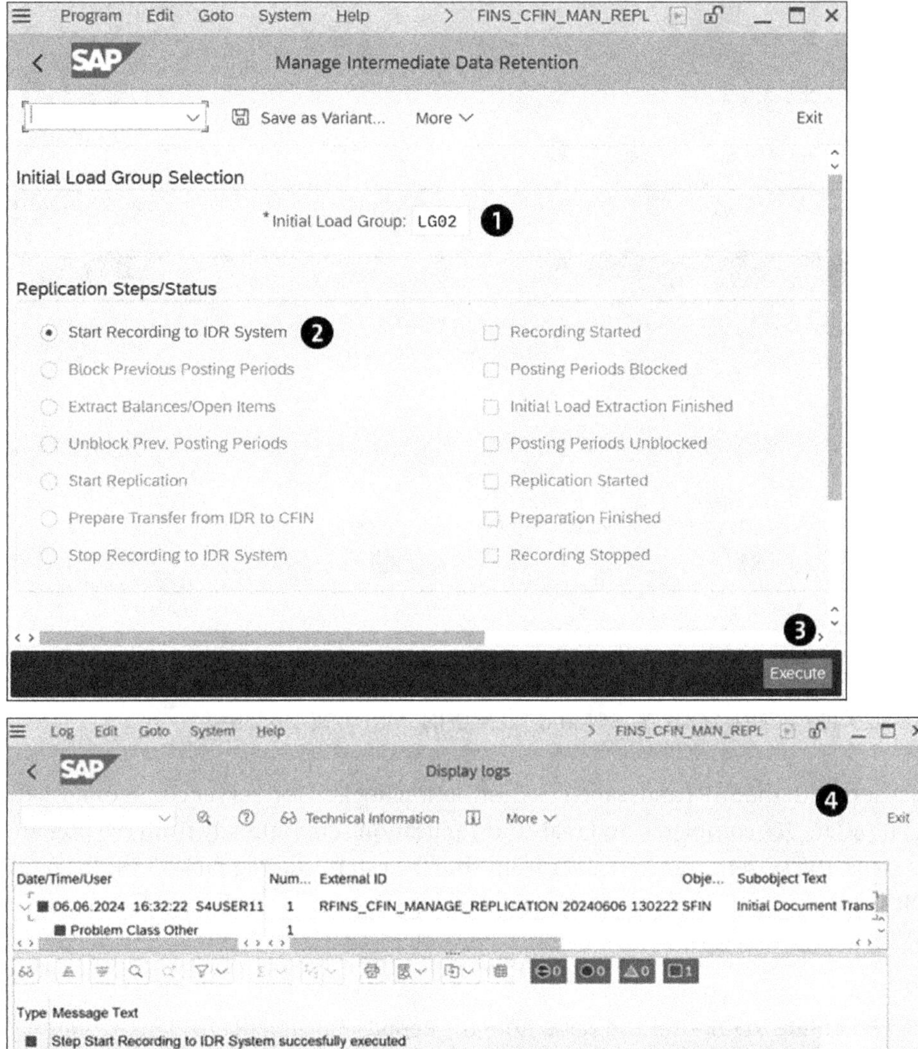

8

Figure 8.28 Manage Intermediate Data Retention

After completing the start recording step, the source system table CFIN_MIG_LOG is updated with an entry for the logical system, company code, and timestamp based on VCFIN_SOURCE_SET **Period - Documents**, as shown in Figure 8.29 ❶. As you can see, fields START_RECORDING_TIME ❷ and START_POSTING_DATE ❸ are updated in the CFIN_MIG_LOG table. START_RECORDING_TIME confirms when the IDR recording starts in the source system. In our example, START_POSTING_DATE in table CFIN_MIG_LOG is **01.07.2024** (DD.MM.YYYY format). Any new posting in the source system with a posting date from this date onward will start capturing data in source staging table CFIN_ACCHD_IDR.

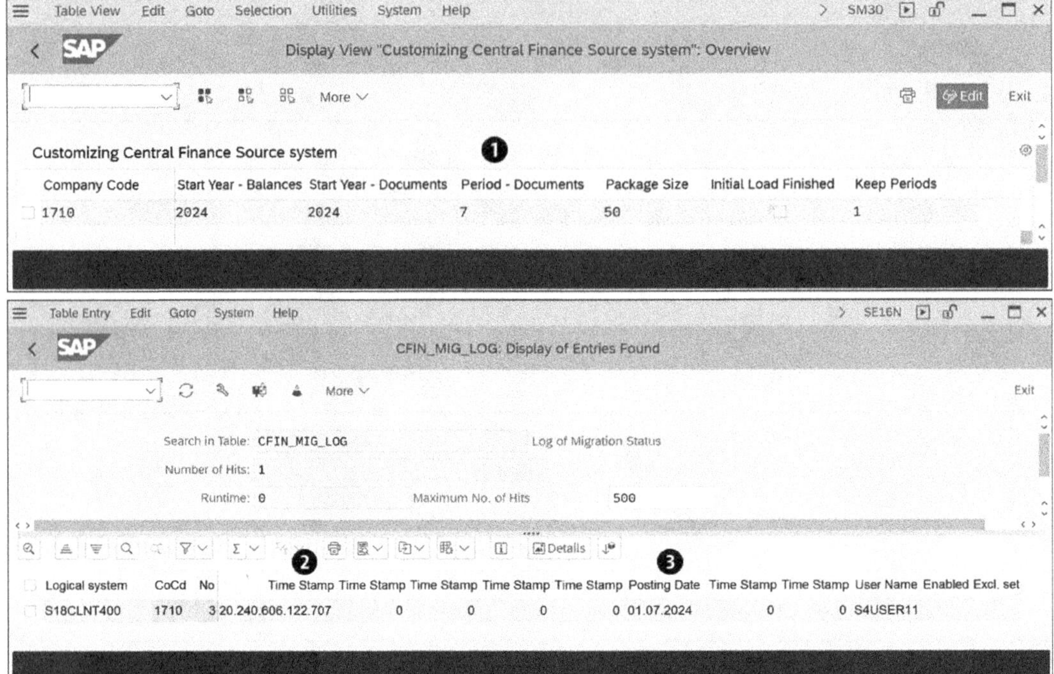

Figure 8.29 IDR VCFIN_SOURCE_SET and Table CFIN_MIG_LOG Updates

In addition, with the SAP Landscape Transformation Replication Server configuration for the CFIN_ACCHD_IDR completed and real-time replication active already from the preparations steps, the system transfers data from source system staging table CFIN_ACCHD_IDR to IDR staging table CFIN_ACCHD_IDR and dependent tables in the dedicated IDR system.

Tip

Start Recording to IDR uses the value from the **Period - Documents** field in the source system VCFIN_SOURCE_SET view. The start date of recording must occur at least one day before the end of the month prior to the period maintained in the **Period - Documents** column in the **General Replication Setting** VCFIN_SOURCE_SET view. Otherwise, the system issues the message **Start Document Transfer date for company code 1710 is in the past**. For example, if the **Period - Documents** field is set up with 7, then start recording must be started by June 30th at the latest.

Block Previous Posting Periods

As you observed from Figure 8.30, once the **Start Recording to IDR System** step is completed, the system enables the **Recording Started** checkbox. As a result, any changes to source table VCFIN_SOURCE_SET is prevented for the company codes setup within the initial load group. In the next step, on the **Manage Intermediate Data Retention** screen, choose the **Block Previous Posting Periods** radio button ❶, and click the **Execute** button to continue and see the message **Block Previous Posting Periods successfully executed** ❷. By starting this step, the system prevents posting any documents into previous open item periods in the source system. SAP recommends you initiate this step several hours before starting the subsequent step to avoid no fresh transactions being posted, which could cause posting into an open-item period. In other words, the system will not allow you to post any financial documents in the posting period prior to the period maintained in **Period - Documents** of **General Replication Settings** (VCFIN_SOURCE_SET) table. As a result, upon posting the document, the system displays the **Posting Dates before 07.01.2024 are blocked due to ongoing IDR** error message. In our example, **Period - Documents** of the **General Replication Settings** (VCFIN_SOURCE_SET) table is set with **7**.

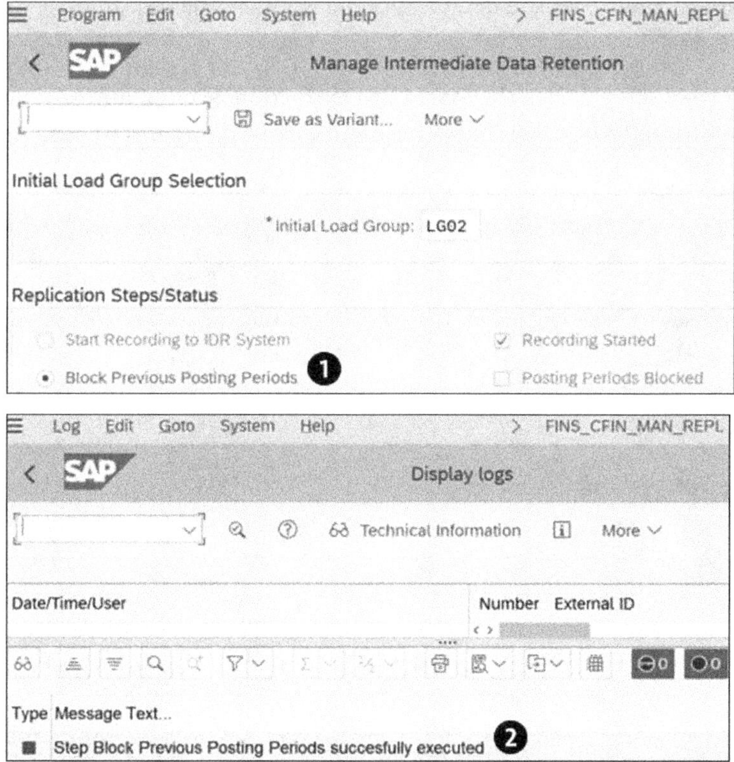

Figure 8.30 Block Previous Posting Periods

Extract Balances/Open Items

Next, on the **Manage Intermediate Data Retention** screen, select the **Extract Balances/ Open Items** radio button, as shown in Figure 8.31 ❶. When you choose this radio button, as you've observed, the system displays two buttons: **Start Initial Load** and **Monitor Initial Load**. Choose the **Start Initial Load** button, and click the **Execute** button to begin extracting data for the initial load. This brings up the **Execute Initial Load** screen, which is the same screen you've seen with the classic initial load earlier. Blocking the periods is a prerequisite to starting the extraction of balances step. Recall, there is no difference in how you process balances and open items between the classic initial load and IDR. However, for documents replication, you'll follow changes in the process as follows:

- During the initial load extraction, detailed documents aren't extracted but are loaded from IDR tables.
- SAP recommends recording IDR as early as possible to have a better quality of IDR recorded data. Documents extracted through the classic initial load sometimes don't contain all details due to summarization or substitution configurations. In such cases, you won't have a better quality of documents as original entries will be lost.
- As you're blocking periods before the document load period, a delta load isn't required.

On the **Execute Initial Load** screen, click the **Start Data Extraction** ❷ radio button, and then click the **Execute** button to begin extracting the initial load data from the source system.

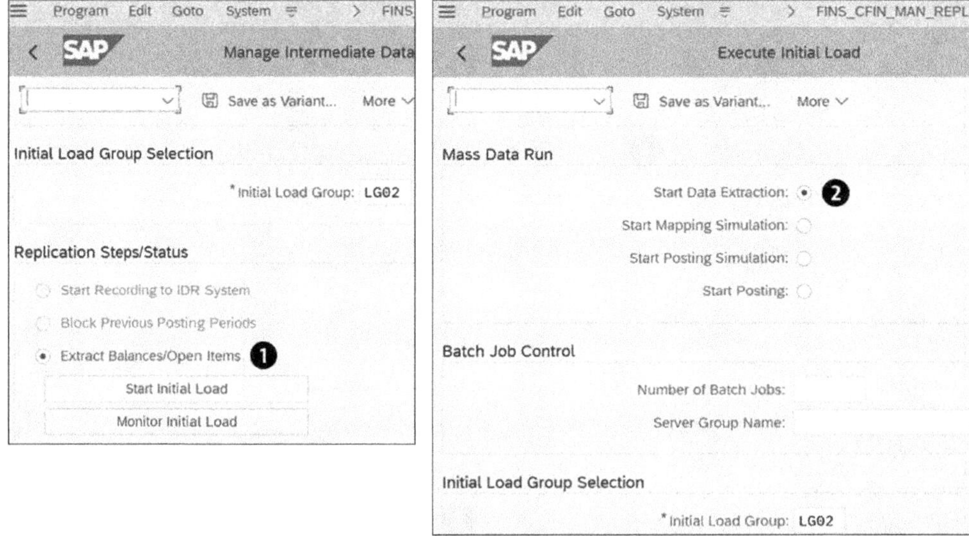

Figure 8.31 Extract Balances and Open Items

The initial load extraction can be monitored using the **Monitor Initial Load** button, as shown previously on the left side of Figure 8.31. After completing the extraction of balances and open items, the system transfers data to Central Finance staging tables (CFIN_AC*). In addition, a new entry is created in table CFIN_MIG_LOG for the company codes in scope, as shown in Figure 8.32.

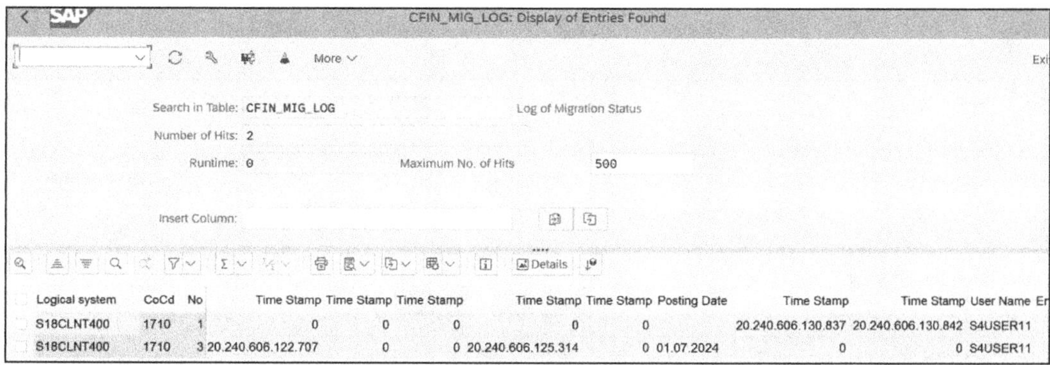

Figure 8.32 CFIN_MIG_LOG_IDR Balances and Open Items Step

As you can see, values in the posting date, start time, and end time fields are recorded in the log table. Immediately after this entry in the table, any new financial posting from the **Posting Date** (in table CFIN_MIG_LOG) onward won't be captured in staging table CFIN_ACCHD; instead, rather transactional data will be captured in IDR staging table CFIN_ACCHD_IDR. Subsequently, these new documents will be transferred via the IDR document replication route. In addition, the delta load impacts only documents, so there is no impact on balances and open items. Therefore, no delta load is required with the IDR approach.

Unblock Previous Posting Periods

Next, on the **Manage Intermediate Data Retention** screen, select the **Unblock Prev. Posting Periods** ❶ radio button to unblock posting into previous open-item periods in the source system (Figure 8.33). Before executing this step, be sure to unblock the previous posting periods in the source system for the corresponding company code. Click the **Execute** button to continue. The system prompts with a **Caution** window with the message **Do you want to activate Unblock Prev. Posting periods for Initial Load Group L02?** Click the **Yes** button to continue. On the resulting **Display Logs** window, the system shows the unblocking of posting periods executed ❷.

Start Replication

As shown in Figure 8.34, on the **Manage Intermediate Data Retention** screen, select the **Start Replication** ❶ radio button to begin real-time replication from the source system to the Central Finance system. Click the **Execute** button to continue. The system prompts with a **Caution** window with the message **This step will trigger the standard replication. Did you stop the AIF processing to Avoid Sequencing issues in process chains?** Click the **Yes** button to continue. On the resulting **Display Logs** window, the system shows that the start replication was executed successfully ❷.

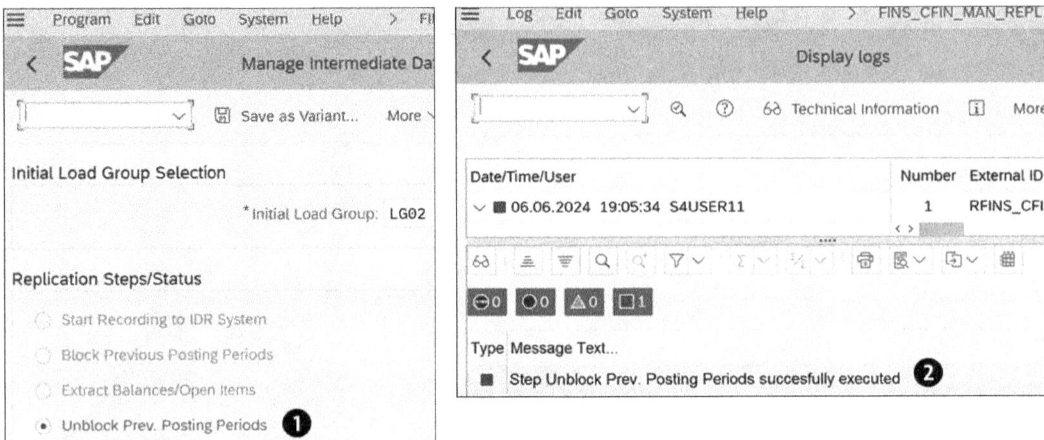

Figure 8.33 Unblock Previous Posting Periods

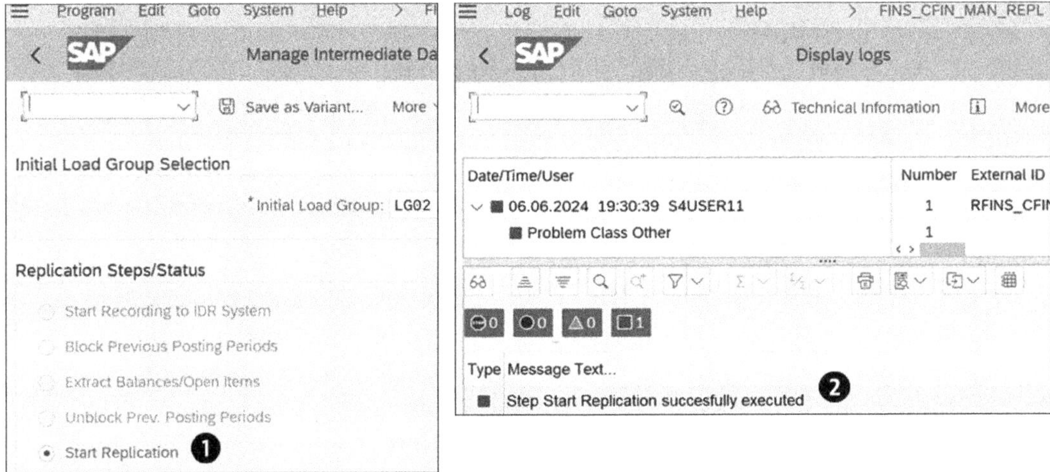

Figure 8.34 Start Replication for IDR

Note

SAP recommends not activating SAP Application Interface Framework until transferring all data from IDR to the Central Finance system to prevent sequencing issues. The system also displays the **Caution** window to alert that SAP Application Interface Framework processing should be done prior to starting this step. SAP recommends setting up a separate SAP Application Interface Framework runtime configuration for each initial load group to avoid SAP Application Interface Framework message sequencing problems.

Once the system completes this step, the system updates the value in the `CFIN_MIG_LOG-START_REPLICATION_TIME` field. After this step, the system captures data in both source staging tables CFIN_ACCHD_IDR and CFIN_ACCHD for new FI postings in the source system. From this point onward, any FI postings are routed for online replication. The main difference between the data captured before and after the **Start Replication** step is what is recorded in table CFIN_ACCHD_IDR. The data captured in staging table CFIN_ACCHD_IDR earlier is relevant for the initial load and used if you need to repeat the initial load. The TRANSFERRED field in table CFIN_ACCHD_IDR explains if the document is captured after the **Start Replication** step started. The TRANSFERRED field value updated with X says the system captured new FI postings should be updated in source staging table CFIN_ACCHD as well.

Prepare Transfer from Intermediate Data Recording to Central Finance

In the last step, as shown in Figure 8.35, on the **Manage Intermediate Data Retention** screen, select the **Prepare Transfer from IDR to CFIN** ❶ radio button. Once you choose this option, as you've observed, the system brings up two buttons: **Start Preparation** and **Monitor Preparation**.

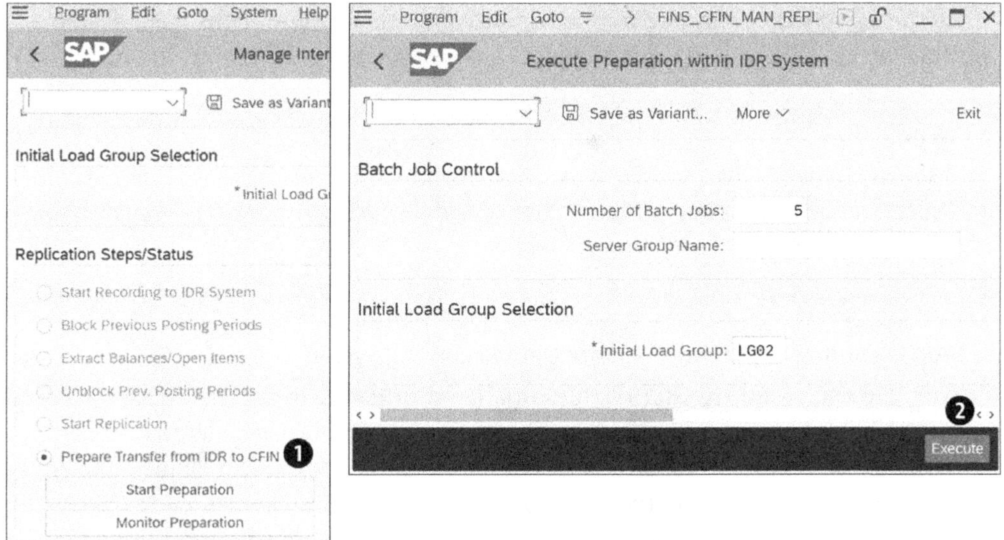

Figure 8.35 Execute Preparation within the IDR System

As a precheck, verify these documents aren't directly transferred from the source system to Central Finance to prevent duplicates. IDR document load comes in two circumstances:

- Documents accumulated in table CFIN_ACCHD_IDR between the beginning of IDR recording and the start of online replications
- Any documents posted before the **Period - Documents** of **General Replication Settings** (VCFIN_SOURCE_SET) but after the extraction balances/open items and beginning of online replication

As a result, the system moves data from staging table CFIN_ACCHD_IDR to staging table CFIN_ACCHD in the IDR system. Note that prior to executing this step, table CFIN_ACCHD in the IDR system is empty. When the preparation step is executed, the system moves only GUIDS with CFIN_ACCHD_IDR-TRUE(TRANSFERRED) = Blank to CFIN_ACCHD in the dedicated IDR system. On the other hand, GUIDs with CFIN_ACCHD_IDR-TRUE(TRANSFERRED)= X are routed to CFIN_ACCHD in the source system for real-time replication through the standard SAP Landscape Transformation Replication Server route when it's turned on from SAP Landscape Transformation Replication Server to load/replicate the data after the initial load step is completed. In addition, the CFIN_ACCHD-LOAD_GROUP ID field is populated with initial load group **LG02** for which the preparation step was executed, and the CFIN_ACCHD-TRUE (IDR_SOURCE) field is also set as X in table CFIN_ACCHD. This represents that the data transferred is captured in the dedicated IDR system when the preparation step is executed.

Click the **Start Preparation** button to prepare documents to be transferred from the dedicated IDR system to the Central Finance system. Once the **Start Preparation** button is selected, the system takes you to the **Execute Preparation within IDR system** screen. On this screen, choose **Number of Batch Jobs** and **Server Group Name** if applicable, and then enter a value in **Initial Load Group**. Click the **Execute** ❷ button to finish.

Tip

SAP recommends deactivating SAP Application Interface Framework processing via runtime configuration group to ensure the right sequencing is followed for the follow-on documents to prevent serialization problems.

Stop Recording to the Intermediate Data Recording System

After the successful transfer of the initial load of IDR with no the initial load reset required, then IDR recording can be stopped. In addition, you can delete all the data you transferred in the IDR persistency. To stop recording, as shown in Figure 8.36, on the **Manage Intermediate Data Retention** screen, select the **Stop Recording to IDR System** ❶ button. Click the **Execute** button to continue. The system prompts with a **Caution** window with the message **Do you want to activate stop Recording to IDR System for Initial Load Group LG02?** Click the **Yes** button to continue. As a result of the execution, the system updates table CFIN_MIG_LOG-END-RECORDING_TIME. After this step, any new posting won't be recorded in table CFIN_ACCHD_IDR, but FI postings will be captured in table CFIN_ACCHD for continuous ongoing replication. Once the IDR recording is stopped, you can start the load of FI-CO postings from the dedicated IDR system to the Central Finance system. Table CFIN_ACCHD and associated staging tables data in the dedicated IDR system is then transferred to the Central Finance system via the standard SAP Landscape Transformation route. The main advantage is that you can start working on the IDR load errors in parallel much earlier before activating standard SAP Landscape Transformation. The IDR data from the dedicated IDR system to the Central Finance system can be seen in the SAP Application Interface Framework interface AC_DOC version 2. In the resulting **Display Logs** window, the system shows the results of execution ❷.

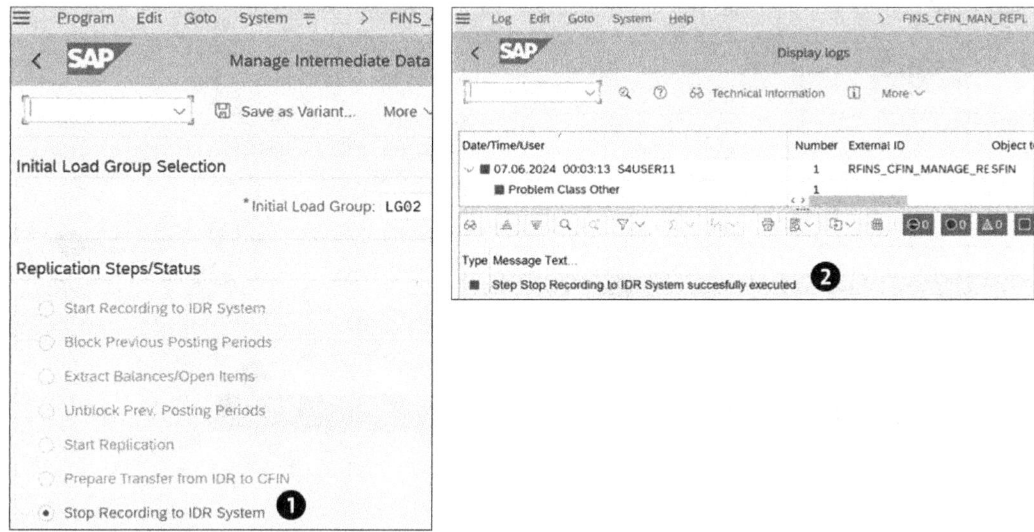

Figure 8.36 Stop Recording to the IDR System

Resetting Source Staging Tables for Intermediate Data Recording

SAP provides an IDR reset report to delete online transfer table and migration log tables in the source system. To reset data in the source system, navigate from Central Finance Customizing Transaction CFINIMG at **Central Finance • Central Finance Source System Settings • Delete Local Initial Load Data** or use Transaction CFIN_DEL_MIG (report RCFIN_DEL_MIG). This report deletes the online transfer tables and table CFIN_MIG_LOG entries created during the IDR process. Upon entering the transaction, it provides the following options:

- **Keep IDR Recording**
  This option enables you to repeat the initial load without resetting the IDR recording in IDR table CFIN_ACCHD. However, migration log table CFIN_MIG_LOG data will get deleted for the selected company code.
- **Full Reset**
  Select this option if you want to do the full reset. This deletes the entries in online persistent table CFIN_ACCHD, IDR persistent table CFIN_ACCHD_IDR, other source staging associated transfer tables and table CFIN_MIG_LOG in the source system. The **Start Replication** step as part of the IDR will record entries in table CFIN_ACCHD.
- **Delete IDR recording**
  This option deletes entries in IDR persistency tables only.

In our examplein Figure 8.37, we've chosen the **Full Reset** ❶ radio button. Upon clicking the **Execute** ❷ button, the system displays the **CAUTION: Deletion of All IDR Data** window with the message **Do you really want to reset the initial load and delete all IDR data (cannot be restored)?** Click the **Yes** button to continue. Upon deletion of data in tables, the system displays an output screen ❸ with results.

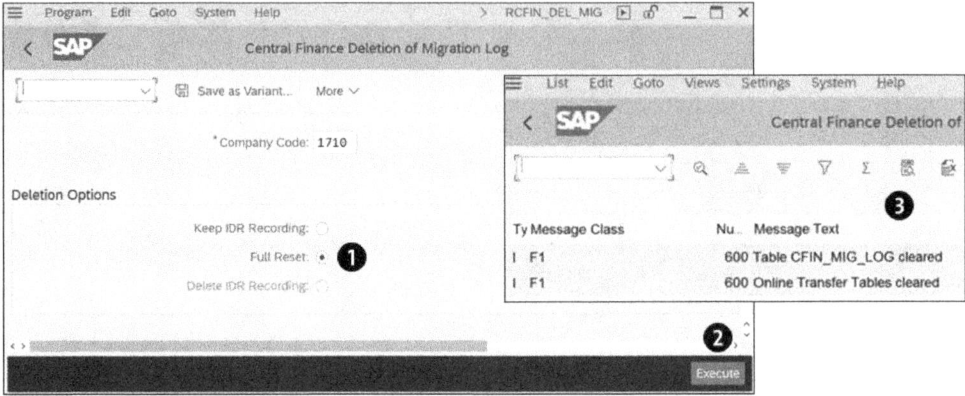

Figure 8.37 Source Staging Tables Reset for IDR

Delete Intermediate Data Recording Persistent, Online Persistent, and Other Associated Tables

You can execute the same Transaction CFIN_DEL_MIG (report RCFIN_DEL_MIG) in the dedicated IDR system to delete the staging table in that system. Note that the table CFIN_MIG_LOG entry is required in the dedicated IDR system to run the report RCFIN_ DEL_MIG. Recall that the system used table CFIN_MIG_LOG to load the dedicated IDR system from the source system, so an entry is required in that table to reset data in the dedicated IDR system.

Delete Initial Load Data in Central Finance

Likewise, a similar exercise can be done to reset the IDR initial load data in the Central Finance system. To reset, go to the transaction by navigating from the Central Finance Customizing Transaction CFINIMG at **Central Finance • Central Finance Target System Settings • Initial Load • Initial Load Execution for Financial Accounting • Initial Load Execution for Selected Company Codes • Delete Initial Load Data** or use Transaction FINS_ CFIN_LOAD_DEL (program RFINS_CFIN_CLEAR_INIT_LOAD). Upon entering the transaction, you'll arrive at the **Delete Initial Load Data** screen. On this screen, enter a value in the **Initial Load** Group field, and then choose the appropriate reset radio button. In our example, we've chosen **IDR Initial Load – Full Reset**, as shown in Figure 8.38 ❶. As you can see, the classic initial load reset functions (first two radio buttons) are disabled. Click the **Execute** button. After completing the reset, the system displays an output screen ❷.

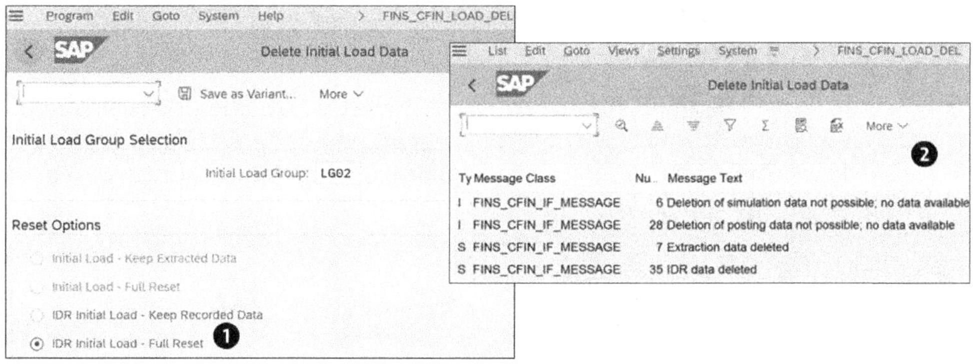

Figure 8.38 Delete the IDR Persistent, Online Persistent, and Other Associated Tables in the Dedicated IDR System

8.6 Initial Load of Controlling Internal Postings

Central Finance supports replication of CO internal postings from the source system to Central Finance. Recall that primary CO postings (business transaction type **COIN**) are replicated along with FI documents. On the other hand, all CO internal postings are transferred with CO replication. For SAP ERP as the source system, if the **GL Reconciliation Postings Transferred** flag is set in VCFIN_SOURCE_SET, CO-FI reconciliation postings are transferred with FI document replication. SAP recommends this flag be set only when CO replication isn't in scope. Otherwise, enabling both the **GL Reconciliation Postings Transferred** flag in VCFIN_SOURCE_SET and CO replication will result in duplicate records. Conversely, if SAP S/4HANA is set up as the source system, there is no **GL Reconciliation Postings Transferred** flag in VCFIN_SOURCE_SET as the FI-CO postings are no longer applicable. In general, from a sequencing standpoint, CO internal postings are executed after the FI initial load is performed.

This section is divided into subtopics to cover the initial load and replication of CO internal postings:

- Business transaction types supported for CO secondary postings
- Overview of the CO initial load and replication of CO secondary postings
- Preparation of the initial load of CO secondary postings
- Smoke test for CO document replication
- Simulation of the initial load of CO secondary postings
- Enabling real-time replication
- Resetting the CO initial load
- CO staging tables cleanup

Let's dive deeper into each of these topics next.

8.6.1 Business Transaction Types Supported for Controlling Secondary Postings Replication

At the time of writing, most of the business transaction types in CO are supported. However, there are still limitations that apply to different SAP S/4HANA release levels. Table 8.14 lists business transaction types supported for CO internal postings replication by different SAP S/4HANA releases.

Business Transaction Types (COBK-VRGNG)	Business Transaction Type Description	Supported from SAP S/4HANA
KAMV	Manual Cost Allocation	On-premise 1503
KOAL	Actual Settlement of Internal Activity Allocation	
KOAO	Actual Settlement (not for the costing-based profitability analysis scenario)	
KSI0	Actual Cost Center Split	
KSI1	Actual Split Costs (Primary)	
KSI2	Actual Split Costs (Secondary)	
KSI3	Actual Split Costs Internal Activity Allocation	
KSII	Actual Price Calculation	
KZPI	Actual Overhead (Periodic)	
RKIU	Actual Overhead Assessment	
RKIV	Actual Overhead Distribution	
RKL	Actual Activity Allocation	
RKLN	Revaluation at Actual Price	
RKLT	Actual Template Allocation	
RKU1	Repost Costs	
KZPR	Actual Overhead (Real Time)	On-premise 1503 SP01
RKIL	Actual Inverse Activity Allocation	
RKS	Enter Statistical Key Figures	
RKLX	KAZI	On-premise 1503 SP02

Table 8.14 Central Finance: Supported Business Transactions for CO Secondary Postings

Business Transaction Types (COBK-VRGNG)	Business Transaction Type Description	Supported from SAP S/4HANA
RKIB	Actual Periodic Reporting	
RKU2	Repost Revenue	
KTDA	COPA: Top-Down: Actuals	
RRIU	Segment Adjust.: Act. Assess.	
RRIV	Segment Adjust.: Act. Distrib	
RKN	Actual Non-Alloc. Activities	
KOLI	Collective Order Delivery	
RRIB	Seg. Adjust.: Act.Per.Repost.	
RKU3	Repost CO Line Items	On-premise 1503 SP04
GPDP	Distribution Primary Costs	On-premise 1503 SP04
GPDS	Distribution Secondary Costs	
JVIU	JV Actual Assessment	On-premise 2020
JVIV	JV Actual Distribution	
JRIU	JV-Seg.Adjustm.Assessment	
JRIV	JV-Seg.Adjustm.Distribution	

Table 8.14 Central Finance: Supported Business Transactions for CO Secondary Postings

Note

To find the latest updates on supported business transaction types, consult SAP Note 2103482.

8.6.2 Overview of the Initial Load of Controlling Secondary Postings

First, let's first understand the activities involved with CO secondary posting in this section. You begin preparation of the initial load of CO secondary posting from Central Finance. Then, you'll run both the simulation and final initial load CO postings from SAP System Landscape Replication Server and monitor from SAP Application Interface Framework. During the initial load of CO internal postings, the system reads data from table COBK, table COEP, and dependent staging tables from the source system and then replicates the data to Central Finance. Before you begin CO preparation, simulation, or the initial load of CO internal postings, you'll carry out **Define COPA Mapping** configuration, as described in Chapter 10, Section 10.8.2.

Following are the steps involved with the initial load of the CO internal postings process:

1. **Preparation of Initial Load of CO Secondary Posting**
   From a sequencing standpoint, first you'll run this step prior to the simulation or real initial load of CO internal postings from the Central Finance system. This step establishes prerequisites such as costing-based profitability analysis characteristics required for actual CO internal postings. For the preparation step of the initial load of CO internal postings, the system considers VCFIN_SOURCE_SET table in the source system for selection criteria. Table 8.15 shows the source system tables updated during the preparation phase of the initial load of CO secondary postings.

Table Name	Where Is the Table Located?	What's Updated?
CFIN_CO_ADD	Source system	Cost object information related to CO document replication and during the initial load.
CFIN_COPA	Source system	Costing-based profitability analysis information related to CO document replication and the initial load. The system stores profitability segment numbers in this table.
CFIN_JOB_LOG_CO	Source system	CO initial load preparation related logs.
CFIN_MIG_LOG_CO	Source system	CO internal postings initial postings.
CFIN_COEP_ADD	SAP S/4HANA as source system	CO line-item extended fields.

Table 8.15 Source System Tables Updated during the CO Initial Load Preparation Step

[+]

Tip

If you don't want to replicate data related to costing-based profitability analysis from the source system to the Central Finance system, then you need to apply the skip rule by setting BAdI FIN_CFIN_DOCUMENT~SKIP_COPA with the flag rv_skip_copa as abap_true in the source system. By setting this flag, the system skips saving costing-based profitability analysis information in table CFIN_COPA in the source system.

2. **Smoke Test of Secondary Postings**
   You'll carry out this step from Central Finance, which also reads tables COBK and COEP in the source system using an RFC. The monitoring is done in the application log screen, which is built within the smoke test report itself. In addition, you're limited

to a maximum of 999 records to simulate. Furthermore, the smoke test doesn't update tables.

3. **Simulation of CO Secondary Postings**
   You'll initiate this step from SAP Landscape Transformation Replication Server. In addition, you'll set any filters in SAP Landscape Transformation Replication Server to restrict the scope documents to be replicated to Central Finance. The simulation of CO secondary postings is monitored from SAP Application Interface Framework using interface `CO_DOC_SIM` (CO document simulation) version 1. This step identifies any potential errors and assesses performance problems prior to the initial load of CO internal posting.

4. **Initial Load of CO Secondary Postings**
   Once both the smoke test and simulation of CO secondary postings are completed, you'll proceed with the actual initial load of CO secondary postings in this step. As a result, the system updates the Universal Journal for CO documents in the Central Finance system. During CO internal postings replication, the system calls function module `FINS_CFIN_CO_POSTING_BULK` to post CO internal postings to the Universal Journal.

8.6.3 Preparation of the Initial Load of Controlling Secondary Postings

In this section, you'll learn how to prepare the initial load of CO secondary postings required prior to loading the actual CO secondary postings. You'll execute this step from the Central Finance system. Upon execution of this step, the system does the following:

1. Reads the costing-based profitability analysis line-item characteristics and converts them into a key-value pair structure. As a prerequisite to complete this step, you must use the **Define COPA Mapping** configuration discussed in Chapter 10, Section 10.8.2.
2. Updates the converted pair values in `CFIN_COPA` tables. This table stores data relevant to costing-based profitability analysis from the CO document replication.
3. Stores references of the original documents, which are used later for reposting activity using Transaction RKU3.
4. Stores additional information in table `CFIN_CO_ADD`, which is used to store CO document additional data, mainly the cost object information. In addition, the system generates the CO key subnumber field in table `COEP`. Moreover, the system updates `CFIN_JOB_LOG_CO` and `CFIN_MIG_LOG_CO` for storing the CO initial load preparation steps and the CO initial load execution logs, respectively.
5. Stores line-item extended fields in table `CFIN_COEP_ADD`, if SAP S/4HANA is set up as a source system.

To begin the preparation of the initial load of CO secondary postings, navigate to IMG Customizing menu **Financial Accounting • Central Finance • Central Finance: Target**

System Settings • Initial Load • Initial Load Preparation for Management Accounting • Prepare for and Monitor the Initial Load of CO Postings or use Transaction CFIN_CO_INIT_PREP (program RFIN_CFIN_CO_PREPARE). After entering the transaction, you'll arrive at the **Preparation of Initial Load of CO Secondary Posting: Summary** screen, as shown in Figure 8.39.

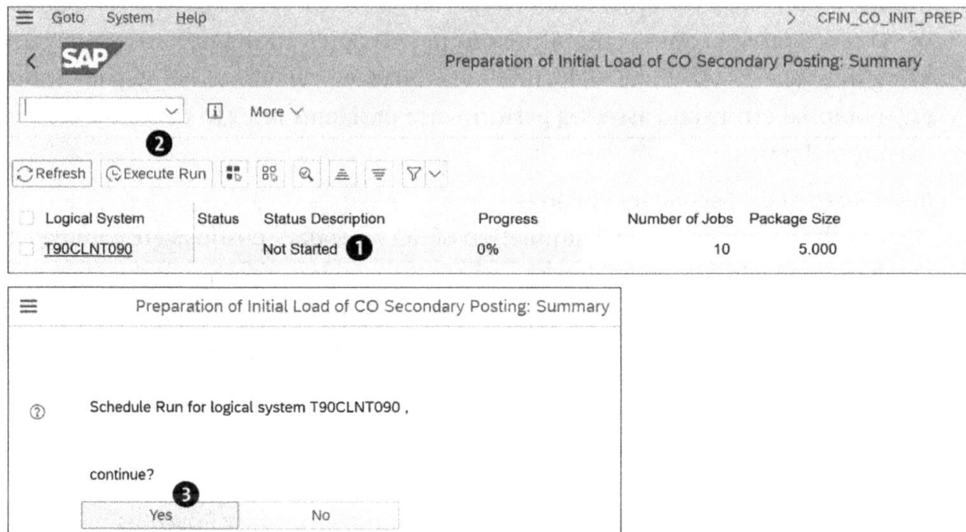

Figure 8.39 Preparation of the Initial Load of CO Secondary Posting Summary

On this screen, you'll notice a list of logical systems for source systems and **Status Description** showing **Not Started** ❶. Status represents the system activity status of preparation of the initial load. Table 8.16 lists the various statuses the system progresses along for the preparation of the initial load of CO postings.

Status	Description of Status
Not Started	This is the default status prior to running preparation of the initial load of CO secondary posting.
Aborted	If the job fails for any reason, the run will end in **Aborted** status. You must restart the aborted jobs.
Finished and Delta Run Required	After the first run, the system status is set to **Finished and Delta Run Required**. You'll restart to complete the delta run.

Table 8.16 Preparation Job Status

Status	Description of Status
Finished	Once the preparation job is completed, the system sets the status of the job to **Finished**.
In Process	During the execution, the system displays **In Process** status. You must wait until the job changes the status to **Finished**. The time taken to complete the job depends on volume and system landscape parameters such as memory and number of processes available.

Table 8.16 Preparation Job Status (Cont.)

Next, validate values in **Number of Jobs** for background jobs and **Package Size** to be entered based on server capacity. By default, the system uses **10** and **5000** for **Number of Jobs** and **Package Size** fields, respectively. Once the values are confirmed, select the logical system of the source system that you want to run preparation of the initial load of CO postings, and click the **Execute Run** ❷ button. This brings up the **Preparation of Initial Load of CO Secondary Posting: Summary** screen to confirm the job will be the schedule job for the chosen logical system. Click the **Yes** ❸ button to continue. At this point, the system kicks off a background job, which is executed in the source system.

Tip

An entry of logical system shows up in this configuration only when the following are true:

- You added the logical system in the **Choose logical System** configuration under **Initial Load Settings** located in Central Finance Customizing.
- You added the **9 Extraction Data for Management Accounting** entry under the **RFC Destination** dropdown list in the **Maintain RFC Assignments and Settings for Source Systems** Customizing activity.

If the **General Replication Settings** view VCFIN_SOURCE_SET is set up with an empty initial load in the source system, you won't see a logical system for that company code.

Click the **Refresh** button to view the job status. Upon completion of preparation of initial load CO secondary postings the first time, you'll notice **Run Mode** with a value of **F**, which means the system has completed the first run successfully, as shown in Figure 8.40. In addition, the **Status** and **Progress** columns display completion status ❶. In addition, you'll be able to confirm an entry in the source system logging table CFIN_MIG_LOG_CO ❷ for the company code in scope.

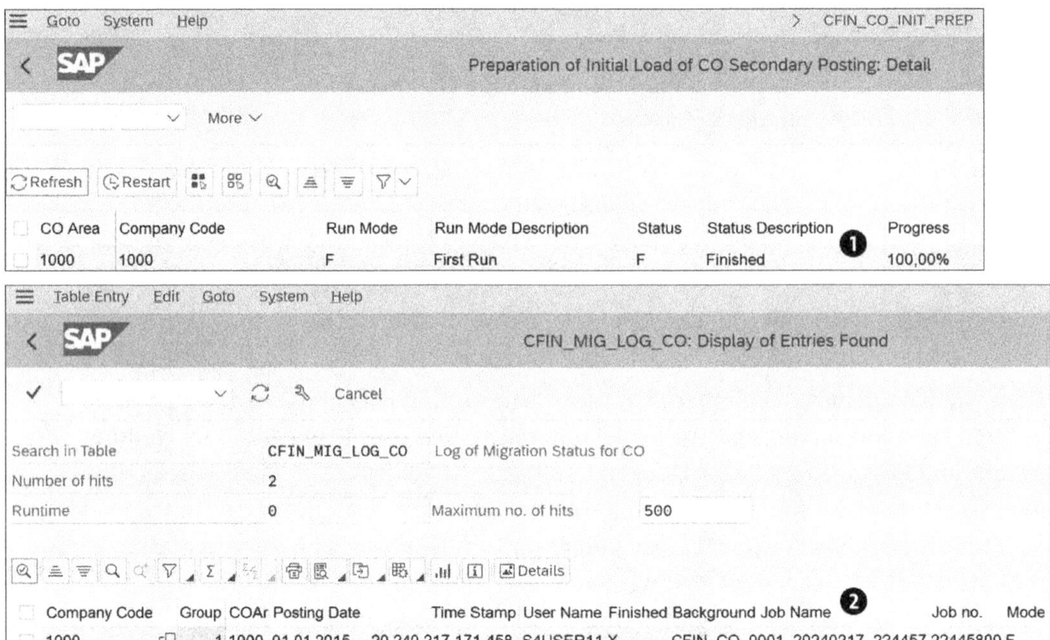

Figure 8.40 Preparation of the Initial Load of CO Secondary Posting: First Run

Now that you've completed the first run, let's run the delta run next. As shown in Figure 8.41, to begin the delta run, choose the logical system ❶, and click the **Execute Run** ❷ button to continue. As you've observed, the system prompts with a confirmation screen regarding whether to continue the delta run. Click the **Yes** ❸ button to continue with the delta run.

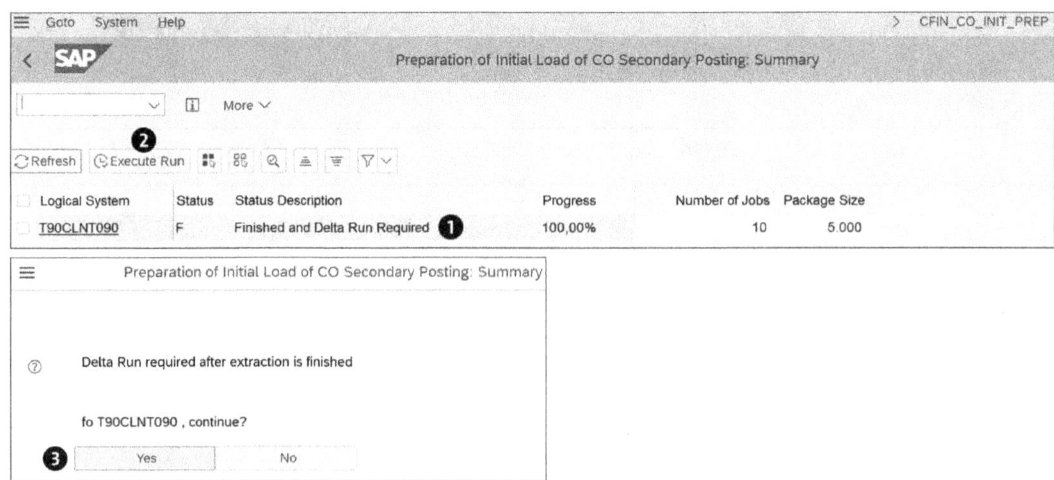

Figure 8.41 Preparation of the Initial Load of the CO Secondary Posting: Delta Run

After the delta run is completed, as shown in Figure 8.42, the system sets **Run Mode** to **D** along with **Status** and **Progress** ❶ columns that confirm the delta run completion. The bottom of the screen shows source system table CFIN_MIG_LOG_CO entries for both the first run and delta runs in completion status as shown in the **Mode** ❷ column.

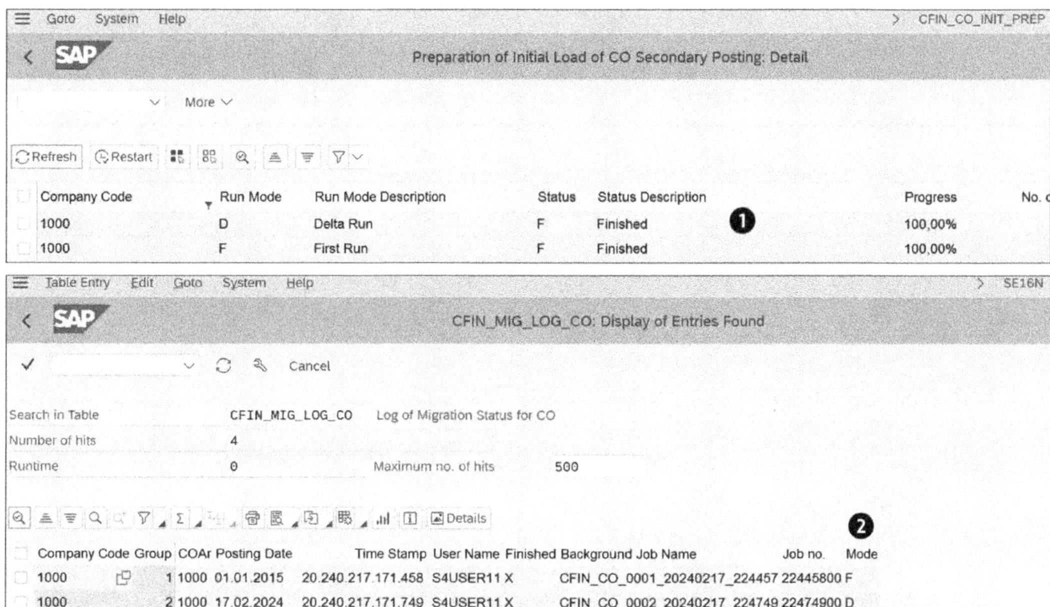

Figure 8.42 Preparation of the Initial Load of CO Secondary Posting: Delta Run Status

Upon completion of both the first and delta runs, as shown in Figure 8.43, you'll notice tables CFIN_JOB_LOG_CO ❶ and CFIN_CO_ADD ❷ updated with additional CO information and costing-based profitability analysis line-item characteristics, respectively. Table CFIN_COPA is updated only when you have costing-based profitability analysis data in the source system.

[!]

Warning

SAP recommends adding a new company code in the existing logical system only after completing the current scheduled preparation of the initial load of CO posting for which both first and delta runs are completed.

Now that you've completed the preparation steps, let's run the smoke test to identify any missing configuration or master data.

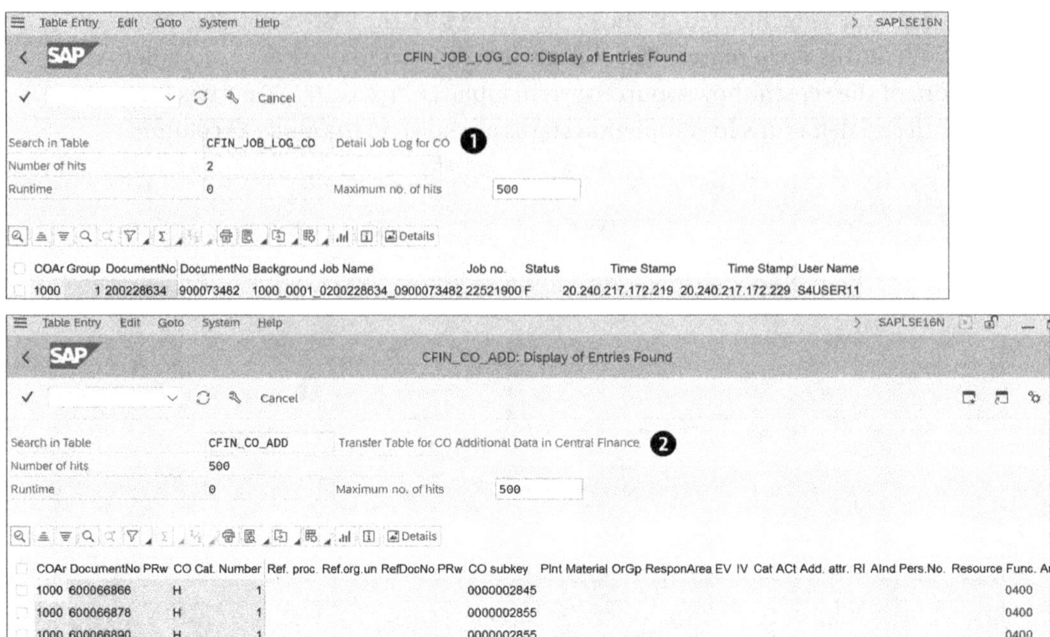

Figure 8.43 Job Log Entries for CO

8.6.4 Smoke Test for Controlling Document Replication

As you recall, in Central Finance, the smoke test program is available for both cost objects and secondary CO postings replication. The program is executed from the Central Finance system, which makes an RFC to the source system to read tables COBK and COEP from the source system. SAP Landscape Transformation Replication Server is *not* used for smoke tests of CO document replication. As the name implies, the smoke test is used primarily to check configuration and missing master data prior to performing the actual CO initial load. During the smoke test, the system doesn't create any CO secondary postings; instead, the check is performed as a sanity check to ensure all Customizing and master data exists. Although the smoke test is an optional step, as a best practice, SAP highly recommends this step to identify any missing configuration and master data.

To perform the smoke test for CO internal postings (table COBK), navigate to Customizing menu path **Financial Accounting • Central Finance • Central Finance: Target System Settings • Initial Load • Initial Load Preparation for Management Accounting • Smoke Test for Cost Object Mapping and CO Document Replication** or use the Transaction CFIN_CO_SIMU. After entering the transaction, you'll arrive at the **Smoke Test for Cost Object Mapping and CO Document Replication** screen, as shown in Figure 8.44. On this screen, enter values in the following fields:

❶ **Technical Setting**
Enter the source system value in **Logical System ID** from which CO secondary postings will be replicated to Central Finance. **Max No. of Records Simulated** can be adjusted from the default value of 200. Enter a numeric value to represent the number of records to be simulated. As you recall, the program allows you to enter a maximum of 999 records in this field.

❷ **Options for Smoke Test**
As you've noticed, this program is used for both **CO Document Replication** and **Cost Object Mapping**. Choose the **CO Document Replication** radio button for the CO secondary postings smoke test. Note if you choose this option, the system expands the selection criteria to include the source controlling area.

❸ **CO Document Selection Criteria**
Enter values in the **Source Document Number** or **Source Controlling Area** fields.

❹ **Administrative Data**
Under this section, you have an option to choose the user ID of the person who created internal order master data or the date when it was created.

Upon entering the necessary data in the selection screen, click the **Execute** ❺ button. This action takes you to the **Selected CO Objects** screen displaying CO secondary documents based on chosen criteria from the previous screen. On this resulting screen, choose the row ❻, and click the **Post** ❼ button to perform the smoke test. The system displays the **Smoke Test Result** screen ❽.

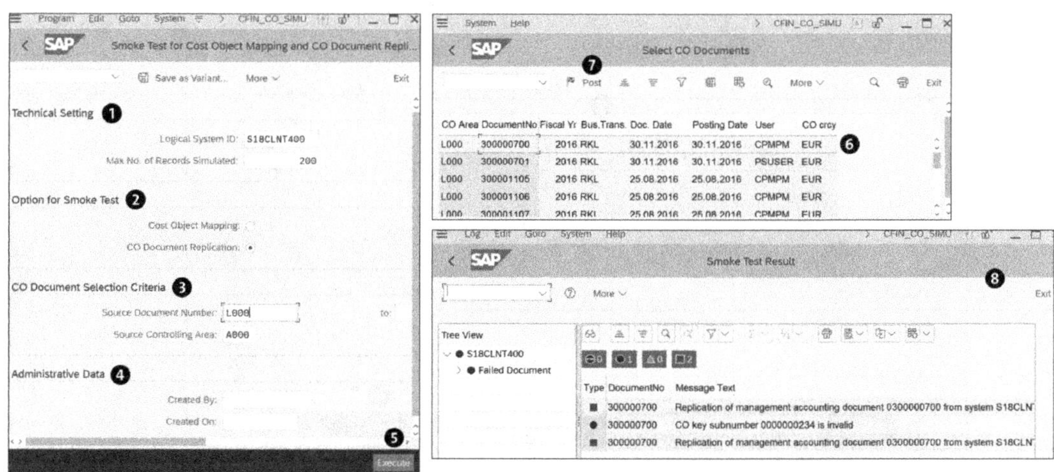

Figure 8.44 Smoke Test for CO Secondary Cost Replication

8.6.5 Simulation of the Initial Load of Controlling Secondary Documents

After the smoke test of CO internal postings is completed, next you'll carry out the simulation of the initial load of CO secondary documents to check missing configuration and master data, but with larger data volume. As the name implies, the simulation of the initial load of CO internal posting doesn't update tables, and it's used primarily to check for any inconsistency in configuration and to find the load time for cutover purpose. This step is carried out in SAP Landscape Transformation Replication Server and monitored from SAP Application Interface Framework with interface `CO_DOC_SIM`. Although smoke test and simulation of the initial load of CO internal postings are optional steps, SAP recommends carrying out these steps prior to performing the real initial load.

The prerequisite to this simulation step is that you must configure the MT ID for CO internal document replication (table `COBK`) with predefined initial load object `CFI_SIM_COBK_L` as covered in Chapter 5, Section 5.2.3. In addition, appropriate filters set up in SAP Landscape Transformation are covered in Chapter 5, Section 5.3.1. In this section, we'll focus on simulation of the initial load of CO internal postings.

To perform simulation of the initial load of CO internal postings, access the SAP Landscape Transformation Replication Server cockpit using Transaction LTRC in the SAP Landscape Transformation client. Choose MT ID **006**, and click on the **Participating Objects** tab, as shown in Figure 8.45 ❶.

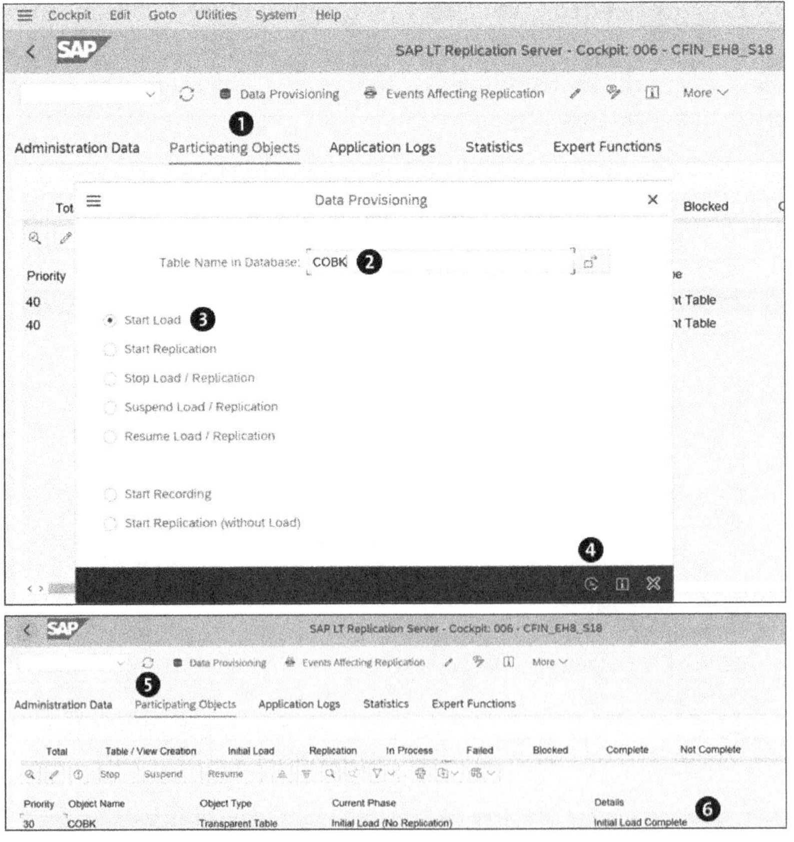

Figure 8.45 Simulation of CO Secondary Posting

This brings up the **Data Provisioning** dialog box. To perform simulation of the initial load of CO internal postings, enter "COBK" in the **Table Name in Database** ❷ field, choose the **Start Load** ❸ radio button, and click the **Execute** ❹ icon. This action starts simulation of the initial load of CO internal postings. To check the status of replication, from the SAP Landscape Transformation Replication Server cockpit, click the **Refresh** ❺ icon to see the simulation run status ❻.

A successful simulation of the initial load of CO internal posting will show up in SAP Application Interface Framework under interface `CO_DOC_SIM`.

8.6.6 Initial Load and Enabling Real-Time Replication of Controlling Secondary Documents

To perform the real initial load and replication of CO internal postings, access the SAP Landscape Transformation Replication Server cockpit using Transaction LTRC in the SAP Landscape Transformation Replication Server. Choose MT ID **006**, and click on the **Participating Objects** tab, as shown earlier in Figure 8.45 ❶. This brings up the **Data Provisioning** dialog box. To perform the initial load of CO internal postings, enter "COBK" in the **Table Name in Database** ❷ field, and choose the **Start Replication** radio button this time. Click the **Execute** ❹ icon. This action starts the initial load of CO postings. Upon completion of the initial load, the online replication of CO internal postings starts automatically. To check the status of the replication from the SAP Landscape Transformation Replication Server cockpit, click the **Refresh** ❺ icon to see the replication status ❻.

[«]

Note

SAP also provides report FINS_CFIN_REPL_CO_DOC to transfer CO individual documents from the source system to the Central Finance. This may be necessary when a CO document is missing or a relevant SAP Application Interface Framework message canceled by mistake or due to some errors. SAP recommends this transaction for a low volume of data.

8.6.7 Resetting the Controlling Initial Load

There are situations where you'll come across to reset the preparation of the CO initial load data especially for mock loads or if there is an error in the load in lower environments. SAP provides two reports to reset the CO initial load data:

- Using the source system report RCFIN_DEL_MIG_CO that deletes entries of preparation data of CO internal postings.
- Using Transaction FINS_CFIN_CO_DEL_SRC (report FINS_CFIN_CO_DEL_MIG_SRC) accessed from the Central Finance system. This report uses RFCs. To access the transaction, follow the SAP Customizing menu path **Financial Accounting • Central**

Finance: Target System Settings • Initial Load • Initial Load Preparation for Management Accounting • Undo Preparation of CO Initial Load or navigate to it from Transaction CFINIMG. Note this feature is available since the release of SAP S/4HANA 1809. If you're on an earlier release, implement SAP Note 2748263 to enable the transaction.

Let's discuss how to execute these reports next.

Reset the Initial Load Data in Source System

After reaching report RCFIN_DEL_MIG_CO, you'll arrive at the **Reset Initial Load CO Preparation Data** screen, as shown at the top of Figure 8.46. On this screen, enter values in the **Company Code** ❶ and **Package Size** ❷ fields. Click the **Execute** ❸ button. On the resulting screen, the system displays results of the reset ❹. Click the **Exit** ❺ button to return to the previous screen.

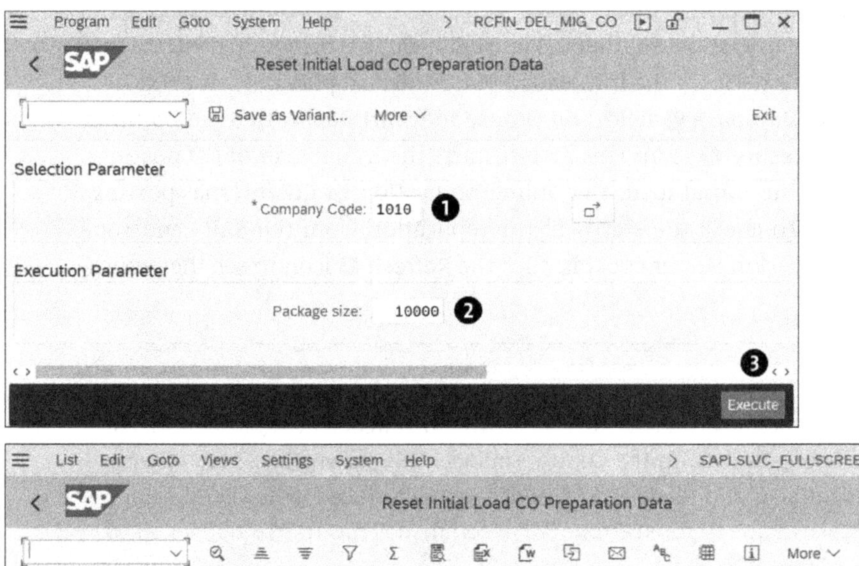

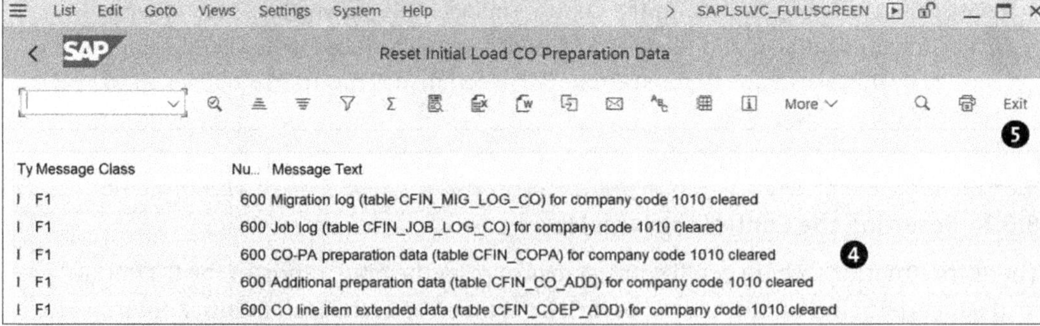

Figure 8.46 Reset Preparation of the CO Initial Load

[!]

Warning

SAP provides Transactions CFIN_CO_DOC_DEL to delete a single CO document and whereas Transaction CFIN_CO_INIT_DEL to delete all CO documents for a selected controlling area or a source company code. Caution should be exercised when resetting the CO initial load or deleting CO documents as these activities shouldn't be done in a live system. Typically, this activity is carried out in mock loads in lower environments. In addition, Transaction CFIN_CO_DOC_CRCT can be used to reverse and repost to make corrections for CO documents posted with incorrect mapping or configuration.

Undo Preparation of the Initial Load Data

Upon executing Transaction FINS_CFIN_CO_DEL_SRC in Central Finance, you'll arrive at the **Central Finance: Undo Preparation of CO Initial Load** screen, as shown in Figure 8.47. On this screen, enter values in **Logical System** ❶ and **Source Company Code** ❷, and then click the **Execute** ❸ button. Upon completion of the report, the system displays **The Job RCFIN_DEL_MIG_CO has run and completed with job number 00145800** ❹ status message at bottom of the screen for the selected company code. Click the **Exit** ❺ button to return to the previous screen.

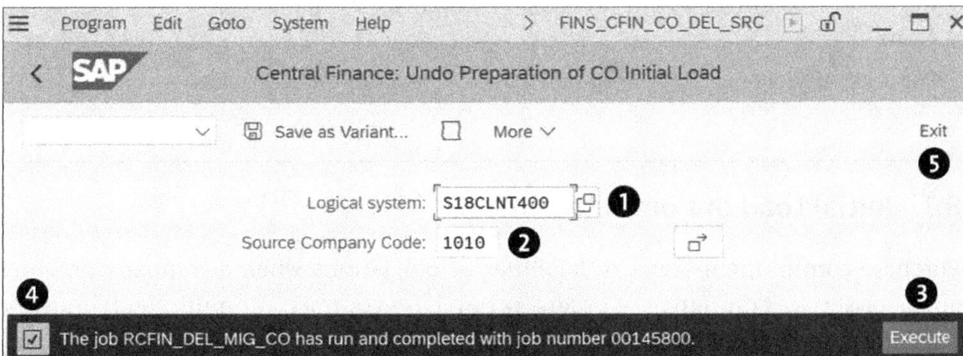

Figure 8.47 Reset Preparation of the CO Initial Load from Central Finance

8.6.8 Controlling Staging Tables Cleanup

After performing the initial load and enabling CO replication, over a period of time, source tables CFIN_COPA and CFIN_CO_ADD may fill up quickly. To improve overall performance and to support ongoing maintenance, SAP provides report FIN_CFIN_CO_CLEANUP, which deletes entries from these tables. Based on the SAP Landscape Transformation Replication Server log entries, the report identifies CO documents that transferred via SAP Landscape Transformation Replication Server. In addition, the system determines the retention period from the **Keep Periods** field in table VCFIN_SOURCE_SET. Upon calling up the report, you'll arrive at the **CO and Commitment Transfer Data Cleanup Tool** screen, as shown in Figure 8.48. On this screen, choose the **Cleanup CO Preparation Data** ❶ checkbox. Next, enter values in the **Company Codes** ❷ and **Package Size** ❸ fields. Click the **Execute** ❹ button. The report deletes entries from tables CFIN_COPA and CFIN_CO_ADD for the selected company code.

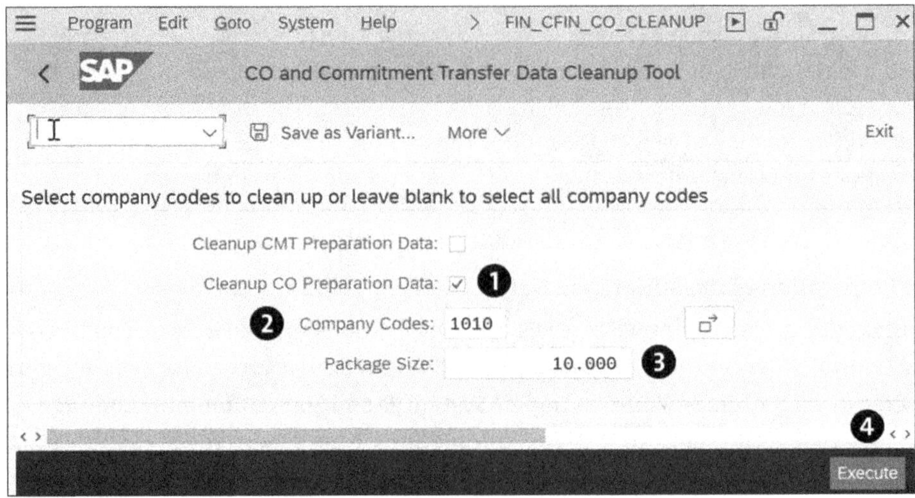

Figure 8.48 CO and Commitment Cleanup

Note

Refer to SAP Note 295532 for frequently asked questions on CO replication.

8.7 Initial Load of Commitments

Purchase commitments refer to liabilities or obligations when a company procures goods or services on purchase orders or purchase contracts. Commitments must be tracked accurately to reflect correct general ledger account balances based on accrual accounting. In addition, the commitments are used in financial planning, budgeting, and reporting to assist managers with monitoring actual spending against budgets. SAP released a new commitment management solution in SAP S/4HANA 1809 release, which is integrated with the Universal Journal (table ACDOCA) for predictive accounting. This requires a predictive ledger maintained in Central Finance. On the other hand, classic commitment management stores data in tables COOI and FINS_CFINT_CMTRL. Central Finance supports real cost object orders, cost centers, WBS/projects, and network and sales orders for commitments. Figure 8.49 illustrates replication of the commitment process in Central Finance. Let's understand how commitment data flows from the source system to Central Finance:

❶ **Purchase processes**

This begins with creating purchase requisitions or purchase orders by identifying a need for goods and services, and it ends with paying suppliers for the goods and services received. SAP supports reference procedure AWTYP-PREQ for purchase requisitions and AWTYP-PORD for purchase orders.

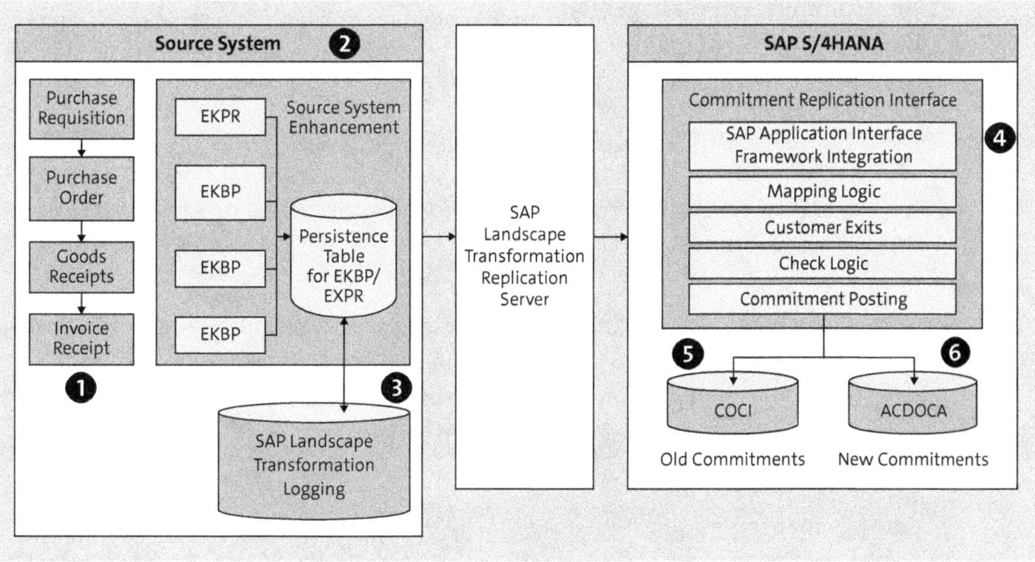

Figure 8.49 Replication of Commitments Process

- ❷ **Commitment updates**
  During execution of purchase processes, commitment management involves creation, updating, and closing commitments as goods or services received. As a result, commitment posting is updated in tables EKPR and EKBP. Commitment triggers during the following:
  - New commitments created through purchase requisitions or purchase orders
  - Commitment updates when changes are made to quantities, prices, indicators, and so on
  - Good receipts or services received
  - Invoice receipts
  - Reversals of goods or services receipts or invoice receipts (reversal of GR/IR)
- ❸ **SAP Landscape Transformation Replication Server**
  The initial load and online replication performed through SAP Landscape Transformation Replication Server.
- ❹ **Application interface framework/mapping**
  During transfer of commitments data, the system uses SAP Application Interface Framework and SAP Master Data Governance mapping prior to posting commitments in Central Finance.
- ❺ **Old commitments**
  During commitments posting, the system posts to tables COOI and FINS_CFINT_CMTRL for old commitments in the Central Finance system.

❻ **New commitment management**
If new commitment management is configured in the Central Finance system, then the system posts to the Universal Journal (table ACDOCA) as well.

Tip
Consult SAP Note 3281627 for answers to frequently asked questions. SAP Note 2549128 provides further guidance on CO commitment replication.

Now that you understand commitment data flow in Central Finance, let's discuss configuration requirements to enable the commitment initial load and replication functionality. The configuration is divided into seven sections for commitment management replication in the order mentioned here:

1. Source system requirements
2. Central Finance configuration
3. SAP Landscape Transformation configuration
4. Reset commitments
5. Replicate individual commitments
6. Delete replicated commitment documents
7. Clean up commitments

Let's explore the configuration for each system with step-by-step instructions.

8.7.1 Source Configuration

The commitments initial load uses **General Replication Settings** maintained in view VCFIN_SOURCE_SET. The company code in scope must be maintained in this view. The system uses this view to load the open commitments initial load and open commitments of purchase requisitions or purchase orders for the company maintained.

SAP doesn't support commitments replication when the source system is set up with a third-party database with a runtime database license.

8.7.2 Central Finance Configuration

Let's begin with the Central Finance configuration activities first. We'll describe these in the following sections.

Business Configuration Set for Commitments

As a first step, you'll install BC sets for commitment replication in the Central Finance system. To apply the BC sets for commitment replication, navigate to **SAP Easy Access** menu path **Tools • Customizing • Business Configuration Sets • Display and Maintain**

BC Sets or use Transaction SCPR20. Upon entering the transaction, you'll arrive at the **Business Configuration Sets: Activation** screen, as shown in Figure 8.50. On this screen, enter value "FINS_CFIN_AIF_CMT" ❶ in **BC Set**, and click **Enter** ❷. Click the **Activation** ❸ icon from the submenu. The system opens the **Activations Option** window to record transport information. Leave default values on this screen, and then click the **Continue** button. Upon activation, the system displays activation logs ❹. If the activation is successful, the system enables SAP Application Interface Framework content for interfaces `CMT_SIM` and `CMT_DOC` version `1` under namespace `/FINCF` in SAP Application Interface Framework.

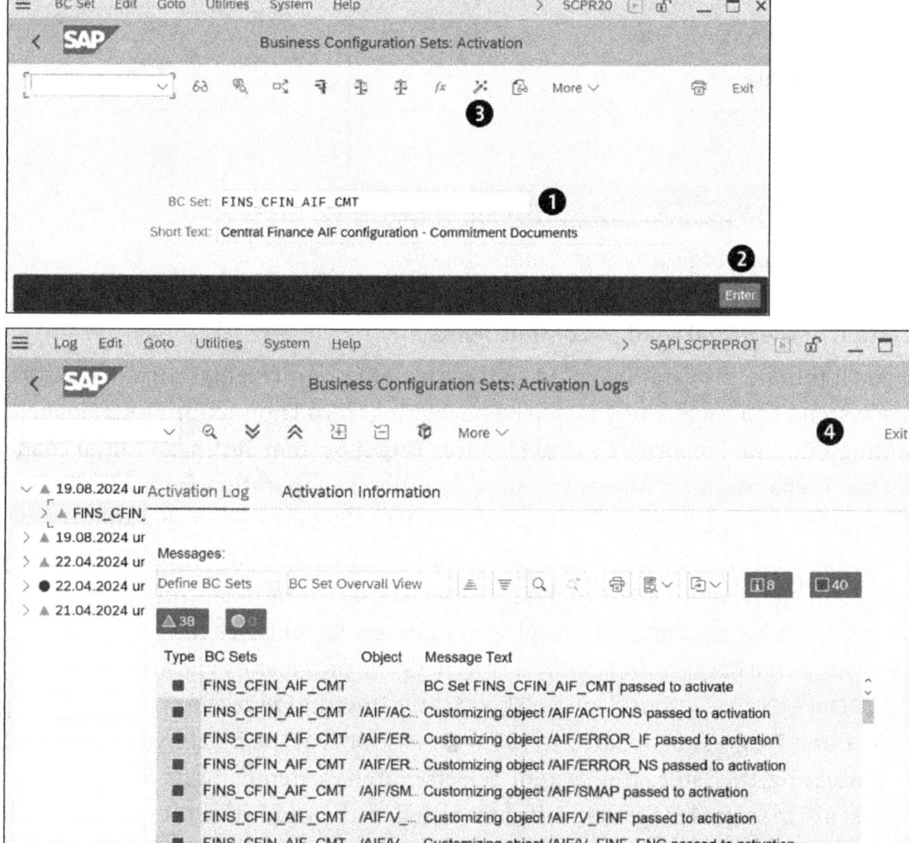

Figure 8.50 BC Set for Commitment Replication

[+]

Tip

Commitment version 2 is recommended as several bugs were resolved in version 2. Refer to SAP Note 2973121 on how to enable the version 2 interface for commitment load/replication and simulation.

Initial Load Logical System

To carry out the commitments initial load, you need to maintain the source logical system in Central Finance. To maintain the source logical system, navigate from Transaction SPRO **Financial Accounting • Central Finance • Central Finance: Target System Settings • Initial Load • Initial Load Settings • Choose Logical System**. After entering the transaction, click the **New Entries** button to maintain a new logical system entry. As shown in Figure 8.51, **Logical system** and **Size** are maintained.

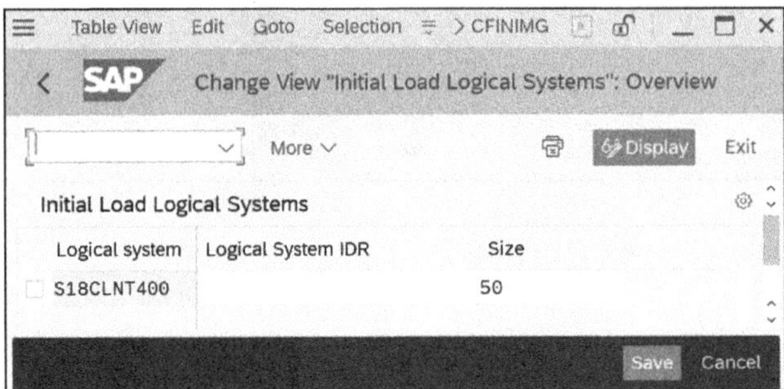

Figure 8.51 Initial Load Logical System: Commitments

Preparation of the Initial Load of Commitments

Next, you'll initiate preparation of an initial load of commitments from the Central Finance system. You'll execute this step by navigating from Transaction SPRO **Financial Accounting • Central Finance • Central Finance: Target System Settings • Initial Load • Initial Load Preparation for Management Accounting • Preparation for Initial Load of Commitments** or using Transaction CFIN_CMT_INT_PREP. After entering the transaction, the system displays the **Preparation of Initial Load of Commitment: Summary** screen, as shown in Figure 8.52. **Progress** is shown as **0%** prior to starting the preparation step. On this screen, choose the logical system row ❶, and click the **Execute** ❷ button. This will start a background job. You can check the progress by clicking the **Refresh** ❸ button. Once the background job finishes, **Status Description** of the preparation step is **Finished**, and **Progress** is updated to **100%** ❹, as shown at the bottom of Figure 8.52. While processing this step, the system selects purchase requisitions and purchase orders that are in scope for the commitment initial load. In addition, the system builds global transfer structures `CFIN_CMT_H` and `CFIN_CMT_I`, which are later used in SAP Landscape Transformation Replication Server to perform the commitment initial load. Note, if interface version 2 functionality is enabled, then source staging tables `CFIN_CMT_HT` and `CFIN_CMT_IT` will transfer the data from the source system to the Central Finance system through SAP Landscape Transformation.

From Figure 8.52, click on the underlined **Logical System** field, which will navigate to the **Preparation of Initial Load of Commitments: Detail** screen. On this screen, you can see the company code with progress as you've seen in the summary screen. As shown in Figure 8.53, upon successful completion of the preparation of commitments step, the system updates entries in tables CFIN_MIG_LOG_CMT ❶ and CFIN_JOB_LOG_CMT ❷ in the source system.

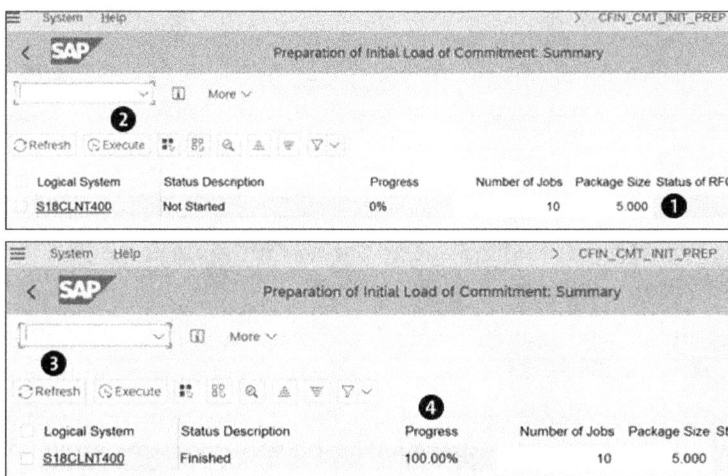

Figure 8.52 Preparation of the Initial Load of Commitments

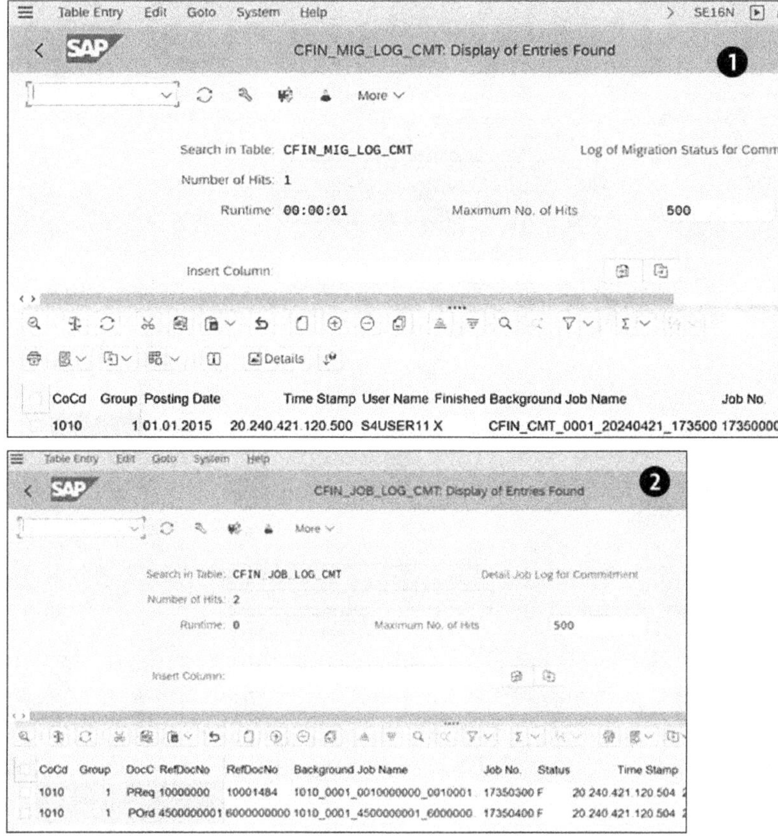

Figure 8.53 Log Tables for Commitments

[+]

Tip

With report FINS_CFIN_CMT_FILTER_ACTIVATE in Central Finance, you can now set filter purchase requisitions and purchase orders to exclude noncommitment line items replicated to Central Finance. To enable the report, consult SAP Notes 3140327 and 3298171 for the Central Finance system and the source system, respectively. One restriction of activating filters is that you can't deactivate them once activated.

8.7.3 SAP Landscape Transformation Replication Server Configuration Requirements for Commitments

For the commitments initial load and replication, you need to define replication objects for table CFIN_CMT_H (interface CMT_DOC version 1) or table CFIN_CMT_HT (interface CMT_DOC version 2). Table 8.17 provides a list of load and replication objects you need to create in SAP Landscape Transformation Replication Server for commitment replication. These predefined load and replication objects are part of project REPL_CFIN and subproject REPL_CFIN. Before you start the actual load and replication of commitments, the system provides an option to perform the simulation of commitments.

Type of Replication	Name	Comment
Predefined simulation load object	CFI_CFIN_CMT_SIM	Simulation of commitment
Predefined load object	CFI_CFIN_CMT_H_L	Transfer of commitment
Predefined replication object	CFI_CFIN_CMT_H_R	Replication of commitment
Predefined simulation load object	CFI_CFIN_CMT2_SM	Simulation of commitment version 2
Predefined load object	CFI_CFIN_CMT2_HL	Transfer of commitment version 2
Predefined replication object	CFI_CFIN_CMT2_HR	Replication of commitment version 2

Table 8.17 SAP Landscape Transformation Replication Server Predefined and Load and Replication Objects for Commitments

8.7.4 Reset Commitment Preparation in Source System

SAP provides report FIN_CFIN_CMT_RESET to reset the commitment preparation data in the source system. This report may be used in mock cycles when you want to reset and restart new commitment replications. This tool can delete relevant commitment data at the company code level. Upon reaching the configuration, you'll arrive at the

Reset Tool for Commitment Replication screen, as shown in Figure 8.54. On this screen, enter a value in the **Company Code** ❶ field, and click the **Execute** ❷ button. You'll then see the output screen ❸ after the report execution. Once the report is executed, the system deletes data from staging tables CFIN_CMT_H and CFIN_CMT_I in case of version 1 or from staging tables CFIN_CMT_HT and CFIN_CMT_IT in case of version 2 of commitment depending on your circumstances. In addition, the system deletes data from tables CFIN_MIG_LOG_CMT and CFIN_JOB_LOG_CMT in the source system.

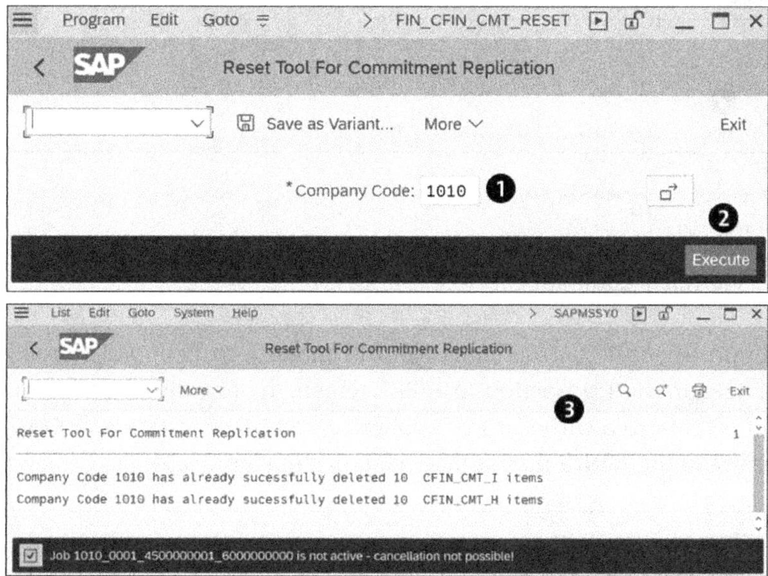

Figure 8.54 Reset Commitment Preparation in the Source System

8.7.5 Replicate Individual Commitments

If you want to replication individual commitments from the system to Central Finance, SAP provides Transaction CFIN_CMT_DOC_REPL (report FINS_CFIN_CMT_REPL) to fulfill this requirement. After reaching the transaction, you'll arrive at the **Replicate Individual Commitment** screen, as shown in Figure 8.55. Next, enter the selection criteria values ❶ in **Logical System** for the source system, which is a mandatory field on this screen. Enter a value in **Purchase Requisition Number in Source System** or **Purchasing Order Number in Source System** to replicate the associated commitments from the source system to the Central Finance system. Click the **Execute** ❷ button. The system displays **Pur. Doc.** and **Replication Status** showing **Not Replicated** based on the selection criteria. Highlight the row, and click the **Simulation** button to check for any errors prior to posting. After a satisfactory **Simulation** step, click **Posting** ❸ to replicate commitments associated with the purchase order.

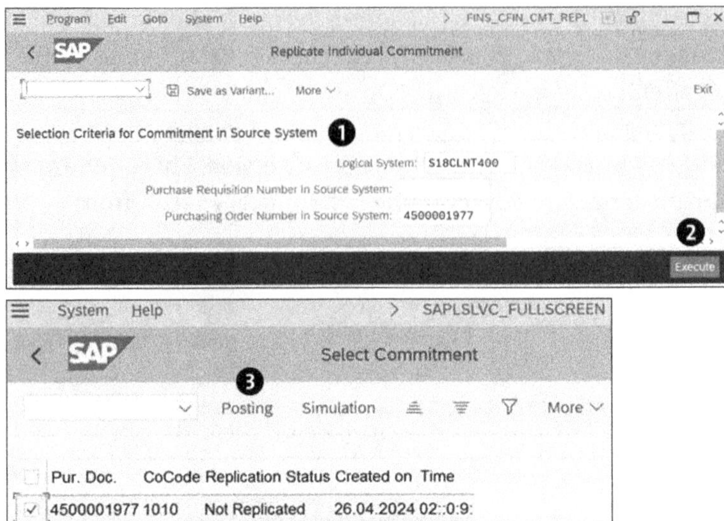

Figure 8.55 Replicate Individual Commitments

Once posting of commitment is successful, the commitment document will be replicated, and it can be seen in the Transaction /AIF/ERR screen. In our example, the commitment document is replicated with the SAP Application Interface Framework **CMT_DOC** version **1** interface of the **/FINCF** namespace, as shown in Figure 8.56.

Figure 8.56 Commitment Replication in SAP Application Interface Framework

8.7.6 Delete Replicated Commitments Documents

SAP provides Transaction CFIN_CMT_DO_DEL (report FINS_CFIN_CMT_DOC_DEL) to delete the replicated commitment documents for specific purchase orders or purchase requisitions in the Central Finance system. After entering the transaction, you'll arrive at the **Delete Replicated Commitment Documents** screen, as shown in Figure 8.57. Next, enter the selection criteria values ❶ in the **Logical System** field of the source system, which is a mandatory field on this screen. Enter a value in **Purchase Requisition Number** or **Purchase Order Number** to delete the associated commitments from the Central

Finance system. Click the **Execute** ❷ button. Based on the selection criteria, on the resulting screen ❸, the system displays the values in the **Reference Document Category** and **Reference Document Number** columns showing commitment documents replicated already in the Central Finance system. Highlight the row and click the **Delete** button to delete the commitment document. The system displays the **Log Display** window showing whether the deletion was successful.

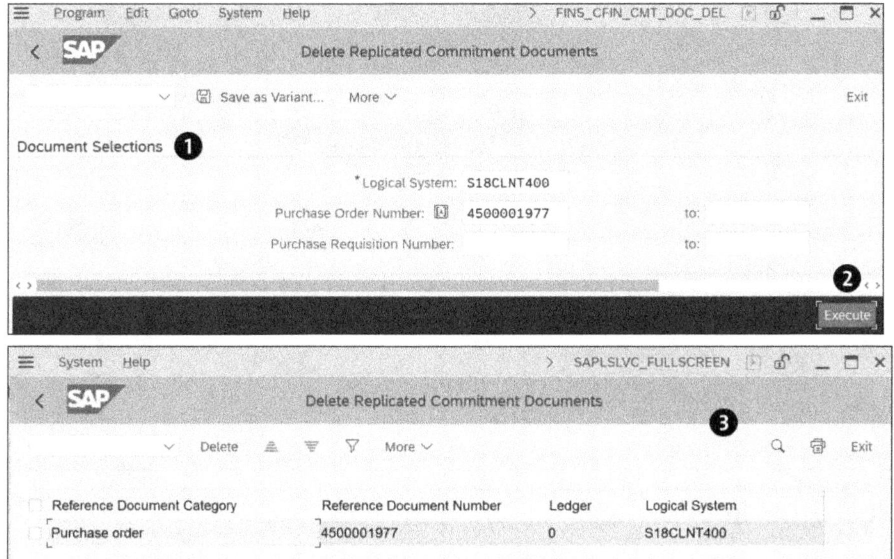

Figure 8.57 Delete Replicated Commitment Document

8.7.7 Clean Up Commitments Report

With cleanup report FINS_CFIN_CMT_DOC_CLEANUP, the system allows you to delete the replicated and posted commitment information of a purchase requisition or purchase order in the Central Finance system. This is especially useful when used as part of housekeeping to clean up entries. In addition, the report enables you to delete the posted commitment documents in bulk with the controlling area and company code as selection criteria. After reaching the report, you'll arrive at the **Delete Replicated Commitment Documents** screen, as shown at top of Figure 8.58. Enter the selection criteria values ❶ in the **Logical System of Source System**, **Controlling Area**, and **Company Code** fields. Enter the value in the **Number of Parallel Jobs** field as needed. Click the **Execute** ❷ button. Based on the selection criteria, on the resulting screen ❸, the system displays whether the job is scheduled. The system deletes entries from tables `COOI`, `FINS_CMT_CTRL`, and `ACDOCA` (if a new commitment is active). In addition, the system uses entries in table `FINS_CMT_CTRL` as a reference to delete commitment data through this report.

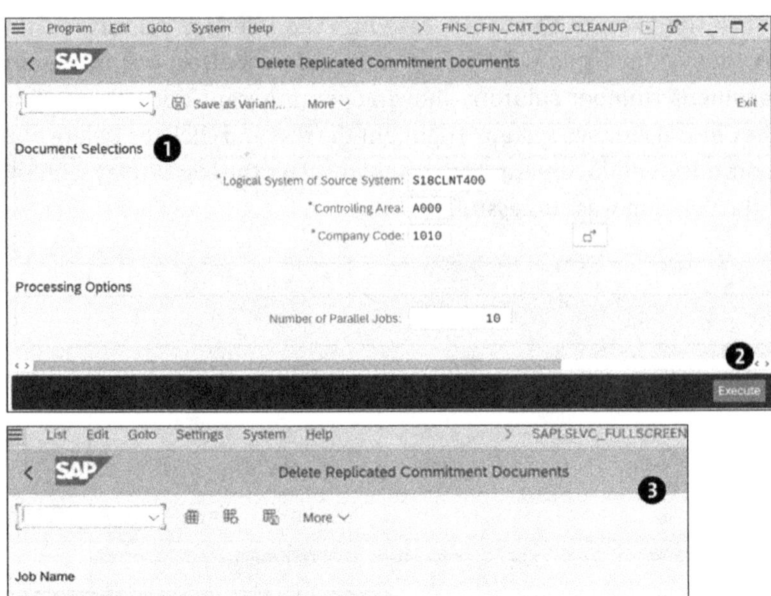

Figure 8.58 Delete Replicated Commitment Documents

8.8 Real-Time Replication

In this section, we'll explore a few important real-time replication-related topics for FI.

8.8.1 Change Documents

Central Finance supports replication of changes made to FI postings in the source system to Central Finance. When an FI document changes on the header or item level in the source system, the system transfers changes to the Central Finance system. Central Finance provides SAP Application Interface Framework **AC_DOC_CHG** versions **2** and **10** (**bfRFC**) interfaces under Central Finance namespace **/FINCF** used for monitoring FI-related changes. Central Finance supports most changes made to financial documents. Table 8.18 provides the supported field changes made to financial documents for SAP-based source systems considering SAP ERP as a source system.

Field	Field Description
BKPF-BKTXT	Header text
BKPF-XBLNR	Reference number
TBD	Document header

Table 8.18 SAP-Based Source System

Field	Field Description
BSEG-SGTXT	Line-item text
BSEG-ZUONR	Assignment
BSEG-VBUND	Trading partner
BSEG-ZTERM	Payment terms
BSEG-ZFBDT	Baseline date
BSEG-MABER	Dunning area
BSEG-FWBAS/HWBAS	Tax base

Table 8.18 SAP-Based Source System (Cont.)

When a third-party system is configured as a source system, SAP supports changes made to financial documents such as general ledger documents, AR documents, and AP documents. The third-party interface also supports changes made to reversals, cleared documents, and withholding tax information.

Tip

If you observe document changes that aren't correctly transferred, refer to source system SAP Note 2147776 for further information.

SAP provides BAdI BADI_FINS_CFIN_AC_DOC_CHG_INT to enhance above and beyond the standard FI document change replication.

8.8.2 Costing-Based Profitability Analysis Characteristics

If source systems are implemented with costing-based profitability analysis and the Central Finance system is configured with margin analysis, Central Finance transfers costing-based profitability analysis characteristics and cost components splits to the Universal Journal along with FI or CO postings replication. The source system FI document line items characteristics posted to a profitability segment (PAOBJNR) and transferred to staging table CFIN_ACCPA_CHAR are then replicated to a profitability segment in Central Finance. If no costing-based profitability analysis is active in the source system, profitability segment derives information from the Central Finance ACCIT structure based on profitability analysis derivation rules and then persists that information in table ACDOCA. If source postings don't contain a profitability segment, for example, in the case of goods delivery, characteristics may not be transferred to Central Finance. For full coverage of margin analysis, refer to Chapter 10, Section 10.8.

Note

SAP provides source system BAdIs to implement custom logic to transfer customer-specific costing-based profitability analysis data to Central Finance. Refer to SAP Note 2543899, which provides source system BAdIs to fill in missing characteristics in line items.

8.8.3 Clearing Information

If clearing transfer is enabled, the system transfers the clearing status, which is a prerequisite for central payment activation. On the other hand, if clearing transfer isn't activated in the Central Finance system, the system sets the **Technically Cleared** (`ALE-Extern`) status on clearing documents and all open items for the initial load and online replication. In addition, you'll be able to replicate changes made to ordinary cleared items and technically cleared items (based on release) to Central Finance. For historical items in lower releases, when Transaction FINS_MIG_CJ3 is run in Central Finance, the system reopens open items if the corresponding open items are open in the source system. If clearing transfer is activated prior to the initial load extraction, execution of Transaction FINS_MIG_CJ3 is no longer needed because the status of the initial load open items reflects a clearing transfer.

8.8.4 Replicating Withholding Tax

Withholding taxes allows a buyer to withhold a certain amount from the payment made to a supplier. This is required in certain countries to comply with local and regulatory requirements. In general, withholding tax is calculated as a percentage of the net or base amount. In SAP, withholding tax code is captured in the invoice, and the system calculates the withholding tax based on a predefined percentage and base amount. In addition, withheld tax can be recorded in a separate tax general ledger account. For Central Finance, extended withholding tax replication has been supported since the SAP S/4HANA 1709 release by applying relevant SAP Notes. Later, SAP supported extended withholding tax officially from SAP S/4HANA 1809 FPS0 onward. Classic withholding tax isn't supported, so source systems running with classic withholding taxes must be migrated to extended withholding taxes. Table 8.19 shows SAP-supported withholding taxes online replication scenarios. For replication of withholding taxes, SAP recommends enabling withholding tax tables available in Central Finance to carry out the checks, as discussed in Chapter 5, Section 5.2.2.

Scenario	Central Finance Support
Non-central payment scenario: Payment is posted in the source system.	■ Invoice includes withholding tax code that is posted in the source system and supported as part of FI document replication. ■ Payment data along with withholding tax data is also replicated to the Central Finance system. ■ Reversal of invoice or payment in the source system is transferred to Central Finance and posts reversals in the Central Finance system. ■ Withholding tax reporting is performed in the source system. ■ Accumulated withholding tax is carried out in the source system for both invoices and payments. This information isn't available in the Central Finance system.
Central payment scenario: All payments are processed in Central Finance.	■ The system calculates and posts invoices from Central Finance only. ■ Reversals of invoices are allowed based on clearing status. ■ After central payment activation, all withholding tax reporting is carried out in the Central Finance system. ■ Accumulated withholding tax is carried out in the source system for invoices as this information isn't available in Central Finance. On the other hand, when payments are processed, accumulated withholding tax information is available in the Central Finance system after activation of central payment at the beginning of the month or year. SAP recommends activating central payment at the beginning of the period or year rather than activating mid-month or mid-year to avoid missing information or no information for accumulated withholding tax.
Changes to withholding taxes	■ Withholding tax changes in the source system, for example, the withholding tax base amount, are transferred as part of change document replication.

Table 8.19 Central Finance Withholding Supported Scenarios

Tip

Consult SAP Note 2623514 to understand prerequisites, functionality delivered, restrictions, and SAP-delivered documentation.

8.8.5 Deferred Tax

Companies measure and recognize deferred taxes, which are future obligations to pay taxes. In an accounting context, when business transactions are recorded, temporary and permanent differences arise. Temporary differences arise when a company is following one method of accounting for books, but another method of accounting on tax returns. This will result in two types of temporary differences:

- Revenue expense items that belong on the income statement now, but won't appear on tax returns until later
- Revenue and expense items that belong on the tax return now, but won't appear on the income statement until later

These situations lead to recording future taxable and future deductible deferred amounts in the book to reflect on financial statements.

On the other hand, permanent differences never reverse themselves. In SAP, deferred taxes must be accrued and posted on a separate deferred tax general ledger account. In the Central Finance context, deferred tax functionality was initially released in SAP S/4HANA 1900 FPS1, but additional scenarios were released in SAP S/4HANA 2020. From an online replication standpoint, Central Finance supports deferred tax replication of deferred tax for invoices, clearing documents, and tax transfer documents.

8.8.6 Specific Ledger Group Postings

Upon replication, FI postings from the source system post to all ledgers defined in the Central Finance system. However, if the source system contains FI postings that are specific to a ledger group, the system replicates postings to the corresponding ledger group in Central Finance. This is the standard behavior regardless of the number of ledgers between the source system and the Central Finance system.

8.8.7 Manage Pending Journal and Manage Temporary Postings Apps

During the period close to the Central Finance system, many documents won't be posted due to master data or configurations errors. Often, these errors couldn't be fixed on time, which could delay the period-end close process. As part of the period-end close process, business users analyze the financial impact before posting any adjustment entries in the Central Finance system or wait for the SAP Application Interface Framework error resolution. To overcome these issues and support period-end activities in Central Finance, if an SAP Application Interface Framework error can't be resolved on time, SAP provides the following SAP Fiori apps that post temporary journal entries:

- **Manage Temporary Postings (App ID F6062)**
  For temporary postings and reversal solutions.

- **Manage Pending Journal Entries (App ID F5482)**
  To display journal entries that contain errors.
- **Manage Pending Journal Entries - Message View (App ID F7149)**
  To display all pending journal entries, including financial data.

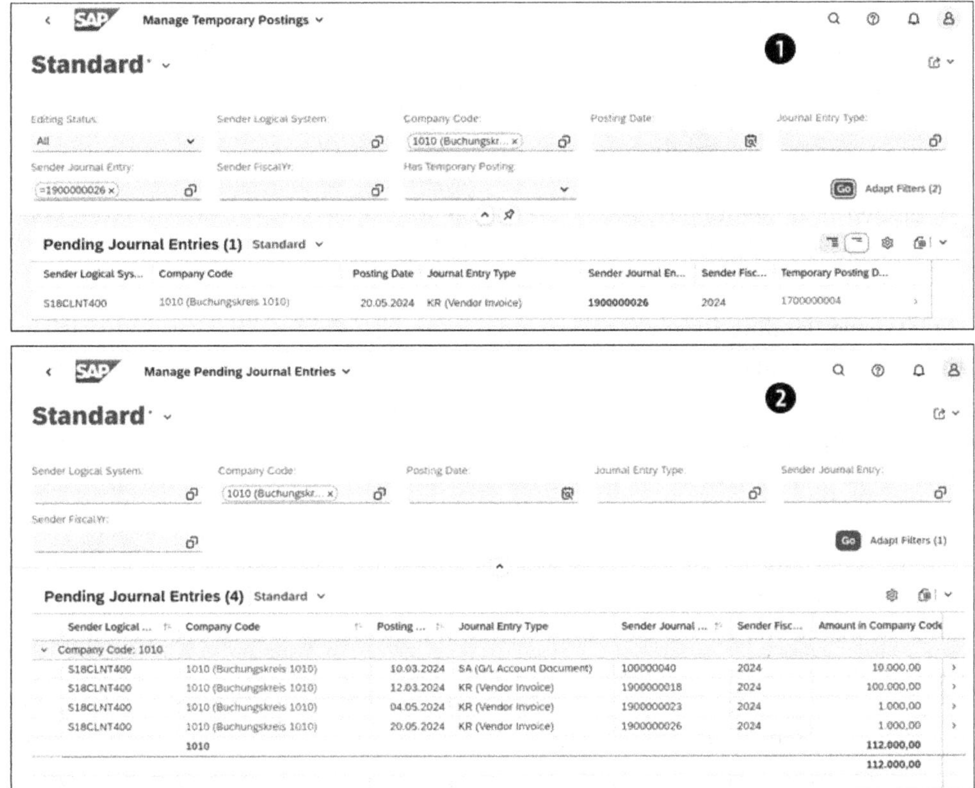

Figure 8.59 Manage Temporary Postings App and Manage Pending Journal Entries App

With the Manage Temporary Postings app shown in Figure 8.59 ❶, you can post temporary journal entries for documents stuck in SAP Application Interface Framework with errors that haven't been resolved on time. The app displays pending journal entries from SAP Application Interface Framework. You can create temporary journal entries, including references using pending items from SAP Application Interface Framework. These entries are outstanding or pending journal entries that require further troubleshooting effort or action by a specific team. However, the journal entries are required from an accrual accounting standpoint before the close and can't wait for a resolution on time. Upon entering a temporary posting, the system updates table `CFIN_TMP_IT`. The system achieves temporary posting functionality with two configurations: document types and reversal/payment block reason codes. If the SAP Application Interface Framework errors are resolved and posted, the system automatically reverses the temporary posting, thereby ensuring the correct information is reflected in reporting. Alternatively, the system provides you an option to reverse a temporary journal manually. When temporary postings are processed or cancelled successfully, the system removes the entries in table CFIN_TMP_IT automatically. In addition, temporary documents posted with open item–managed general ledger accounts must be cleared with the Manage Temporary Postings app. Journal entries can be exported to a Microsoft Excel spreadsheet. The Central Finance comparison reports for FI provide a checkbox to include temporary postings.

Tip

If discrepancies are found, you can check table for status CFIN_AIF_IDX_TBL with key MSGGUID which is also referenced with the same key in table CFIN_TMP_IT.

The Manage Pending Journal Entries app (Figure 8.59 ❷) also displays journals that aren't yet posted in the system. This app gives users an opportunity to review and take the necessary action prior to posting. The app has a filter capability to include or exclude temporary postings. Moreover, you can navigate to SAP Application Interface Framework messages to view pending journal entries.

As you see from Figure 8.60, you can check the temporary postings in SAP Application Interface Framework interface **AC_DOC_TMP**. Once the original FI document is posted successfully in SAP Application Interface Framework, the system posts two SAP Application Interface Framework messages—temporary postings and in-process messages. The in-process message was created for reversal of temporary posting. Once the original accounting document is posted successfully, the in-process SAP Application Interface Framework message (reversal SAP Application Interface Framework GUID) will be reprocessed and posted successfully. The system allows you to manually reverse a temporary posting if automatic reversal doesn't succeed.

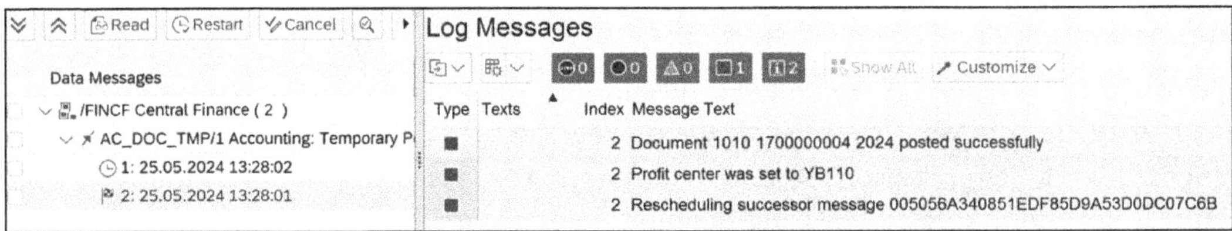

Figure 8.60 Temporary Postings on SAP Application Interface Framework

Figure 8.61 displays the initial screen of the Manage Pending Journal Entries – Message View app, which displays all pending journal entries errors along with financial data. The header screen provides a dashboard count of error messages. You can navigate to pending journal entries from this app.

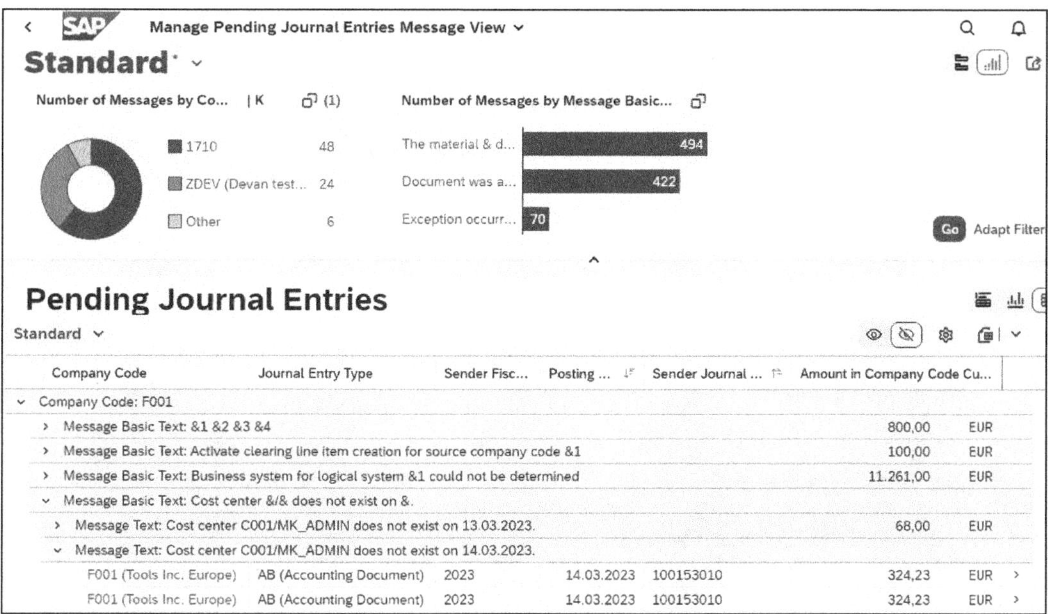

Figure 8.61 Manage Pending Journal Entries - Message View App

Table 8.20 shows two SAP-provided configurations that are used to support temporary postings.

Customizing Activity	Central Finance Customizing Menu Path (Transaction CFINIMG)	Purpose
Maintain Document Types for Temporary Postings	**Central Finance: Target System Settings • Temporary Postings • Maintain Document Types for Temporary Postings**	Document types maintained in this setting are used for the temporary postings and the reversal of the temporary postings, as shown in Figure 8.62 ❶. Document types need to be maintained for each account type (**Customer**, **Vendor**, and **Other**). SAP recommends setting up a unique document type to identify the temporary postings.

Table 8.20 Central Finance Configurations: Temporary Postings

Customizing Activity	Central Finance Customizing Menu Path (Transaction CFINIMG)	Purpose
Maintain Reversal and Payment Block Reasons for Temporary Postings	**Central Finance: Target System Settings • Temporary Postings • Maintain Reversal and Payment Block Reasons for Temporary Postings**	Maintain reversal reason for automatic and manual reversals for the temporary postings, as shown in Figure 8.62 ❷. In addition, maintain the payment block reasons for the temporary postings.

Table 8.20 Central Finance Configurations: Temporary Postings (Cont.)

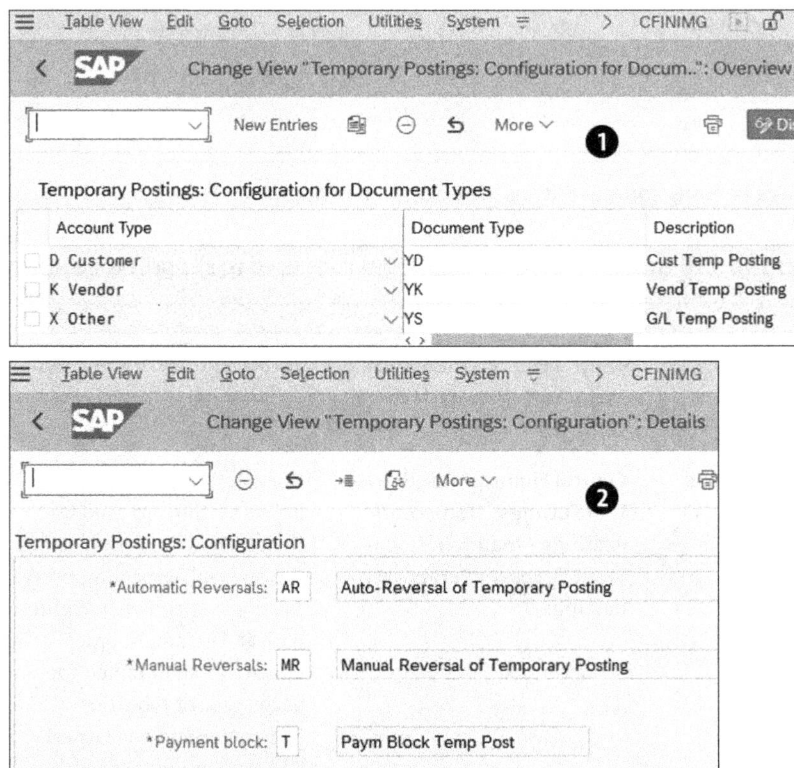

Figure 8.62 Temporary Postings Configuration

[+]

Tip

If the central payments and payment block are active, the system excludes those items for temporary posting. SAP recommends excluding central follow-on process-related documents in dunning and collection management. In addition, if open item-managed accounts are used in the temporary posting, the system doesn't clear or reverse the document. In such cases, you must do it manually.

8.8.8 Cross-Company Code Postings

Central Finance supports replicating cross-company code postings from the source system to Central Finance. For replicating cross-company code postings, SAP recommends reviewing several settings and recommendation as follows:

- All involved company codes transact intercompany transactions are set up in automatic posting and clearing configuration (Transaction OBYA) correctly. If not, only active company codes documents are replicated to Central Finance, which results in reconciliation differences. This is especially important if document splitting is activated in Central Finance.
- A chart of accounts must be set up to support intercompany transactions.
- Intercompany postings (T003-XGSUB) settings on document type configuration must be identical between the source system and Central Finance system. This might lead to the **F5 080 Consolidated companies & and & are different** error in SAP Application Interface Framework. This indicator on the FI document type needs to be enabled at least in the target system to avoid the error.
- If company codes are merged (M:1), this scenario isn't supported out of the box. In such cases, a custom solution must be adopted to prevent cross-company code clearing related errors during online replication.

8.8.9 Foreign Currency Transactions

Replicating foreign currency transactions depends on whether your source system is using the classic general ledger or new general ledger. If the classic general ledger is set up as the source system, foreign currency valuation differences posted with Transaction F.05 (table BSEG-BUZID = U) aren't transferred correctly to Central Finance due to how SAP S/4HANA open items store foreign currency transactions in Central Finance compared to the source system process. This is due to the logic of how foreign currency differences can be posted. In the classic general ledger, foreign currency differences can be posted to open items (table BSEG-BDIF*), whereas in SAP S/4HANA, foreign currency valuation differences are posted to an adjustment account separately. Likewise, reversal of posted foreign currency transactions may contain fewer numbers of lines in Central Finance when compared to the corresponding document in the source system. These transactions may result in reconciliation of differences. If Central Finance is the book of record, SAP recommends foreign current valuation processes run from Central Finance.

8.8.10 Drill-Back to the Source System

Central Finance supports displaying the source system document from the corresponding replicated document into the Universal Journal with the drill-back functionality. This feature enables seamless navigation from Central Finance to the source system without needing to log on separately. As a prerequisite, SAP recommends that the RFC user has been assigned as the current user with the appropriate security access in the source system.

To execute the drill back to a source system document, you can display it using Transaction FB03, Transaction KSB5, or the Manage Journal Entries app from the Central Finance system. Once a document is displayed using Transaction FB03, there are two ways to drill back to the source system. In the first way, go to the document header, and double-click the document number in the **Sender Doc.No** field, as shown in Figure 8.63 ❶. This takes you to the source system document.

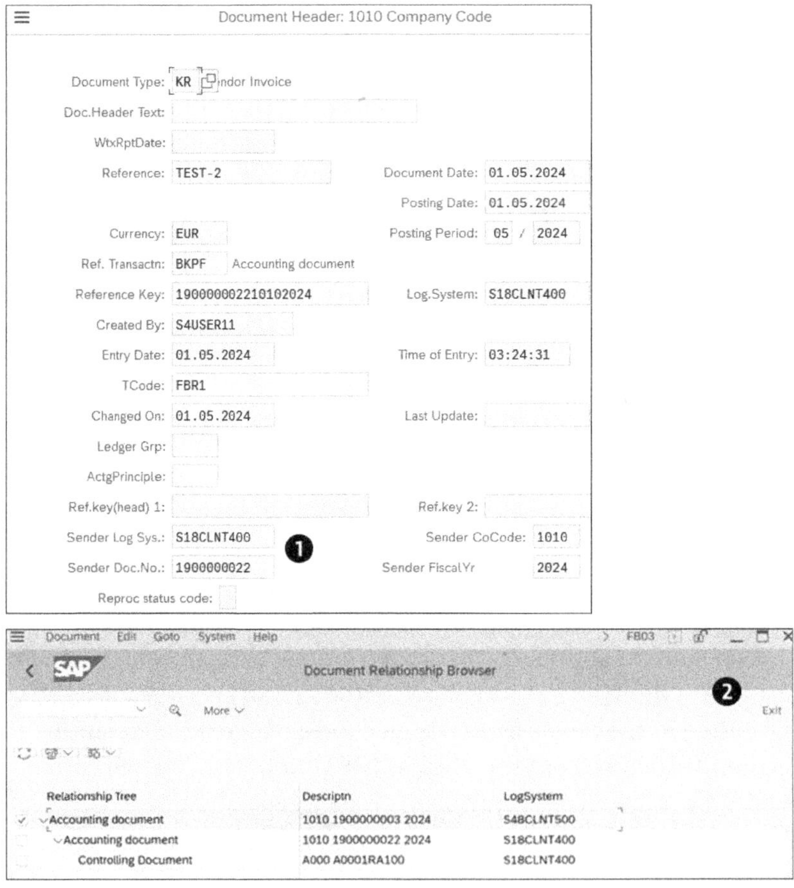

Figure 8.63 Drill Back to the Source System

In the second way, navigate to **Document Relationship Browser** from the menu **Goto • Environment • Document Environment • Relationship Browser**. The system displays documents with their relationship in a hierarchy under the **Relationship Tree** column ❷. This screen also shows the logical system assigned, which helps show where the document is replicated from. In our example, as you see from the figure, document number **1900000003** has the logical system **S48CLNT500**, which represents the logical system of the Central Finance system. On the other hand, the other two documents have the logical system **S18CLNT400**, which were replicated from the source system. When you double-click either the **Accounting document** or **Controlling Document**, the system drills back to the source system and displays the document in its respective transaction, that is, Transaction FB03 (Display Document) or Transaction KSB5 (Display Actual Cost Document).

[+]

Tip

For drill-back to work, you must have configured **Assign RFC destination for Displaying Objects from Source Systems**, as described in Chapter 7, Section 7.4.2, with the source system's RFC destination for standard dialog calls. In addition, RFC connection configuration is setup, and RFC user logon information is maintained in the **Security** tab of Transaction SM59.

8.8.11 Document Flow Analyzer

Central Finance provides the Transaction FINS_CFIN_DFA (report FINS_CFIN_DOCFLOW_ANALYZER), which displays the document flow of all related documents by inputting either the source document or Central Finance document. This transaction is available from the SAP S/4HANA 1909 release onward.

After reaching the transaction initial screen, as shown in Figure 8.64, enter values for the fields under **Selection Option** ❶, which provide options to enter either the **Source Document** number or **Central Finance Document** number to trace the document flow. In our example, we've chosen **Source Document** number. Choose the source **Logical System** under the **Selection for Source System** ❷ section. Next, enter values for **Company Code**, **Document Number**, and **Fiscal Year** under the **Keys for Document Selection** ❸ section. Optionally, you can enter **Cross-Comp Code No** under the **Cross-Company Code Document Selection** ❹ section if applicable. Finally, choose the appropriate radio button under the **Document Flow Analysis Mode** ❺ section. The transaction provides **Upstream**, **Downstream**, **Overall**, and **Include Recent Transactions** options under this section. Choose the **Overall** radio button. Click the **Execute** ❻ button to continue.

On the resulting **Document Flow Analyzer** screen, the output screen provides a split screen showing both documents in the source system ❼ and Central Finance documents ❽ in a tree structure with related documents. On the header section of the screen, the transaction shows several buttons to navigate to further details. When you click the **Display Items** ❾ button, the system navigates you to the top of Figure 8.65. When you click **Compare Items** ❿, the system navigates you to the bottom of Figure 8.65.

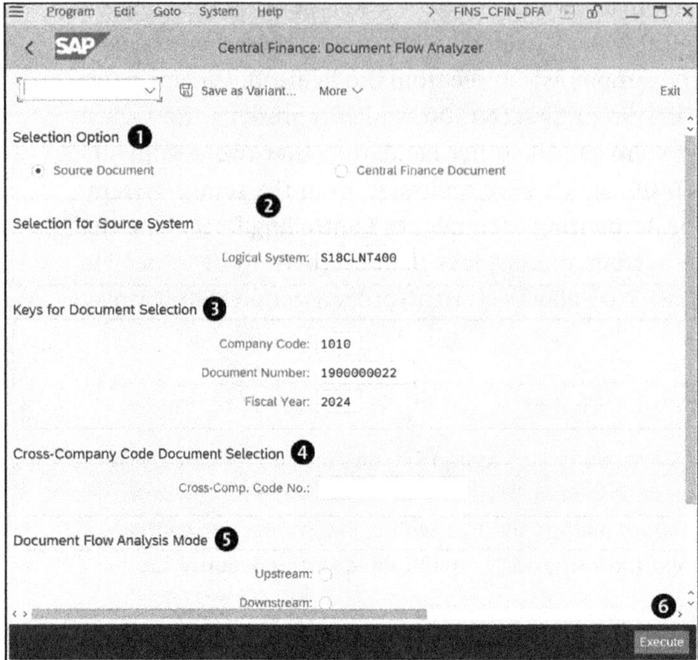

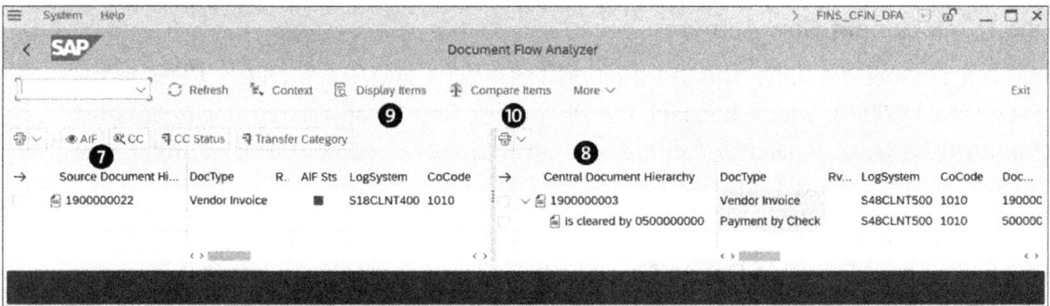

Figure 8.64 Document Flow Analyzer

When the **Display Items** button is chosen, transactions are displaced at the bottom of the screen with full details. As you can see, the succeeding clearing document is also shown. In addition, source document **1900000022** is cleared technically, and the clearing document is set with value `ALE-extern` as central payment is activated in this system. Corresponding Central Finance document **1900000003** is also cleared by **0500000000** in the Central Finance system on the right side of the screen, as shown in Figure 8.65 ❶.

When the **Compare Items** button is chosen, both source system and Central Finance fields are displayed side-by-side for comparison ❷. You can change the layout to show the most important fields as necessary.

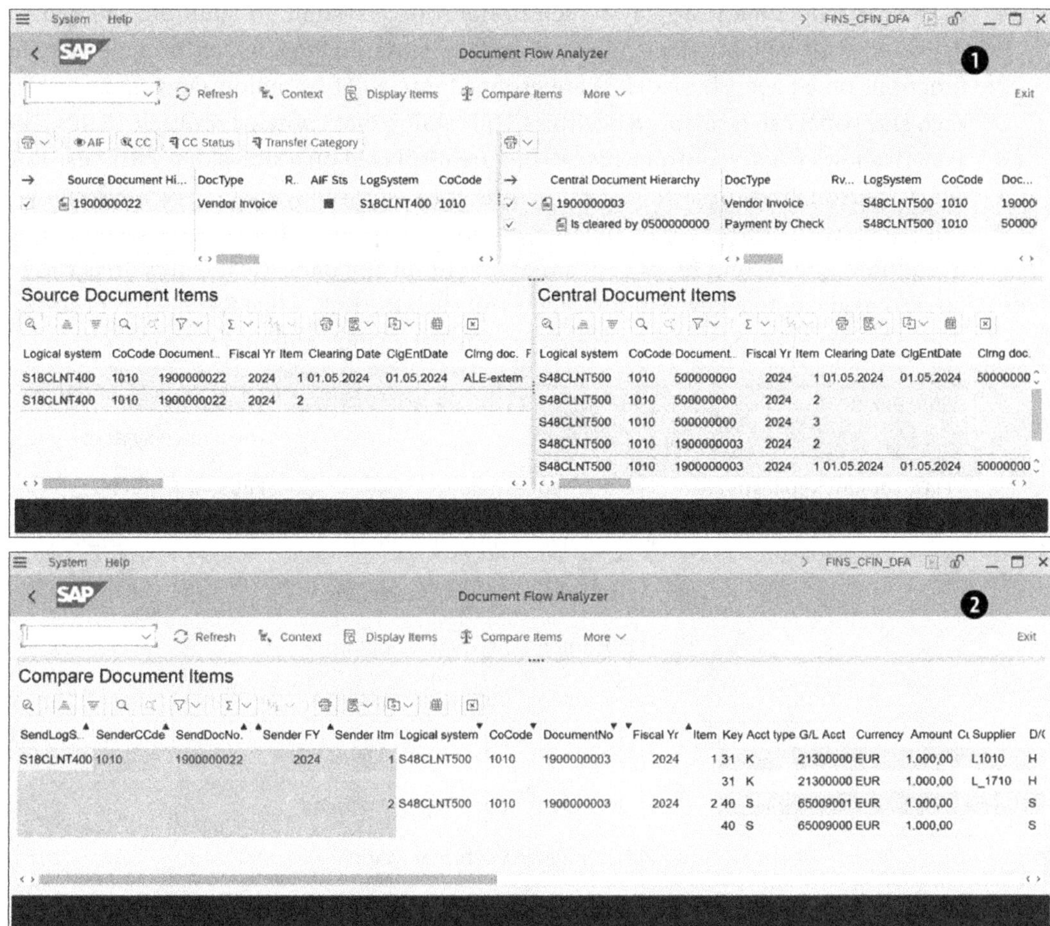

Figure 8.65 Document Flow Analyzer: Display and Compare Items

[+]

Tip

If the transaction is missing in your system, consult SAP Note 3152812 to apply missing SAP Notes to enable this transaction.

8.8.12 Side Panel

Like the drill-back functionality where you can drill back to a document from Central Finance to the source system document, SAP provides the side panel functionality, which allows users to navigate to Central Finance documents from the source system to Central Finance. The side panel, based on the Web Dynpro ABAP page builder, provides additional contextual information related to Central Finance documents next to the main document display screen in the source system. In addition, side panels delivered with collaborative human interface parts (CHIPs), which provide specific functions on a page. These CHIPs are grouped into a CHIP Catalog. On a main screen, each side panel can be displayed with a CHIP catalog that contains multiple CHIPs. The system allows you to control which CHIPs are displayed on a side panel with Customizing or personalization settings. Besides this, the system allows you to create your own side panels with desired CHIPs and then combine them with the Central Finance CHIPs. In addition, the system hides CHIPs that don't pull any data. Most CHIPs links can be used to navigate to a new window when you click the link. Table 8.21 displays transactions supported in the source systems for side panel and CHIP functionalities.

Side Panel	CHIP	Transactions in the Source System
FI Document Replicated to Central Finance	FI Document - Header	FB03, MIR4, VF03
	FI Document - Entry View Items	
	FI Document - General Ledger View	
	FI Document - Account Mapping	
	FI Document - Clearing Status	
	FI Document - Cost Center Mapping	
	FI Document - Cost Object Mapping	
	FI Document - Customer Mapping	
	FI Document - Vendor Mapping	
Central Finance Master Data	Master Data - Cost Center	KS03
	Master data - Customer	XD03
	Master Data - General Ledger Account	FS00
	Master Data - Vendor	XK03
Central Finance Credit Management	Master Data - Customer (available on Central Finance master data panel)	VA01, VA02, VA03, XD03
	Credit Management - Profile	
	Credit Management - Segment	

Table 8.21 SAP-Supported Side Panels and CHIPs

The side panels functionality works only with the SAP NetWeaver Business Client (Transaction NWBC), which requires extensive configuration to enable the feature. Table 8.22 provides configuration activities you need to perform to enable the side panel functionality.

Activity	System	Task
SAP NetWeaver setup	SAP Business Client	SAP recommends installing and setting up SAP NetWeaver for desktop version 3.5PL5 or version 4.0 or higher.
Security roles	Source system	Setup and assign roles.
	Central Finance system	SAP recommends two possible ways to create the role to provide access to side panels: ■ Copy template role SAP_SFIN_CFIN_SP_MENU in the Central Finance system to the roles in the source system using RFC connection type 3 from Transaction PFCG. ■ Copy template role SAP_SFIN_CFIN_SP_MENU in a customer namespace starting with "Z" as the prefix. Later, you load the role in the source system. In this case, you also set a list of transactions in the **Folder** option. Finally, assign users to the role who need side panel access in the source system. In addition, add Web Dynpro application configuration authorization objects for the following: ■ FINS_CFIN_SP_ACDOC to use the Central Finance FI Document side panel ■ INS_CFIN_SP_CM to use the Central Finance Credit Management side panel ■ FINS_CFIN_SP_MD to use the Central Finance Master Data side panel Grant access to Transaction FINS_CFIN_SP_AIF to open the SAP Application Interface Framework cockpit from the Central Finance FI Document side panel.
Basic connections	Central Finance	SAP recommends setting up a Single Sign-On (SSO) scenario using client certificates, for example, X.509 certificates, or using logon tickets for user authorization and SSO if used.

Table 8.22 Side Panel Setup

Activity	System	Task
RFC connections	Source system	▪ Maintain an RFC connection of HTTP type 3. ▪ Maintain RFC assignments and settings for source systems.
Enabling side panels	Source system	For all source systems connected to Central Finance, you need to enable SAP GUI scripting (sapgui/user_scripting_set_readonly) with Transaction RZ11.
	Central Finance system	▪ Enable parameters WDSIDEPANELREMOTE-CONSUMER and WDSIDEPANELREMOTEPRO-DUCER in Web Dynpro application WD_GLOBAL_SETTING. ▪ To use other side panels besides Central Finance side panels, SAP recommends activating parameters WDSIDEPANEL-REMOTECONSUMER and WDSIDEPANEL-REMOTEPRODUCER in Web Dynpro application WD_GLOBAL_SETTING in the Central Finance system.
Customizing and personalization	Source system	Assign Customizing access to key users, administrators, and end users with personalization. Customizing access provides the option to change the default settings for all users within a client. End users can override default settings or Customizing settings.

Table 8.22 Side Panel Setup (Cont.)

8.8.13 Back Posting

After financial documents are posted from the source system to Central Finance, write-back capability isn't available. In other words, Central Finance data flow happens from the source system to Central Finance one way, and bidirectional data transfer isn't supported. If source systems, for example, with a third-party system, rely heavily on data to support third-party system processes such as credit management, collection, and dispute management functions, then you can build a custom interface to sync information back to the source systems. If central payment is active, based on project requirements, the clearing information can be sent back to the source system with the custom interface. Alternatively, third-party tools, such as Insightsoftware's SourceConnect, provide sync-back capabilities that can be evaluated based on your requirements.

8.9 Summary

Central Finance offers customers the replication of financial postings from multiple source systems into central SAP S/4HANA systems to achieve centralized consolidated reporting and centralized processes without disturbing existing systems. To realize these objectives, Central Finance FI initial loads are critical tasks required before beginning real-time replication. The initial load of FI postings is a mandatory step that lays the foundation for online replication in Central Finance. In this chapter, you learned about the initial load, initial load sequence, and resetting an initial load. Then, we delved into the step-by-instructions to carry out the initial load for cost objects, the classic FI-CO initial load, IDR, the initial load of CO secondary postings, and the initial load of commitments. We highlighted important performance-related considerations for the classic FI-CO initial load. Then, we transitioned into important topics in real-time replication FI-CO postings. In the next chapter, you'll learn about central payments, which are an important centralized process in Central Finance.

Chapter 9
Central Payment

Several large companies use decentralized payment systems that require significant resources and are often inefficient. As a result, companies build local processes and controls that lack visibility and transparency across enterprise divisions. This leads to additional communications and delays in reporting cash flow on time. With the central payment functionality offered by Central Finance, customers eliminate redundant controls and processes, thereby improving cash flow reporting seamlessly.

In a distributed system landscape, enterprises make or receive payments in an either decentralized or centralized process. This means the payments can be processed from a local decentralized system or from one centralized global location. One major use case of Central Finance is to leverage process centralization, or shared services solutions offered to customers. Central Finance offers the central payments functionality to centralize accounts receivable and accounts payable processes and relevant clearing activities from a single place using the Universal Journal in SAP S/4HANA.

In this chapter, you'll learn about the central payment processes and configuration requirements. Then, we'll discuss the functional processes across different workstreams after implementing central payments.

9.1 Overview of Central Payment

Central Finance evolved over time with several new features. Initially, SAP introduced the central payment functionality in SAP S/4HANA 1709 with pilot SAP Note 2346233 released to only customers who requested it at that time. When the initial central payment was released, SAP listed several restrictions in SAP Note 2827364. Over time, the central payment functionality evolved with additional features with different on-premise SAP S/4HANA releases. Figure 9.1 shows the central payment features released in SAP S/4HANA release cycles.

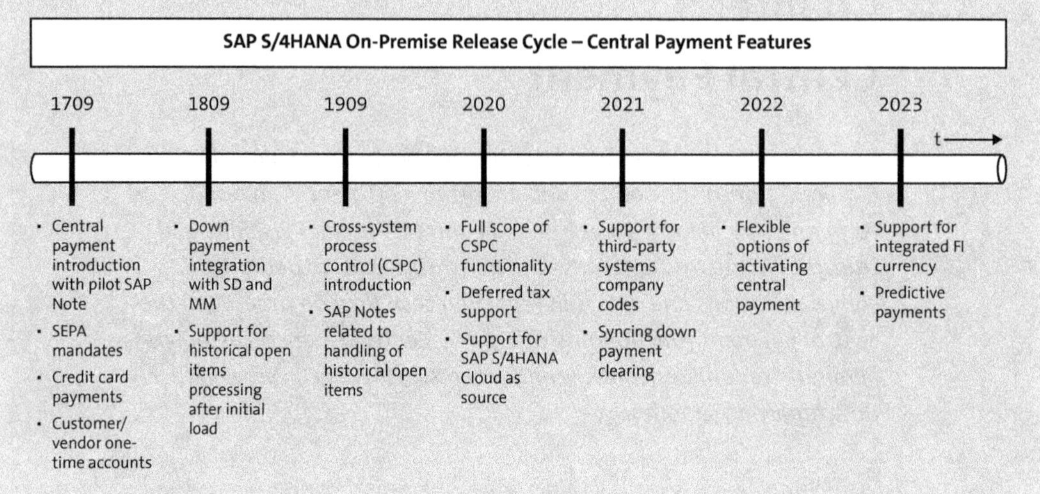

Figure 9.1 Central Payment Functionality by SAP S/4HANA On-Premise Release

Central payment allows companies to manage their payment processes from a single SAP S/4HANA system by replicating customer and vendor open items from different source systems. As a result, you'll see increased financial management, operational efficiency, and control and compliance. Let's explore other important benefits of central payment in Central Finance for customers:

- Improves visibility on cash flow across enterprises as you have better control over cash flow activities
- Standardizes the payment process across all companies because the payment process is run from a single place
- Eliminates duplicate or redundant processes around cash management and treasury functions
- Enhances controls and streamlines audit processes
- Helps to avoid duplicate payments and improve reconciliation as the payment process is consistently followed
- Allows companies to leverage economies of scale to streamline payment processes, which leads to negotiations with suppliers for better terms of conditions
- Allocates resources efficiently as you have better visibility of payment processes
- Increases transparency and cross-system views of customer and supplier accounts
- Centralizes accounts receivable (AR) processes such as credit exposure, dunning, dispute, and collection processes

- Lowers cost of maintaining bank accounts as the number of accounts required is greatly reduced
- Integrates digital payments seamlessly
- Improves bank communication with built-in SAP S/4HANA workflows and monitoring tools
- Integrates with bank communication management and in-house cash management
- Combines payments to suppliers involved in cross-border transactions
- Uploads bank statement with standard interface and reconciliation
- Leverages SAP S/4HANA reporting tools and user experience with SAP Fiori, SAP Analysis for Microsoft Office, SAP Analytics Cloud, and so on
- Integrates central payments with SAP Business Technology Platform (SAP BTP) processes
- Optimizes SAP HANA payment proposal generation in SAP S/4HANA

Now that you understand the important benefits of central payment offered with Central Finance, let's explore the use cases of central payments. Figure 9.2 illustrates important use cases of central payments used in an organization:

❶ **External payments**
Payments are made to external suppliers, debtors, or any other parties involved in the company's day-to-day transactions. An external invoice is created from the core or source system, and the invoice open item is technically cleared in the source system. The invoice document is then replicated to the Central Finance system for further processing of payments from your company's bank account. This improves centralized SAP cash operations. In addition, you'll leverage centralized SAP Treasury and Risk Management features and innovations.

❷ **Intercompany payments**
An intercompany invoice is created and is technically cleared in the source system. The invoice document is then replicated to the Central Finance system for further processing of the payment. Besides seamless integration with the cash management tool, you'll be able to leverage intercompany reconciliation tools offered with the SAP S/4HANA system.

❸ **Incoming payments**
Central payments can be used to process incoming payments from customers or other stakeholders related to a company. Transactions related to incoming payments are centralized in a shared services model.

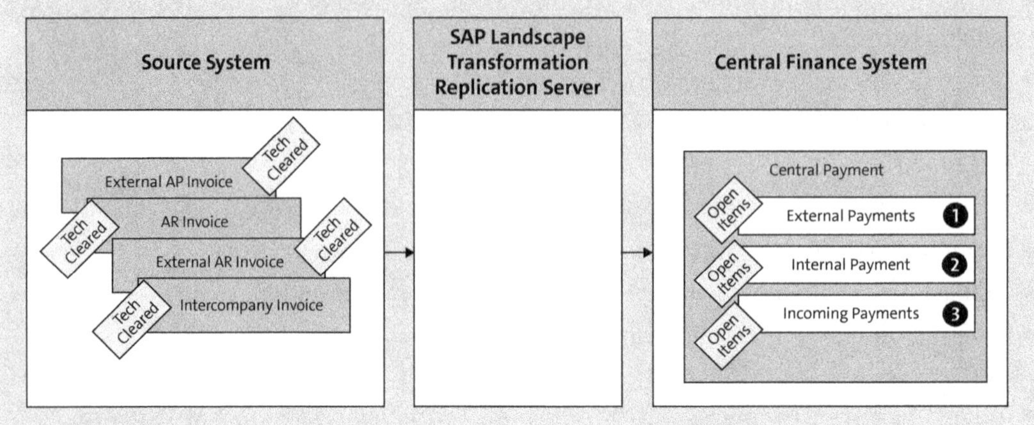

Figure 9.2 Central Payments: Overview

9.2 Getting Started with Central Payment

As we just learned, the central payment functionality offers customers the chance to centralize payments across enterprises with multiple source systems. This enables customers to streamline payment processes in a single location using the Universal Journal features in SAP S/4HANA, thereby improving an entity's cash flow. Corporate controllers and treasurers like the central payment functionality offered with Central Finance because it provides greater transparency into cash flows and integration with treasury and working capital management. Before we discuss central payment prerequisites and relevant configurations, let's see how data flows in procure-to-pay and order-to-cash process cycles.

9.2.1 Procure-to-Pay Process

When central payment is activated in the Central Finance system by company code, supplier open items are technically cleared when they are posted in the source system automatically. Then, the corresponding open items are replicated over to Central Finance. You'll then manage clearing activities centrally using SAP S/4HANA's cash management features. Figure 9.3 illustrates a typical central payment scenario with the procure-to-pay cycle:

❶ **Business processes**
Core business processes related to procurement activities such as creating purchase requisitions and purchase orders, receiving goods, and posting invoices are carried out in source systems. As shown in Figure 9.3, goods receipt/invoice receipt (GR/IR) accounting documents are replicated to the Central Finance system in real time.

Upon the invoice receipt being posted in the local or source system, the system clears the invoice technically in the source system.

❷ Open items in Central Finance
Upon invoice replication from the source system, open items are reposted in Central Finance, thereby leaving the invoice open in Central Finance for further processing.

❸ Central payment
All payments are processed in Central Finance using the centralized open item clearing offered with the central payment functionality.

❹ Follow-on processes
Any follow-on processes, such as reversals, happening in the source system are carried out with a caution as clearing and reset clearing processes occur in Central Finance. Reversals are allowed only if open items aren't cleared in the Central Finance system. Similarly, resets are allowed only in Central Finance.

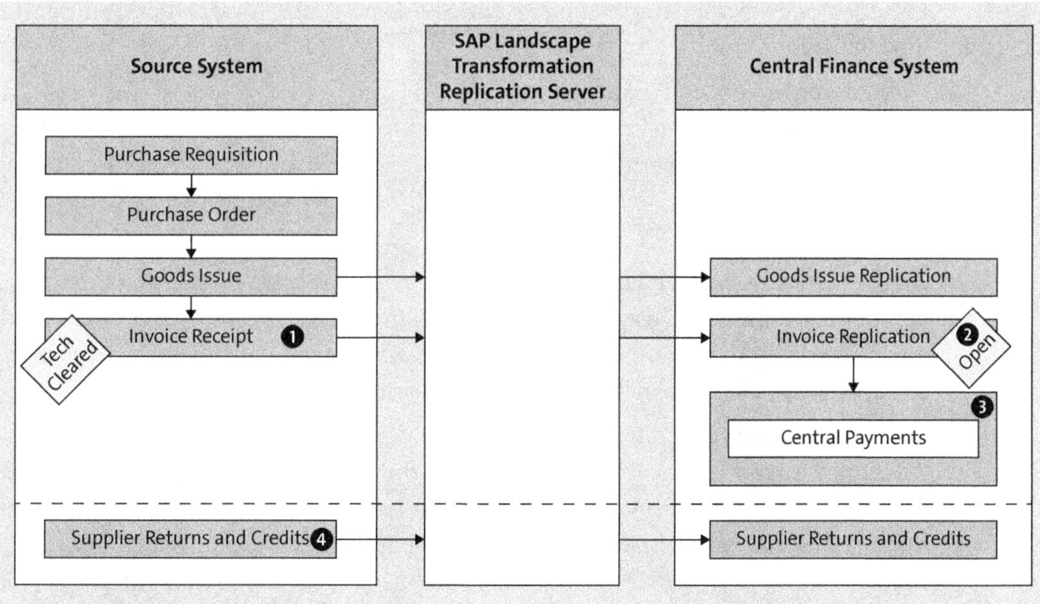

Figure 9.3 Central Payment: Procure-to-Pay Process

9.2.2 Order-to-Cash Process

This process contains a full lifecycle that starts with a customer quote or customer order and ends when payment is received and recorded in the accounting books. The order-to-cash process integrates with other components to support a variety of interconnected processes such as credit management, dispute resolution, and collections.

Figure 9.4 illustrates the central payment scenario with a typical order-to-cash cycle. The flow diagram highlights important processes:

❶ **Sales order credit check**
Core business processes related to the order-to-cash cycle are carried out in the source system, whereas clearing activities related to incoming payments activities are performed in Central Finance. The order-to-cash processes involving sales requisition, sales order, delivery of goods, and process billing are carried out in a source system. During creation of the sales order delivery, a credit check on a business partner can be carried out with a web services call. The system uses check rules assigned to the credit segment on the business partner.

❷ **Technically clearing in the source system**
Accounting documents are related to both goods delivery and billing documents replicated to Central Finance in real time. Upon billing document posting in the local system, the system clears the accounting invoice document related to the billing document in the source system, thereby leaving the invoice open in Central Finance only.

❸ **Open items in Central Finance**
Upon invoice replication from the source system, open items are reposted in Central Finance, thereby leaving the invoice open in Central Finance for further processing by central payment.

❹ **Central payment**
All payments are received in Central Finance using the centralized open item management functionality. In addition, Central Finance integrates centralized cash management, credit management, dispute management, and collections management to support customer open items clearing activities.

❺ **Follow-on processes**
Any follow-on processes, such as reversals, in the source system are carried out with caution. Clearing and reset clearing processes occur in Central Finance due to cross-system process controls (CSPC). Reversals related to sales and distribution (SD) invoices, down payments, or FI invoices are allowed only if open items aren't cleared in the Central Finance system.

Now that you've learned the high-level process of how central payment functions work with examples of procure-to-pay and order-to-cash cycles, let's explore how open item management functions in Central Finance.

Note

Once the central payment scenario is activated for a company, you'll always refer to the Central Finance system for accounts receivable (AR) and accounts payable (AP) open items and related reporting.

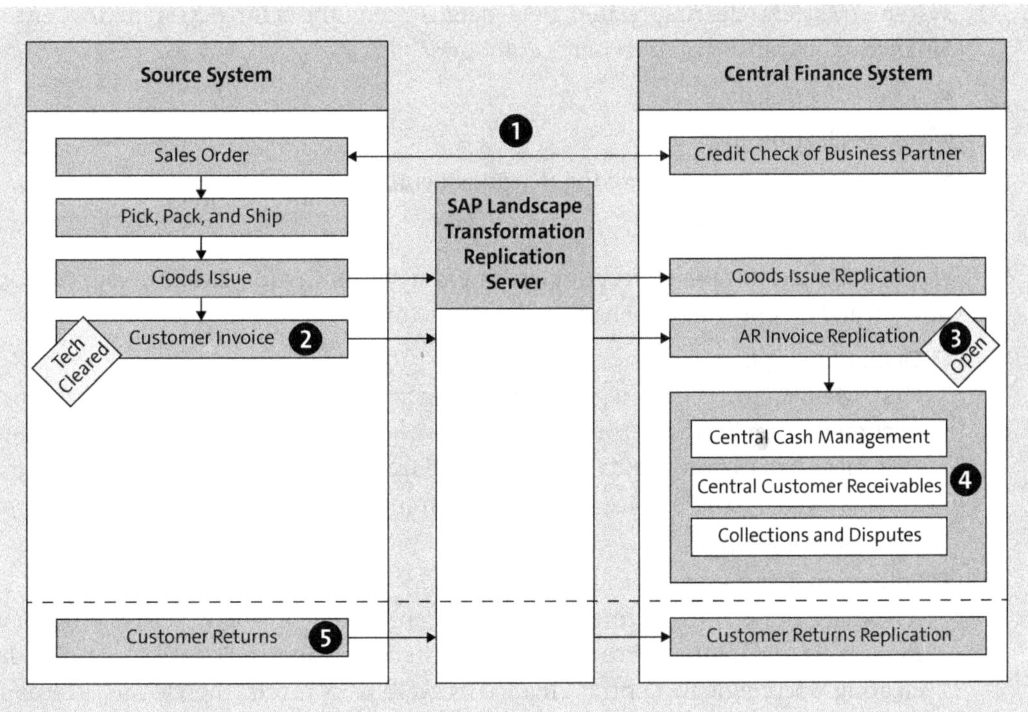

Figure 9.4 Central Payment Order-to-Cash Process

9.2.3 Managing Accounts Receivable and Accounts Payable Open Items in Central Finance

With the end goal of centralizing processes in Central Finance, the central payment functionality offers customers the chance to process the open item clearing functionality in one single central location. Open AR and AP items from source systems are replicated to Central Finance to have a consistent process across the enterprise. To support open item management, especially for customer and vendor open items, SAP provides two scenarios to manage open items in Central Finance: *clearing transfer* and *central payments*. Let's understand each scenario next.

Clearing Transfer

As the name implies, clearing transfer reflects the status of open items across a distributed landscape. Clearing transfer is a prerequisite for central payments activation. In addition, it's an essential step for document splitting to work correctly in Central Finance, especially when freely defined currencies are enabled. Relevant prerequisite notes must be implemented in the source system to enable this functionality. As a result of enabling clearing transfer activation functionality, this helps users to monitor the status of open items replicated from various source systems to Central Finance. The system transfers clearing-related field details from the source system to Central Finance upon transfer of open items as follows:

- AUGDT: Clearing date
- AUGCP: Clearing entry date
- AUGBL: Document number of the clearing document
- AUGGJ: Clearing fiscal year

Figure 9.5 illustrates the concept of clearing transfer in Central Finance. Let's explore how AR and AP open items behave in a clearing transfer scenario:

❶ **Open item replication**
First, when an invoice open item is posted in a source system, the relevant account document is replicated to the Central Finance system along with its clearing status. This means all clearing fields are transferred from the source system to the Central Finance system. The invoice is open in both the source system and target system, reflecting clearing status.

❷ **Follow-on document replication**
Assuming the central payment activation isn't enabled, when the clearing document is cleared in the source system, the system transfers the clearing status of the clearing document to Central Finance as well. As a result, the clearing status is reflected in both the source and target systems. Likewise, reset or reversal is carried out in the source system and the relevant status is reflected in Central Finance.

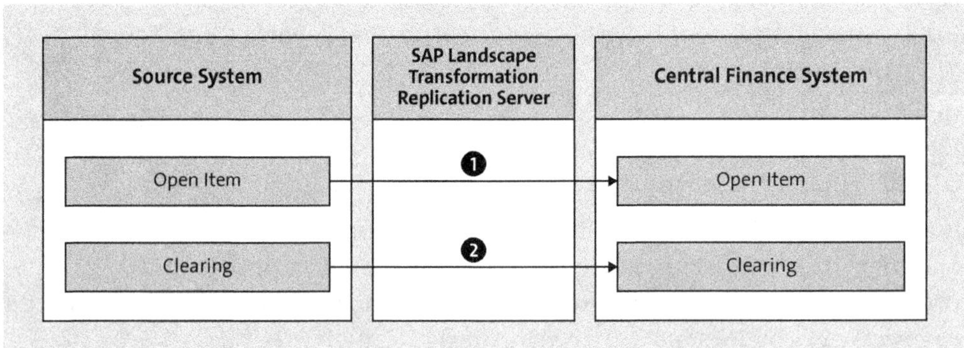

Figure 9.5 Central Payment: Clearing Transfer Scenario

As you can see in Figure 9.5, the open item is cleared in the source system as the payment process resides within the source system. Therefore, the source system is considered a book of record in this case. The Central Finance system is used merely as a reporting platform for centralized reporting. You'll gain a cross-system view of customers and vendor AR and AP open items. In addition, you can integrate with central dispute management, collections worklist, and credit evaluation processes to enable shared services for certain invoice management functions. Clearing transfer, however, prepares Central Finance to enable the central payment functionality at a later point.

Clearing transfer is a required prerequisite for central payment activation. When clearing transfer is activated, the system updates table FINS_CFIN_CTRL with key **Char** as **DISABLE_TECH_CLEARING_OF_OPEN_ITEMS** and the **X** flag set for **Char80**, as shown in Figure 9.6.

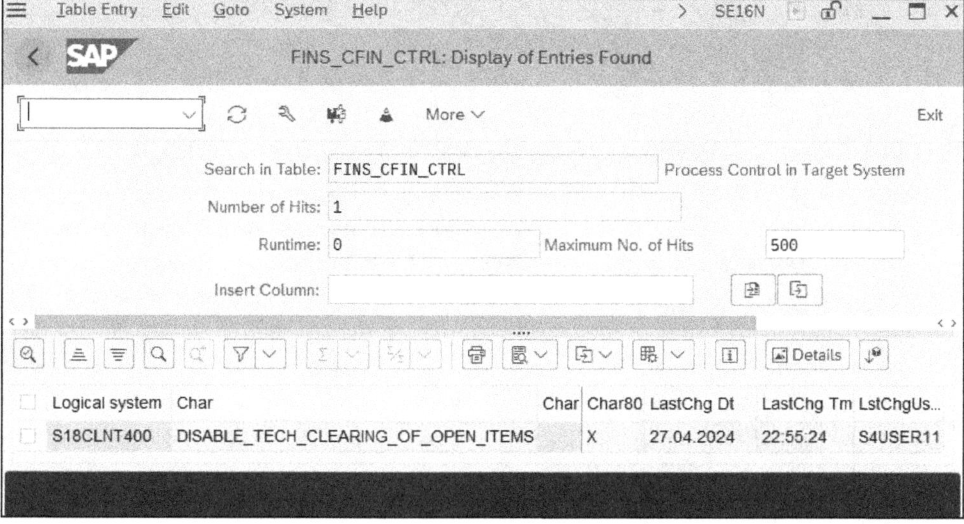

Figure 9.6 Clearing Transfer Update in Table FINS_CFIN_CTRL

Restrictions

Although the clearing transfer functionality has existed since SAP S/4HANA version 1809, SAP recommends reviewing several restrictions before enabling the clearing transfer functionality in Central Finance:

- Allowed currency configurations specified in SAP Note 2863836 also apply to clearing transfers. The SAP Note provides additional guidelines and configurations for handling currencies within Central Finance.
- In SAP, foreign currency revaluation postings logic is different between classic general ledger and SAP S/4HANA general ledger or new general ledger. As a result, any differences that arise have persisted to open items table BSEG-BDIF*. SAP recommends considering the balance in corresponding balance sheet correction accounts when reconciling account balances. On the other hand, in SAP S/4HANA, foreign currency valuation differences aren't stored in open items. As a result, the foreign currency valuation coming in from a classic general ledger system isn't transferred to the Central Finance system. This must be accounted for in the replication. The same behavior applies to exchange rate gain/loss recognized (table BSEG-RDIF*), realized exchange rate gain/loss for partial payments, and penalty charge amounts (table BSEG-PENC*).
- To avoid rounding differences, source systems currency decimals must comply with the Central Finance system. If currency decimals exist when connecting to the Central

Finance system, SAP recommends having more decimals defined in Central Finance. Inconsistent decimals lead to errors during follow-on document replication.

- Cross-company postings from the reconciliation ledger using Transaction KALC aren't supported.
- Document splitting configuration must be identical between source systems and Central Finance systems. SAP doesn't support document splitting scenario, not being active in the source system but being active in Central Finance. This may lead to errors during follow-on processes such as with the clearing transaction as these processes rely on characteristics derived from the clearing transaction.
- Document summarization settings must be identical between the source system and the Central Finance system. Inconsistent summarization settings may lead to errors in SAP Application Interface Framework. Harmonization consideration must be given when multiple source systems are configured with FI summarization with Transaction OBCY, but enhanced summarization is configured with Transaction OBCYX in the Central Finance system. In addition, customer-specific fields (coding-block) should be added to table ACDOCA only to support line-item reporting in SAP S/4HANA, but not to accounting entry view table BSEG. In other words, custom fields have been included in ACDOA table not extended to BSEG to prevent document summarization errors.
- Clearing postings from real estate management settlement transactions such as RESCE, RESCBC, RESRBCSINGLE, RESCSECO, and RESCSETN are transferred to the Central Finance system, but the clearing isn't executed.
- The system doesn't transfer open items to managed general ledger line items, which fall within the initial load balance. SAP recommends implementing SAP Notes 2396399 and 2397166 for the items that fell within the current period, but the system failed to transfer the clearing. After applying the SAP Notes, you must extract the initial load data again.
- SAP recommends applying SAP Note 2210341 to transfer clearing open items along with the initial load. If the SAP Note isn't applied, the transferred open items will be missing the references to predecessors. As a result, clearing can't be performed in the Central Finance system.
- The system transfers clearing postings from the loans accounting module (TR-LO-AC), but the clearing isn't executed in the Central Finance system.
- For the clearing transfer to function properly for open-item managed GL accounts, the open-item managed general ledger must be mapped to the corresponding open-item general ledger in the Central Finance system.
- Invoice withholding tax that is calculated during the payment clearing in the source system isn't updated on the corresponding invoice in the target system.

As you recall, clearing transfer is a prerequisite for central payment activation in Central Finance. SAP provides two ways of activating clearing transfer in Central Finance: *before initial load* and *after initial load*. Let's understand each method next.

Activating Clearing Transfer Before Initial Load

Figure 9.7 shows activating clearing transfer before initial load. As you can see, clearing transfer is activated ahead of starting the initial load extraction.

SAP recommends activating before extracting the initial load for any SAP S/4HANA release 1809 FPS0 or higher. However, if clearing transfer is already activated in higher releases, this step isn't relevant and can be ignored.

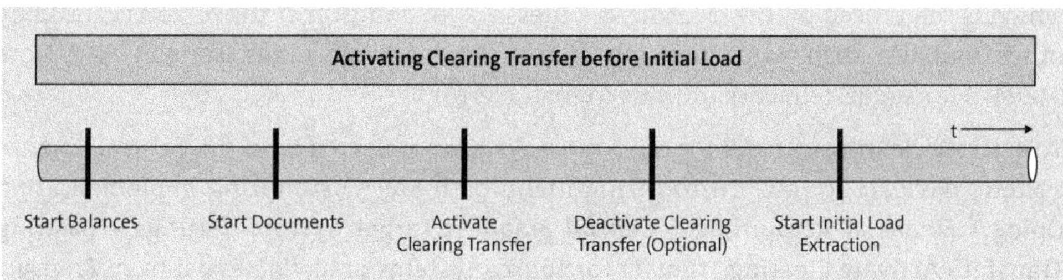

Figure 9.7 Transfer Clearing Activation Before Initial Load

[+]

Tip

SAP allows you to deactivate clearing using Transaction FINS_CFIN_DEACT_CLR if there are no open items in the company code in Central Finance. As such, you'll perform this task generally before extracting for the initial load.

Activating Clearing Transfer After Initial Load

If the initial load is executed without clearing transfer activation, especially when the Central Finance release is lower than SAP S/4HANA 1809, then this option can be relevant. Moreover, this approach is relevant only for SAP S/4HANA release 1610 through SAP S/4HANA 1709 FPS2. With this approach, as shown in Figure 9.8, clearing transfer isn't activated initially, but the initial load is already executed. In this case, you'll activate clearing transfer after the initial load. In addition, you'll reopen technical clearing with Transaction FINS_MIG_CJ3 (Reopen Technical Clearing). Consult SAP Note 2292043 on transferring clearing data to Central Finance, especially for historical open items on prior releases.

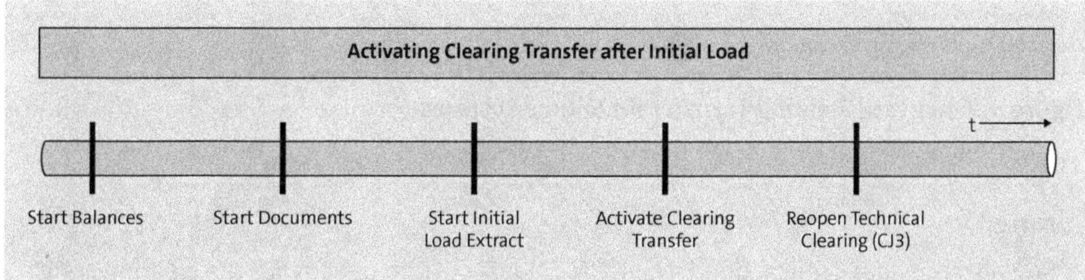

Figure 9.8 Transfer Clearing Activation After the Initial Load

In some instances, reset clearing is stuck in SAP Application Interface Framework with the error message **Ref. open item already cleared** (message **FINS_CFIN_AC_DOC 056**). This happens when a reset clearing is missing as it was processed before clearing transfer was activated. In other words, the SAP Application Interface Framework message that failed refers to the clearing message in SAP Application Interface Framework that has been technically cleared. In such situations, SAP recommends running report RFINS_CFIN_REOPN_TECH_CLRD_DOC to reopen the technically cleared open item which is referenced by the clearing AIF message. In addition, if the error is caused by other problem such as sequencing or duplicate AIF messages, consult SAP Note 3489479 for suggested workarounds to fix the error.

To activate clearing transfer for the source system connected to in the Central Finance system, navigate to IMG Customizing menu path **SAP Customizing Implementation Guide • Financial Accounting • Central Finance: Target System Settings • Clearing Transfer • Activate Clearing Transfer for Source Systems** or navigate to it from Transaction CFINIMG. Upon opening the configuration, you'll arrive at the **Activate Clearing Transfer for Logical Sender System** screen, as shown in Figure 9.9. On this screen, enter a value in the **Logical Sender System** ❶ field, and then click **Execute** ❷. If the clearing transfer is activated for the sender logical system, then the system displays a successful message ❸. For the example logical system, the system displays **Clearing transfer is already active**. Notice the **Test Mode** option available prior to executing update mode.

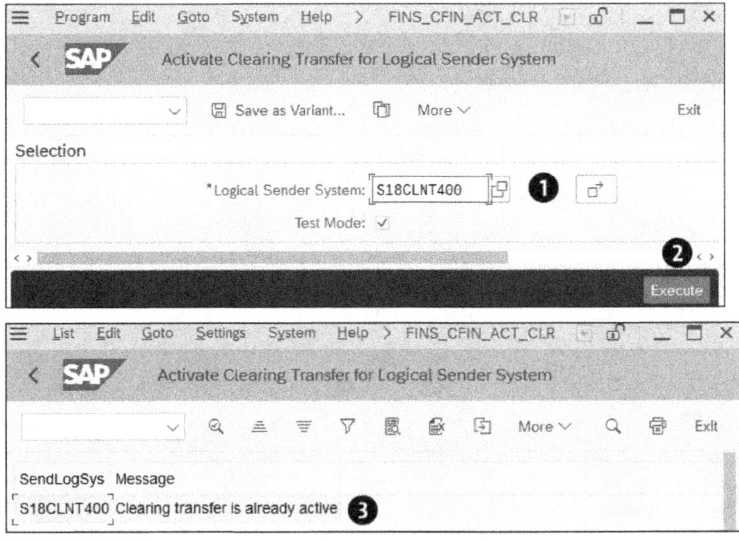

Figure 9.9 Activate Clearing Transfer for Source Systems

[+]

Note

SAP also provides Transaction FINS_CFIN_DEACT_CLR to deactivate the clearing transfer if it was mistakenly activated.

Confirm whether the clearing transfer is activated successfully by going to the **Maintain RFC Assignments Settings for Source System** configuration as shown in Figure 9.10. As you've noticed, the **Clearing Transfer** indicator shows whether the clearing is active.

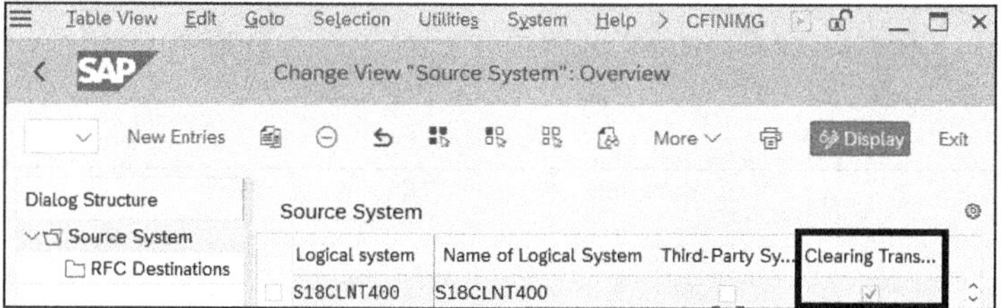

Figure 9.10 Clearing Transfer Indicator When Maintaining RFC Assignments

When clearing transfer is activated, the system updates the source system control table `CFIN_CTRL` as shown in Figure 9.11 ❶. In addition, the system updates central finance control table `FINS_CFIN_CTRL` ❷ with key **Char** as **DISABLE_TECH_CLEARING_OF_OPEN_ITEMS** ❸ and the **X** flag set for **Char80**.

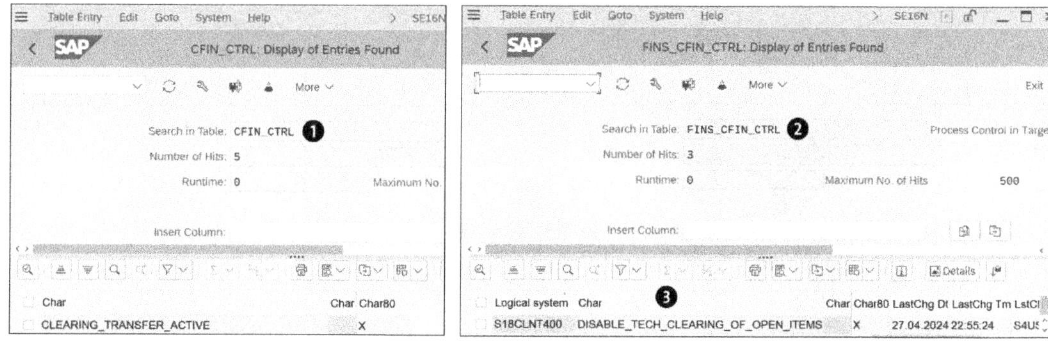

Figure 9.11 Clearing Transfer Update in Tables CFIN_CTRL and FINS_CFIN_CTRL

Central Payment Scenario

With the central payment scenario, you'll activate central payment at the company code level in Central Finance. As a result, when AP or AR open items are created in the source system, the system clears open line items automatically. In addition, the system clears credit card settlement entries for deferred tax line item, net cash discount line item and bank clearing line items. This is referred to as technically cleared in the system update clearing document field with value `ALE-Extern`. Figure 9.12 illustrates the central payment scenario. As a result of technical clearing of open items in the source system, replicated invoice open items are cleared with the central payment functionality in Central Finance. In addition, as shown in Figure 9.12, clearing status isn't updated back to the source system. This is standard behavior in Central Finance. Any sync back to the source system isn't delivered as an out-of-the-box Central Finance solution. In addition, ensure replicated documents processed in Central Finance prior to archiving the source system documents to avoid errors in SAP Application Interface Framework.

Technically, when you activate central payment in Central Finance, the system updates table FINS_CFIN_CTRL with key ENABLE_CENTRAL_PAYMENT = X for all logical systems in the Central Finance system, as shown in Figure 9.13 ❶. Upon central payment activation, the system adds an entry in SETTING_KEY = LEADING_CENTRAL_PAYMENT_SYSTEM ❷ in control parameter table CFIN_CTRL in the Central Finance system.

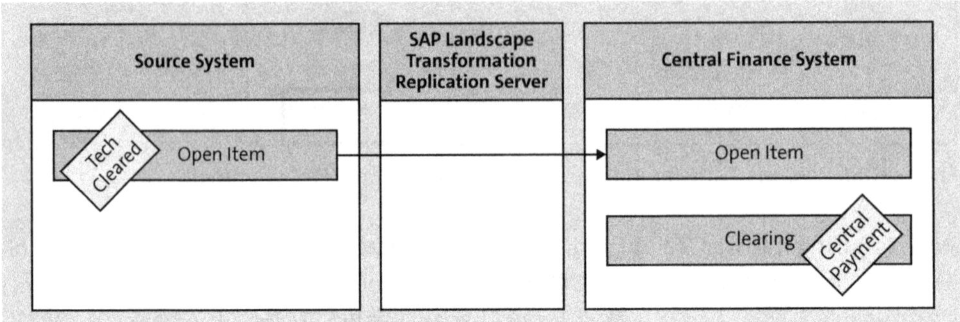

Figure 9.12 Central Finance: Central Payment Scenario

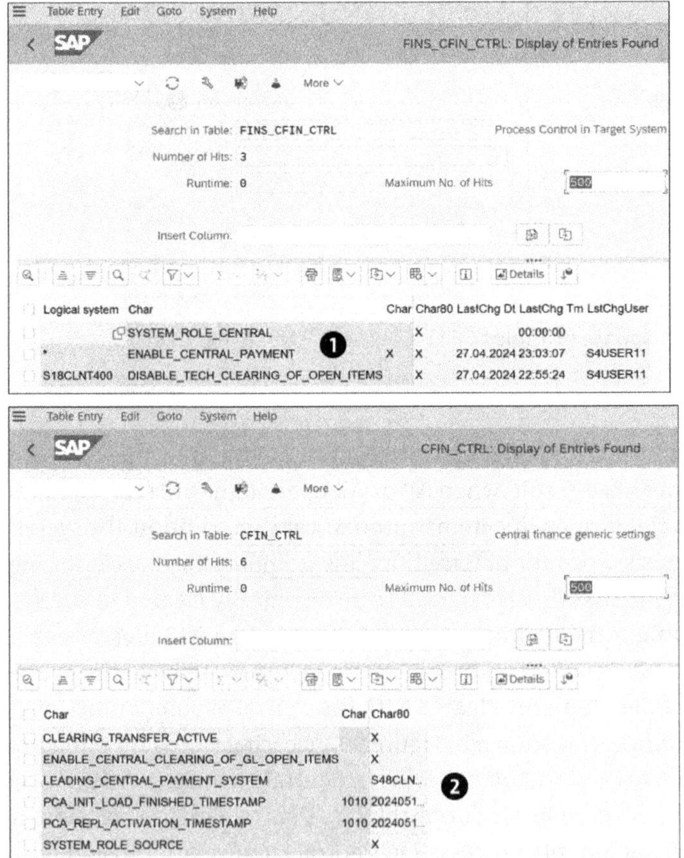

Figure 9.13 Enabling Central Payment: Tables FINS_CFIN_CTRL and CFIN_CTRL

[«]

Note

If you're on a lower release level, you'll apply several SAP Notes to enable central payment functionality to work correctly. To check SAP Notes to switch on central payment in Central Finance, consult SAP Note 2346233. Likewise, refer to SAP Note 3063148 for enabling central payment involved with a third-party source system.

9.2.4 Cross System Process Control

Central payment activated in a distributed landscape faces multiple challenges. As the end-to-end business processes split between different systems in a distributed landscape, this will lead to several issues, as shown in Figure 9.14 and described here:

- **Actual clearing**
  Actual clearing status availability is only in the Central Finance system, not in the source system. There is no write-back to the source system functionality available in Central Finance. Any such requirement is handled as a custom enhancement project.
- **Dependent processes**
  Dependent processes in the source system rely on the actual clearing status of open items, which isn't available because open items are technically cleared due to central payment activation.
- **Lock and clearing status**
  The mechanism used for the lock and validate clearing status isn't possible in a standalone system in a cross-system situation.
- **Unintended reversals**
  If a payment is already processed, for example, clearing has already occurred in the Central Finance system, you want the system to prevent accidental reversals in the source system.
- **Inconsistent process**
  In a heterogeneous landscape, following inconsistent processes leads to workarounds, and manual adjustments must be prevented.
- **Compliance**
  Preventing or reducing compliance-related risks when local and statutory reporting is involved is challenging in a distributed system where data is spread between different systems.
- **Data integrity**
  Without following consistent processes for payments, it's impossible to achieve the highest data integrity.

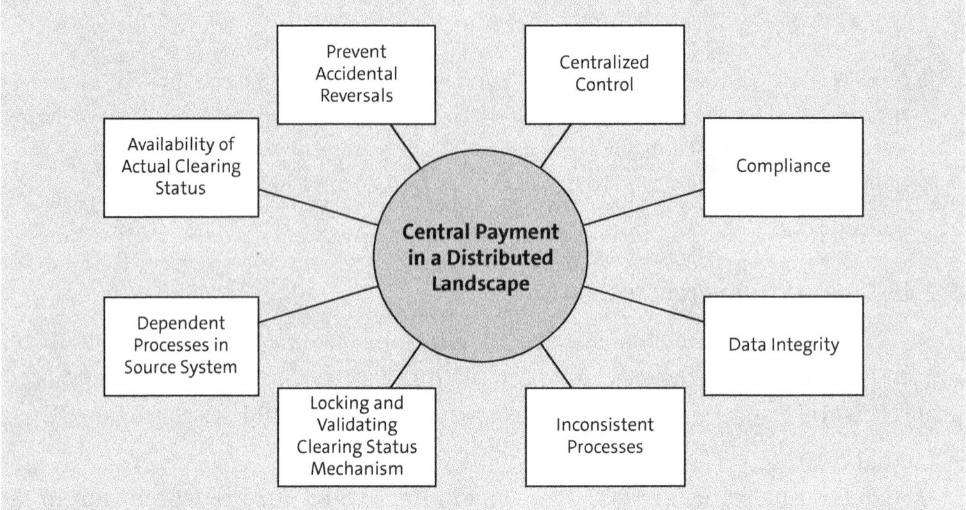

Figure 9.14 Challenges of Central Payments

To mitigate some of the challenges we've discussed, SAP introduced the CSPC framework in Central Finance to have better controls over business processes for documents replicating in a distributed landscape. Although CSPC was introduced in SAP S/4HANA 1909, full functionality became available from SAP S/4HANA 2020 onward. SAP downported this functionality to SAP S/4HANA 1709 to support central payment in that release. In addition, SAP made CSPC mandatory for central payment since SAP S/4HANA 2020 to ensure data integrity. With central payment and CSPC activated, the framework ensures AP and AR open item clearing activities happen only in Central Finance. The CSPC works with a locking mechanism using a virtual token to control business transactions. The virtual token determines whether to allow or prevent a transaction and is stored in table CFIN_CSPC_TOKEN in the source system and CSPC_TOKEN in the target systems. For every open item, the system creates a virtual token and is active in one system at a time so that users can execute a specific process in one system only. Without a token, the system prevents process steps for processing open items. In addition, if a token is inactive in one system, the system will do a RFC call to check if the token is active in the other system to prevent or allow a process to carry on. As a prerequisite, in the CSPC configuration, you must activate process type CPAY, which facilitates defining a group of process steps for CSPC to benefit from CSPC functionality.

Let's discuss CSPC controls and how they work when an open item is replicated and cleared in Central Finance. In our example, we've chosen the reversal Transaction MR8M to explain the concept, as shown in Figure 9.16:

❶ **Vendor invoice technically cleared**
When a supplier invoice is created in the source system, the system creates a virtual token for the open item. As shown in Figure 9.15, the system creates a token in source system table CFIN_CSPC_TOKEN with a **Token Transfer Counter** (CFIN_CSPC_TOKEN-

`TRANSF_CNTR`) of value **1** for an invoice in **Controlled Object** (`CFIN_CSPC_TOKEN-CTRLD_OBJ_KEY`). The controlled object is the combined value of company code, document number, fiscal year, and line item. Then the system writes the token to target system table `CSPC_TOKEN` with a value of **Token Transfer Counter** = **1** during the open item replication.

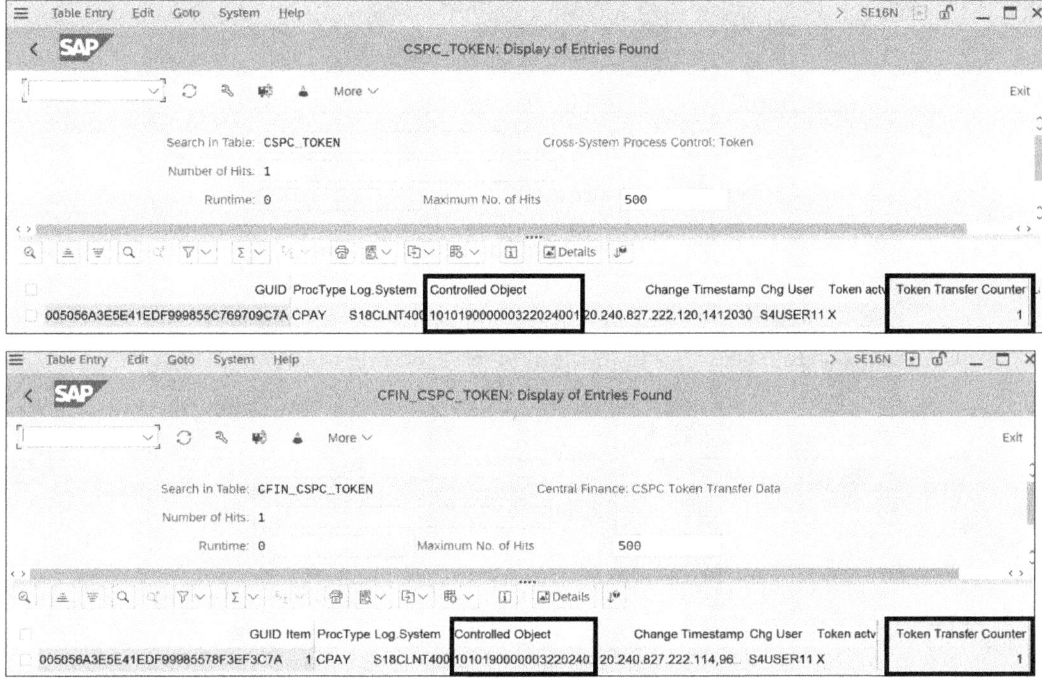

Figure 9.15 CSPC Tables CFIN_CSPC_TOKEN and CSPC_TOKEN – invoice

❷ Open item active in Central Finance

CSPC locks payment/clearing processes in the Central Finance system to process the open items.

❸ Follow-on reversals not allowed

Upon initiating reversal Transaction MR8M in source system, the system performs a remote clearing status check in Central Finance, whether the invoice is open or already cleared. The system then responds back whether to allow the transaction in source system table `CFIN_CSPC_TOKEN`. Then, the system checks the following scenarios and prevents invoice reversal (e.g., if a reversal happens with the materials management [MM] Transaction MR8M) in the source system if the following are true:

- The open item is already cleared in the Central Finance system.
- The open item is already included in the successful payment proposal.
- The replicated document is locked.
- Customer or vendor accounts are locked by the payment clearing process.

- Reversals or replicated change documents are stuck in SAP Application Interface Framework.
- Open items are created prior to CSPC initialization.

❹ **Follow-on reversals allowed**
The system checks the following scenarios and allows reversal of the invoice with Transaction MR8M in the source system if the following are true (see Figure 9.16):
- The open item isn't cleared in the Central Finance system.
- The open item isn't included in a successful payment proposal.
- A replicated document isn't locked.
- Customer or vendor accounts aren't locked by the payment clearing process.
- The CSPC initialization is carried out.
- Transaction MR8M is allowed in the source system for invoices that only have general ledger line items. On the other hand, the AP line items are still open in Central Finance. This will result in inverse posting in Central Finance.

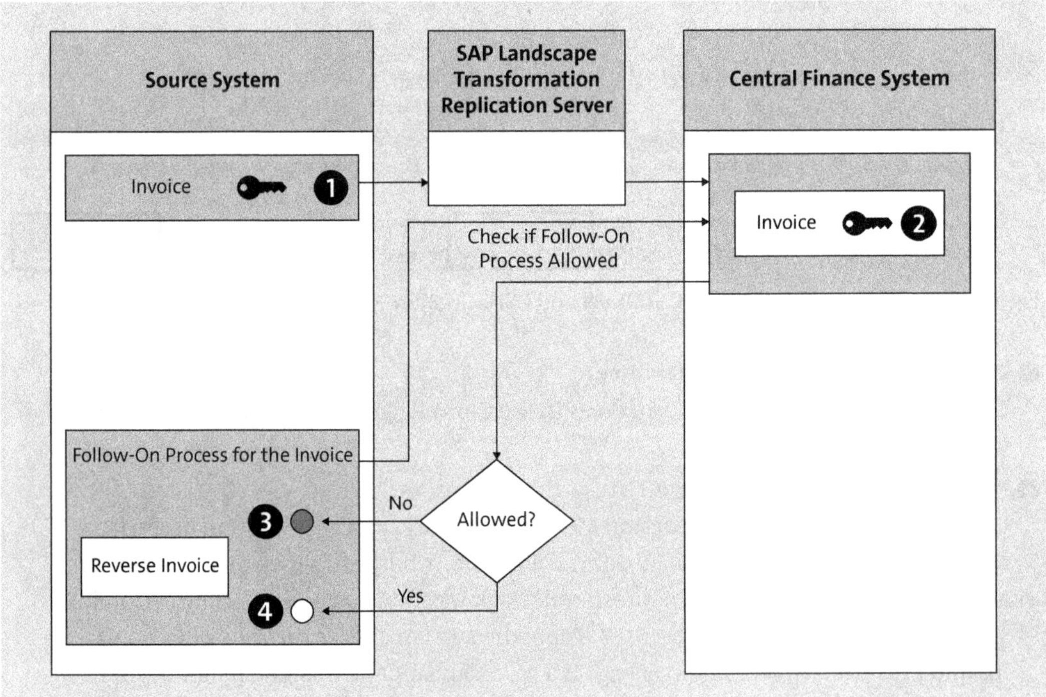

Figure 9.16 Central Payment: Cross-System Process Control

When a reversal document is successfully posted in the source system, a new record with the token value is set to **Token Transfer Counter** = **2** and is written in the source system table `CFIN_CSPC_TOKEN` for the invoice. On the other hand, the target system table CSPC_TOKEN is updated in the same invoice record with **Token Transfer Counter** = **2**, as shown in Figure 9.17. In addition, a new token record with value **Token Transfer Counter** = **1** is created in both the source system and Central Finance for the reversal document.

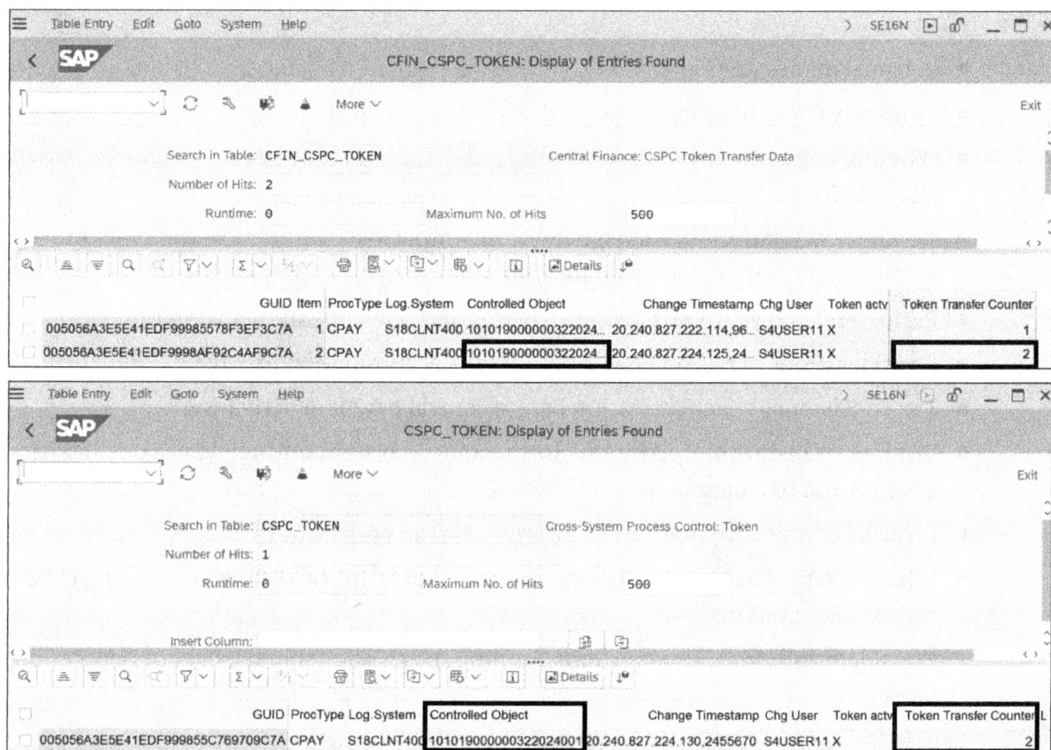

Figure 9.17 CSPC Tables CFIN_CSPC_TOKEN, and CSPC_TOKEN – Reversals

These controls apply to individual FI AP, AR, or general ledger line-item document reversals using Transaction FB08 or when an intercompany reversal is carried out with Transaction FBU8. In addition, CSPC controls apply for an inverse posting triggered through Transaction VF11 or MR8M to prevent inconsistencies. In such circumstances, SAP recommends clearing AR or AR line items and inverse documents in Central Finance.

> **Note**
>
> Consult SAP Notes 3117154 and 2985804 for additional tips and recommendations on CSPC for central payment and SAP Landscape Transformation Replication Server for Central Finance.

9.3 Configuring Central Payment

Prior to central payment activation, SAP recommends you check several considerations and gather information required for central payment activation to assist you with a successful central payment activation with minimal interruptions and minimal errors:

- Central Finance release
- Company codes in scope for central payment activation
- Whether any company codes were merged in the source system and mapped in Central Finance
- Timing of activation such as right after the initial load completion and real time replication activated or at a later date after real-time replication is active
- Period-end and year-end closing activities and relevant cutover activities
- Considerations of period of withholding tax accumulation involved
- Tax consistency check activation before central payment activation
- Volume of historical open items in the source system and relevant consideration after central payment activation
- Clearing of any SAP Application Interface Framework error messages
- Where foreign exchange valuation is done, that is, in the source system or the Central Finance system
- Considerations for source systems running the classic general ledger or new general leger

To make centralized payments and perform centralized clearing activities in Central Finance, you'll configure several steps involved in both the source system and the Central Finance system. SAP recommends seven steps to enable central payment functionality fully. A high-level overview of these steps is discussed next:

1. **Activating clearing transfer**
   We discussed activating clearing transfer before the initial load and after the initial load scenarios in Section 9.2.3. Evaluate the clearing transfer activation requirements along with SAP recommended restrictions per your project needs with relevant stakeholders.
2. **SEPA mandates**
   If SEPA mandates are in scope for Central Finance, you must evaluate configuration requirements for SEPA mandates replication in both the source system and the Central Finance system. In Central Finance, you'll begin with Business Configuration Sets (BC Sets). In addition, consideration must be given to initial SEPA mandates migration.

3. **Consistency check**
   SAP recommends running consistency checks using Transaction FINS_CFIN_CC (Central Finance Configuration Consistency Checks). You'll choose the Central Payment check group on the transaction. The transaction identifies any inconsistencies in the configuration and reveals any corrections required prior to activating central payment.
4. **Activate CSPC**
   SAP offers CSPC to place controls over business processes for the open items replicated from the source system to Central Finance systems. With CSPC activation, the system creates a virtual token for each open item. If the system finds any issues during the clearing process, the relevant process will be blocked.
5. **Source system settings for central payment**
   You must carry out central payment configurations in the source system, which includes setting up the following activities:
   - **Defining Logical System for Central Finance System**
   - **Set Up RFC Destination for Central Finance System**
   - **Assign RFC Destination for Displaying Object from Central Finance System**
   - **Maintain RFC Assignments and Settings for the Central Finance System**

   These activities are covered in Section 9.3.2.
6. **Switching and activating central payment**
   In this step, you'll be switching central payment and activating it by company code. SAP offers a flexible option to activate central payment from the SAP S/4HANA 2022 release onward.
7. **Handling historical open items**
   In this step, to avoid duplicate payments, you'll set open items to technically clear the open items created before central payment activation. Furthermore, you'll run CSPC initialization reports in the Central Finance system. In addition, you'll run historical down payments reports to update downpayments with sales order information and relationships between historical open down payments and sales orders.

Note

For additional guidance on central payment functionality, review SAP Notes 2827364 and 2597587 completely on the central payment functionality for SAP-based source systems and third-party source systems, respectively.

Now that you've learned about the high-level requirements and process steps of activating the central payment functionality in Central Finance, let's go over the prerequisites and step-by-step configuration in both the source system and the Central Finance system.

9.3.1 Prerequisites

SAP recommends several prerequisites to be considered prior to activating central payment in Central Finance, as described in Table 9.1.

Prerequisite	Recommended action
Company code	Identify company codes that are in scope for central payment activation.
Timing of activation	Central payment can be activated at different times in the Central Finance implementation: ■ When Central Finance goes live ■ At a later date after Central Finance replication is turned on In both cases, you must plan for relevant close activities such as period-end or year-end activities. These activities must be captured in a cutover plan with all details.
Withholding tax	You must understand the period of withholding tax accumulation for local and statutory reporting purposes. Consult SAP Noe 2623514 for additional guidance on withholding tax functionality and prerequisites.
Historical open items	Dependent processes rely on historical open items. You must evaluate such dependent processes prior to activating central payment.
SAP Application Interface Framework messages	SAP recommends activating central payment after clearing messages in the SAP Application Interface Framework queue.
Dependent processes	All dependent processes must be evaluated such as the cash cycle and procure payment cycle. For example, in the order-to-cash cycle, you'll review processes such as credit management, collection management, dispute management, dunning, and intercompany reconciliation, as well as their impact on central payments.

Table 9.1 Central Payment Prerequisites

In the next section, you'll learn step-by-step configuration for central payment to be carried out in both the source system and the Central Finance system. Let's explore configuration requirements in the source system first.

9.3.2 Source System Settings for Central Payments

SAP recommends you perform central payment configurations in the source system. The activities for central payment are divided into two configuration sections:

- General configurations located under the Transaction CFINIMG menu path **Central Finance: Source System Settings • Set up Connection to Central Finance System**:
  - **Define Logical System for Central Finance System**
  - **Set Up RFC Destination for Central Finance System**

- **Assign RFC Destination for Displaying Object from Central Finance System**
- **Maintain RFC Assignments and Settings for the Central Finance System**

- CSPC configurations located under the Transaction CFINIMG menu path **Central Finance: Source System Settings • Cross System Process Control**:
  - **Setup RFC Connections for Cross-System Process Control**
  - **Configure Remote Systems for Cross System Process Control**

> **Tip**
>
> Consult SAP Notes 2346233 and 3220340 before you begin the source system configurations. These SAP Notes provide prerequisites information and recommendations to support central payment functionality and flexible options to activate central payment. In addition, SAP Note 3135178 provides prerequisites SAP Notes required in the source system to enable flexible options of central payment activation.

Let's dive deeper into each of these source system configurations with the step-by-step instructions in the following sections.

Define Logical System for Central Finance

To set up the connection to the Central Finance system for central payment and CSPC communication, a logical system is required. To define a logical system, navigate to IMG menu path **SAP Customizing Implementation Guide • Financial Accounting • Central Finance: Source System Settings • Set Up Connection to Central Finance System • Define Logical System for Central Finance System** or navigate to it from Central Finance IMG Transaction CFINIMG. After reaching the configuration, the system displays a popup window with a warning **Caution: The table is cross-client**. Click the **Continue** icon to enter. On the **Change View: "Logical Systems:" Overview** screen, click the **New Entries** button. This will lead to the **New Entries: Overview of Added Entries** screen, as shown in Figure 9.18.

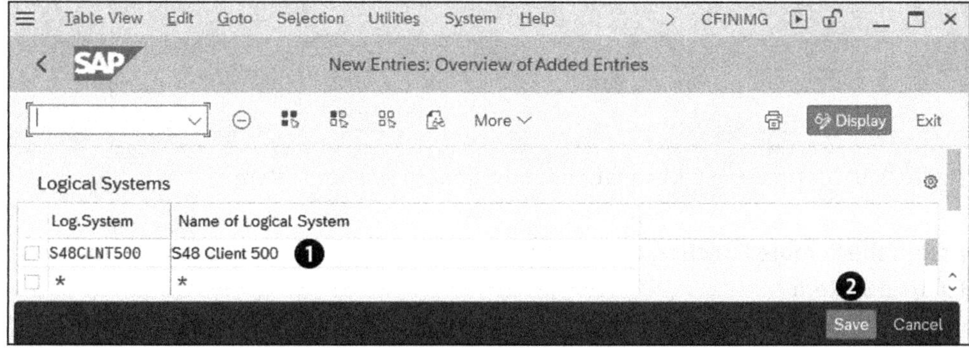

Figure 9.18 Define the Logical System for the Central Finance System

On this screen, maintain values in the **Log.System** and **Name of Logical System** ❶ fields to define the logical system for the target Central Finance system. Click the **Save** ❷ button to save the configuration entries.

Alternatively, this configuration can be maintained in view V_TBDLS via Transaction SE54 or SM30. Note this table is a cross-client activity.

Set Up the Remote Function Call Destination for the Central Finance System

In this configuration, you'll set up the RFC destination for the Central Finance system to receive RFCs from the source system to the Central Finance system. This is no different from the logical system we walked you through in Chapter 7, Section 7.4.2, so we'll provide only the menu path to the configuration activity here. To set up the RFC destination for the Central Finance system, navigate to IMG menu path **SAP Customizing Implementation Guide • Financial Accounting • Central Finance: Source System Settings • Set up Connection to Central Finance System • Set Up RFC Destination for Central Finance System** or navigate to it from Central Finance IMG Transaction CFINIMG. Figure 9.19 shows the RFC destination defined for the Central Finance system.

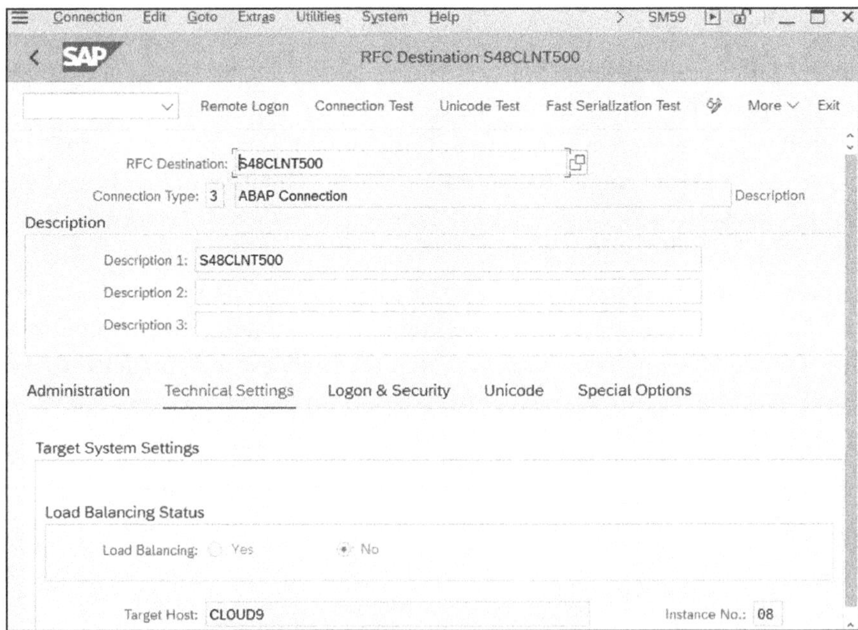

Figure 9.19 Set Up the RFC Destination for the Central Finance System

Assign the Remote Function Call Destination for Displaying Objects from the Central Finance System

To display objects or read data from the source system, you must assign an RFC destination to the logical system you've created for Central Finance. In this section, we'll highlight the menu path only as this is covered in Chapter 7.

To assign RFC destinations to the logical systems for business application programming interfaces (BAPIs), dialogs, and special methods, refer to Chapter 7, Section 7.4.2. To reach the configuration, navigate to IMG menu path **SAP Implementation Guide • Financial Accounting • Central Finance: Source System Settings • Set Up Connection to Central Finance System • Assign RFC Destination for Displaying Objects from Central Finance System** or navigate to it from Central Finance IMG Transaction CFINIMG. You can also reach the settings using Transaction BD97. Figure 9.20 shows the RFC destination defined for the Central Finance system.

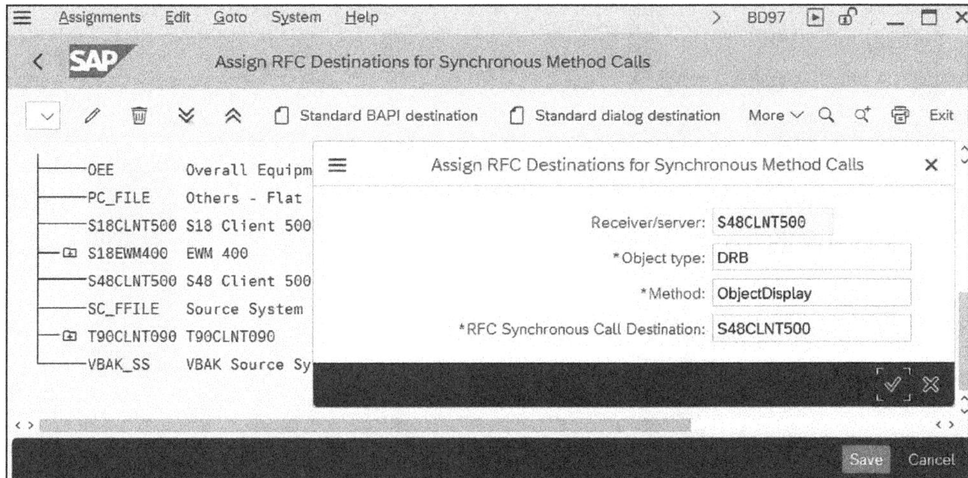

Figure 9.20 Assign an RFC Destination for Displaying Objects from the Central Finance System

Maintain Remote Function Call Assignments and Settings for the Central Finance System

For each logical system, RFC usage and RFC destinations must be maintained for certain central payment scenarios. If central payment is activated in the target system, this configuration specifies one leading Central Finance system on which you use central payment. We explained this with detailed steps in Chapter 7, Section 7.3.4 from the Central Finance configuration side. You'll repeat similar steps from the source system side in this configuration for Central Finance. To reach the configuration, navigate to IMG menu path **SAP Customizing Implementation Guide • Financial Accounting • Central Finance: Source System Settings • Set Up Connection to Central Finance System • Maintain RFC Assignments and Settings for the Central Finance System** or navigate to it from Central Finance IMG Transaction CFINIMG. Figure 9.21 shows the source system configuration of **RFC Usage** and **RFC Destination** settings for the Central Finance target system.

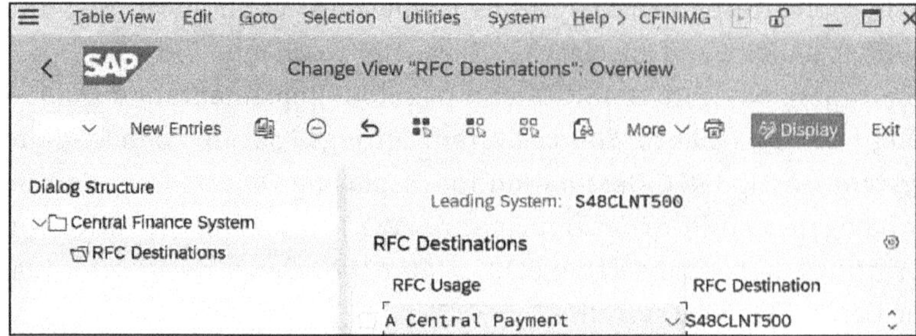

Figure 9.21 Maintain RFC Assignments and Settings for the Central Finance System

Alternatively, if the Central Finance customizing menu is not available in your source system, you can maintain it in view cluster VC_FIN_CFIN_LSYS via Transaction SM34.

The system updates source system control table CFIN_CTRL with the technical field SETTING_KEY value LEADING_CENTRAL_PAYMENT_SYSTEM, as shown in Figure 9.22. The settings in this table are used for RFCs from the source system into the Central Finance system for central payment-related activities.

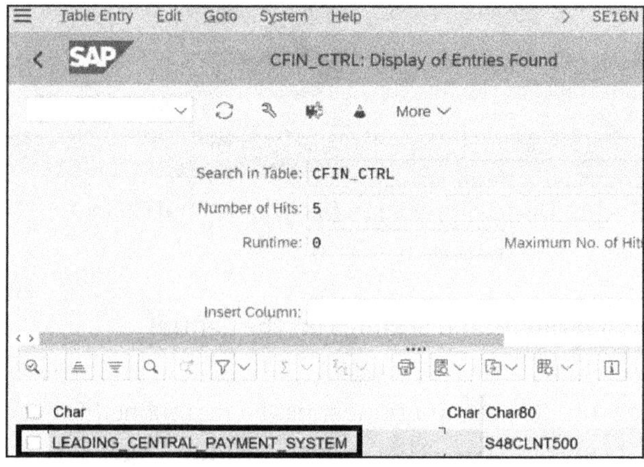

Figure 9.22 Set Leading Central Payment System Table CFIN_CTRL

Set Up Remote Function Call Connections for Cross-System Process Control

In this configuration, you'll define parameters for RFC destinations for CSPC. The system uses these parameters during RFCs. Detailed steps are covered in Chapter 7, Section 7.4.2, under "Set Up RFC Destination for Source System." However, a similar configuration is required for CSPC in the source system. To reach the configuration settings, navigate to IMG menu path **SAP Customizing Implementation Guide • Financial Accounting • Central Finance: Source System Settings • Cross System Process Control (Source System) • Set Up RFC Connections for Cross System Process Control** or navigate to it from Central Finance IMG Transaction CFINIMG.

The SAP-recommended RFC destination settings for the CSPC functionality are provided in Table 9.2 with the columns described here:

- `USER_TYPE`
  Tells whether the type is communication or dialog.
- `RFC_TYPE`
  Identifies the type of RFC authorization object. A value in **FUGR** indicates function group, and **FUNC** refers to function module.
- `RFC_NAME`
  Indicates the name of the RFC object or function group.
- `ACTVT`
  Indicates the activity being performed. For example, 16 means execute, 03 means to display, and so on.
- `CF_PURPOSE`
  Indicates the usage of activity in the source system.
- `AUTHORIZATION OBJECT`
  Authorization object in SAP grants users access to perform certain functions or actions.

USER_TYPE	RFC Type	RFC_NAME	ACTVT	CF_Purpose	Authorization Object
C - Communication	**FUGR** - Function Group	FIN_CSPC_INBOUND_REQUEST FIN_CSPC_ACTIVATION_SYNC	■ **16 – Execute**	**01**	S_RFC
C - Communication		CSPC_PRCTY = *	■ **CSPC_PRCTY = *** ■ **49 (Request)** ■ **63 (Activate)** ■ **H1 (Deactivate**		F_CSPC_TKN

Table 9.2 Set Up RFC Connections for CSPC

[+]

Tip

CSPC communicates in a bidirectional way between source system and central finance system. Therefore, you'll ensure authorization granted for CSPC specific users in both the systems. In addition, The remote function call user must be assigned with authorization object F_CSPC_TKN.

Configuring Remote Systems for Cross-System Process Control

In this configuration, you'll define the *process type*, which is used in CSPC to control business and technical checks for the central payment scenario. For example, the configuration determines whether a reversal is allowed in the source system based on the checks performed in the target Central Finance system. If an open item is already cleared in the Central Finance system, the system prevents a reversal in the source system. This is a required configuration activity in both the source system and the Central Finance system. Thus, CSPC functions in a bi-directional between source system and target system.

To define process types for the source system involved in the central payment scenario, navigate to IMG menu path **SAP Customizing Implementation Guide • Financial Accounting • Central Finance: Source System Settings • Cross System Process Control (Source System) • Configure Remote Systems for Cross-System Process Control** or navigate to it from Central Finance IMG Transaction CFINIMG. After entering the configuration, you'll arrive at the **Change View "CSPC Remote Systems": Overview** screen and click the **New Entries** button. On the resulting **New Entries: Overview of Added Entries** screen, as shown in Figure 9.23, maintain values in the target **Logical System** and **RFC Destination** ❶ fields for CSPC remote systems.

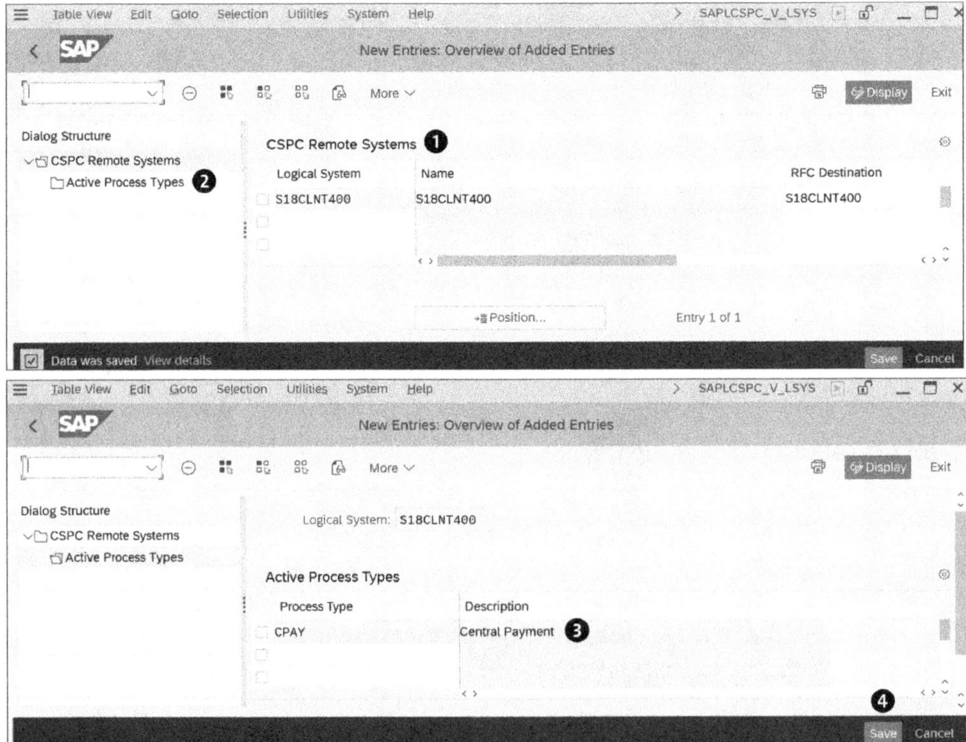

Figure 9.23 CSPC Central Payment Configuration

Next, select the logical system row you've just added, and double-click on the **Active Process Types** folder icon located in the **Dialog Structure** ❷ section and click the **New Entries** button to add process types. On the resulting **New Entries: Overview of Added Entries** screen, select **CPAY** ❸ for the **Process Type** column. Click the **Save** ❹ button to save the entry.

Once the configuration is saved, as you'll see, when adding or removing **Process Type** in the source system, the system synchronizes the process type configuration in the corresponding activity in the Central Finance system. If this behavior isn't occurring, consult SAP Note 2889748 for lower releases.

Alternatively, this configuration is also carried out with view VC_CSPC_LSYS using Transaction SM34 in the source system if the Transaction CFINIMG menu isn't available in your source system.

Note

SAP recommends this configuration as a client-dependent activity carried out in the production system as a manual activity. SAP doesn't support transporting this configuration to other systems. Therefore, this activity should be included in your project cutover activity.

9.3.3 Central Finance Configuration

In this section, you'll learn central payment configurations carried out in the Central Finance system. Before you begin the Central Finance configuration, ensure you've completed the source system configuration for central payment. In addition, consult SAP Note 2827364 to understand the functional restrictions for central payment. Next, you must complete the FI initial load prior to activating central payment.

Central Payment Switch

As a first step in enabling the central payment functionality in Central Finance, you must turn on the central payment switch using this configuration. As you recall, the central payment functionality was first introduced with a pilot SAP Note in SAP S/4HANA 1709, so this configuration is available from that release onward. To enable the central payment switch in the Central Finance system, navigate to IMG menu path **SAP Customizing Implementation Guide • Financial Accounting • Central Finance: Source System Settings • Cross System Process Control (Target System) • Central Payment • Switch On Central Payment** or navigate to it from Central Finance IMG Transaction CFINIMG. After entering the program, you'll arrive at the **Switch for Enabling Central Payment** screen, as shown at the top of Figure 9.24.

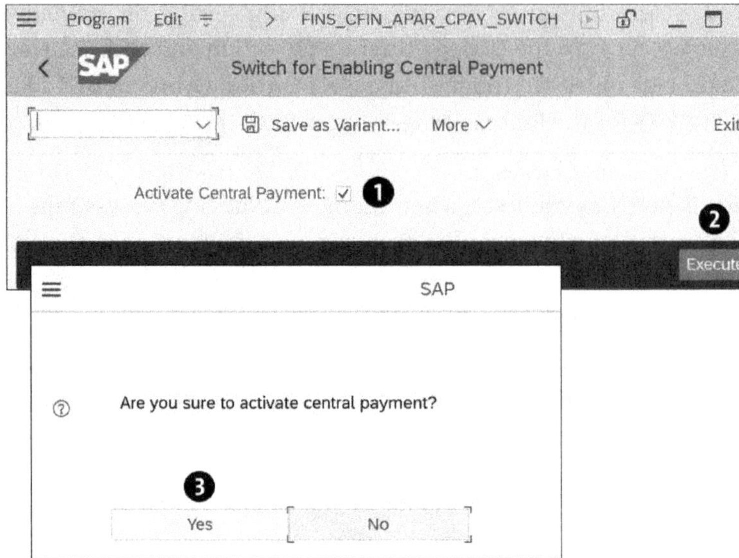

Figure 9.24 Switch for Enabling Central Payment

On this screen, choose the **Activate Central Payment** ❶ checkbox, and click the **Execute** ❷ button. As shown at the bottom of Figure 9.24, the system prompts with a popup screen message asking **Are you sure to activate central payment?** If the central payment switch is already active in the system, then the system issues the message **Central Payment has already been enabled.** Click the **Yes** ❸ button to confirm.

After the successful activation, the system displays the confirmation popup window, as shown in Figure 9.25.

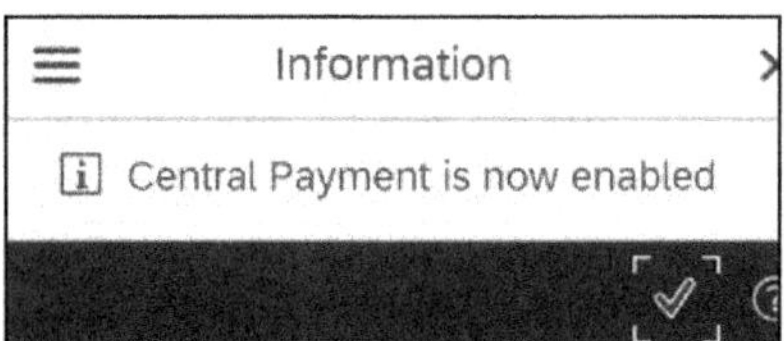

Figure 9.25 Central Payment Enabled Confirmation

As a result of executing this configuration, the system updates central finance table `FINS_CFIN_CTRL` with the SETTING_KEY with value "ENABLE_CENTRAL_PAYMENT" = x, as shown in Figure 9.26.

Note

This configuration switch is non-reversable. Consult SAP Note 2346233, which provides guidance on switching on central payment for lower Central Finance releases.

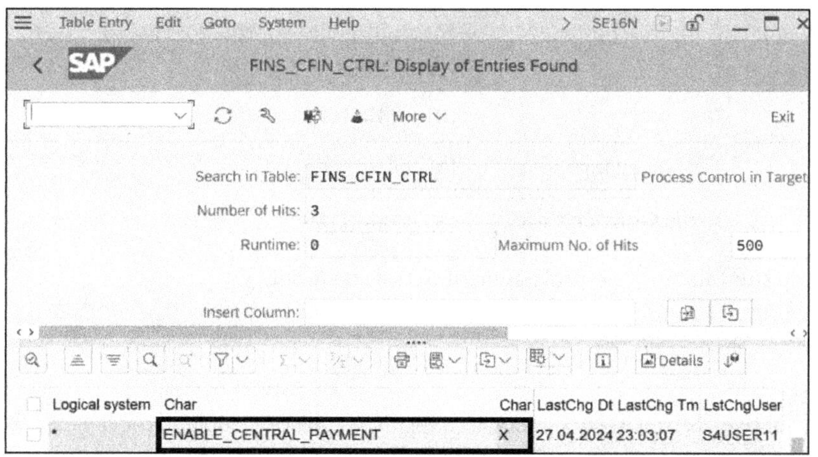

Figure 9.26 Enabling Central Payment - Table FINS_CFIN_CTRL

Central Payment Activation

In this configuration, you'll activate the tax consistency check and central payment with flexible options in the Central Finance system. This is a one-time activity that isn't reversible. Prior to activation, you must ensure that you've configured using navigating it necessary prerequisites both in the source system and the Central Finance system. In addition, you must understand restrictions of activating central payment on dependent processes covered in Section 9.6. The following scope options are available when activating the central payment functionality in the Central Finance system, as shown in Figure 9.27:

❶ **Tax Consistency Check Active (activation scope 1)**
SAP recommends enabling this check before initial load extraction to have consistent tax data in the initial load and online replication to support central tax reporting. Besides reviewing dependent processes and prerequisite configurations, such as CSPC, to activate the tax consistency check, you must configure relevant tax tables in SAP Landscape Transformation Replication Server, as shown in Table 9.3. These tables are validated before performing the initial load and online replication of FI documents.

Source Table Name	Table Description	Replicated Table Name in the Central Finance Table
T000F	Cross-Client FI Config.	FINS_CFIN_T000F
T001	Company Code	FINS_CFIN_T001
T003	Document Types	FINS_CFIN_T003 (available in SAP S/4HANA 2022 onward)
T005	Country Code	FINS_CFIN_T005

Table 9.3 Prerequisite Tables for Central Payment Activation

Source Table Name	Table Description	Replicated Table Name in the Central Finance Table
T007A	Tax Code	FINS_CFIN_T007A
T007B	Tax Processing Key	FINS_CFIN_T007B
TTXD	Tax Jurisdiction Code	FINS_CFIN_TTXD

Table 9.3 Prerequisite Tables for Central Payment Activation (Cont.)

Tip

If you deactivate the tax consistency check, as a result, you'll not be able to process central payment for the company code. This will result in inconsistent tax data between the source system and central finance system In addition, central tax reporting is not feasible in Central Finance. In that case, you'll run tax reporting either in the source system or external tax provider.

Setting up tax tables is covered extensively in Chapter 5, Section 5.2.1.

❷ **Central Payment Active with Tax Consistency Check (activation scope 0)**
This option has been available since the introduction of the central payment functionality from the SAP S/4HANA 1709 release onward. Once central payment is activated by company code in the Central Finance system, to avoid duplicate payments, the system technically clears line items with `ALE-extern`:

- AP and AR open items in the source system to facilitate centralized payments in the Central Finance system.
- Deferred tax line item, net cash discount line item, and bank clearing line item as part of credit card settlement posting in the source system

If central payment isn't activated and clearing transfer is activated, the source system processes invoice clearing. In this case, replication of financial documents would be used only for reporting purposes in the Central Finance system. When central payment is activated, tax consistency check is also activated.

❸ **CPAY Active for Intercompany APAR with Tax Consistency Check (activation scope 2)**
This option has been available since the release of SAP S/4HANA 2022. As the name implies, SAP technically clears only intercompany AP and AR open items. The system determines intercompany open items by using the trading partner on the open item line item (table VBUND). Open items for external customers and vendors are still processed in the source system. Upon activation of this option, the system technically clears (`ALE-extern`) open items in the source system to facilitate the centralized payment process in the Central Finance system for the following types of transactions:

- Intercompany AP line items populated with trading partner (BSEG-VBUND)
- Intercompany AR line items populated with trading partner (BSEG-VBUND)

- Deferred tax line item with AP and AR line items and populated with trading partner (BSEG-VBUND)
- Net cash discount line item posted with AP and AR line items and populated with trading partner (BSEG-VBUND)

SAP allows switching from activation scope **2** to scope **0** when the business transitions to full central payment functionality in Central Finance. This option gives businesses the flexibility to transition to central payment when ready to do so.

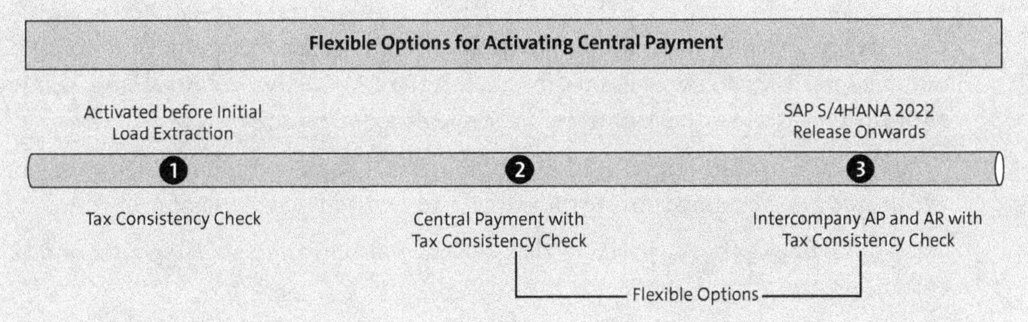

Figure 9.27 Flexible Options for Activating Central Payment

[«]

Note

Prior to choosing the intercompany AP and AR with tax consistency check option, SAP recommends considering the following functionalities:

- Trading partner consistency check is necessary for central payment for intercompany AP and AR transactions. Consult SAP Note 3220340 on minimal SAP Notes required for the source system and the Central Finance system. Refer to the section for trading partner consistency check (see Chapter 4, Section 4.4.8).
- Evaluate dependent processes relevant for intercompany AP and AR open items and the transition to Central Finance. This will involve additional regression testing to understand any impact on the new process. In addition, SAP recommends testing all relevant local statutory reports both in the source system and the Central Finance system.
- Cross-process control features work only for intercompany AP and AR line items for the company codes that are in scope for central payment.
- As of writing this book, SAP doesn't support the following processes for the intercompany AP and AR with tax consistency check option:
  - SD Down Payment Status Transfer
  - Down Payment Integration with SD
  - Down Payment Integration with MM
- You must activate trading partners for the company code in scope.

Now that you understand the central payment options and relevant concepts, let's activate central payment in the system. To reach the configuration settings, navigate to IMG menu path **SAP Customizing Implementation Guide • Financial Accounting • Central Finance: Target System Settings • Central Payment • Activate Central Payment with Flexible Options** or use Transaction CFIN_CPAY_CUST. After entering the customizing activity, you'll arrive at the **Change View "CPAY: Central Payment Activation": Overview** screen, as shown in Figure 9.28. Our focus here is on central payment activation assuming tax consistency is already active. On this screen, if a full central payment functionality is in scope for activation, then choose **0 Central Payment Active with Tax Consistency Check** ❷ from **the Scope** dropdown list. If the central payment scope is only for intercompany open items, then select the **CPAY Active for Intercompany APAR with Tax Consistency Check** ❸ from the **Scope** dropdown list.

Note that you're switching the **Scope** option from **1 Tax Consistency Check Active** ❶, which is activated before starting the initial load, as discussed in Chapter 5, Section 5.2.1.

Once the **Scope** option is selected, click the **Save** ❹ button to save the configuration entries.

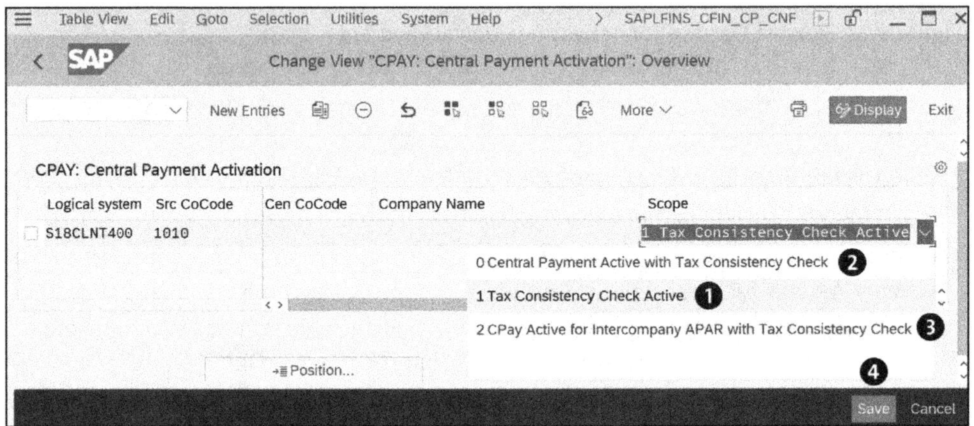

Figure 9.28 Activating Central Payment

[»]

Note

Once central payment is activated for a company code, the system doesn't allow you to deactivate it. This is to prevent inconsistencies, such as if payments are already processed in the Central Finance system, that will lead to inconsistencies. SAP provides program CPAY FINS_CFIN_APAR_CPDEL to deactivate central payment activation. However, SAP doesn't recommend using this program in a live system. The program may be used for testing purposes in a lower system, especially when you're running mock cycles where a reset is required.

> **Tip**
>
> SAP recommends you complete the FI initial load prior to activating central payment.

Technically, as highlighted in Figure 9.29, after you've saved central payment for the company code, the system updates source system table `CFIN_CPCTL` ❶ with a timestamp via an RFC call. AR and AP open items posted in the source system from this timestamp onward are technically cleared prior to replication. Likewise, table `FINS_CFIN_CPCTL` ❷ is updated for relevant logical system and company code combinations on the Central Finance side.

When you choose **Reconcile** button from Figure 9.28 ❺, the system compares mapping relationships in the source system and Central Finance for the company code in scope for central payment and then displays any inconsistencies. In our example in Figure 9.30, the system displays no exceptions with a successful activation status. If any inconsistencies are found during activation, the system displays those in a red status. After fixing inconsistencies, you'll use the **Sync** ❶ button to check the status. In addition, the **Refresh** ❷ button can be used to display the status.

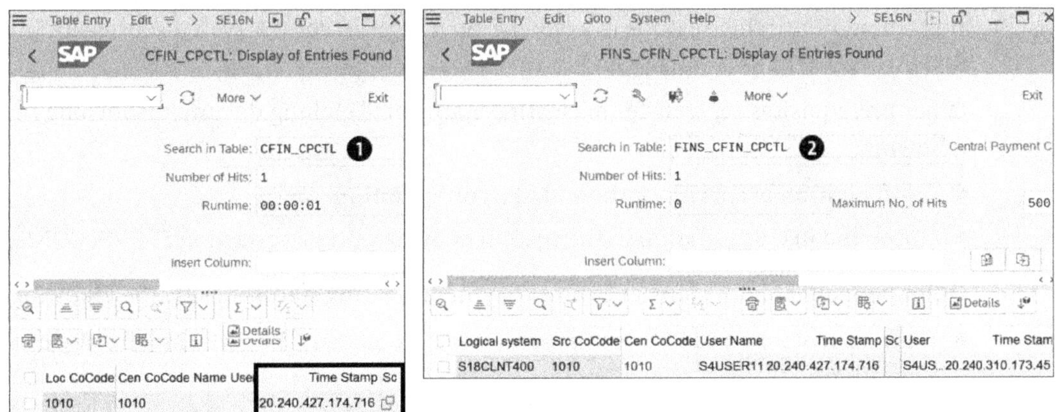

Figure 9.29 Central Payment Activation (Table CFIN_CPCTL)

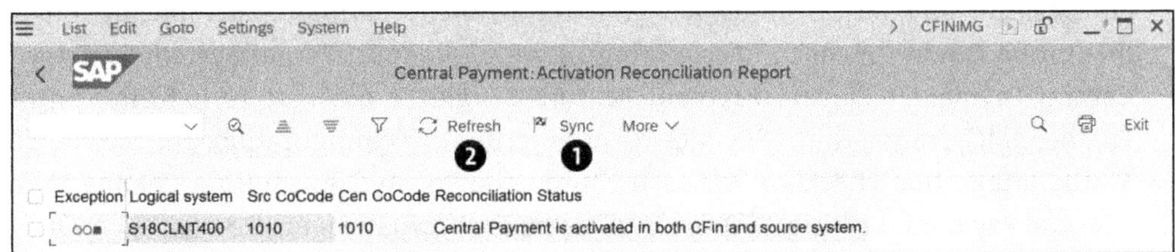

Figure 9.30 Activating Central Payment – Reconcile

For environments below the SAP S/4HANA 2022 release, the **Reconcile** button is not available. The only way to check the activation in those lower instances is to run the reconciliation report to synchronize the setting with the source system for any inconsistencies.

Note

Central Finance does not support central payment activation if you're connected SAP S/4HANA Cloud (Public) as a source system. As a result, central payment, cross-system process control, and SD processes are not available. Consult SAP Note 3398470 for further guidance on restrictions when SAP S/4HANA cloud as a source system connect to central finance.

Central Payment Feature Toggles

In this configuration, you'll toggle the SD integration system status with the Central Finance system. As a result of activating this check, the system transfers the SD billing document clearing status and SD downpayment status from the source system to the Central Finance system. Activating central payment is a prerequisite for this configuration. In addition, this feature is available only for the source system either in SAP S/4HANA 2020 or SAP S/4HANA Cloud systems.

Other prerequisites for transfer of clearing status of SD billing documents include the following:

- Configuration of specific outbound RFC configuration
- Outbound RFC destination in the `bgRFC` scheduler using Transaction SBGRFCCONF

Other prerequisites for transfer of downpayment clearing status include the following:

- Definition of inbound RFC destination `FINS_CFIN_SD_DPST_INB_DEST` in the `bgRFC` configuration
- Addition of prefix `CFIN_SD_DPST` for destination `FINS_CFIN_SD_DPST_INB_DEST` in the `bgRFC` configuration

To toggle SD integration status, navigate to IMG menu path **SAP Customizing Implementation Guide • Financial Accounting • Central Finance: Target System Settings • Central Payment • Central Payment Feature Toggle** or navigate to it from Central Finance IMG Transaction CFINIMG. After entering the configuration, choose the **SD Status Integration** checkbox for each source system on the resulting **Change View "Central Payment: Features": Overview** screen, as shown in Figure 9.31.

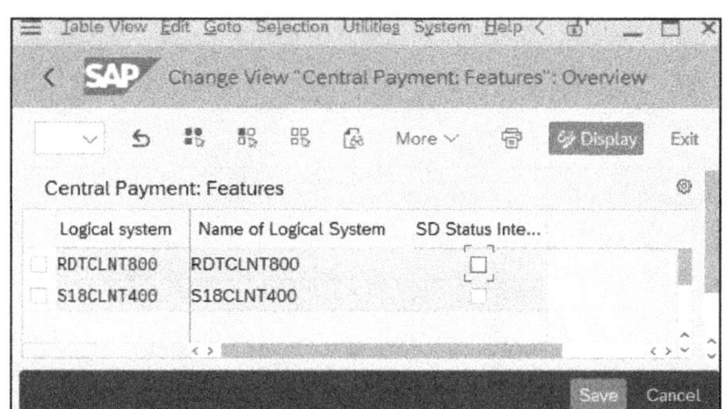

Figure 9.31 Central Payment Feature Toggles

9

Cross-System Process Control for the Central Finance System

Like the source CSPC setting we discussed in Section 9.3.2, in this configuration, you'll define the remote system used in CSPC to control the business and technical checks for the central payment scenario. As discussed earlier, the process type is synchronized between the systems. However, as a perquisite, the remote system creation part requires configuration activity in both the source system and the Central Finance system. In addition, SAP recommends implementing SAP Note 2889748 in both the source system and the Central Finance system. SAP Note included with SAP S/4HANA software component 105 and above.

To configure the remote system for the source system involved in the central payment scenario, navigate to IMG menu path **SAP Customizing Implementation Guide • Financial Accounting • Central Finance: Target System Settings • Cross System Process Control (Target System) • Configure Remote Systems for Cross-System Process Control** or navigate to it from Central Finance IMG Transaction CFINIMG. After entering the configuration, you'll arrive at the **Change View "CSPC Remote Systems": Overview** screen, as shown in Figure 9.32. On the resulting **New Entries: Overview of Added Entries** screen, click the **New Entries** button to maintain target **Logical System** and **RFC Destination** ❶ for CSPC remote systems. As discussed in Section 9.3.2, the **Process Types** ❷ are maintained automatically when the process types are added in the source system configuration. If the system displays an error upon synchronization, you'll able to maintain the changes in the respective system directly.

[«]

Note

SAP recommends performing this activity in a production system as a manual activity, so it needs to be tracked in the Central Finance cutover activity.

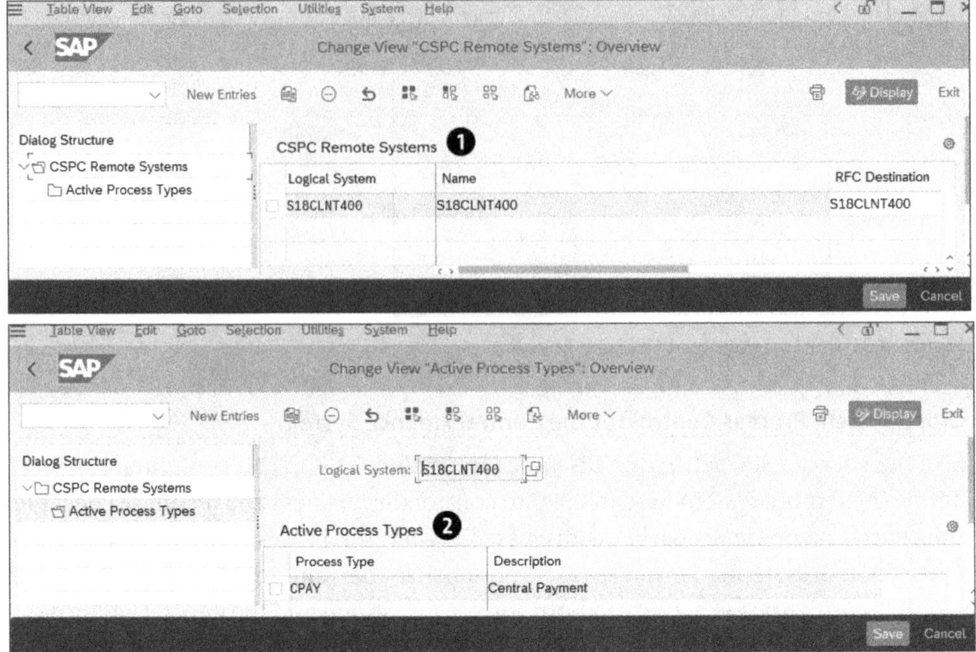

Figure 9.32 CSPC Configuration – Target

9.3.4 Central Payment Activation with a Third-Party System

An important prerequisite to activate central payment for a third-party system is to maintain the third-party logical system in the **Maintain RFC Assignments and Settings for Source Systems** configuration located under IMG menu path SAP **Customizing Implementation Guide • Financial Accounting • Central Finance: Target System Settings • Set Up Systems • Maintain RFC Assignments and Settings for the Source System** or navigate to it from Central Finance IMG Transaction CFINIMG. As shown in Figure 9.33, you'll maintain the **Third-Party System** ❷ checkbox for the third-party logical system under the **RFC Destination** ❶ submenu.

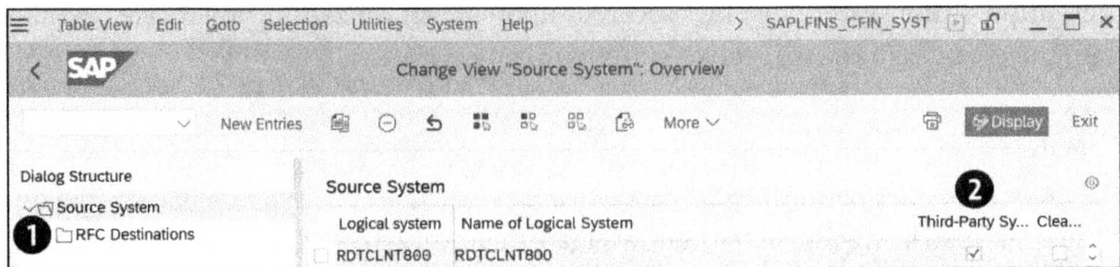

Figure 9.33 Maintain RFC Assignments and Settings for the Source Systems: Third-Party System

[+]

Tip

Once you've configured **Maintain RFC Assignments and Settings for Source Systems** for a third-party system, the system makes the logical system available under the central payment configuration automatically.

Next, you'll activate central payment for the third-party system under IMG menu path **SAP Customizing Implementation Guide • Financial Accounting • Central Finance: Target System Settings • Central Payment • Activate Central Payment for the Third-Party Source System** or navigate to it from Central Finance IMG Transaction CFINIMG. After reaching the configuration, you'll arrive at the **Change View "Activate Central Payment for Third Party Sy..": Overview** screen, as shown in Figure 9.34. On this screen, maintain the system-provided **Activation Option** ❷ dropdown list under the **Company Code** ❶ subscreen:

- **Activate Central Payment for all company codes**
  This option activates central payment for all company codes, thereby reducing maintenance of individual company code maintenance. If you've made this choice, the change is irreversible. This means you won't be able to activate central payment at an individual company code level.
- **Activate Central Payment for individual company codes**
  With this option, you'll maintain individual company codes. Likewise, once you make this choice, the system doesn't allow you to switch to the **Activate Central Payment for all company codes** option.

Once you've selected the dropdown choice, click the **Save** button to save the configuration entry.

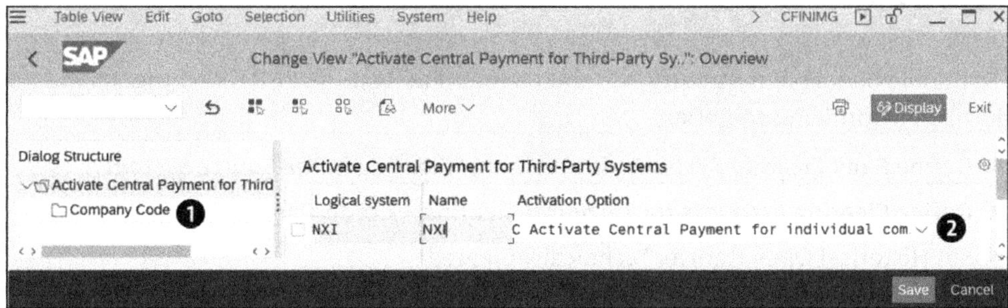

Figure 9.34 Activate Central Payment for Third-Party Systems

[«]

Note

For additional guidance, consult SAP Note 2597587 on enabling the central payment functionality for a third-party system.

9.4 Handling Open Items with Central Payments

Once the central payment functionality is activated, all payments are processed in the Central Finance system. To avoid duplicate payments, any historical open items that are open in the source system prior to central payment activation must be technically cleared. However, in some situations, clearing a large volume of open items isn't feasible in the source system. SAP addresses handling of historical open items in the source system to ensure payment and clearing takes place in the Central Finance system. SAP supports historical open items handling of the following categories:

- AR
- AP
- Cash discount clearing accounts
- Deferred tax accounts
- Bank clearing accounts
- Credit card clearing accounts for scope 00 on central payment

Note

On lower SAP S/4HANA releases, SAP provided **Ensuring Payment and Clearing Takes Place in the Source System** solution to process historical open items in the source system. With this approach, payments of historical open items are processed in the source system only by applying a specific payment block in the Central Finance system. SAP recommends switching to the **Ensuring Payment and Clearing Takes Place in the Central Finance System** solution by following all tasks outlines in the recommended approach. If the earlier approach is used, SAP recommends removing the payment block from the historical items in the Central Finance using report FINS_CFIN_APAR_HIST_OI_PRC.

To handle historical items after central payment activation, SAP recommends the following customizing activities:

- **Define Bank Clearing Accounts for Technical Clearing in the Source System**
- **Define Clearing Accounts for Technical Clearing – Credit Cards**
- **Set Historical Open Item as Technically Cleared**
- **Initializing CSPC**
- **Reset Historical Cleared AP/AR**
- **Correcting Tokens**

Let's discuss the customizing steps for these activities next: Handling of open items required for the full central payment activation with activation scope 00, central payment for intercompany AP and AR line items with activation scope 02, and switching from activation scope from 02 to 03.

9.4.1 Define Bank Clearing Accounts for Technical Clearing in the Source System

In this customizing activity, you'll maintain a bank clearing account that is used for historical open items technical clearing. For example, if you have historical outgoing checks created before central payment activation, the bank clearing account setup in this configuration is used for migrated checks to technically clear in the source system. You'll maintain a bank clearing account for each company code that is active for central payment. In addition, the general ledger account must be an open-item managed account.

To reach the settings, navigate to IMG menu path **SAP Customizing Implementation Guide • Financial Accounting • Central Finance • Central Finance: Source System Settings • Settings for Central Payment • Handling of Historical Open Items • Define Bank Clearing Accounts for Technical Clearing**. After reaching the configuration, as shown at the top of Figure 9.35, click **New Entries** ❶ on the **Change View "G/L Clearing Account for Technical Clearing": Overview** screen. On the resulting **New Entries: Overview of Added Entries** screen, maintain values in the **CoCD** and **G/L Account** ❷ fields. Click the **Save** ❸ button to save the configuration entry.

9

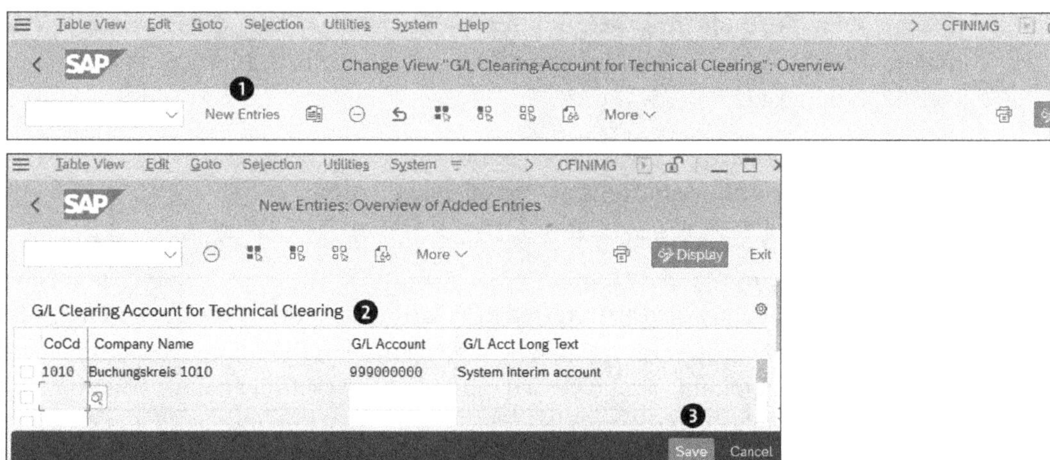

Figure 9.35 Define Bank Clearing Accounts for Technical Clearing

9.4.2 Define Clearing Accounts for Technical Clearing: Credit Card

In this customizing activity, you'll maintain a credit card clearing account that is used for historical open items settlement processes in the source system. If you've maintained a bank clearing account for central payment as we discussed in Section 9.4.1, the system defaults to the same account in this configuration. Nevertheless, the system provides the configuration under the credit card node if additional entries are required to be maintained. One prerequisite is that the general ledger account must be an open-item managed general ledger account.

To reach the settings for the SAP S/4HANA, SAP Simple Finance, or SAP ERP source systems, navigate to IMG menu path **SAP Customizing Implementation Guide • Financial Accounting • Central Finance • Central Finance: Source System Settings • Settings for Central Payment • Credit Card • Define Clearing Accounts for Technical Clearing.**

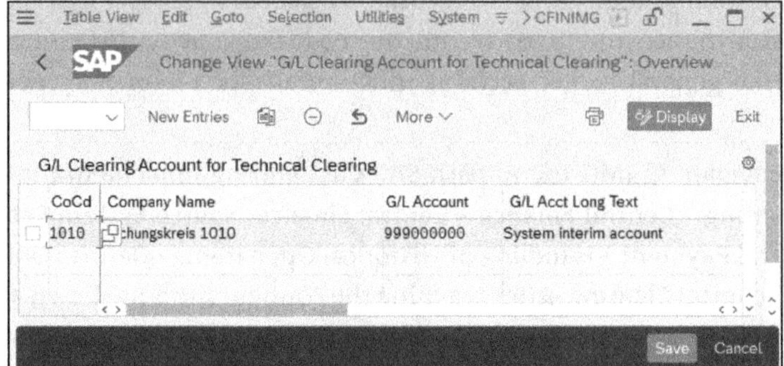

Figure 9.36 Define Clearing Accounts for Technical Clearing

After reaching the configuration, as shown in Figure 9.36, click **New Entries** on the **Change View "G/L Clearing Account for Technical Clearing": Overview** screen to maintain additional accounts. The configuration is similar to the configuration we covered in Section 9.4.1.

9.4.3 Set Historical Open Items as Technically Cleared

With central payment activation, all open items are processed only from the Central Finance system. Prior to activation of central payment, the source system may have contained open items, which are referred to as historical items. These open items must be technically cleared after central payment activation for the company codes that are in scope, so all open items are processed only from the Central Finance system going forward. SAP provides Transaction FINS_CFIN_HISOI_CLR (program FINS_CFIN_CPAY_HISTL_OI_MASTER) to set historical open items with technically cleared status. Upon execution of the transaction, the system marks all open historical open items with clearing document (BSEG-AUGBL) and clearing date (BSEG-AUGDT) fields with values of **ALE-extern** and the posting date, respectively, as shown in Figure 9.37. Note Transaction FINS_CFIN_TECH_CLR is available only in lower environments and it's replaced with Transaction FINS_CFIN_HISOI_CLR from SAP S/4HANA 2022 release onward.

As a result, you'll be able to process open items from the Central Finance system only. The transaction must be run interactively in Central Finance, which kicks off multiple background jobs. Ensure you have execution access in the source system with authorization object F_CFIN_SRC and Central Finance with authorization object F_CFIN_TRG. In addition, based on the historical open item volume, you'll have to adjust the timeout parameter for the job to be run in the Central Finance system interactively. SAP limits a maximum processing of 5 million open items per account activity for each run. If the volume exceeds this limit, the system issues warning message **Remaining open items may exist under account &1 (&2) in system &3**. If the volume is higher than 5 million per each account, then you'll have to repeat the transaction multiple times until all historical open items are cleared. Consult with the Basis team on the number of batch work processes available to execute the transaction. Figure 9.37 highlights an example of invoice technically cleared document (**ALE-extern**) in the source system.

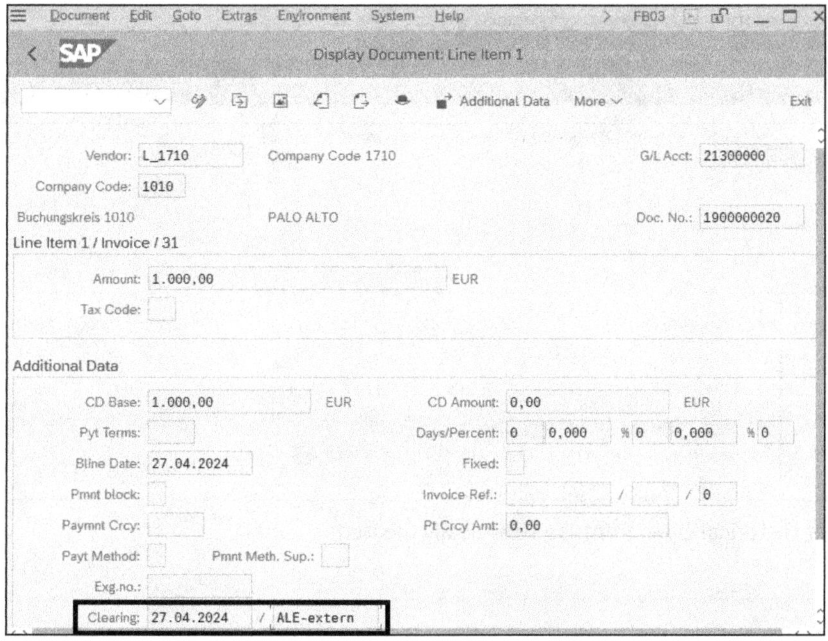

Figure 9.37 Technically Cleared Documents

To run the transaction, navigate to IMG menu path **SAP Customizing Implementation Guide • Financial Accounting • Central Finance • Central Finance: Target System Settings • Central Payment • Handling of Historical Open Items • Set Historical Open Items as Technically Cleared** or use Transaction FINS_CFIN_HISOI_CLR. After running the transaction, you'll arrive at the **Set Historical Open Items as Technically Cleared** screen, as shown at the top of Figure 9.38.

On this screen, enter values in **Number of Batch Jobs** and **Server Group** in the **Batch Job Control** section ❶ based on your Basis team's recommendation on parallel processes capacity. Click the **Execute** ❷ button to continue. Note this report should be run in online mode. The system sets clearing document field (BSEG-AUGBL) with ALE-Extern along with posting time populated on field clearing date (BSEG-AUGDT) to facilitate these open items processed only in Central Finance. Upon completion of the report, the job log displays how many packages ❸ have been processed by the system, as shown at the bottom of Figure 9.38.

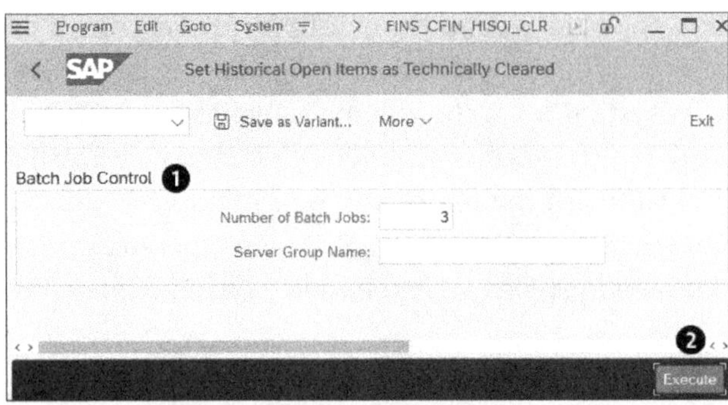

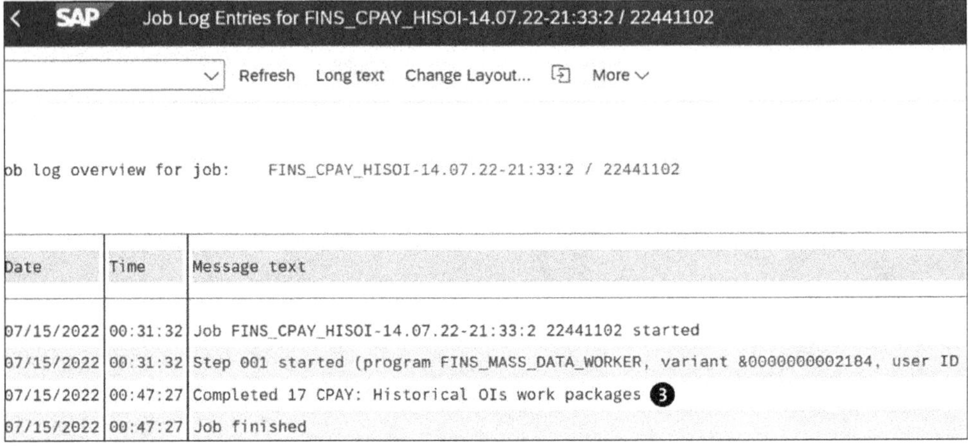

Date	Time	Message text
07/15/2022	00:31:32	Job FINS_CPAY_HISOI-14.07.22-21:33:2 22441102 started
07/15/2022	00:31:32	Step 001 started (program FINS_MASS_DATA_WORKER, variant &0000000002184, user ID
07/15/2022	00:47:27	Completed 17 CPAY: Historical OIs work packages 3
07/15/2022	00:47:27	Job finished

Figure 9.38 Set Historical Open Items as Technically Cleared

[!]

Warning

The system processes a maximum of 5 million open items per account for each run. If the open items exceed 5 million, you'll repeat the transaction until all open items are set to technically cleared (ALE-Extern).

Once the report is executed, the system provides an application log with the activity performed by the report. Additional information on the background job can also be found in Transaction SLG1 with **Object** and **Subobject** fields having values of **FINS** and **FINS_CFIN_HISTOI_CLR**, respectively. SAP also provides monitoring Transaction FINS_CFIN_HISOI_MON (Display Technical Clearing Status of Historical Open Items) to display the status of the job.

[»]

Note

The system considers both full central payment activation (scope **0**) and CPAY active for intercompany AP and AR (scope **2**) historical open items, depending on what's configured in the system. In the scope 2 scenario, the system technically clears only intercompany AP and AR open items where the trading partner is entered (BSEG-VBUND).

9.4.4 Initializing Cross-System Process Control

For SAP S/4HANA 2020 or higher releases, SAP provides Transaction FINS_CFIN_CPAYCSPC_I (program FINS_CFIN_CSPC_CPAY_MASTER) to create initial tokens for historical AR and AP open items. This is to ensure the system follows a consistent flow of business processes per CSPC controls. This is a required step. Otherwise, the system doesn't allow payment clearing in the Central Finance system. One prerequisite is that real-time replication is turned on. You'll execute the transaction in three scenarios:

- **Central payment is activated first before CSPC activation**
  You'll execute Transaction FINS_CFIN_CPAYCSPC_I for the historical open items before activating CSPC.
- **CSPC is activated first before central payment activation**
  You'll run Transaction FINS_CFIN_CPAYCSPC_I for the historical open items created prior to central payment activation.
- **A new company code activated for central payment**
  Whenever you activate a new company code, you'll run Transaction FINS_CFIN_CPAYCSPC_I to initialize all historical open items on AP and AR open items.

To run the transaction for these scenarios, navigate to IMG menu path **SAP Customizing Implementation Guide • Financial Accounting • Central Finance • Central Finance: Target System Settings • Central Payment • Cross-System Process Control for Central Payment • Initialization of Cross-System Process Control** or use Transaction FINS_CFIN_CPAYCSPC_I. After running the transaction, you'll arrive at the **Cross-System Process Control for Central Payment Initialization Report** screen, as shown in Figure 9.39.

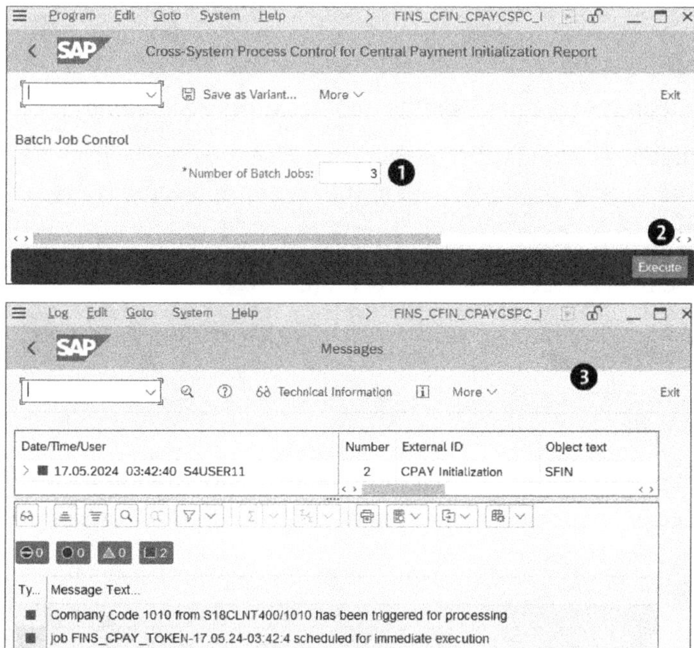

Figure 9.39 Initialization of CSPC

On this screen, enter **Number of Batch Jobs** ❶ located in the **Batch Job Control** section based on your Basis team's recommendation on the parallel processes capacity of the server. Click the **Execute** ❷ button to continue. You'll observe the system displays a confirmation message ❸ that the job has been triggered for processing.

Next, you'll run Transaction FINS_CFIN_CPAYCSPC_M (Initialization Status Monitor) to check the status of the submitted background job, as shown in Figure 9.40. Fix any errors and rerun until you have a successful run.

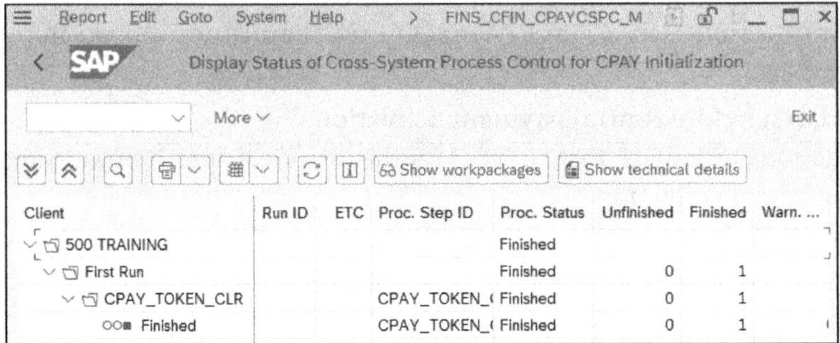

Figure 9.40 CSPC Initialization Status Monitor

> **Prerequisites**
>
> To run the transaction, you'll need authorization object F_BKPF_BUK for company codes and F_CFIN_TR for central payment. In addition, the RFC user must be assigned with authorization object F_CSPC_TKN.

9.4.5 Reset Historical Cleared Accounts Payable and Accounts Receivable

Historical AP and AR open items cleared with the following scenarios can be reset with Transaction FBRA to facilitate centralized payment in the Central Finance system:

- Central payment is activated first before CSPC activation.
- CSPC is activated first before central payment activation.

In both instances, CSPC will create a token for reopened items via Transaction FBRA.

9.4.6 Correcting Token Issues

While executing the payment process after activating CSPC, as shown in Figure 9.41, the system may raise a token error. In such cases, the system won't allow you to process the payment in the Central Finance system.

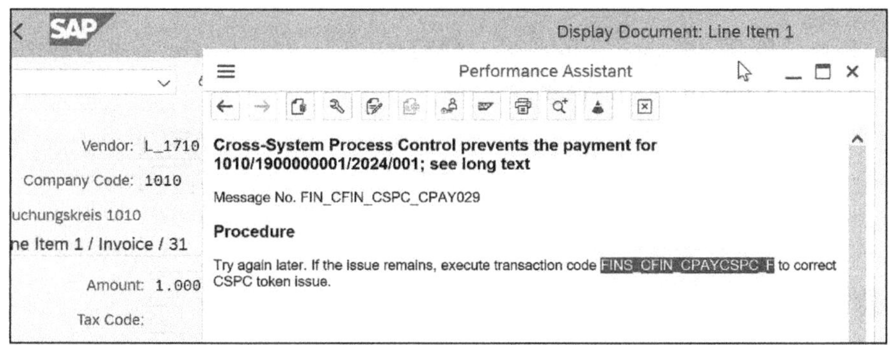

Figure 9.41 CSPC Preventing an Invoice

9

The system raises common errors in situations identified in Table 9.4.

Type of Error	Scenario
Configuration-related errors	RFC connection failure
Locking of objects	An account locked by a user
SAP Application Interface Framework errors	SAP Application Interface Framework errors such as **A predecessor exists with MGGUID xxx**
Semantic errors	**Document is cleared/reversal in System xxx**

Table 9.4 Common Token Issues

SAP provides the token correction Transaction FINS_CFIN_CPAYCSPC_F to fix such errors in the Central Finance system. Transaction corrects token only if no issues are found with the SAP Application Interface Framework message. In addition, prior to executing this transaction, ensure you've cleared all predecessor messages in SAP Application Interface Framework before the token can be corrected. Otherwise, the system would return an error message stating a predecessor exists upon executing this transaction (message ID FIN_CFIN_CSPC_CPAY015). To run the token correction transaction, navigate to IMG menu path **SAP Customizing Implementation Guide • Financial Accounting • Central Finance • Central Finance: Target System Settings • Central Payment • Cross-System Process Control for Central Payment • Correcting Token Issues for Cross-Process Control** or use Transaction FINS_CFIN_CPAYCSPC_F. After reaching the transaction, you'll arrive at the **Correcting Token Issues for Cross-System Process Control** screen, as shown in Figure 9.42. On this screen, you'll see that the system provides two options to choose under the **Options** ❶ section:

- **Execution**
  If you have a specific token issue, choose this radio button to fix token issues for open items.

- **Display Logs**
  If you choose this radio button, the system displays logs of activities carried out for payment run or automatic clearing.

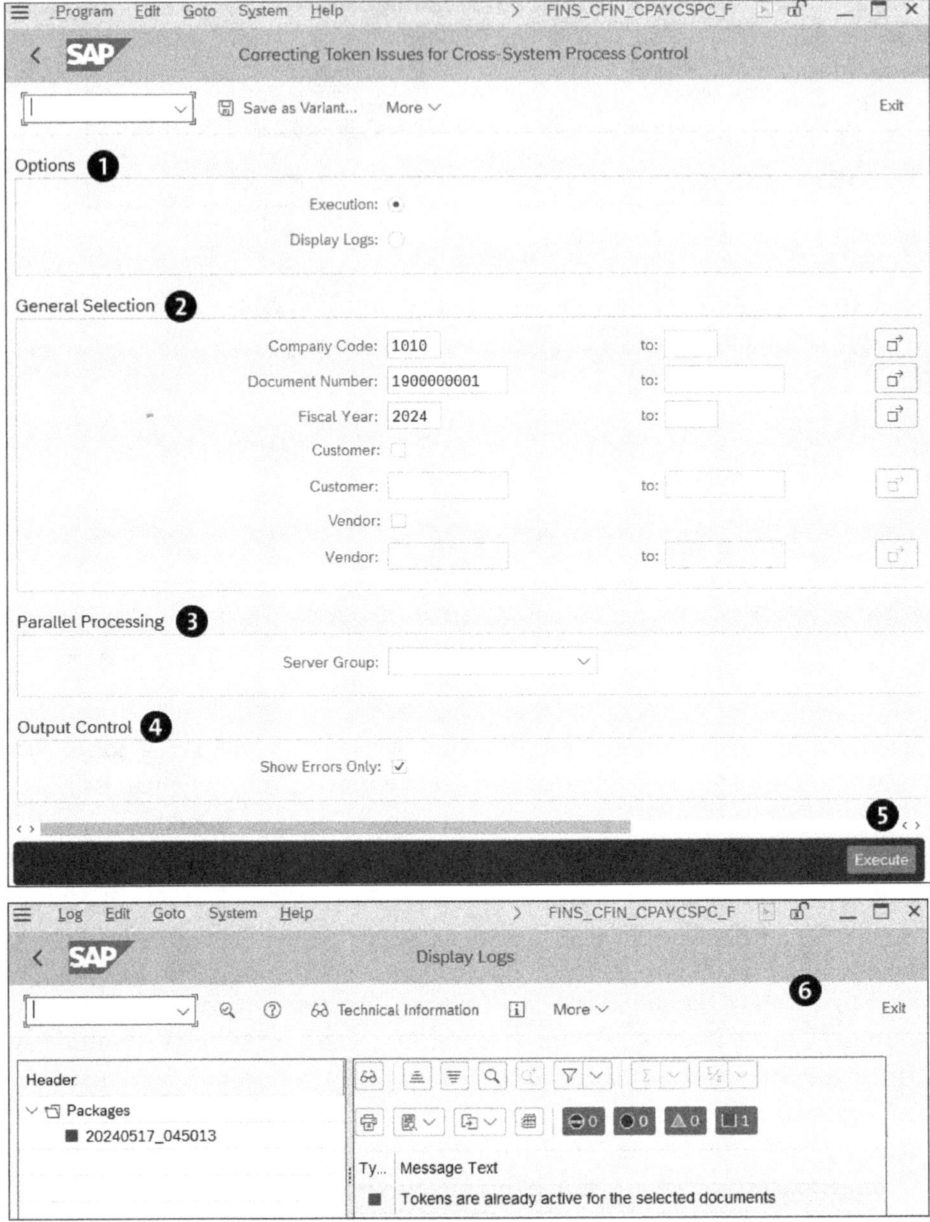

Figure 9.42 Correcting Token Issues for CSPC

On this screen, enter selection criteria under **General Selection** ❷. In our example, we've selected values for **Company Code**, **Document Number**, and **Fiscal Year** fields. Next, choose a **Server Group** under the **Parallel Processing** ❸ section based on your Basis team's recommendation. In addition, if you want to display only errors, choose the **Show Errors Only** checkbox under the **Output Control** ❹ section. Click the **Execute** ❺ button to correct the token error. You'll observe the system displays a confirmation message on the **Display Logs** ❻ output screen:

- A successful message that the token has been corrected or an error message with specific reasons
- A notification that the token issue is already corrected for the selected document

Prerequisites

To run the transaction, ensure the following:

- You have authorization object F_BKPF_BUK for company codes and F_CFIN_TR for central payment.
- SAP Landscape Transformation Replication Server is actively running.
- Both central payment and CSPC are active.

SAP provides Transaction FINS_CFIN_TRWPR_REG (report RFINS_CFIN_TRWPR_REGISTRATION) in the Central Finance system to display, register, or deregister TRWPR function modules, which are used in the central payment scenario. This transaction enables FI-relevant transactions to be included for the central payment process. By running this transaction, the system ensures the following:

- Register document types that are pertinent for the central payment process, especially to identify when invoices, credit memos, and other similar documents are replicated over to Central Finance.
- Include transactions with document types registered for the central payment process.
- Perform mapping.
- Maintain consistency and reduce risk of errors as specific documents are registered for the central payment process.

After running the transaction, you'll arrive at the **Program RFINS_CFIN_TRWPR_REGISTRATION** screen, as shown in Figure 9.43. Choose the **Display TRWPR Modules** ❶ radio button, and then click the **Execute** ❷ button. On the resulting screen, the system displays the registered transactions ❸.

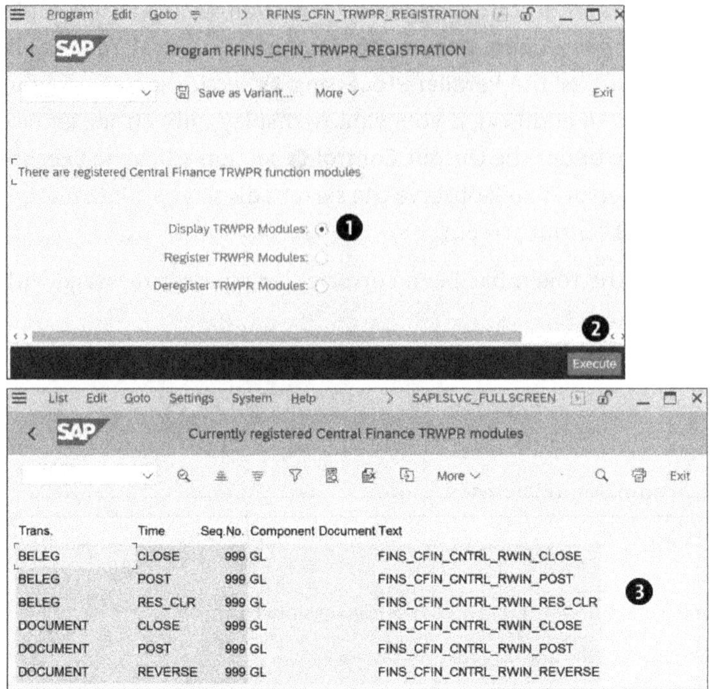

Figure 9.43 Register-Deregister Central Finance TRWPRs

[+]

Tip

Consult SAP Note 3086334 if TRWPR modules can't be registered.

SAP provides Transaction FINS_CFIN_LOG_DEL (report FINS_CFIN_APAR_APPL_LOGS_DEL) to delete the central payment application logs created based on the selection criteria. The transaction deletes entries from application log table FINS_CFIN_EXECLG. You may consider running this report periodically to reduce the redundant data to improve the performance of the system.

9.5 Down Payment Integration

A down payment is a type of partial payment made or received in advance for exchange of goods or services from customers and vendors. This occurs commonly on large transactions to secure orders or commit to buy or purchase. To handle down payments consistently and centrally across an enterprise, SAP supports replicating down payments from various source systems to the Central Finance system. As a result, you'll gain centralized control over transactions, better system integration among integrated systems, and improved cash flow. In addition, this brings enhanced reporting capabilities by leveraging SAP S/4HANA's latest innovations.

In SAP, down payments are considered open items until they are cleared. Central Finance supports replicating down payment postings from the source systems for both SD and MM components. In this section, you'll learn about down payment replication requirements, configurations, and restrictions to be noted during implementation. Let's discuss down payment integration with SD first, and then we'll transition to the MM side.

9.5.1 Down Payment Integration with Sales and Distribution

In Central Finance, you'll run the down payment process in the Central Finance system with the central payment functionality. Central Finance supports two types of down payment replication and clearing activities, as well as invoices for SD components:

9

- **Request-based**
  Down payments are posted in the source system with a reference to a down payment request. Sales orders are created with item category **TAO**, which is normally used in milestone billing plans.
- **Condition-based**
  Down payments are posted in the source system with a reference to a sales order of item category **TAN**.

Let's go over each type of down payment replication next.

Request-Based Down Payment Replication

In SAP, SD-related down payment requests can be initiated either from milestone billing or financials for a customer before delivery of goods or services in the source system. A down payment request begins with a sales order that is created with a milestone billing plan with item category **TAO**. A down payment agreement was created in the billing plan to reflect a schedule of payments based on agreed terms. The value of the agreed payment can either be an amount or a percentage value of the item or service. The Central Finance process for request-based down payment replication is illustrated in Figure 9.44, and the data flow is explained here:

❶ **Down payment request (DPR) replication**
As soon as a down payment request is created in the source system, the DPR line gets cleared technically with `ALE-Extern` and replicated to the Central Finance system.

❷ **Remote read of DP**
During the billing cycle run, relevant down payment request information is retrieved from the Central Finance system. The system pulls the important fields such as the following:

- **Company Code**
- **Customer Number**
- **G/L Account**
- **Special G/L indicator**

- Assignment
- Tax code
- Business Area
- Profit Center

❸ **Replicate receivable item**
The receivable item is technically cleared in the source system and replicated to the Central Finance system. The down payments clearing process is carried out in the Central Finance system.

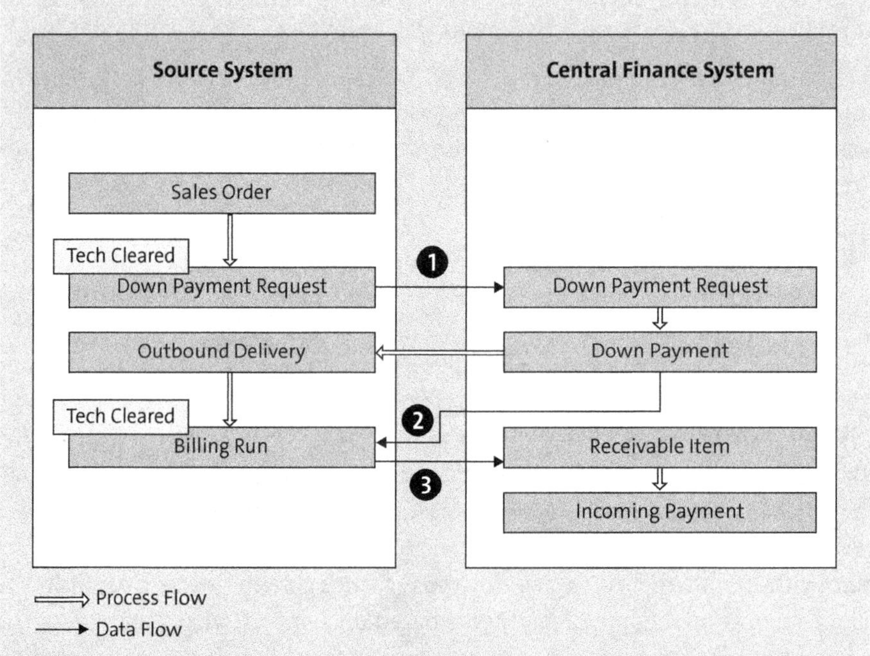

Figure 9.44 Down Payment Integration with SD: Request-Based

Condition-Based Down Payment Replication

As shown in Figure 9.45, the condition-based down payment posted directly without any reference to the down payment request:

❶ **Down payment request (DPR) replication**
When a sales order is created using conditions for down payment in the source system, a down payment is sent to the Central Finance system.

❷ **Remote read of DP**
In the subsequent step, the following down payment fields are pulled from the Central Finance system.

- Company Code
- Customer Number

- G/L Account
- Special G/L indicator
- Assignment
- Tax code
- Business Area
- Profit Center

❸ **Replicate receivable item**
The receivable item is technically cleared in the source system and replicated to the Central Finance system. Down payment clearing activities are then carried out in the Central Finance system.

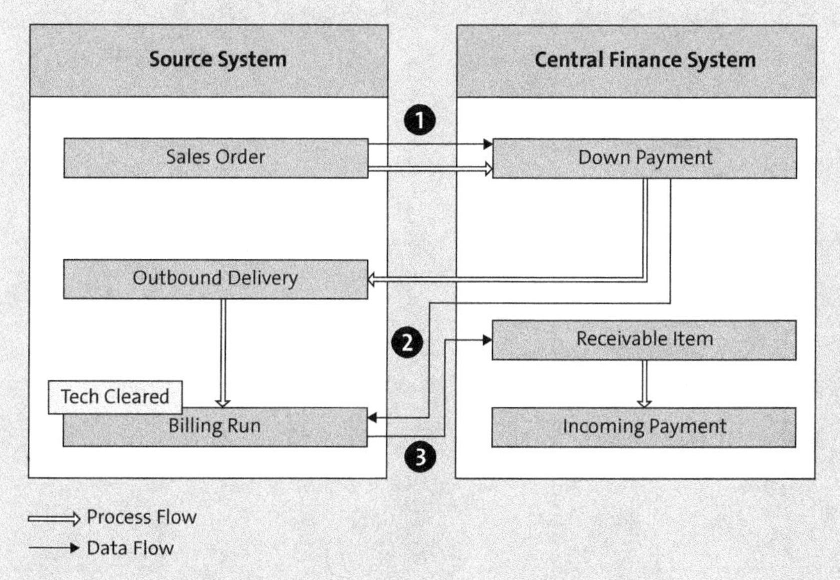

Figure 9.45 Down Payment Integration with SD: Condition-Based

[«]

Note

The system only pulls information about unique values from the source system. In other words, if field mapping in Transaction FINS_CFIN_MAP_REL is set with an N:1 relationship, the system won't retrieve a unique value that applies in the source system.

In addition, SAP recommends considering the following:

- When you have an N:1 mapping scenario, define a default value in value mapping. In key mapping, you'll select the value of index 1 as the default value.
- SAP also offers two BadIs that can be used for additional customer-specific enhancements: `BADI_FIN_CFIN_DP_MAP` for the source system and `BADI_FINS_CFIN_DP_MAP` for Central Finance.

Prior to central payment activation, down payments are replicated from the source system to Central Finance. However, the system removes sales order information from the down payments as this information doesn't exist in the target system. When you activate central payment at a specific point, any open items in the source system are considered as historical open items. You'll need to update sales order information, for example, sales order number, back on these historical open items, so they can be processed with the central payment functionality. Refer to Section 9.5.3 and Section 9.5.4 regarding customizing activities to update historical down payment open items.

For SD-related down payment processes, Table 9.5 lists SAP Fiori apps and SAP GUI transactions supported in the Central Finance system.

SAP Fiori apps		SAP GUI Transaction	
App ID	**Description**	**Transaction Code**	**Description**
F1689	Manage Customers Down Payment Requests	F-29	Post Customer Down Payment
F1345	Post Incoming Payments	F-28	Post Down Payment/ Post Incoming Payment
F1345	Clear Incoming Payments	F-39	Clear Customer Down Payment
		F-32	Clear Customer
F0770	Manage Automatic Payments	F110	Parameters for Automatic Payment
F1564	Manage Bank Statement	FF67	Manual Account Statement
F1520	Reprocess Bank Statement Items	FEBA_BANK_STATEMENT	Postprocessing Bank Statement
FF68	Manage Check Deposits	FF68	Manual Check Deposit Transaction
FEBA_CHECK_DEPOSIT	Reprocess Incoming Checks	FEBA_CHECK_DEPOSIT	Postprocessing Check Deposit Trans.
F1680	Manage Incoming Payment Files	FLB2	Import Lockbox File
FLB2	Upload Lockbox Files	FLB2	Import Lockbox File
F1681	Manage Lockbox Batches	FLB1	Postprocessing Lockbox Data
F2223	Reset Cleared Items	FBRA	Reset Cleared Items
F0717	Manage Journal Entries	FB08	Reverse Document

Table 9.5 Down Payment Supported Apps and Transactions in Central Finance

[+]

Note

For further guidance on SD down payments in Central Finance with central payment activation, consult SAP Note 2474760.

9.5.2 Down Payment Integration with Materials Management

In MM, down payments are partial advance payments made to suppliers for the purchase of goods and services. This helps the procurement process secure orders or commit to buy goods and services. The down payment process begins with creating a purchase order in the source systems. While creating purchase orders in the MM component, you'll indicate down payment terms and percentage in item conditions in the **Payment Processing** tab. Likewise, a down payment request can also be initiated with Transaction F-47 and posted with Transaction F-48 in the source system. The Central Finance process for down payments is demonstrated in Figure 9.46.

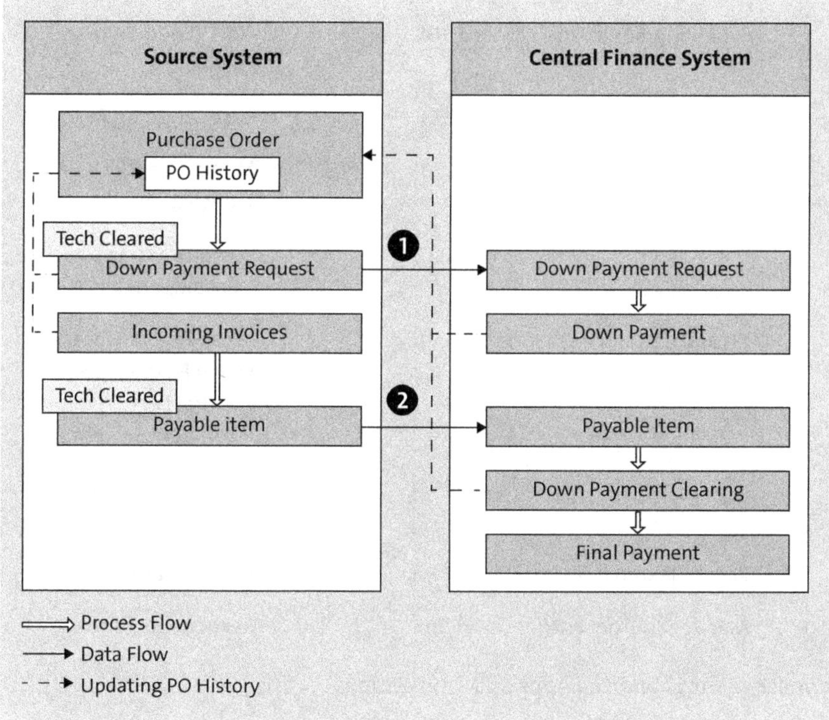

Figure 9.46 Down Payment Integration with MM Overview

Let's explore the down payment data flow architecture next:

❶ **Down payment request (DPR) replication**
Upon posting a down payment request document in the source system, the system marks the DPR request cleared technically with `ALE-Extern` and replicates it to the

Central Finance system. In the Central Finance system, you'll post the down payment to clear the DPR request, which is replicated from the source system using apps or transactions as listed previously in Table 9.5.

❷ **Replicating payable open item**
Upon posting a supplier invoice in the source system, the payable items are replicated to Central Finance for the down payment. As you can see in Figure 9.46, payable items are technically cleared in the source system. SAP recommends clearing down payments in reference to supplier invoices in the Central Finance system.

Table 9.6 provides SAP Fiori apps and SAP GUI apps used to process down payments in Central Finance.

SAP Fiori Apps		SAP GUI Transaction	
App ID	**Description**	**Transaction Code**	**Description**
F0770	Manage Automatic Payments	F110	Parameters for Automatic Payment
F1612	Post Outgoing Payments	F-48	Post Vendor Down Payment
		F-53	Post Outgoing Payment
		F-53	Manual Check Deposit Transaction
		FBZ2	Postprocessing Check Deposit Trans.
F1367	Clear Outgoing Payments	F-44	Clear Vendor (Clear Down Payment: Special General Ledger: A for Open Item Selection)
		F-54	Clear Vendor Down Payment
F2223	Reset Cleared Items	FBRA	Reset Cleared Items
F0717	Manage Journal Entries	FB08	Reverse Document

Table 9.6 Down Payment Supported Apps and Transactions

[»]

Note

SAP doesn't support integrating the clearing down payment functionality with creating supplier invoices in the source system. This requires Transaction MIRO integration through business function activation LOG_MMFI_P2P via Transaction SFW5. Consult SAP Note 2683870 for restrictions regarding central down payment integration with MM.

9.5.3 Update Down Payments with Sales Order Information

Prior to central payment activation, when replicating accounting documents generated from the billing process, the system removes sales order–related information such as sales document number (BSEG-VBEL2), item number (BSEG-POSN2), and schedule line (BSEG-ETEN2) from FI line items. Now that you've activated central payment, replicated down payments require sales order–related information. With Transaction FINS_CFIN_SO_UPDATE (program FINS_CFIN_APAR_SO_UPDATE) in the Central Finance system, you can update down payments with missing information by retrieving them from the source system. Based on the selection criteria in the transaction, the system checks for missing information, pulls relevant information from the source system, and updates missing fields in table BSEG.

To run the transaction, navigate to IMG menu path **SAP Customizing Implementation Guide • Financial Accounting • Central Finance • Central Finance: Target System Settings • Central Payment • Handling of Historical Open Items • Update Down Payments with Sales Order Information** or use Transaction FINS_CFIN_SO_UPDATE. After running the transaction, you'll arrive at the **Update Down Payments with Sales Order Information** screen, as shown in Figure 9.47.

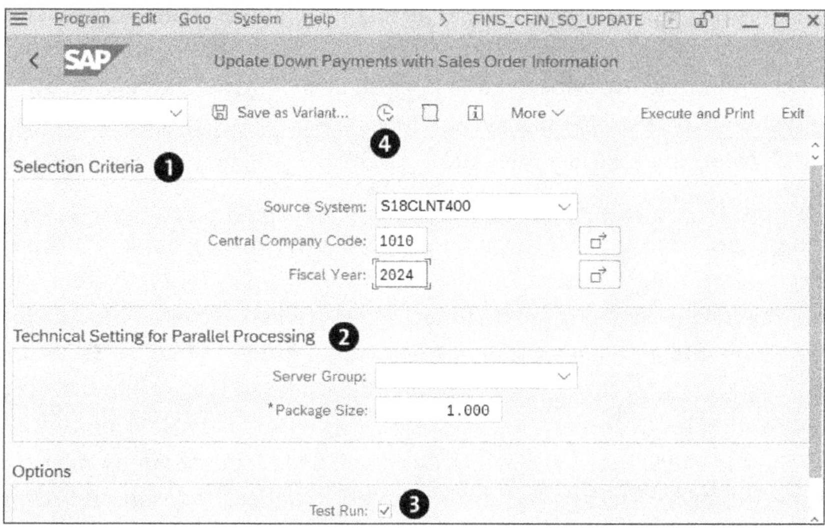

Figure 9.47 Update Down Payments with Sales Order Information

On this screen, enter values in the **Source System**, **Central Company Code**, and **Fiscal Year** fields located under the **Selection Criteria** section ❶. Next, enter **Server Group** located under the **Technical Setting for Parallel Processing** ❷ section based on your Basis team's recommendation. **Package Size** defaults to 1,000 but can be modified based on the parallel processes capacity and volume of transactions from different source systems. The system divides and packages documents by the source system and package size defined in this parameter. The **Test Run** ❸ checkbox is available to preview the results. Click the **Execute** ❹ button to continue. Upon executing the transaction in final mode, the system updates table BSEG with the relevant sales order information. In addition, the resulting screen displays which document line items and fields are updated.

Note

You must have two prerequisites to run Transaction FINS_CFIN_SO_UPDATE in Central Finance:

- You must have the authorization to execute the transaction.
- Central payment must be active for the company code you want to run this report.

9.5.4 Update Relationship between Down Payments and Sales Orders

Before central payment activation, historical open SD down payment items that replicated to the Central Finance system weren't updated with down payment log table FINS_CFIN_SDDPLG. For such historical SD invoice documents, while fetching SD down payment details, the system won't process down payments due to missing the link between the down payment and sales order. This log table contains the relationship between down payments and sales order information. After central payment activation, the down payment relationship log table updates at the time of billing execution in the source system. However, for historical items, SAP provides Transaction FINS_CFIN_SDDPLG_UPD (program FINS_CFIN_APAR_SDDPLOG_UPDATE) to update missing relationship information.

To run the transaction, navigate to IMG menu path **SAP Customizing Implementation Guide • Financial Accounting • Central Finance • Central Finance: Target System Settings • Central Payment • Handling of Historical Open Items • Update Relationship between Down Payments and Sales Orders** or use Transaction FINS_CFIN_SDDPLG_UPD. After reaching the transaction, you'll arrive at the **Update Relationship Between Down Payments and Sales Orders** screen, as shown at the top of Figure 9.48. On this screen, enter values in the **Source System** and **Company Code** fields located under the **Selection Criteria** section ❶. **Package Size** defaults to 1,000 under the **Technical Setting for Parallel Processing** ❷ section, which can be modified based on parallel processes capacity and volume of transactions from different source systems. The system divides and packages documents by the source system and package size defined in this parameter. The **Test Run** ❸ checkbox is available to preview the results. Click the **Execute** ❹ button to continue. Upon executing the transaction in final mode, the system updates table FINS_CFIN_SDDPLG with the relevant sales order relationship information for down payments. In addition, the resulting screen displays the total number of records updated for the selected company code and fiscal year combinations. In addition, the system has an option to display and view detailed output.

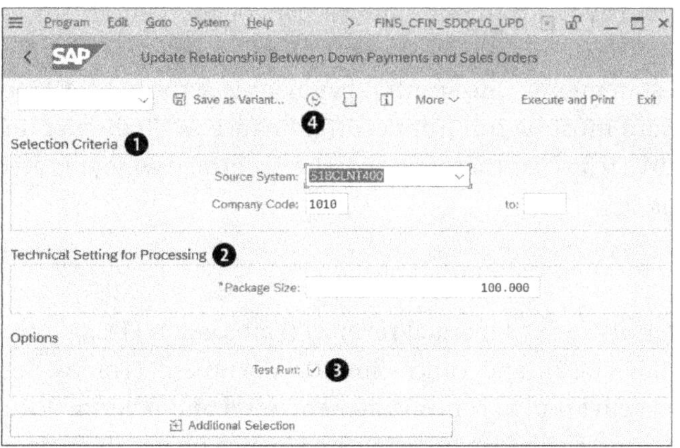

Figure 9.48 Update Relationship between Down Payments and Sales Orders

Note

As a prerequisite, you must have access to the transaction. In addition, central payment must be activated for the company code that is missing the sales order to down payment relationship in log table FINS_CFIN_SDDPLG.

Over a period as the volume grows in the log table, consider archiving the table based on your company's retention policy. The data in this table is included with the archiving of accounting documents.

9.6 Impact on Dependent Process

After activating central payment for the company codes that are in scope, all open items are processed from the Central Finance system. SAP provides guidance on several dependent processes related to centralized payment in the Central Finance system. We've categorized these dependent processes into the following broader categories:

- General ledger
- Asset Accounting
- Financial supply chain management (FSCM)
- Cash operations
- Intercompany reconciliations
- Tax

Let's consider the impact of these dependent processes as a result of central payment activation.

9.6.1 General Ledger

The general ledger in SAP is a critical component in FI that captures all financial transactions. In Central Finance, you'll assess which general ledger processes may have an impact after activating central payment. In addition, as part of the period closing process, you'll make manual corrections and adjustments to reflect balances, thereby producing accurate financial statements. Once central payment is activated, SAP recommends balance carry-forward must be run from central finance. We'll discuss important general ledger relatedWe'll discuss these actions when central payment is activated in the following sections.

Reset and Reversal

In SAP, different types of reversals such as manual reversal (Transaction FB08), intercompany reversal (Transaction FBU8), and cancel invoice document (Transaction MR8M) are processed. Prior to activation of central payment, reset and reversal documents are replicated from the source system to the Central Finance system. If an open item is cleared already, a reset document must arrive prior to the reversal document. During posting, the system checks the timestamp and logical sequence of events occurring with reset and reversals. In addition, a clearing is posted without line items (BSTAT= A), the clearing document is reset in the source system, and the clearing document and the corresponding reset document of reversal are also transferred to the Central Finance system.

After activating the central payment functionality, if an open item has already been cleared through central payment in the Central Finance system, reversal isn't allowed in the source system due to the CSPC controls in place. Figure 9.49 displays a typical reversal sequence in a central payment scenario. In a central payment activation scenario, AP and AR open items are technically cleared in the source system with ALE-extern. As a result of central clearing, a reset of the reversal document isn't allowed in the source system anymore. As shown in Figure 9.49, you must reset ❶ the clearing in Central Finance first prior to initiating the reversal ❷ in the source system.

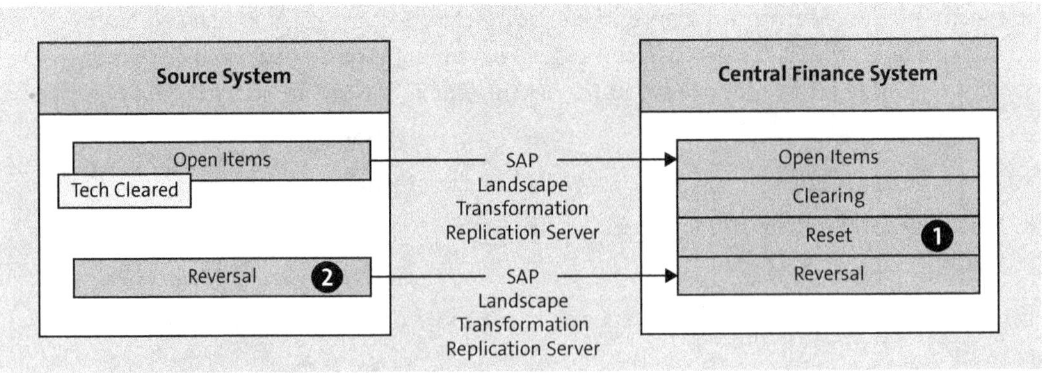

Figure 9.49 Reversal Sequence

After activating central payment, you'll observe the following behavior when transferring reversals and resetting transactions:

- **Transaction FB08 (Individual Reversal)**
  The system prevents reversal in the source system if the corresponding invoice is cleared in the Central Finance system.
- **Transaction F.80 (Mass Reversal)**
  The transaction allows you to perform mass reversal of multiple documents. If open items are already cleared in Central Finance, the system will prevent reversal in the source system.
- **Transaction MR8M (Cancel Invoice document)**
  If a replicated MM invoice is cleared in the Central Finance system, canceling invoices is prevented in the source system.
- **Transaction VF11 (Cancel Billing Document)**
  If a replicated SD invoice is cleared in the Central Finance system, canceling billing documents is prevented in the source system.

In all of these cases, you'll reset using Transaction FBRA (Reset Cleared Items) in the Central Finance system prior to initiating a reversal in the source system. Transaction FBRA has both reset and reversal.

If an open item is posted directly in Central Finance after activation of central payment using transactions such as Transaction FB60, you'll perform the reset and reversal directly in the Central Finance system.

SAP provides Transaction FINS_CFIN_CREV – Central Reversal and Reposting to reverse and repost documents which are replicated with incorrect mapping, incorrect configuration, or errors in the application logic. The transaction reverses and posts a new document in central finance based on original data retrieved from the document and posts as inverse posting. In addition, it skips business mapping of Central Finance but uses FI configuration and accounting interface logic to post revised documents. However, this transaction doesn't work for clearing documents and situations requiring CSPC controls. In addition, it comes with several limitations. Consult SAP Note 2393791 prior to using this transaction.

Tip

If a document reversal falls within the open period of initial load document period, the system displays error message **FINS_CFIN_AC_DOC051 reference document not yet posted**.

Clearing was done before the initial load, reset reversal was done after the initial load, and then the replicated reset document failed in SAP Application Interface Framework with the reset clearing failing with the following error message. **FINS_CFIN_AC_DOC051 reference document not yet posted.**

For these errors, you'll follow solution discussed in Chapter 6, Section 6.5. In addition, consult SAP Note 2710447 - Automatic Reload of Missing Cleared Items (see Chapter 6, Section 6.5).

Refer to SAP Note 3345175, which provides information on the scope of the reversal process to documents mixed with general ledger and AP and AR open item postings.

Foreign Currency Revaluation

US GAAP ASC 830 and IFRS IAS 21 require companies to have more than one currency when preparing financial statements to recognize and measure the changes in relationship to all currencies involved. Accounting standards require transaction gains or losses to be measured and recognized due to exchange rate fluctuations for all currencies in an entity.

From a central payment context, SAP recommends you carry out the foreign currency revaluation process in the Central Finance system. This depends on whether you're using delta valuation. Delta valuation is the difference between the book value on the day a transaction is recorded in the financial statement and the current valuation based on the current exchange rate at the valuation date. If no delta valuation is used, then SAP recommends running the foreign currency valuation in Central Finance.

When delta valuation is used, SAP provides the following guidance if central payment is activated regardless of whether the source system is using the classic general ledger or new general ledger (this is important to reflect accurate financial reporting and to comply with local and statutory requirements in Central Finance):

- During the initial load or online replication, ensure that all valuated documents are replicated to the Central Finance system.
- Reset the foreign currency valuation in the source system for the companies that are in scope for central payment via the following:
  - Classic general ledger Transaction F.05
  - New general ledger Transaction FAGL_FCV (Foreign Currency (New))
- Re-create foreign currency valuation entries in the Central Finance system with the exchange rates using Transaction FAGL_FCV or the Post Currency Adjustments app.

Tip

If the initial load didn't extract foreign currency valuations posted in the source system, consult SAP Note 2666114.

Balance Sheet Adjustment

To maintain the integrity and accuracy of financial statements, balance sheet adjustments are required, especially during period closing. In the Central Finance scenario, consider balance sheet adjustments before central payment activation and after central payment activation. Table 9.7 lists SAP recommendations on balance sheet adjustments for Business Area and Profit Center.

Balance Sheet Adjustments		
Business Scenario	**Before Central Payment Activation**	**After Central Payment Activation**
Business Area	Balance adjustments posted in the source system are replicated to the Central Finance system.	After central payment activation, balance adjustments must be processed in Central Finance only.
	Balance sheet adjustments are run in the Central Finance system only.	
Profit Center	Balance adjustments are posted in both the source system and the Central Finance system separately. Replication isn't supported.	
	Run balance adjustments in the Central Finance system only.	
	Other scenarios are possible.	

Table 9.7 Balance Sheet Adjustments in the Central Finance Scenario

> **Tip**
> After central payment activation, SAP recommends deactivating the balance sheet adjustments program in the source system and activating it in the Central Finance system to prevent inadvertent posting in the source system.

Cash Journal

In SAP, the cash journal is treated as a subledger of bank accounting, which includes general ledger accounts, customer accounts, and vendor accounts. The subledger helps to manage the cash transactions of an enterprise. The cash journal calculates and displays opening and closing balances of both cash receipts and cash payments accounts and their totals. It provides a cash journal view at a company code level. From a Central Finance context, SAP recommends certain steps for the cash journal closing process, especially when central payment is in scope. This includes executing these steps in the source system to transfer cash journals accurately to Central Finance prior to central payment activation. The steps are as follows:

1. All cash journal postings must be in **Posted**, **Reversed**, or **Deleted** status. Ensure you don't have any remaining documents with **Saved** status. The system shows the **Saved** status of document with a yellow traffic light.

2. Ensure you've submitted all checks, and all check logs are posted.
3. Run report RFCJ14 (Analysis Report for Cash Journal Balance Determination) to ensure that balances in the cash journal and general ledger accounts are the same. For additional guidance, consult SAP Note 2515460.
4. Establish a zero balance in the cash journal by using a closing entry in the source system. This implies that the money that is still owed must be transferred from the source system's cash journal to the Central Finance system's new cash journal.
5. Close the cash journal in the source system with Transaction FBCJCO. Ensure that no posting is carried out on the cash journal after closing.

Year-End Carryforward

In this section, we'll discuss important considerations for the year-end carry-forward-carryforward process in the central payment scenario. Table 9.8 provides guidance on before and after central payment activation for the year-end process.

Business Scenario	Before Central Payment Activation	After Central Payment Activation
Year-End Carry-forward of Balances	Run balance carryforward as part of normal business processes.	▪ Execute the general ledger balance carryforward (Transaction FAGLGVTR or SAP Fiori app in the Central Finance system). ▪ Balances may not match between the source system and the central system due to varying factors such as mapping accounts M:1.

Table 9.8 Year-End Balance Carryforward in the Central Payment Scenario

9.6.2 Asset Accounting

You must be aware of restrictions related to asset under construction (AuC) postings in a central payment scenario, as described in Table 9.9.

Before Central Payment Activation	After Central Payment Activation
If you've posted the down payment and capitalized to an AuC, SAP recommends clearing the down payment in the source system.	▪ Down payment and down payment clearing are posted in Central Finance. However, there is no sync back to AuC in the source system. ▪ Central payment is available only for those assets where no reactivation is required when cash discount is taken or no cash discount is available. ▪ No impact on central asset accounting processes.

Table 9.9 Asset Accounting Impact on Central Payment Activation

9.6.3 Accounts Receivable and Accounts Payable

In this section, you'll learn about important considerations during central payment activation for AR and AP processes.

Individual Value Adjustment

After central payment activation, you'll run individual value adjustments only in Central Finance. For example, individual value adjustment is done through Transaction F-21, which is used to capture transfer postings using the **Special G/L** indicator **E** to differentiate from regular postings. You'll access the transaction via **SAP Easy access** menu path **Accounting • Financial Accounting • Accounts Receivable • Document Entry • Other • Transfer without Clearing** or using Transaction F-21. A typical use case of this transaction is to record an installment payment or down payment received from a customer. The transaction can also be used if an installment payment or down payment is made to a vendor.

Flat-Rate Adjustment

In SAP, a flat-rate adjustment is performed on provisions, depreciation, or across other accounts to post adjustments. SAP program SAPF107V enables you to perform automatic flat-rate individual adjustments for overdue receivables. Table 9.10 lists flat-rate adjustment scenarios after activating central payment.

Source System	Central Finance System
■ If the flat rate adjustment valuation is generated in the source system, the adjustments are available in the source system only. ■ Valuation adjustments aren't transferred to the Central Finance system.	If flat-rate adjustments are generated only in the Central Finance system, the valuation adjustments are available in Central Finance only.

Table 9.10 Flat-Rate Adjustments in the Central Payment Scenario

Provisions for Doubtful Receivables

Per GAAP, provisions for doubtful receivables are required for customer open items that are irrecoverable. Generally, you'll carry out doubtful receivables as individual value adjustments in SAP during year-end closing. In the Central Finance scenario, you'll process doubtful receivables for historical open items in Central Finance for both the classic general ledger and new general ledger. SAP provides the following guidance on handling provisions for doubtful receivables:

- During the initial load or online replication, ensure that all provisions for receivables postings are replicated to the Central Finance system.
- Prior to activation of central payments, reverse all provision documents in the source system.
- All historical provision items are reversed prior to central payment activation in the source system. Reversals are carried out with an offsetting account via Transaction FB01 to bring the provision of the doubtful account balance to zero. The reversals must happen before running Transaction FINS_CFIN_HISOI_CLR, which sets all historical open items technically cleared (`ALE-extern`).
- SAP recommends redoing the provision of doubtful receivable open items in Central Finance prior to central payment activation.

Tip

SAP recommends keeping the key date and open items at key date the same from the last run in the source system.

Bill of Exchange

Bill of exchange (BoE) is a type of payment medium for one party to pay another party with a specified amount and defined terms. In general, BoEs can be created in AR or AP. In AR, BoE receivables involve a commitment by a customer to pay a specified amount based on specific terms. On the other hand, a BoE can be used as a payment medium for vendor accounts in AP.

Table 9.11 displays key points to be considered regarding BoEs when activating central payment in the Central Finance system.

Scenario	Before Central Payment Activation	After Central Payment Activation
BoE process	BoE processes are carried out in the source system, and relevant documents are replicated to the Central Finance system.	■ You must run all BoE processes in Central Finance. ■ Map special general ledger transaction type W in Central Finance with guidance from SAP Note 2288901.

Table 9.11 BoE Considerations in the Central Payment Scenario

Scenario	Before Central Payment Activation	After Central Payment Activation
Historical open items	Historical open items open in the source system.	You'll process historical open items with Transaction F-36 and Transaction F110.
Other BoE processes: ■ BoE presentation (Transaction FBWE) ■ BoE collection (Transaction F-34) ■ Discount (Transaction F-33) ■ Forfaiting transaction (Transaction F35) ■ Reversal of liability (Transaction F-20)	After implementing SAP Note 2288901, carry out these processes in the source system.	If these processes need to be transferred to the Central Finance system, you must reverse the transactions in the source system and report directly in the Central Finance system.

Table 9.11 BoE Considerations in the Central Payment Scenario (Cont.)

[+]

Tip

In the case of central payment scenario, special general ledgers are posted in Central Finance. However, if an invoice is created with a special general ledger in the source system, the system allows you to remotely check existence of special general ledger transactions in Central Finance via an RFC call. To enable this functionality, consult SAP Notes 3090910 and 3090911 for prerequisites and necessary authorization objects required for RFC user respectively.

Consider reviewing SAP Notes 2288901 and 3132166, which provide additional guidance on enabling the **Special G/L** indicator in the Central Finance system.

Outgoing Checks

Once central payment is activated, any outgoing checks in the source system are considered historical outgoing checks. To ensure these historical outgoing checks are processed correctly with the new central payment process, you must move all historical outgoing checks data to Central Finance. This ensures the continuity check operations function as expected in Central Finance. Table 9.12 displays key points to be considered before and after central payment activation.

Scenario	Before Central Payment Activation	After Central Payment Activation
Regular outgoing checks	▪ Set bank clearing accounts to be technically cleared in the source system.	▪ Process all outgoing checks in the Central Finance system. ▪ Maintain check lots with Transaction FCHI.
Historical outgoing checks		▪ SAP recommends migrating all historical outgoing checks to the Central Finance system. ▪ To cancel a payment, reverse the historical outgoing check payment document in the source system by voiding the check with Transaction FCH8 in the source system. Then void the check in the Central Finance system with Transaction FCH9. ▪ During the migration of check data, if an error occurs, SAP recommends running the deletion program. To navigate to the migration program, follow Central Finance customizing menu path **Central Finance: Target System Settings • Central Payment • Handling of Historical Outgoing Checks • Delete Migrated Historical Outgoing Checks**.
Special handling – migrated check reversal	▪ Because historical payment documents can only be reversed in the source system, you must take the following actions to cancel payment for migrated checks with associated payment documents that need to be reversed: – Void the check by canceling payment in the source system with Transaction FCH8. – Void the check in Central finance with Transaction FCH9.	▪ Void the check in Central Finance with Transaction FCH9.

Table 9.12 Outgoing Checks Consideration in the Central Payment Scenario

Credit Card Payments

After central payment is activated, SAP recommends executing credit card settlement in the source system where card authorization occurs, however, corresponding FI documents are replicated to Central Finance for clearing. Open Items are technically cleared in the Central Finance system. However, if open items were created earlier than central payment activation, you run report FINS_CFIN_HISOI_CLR to set technically cleared (ALE-Extern). In addition, the Digital Payment app is not supported in credit card payment.

Foreign Trade Reporting

SAP provides the flexibility of running foreign trade reporting either in the source system or target system depending on the point of invoice or point of payment per the contract terms with the central bank. It can remain in the source system if reporting is at the point of invoice. On the other hand, it can be in Central Finance if reporting is at point of payment.

Intercompany Reconciliation Process

SAP provides the following guidance on the intercompany reconciliation process after initial load and after central payment activation in the Central Finance system:

- After completing the financial accounting/Controlling (FI-CO) initial load, as a leading practice, both technical and business reconciliations must be completed before activating central payment.
- After central payment activation, SAP recommends running the intercompany reconciliation process in Central Finance going forward.
- Central payment activation must be done for all company codes involved in the intercompany transactional process.

Item Interest Calculation

SAP provides Transaction FINTAP to calculate interest on vendor open items in AP. This is helpful especially when you want to calculate interest on vendor line items to assess late payment interest. You'll configure interest calculation rules in the system according to business requirements. For Central Finance, Table 9.13 provides important considerations during central payment activation.

Scenario	Before Central Payment Activation	After Central Payment Activation
Interest calculation	The interest calculation process is carried out in the source system.	You must carry out item interest calculation in the Central Finance system.

Table 9.13 Interest Calculation

SAP recommends steps highlighted in this section to make sure you have the correct interest calculation in Central Finance. SAP provides Transaction FINTAP to display the item interest calculation. However, before running the transaction, you must turn off the **Post Interest** indicator in the configuration. To turn off the indicator, follow the customizing IMG menu path **Financial Accounting • Accounts Receivable and Accounts Payable • Business Transactions • Interest Calculation • Interest Calculation Global Settings • Prepare Item Interest Calculation**. After reaching the configuration, on the **Change View "Prepare Item Interest Calculation": Details** screen, unselect the **Post interest** ❶ checkbox, as shown in Figure 9.50. Click the **Save** ❷ button. By not selecting this indicator, interest isn't posted on the document.

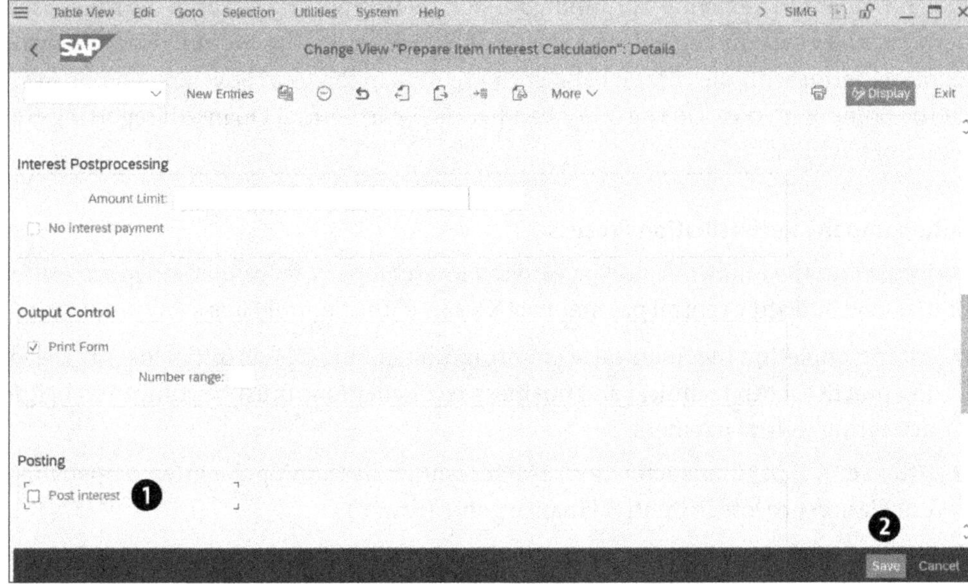

Figure 9.50 Prepare Item Interest Calculation

Now, you'll run the item interest calculation Transaction FINTAP in the Central Finance system. SAP recommends keeping the same date as the source system run date. To run this transaction, navigate to **SAP Easy Access** menu path **Accounting • Financial Accounting • Accounts Payable • Periodic Processing • Interest Calculation** using Transaction FINTAP. After entering the transaction, on the **Item Interest Calculation** screen, enter the relevant information to post interest calculation to vendor accounts, as shown in Figure 9.51. After a successful execution run, you'll be able to display the run results using Transaction FINTSHOW. Now, go back to the **Prepare Item Interest Calculation** screen, as shown in Figure 9.50. Choose the **Post interest** indicator, and save the configuration.

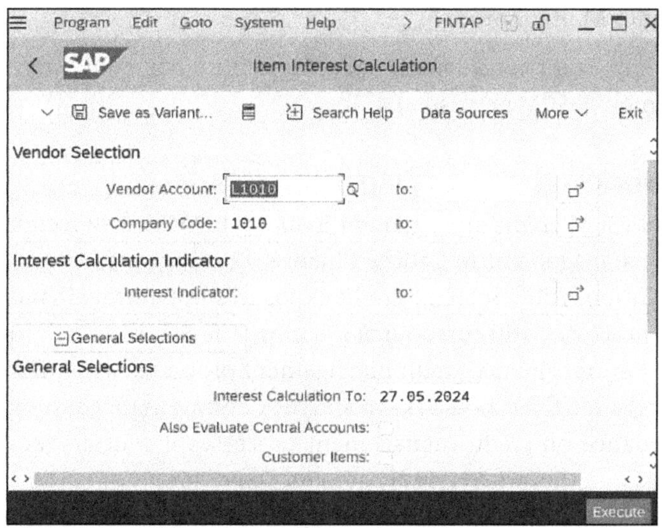

Figure 9.51 Item Interest Calculation

With this exercise, the system is ready to process item interest calculations accurately in Central Finance going forward.

> **Tip**
> SAP provides additional guidance through SAP Note 2184320 for item interest calculation in the Central Finance system.

Special General Ledger Indicator

If central payment is active, during invoice posting in the source system, SAP now supports checking whether the **Special G/L** indicator exists in the Central Finance system through an RFC from the source system. This is relevant for down payment or advance payment scenarios and facilitates all special general ledger postings being processed in Central Finance after central payment activation.

Loans Management

Loans management (FS-CML) facilities structuring loans of consumers and complex mortgages loans. From a central payment scenario, if source system is enabled with FS-CML, this component is not supported for central payments.

Contracts Accounts Receivable and Payable

Central finance does support replicating contracts accounts receivable and payable (FI-CA) documents from the source system to Central Finance. As a result, central payment activation is not relevant for FI-CA component.

9.6.4 Financial Supply Chain Management

In this section, you'll learn important considerations during central payment activation for receivables management (FSCM processes).

Credit Management Considerations

With an improved version of FSCM credit management in SAP S/4HANA, SAP recommends implementing credit management in Central Finance. The improved version replaces the classic SD-FI solution. You'll manage credit exposures in Central Finance while relevant SD business processes still run source systems. The SD business processes must be aligned with Central Finance credit management processes with information sent back and forth between the source system and the Central Finance system. Table 9.14 provides SAP guidance on credit management processes if central credit management is implemented together with central payment activation.

Scenario	Source System	Central Finance System
Before Central Payment Activation	FI documents are posted in the source system.	Both FI documents replicated and credit exposure open invoice updates in FSCM result in duplicate invoices in the Central Finance system. To avoid duplicates, SAP recommends filtering company code using BAdI UKM_FILL (methods IF_EX_UKM_FILL~FILL_FIELDS and IF_EX_UKM_FILL~FILL_VECTOR) per guidance from SAP Notes 2839326 and 2833145. Rebuild the credit exposure with type 200 open invoices to fix incorrect values in credit exposure values. SAP recommends running Transaction UKM_COMMITMENTS to delete incorrect credit exposure values and reconstruct incorrect credit exposure values with report UKM_TRANSFER_ITEMS with guidance from SAP Notes 2576445 and 1124271.
After Central Payment Activation	Customer open items are set to technically cleared (ALE-Extern) upon posting FI documents in the source system.	FI documents are replicated to Central Finance. Replicated documents don't update the credit exposure with type **200 Open Invoice** in FSCM.

Table 9.14 Credit Management Considerations for Central Payment

Scenario	Source System	Central Finance System
After Central Payment Activation (Cont.)	Historical open items created prior to central payment activation are set to technically cleared with ALE-Extern.	Historical open documents are replicated to Central Finance. SAP recommends running Transaction FINS_CFIN_HISOI_CLR in Central Finance to technically clear the items. BAdI UKM_FILL (methods IF_EX_UKM_FILL~FILL_FIELDS and IF_EX_UKM_FILL~FILL_ VECTOR) is implemented to filter out company codes in Central Finance. It must be switched off, or code is enhanced to remove filtering logic.

Table 9.14 Credit Management Considerations for Central Payment (Cont.)

Collection Management Considerations

In SAP, collection management helps businesses perform cash collection activities for customer accounts as part of FSCM. It provides several capabilities:

- Classifies customers with various criteria
- Determines risk of bad debt
- Automates collection strategies
- Integrates with dispute management
- Corresponds with customers
- Monitors and reports

From a Central Finance context, you'll determine handling of collection management based on whether you've activated central payment in the Central Finance system. Table 9.15 identifies different collection scenarios for the central payment scenario.

Scenario	Before Central Payment Activation	After Central Payment Activation
Central Finance and collection management are in a stand-alone system (same system).	No collection management is carried out in the source system.	■ Collection management is performed in the Central Finance system. ■ No connection is required to the collection management system.
Collection management runs on different systems.	Collection management is performed in the source system.	■ Connecting Central Finance to the collection management system is required. ■ Disconnect the source system.

Table 9.15 Collections Management with the Central Payment Scenario

Scenario	Before Central Payment Activation	After Central Payment Activation
Multiple source systems are involved.	No collection management is carried out in any source systems.	■ Collection management is performed in the Central Finance system.
	For historical collection processes, SAP recommends leaving historical collection objects—promises to pay, dispute cases, dunning history, notes, resubmissions, and customer contacts—in the source systems to process.	■ If processing historical collection processes in Central Finance, you'll require a migration project to transfer the historical data to the Central Finance system.

Table 9.15 Collections Management with the Central Payment Scenario (Cont.)

Tip

SAP provides the following restrictions to consider for central payment activation:

- Dunning history (table KNB5) isn't accessible in Central Finance if previous dunning runs have been carried out in the source system.
- Printing out billing documents from the worklist in Central Finance isn't supported.
- Payment history (table KNB4) isn't available unless the **Record Payment History** indicator is activated prior to central payment activation.

In these circumstances, a custom solution may be evaluated and adopted per project requirements.

Dispute Management Considerations

In AR and contract accounting components, you'll create dispute cases for some open items where further action is required to resolve an issue for an open item. SAP adds selected open items to disputed objects to customers under the business partner in the dispute case. In the Central Finance scenario, SAP provides guidance on both prior to central payment activation and after central payment activation scenarios, as listed in Table 9.16.

Scenario	Source System	Central Finance System
Before Central Payment Activation	Dispute management can be run in the source system.	■ SAP recommends Dispute management must be run in Central Finance. ■ Activation of the clearing transfer is required.

Table 9.16 Dispute Management in the Central Payment Scenario

Scenario	Source System	Central Finance System
After Central Payment Activation	No disputes are created in the source system.	▪ Dispute management must be implemented in the Central Finance system. ▪ Billing documents are linked to dispute cases in the Central Finance system.
	Historical open dispute cases exist prior to central payment activation.	▪ Historical open disputes must be migrated to the Central Finance system with a custom approach.

Table 9.16 Dispute Management in the Central Payment Scenario (Cont.)

SAP provides the **Define Derivation of Element Types** configuration activity to link dispute cases to billing documents in the source system automatically. To accomplish this, the system must determine the element type with which the object is to be added. To define derivation of element types, navigate to IMG menu path **Financial Supply Chain Management • Dispute Management • Dispute Case Processing • Process Integration • Define Derivation of Element Types**. As shown in Figure 9.52, SAP predelivered values **VBRK** and **UDM_SPS_VBRK** for the **Object Type** ❶ and **Element Type ID** ❸. An entry can be maintained for a specific source's logical system ❷ as well. As you can see in Figure 9.52, the system delivers element types for both AR and contract accounting object types.

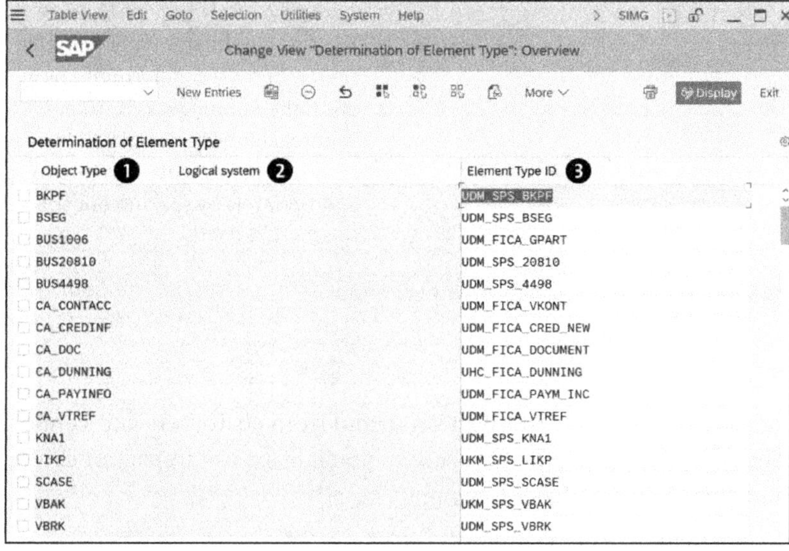

Figure 9.52 Define Derivation of Element Types

Dunning

SAP recommends activating the dunning process once central payment is activated in Central Finance. Table 9.17 provides recommendations when considering dunning and central payment activation.

Scenario	Before Central Payment Activation	After Central Payment Activation
Regular dunning process	The dunning process is handled in the source system.	■ The dunning process must be run in the Central Finance system. ■ No connection is required to the collection management system.
Historical open items		■ Manual correction may be required for historical open items that incorrectly contain values due to different dunning procedures used in the source system. You'll correct the dunning level and dunning key at the item level. ■ **Dunning Area** and **Dunning Block** are mapped at the item level between the source system and the Central Finance system. Corrections are necessary only if the configuration is changed. ■ If **Highest Dunning Level** in the customer master is blank, you'll populate it manually by checking the corresponding field in the source system. ■ The **Last Dunning Date** field in the customer master determines how often the dunning is run. To prevent reaching the dunning level early, plan to execute dunning appropriately by spacing out the runs.

Table 9.17 Dunning Process in the Central Payment Scenario

Correspondence

In SAP, correspondence refers to communication to and from customers and vendors to check the balances of AR and AP. In this section, you'll learn the impact of central payment activation as it relates to the following:

- **Payment advice notes**
  These are generated to communicate with customers and vendors related to open items. Companies follow payment advice notes to speed up the payment process. SAP provides a few transactions to process payment advice note activities.
  - Transactions FBE1, FBE2, FBE3, and FBE6: Payment advice (program SAPMF06A).
  - Transaction FB12: Request correspondence (program SAPMF05M).
- **Account statements**
  These provide summarized open items transactions for a specific customer or supplier or for a group of customers or vendors. With account statements, you'll be able to track outstanding balances. SAP provides correspondence printing transactions:
  - Transaction F.27: Periodic account statements (program RFKORK00). For example, this transaction can be used to generate a customer account statement.
  - Transaction F.61: Per requests (program SAPF140). This is used to trigger customer correspondence.
  - Transaction F-62: Internal documents (program RFKORB00). This is used to generate customer statements detailing invoices, payments and other information.
  - Transaction F.63: Delete requests (program SAPF140D). This is used to delete correspondence requests.
- **Balance confirmation**
  SAP provides balance confirmation activities to communicate with customers and vendors to check receivable and payable balances. With the reports, you'll be able to send balance confirmations, balance notifications, and balance requests about open items, account balances, and any other information required. These reports also display a checklist and error list for each company code during evaluation:
  - Transaction F.17: Print letters (report SAPF130D). This is used to create customer balance confirmations.
  - Transaction F.18: Create Index (report SAPF130Z). Similar to customer balance, this transaction generates vendor balance confirmation.
  - Transaction F.1A: Grouping Customer/Vendor Master Records (report SAPF130P). This is used to generate balance confirmations for group master records.
  - Transaction FB12: Request Correspondence (program SAPMF05M). This transaction is used to generate correspondence notices.
  - Transaction F.64: Maintain Correspondence Requests (program SAPF140P)

From a Central Finance context, Table 9.18 shows the impact of correspondence-related activities in the central payment scenario.

Scenario	Before Central Payment Activation	After Central Payment Activation
Payment advice notes	■ Payment advice notes are processed in the source system.	■ You must create payment advice notes for incoming payments in Central Finance. ■ You must process payment advice notes along with outgoing payments in the Central Finance system.
Account statements	■ Customer and supplier open items are posted in the source system. ■ Account statements are generated in the Central Finance system.	■ You'll ensure customer and supplier open items are replicated to Central Finance without errors. ■ The **Bank statement** field on the customer business partner is maintained in the Central Finance system. ■ The account statement process is managed in the Central Finance system.
Balance confirmation	■ Down payment requests and payment requests are posted in the source system.	■ The balance confirmation is executed from the Central Finance system. ■ Down payment requests and payment requests are replicated to the Central Finance system. ■ Noted items aren't included in the balance confirmation.

Table 9.18 Correspondence in the Central Payment Scenario

Bank Account Management

In the Central Finance context, house banks and house bank accounts should be identical between the source system and the Central Finance system. However, if they differ in Central Finance, you'll maintain mapping entities of house bank (HBKID) and account details (HKTID) between the source system and the Central Finance system. Likewise, if head office and branch office are involved in a transaction (FILKD), you must consider settings for partner bank type (BVTYP). If source system data sends inconsistent data, you'll use central finance BAdI to clear data or fill in with missing information. Table 9.19 provides important considerations in the central payment activation scenario for bank account management.

Scenario	Before Central Payment Activation	After Central Payment Activation
Bank account management	All bank statements must be posted in the source system before activation of central payment. This means you'll ensure there are no outstanding batch input sessions.	▪ You must maintain bank relationship management and house bank–related configuration in the Central Finance system. ▪ All payments are generated for the company codes that are in scope for Central Finance, including bank statements.
Bank account master data	The bank account master is used in the source system to assign it to documents.	▪ Run Transaction DRFOUT to replicate house banks, house bank accounts, and bank accounts from the Central Finance system to the source system. The transaction is flexible to send selected accounts or mass transfer of accounts, as well as specific changes.
Historical bank statements	Historical bank statements are accessed only from the source system.	▪ Historical bank statements aren't available in Central Finance.

Table 9.19 Bank Account Management Considerations in the Central Payment Scenario

Tip

Refer to SAP Note 2165520 for additional guidance on feature scopes in bank account management.

9.6.5 Cash Management

SAP recommends running cash management in Central Finance after activating central payment. Prior to central payment activation, cash management is run in the source system. Clearing transfer is a prerequisite. SAP recommends using IDoc message types CMSEND and CMREQU to transfer classic cash management data, for example, purchase orders and sales orders. To avoid double posting of cash management items for FI posting in One Exposure (FQM_FLOW), SAP recommends applying filters in the IDoc distribution model. Table 9.20 provides important considerations for the central payment activation scenario.

Scenario	Before Central Payment Activation	After Central Payment Activation
Cash management process	▪ Cash management is run in the source system.	▪ Cash management is run from the Central Finance system. ▪ If the planning date is blank, the system redetermines the planning date. ▪ SAP provides BADI_FINS_CFIN_AC_INTERFACE to clear the planning date on replicated FI documents, so the planning date can be redetermined.
Payment history	▪ The **Payment History Record indicator** can be activated on the customer master (KNB4), which could facilitate sending the payment history to Central Finance. ▪ After central payment activation, the planning date may not calculate correctly.	▪ The payment history is updated considering the planning date calculation in Central Finance. ▪ Consult SAP Note 24883 for additional guidance on payment planning date determination.
Check cashing	▪ The check cashing for vendors planning date may not calculate correctly after central payment activation.	▪ Check cashing is applicable if the payment method chosen is check when the planning date is calculated for suppliers. SAP recommends maintaining check cashing time manually or using report RFSRUE10 (outstanding check analysis for general ledger acct supplier, house bank, and account ID).

Table 9.20 Cash Management for the Central Payment Activation Scenario

9.6.6 Tax

In this section, we'll discuss what aspects are to be considered for tax when activating central payment in Central Finance. Let's explore the processes that may have an impact for value-added tax, withholding taxes, and deferred tax.

Value-Added Tax

Table 9.21 provides guidance from SAP on value-added tax considerations in the central payment activation scenario.

Scenario	Before Central Payment Activation	After Central Payment Activation
Running tax return and payment	▪ You'll always run **Advanced Return for Tax on Sales/Purchases** and payment to the tax authority in the source system prior to central payment activation. ▪ If short payments, for example, discounts, don't result in tax adjustments, **Advanced Return for Tax on Sales/Purchases** can be run in the Central Finance system or the source system.	▪ You'll determine the process of running **Advanced Return for Tax on Sales/Purchases** depending on the country. ▪ If short payments, for example, discounts, don't result in tax adjustments, **Advanced Return for Tax on Sales/Purchases** can be run in the Central Finance system or the source system. However, you'll process payments to the tax authority in the Central Finance system. ▪ If the central payment is active and short payments are made, for example, discounts, resulting in tax adjustments, **Advanced Return for Tax on Sales/Purchases** and payments to the tax authority must be run only in the Central Finance system.

Table 9.21 Handling of Value-Added Tax in the Central Payment Scenario

Withholding Tax

In Central Finance, extended withholding tax replication is supported, but with a few restrictions. Table 9.22 displays what you'll consider for withholding tax in the central payment scenario.

Scenario	Before Central Payment Activation	After Central Payment Activation
Invoice posting and payment posting	No collection management is carried out in the source system.	▪ If the country in scope includes withholding tax at the time of invoice posting and payment posting, there are no restrictions on withholding tax replication and relevant processes.

Table 9.22 Withholding Tax in the Central Payment Scenario

Scenario	Before Central Payment Activation	After Central Payment Activation
Accumulated withholding tax		■ SAP recommends activating central payment before the next accumulation period begins. This means it could be a month, a calendar year, or a 12-month period starting on a specific month to calculate accumulated withholding tax. ■ Accumulated data for withholding tax is accessible in Central Finance only at the time of payment and not available at the time of invoice.

Table 9.22 Withholding Tax in the Central Payment Scenario (Cont.)

Deferred Tax

In certain countries, reporting deferred tax is required. Table 9.23 provides guidance from SAP on deferred tax considerations in the central payment activation scenario.

Scenario	Before Central Payment Activation	After Central Payment Activation
Running tax return and payment	■ You'll run deferred tax processes in the source system prior to central payment activation. ■ Once central payment is activated, SAP recommends not activating deferred tax in the source system.	■ Deferred tax must be processed in the Central Finance system.

Table 9.23 Deferred Tax in the Central Payment Activation Scenario

Tip

Central payment activation is supported for company codes for those countries where central tax reporting is supported by SAP. To understand country- or region-specific central payment restrictions, consult SAP Note 3290988. This SAP Note provides guidance on central payment restrictions for Brazil, India, Italy, Mexico, Poland, and Ukraine. Prior to activating central payment for these countries, review the restrictions.

9.7 SEPA-Mandated Replication

Single Euro Payment Area (SEPA) helps companies collect payments from a customer's bank account directly without the customer's intervention. Customers must grant companies a signed mandate. The mandate is a legally binding agreement that allows companies to debit money from a customer's bank account. The SEPA mandated payment method is a standardized way to process payments offered to customers across most European countries. In SAP, SEPA mandates are created from **SAP Easy Access** menu **Accounting • Financial Accounting • Accounts Receivable • Master Records • SEPA Mandates** or by using Transaction FSEPA_M1. Using this transaction, you'll be able to create a SEPA mandate for customers or vendors. The transaction captures all required mandate information. Central Finance supports SEPA mandate replication from the source systems to the Central Finance system to support the centralized SEPA direct debit process. SEPA mandate handling was introduced in the SAP S/4HANA 1709 on-premise release. Once SEPA mandates are replicated to Central Finance and central payment is active for a company code, you'll be able to run an automatic payment program (Transaction F110) or any manual clearing transactions to clear open items based on parameters and configuration settings in Central Finance.

In this section, you'll learn about SEPA mandate replication and central payment considerations. SAP supports the following:

- SEPA mandates replication from an SAP source system to the Central Finance system
- SEPA mandates replication from the Central Finance system to an SAP source system

Let's explore each of the scenarios with a process flow diagram next.

9.7.1 SEPA Mandates form an SAP Source System to Central Finance

Central Finance supports SEPA mandate replication from an SAP source system to the Central Finance system. Figure 9.53 illustrates this scenario as follows:

❶ SEPA mandates are posted in the source system transfers to the Central Finance system via IDoc. The standard core layer function calls such as BAdI `IDOC_DATA_MAPPER` and BAPI calls are made in Central Finance.

❷ The system performs business mapping using SAP Master Data Governance mapping from Central Finance.

Once business mapping is completed, the system posts SEPA mandates in the Central Finance system. SAP Application Interface Framework is used to review posting errors in interface `ISEPA_CR` under namespace `/FINCF`.

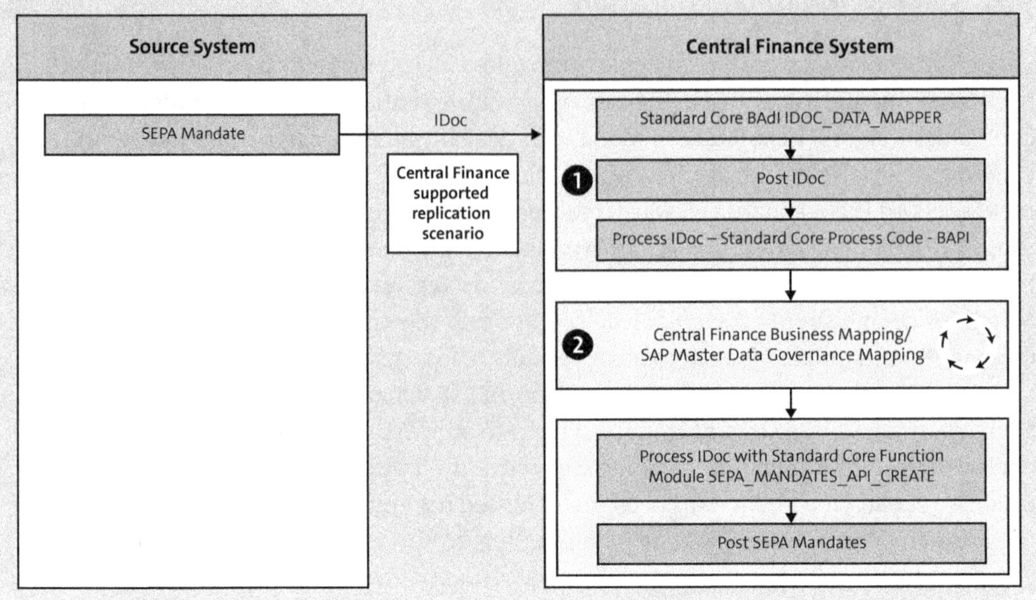

Figure 9.53 SEPA Mandate Replication: SAP Source System to Central Finance System

For company codes that are activated for central payments, you can migrate SEPA mandates for those company codes from the source system to Central Finance. To migrate SEPA mandates, navigate from **SAP Easy access** menu path **Accounting • Financial Accounting • Accounts Receivable • Master Records • SEPA Mandates • List** or use Transaction FSEPA_M4. Upon executing the transaction, it displays a **List of Mandates** to migrate from the source system to the Central Finance system based on your selection criteria. After choosing mandates, the system transfers the mandates to Central Finance if they aren't already created in Central Finance with a click of the **Migrate** button. If changes are distributed, the changes are transferred over to Central Finance when you click the **Migrate** button again.

9.7.2 SEPA Mandates from Central Finance to an SAP Source System

Central Finance also supports SEPA mandate replication from the Central Finance system to an SAP source system. Figure 9.54 illustrates this scenario as follows:

❶ SEPA mandates are posted and mapped with SAP Master Data Governance reverse mapping in the Central Finance system. The mapped data is then used to create IDocs for distribution to the source system.

❷ The system transfers the mapped data to the source system using IDocs. In the source system, the IDoc standard core process code is executed. The system calls function module `SEPA_MANDATES_API_CREATE` to post SEPA mandates in the source system.

Posted SEPA mandate data can be viewed using Transaction SEPA_M3 in the source system.

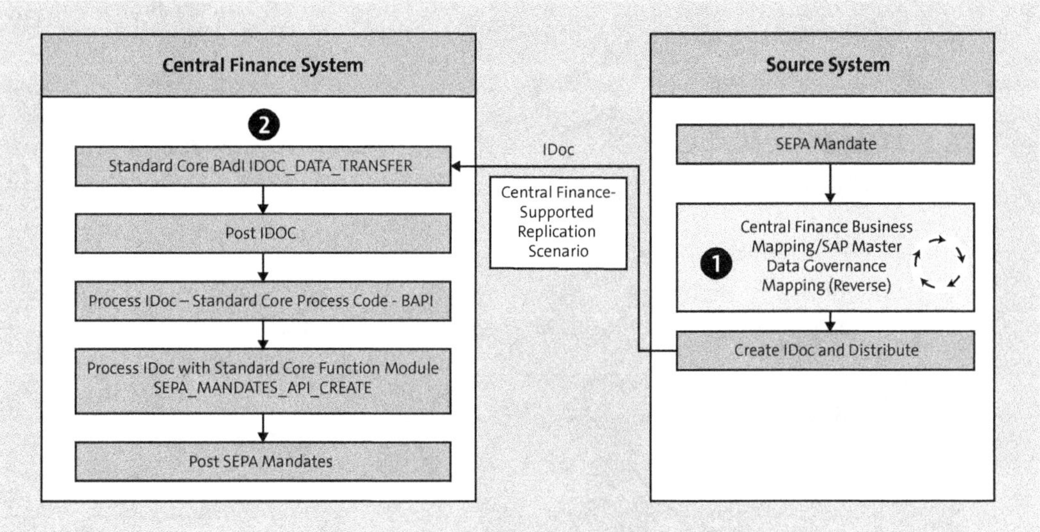

Figure 9.54 SEPA Mandate Replication: Central Finance to SAP Source System

9.7.3 Configuring SEPA Mandates in the Central Finance System

In this section, you'll learn SEPA mandate replication configuration in the Central Finance system. Before you begin SEPA mandate configuration, SAP recommends that you review SAP Notes 1887572 and 3220340.

Enable SEPA for Applications

The first step in using SEPA mandates is to activate SEPA mandate management in the Central Finance system. To activate SEPA mandate management, navigate to the configuration from IMG menu path **SAP Customizing Implementation Guide • Financial Accounting • Central Finance • Central Finance: Target System Settings • Central Payment • SEPA • Enable SEPA for Application**. After reaching the configuration, you'll arrive at the **Change View "Mandates Management: General Settings": Overview** screen. Click the **New Entries** button to enable a SEPA application. On the **New Entries: Details of Added Entries** screen, choose an application from the **Applic** field dropdown, as shown in Figure 9.55. Select **Active** ❶ under the **Mandate Management: General Settings** section. Optionally, you can register specific function modules to make data enhancements and additional checks. The screen allows you to define a **Form Name** for printing SEPA mandates. Fill in the required fields under **Val.** ❷ to support SEPA Mandates as specified in Table 9.24.

Parameter	Val.
ALE: Logical Name of the Consolidation System	S48CLNT500 (in this example)
ALE: Load Additional Data	FIN_CFIN_APAR_SEPA_LOAD_SRC
ALE: Unload Additional Data	FIN_CFIN_APAR_SEPA_UNLOAD_SRC
ALE: Filter Data	FIN_CFIN_APAR_SEPA_ALE_FILTER
Function Module for Settable Button 2	FIN_CFIN_SEPA_MIGRATE_BUTTON

Table 9.24 Enable SEPA for Applications

After filling in the relevant fields, click the **Save** ❸ button to save the entries.

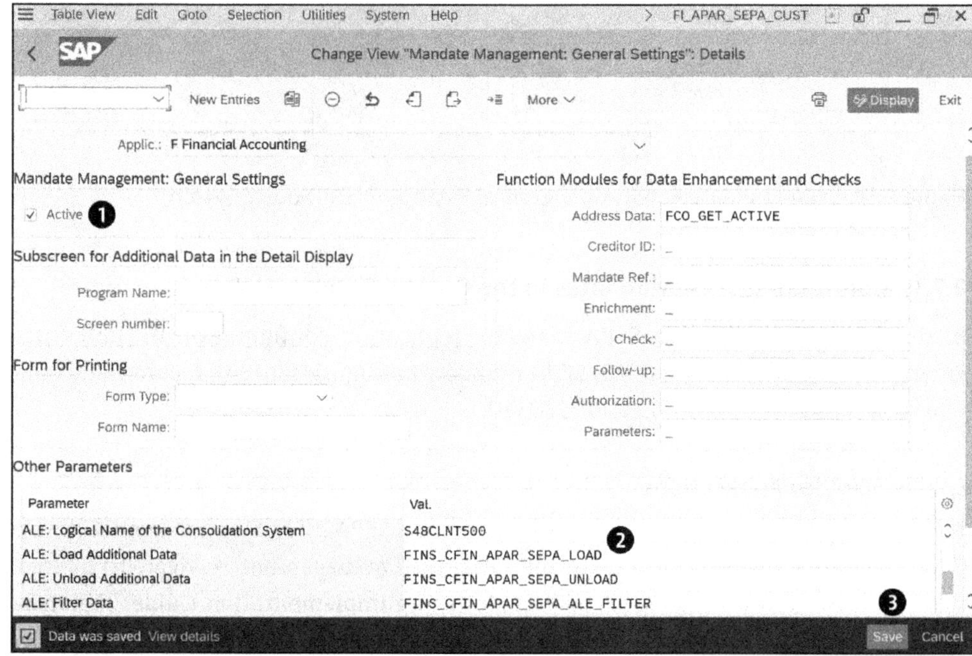

Figure 9.55 Enable SEPA for Applications

Define ALE Filter Object Types

SAP defines filter objects as attributes of a message in the distribution model that are assigned to BAPIs during IDoc calls. ALE filter object types are used in Application Link Enabling (ALE) during inbound or outbound processing of data in a distributed network. In this configuration, you'll specify the types of filter objects required for SEPA mandate replication. Filter objects are characteristics of a message that you define in a distribution model. To define ALE filter object types, navigate to the configuration from IMG menu path **SAP Customizing Implementation Guide • Financial Accounting • Central Finance • Central Finance: Target System Settings • Central Payment • SEPA •**

Define ALE Filter Object Types. As shown in Figure 9.56, the system predelivered values **RECEIVER** ❶, **TBD06** ❷, and **RCVSYSTEM** ❸ for **ALE Object Type**, **Table Name**, and **Field name** fields, respectively, that can be used for SEPA mandate replication. Nevertheless, you can define a new **ALE Object Type** in this configuration.

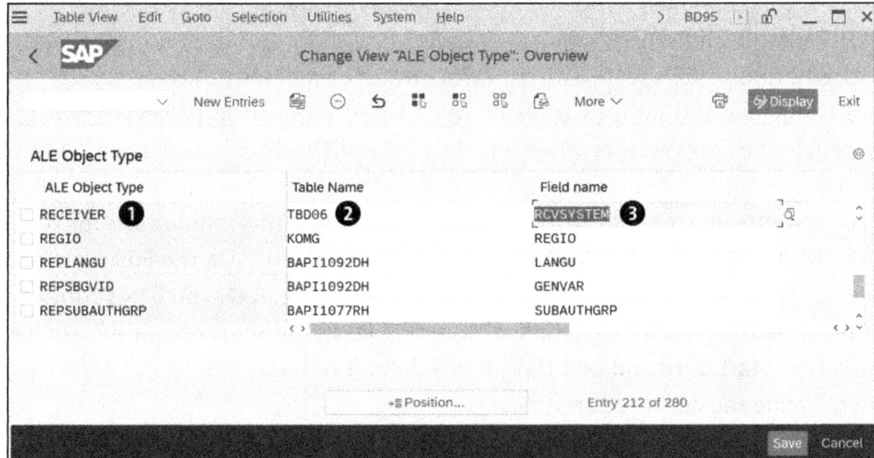

Figure 9.56 Define ALE Filter Object Types

Assign ALE Filter Object Types

In this configuration, you'll assign ALE filter object types you've created from the previous configuration. To assign the ALE filter object types, navigate to the configuration from IMG menu path **SAP Customizing Implementation Guide • Financial Accounting • Central Finance • Central Finance: Target System Settings • Central Payment • SEPA • Assign ALE Filter Object Types**. After reaching the configuration, on the **Change View "Filter Objects for Receiver Determination o: "Overview** screen, click the **New Entries** button, as shown in Figure 9.57.

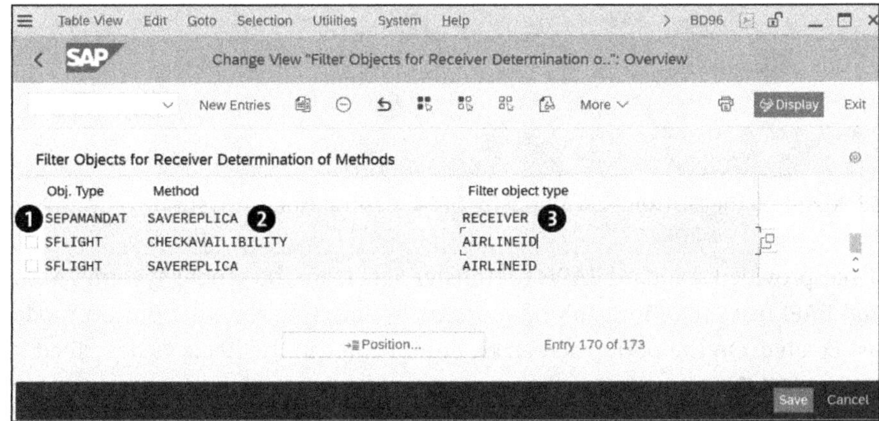

Figure 9.57 Assign ALE Filter Object Types

On the resulting screen, enter values in **Obj. Type**, **Method**, and **Filter object type** fields. Click the **Save** button to save the configuration. The system displays saved information **SEPAMANDAT** ❶, **SAVEREPLICA** ❷, and **RECEIVER** ❸ values for the **Obj. Type**, **Method**, and **Filter object type** fields, respectively.

Maintain the Distribution Model

In this configuration, you'll create a distribution model that determines the sender and receiver of the IDocs and defines transfer rules when replicating projects from the source system to the Central Finance system. In addition, the distribution model allows you to define a system-wide message flow. To create a distribution model for replication project systems master data, go to Transaction BD64 in the source system. You'll arrive at the **Display Distribution Model** screen. On this screen, click the **Edit** icon, and then click the **Create Model View** button, as shown in Figure 9.58 ❶. You'll be prompted with the **Create Model View** dialog window. Enter values in the **Short text** and **Technical Name** fields. The **Start date** and **End Date** fields default to today's date. Click the **Continue** icon to create the distribution model.

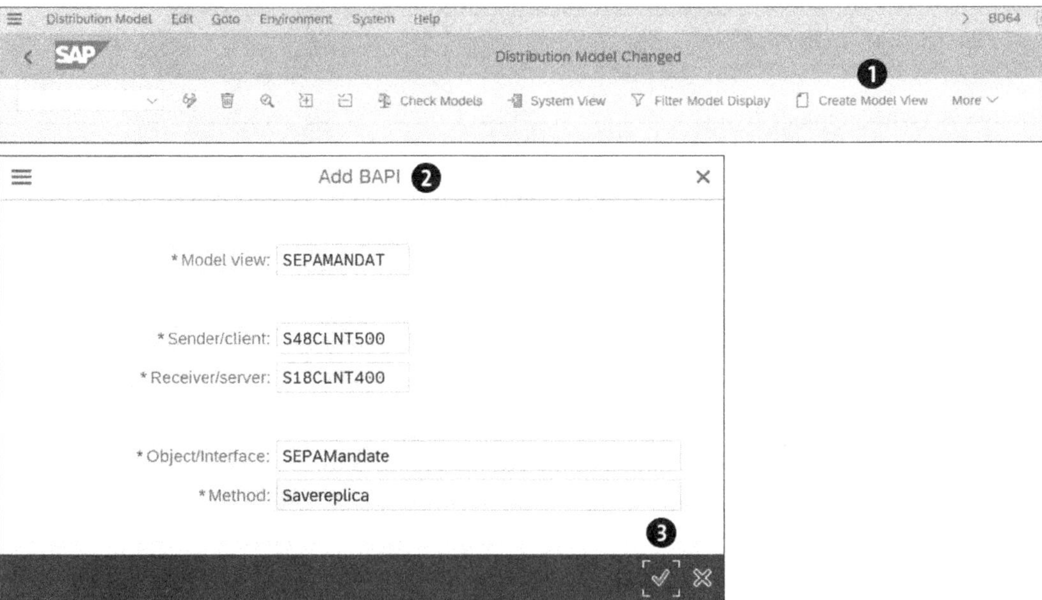

Figure 9.58 Maintain the Distribution Model for SEPA: Central Finance System

Next, SAP has provided standard BAPIs to transfer data from the sender systems. Navigate to **Add BAPI** from the **More** submenu items by selecting the distribution model you've just created. On the popup screen ❷, enter values in the fields as described in Table 9.25. Click **OK** ❸ to continue.

Field	Value
Sender/Client	**S48CLNT500** (Central Finance logical system)
Receiver/Server	**S18CLNT400** (source logical system)
Object/Interface	**SEPAMandate**
Method	**Savereplica**

Table 9.25 Add BAPI for Maintenance of Distribution Model:Central Finance System

Next, you'll generate partner profiles for the model view you've just created in the source system. To generate partner profiles, select the **SEPAMANDAT** distribution model you created in the previous step, and then navigate to **Generate Partner Profiles** from the **Environment** submenu item in Transaction BD64. Alternatively, you can generate the partner profiles by reaching Transaction BD82 directly. Upon reaching the transaction, maintain the following fields:

❶ **Model View and Partner System**
Enter the model view name, for example, "SEPAMandate" and "S48CLNT500" for **Partner System**.

❷ **TY and ID**
The system defaults the type of individual (**US**) and user ID in the **Postprocessing Authorized Users** section for IDoc error notification.

❸ **Version and Pack.size**
The system defaults to **3** for IDoc types from version 4.0 onward. For older versions, version **2** is used.

❹ **Trigger immediately**
The system defaults to the **Trigger immediately** radio button.

Once all fields are filled, click the **Execute** button to generate the partner profile. As shown in Figure 9.59, the system displays logs if partner profiles are generated ❶ and the distribution model ❷ is created successfully.

Activate Business Configuration Sets

For SEPA mandate replication, you must install the relevant BC set. It contains configuration settings by grouping table values for a specific business process. To apply the BC sets for SEPA mandate replication, navigate to **SAP Easy Access** menu path **Tools • Customizing • Business Configuration Sets • Display and Maintain BC Sets** or use Transaction SCPR20. Upon entering the transaction, you'll arrive at the **Business Configuration Sets: Activation** screen, as shown in Figure 9.60. On this screen, enter the value "FINS_CFIN_AIF_SEPA" in the **BC Set** ❶ field. Click the **Activation** ❸ icon to activate the BC set. Upon activation, the system displays the activation log ❷ screen of the status. You need to make changes if the system displays any errors.

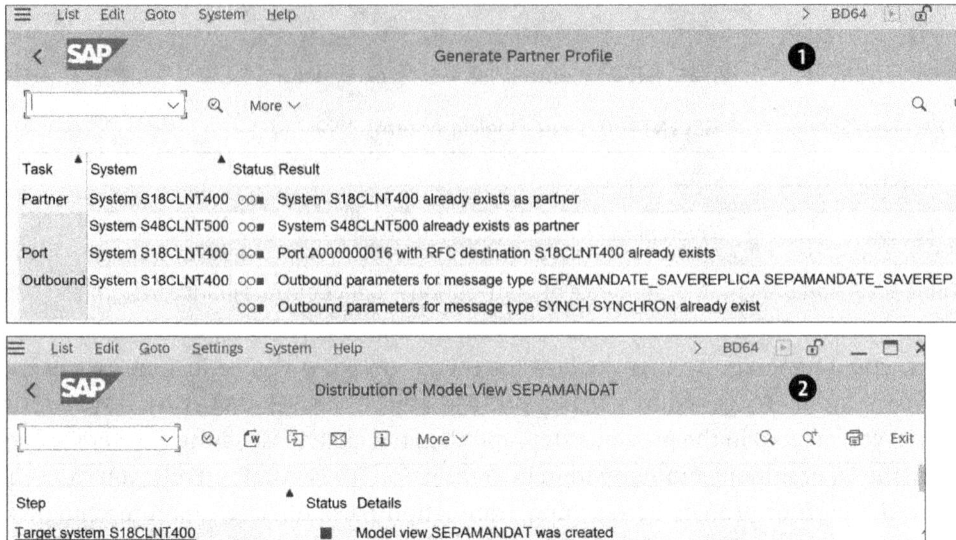

Figure 9.59 Distribution Model View and Generating the Partner Profile

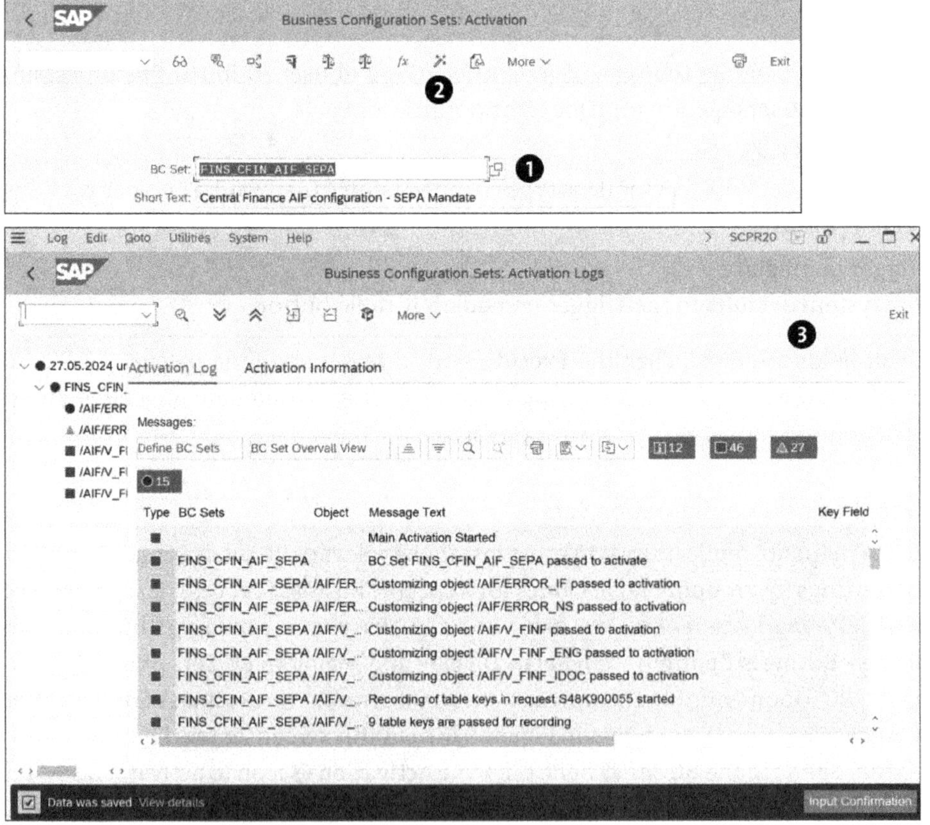

Figure 9.60 Activate BC Sets for SEPA

Define SAP Application Interface Framework Reprocessing Action

Next, in this step, you'll configure **Define Reprocessing Action** for reprocessing action definition **RECREATE**. To configure this, go to **SAP Easy Access** menu and choose **Cross Applications Components • SAP Application Interface Framework • Administration • Configuration • Define Reprocessing Action** or use Transaction /AIF/REP_AC_DEF. After entering the transaction, the system prompts you with a popup screen to enter a **Namespace**. Enter the Central Finance namespace as "/FINCF", and then click the **Continue** button. The system brings up another warning screen with the **All Fields which define the area have initial values. Check Input Values** message. Click **Continue** on this screen. Next, you'll arrive at **the Change View "Namespace Lock": Overview** screen, as shown in Figure 9.61. On this screen, click the **New Entries** button ❶. On the resulting 9
New Entries: Details of Added Entries screen, enter values in the following fields:

- **Repr. Action Name**
  Enter the name of the automatic reprocessing action as "RECREATE" ❷.
- **Function Module**
  Enter the SAP-delivered function module "/AIF/RESTART_MSG" ❸.
- **Namespace**
  Enter the Central Finance namespace as "/FINCF".

Once all fields are filled, click the **Save** ❹ button to save the entries.

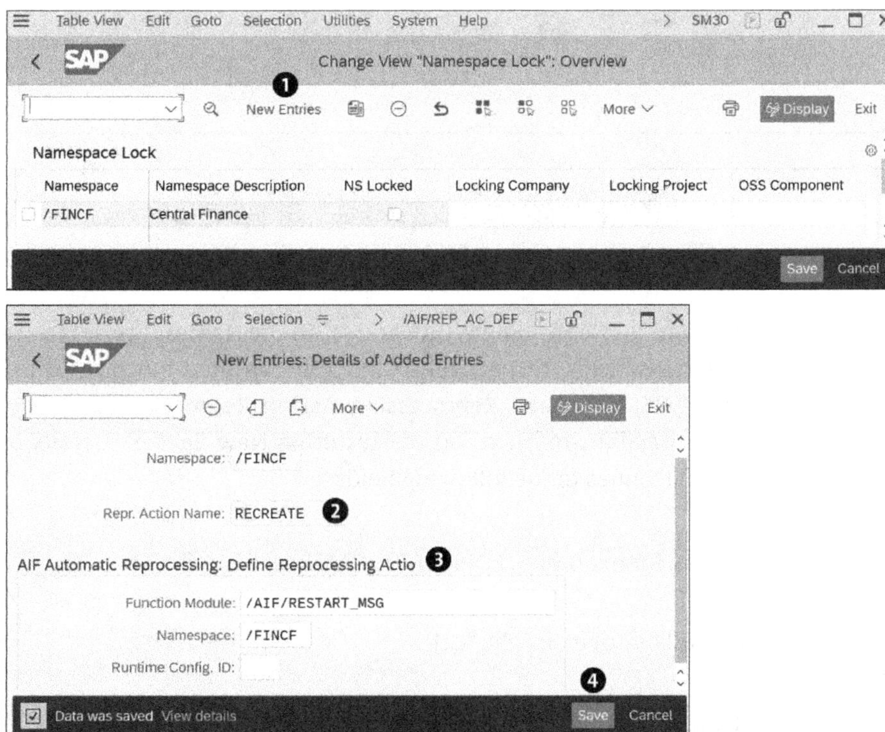

Figure 9.61 Define SAP Application Interface Framework Reprocessing Action

Assign SAP Application Interface Framework Reprocessing Action

Next, you'll configure **Assign Reprocessing Action** for interface ISEPA_CR. To assign the reprocessing action, navigate to the configuration from the **SAP Easy Access** menu and choose **Cross Applications Components • SAP Application Interface Framework • Administration • Configuration • Assing Reprocessing Action** or use Transaction /AIF/REP_AC_ASGN. After reaching the transaction, the system prompts with the **Determine Work Area: Entry** screen, as shown in Figure 9.62 ❶.

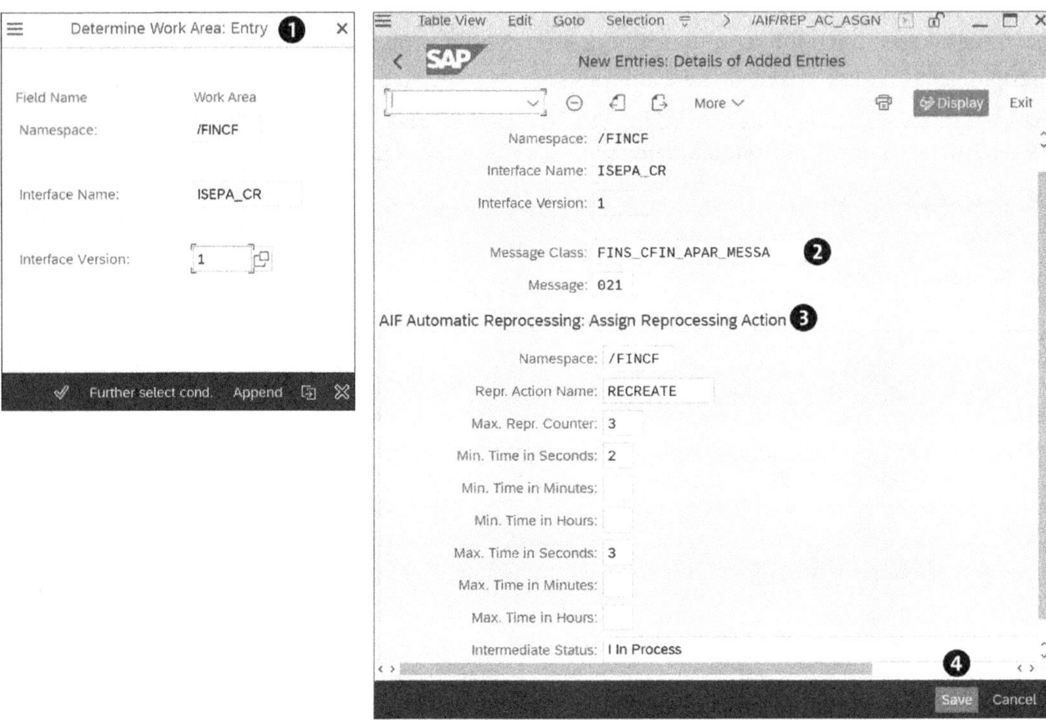

Figure 9.62 Assign SAP Application Interface Framework Reprocessing Action

On this screen, enter values "/FINCF", "ISEPA_CR", and "1" in the **Namespace**, **Interface Name**, and **Interface Version** fields, respectively. Click the **Continue** button. On the resulting **Change View "AIF Automatic Reprocessing Assing Reprocess: "Overview** screen, click the **New Entries** button. Next, on the resulting **New Entries: Details of Added Entries** screen, enter values in the following fields:

- **Message Class**
  Enter the message class, for example, "FINS_CFIN_APAR_MESSA" ❷.
- **Message**
  Enter the message number, for example, "011".
- **Namespace**
  Enter the Central Finance namespace "/FINCF" ❸.

- **Repr. Action Name**
  Enter the reprocessing action name you've created in the previous step: "RECREATE".
- **Max. Repr.** counter
  Enter a maximum number of reprocessing trials: "3".
- **Min Time in Seconds**
  Enter the minimum value in seconds: "2".
- **Max. Time in seconds**
  Maintain the maximum time in seconds: "3".
- **Intermediate Status**
  Choose **In Process** from the dropdown list.

When finished, click the **Save** ❹ button to save the entries.

Define the IDoc Interface

To support SEPA mandate replication in SAP Application Interface Framework, you must define the interface for IDoc interfaces. This is done based on the fields whose values are required to identify the appropriate interface using IDoc basic type and message type. These definitions are required for handling the SEPA mandate incoming data messages.

To define the IDoc interface, navigate from SAP Application Interface Framework customizing (Transaction /AIF/CUST) menu path **SAP Application Interface Framework • System Configuration • Interface Determination • Define Determination for IDoc Interface**. After reaching the customizing activity, you'll arrive at the **Change View "Define Determination Key": Overview** screen, as shown at the top of Figure 9.63. On this screen, click the **New Entries** button. Next, on the resulting **Change View "Assing Interfaces": Details** screen, maintain values in the **Basic Type** and **Message Type** fields, as shown in Figure 9.63 ❶ and described in Table 9.26.

Basic Type	Message Type
SEPAMANDATE_CHANGE_ALE01	SEPAMANDATE_CHANGE_ALE
SEPAMANDATE_CREATE_ALE01	SEPAMANDATE_CREATE_ALE
SEPAMANDATE_SAVEREPLICA01	SEPAMANDATE_SAVEREPLICA

Table 9.26 Define the Determination Key

Next, choose each row, and double-click the **Assign Interface** folder option located under the **Dialog Structure** subscreen ❷.

On the resulting **Change View "Assing Interfaces": Details** screen, enter the values for all three **Basic Types** ❸ from Table 9.27.

Basic Type	Value Number	Namespace	Interface Name	Interface Version
SEPAMANDATE_CHANGE_ALE01	0	/FINCF	ISEPA_CH	01
SEPAMANDATE_CREATE_ALE01	0	/FINCF	ISEPA_CR	01
SEPAMANDATE_SAVEREPLICA01	0	/FINCF	ISEPA_SR	01

Table 9.27 Assign Interfaces

Once each **Basic Type** is populated with interface information, click the **Save** ❹ button to save the entry.

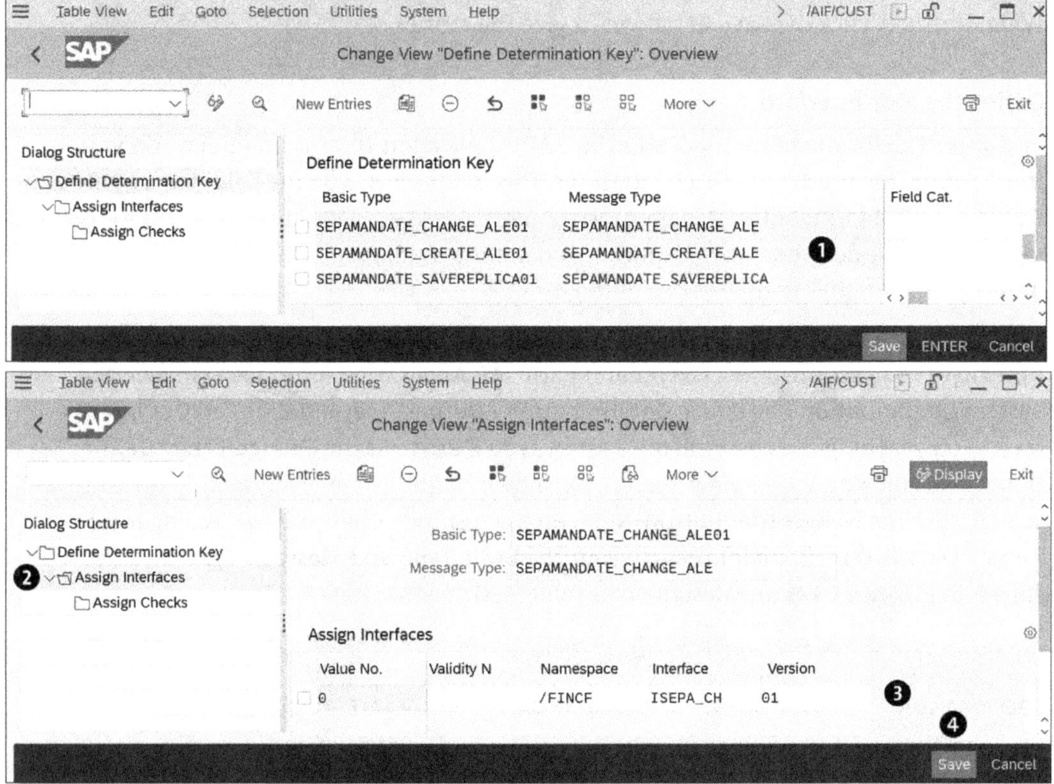

Figure 9.63 Interface Determination: Assign Interfaces

9.7.4 Configuring SEPA Mandates in the Source System

In this section, let's discuss the source system configuration requirements to support SEPA mandate replication. Primarily, there are four configurations you'll maintain in the source system:

- Maintain entries in view V_SEPA_CUST.
- Specify ALE object types.
- Assign ALE filter object types.
- Maintain distribution model.

Let's explore each configuration with step-by-step instructions.

Maintain Entries V_SEPA_CUST

Maintenance view V_SEPA_CUST stores ALE parameters for SEPA mandate replication. To maintain values in this view, call up Transaction SM30 (Call View Maintenance). After entering the transaction, enter the value "V_SEPA_CUST" in the **Table View** field, and click the **Edit** icon. On the resulting **Change View "Mandate Management: General Settings": Overview** screen, choose the **Financial Accounting Application** row, and then click the **Details** icon. Next, you'll land at the **Change View "Mandate Management: General Settings": Details** screen, as shown in Figure 9.64.

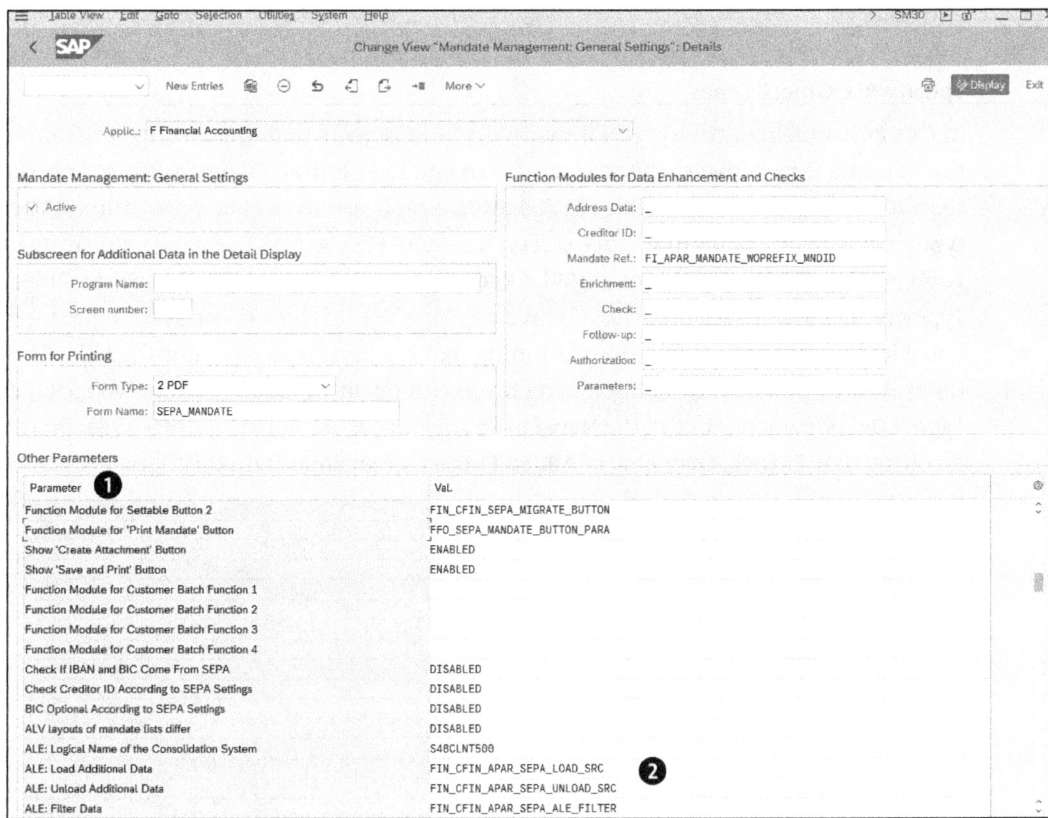

Figure 9.64 Maintain Table V_SEPA_CUST

On this screen, maintain values for **Parameter** ❶ and **Val.** ❷ fields, which are described in Table 9.28.

Field	
Parameter	**Val.**
Function Module for Settable Button 2	FIN_CFIN_SEPA_MIGRATE_BUTTON
ALE: Logical Name of the Consolidation System	S48CLNT500
ALE: Load Additional Data	FIN_CFIN_APAR_SEPA_LOAD_SRC
ALE: Unload Additional Data	FIN_CFIN_APAR_SEPA_UNLOAD_SRC
ALE: Filter Data	FIN_CFIN_APAR_SEPA_ALE_FILTER

Table 9.28 Maintain Parameters in Table V_SEPA_CUST

Once the parameters are maintained, click the **Save** button to save the configuration entries.

Specify ALE Object Types

In this customizing activity, you'll maintain filter objects that are used in controlling the ALE data flow between the source system and the Central Finance system. This is a required configuration to enable SEPA mandate replication. To specify ALE filter object types, navigate to the maintenance screen from **SAP Easy Access** menu by navigating to **Tools • ALE • ALE Development • BAPI • Receiver Determination • Define Filter Object Types** or using Transaction BD95. Upon entering the transaction, the system prompts you with an **Information** window cautioning that the table is cross-client. Click the **Continue** icon to proceed to the next screen. On the resulting **Change View "ALE Object Type": Overview** screen, click the **New Entries** button, as shown in Figure 9.65 ❶. On the resulting **New Entries: Overview of Added Entries** screen, maintain the following values in the fields ❷:

- **ALE Object Type**
  Enter value "RECEIVER".
- **Table Name**
  Enter "TBD06" as the table name.
- **Field name**
  Enter "RCVSYSTEM".

Click the **Save** ❸ button to save the entries.

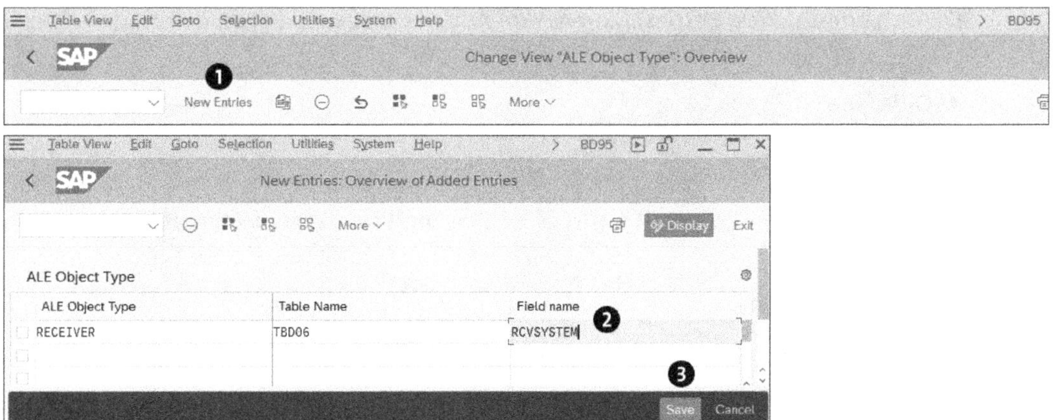

Figure 9.65 Specify ALE Object Types

Assigning Filter Object Type to Business Application Programming Interface

In this customizing activity, you'll assign filter object types, which you've just created from the previous configuration, to a BAPI. To assign filter object types to BAPI ALE filter object types, navigate to the maintenance screen from the **SAP Easy Access** menu by navigating to **Tools • ALE • ALE Development • BAPI • Receiver Determination • Assign Filter Object Types to BAPI** or using Transaction BD96. Upon entering the transaction, the system prompts you with an **Information** window cautioning that the table is cross-client. Click the **Continue** icon to proceed to the next screen. On the resulting **Change View "Filter View "Filter Objects for Receiver Determination o..": Overview** screen, click the **New Entries** button, as shown in Figure 9.66 ❶.

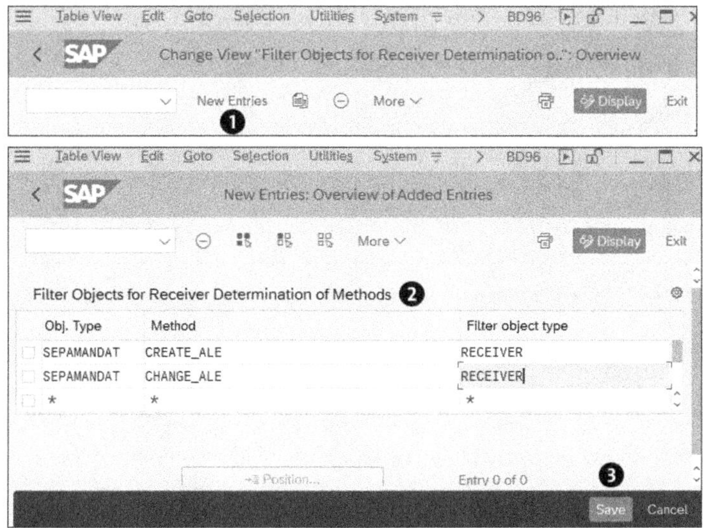

Figure 9.66 Filter Objects of Receiver Determination

On the resulting **New Entries: Overview of Added Entries** screen, maintain the values in the fields ❷, as follows:

- **Obj. Type**
  Enter value "SEPAMANDAT".
- **Method**
  Enter "CREATE_ALE" and "CHANGE_ALE" as method names, as shown in Figure 9.66.
- **Filter object type**:
  Enter "RECEIVER".

Click the **Save** button to save the entries.

Maintain the Distribution Model

In this configuration, you'll create a distribution model that determines the sender and receiver of the IDocs and defines the transfer rules when SEPA mandates go from the source system to the Central Finance system. In addition, the distribution model allows you to define the system-wide message flow. To create the distribution model for the replication project systems master data, go to Transaction BD64 in the source system. Refer to the steps we've covered in Section 9.7.3, subsection "Maintain the Distribution Model," for Central Finance configuration. Follow the same steps for the source system. However, there are a couple of differences when maintaining details in the **Add BAPI** popup screen, as shown earlier in Figure 9.58. On this popup screen, maintain the details as shown in Table 9.29 for the source system configuration.

Field	Value
Sender/Client	S18CLNT400 (source logical system)
Receiver/Server	S18CLNT 500 (Central Finance logical system)
Obj. name/interface	SEPAMANDATE
Method	CREATE_ALE

Table 9.29 Add a BAPI for Maintenance of the Distribution Model: Source System

Tip

Consult SAP Note 1887572 for additional trips and tricks for troubleshooting SEPA mandates through ALE distribution.

9.8 Summary

The central payment functionality offered with Central Finance allows companies to process payments centrally in one location from the Central Finance system. This is helpful especially for companies with multiple ERP systems where invoices from several source systems require central processing in the Central Finance system to improve cash flow. In this chapter, first you learned the basic concepts of central payments. Then, we transitioned to review configuration requirements in the source system and the Central Finance system. Addressing issues around handling open items with central payments is a critical step during the cutover phase of Central Finance. The chapter delved into each dependent process consideration during central payment activation. In addition, you've learned SEPA mandate configuration requirements are required for Central Finance to process SEPA direct debit mandates centrally. In the next chapter, you'll learn about additional replication scenarios in Controlling from a Central Finance perspective.

Chapter 10
Controlling

Controlling provides management users with information that helps them to make decisions on day-to-day, short term, and long-term operational activities. As a result of financial documents replication in Central Finance, relevant cost information is available from the Universal Journal for management reporting. In addition, Central Finance offers several replication scenarios in the controlling space. This chapter provides information you need to understand how various controlling objects and their relevant transactions transfer to Central Finance.

The goal of financial accounting is to provide relevant and useful information to external users such as investors, creditors, employees, and regulatory bodies. The relevant information, in this case, is provided in the form of financial statements primarily used to make informed decisions on the allocation of company resources efficiently. The financial accounting (FI) component within SAP accomplishes this goal. On the other hand, management accounting focuses on internal users of an organization. Examples of internal users in an organization include production managers, marketing managers, finance directors, and controllers. The internal users also need detailed information to make the right decisions. The Controlling (CO) module in SAP fulfills management accounting needs. Nevertheless, both accounting functions are aimed at maximizing the company's future profits. In SAP, FI and CO components are seamlessly integrated to provide the necessary information to stakeholders.

From a Central Finance perspective, SAP has delivered several CO-related replication scenarios to transfer data from the source system to the Central Finance system. When a financial document is posted, the CO document is also reposted along with financial document, capturing cost object information in the Universal Journal for reporting. We've covered a few CO topics in Chapters 4 and 8. In this chapter, you'll learn about additional features from the CO module perspective:

- **Projects and work breakdown structure (WBS) replication**
  This supports replication of project definitions and breakdown structure (WBS) elements from the source system to the Central Finance system. When FI postings are replicated, the project or WBS master data is retrieved from mapping prior to posting to the Universal Journal. See Section 10.1.

- **Central budgeting**
  This is a centralized approach in Central Finance to manage budgets in a single place across an enterprise. See Section 10.2.
- **Classic profit center replication**
  This enables replicating of EC-PCA transactional postings from the source system to the Central Finance system. See Section 10.3.
- **Material cost estimate**
  This replication is supported for a released material cost estimate with or without quality structure. See Section 10.4.
- **Cost rate replication**
  Central Finance supports replicating source system cost rates, which are prices for internal activities in CO, to the Central Finance system. See Section 10.5.
- **Activity type replication**
  This supports transferring activity types used in cost centers in product cost controlling. See Section 10.6.
- **Transfer of plan data**
  This is supported from SAP S/4HANA 2023 release onward. See Section 10.7.
- **Margin analysis**
  This supports replicating profitability analysis characteristics from the source system to the Central Finance system. See Section 10.8.
- **Cost of goods sold split (COGS split)**
  This provides the relevant cost component structure of material cost estimates. Central Finance supports different scenarios of COGS split. See Section 10.9.
- **Price difference split**
  This is supported in Central Finance based on the source system data received from the source system. See Section 10.10.

Let's now discuss each CO feature in the context of Central Finance. We'll begin with WBS element replication and then transition into other CO-relevant innovations topics in Central Finance.

10.1 Projects and Work Breakdown Structure Replication

One of the primary use cases of Central Finance is central project reporting. The project structure, including WBS master data, is created and managed in the source system. Often project systems processes are integrated with other processes, for example, logistics, in the source system. Subsequent project plan and budget information is also maintained in the source system. Central Finance supports projects and WBS master data replication from the source system to Central Finance. Subsequent FI and CO postings assigned with WBS as account assignment objects in the source system are replicated to the Central Finance system. Before posting, the system uses SAP Master Data Governance mapping for source WBS to get corresponding WBS mapping from Central Finance with the collected actual data in the Universal Journal facilities to achieve central project reporting. In addition, Central Finance supports asset settlement and central budgeting scenarios. To support projects and WBS master data replication, SAP uses application link enabling (ALE) and intermediate documents (IDocs) technology to replicate project systems data from the source system to the Central Finance system.

Figure 10.1 shows an overview of the projects and WBS master data replication process in Central Finance.

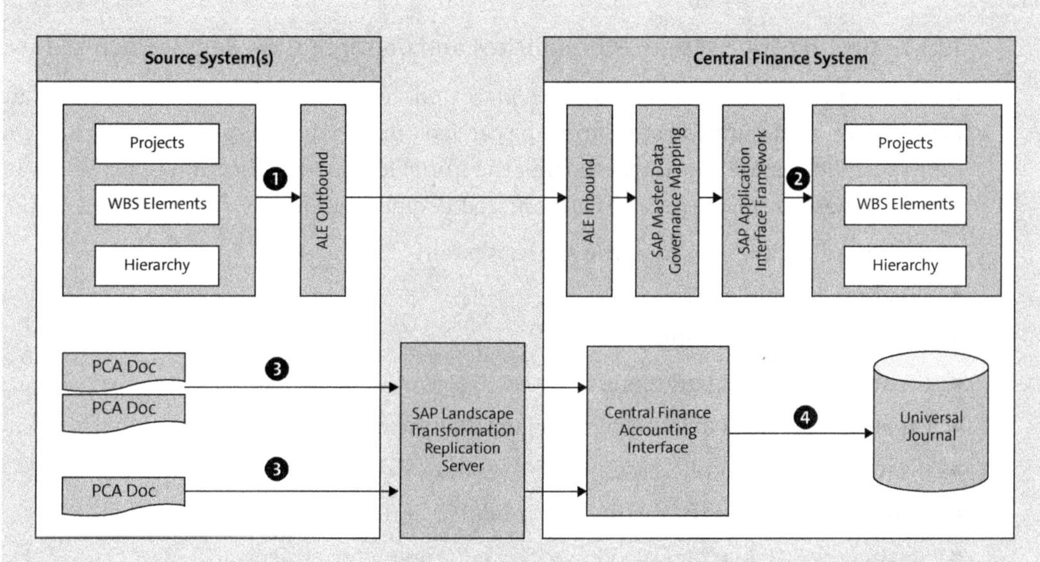

Figure 10.1 Projects Replication Overview

The data flow is as follows:

❶ **Project master data**
First, project definition and WBS master data is created in the source systems. Later, the WBS cost objects are then used in FI and CO transactional posting in the source system.

❷ **ALE inbound and outbound interfaces**
The project systems master data is then replicated via ALE IDoc technology using ALE outbound and inbound configurations setup in the source system and the Central Finance system. Upon the master data creation in the source system, the system creates an IDoc that is transferred to the Central Finance system. The replicated master data goes through SAP Master Data Governance mapping before the maser data created in the project systems master data table (PRPS) and dependent tables in the

Central Finance system. SAP Application Interface Framework is used to monitor and troubleshoot projects and WBS master data replication-related errors.

❸ **Transactional posting in the source system**
WBS cost objects created in the earlier step are used in FI and CO transactional postings in the source system.

❹ **Financial postings in the Central Finance system**
When the FI and CO transactional postings are transferred from the source system to the Central Finance system, the system retrieves project master data and WBS elements from SAP Master Data Governance mapping in Central Finance to post documents to the Universal Journal.

10.1.1 Getting Started with Prerequisites and Configuration Requirements

Besides applying the necessary service pack updates and required SAP Notes, certain prerequisite configuration steps must be carried out in both the source system and the Central Finance system prior to starting the initial load and replication of project master data. These prerequisite steps are covered in eight sections, as follows:

- Creating a logical system in the source system
- Creating the tRFC port
- Creating a partner profile
- Creating outbound parameters
- Creating the ALE distribution model in the source system
- Configuring SAP Application Interface Framework in Central Finance
- Configurating the ALE Distribution model in Central Finance
- Setting the SAP Master Data Governance mapping for projects

Next, you'll learn about each of the configuration activities with step-by-step instructions.

Configuring a Logical System in the Source System

For the ALE integration scenario to work for a project's master data replication, every source system connected to Central Finance must be assigned as a logical system. This logical system acts as a communication bridge between the sender system and receiver system. To define a new logical system, navigate to Transaction BD54. Alternatively, navigate to the transaction from Transaction SALE via **IDoc Interface/Application Link Enabling (ALE) • Logical Systems • Define Logical Systems**. Upon executing the transaction, the system displays the warning message that the table is cross-client. Click the **Continue** icon to go to the **Change View "Logical Systems": Overview** screen. On this screen, click the **New Entries** button. You'll be taken to the **New Entries: Overview of Added Entries** screen, as shown in Figure 10.2. On this resulting entry screen, maintain logical system name and description values in the **Log.System** and **Name of Logical System** ❶ fields. Click the **Save** ❷ button to save the entries.

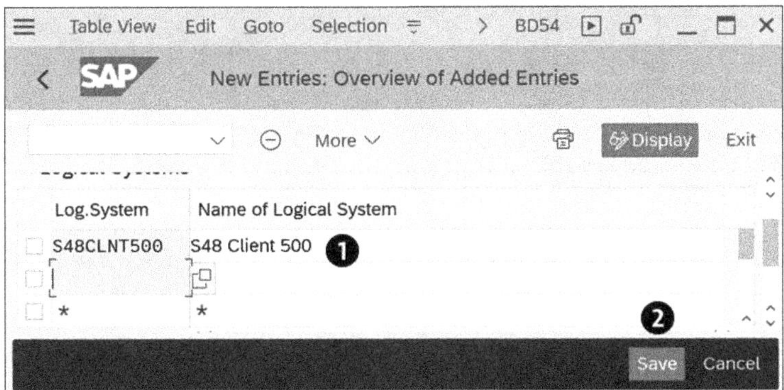

Figure 10.2 Create the Logical System ALE Distribution

[«]

Note

This is like the logical system you've created in configuration in Chapter 7, Section 7.4.2. If you've already done this earlier for the other replication scenarios, this step can be ignored.

You'll also define necessary remote function call (RFC) connections from the source system to Central Finance using Transaction SM59 as we've discussed in Chapter 7, Section 7.4.2.

Creating a Transactional Remote Function Call Port

To establish communication between the source system and the Central Finance system, SAP uses RFCs. The system executes the asynchronous communication method using transactional RFC (tRFC). SAP recommends using the tRFC communication method to transfer data between systems in a distributed landscape. To configure a tRFC port, navigate to the transaction from **SAP Easy Access** menu path **Tools • ALE • ALE Administration • Runtime Settings • Port Maintenance** or use Transaction WE21. After entering the transaction, navigate to **Ports • Transaction RFC** from the **Ports in IDoc processing** initial screen, as shown in Figure 10.3. On the resulting screen, choose **Transaction RFC** and then click the **Create** ❶ icon. This brings up **Ports in IDoc processing** dialog screen,. Choose the **own port name** radio button and enter the **Name** of port and Click the **Continue** icon, On the resulting screen, enter the **Description** ❷ of port. Next, choose **IDoc record types SAP release 4.x** radio button under **Version** section ❸. Next, enter the name in **RFC destination** ❹ you've created earlier. Finally, you can configure **Port Options** ❺ optionally. Click the **Save** ❺ icon to save the entries.

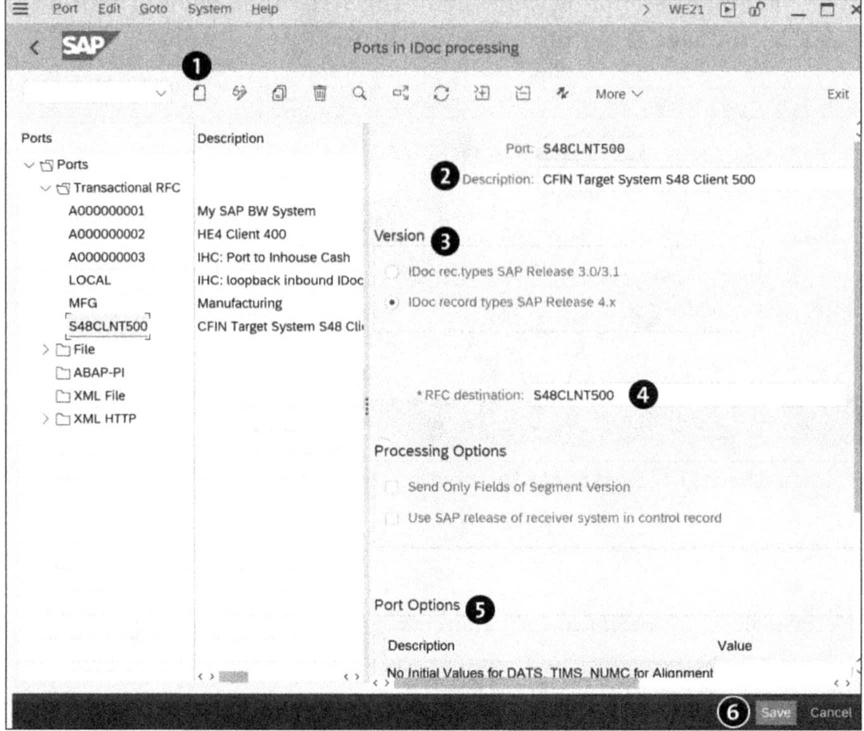

Figure 10.3 Creating the tRFC Port

Note

As you've noticed, in our example, we've shown you how to configure the tRFC port. SAP released background RFC (bgRFC), which is a successor to tRFC and queued RFC (qRFC) communication methods. bgRFC brings significant improvements, especially in performance and new features.

Creating a Partner Profile

Partner profiles contain inbound and outbound parameters that define how data is exchanged between systems. In the source system, you'll create a partner profile and outbound parameters containing the following:

- Basic partner profile information such as data types and asynchronous method
- Outbound parameters to define the project basic type and message type

To define a partner profile, navigate to the transaction from **SAP Easy Access** menu path **Tools • ALE • ALE Administration • Runtime Settings • Port Maintenance** or use Transaction WE20. After reaching the transaction, you'll arrive at the **Partner profiles** screen shown in Figure 10.4. On this screen, choose **Partner Type LS**, and click the **Create** icon ❶.

On the resulting entry screen, maintain values in **Partner No.** and **Partn.Type** for the logical system and LS partner type, respectively. Next, maintain the values in the **Ty**, **Agent**, and **Lang** fields located under the **Post Processing: Valid Processors** ❷ tab. Then, go to the **Classification** ❸ tab to maintain **Partner Class** as **ALE** and **Partner Status** as **A** for active status. Click the **Outbound Parameter** icon ❹ to define the project basic type and message type.

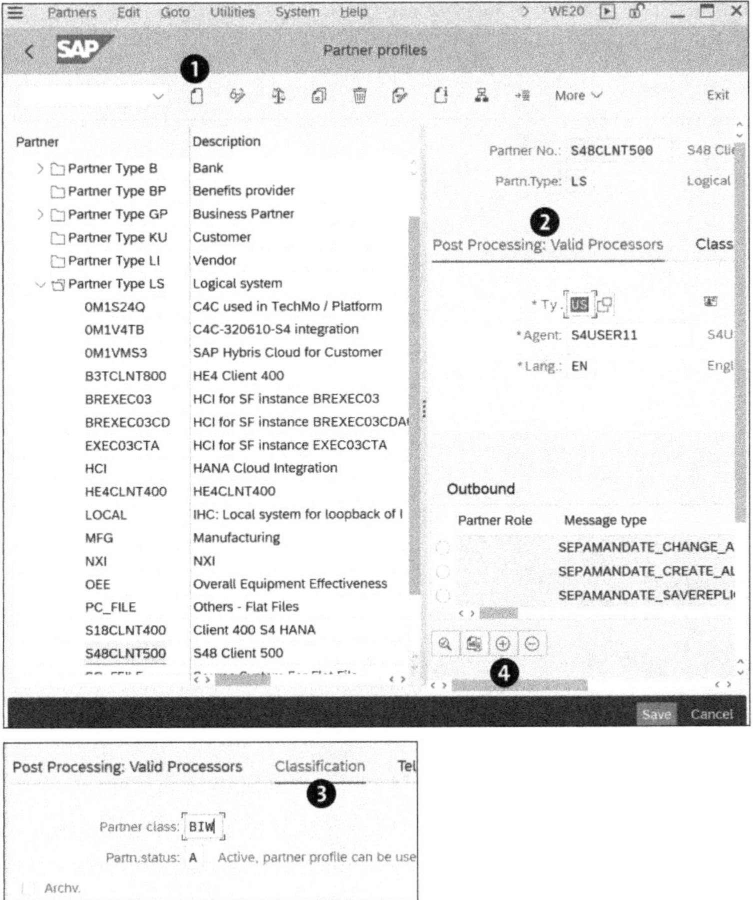

10

Figure 10.4 Create a Partner Profile

Creating Outbound Parameters

In this configuration, you'll create outbound parameters such as IDoc message type, IDoc basic type, and port number for the partner profile you've just created in the previous step. IDocs use outbound parameters to transfer data from the source system to the Central Finance system.

To create outbound parameters, click the **Create outbound parameter** icon from the **Partner Profiles** screen, as shown in Figure 10.4 ❹. This will lead you to the **Partner**

Profiles: Outbound Parameters screen shown in Figure 10.5. On this screen, enter "PROJECT" in the **Message Type** field ❶ in the header section. In the **Outbound Options** tab, choose the **Pass IDoc Immediately** ❷ radio button under the **Output Mode** section. Then enter **PPROJECT02** as a value in **Basic Type** ❸ field under **IDoc Type** section, and then choose the tRFC port number under **Receiver port** that you created in the previous step. Click the **Save** ❹ icon to save the entries.

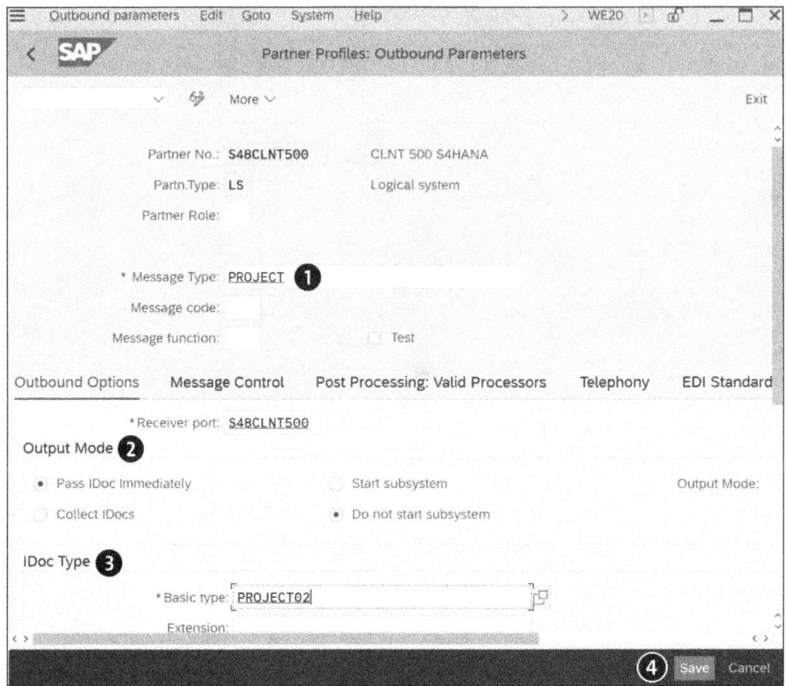

Figure 10.5 Partner Profile Outbound Parameters

Create an Application Link Enabling Distribution Model in the Source System

In this configuration, you'll create a distribution model that determines sender and receiver of the IDocs and defines transfer rules when replicating projects and WBS master data from the source system to the Central Finance system. In addition, the distribution model allows you to define a system-wide message flow. To create a distribution model for project systems master data replication, go to Transaction BD64 in the source system. You'll arrive at the **Display Distribution Model** screen. On this screen, as shown in Figure 10.6, click the **Edit** ❶ button and then click the **Create Model View** ❷ button. As observed, you'll be prompted with the **Create Model View** ❸ dialog window. On this screen, maintain values in **Short text** and **Technical name** fields. As you've also noticed, the **Start date** and **End Date** fields default to today's date. Click the **Continue** ❸ icon to create the distribution model. Next, SAP has provided a standard business application programming interface (BAPI) to transfer data from sender systems. On the **Change Distribution Model** screen, choose the **Project WBS Replication from S18 S48** ❹ distribution model that you've just created, and navigate to **Add BAPI** ❺ from the **More** submenu item.

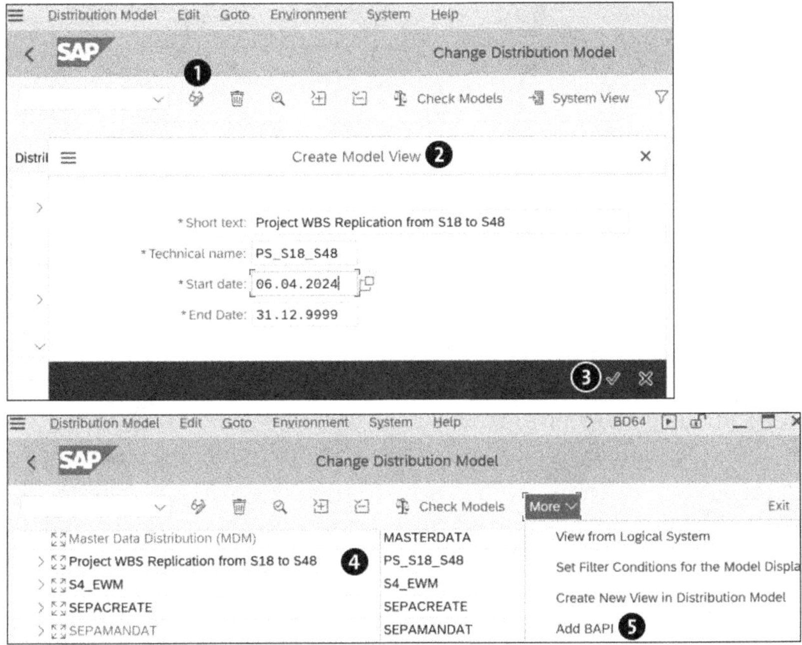

Figure 10.6 Create a Distribution Model

Next, on the resulting **Add BAPI** dialog box, as shown in Figure 10.7, maintain values as follows:

❶ **Model view**
Enter the model view name, for example, "PS_S18_S48".

❷ **Sender/client**
Enter the source system client information in this field.

❸ **Receiver/server**
Enter the target Central Finance system as a receiver system.

❹ **Object/Interface**
Enter "WorkBreakDownStruct" for project systems replication.

❺ **Method**
Enter "SaveReplica".

Once all fields are filled, click the **Continue** ❻ icon to save the entries.

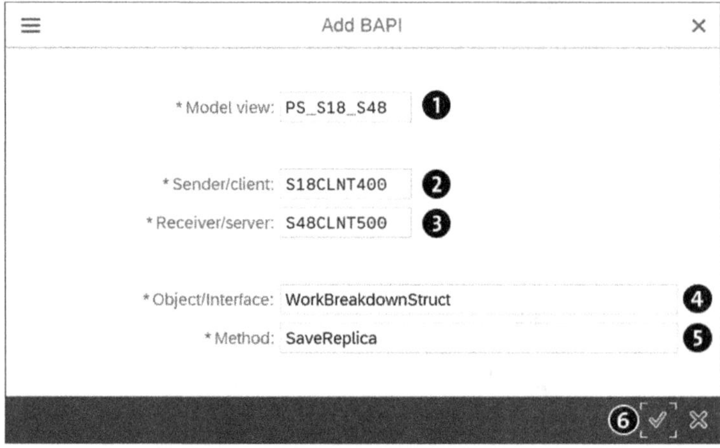

Figure 10.7 Add the BAPI

Next, you'll generate partner profiles for the model view you've just created in the source system. To generate partner profiles, select the project WBS replication distribution model (see Figure 10.8), and then navigate to **Generate Partner Profiles** ❶ from the **Environment** submenu item in Transaction BD64. Alternatively, you can generate the partner profiles by reaching Transaction BD82 directly. Upon reaching the transaction, maintain the following fields:

- **Model View** and **Partner System** ❷
  Enter a model view name and partner system name, for example, "PS_S18_S48", and "S48CLNT500", respectively.
- **TY** and **ID** ❸
  The system defaults the type of individual (**US**) and user ID under **Postprocessing Authorized Users** section for IDoc error notification.
- **Version** and **Pack. Size** ❹
  The system defaults **3** for IDoc types from version 4.0 onward. For older versions, version 2 is used.
- **Pass IDoc immediately** ❺
  The system defaults to trigger immediately.

Once all fields are filled, click the **Execute** ❻ button to generate the partner profile. The system displays logs if the partner profiles are generated successfully. You'll be able to check whether the partner profiles generated successfully by navigating to **Checking Partner Profiles** via Transaction WE20.

Next, you'll create a filter group that is used to control how the system transfers segments of an IDoc based on certain filter conditions. This is an optional step, but it might be necessary considering the volume of data and other selection criteria. From the **Display Distribution Model** screen, navigate to and double-click on **Receiver Determination: Filter Active**, as shown in Figure 10.9 ❶.

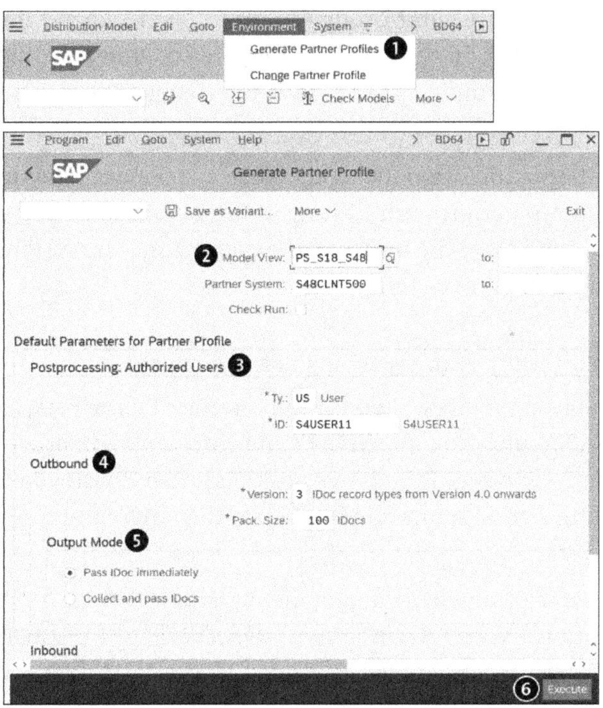

Figure 10.8 Generate a Partner Profile

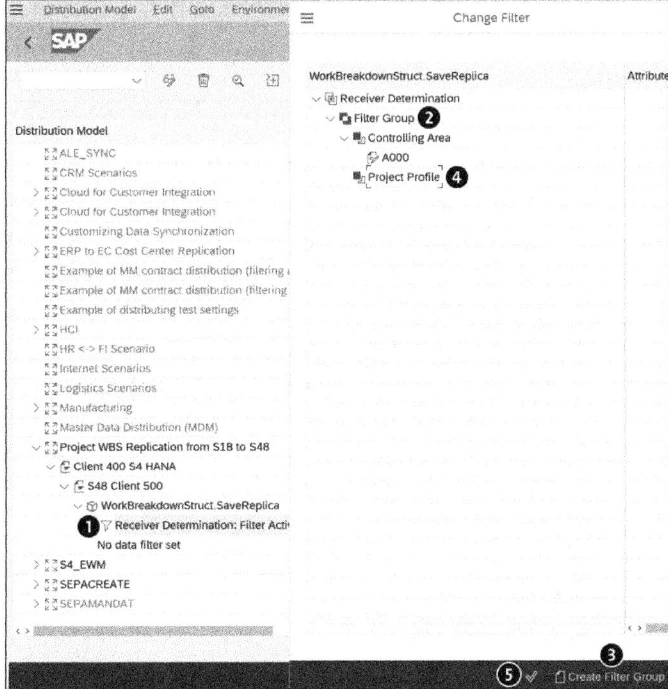

Figure 10.9 Create filter group

On the **Change Filter** screen, click the **Create Filter Group** ❷ button. Next, expand the **Filter Group** ❸ menu item, and double-click on either **Controlling Area** or **Project Profile** to enter the controlling area value, for example, "A000". Save the changes by clicking the **Continue** ❺ icon.

Once IDoc filter objects are created, you can assign the filter objects to message types in Transaction BD59. In addition, you can perform unit testing of filter conditions using Transaction WE02 to ensure filter conditions are working as expected and the system populates data in the respective segments correctly.

Configuration of SAP Application Interface Framework in Central Finance

To monitor project master data replication in SAP Application Interface Framework, as a prerequisite, you'll enable the SAP Application Interface Framework deployment scenario and relevant business configuration sets (BC sets) for project systems master data replication. Following are the configuration items you'll perform that are relevant for SAP Application Interface Framework:

- To activate SAP Application Interface Framework deployment scenario SAP_AIF_0045 for projects master data replication, execute Transaction /AIF/CONTENT_EXTRACT. Refer to the step-by-step instructions in Chapter 6, Section 6.1.2.
- To activate BC set FINS_CFIN_ALE_PS for projects master data replication, use Transaction SCPR20. Refer to the step-by-step instructions in Chapter 6, Section 6.1.2.

Once these Central Finance configurations are completed, you'll see the PS_OBJ interface available within the /FINCF Central Finance namespace in the Transaction /AIF/ERR monitoring and error handling.

Configuration of an Application Link Enabling Distribution Model in Central Finance

Like source system configuration ALE and IDocs, you'll also create an ALE distribution model in Central Finance to support projects system master data replication. First, you'll create the ALE distribution model in Central Finance using Transaction BD64. Next, you'll assign the process code /FINS_CFIN_PS_PROJECT in the inbound parameter of the partner profile created in the Central Finance system using Transaction WE20. Repeat the same steps you followed in the beginning of the section for the source system. Figure 10.10 shows the ALE distribution created and assigned a message type with process code **/FINS_CFIN_PS_PROJECT** ❶. Once you've configured it, click the **Save** ❷ button to save the entry.

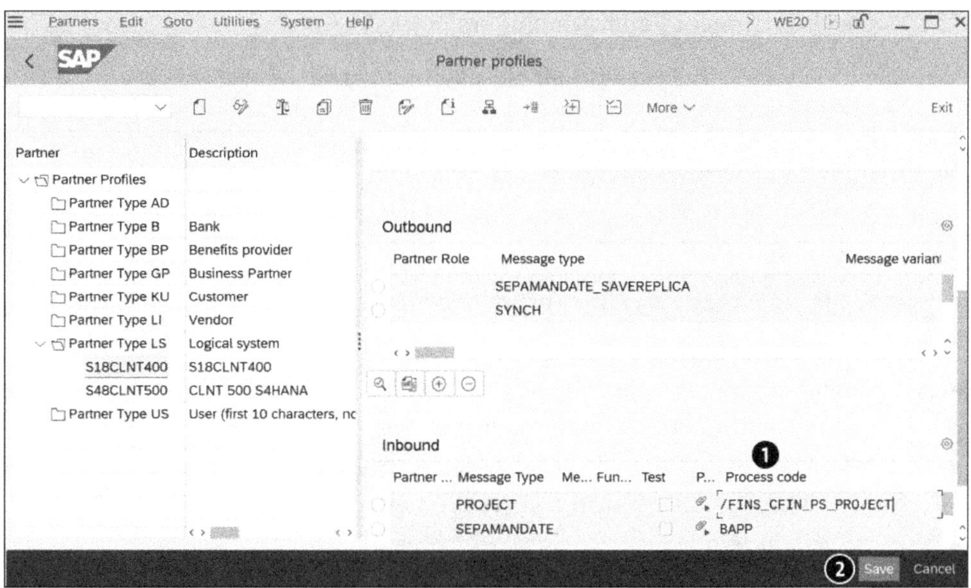

Figure 10.10 Create an ALE Distribution Model in Central Finance

SAP Master Data Governance Mapping for Projects

To replicate project system master data, you need to set **Mapping Action** to **1 Mapping Obligatory** in the **Define Mapping Action for Mapping Entities** customizing activity for the **PROJECT_DEFINITION** and **WBS_ELEMENT** mapping entities. Figure 10.11 shows the **Mapping Action** is for **WBS_ELEMENT** mapping entities. Refer to Chapter 4, Section 4.4.2, for step-by-step instructions for this customizing activity.

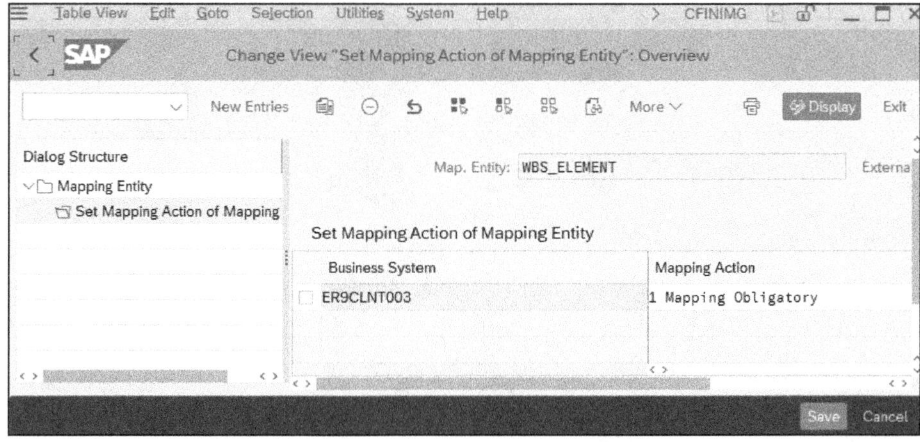

Figure 10.11 Mapping Action for WBS Replication

10.1.2 Check Setting for Replication of Projects

Once you've completed all required configuration, SAP provides report FINS_CFIN_PS_CHECK to check any inconsistencies in project systems replication-related configuration settings. **Maintain Source Systems and RFC Assignments** is required as a prerequisite to run this report. SAP strongly recommends you run this report prior to project master data replication. The report validates the following activities and displays a log report of the same:

- SAP Application Interface Framework deployment scenario `SAP_AIF_0045` activation
- Activation of BC set `FINS_CFIN_ALE_PS`
- Assignment of process code `/FINS_CFIN_PS_PROJECT` to message type `PROJECT` in the ALE inbound partner profile for the source system
- Mapping action configuration set to **Mapping Obligatory** for the connected source system for **PROJECT_DEFINITION** and **WBS_ELEMENT** mapping entities
- No coding mask defined
- Interface determination in SAP Application Interface Framework for projects replication

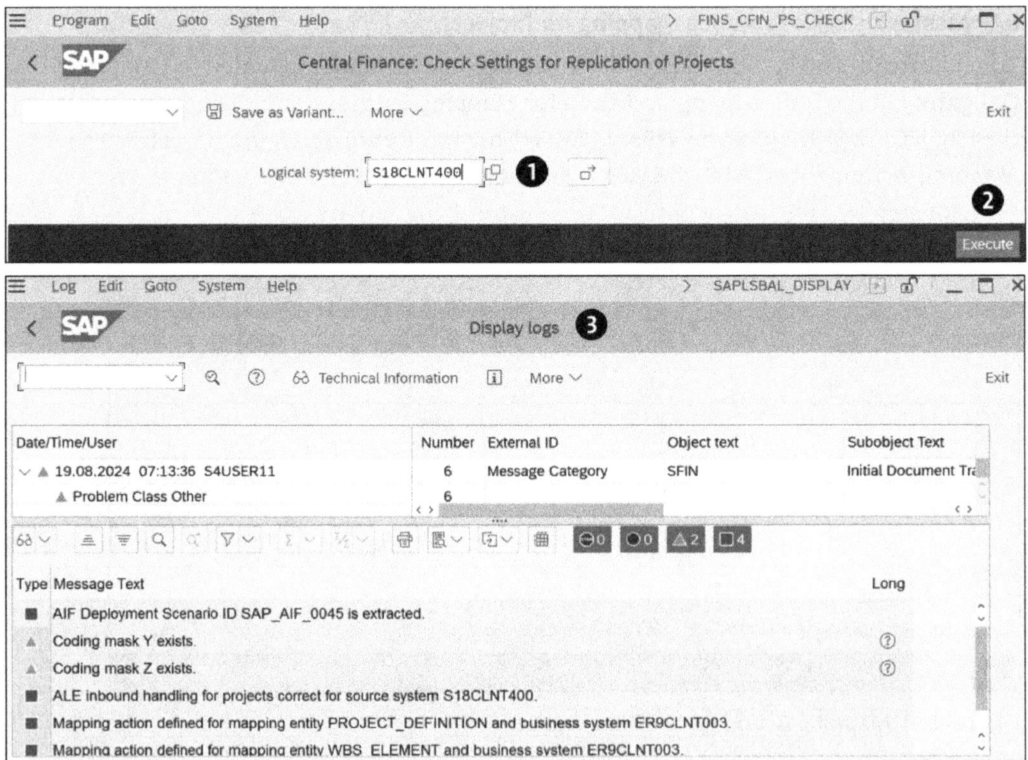

Figure 10.12 Check Settings for Replication of Projects

To run the report, execute it from Transaction SA38 or Transaction SE38. Upon reaching the report, on the resulting **Central Finance: Check Settings for Replication of Projects** screen, as shown at the top of Figure 10.12, enter the source **Logical System** ❶ value, and click the **Execute** ❷ button. The system **Displays logs** ❸ with success, warning, and error messages.

10.1.3 Initial Load and Replication of Projects

Once all errors and inconsistencies are addressed, you'll begin the initial load of projects master data by executing Transaction CJAL from the source systems. This transaction takes you to the **Report Program for Directed Transmission of WBSs** screen, as shown in Figure 10.13.

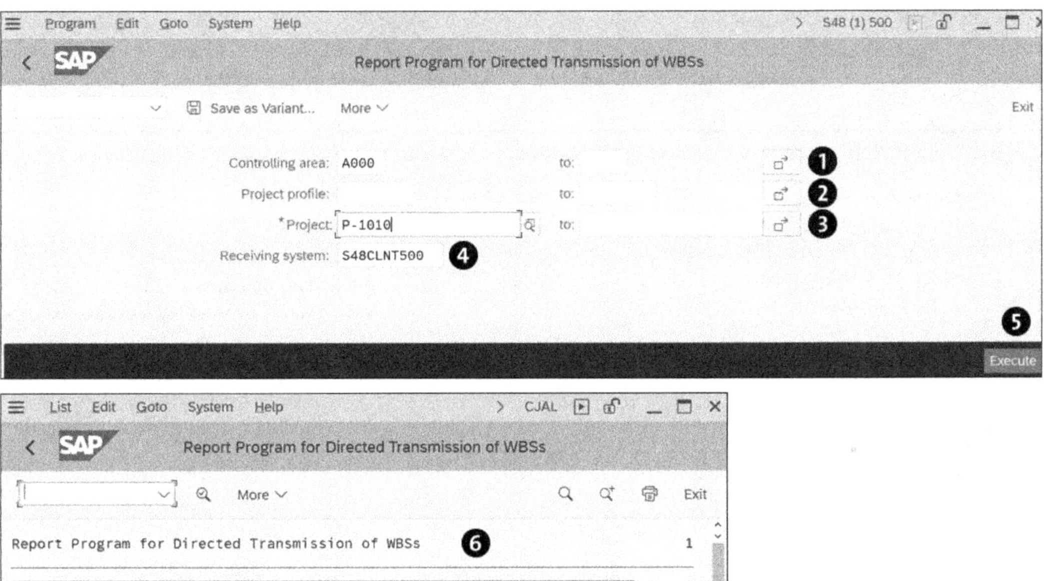

Figure 10.13 Initial Load of Projects Master Data

On this screen, enter values in the following fields:

❶ **Controlling area**
Enter a controlling area, for example, "A000".

❷ **Project profile**
Enter a project profile or range of project files you want to replicate from the source system to the Central Finance system. This is an optional field.

❸ **Project**
Enter a project or range of projects to replicate.

❹ **Receiving system**
Enter the target Central Finance system as a receiver system.

Click the **Execute** ❺ button to begin the initial load of projects. The resulting output screen ❻ shows the projects that came through the initial load.

A successful initial load transfers project master data from the source system to the Central Finance system. You'll able to view replicated project master data in SAP Application Interface Framework under namespace **/FINCF**, interface **PS_OBJ**, and version **1**, as shown in Figure 10.14.

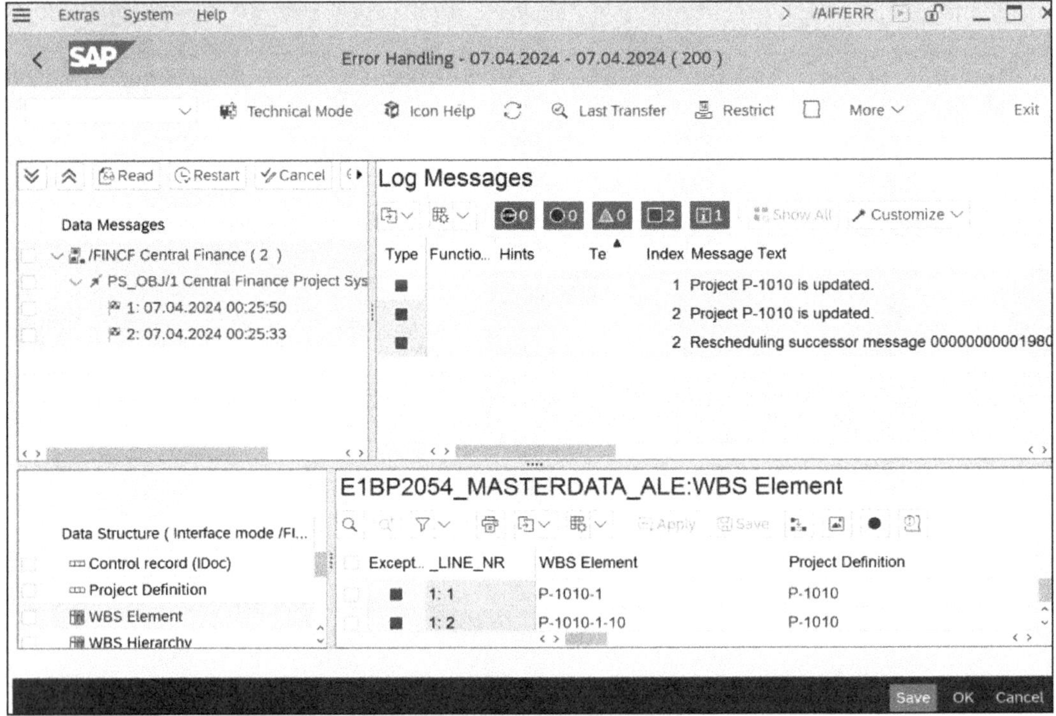

Figure 10.14 Projects Replication in SAP Application Interface Framework

As shown in Figure 10.15, besides project master data replication, the system also generates mapping for project definition ❶ (mapping entity **PROJECT_DEFINITION**) and WBS element ❷ (mapping entity **WBS_ELEMENT**) in SAP Master Data Governance key mapping in Central Finance. The system adds the key mapping for the first time or when no key mapping exists. If there is a key mapping that already exists, the system doesn't update. At the time of financial documents posting with WBS as an account assignment cost object on a line item, the system uses the SAP Master Data Governance mapping in Central Finance.

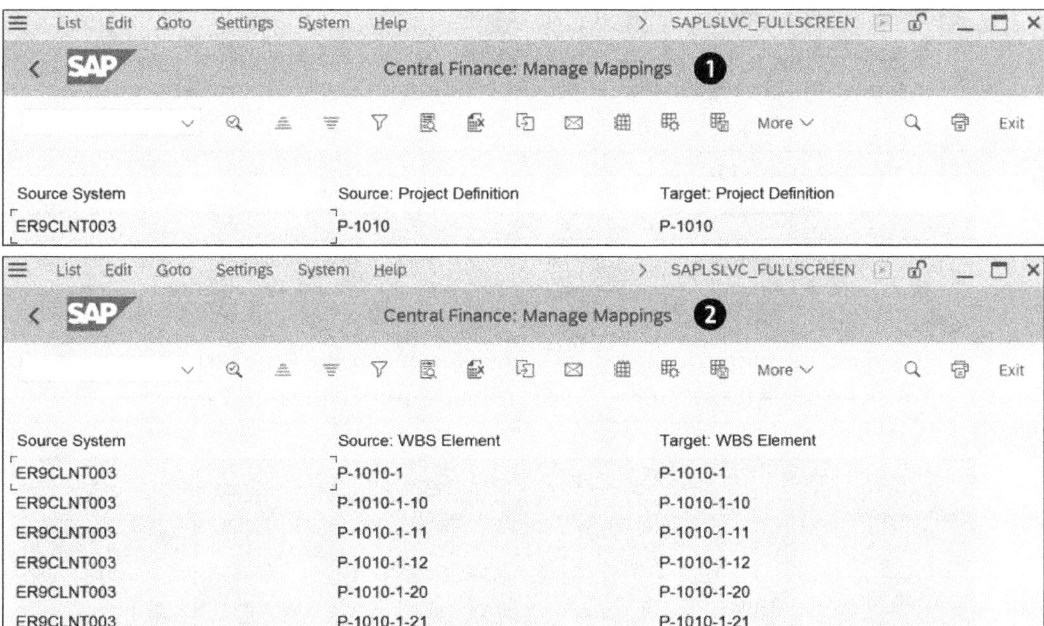

Figure 10.15 SAP Master Data Governance Mapping for Projects

10.1.4 Central Projects: Asset Settlement Scenario

With Central Finance asset settlement scenario, the system creates an asset under construction (AuC) automatically. Accumulated cost on WBS can be settled to the AuC, and then you can settle the cost to final fixed assets in Central Finance based on asset accounting policies. This is useful especially when fixed assets processes are moved to Central Finance. Figure 10.16 illustrates the data flow between the source system and the Central Finance system for the Central Finance projects asset settlement scenario:

❶ **Investment profile and project profiles**
Besides Central Finance configuration requirements for the asset settlement scenario, you'll activate the **Manage AuC** checkbox on investment profile configuration in Central Finance only, not in the source system. As a result of this configuration, the system generates an AuC automatically for a released WBS in the Central Finance system. No AuC is generated in the source system. In addition, SAP recommends creating new project profiles for asset settlement scenarios.

❷ **Project systems master data replication**
Project definition and WBS master data is created in the source system and replicated to Central Finance. When the WBS system status is set to **REL**, the WBS master data is replicated to Central Finance. In addition, an AuC is generated for the WBS in Central Finance.

❸ **Transactional posting**
FI documents and CO documents that use project system master data as account assignment cost objects to collect cost and then relevant financial documents are replicated to Central Finance.

❹ **Periodic settlement**
With the asset settlement scenario, you'll run periodic settlements in the Central Finance system. Periodic settlements move costs from WBS to the AuC generated from step 2.

❺ **Final settlement**
Based on asset accounting policies, you'll carry out the final settlement in Central Finance. Full settlement moves costs from the AuC to the final asset.

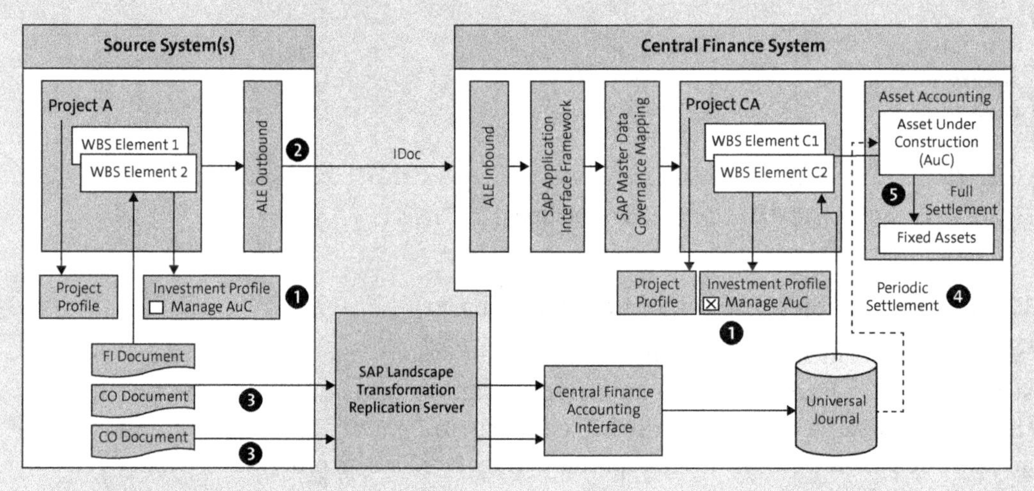

Figure 10.16 Central Asset Settlement

Now that you understand the data flow in the central asset settlement scenario, let's walk through the configuration requirements in the source system and the Central Finance system central asset settlement scenario.

Configuration Activities in the Source System

To enable the central projects assets settlement scenario, you'll have to carry out the required configuration in the source system and the Central Finance system. Table 10.1 provides customizing activities required to enable projects master data replication in the source system. Some configurations are already covered in other chapters, so reference links are provided.

Customizing Activities	IMG Menu Path	Activity
Define Logical System for Central Finance	**Financial Accounting • Central Finance: Source System Settings • Set Up Connection to Central Finance System • Define Logical System for Central Finance Systems.**	For step-by-step instructions, see Chapter 7, Section 7.4.2.
Set Up RFC Destination for Central Finance System	**Financial Accounting • Central Finance: Source System Settings • Set Up Connection to Central Finance System • Set Up RFC Destination for Central Finance System.**	For step-by-step instructions, see Chapter 7, Section 7.4.2.
Assign RFC Destination for Displaying Objects from Central Finance System	**Financial Accounting • Central Finance: Source System Settings • Set Up Connection to Central Finance System • Assing RFC Destination for Displaying Objects from Central Finance.**	For step-by-step instructions, see Chapter 7, Section 7.4.2.
Maintain RFC Assignments and Settings for Central Finance System	**Financial Accounting • Central Finance: Source System Settings • Set Up Connection to Central Finance System • Maintain RFC Assignments and Settings for the Central Finance System.**	For step-by-step instructions, see Chapter 7, Section 7.4.2.
Define Investment Profile	**Investment Management • Projects as Investment Measures • Master Data • Define Investment Profile**	In this configuration, leave the **Manage AuC** checkbox unchecked, so the AuCs aren't generated in the source system but are created in the Central Finance system. Figure 10.16 shows this checkbox enabled in Central Finance.

Table 10.1 Source System Customizing Activities

Customizing Activities	IMG Menu Path	Activity
Define Settlement Profile	**Investment Management • Projects as Investment Measures • Settlement • Maintain Settlement Profiles**	In this configuration, you'll choose the **Not for Settlement** radio button on settlement profile configuration in source system. You'll maintain the **To Be Settled in Full** radio button on settlement profile in central finance system. As a result of these configurations, you'll settle costs in Central Finance only.
Define Project Profile	**Investment Management • Projects as Investment Measures • Master Data • WBS Elements • Define Project Profile**	The project profile stores default values to be populated when creating the project definition. You'll assign an investment profile and settlement profile in the project profile. SAP recommends creating a new project profile for the Central Finance asset settlement scenario.

Table 10.1 Source System Customizing Activities (Cont.)

Tip

SAP recommends implementing source system SAP Note 2994270 to support central projects asset settlement. Check if the SAP Note is relevant in your source system.

Configuration Activities in Central Finance System

Table 10.2 lists configuration activities in the Central Finance system.

Customizing Activities	Menu Path	Activity
Define Investment Profile	**Investment Management • Projects as Investment Measures • Master Data • Define Investment Profile**	Define an investment profile with the **Manage AuC** checkbox chosen, as shown in Figure 10.16. The system reads the settings, and along with if the asset settlement scenario is active, the AuC asset is generated in the Central Finance system.

Table 10.2 Customizing Activity in Central Finance for the Central Projects Asset Settlement Scenario

Customizing Activities	Menu Path	Activity
Define Settlement Profile	**Investment Management • Projects as Investment Measures • Settlement • Maintain Settlement Profiles**	Choose the **To be Settled in full** radio button. You'll execute both periodic and full settlement in Central Finance.
Define Project Profile	**Investment Management • Projects as Investment Measures • Master Data • WBS Elements • Define Project Profile**	Assign both the investment profile and settlement profile in the project profile. When a new project is created and released, the system generates the AuC automatically in the Central Finance system.
Assign Project Profile to Company Code for Asset Settlement	**Financial Accounting • Central Finance • Central Finance: Target System Settings • Central Projects • Assign Project Profile to Company Code for Asset Settlement**	This configuration activates the central projects (WBS) asset settlement scenario by project profile and company code combination. SAP recommends creating a new project profile for the asset settlement scenario.

Table 10.2 Customizing Activity in Central Finance for the Central Projects Asset Settlement Scenario (Cont.)

10

The system considers only the company codes and project profiles defined in **Assign Project Profile to Company Code for Asset Settlement** configuration in Central Finance. In addition, you must set up project profiles and investment profiles mapped in Central Finance SAP Master Data Governance mapping correctly. To assign a company code and project profile to enable the central projects asset settlement scenario, navigate to IMG menu path **Financial Accounting • Central Finance • Central Finance: Target System Settings • Central Projects • Assign Project Profile to Company Code for Asset Settlement**. You'll arrive at the **Change View "Assign Project Profile to Company Codes for..": Overview** screen, as shown in Figure 10.17. On this screen, click the **New Entries** ❶ icon to maintain the company code and project profile. On the resulting **New Entries: Overview of Added Entries** screen, maintain values in the **CoCd** and **Prj.Prf** fields ❷. Click the **Save** ❸ button to save the entries.

As shown in Figure 10.18, select the **Manage AuC** ❶ checkbox in Central Finance. The corresponding checkbox in the source system investment profile must be unchecked. Click the **Save** ❷ button to save the investment profile.

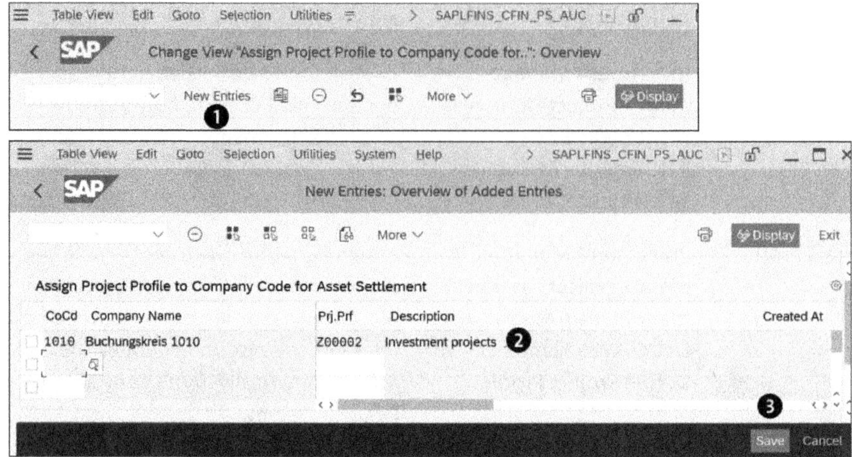

Figure 10.17 Assign a Project Profile to a Company Code for Asset Settlement

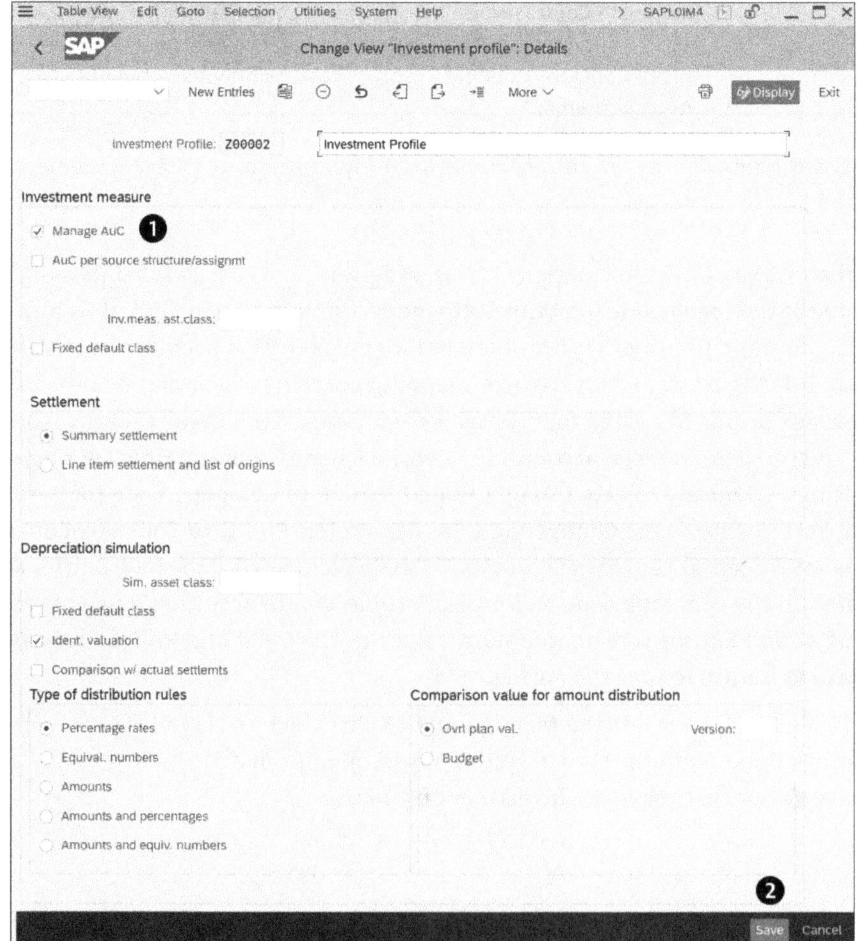

Figure 10.18 Investment Profile

Restrictions for Using Central Projects: Asset Settlement Scenario

SAP recommends reviewing several constraints when considering the asset settlement scenario in Central Finance. Table 10.3 provides a list of such restrictions you need to review prior to considering the asset settlement scenario in Central Finance.

Scenario	Central Projects Asset Settlement Scenario Restriction
SAP S/4HANA Cloud as source system	■ Not supported.
Project systems networks replication	■ Not supported. If networks are created under a project/WBS, SAP recommends settling the network to WBS, a profitability segment, or any other cost object. Mapping network to an internal order or to a WBS is considered a custom solution that requires configuration, coding, and extensive testing.
Project/WBS plan data	■ Not supported.
Project/WBS budget data	■ Not supported.
Deleting of WBS after release status	■ The system won't allow you to delete a WBS. The deletion flag is set in the source system for a WBS. If an AuC exists, you'll first set the deletion flag on the WBS in Central Finance prior to setting the deletion flag in the source system.
Changing IDs of the project and WBS elements	■ The system won't allow changing IDs.
Settlement rule	■ Restricted to **FXA** (fixed assets) and **PSG** (profitability segment) receivers only.
Project master data changes	■ All changes are made in the source system. The system allows only **CLSD** (close) and **DEL** (deletion) statuses in Central Finance.
Deleting investment profiles	■ This is only possible in the source system. One prerequisite is no AuC is capitalized in Central Finance. If no AuC is capitalized and you delete the investment profile in the source system, SAP recommends running program RFINS_CFIN_PS_AUC_DEL_IMPRF, which deletes generated the AuC asset shell in Central Finance.
Clearing account assignment indicator on WBS in the source system	■ The system doesn't delete the AuC automatically. SAP recommends running program RFINS_CFIN_PS_AUC_DEL_IMPRF, which deletes the generated AuC asset shell in Central Finance.
Statistical indicator on WBS	■ Changing statistical indicators on a WBS is possible only if no AuC is in Central Finance. Therefore, you must run program RFINS_CFIN_PS_AUC_DEL_IMPRF, which deletes the generated AuC asset shell in Central Finance.

Table 10.3 Central Finance Asset Settlement Scenario Restrictions

[»]

Note

Refer to SAP Note 2994285, which comes with the "Central Projects (WBS)" document, for further guidance.

10.2 Centralized Budgeting

With Central Finance release 2020 and higher, SAP offers a centralized budgeting approach to manage budgeting activity in Central Finance for the processes operated from the source system. During posting documents, the system checks the budget information in Central Finance via RFC from the source system. To facilitate a remote budget check, the relevant configuration information such as budget profiles and tolerance limits should be maintained in Central Finance. In addition, the relevant central budgeting settings must be configured in the source system located under the Central Finance source system settings. Once the central budgeting functionality is enabled, the system allows you to manage budgets only in the Central Finance system. As of writing this book, SAP supports two central budgeting scenarios in Central Finance:

- Central budgeting for internal orders
- Central budgeting projects

In Chapter 2, Sections 2.4.4 and 2.4.5, you already learned about the central budgeting processes. In the following sections, we'll walk you through central budgeting configurations for internal orders and projects, respectively.

10.2.1 Settings for Internal Orders

In this configuration, to enable central budgeting for internal orders, you'll maintain the company code and internal order type, which are relevant for budget checks during posting to internal orders in the source system. If enough budget exists, the order posting goes through successfully without any errors or warning messages. Insufficient budget may lead to an error or warning message based on the budget profile and tolerance limit settings.

To reach the configuration for internal order settings for central budgeting, navigate to IMG menu **SAP Customizing Implementation Guide • Financial Accounting • Central Finance: Source System Settings • Central Budgeting • Settings for Internal Orders** or use Transaction CFINIMG. After reaching the settings, you'll arrive at the **Change View "Central Budgeting for Internal Orders: Sour..": Overview** screen, as shown in Figure 10.19. On this screen, click the **New Entries** ❶ button. On the resulting **New Entries: Overview of Added Entries** screen, maintain values for the **CoCd** and **Type** ❷ fields. In addition, enable the **Error Msg** checkbox. Click the **Save** ❸ icon to add the configuration entry.

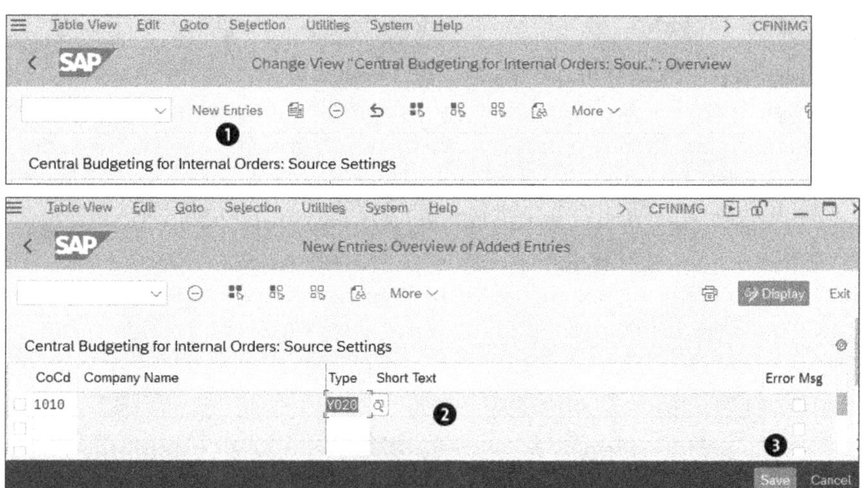

Figure 10.19 Settings for Internal Orders

Note

Several prerequisites must be considered when activating central budgeting for internal orders. First, consult two important SAP Notes for Central Budgeting for internal orders: SAP Note 2955778 and SAP Note 2914565. Second, all prerequisite configurations must be configured both in the source system and the Central Finance system.

In addition, SAP lists the following constraints to be considered prior to implementing central budgeting for internal orders:

- With central budgeting of internal orders, budgets in the source system are no longer maintained. Budget maintenance Transaction KO22 only displays the budget.
- SAP supports both 1:1 and N:1 mapping of internal orders. Due to N:1 mapping of internal orders and budget maintenance allowing only display, budget maintenance is possible in the target Central Finance system. Therefore, SAP recommends internal order reporting in the Central Finance system.
- The scope of central budgeting is limited to internal orders replicated via the cost object framework of master data replication.
- Mapping scenarios defined and mapped within the cost object framework in the Central Finance system are supported.
- Order types mapped in the cost object framework are supported for the central budgeting scenario. Budgets created and maintained in the source system aren't supported.
- During batch processing, if documents are stopped in SAP Application Interface Framework due to errors, the lock concept is limited. As a result of an error, the remote budget doesn't consider documents stuck with errors.
- Central Finance doesn't offer any cutover tools or reports for existing orders conversion.

10.2.2 Settings for Projects

Like central budgeting for internal orders, Central Finance supports central budgeting for projects as well. Likewise, you'll maintain project profiles at the company code level to carry out budget checks in the Central Finance system. During project WBS cost posting in the source system, the system performs budget checks based on budget profile and tolerance limit settings in the Central Finance system. After enabling central budgeting for projects, budget maintenance is performed only in the Central Finance system, and this check is no longer possible in the source system. Budget maintenance Transaction CJ30 is now display mode only. In addition, while configuring company codes and project profiles in this configuration, consistency is maintained in both the source system and the Central Finance system.

To reach the configuration settings, navigate to IMG menu **SAP Customizing Implementation Guide • Financial Accounting • Central Finance: Source System Settings • Central Budgeting • Settings for Projects** or use Transaction CFINIMG. After reaching the settings, you'll arrive at the **Change View "Central Budgeting for Projects: Source Sett..": Overview** screen, as shown in Figure 10.20. On this screen, click the **New Entries** ❶ button. On the resulting **New Entries: Overview of Added Entries** screen, enter values for **CoCd** and **Prj.Prf** ❷. In addition, enable the **Error Msg** checkbox. Click the **Save** ❸ icon to add the configuration entry.

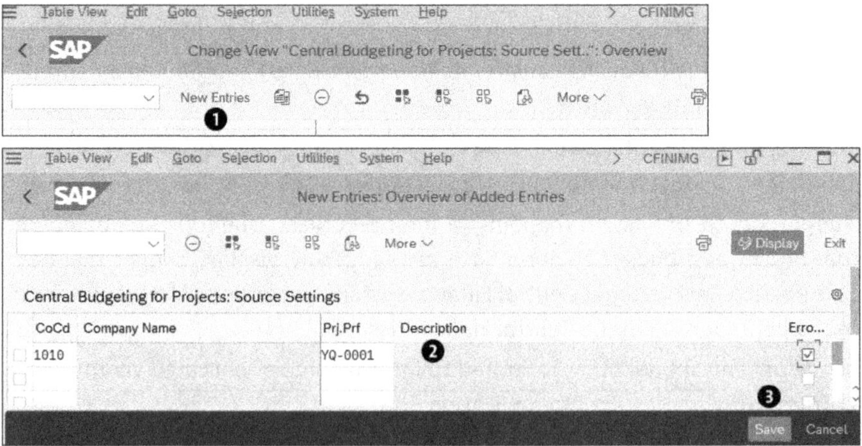

Figure 10.20 Settings for Projects

[»]

> **Note**
>
> The central budgeting for projects is available from SAP S/4HANA release 2022 or higher. In addition to settings for projects, such as central budgeting for internal orders, several prerequisites must be considered to enable central budgeting for projects as well. Otherwise, the system does not execute remote budget checks.

The following prerequisites must be configured in the source system first:

1. **Define Logical System for Central Finance System**
2. **Set Up RFC Destination for Central Finance System**
3. **Assign RFC Destination for Displaying Objects from Central Finance System**
4. **Maintain RFC Assignments and Settings for the Central Finance System** (If Central Finance customizing Transaction CFINIMG isn't available, you can maintain the entry in view VC_FIN_CFIN_LSYS through Transaction SM30.)

Detailed coverage of projects replication and functional restrictions is provided in Section 10.1.1.

10.3 Replicating Classic Profit Center Accounting Documents

In SAP, you'll determine profits using profit center dimensions either with period accounting or the cost-of-sales approach. A profit center is a management-oriented organizational unit used for internal reporting purposes. It determines areas of responsibility for profits to analyze return on investment (ROI), cash flows, or working capital. In addition, profit center accounting (PCA) is used to determine segment reporting to support the accounting standard ASC 280/IFRS 8. SAP provides PCA and profitability analysis tools to capture and analyze profitability data. Different versions of PCA are implemented in the source system depending on the SAP release customers are using. Some SAP ERP systems were implemented with classic enterprise controlling profit center accounting (EC-PCA), which is integrated with the CO component. On the other hand, in SAP's myERP or SAP S/4HANA releases, SAP integrated PCA with the general ledger module.

SAP ERP source systems are enabled with classic EC-PCA to fulfill reporting requirements. In some situations, there is a use case to bring in classic EC-PCA documents to the Central Finance system mostly for the same reporting purposes. Central Finance supports classic EC-PCA replication from SAP S/4HANA release 1709 onward. This means EC-PCA documents posted through Transactions 4KE5, 3KE5, 9KEO, and 1KEL can be replicated to Central Finance in the Universal Journal. Central Finance supports three types of EC-PCA internal posting replications from the source system to Central Finance:

- EC-PCA posting originates from the manual profit center document entry Transactions 9KEO, 9KE9, and 1KEL
- EC-PCA posting comes from period posting for assessment allocations using Transaction 3KE5
- EC-PCA posting come from period posting for distributions using Transaction 4KE5

In the case of replicating PCA assessment and PCA distributions, if source parallel valuation is active, SAP recommends mapping parallel valuation currency types in currency transfer table CFIN_ACCCR to local and group currency in Central Finance.

Central Finance supports classic EC-PCA replication. These transactions are replicated through SAP Landscape Transformation Replication Server and monitored through SAP Application Interface Framework in Central Finance. Figure 10.21 illustrates a high-level data flow of classic EC-PCA in Central Finance.

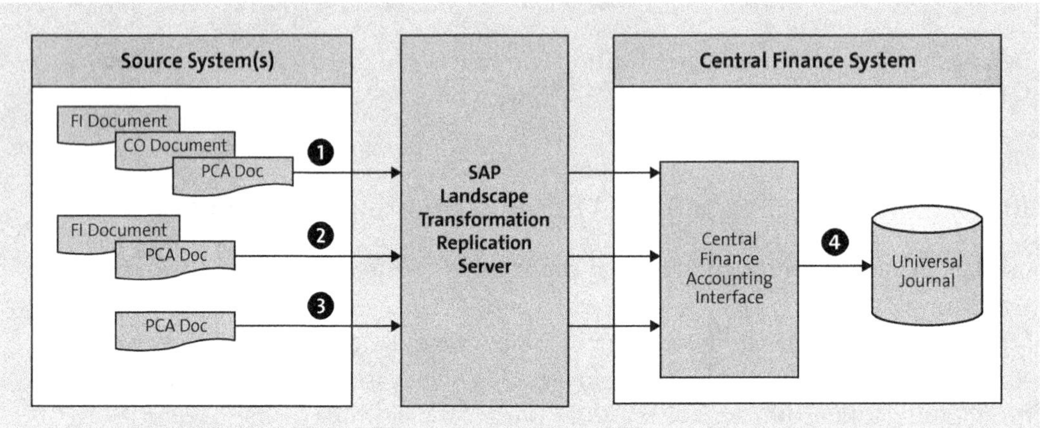

Figure 10.21 Classic Profit Center Replication Data Flow

Let's go over the different scenarios where an EC-PCA document is posted in the source system and how it replicates over to Central Finance:

- ❶ **EC-PCA documents posted with FI postings**
  When EC-PCA documents are posted along with FI documents in the source system, the system automatically transfers profit center information to the Universal Journal.
- ❷ **EC-PCA documents posted with CO secondary postings**
  If a CO document carries a profit center document, the system, during replication, transfers profit center information to line items of replicated document and posts to the Universal Journal.
- ❸ **EC-PCA documents posted with manual transactions**
  With the Central Finance EC-PCA replication functionality, you'll be able to replicate EC-PCA documents to a separate ledger in the Universal Journal (table ACDOCA). Replicated EC-PCA postings are then reconciled between the source system and Central Finance at the profit center level. If source EC-PCA is posted with one-sided entries, especially with the PCA ledger (table GLPCT) with table GLT0, replicated documents may not tie in this case between the source system and the Central Finance system. This requires manual reconciliation effort.

❹ **EC-PCA reporting**
Once all EC-PCA documents are replicated from the source system, you'll be able to report based on the separate ledger posting for EC-PCA in the Universal Journal. Reference fields in the document header table BKPF-AWYP and BKPF-AWSYS are populated during replication, which helps users navigate to the source system EC-PCA document.

In the following sections, you'll learn about simulation and initial load EC-PCA documents from the source system to Central Finance first. Then, we'll cover configuring EC-PCA with step-by-step instructions. The section concludes with EC-PCA comparison reports and restrictions.

10.3.1 Simulation and Initial Load of Enterprise Controlling Profit Center Accounting

SAP provides two transactions to simulate and perform the initial load of EC-PCA: Transaction FINS_CFIN_PCA_SIMU and Transaction FINS_CFIN_PCA_LOAD. Both transactions are initiated from the Central Finance system. Let's explore simulation and the initial load of EC-PCA in the following sections.

Simulation of Enterprise Controlling Profit Center Accounting Replication

Figure 10.22 illustrates the simulation of EC-PCA internal postings in the Central Finance system.

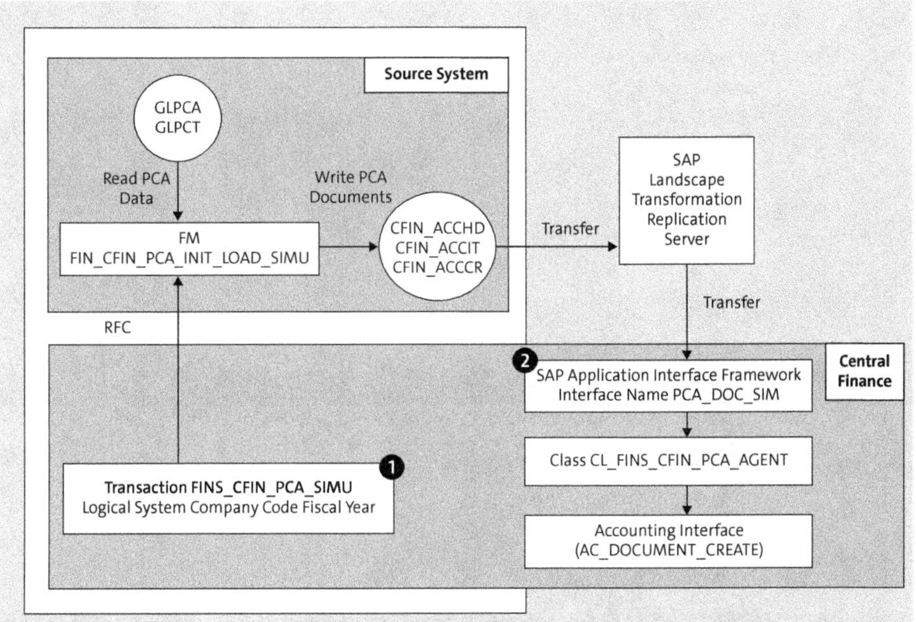

Figure 10.22 Simulation of EC-PCA Internal Postings

As you can see, you initiate the simulation of EC-PCA internal postings from the Central Finance system using Transaction FINS_CFIN_PCA_SIMU ❶. Upon execution of the transaction, the system reads EC-PCA data from the source system tables via RFC and transfers data to the Central Finance system via SAP Landscape Transformation Replication Server. In Central Finance, the system goes through data mapping checks and mapping logic checks. The mapping simulation can be viewed through SAP Application Interface Framework (namespace **/FINCF** and interface **PCA_DOC_SIM**) ❷. Though the simulation doesn't post data into Central Finance, this step is performed for two reasons:

- To perform mapping checks, posting logic checks, and system environment settings
- To simulate the initial load prior to the actual load to understand the time required and to find performance issues

To perform simulation of EC-PCA, navigate to IMG customizing menu path **Financial Accounting • Central Finance • Central Finance – Target System Settings • Profit Center Accounting Documents Replication • EC-PCA: Replication Simulation Load** or use Transaction FINS_CFIN_PCA_SIMU. Upon entering the transaction, you'll arrive at the **Central Finance: EC-PCA Simulation Tool** screen, as shown in Figure 10.23.

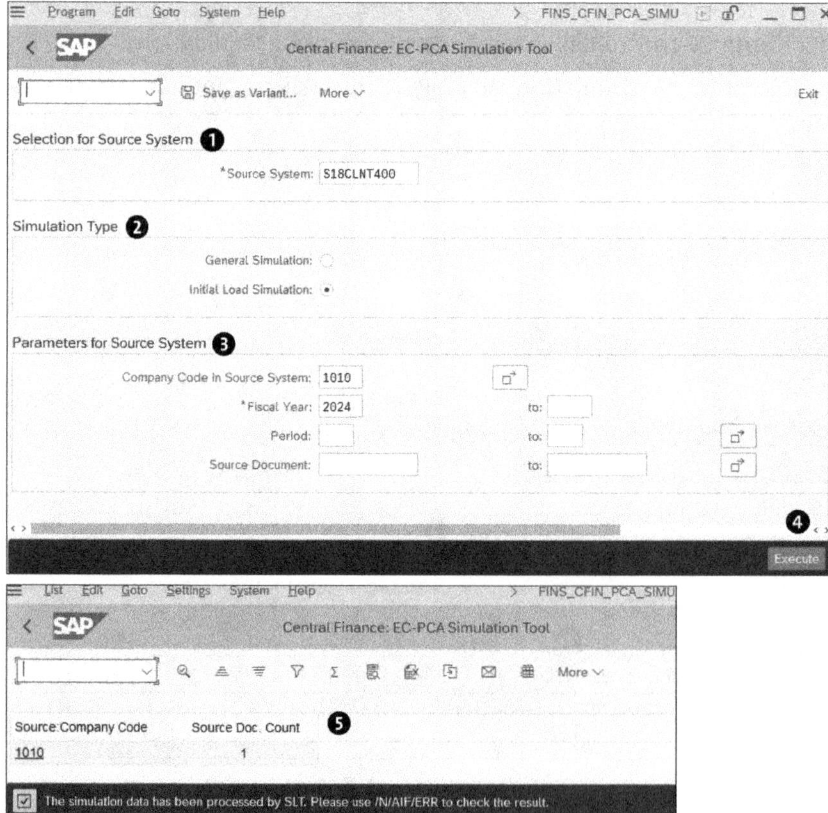

Figure 10.23 EC-PCA Simulation Tool

On this screen, enter the value of the source logical system in **Source System** ❶. The transaction supports two types of simulations: **General Simulation** and **Initial Load Simulation**. Choose the **Initial Load Simulation** radio button ❷. Next, enter the selection parameters ❸ as required. Click the **Execute** ❹ button. The system displays the results of the simulation ❺. In addition, the simulation results are available in the SAP Application Interface Framework PCA_DOC_SIM interface. The simulation can be repeated multiple times until all errors are resolved.

Initial Load of Enterprise Controlling Profit Center Accounting Internal Postings

Like the EC-PCA simulation shown earlier, you'll execute the initial load of EC-PCA for tje selected company code from the Central Finance system using Transaction FINS_CFIN_PCA_LOAD, as shown in Figure 10.24 ❶.

10

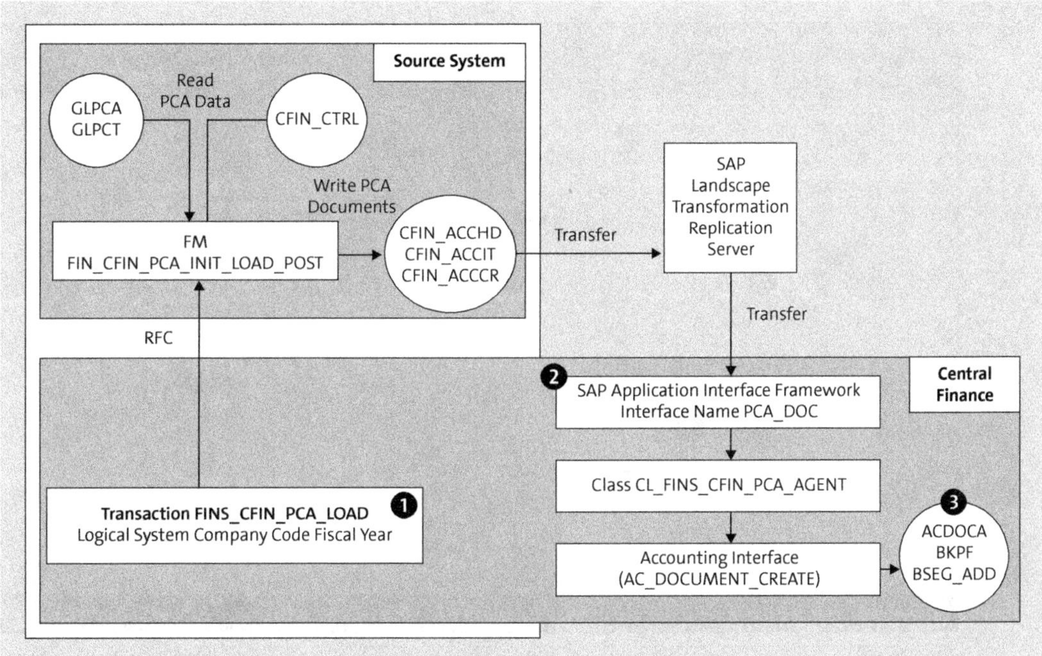

Figure 10.24 Initial Load of EC-PCA Internal Postings

Initial load balances are transferred based on previous **Start** and **Period** fields in **Define Start of Transfer of EC-PCA Postings**. Upon execution, the system reads EC_PCA tables and transfers the EC-PCA documents and balances to the source system staging tables CFIN_ACCHD, CFIN_ACCIT, and CFIN_ACCCR based on criteria specified on the selection screen. In addition, the system also updates with a timestamp in the generic settings control table CFIN_CTRL in Central Finance to confirm the initial load is completed. The timestamp written in this table is used for the system to direct any subsequent EC-PCA documents via online replication mode. The system uses the same initial load configuration for real-time replication, and no separate action is required to activate real-time replication separately. The database triggers in staging tables enable the transfer of data to the Central Finance system via SAP Landscape Transformation Replication Server in an asynchronous process. In Central Finance, the system carries through the data mapping check and mapping logic checks prior to posting. The initial load EC-PCA data can be viewed through SAP Application Interface Framework (PCA_DOC) ❷. Finally, accounting interface is called to post EC-PCA documents to the Universal Journal ❸.

To perform the initial load of EC-PCA, navigate to IMG customizing menu path **Financial Accounting • Central Finance • Central Finance - Target System Settings • Profit Center Accounting Documents Replication • EC-PCA: Execute Initial Load** or use Transaction FINS_CFIN_PCA_LOAD. Upon reaching the transaction, the system displays the **EC-PCA Initial Load Transfer** screen, as shown in Figure 10.25.

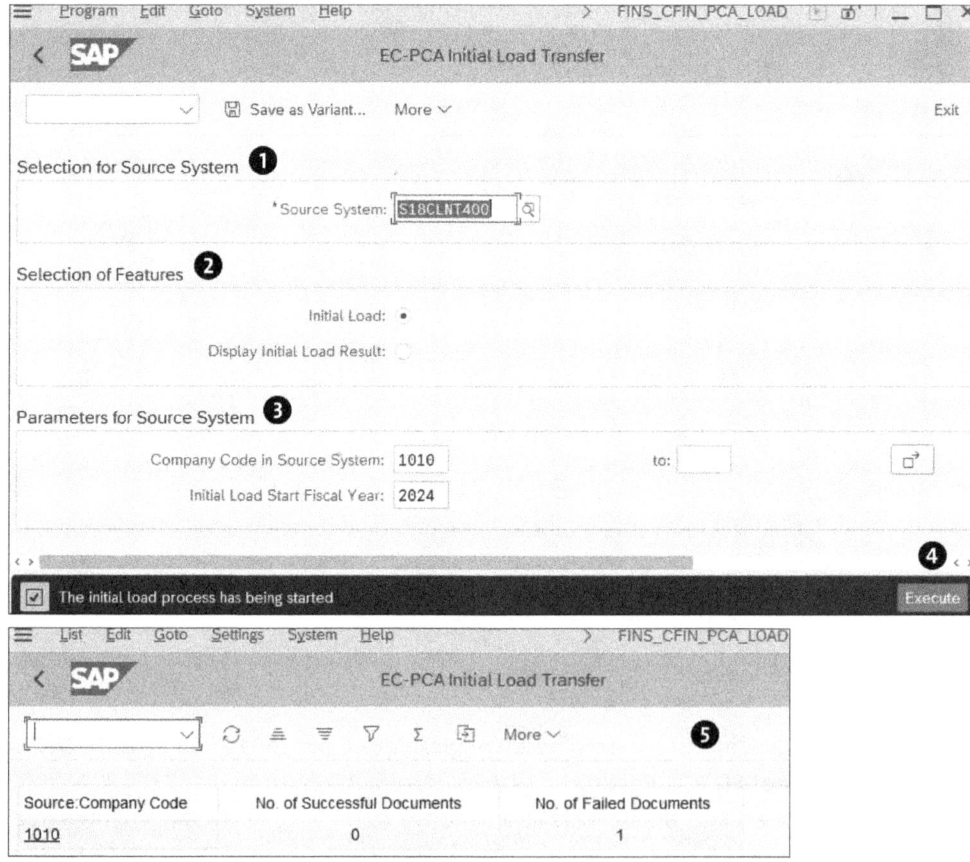

Figure 10.25 EC-PCA Initial Load Transfer

On this screen, enter the value of the source logical system in **Source System** ❶. The transaction supports two types of simulations: **Initial Load** and **Display Initial Load Result**. Choose the **Initial Load** radio button ❷. Next, enter the selection parameters as required under **Parameters for Source System** ❸ such as **Company Code in Source System** and **Initial Load Start Fiscal Year**. Click the **Execute** ❹ button. Unlike simulation of EC-PCA, you'll carry out the initial load only once for each company code. The system displays results of the initial load in the next screen ❺. In addition, to display the initial load results, go to the previous screen by pressing F3, choose the **Display Initial Load Result** radio button, and click the **Execute** ❹ button again.

The initial load results are available in the SAP Application Interface Framework interface **PCA_DOC** version **1**, as shown in Figure 10.26.

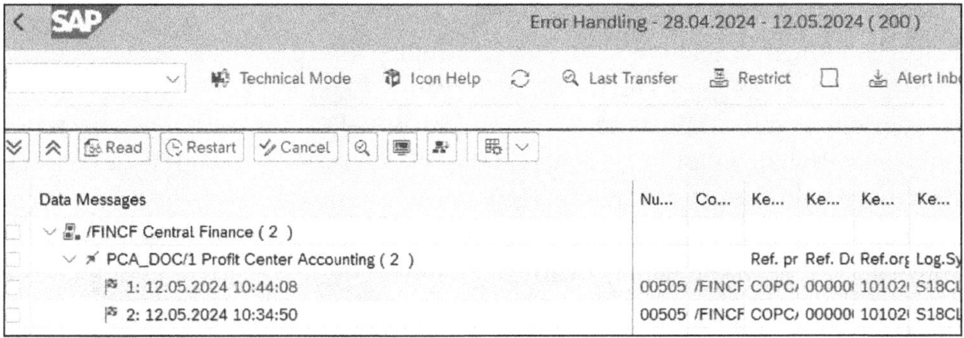

Figure 10.26 EC-PCA Replication in SAP Application Interface Framework

[+]

Tip

EC-PCA replicated data can be deleted with Central Finance report FINS_CFIN_PCA_DOC_DELETE, which deletes information from FI staging tables CFIN_CTRL and CFIN_ACC*. This report isn't recommended in a live system, but it can be used to reset data when mock cycles are executed prior to a system go-live. If you intend to reset staging tables in the source system (CFIN_ACC*), SAP provides Transaction FIN_CFIN_PCA_RESET (report FIN_CFIN_PCA_RESET), which can be executed in the source system.

10.3.2 Configuring for Profit Center Accounting Replication

For classic EC-PCA replication from the source system to the Central Finance system, you need several customizing activities in both the source system and the Central Finance system. In the following sections, we'll walk you through configurations of both systems to enable replication of classic EC-PCA documents.

Source System Customizing Activities

From a source system standpoint, the following steps must be validated and carried out sequentially for EC-PCA replication in Central Finance:

1. Maintain the controlling area settings.
2. Ensure EC-PCA is active in the source system.

3. Perform preparation checks on online items and online transfer control parameters, which are activated for EC-PCA configuration.
4. Define the start of transfer of EC-PCA postings in the Central Finance source system configuration.
5. Activate the start of transfer of EC-PCA postings.
6. Reset Central Finance tables (only when necessary).

We'll explain these configurations next.

Maintain Controlling Area Settings

First, check the maintenance of EC-PCA Controlling area parameters maintained for controlling areas assigned with the company code. To confirm EC-PCA CO parameters by controlling area, navigate to IMG menu path **SAP Customizing Implementation Guide • Controlling • Profit Center Accounting • Basic Settings • Controlling Area Settings • Maintain Controlling Area Settings**. You'll be prompted with the **Change View "EC-PCA: Controlling Area Settings": Overview** screen, as shown in Figure 10.27. The **Active Indicator** checkbox under **Control Indicators** activates the parameters for EC-PA controlling area settings.

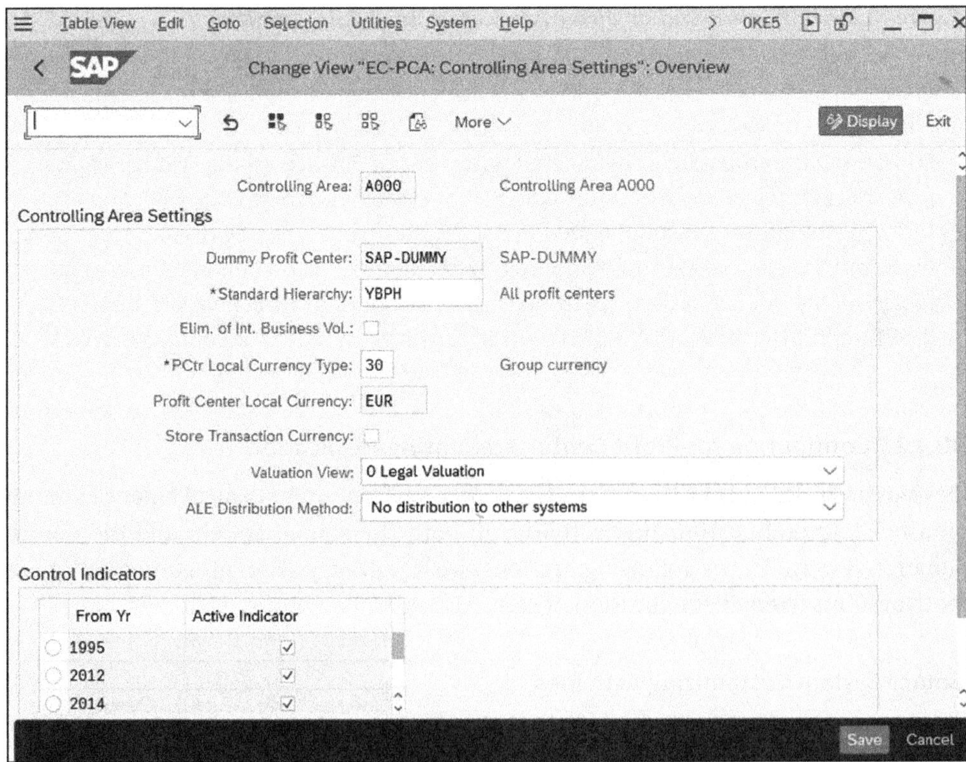

Figure 10.27 Maintain Controlling Area Settings

Set Control Parameters for Actual Data

Next, to check whether the second prerequisite is met, ensure whether the **Line items** and **Online transfer** checkboxes are enabled in EC-PCA control parameters settings. To check these parameters, navigate to IMG menu path **SAP Customizing Implementation Guide • Controlling • Profit Center Accounting • Controlling Area Settings • Set Control Parameters for Actual Data** or use Transaction 1KEL. Upon entering the transaction, the system displays the control parameters for actuals postings in PCA, as shown in Figure 10.28. Make sure that **Line items** ❶ and **Online transfer** ❷ flags are enabled for the years of EC-PCA data extraction and postings for the controlling area in scope for Central Finance scenario. The **Line items** checkbox enables actual line items updated in the ledger. The **Online transfer** control indicator activates simultaneous posting to the EC-PCA ledger. Once these flags are set, click the **Save** ❸ icon to save the configuration.

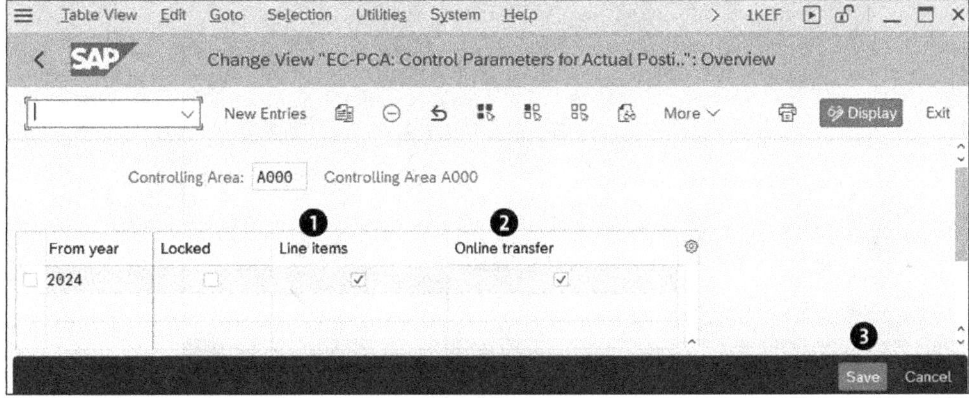

Figure 10.28 Set Control Parameters for Actual Data

Define the Start of Transfer for Enterprise Controlling Profit Center Accounting Postings

From the Central Finance source system settings for EC-PCA perspective, first you define fiscal year and period to start the transfer of EC-PCA postings to Central Finance's Universal Journal. Next, you'll activate postings. In this configuration, you'll define the start date of transfer for EC-PCA replication.

To reach this customizing activity, navigate to IMG menu **SAP Customizing Implementation Guide • Financial Accounting • Central Finance: Source System Settings • Replication Settings for Profit Center Accounting • Define Start of Transfer of EC-PCA Postings** or use Transaction CFINIMG. Upon entering the configuration, on the **Change View "Customizing Central Finance: PCA Document T..": Overview** screen, click the **New Entries** button, as shown in Figure 10.29. On the resulting **New Entries: Overview of Added Entries** screen, maintain the following values:

- **CoCd**
  Maintain company codes in scope for EC-PCA replication.
- **Start**
  Specify the start fiscal year for EC-PCA documents transferred to Central Finance.
- **Period**
  Specify the period within the fiscal year for EC-PCA documents transferred to Central Finance.
- **Period 000**
  Select the period from the dropdown regarding whether to post **Period 000** to the period in the same fiscal year or to post EC-PCA data in the period to the last period of the prior fiscal year.

When all fields have been filled, click the **Save** button to save the entries.

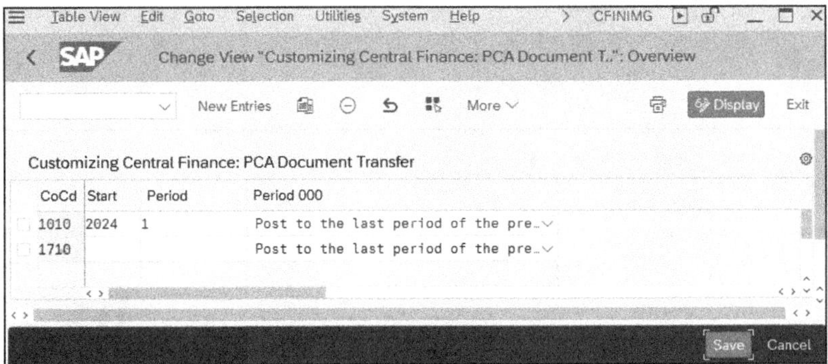

Figure 10.29 Define the Start of Transfer for EC-PCA Postings

Activate Start of Transfer for Enterprise Controlling Profit Center Accounting Postings

In this configuration activity, you'll activate the transfer of EC-PCA postings to transfer data from the source system to the Central Finance system. The activation is done at the company code level.

To activate the transfer of EC-PCA postings, navigate to IMG menu **SAP Customizing Implementation Guide • Financial Accounting • Central Finance: Source System Settings • Replication Settings for Profit Center Accounting • Activate Start of Transfer of EC-PCA Postings** or use Transaction FIN_CFIN_PCA_ACTIVAT. Upon opening the configuration activity, you'll arrive at the **Central Finance: PCA Ongoing Replication Activation** screen, as shown in Figure 10.30. Next, on this screen, enter each company code or range or list of company codes ❶ that are in scope for transfer of EC-PCA postings. Click the **Execute** ❷ icon. The system displays company codes that have been activated successfully ❸.

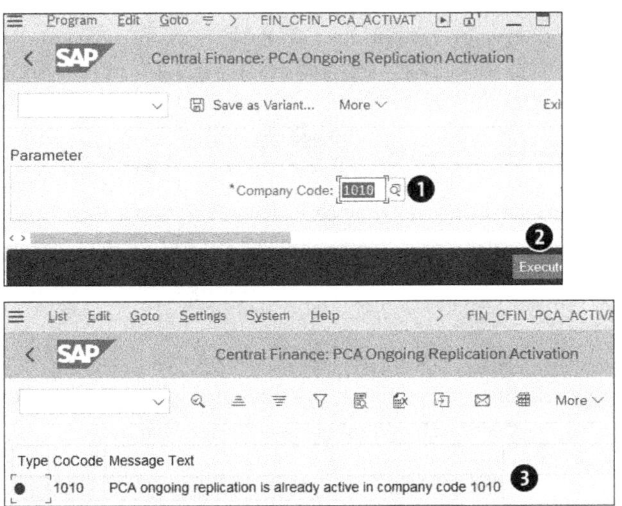

Figure 10.30 Activate Transfer of EC-PCA Postings

Central Finance Customizing Activities

Now, let's look at Central Finance Customizing for EC-PCA replication in this section.

Settings for the Source System

In the Central Finance Customizing, you must assign a target ledger group for the logical system to receive incoming EC-PCA postings from the source system. Note that EC-PCA postings can't be posted to the leading ledger, so the leading ledger shouldn't be included in the ledger group used in EC-PCA postings. To assign a ledger group for EC-PCA replication, navigate to IMG customizing menu path **Financial Accounting • Central Finance • Central Finance - Target System Settings • Profit Center Accounting Documents Replication • Settings for Source Systems.** Upon reaching the customizing activity, the system displays the **Change View "Central Finance: Settings for Source Systems": Overview** screen, as shown in Figure 10.31. On this screen, maintain values in the **Logical system** and **Ledger Grp** fields. The ledger group must exist prior to this configuration to replicate classic EC-PCA document replication.

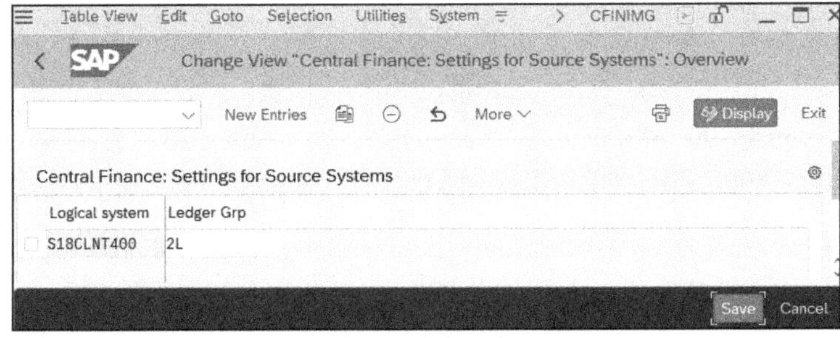

Figure 10.31 EC-PCA Settings for Source Systems

Settings for Company Codes

In this customizing activity, the system allows you to define attributes such as document type, general ledger account and profit center for EC-PCA replicated documents. Upon replication of EC-PCA postings, the system uses these attributes to be assigned. To assign attributes for EC-PCA replication, navigate to IMG customizing menu path **Financial Accounting • Central Finance • Central Finance – Target System Settings • Profit Center Accounting Documents Replication • Settings for Company Codes.** Upon reaching the customizing activity, the system displays the **Change View "Central Finance: Settings for Source Systems": Overview** screen, as shown in Figure 10.32. On the resulting screen, maintain values in the **CoCd**, **Type**, **G/L Account**, and **Profit Center** fields. When EC-PCA postings are replicated from the source system, the system uses the document type (**Z9** in our example) for posting. The general ledger account is used as an offsetting line item if replicated EC-PCA balances don't tie to zero. Likewise, the profit center maintained here is used to populate offsetting additional line items when replicated EC-PCA postings don't tie to zero.

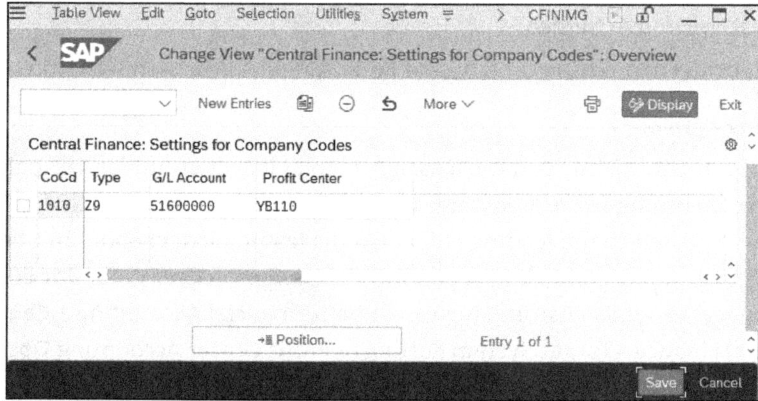

Figure 10.32 EC-PCA Settings for Company Codes

10.3.3 Restrictions

Before you consider activating classic EC-PCA replication in Central Finance, SAP recommends you review several restrictions:

- Only the transactions identified in the beginning of this section are supported.
- The system doesn't transfer the initial load FI balances at the profit center level.
- The segment is rederived based on the configuration in the target Central Finance system.
- Reversals posted with deprecated reversal Transaction GB06 aren't supported.
- Postings from Transaction 1KEK (Reconciliation of Profitability Analysis) aren't supported.

- Currency type 32 must be configured in Central Finance if transfer pricing is activated in the source system.
- Replication of deletion entries posted with Transactions OKEI isn't supported.
- Balance carryforward is performed in the Central Finance system.

10.3.4 Reset Central Finance Enterprise Controlling Profit Center Accounting Tables

When EC-PCA is activated for replication, the system updates Central Finance staging tables such as CFIN_ACCHD, CFIN_ACCIT, and CFIN_ACCCR. There are situations, for example, during mock cycle loads or during system refresh activities, where you'll need to reset EC-PCA data. If such a situation arises, you'll execute the reset program to delete EC-PCA data from the staging tables.

To perform a reset of EC-PCA data, navigate to IMG menu path **SAP Customizing Implementation Guide • Financial Accounting • Central Finance: Source System Settings • Replication Settings for Profit Center Accounting • Reset Central Finance Tables**.

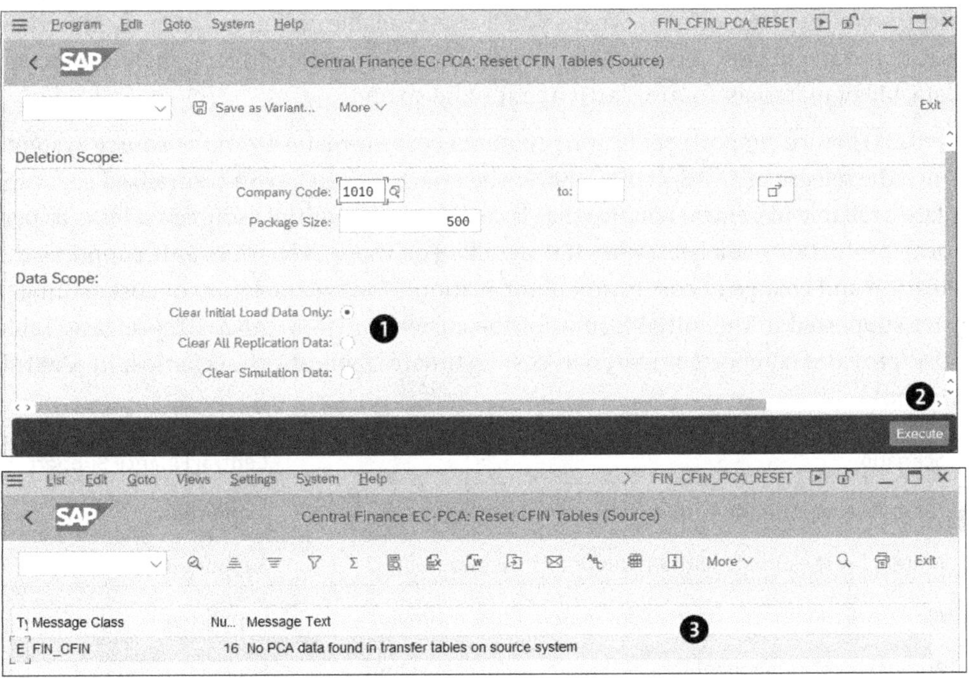

Figure 10.33 Reset Central Finance Tables: EC-PCA

As shown in Figure 10.33, the system allows you to reset EC-PCA data with three **Data Scope** options:

- **Clear Initial Load Data Only**
  Deletes only the initial load data.

- **Clear All Replication Data**
  Deletes any replicated data already cleared.
- **Clear Simulation Only**
  Deletes simulated data before posting.

First, enter values in the **Company Code** and **Package Size** fields in the **Deletion Scope** section. Next, choose the appropriate **Data Scope** ❶ radio button option. The choice is dependent on the project activity and scope of data to be deleted from tables. Choose **Execute** ❷ to execute the reset function for EC-PCA data in staging tables. Upon execution of the report, the system displays the rest of the results on the resulting window ❸.

10.4 Material Cost Estimate Replication

When manufacturing products or providing services to customers, companies determine the appropriate cost for the product or service. For manufacturing products, companies find the cost of an item for both finished and semi-finished materials, which helps in determining cost of sales. Determining the cost of a product or rendering services is an involved process where you'll use available resources such as materials, labor, and overhead costs. In SAP, cost estimates are created using material master data, bill of materials, routing, activity rate, and so on.

Central Finance supports replicating material cost estimates from the source systems since the release of SAP S/4HANA 1909. The end goal is to have a centralized cost estimate available in Central Finance that is connected to multiple source systems to perform profitability analysis with full details. For these scenarios, SAP supports the creation and change of cost estimate replication. However, deletion of cost estimates isn't supported in the initial load as of the current SAP S/4HANA 2023 release. Table 10.4 provides supported material cost estimate replication scenarios in Central Finance.

Scenario	Central Finance Support
Material cost estimate with quantity structure	Supported
Material cost estimate without quantity structure	Supported
Main cost component structure of material cost estimate	Supported
Auxiliary cost component structure of material cost estimate	Supported
Mixed costing and relevant procurement alternatives	Supported
Material estimate for cost collectors	Not supported
Material cost for additive cost	Not supported

Table 10.4 Supported Scenarios for Material Cost Estimate in Central Finance

Let's discuss how data flows in material cost estimate in Central Finance scenario. Figure 10.34 illustrates material cost estimates for replication from different source systems ❶ to the Central Finance system. SAP Landscape Transformation Replication Server is used to replicate cost estimates. As shown in Figure 10.34, you'll have centralized cost estimates available for multiple valuation views ❷ in the Central Finance system. The cost estimate replication supports the following valuation views:

- Standard price – legal valuation
- Group valuation
- Profit center valuation

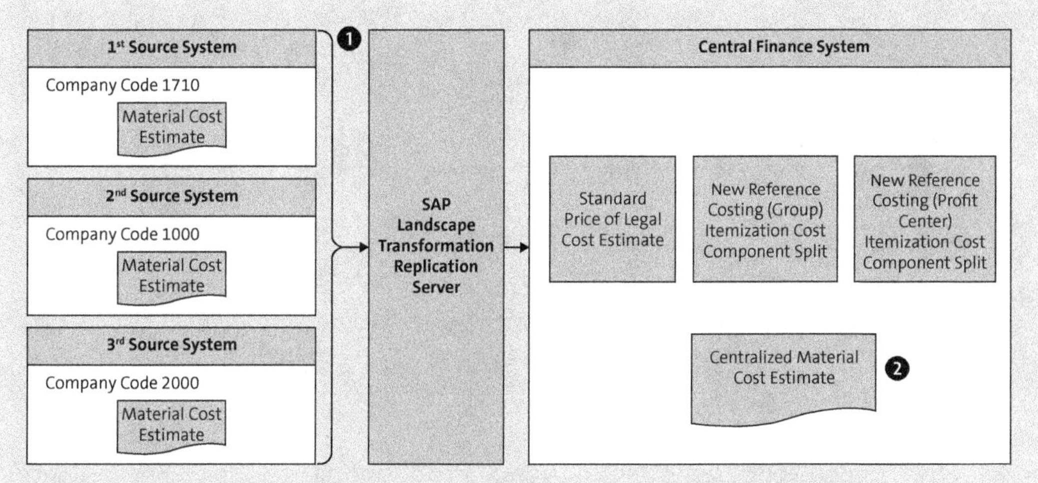

Figure 10.34 Material Cost Estimate Replication Overview

In the following sections, we'll discuss the goals behind replicating cost estimates first. Then, the section walks you through configurations in the source system, SAP Landscape Transformation Replication Server, and Central Finance. The section concludes with the initial load of cost estimates and replicating cost estimates manually.

10.4.1 Replicating Material Cost Estimates Process

When replicating cost estimates for the first time, you'll go through three processes to transfer material cost estimate data from the source system to the Central Finance system:

- **Initial load of material cost estimates**
  Transfer historical material cost estimates.
- **Ongoing replication**
  Once the initial load is completed, activate online replication to continuously replicate material cost estimates in real time.

- **Replicating cost estimates manually**
  Manually replicate material cost estimates for a smaller volume.

Note

SAP supports both historical material cost estimates and online replication via SAP Landscape Transformation Replication Server. However, SAP recommends you review several restrictions that come when replicating material cost estimates from the source system to the Central Finance system. In addition to what's supported as shown earlier in Table 10.4, consider following nonsupported scenarios before considering cost estimate replication:

- Only material cost estimates with value KEKO-BZOBJ=0 are supported for the initial load.
- Cost estimates created for product cost collectors aren't supported, that is, KEKO-KALKA <> 19.
- Cost estimates for configured materials aren't supported, that is, KEKO-CUOBJID <> BLANK.
- Pure additive material costs aren't supported, that is, KEKO-KKZMA <> BLANK.
- Deletion of material cost estimate replication isn't supported. Filter logic for SAP Landscape Transformation Replication Server event-based rule SKIP_KEKO_OPFLG allows only insert and update functions.
- Currency settings of controlling areas must be identical between the source system and the Central Finance system.
- Currency settings for company codes must be identical between the source system and Central Finance system.

10.4.2 Getting Started with Replication of Material Cost Estimates

To replicate the material cost estimate, you'll have to configure both the SAP Landscape Transformation Replication Server and Central Finance systems. In this section, you'll learn configuration requirements of the source system, SAP Landscape Transformation Replication Server, and the Central Finance system. We'll review each system's requirements in the following sections.

Source System

SAP recommends implementing SAP Note 2800883 in all source systems that require material cost estimate replication. It provides any prerequisite SAP Notes to support lower environments. In addition, the SAP Note provides value help for relevant fields when replicating cost estimates from the source system. You also can consult SAP Note 3062966 for the latest content about SAP Landscape Transformation Replication Server.

SAP Landscape Transformation Replication Server Prerequisites and Configuration

From an SAP Landscape Transformation Replication Server standpoint, first you'll check whether you've installed the latest content in SAP Landscape Transformation Replication Server. Consult SAP Note 2154420 for the most up-to-date content. Once the latest content and relevant SAP Notes have been applied, you can configure the SAP Landscape Transformation Replication Server for material cost estimates in the following sequence:

1. Apply advanced filters.
2. Configure predefined load and replication objects.
3. Apply the skip rule.

Let's begin with how to apply advanced filters in SAP Landscape Transformation Replication Server. You can specify the data scope of the initial load of material cost estimates through Transaction LTRS in the SAP Landscape Transformation Replication Server; otherwise, all relevant material cost estimates in the source system will be transferred in the initial load. You can add the advanced filter criteria for the initial load.

To apply filters, go to Transaction LTRS. On this exercise, we're assuming your mass transfer ID (MT ID) has already been created. Refer to Chapter 5, Section 5.2, for more on creating an MT ID. You'll arrive at the **SAP LT Replication Server Change Performance Options (OOC)** screen, as shown in Figure 10.35.

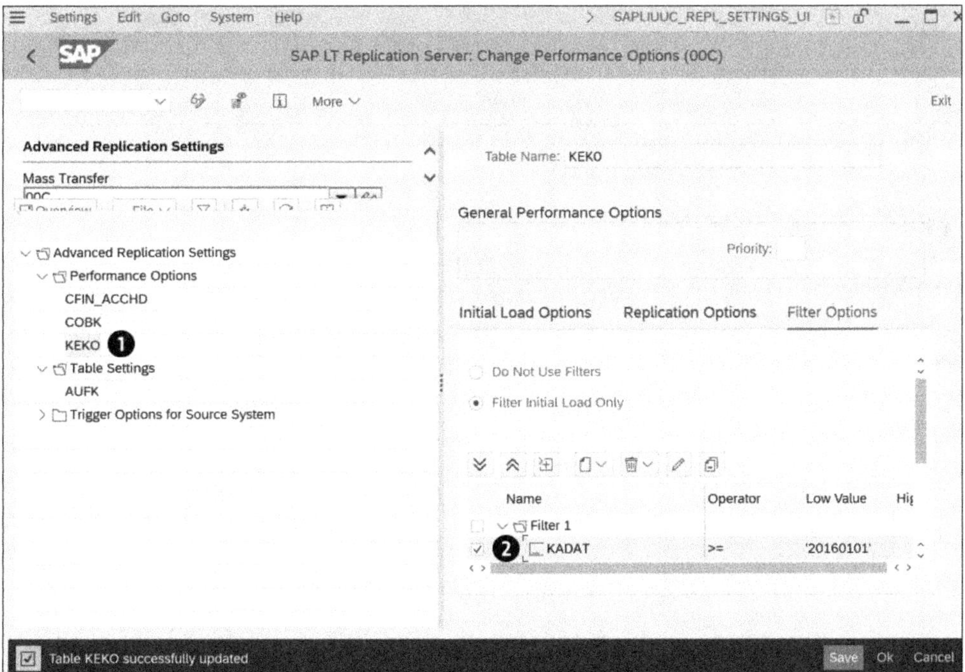

Figure 10.35 Material Cost Estimate: SAP Landscape Transformation Replication Server Filter and Filter Conditions

On the resulting screen, select **Mass Transfer** ID, and navigate to **Advanced Replication Settings**. Under **Performance Options** on the left pane located under **Advanced Replication Settings**, right-click and select **Add Object** to add material cost estimate table **KEKO** ❶. Next, on the **Filter Options** ❷ tab, choose the **Filter Initial Load Only** button, and then add your filter criteria per requirements. In our example, we've chosen costing data element **KADAT >= 20160101**. Click the **Save** button to save the filter and filter conditions.

After applying advanced filters, next you'll create the load and replication objects for material cost estimates, as shown in Table 10.5.

Type	Name
Predefined load object	CFI_KEKO_L
Predefined replication object	CFI_KEKO_R

Table 10.5 Predefined Load and Replication Objects for Cost Estimate Replication

As we've seen earlier, the predefined load and replication objects are part of project REPL_CFIN and subproject REPL_CFIN. Next, you'll apply the skip rule for material cost estimates to transfer cost estimates from the source system to the Central Finance system. As shown in Figure 10.36, SAP provides the SKIP_KEKO event-based skip rule for the predefined load and replication objects to load and replicate the material cost estimates. SAP has added default logic in the SKIP_KEKO rule to restrict material cost estimates.

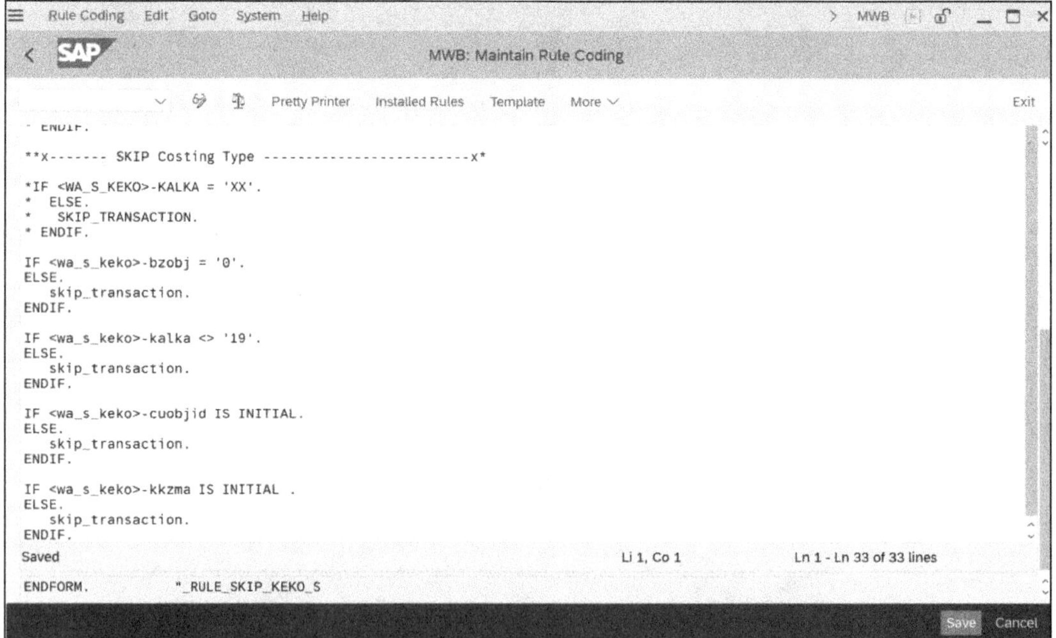

Figure 10.36 Skip Rule: Material Cost Estimate

Central Finance Configuration

In this section, you'll learn Central Finance configuration for material cost estimate replication. To monitor material cost estimates in SAP Application Interface Framework, you must extract SAP Application Interface Framework content using Transaction /AIF/CONTENT_EXTRACT, as shown in Figure 10.37. On the resulting **AIF Content Transport - Deploy** screen, enter "SAP_AFI_0018" ❶ for **Deployment Scenario ID**. Select further selection **Parameters** ❷ to record registration and **Transport recording** ❸, respectively. Click the **Execute** ❹ button to extract content. Once the content extraction is successful, you'll see interface **CE_MAT** available under namespace **/FINCF** in Transaction /AIF/ERR monitoring and error handling.

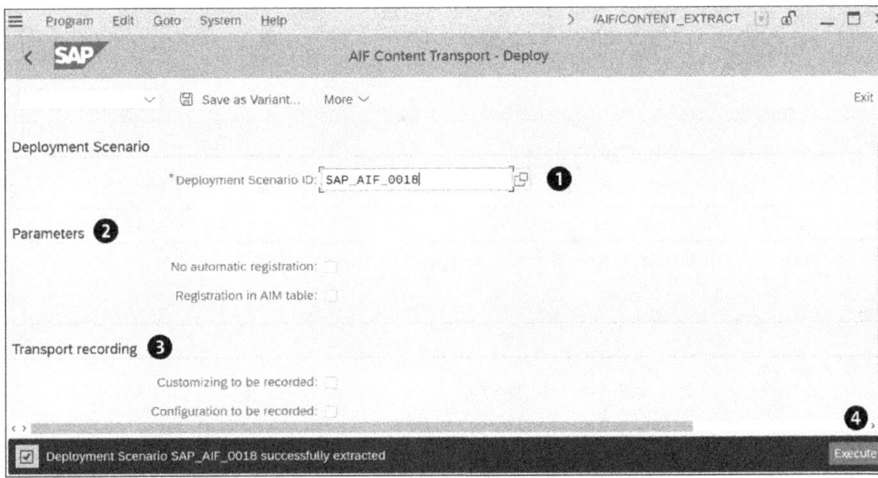

Figure 10.37 SAP Application Interface Framework Content Extract for Material Cost Estimate Replication

Next, you'll configure the transfer rule for material cost estimate replication. This is a mandatory configuration that stores conditions for different source systems to transfer material cost estimate to the Central Finance system. The system uses these transfer rules during replication of cost estimates. If no rules are maintained in this configuration, the system doesn't replicate material cost estimates. To define a transfer rule, navigate to IMG menu path **SAP Customizing Implementation Guide • Financial Accounting • Central Finance: Target System Settings • Cost Estimate Replication Settings • Define Rule for Cost Estimate Replication** or use Central Finance IMG Transaction CFINIMG. After reaching the configuration, as shown in Figure 10.38, click the **New Entries** ❶ button on the **Change View "Cost Estimate Transfer Rule Head": Overview** screen. On this screen, maintain values in the following fields ❷:

- **Logical system**
  Enter the sender logical system (source system) from which you want to transfer the material cost estimate to the Central Finance system. In our example, we used the source client "S18CLNT400" for source system S18.

- **Reference Object**
  Enter the reference object type of the cost estimate transferred, that is, "0 Material".
- **CoAr**
  Enter the source system controlling area here.
- **Plnt**
  Enter a plant or enter an asterisk (*) to select all plants.
- **Cstg Vrnt**
  Enter the costing variant linked for cost estimate replication. During cost estimate creation, you'll always link a cost estimate to a costing variant that contains relevant information.
- **Cstg Date From**
  Pick a date from which you want to transfer the cost estimate.

Next, choose **Cost Estimate Transfer Rule Head** ❸ on the left pane, and then double-click the **Cost Estimate Transfer Rule Item** under **Dialog Structure**. This will take you to the **Change View "Cost Estimate Transfer Rule Item": Overview** screen. On this screen, click **New Entries** ❹, and then maintain values in **Variant** and **Variant Value** ❺ for each transfer rule header. Click the **Save** ❻ button to save the entries.

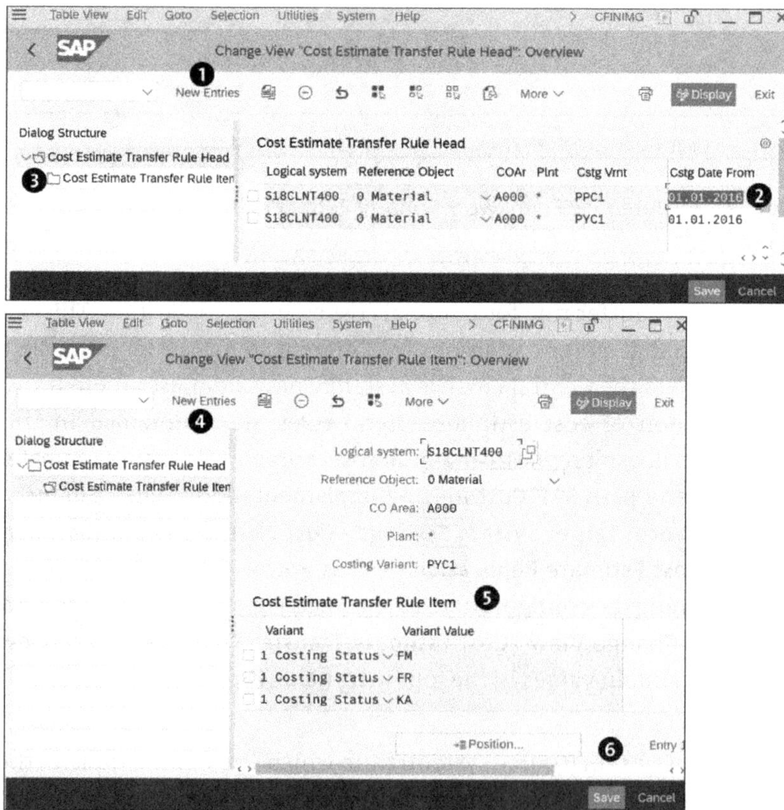

Figure 10.38 Define Rules for Cost Estimate Replication

[«]

Note

If you select one variant and a corresponding value, the system *doesn't* transfer material cost estimates from the source system to the Central Finance system. On the other hand, if you select multiple variants and corresponding variant values, the system considers and transfers the material cost estimate accordingly. Therefore, you need both variants to transfer material cost estimates.

Next, you'll configure SAP Master Data Governance mapping entities and mapping actions for several objects in Central Finance:

- Material number (**MATERIAL_ID**)
- Plant (**PLANT_ID**)
- Controlling area (**KOKRS**)
- Cost center (**COST_CENTRE_ID**)
- Cost element (**COST_ELEMENT_ID**)
- Activity type (**ACTIVITY_TYPE_ID**)
- Cost component structure (**CK_ELESMHK**)
- Cost component number (**CK_ELEMENT**)
- Costing variant (**CK_KLVAR**)

For configuring mapping actions, refer to Chapter 4, Section 4.4.2.

10.4.3 Initial Load of Cost Estimates

Once SAP Landscape Transformation Replication Server and Central Finance configurations are completed, you can trigger the initial load and replication by adding table KEKO in the MT ID in SAP Landscape Transformation Replication Server. You need to select the **Start Replication** radio button after clicking the **Data Provisioning** button. Once the initial load of material cost estimate is finished, data will be loaded in the Central Finance system. During the initial load, the system validates transfer rules you've configured earlier. You can check the initial load material cost estimate data in interface **CE_MAT** under namespace **/FINCF** in Central Finance. For example, Figure 10.39 shows the SAP Application Interface Framework error handling screen after performing the initial load of cost estimate.

[«]

Note

Once the initial load of cost estimate is completed, online replication is turned on automatically.

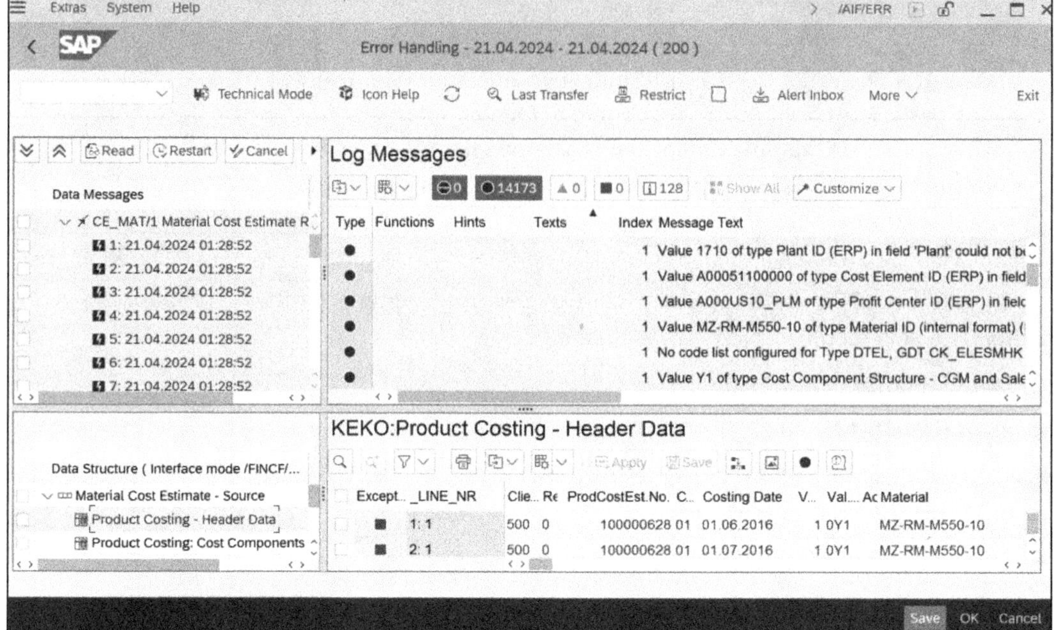

Figure 10.39 Cost Estimate Initial Load

10.4.4 Replicating Material Cost Estimates Manually

SAP provides Transaction FINS_CFIN_MCE_REPL to transfer material cost estimates from the source system to Central Finance manually. You'll use this transaction in the following situations:

- Due to technical errors or SAP Application Interface Framework message that can't be restarted
- An incorrectly configured transfer rule in Central Finance that requires correction and then requires re-transfer of the material cost estimate
- SAP Landscape Transformation Replication Server technical issues causing material cost estimate not transferred to the Central Finance system

If any of these situations occur, you'll execute the transactions by navigating from **SAP Easy Access** menu **Accounting • Central Finance • Manual Transfer Tools • Replicate Material Cost Estimate Manually** or use Transaction FINS_CFIN_MCE_REPL. Upon reaching the transaction, you'll arrive at the **Transfer Material Cost Estimates from Source System to Central Finance** screen, as shown at the top of Figure 10.40. On this screen, enter values in the mandatory fields: **Source System** ❶, **Plant**, and **Costing Date** ❷. The other fields can be chosen further to apply filter criteria as required per business requirements. The report provides options to select **Settings for Parallel Processing** ❸. This option can be used if performance is a consideration factor when executing the transaction. Click the **Execute** ❹ icon to replicate cost estimates manually for the given selection criteria.

As a result of execution, the system brings all material cost estimates, as shown at the bottom of Figure 10.40 based on selection criteria from the previous screen. At this point, the system simply displays all material cost estimate lines waiting to be posted in Central Finance. Next, from the **Transfer Material Cost Estimates from Source System S18CLNT400** screen, choose rows ❺, and click the **Replicate** ❻ button.

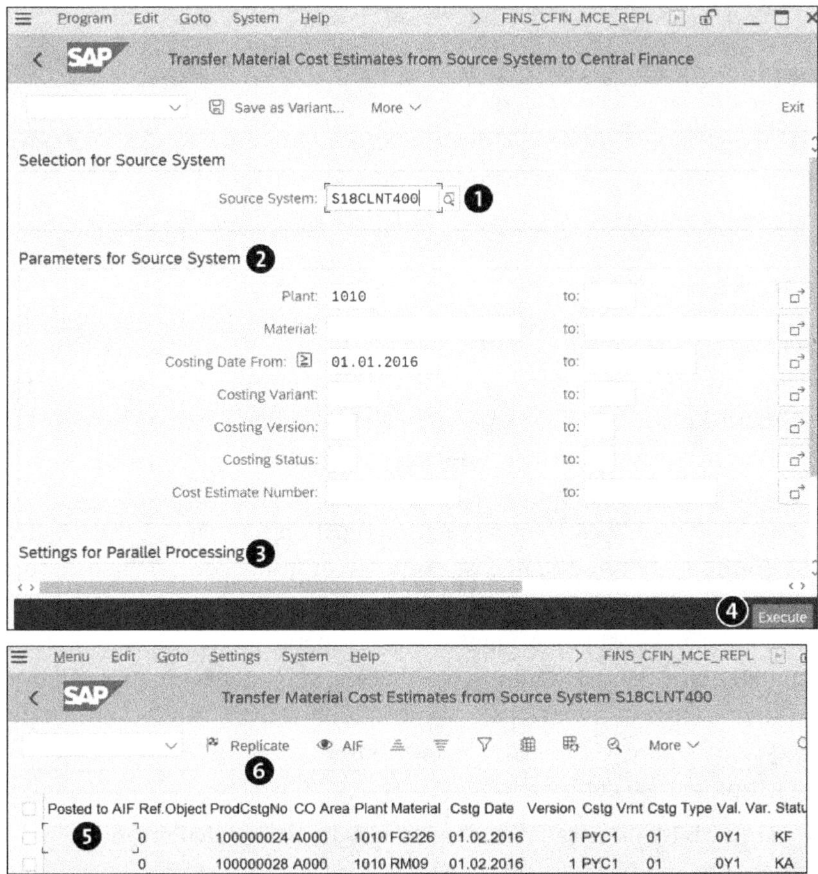

Figure 10.40 Replicating Material Cost Estimates Manually

Once the system processes the chosen rows, the value **N** in the **Posted to AIF** field is updated by the system, as shown in Figure 10.41 ❶. This means the system tried but failed to transfer material cost estimate due to a nonsatisfying transfer rule. In addition, this means the material cost estimate made it to SAP Application Interface Framework as well for troubleshooting.

On the other hand, as you've seen in the bottom of Figure 10.41, value **Y** ❷ in the **Posted to AIF** field confirms the system replicated the material cost estimate to the Central

Finance system successfully. In addition, a value of **Y** in the **Posted to AIF** field also means the material cost estimate transfer satisfied the transfer rule to reach SAP Application Interface Framework. You'll be able to navigate to the SAP Application Interface Framework error handling tool to see the message by clicking the **AIF** icon on submenu from this screen.

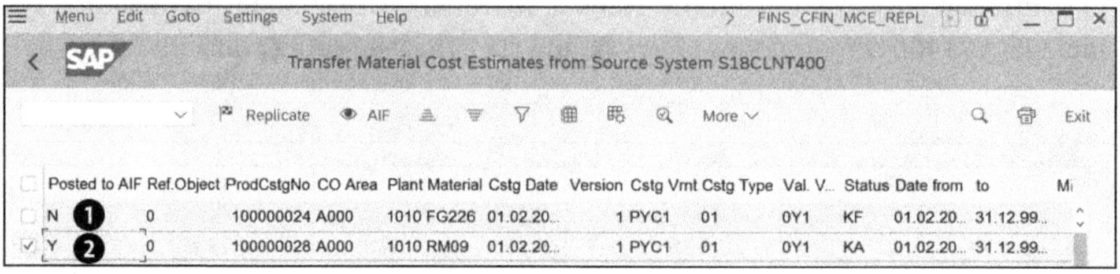

Transfer Material Cost Estimates from Source System S18CLNT400

Posted to AIF	Ref.Object	ProdCstgNo	CO Area	Plant	Material	Cstg Date	Version	Cstg Vrnt	Cstg Type	Val. V...	Status	Date from	to
N ❶	0	100000024	A000	1010	FG226	01.02.20...	1	PYC1	01	0Y1	KF	01.02.20...	31.12.99...
Y ❷	0	100000028	A000	1010	RM09	01.02.20...	1	PYC1	01	0Y1	KA	01.02.20...	31.12.99...

Figure 10.41 Replicating Material Cost Estimates Manually: Report Output

10.5 Cost Rate Replication

To derive production costs from operations, you'll set the activity price or activity rate in cost accounting. The activity price is then considered during routing or confirmation postings in product costing. In SAP, you need an activity rate or cost rate to allocate cost to account assignment objects—cost centers, WBS, internal orders, production orders, service orders, and so on. Cost rates are primarily used in the product costing component of CO to allocate costs to various cost objects. On the other hand, activity price is commonly referred to in cost center accounting. Central Finance supports replicating the activity rate from the source systems to the Central Finance system. The system also brings cost component split information if the cost component structure is maintained in the controlling area. Once activity rates are replicated to Central Finance, they are used for material cost calculation or activity allocation in the Central Finance system. Figure 10.42 illustrates data flow of cost rate replication in Central Finance. Let's explore each component next:

❶ **Activities of plan activity price calculation and activity price calculation**
These are initiated by the business. In the source system, the business carries out plan price planning for the cost objects such as cost center and activity types. Transaction KP26 is used to enter a price for an activity, whereas Transaction KSPI is used to calculate planned activity price.

❷ **SAP Landscape Transformation Replication Server**
This reads cost rate information from table `COST` in the source system and then transfers the cost rate to table `COST` in the Central Finance system.

❸ **SAP Application Interface Framework**
This is used to process cost rate information. In addition, the system applies relevant SAP Master Data Governance key mapping and value mapping in Central Finance.

❹ **Activity rate post service and primary cost splits post service**
These are called to post cost rate information based on relevant mapped values from SAP Master Data Governance key mapping and value mapping in Central Finance. The system updates activity rate tables COST, CSSL, COSL, and COKL. In addition, if the cost component split is active, then the system also updates tables KEKO, KEPH, and COST. Furthermore, the system maintains values in table FINS_CFIN_AR_MAP, which stores the sender system's key fields for reference.

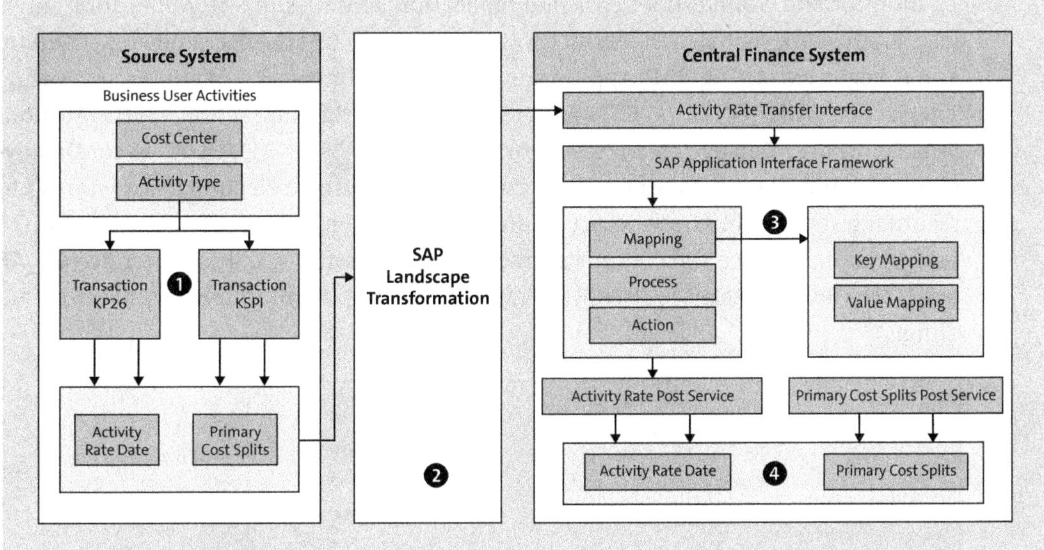

Figure 10.42 Activity Rate Replication Overview

Now that you understand the replication of cost rate, let's discuss the configuration requirements for activity rate replication in the SAP Landscape Transformation Replication Server and Central Finance systems.

10.5.1 Configuring for Cost Rates Replication

Likewise, to support cost rate replication, you need to carry out relevant configuration in the following:

- Source system
- Central Finance system
- SAP Landscape Transformation Replication Server
- Initial load and replication of cost rates

We'll cover the configuration requirements for cost rate replication in each system in the following sections.

Source System

SAP recommends implementing SAP Note 2821651 in the source system. This SAP Note is required if you set up a transfer rule for activity rate replication from the source system to the Central Finance system. In addition, this SAP Note provides search help for several objects such as controlling area, cost centers, versions, and connectivity types.

Central Finance System

To monitor and troubleshoot cost rate replication activities in SAP Application Interface Framework, you'll first extract SAP Application Interface Framework content. You'll begin extracting SAP Application Interface Framework content by executing Transaction /AIF/CONTENT_EXTRACT. Upon entering the transaction, on the **AIF Content Transport - Deploy** screen (see Figure 10.43), enter "SAP_AFI_0023" ❶ for **Deployment Scenario ID** for the cost rate replication. Select further selection parameters ❷ to record registration and transport recording ❸, respectively. Click the **Execute** ❹ icon to extract content. Once the content extraction is successful, you'll see interface **CC_AR** available under namespace **/FINCF** in Transaction /AIF/ERR monitoring and error handling.

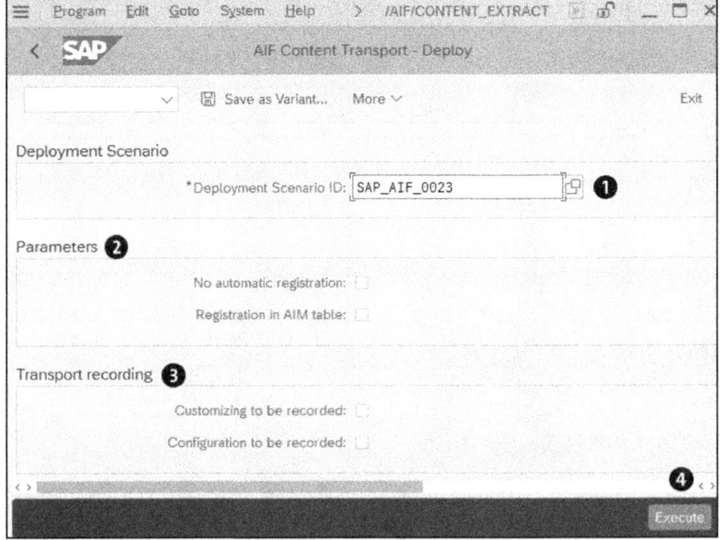

Figure 10.43 SAP Application Interface Framework Content Extract for Material Cost Rate Replication

Next, SAP provides the **Transfer Rule Validation for Activity Replication** report to switch on or off the transfer rule validation for activity rate replication. By default, the system comes with transfer rule validation switched off. To replicate cost rate replication, you must switch on transfer rule validation if you want the system to check the transfer rule configuration. To carry out this configuration, navigate to IMG menu path **SAP Customizing Implementation Guide • Financial Accounting • Central Finance: Target**

System Settings • Activity Rates Replication • Transfer Rate Validation for Activity Rate Replication or use Central Finance IMG Transaction CFINIMG. After reaching the configuration, as shown in Figure 10.44, click the **Switch Off** ❶ button on the **Transfer Rule Validation for Activity Rate Replication** screen. The value in the **Current Status** field changes from **Switched Off** to **Switched On** ❷.

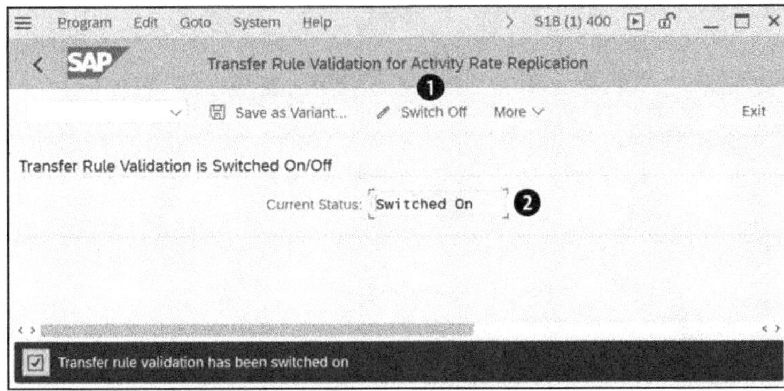

Figure 10.44 Transfer Rule Validation for Activity Rate Replication

Note

The configuration is relevant only if transfer rules are defined in the Central Finance system. If you leave the **Current Status** field with the value of **Switched Off**, the system transfers all data without checking transfer rule configuration.

For cost rate replication in Central Finance, SAP provides transfer rule configuration in both the source system and the Central Finance system to support two-direction replication for activity rate replication. This means you'll be able to replicate activity rates from the source system to the Central Finance system as well as from the Central Finance system to the source system in real time. In our demonstration, we'll walk you through the replication from the source system to the Central Finance system.

To define transfer rules to replicate cost rates from the source system to the Central Finance system, navigate to IMG menu path **SAP Customizing Implementation Guide • Financial Accounting • Central Finance: Target System Settings • Activity Rates Replication • Define Transfer Rules for Activity Rate Replication from Source to Central** or use Central Finance IMG Transaction CFINIMG. Upon entering the configuration, as shown at the top of Figure 10.45, you'll arrive at the **Change View 'Activity Rate Transfer Rule He.. (Source)": Overview** screen. On this screen, click **New Entries** to maintain the source logical system to enable cost rate replication. Select the **Logical System** ❶ row you've just created, and double-click the **Activity Rate Transfer Rule Head (Source)** ❷ option under **Dialog Structure** on the left pane. On the resulting screen, click **New Entries** to maintain the cost rate transfer header parameters. On the resulting screen, maintain values in fields **CoAr**, **Cost Center**, **Value Type**, and **Year From** ❸.

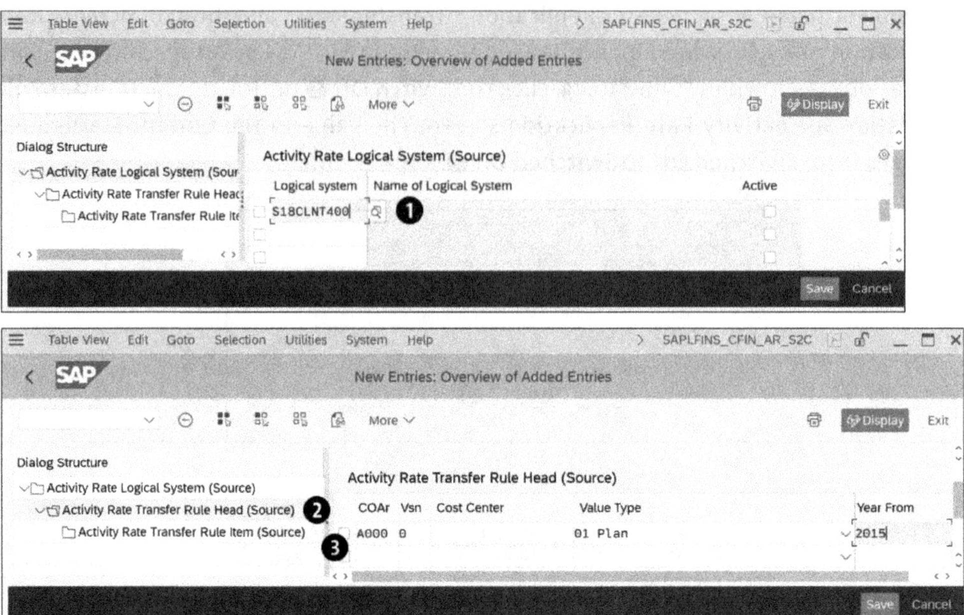

Figure 10.45 Activity Rate Logical System and Activity Rate Transfer Rule Head

Once you've maintained the header information, navigate to the item screen by selecting the header row and double-clicking **Activity Rate Transfer Rule Item (Source)** under **Dialog Structure** located on the left pane, as shown in Figure 10.46 ❶.

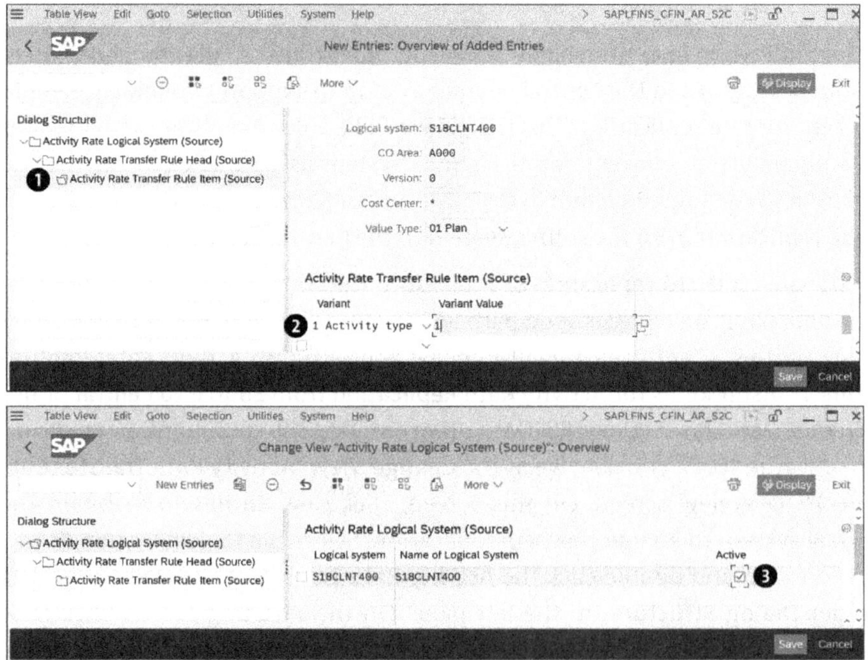

Figure 10.46 Activity Rate Logical System and Activity Rate Transfer Rule Item

Next, click **New Entries** to maintain the activity rate transfer rule item. Choose **Activity Type** from the dropdown on the **Variant** field ❷.

Finally, go to the **Change View "Activity Rate Logical System (Source)": Overview** screen, and click the **Active** ❸ checkbox to activate the **Logical System**.

SAP Landscape Transformation Replication Server Configuration

From the SAP Landscape Transformation Replication Server standpoint, besides downloading the latest SAP Landscape Transformation Replication Server content and applying relevant SAP Notes in SAP Landscape Transformation Replication Server, you must configure SAP Landscape Transformation Replication Server for the cost rate replication. Perform SAP Landscape Transformation Replication Server configuration in the following sequence:

1. Set up table COST to filter data coming in from the source system to the Central Finance system.
2. Create predefined load and replication objects for cost replication.
3. Apply the skip rule as necessary based on your requirements.

To apply filters, go to Transaction LTRS, and select the MT ID you've created for table COST. In our example in Figure 10.47, we selected **MT ID 006** ❶ to set up the COST filter.

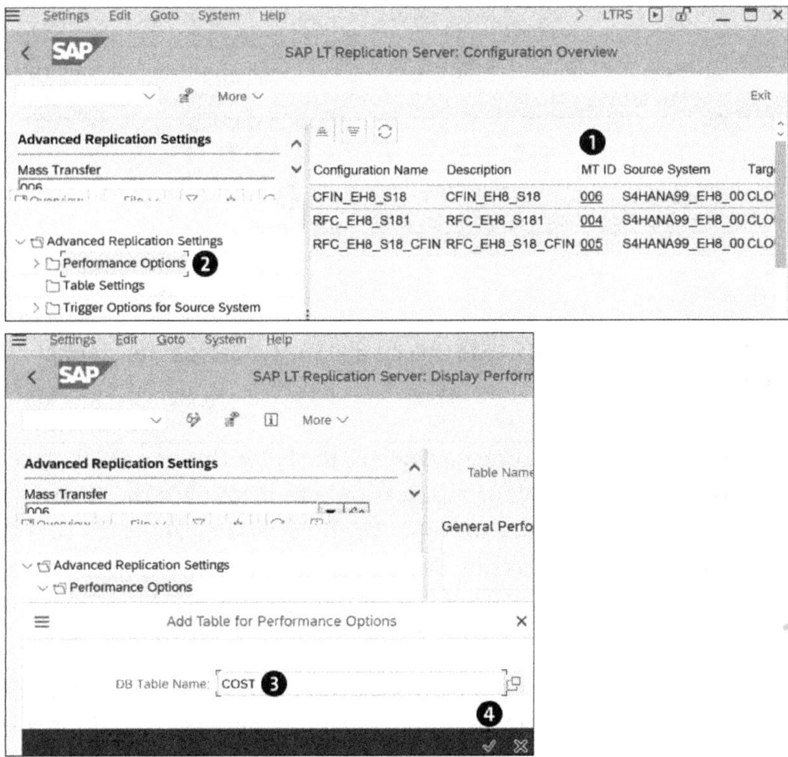

Figure 10.47 Cost Rate Replication: Filters

After reaching the transaction in SAP Landscape Transformation Replication Server, choose the **MT ID** row, and right-click on **Performance Options** ❷ located under **Advanced Replication Settings** on the left pane. On the resulting context menu, click on **Add Object**. This brings up the **Add Table for Performance Options** popup window. Enter "COST" ❸ in the **DB Table Name** field, and click the checkmark icon ❹ to continue.

As shown in Figure 10.48, once table COST is created ❶, on the resulting screen, choose **5 Sender Queue** from the **Reading Type** dropdown list in the **Initial Load Options** ❷ tab. Click the **Save** button to update the configuration.

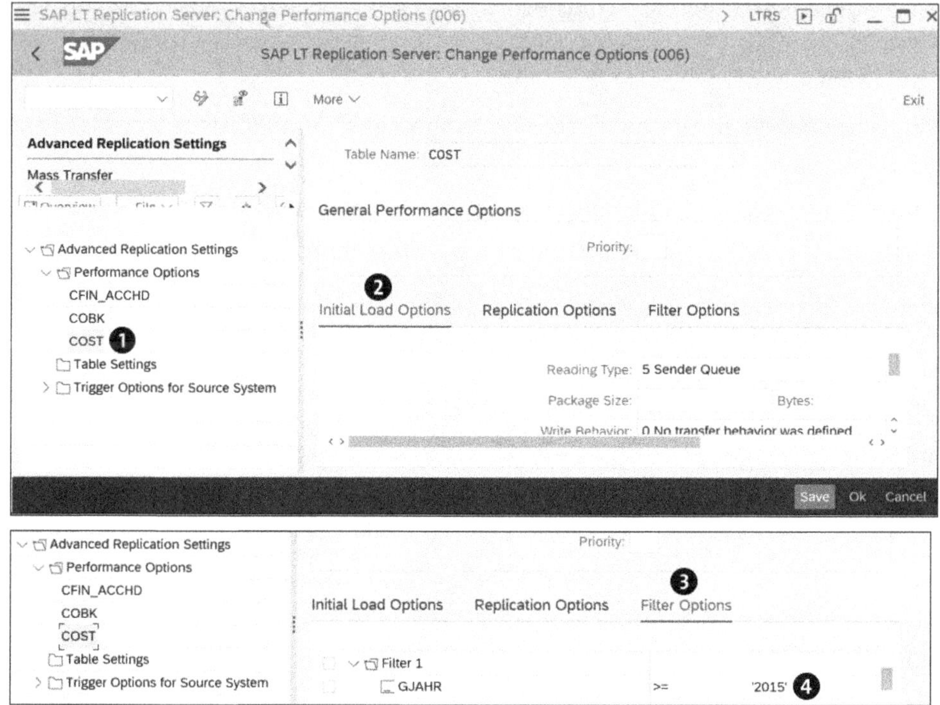

Figure 10.48 Activity Rate Replication: Initial Load Objects and SAP Landscape Transformation Replication Server Filter Options

Next, as shown at the bottom of the figure, navigate to the **Filter Options** ❸ tab to set the filter criteria to load cost rates per your business requirement. Click the **Save** icon to save the configuration. In our example, we've set the fiscal year greater than **2015** ❹ as a filter criterion. This configuration setting determines the activity rate data loaded from fiscal year 2015 onward for the initial load when transferring cost rates from the source system to the Central Finance system.

Once filters are applied, you'll move on to create predefined load and replication objects for table COST replication. Table 10.6 lists the predefined objects created for cost rate replication.

Type of Object	SAP-Delivered Predefined Load and Replication Object
Predefined load object	CFI_COST_L
Predefined replication object	CFI_COST_R

Table 10.6 Predefined Load and Replication Objects for Cost Rate Replication

To create predefined load and replication objects, go to Transaction LTRC. On the **SAP LT Replication Server-Cockpit Overview** screen, choose **MT ID**, and navigate to **Expert Functions • Processing Steps • Preparatory Functions • Create Predefined Load and Replication Objects**. Once you've arrived at the **SAP LT Replication Server – Cockpit: 006** screen, choose **Copy Predefined Object**. This will bring up the **Copy existing Predefined Object** screen, as shown in Figure 10.49. On this screen, maintain predefined load objects and replication objects **CFI_COST_L** ❶ and **CFI_COST_R** ❷, respectively. As a result, the system creates templates for predefined load and replication objects for table COST. The system copies from standard migration objects CFI_COST_L and COST_R listed in Table 10.6. Moreover, the template migration load object Z_COST_006 for table COST was also created. In addition, mapping rules and associated structures are copied from the standard predefined load and replication objects.

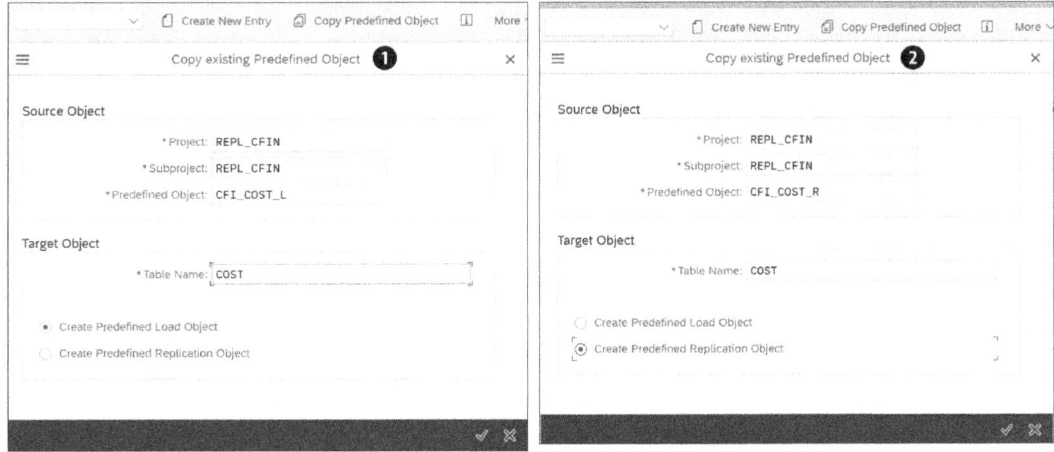

Figure 10.49 Predefined Load and Replication Objects for Activity Rate Replication

After creating predefined load and replication objects by copying from the standard objects, the system defaults to the **Inactive** status. Change the status to **Active** to activate cost rate replication. Like other replication objects, SAP has provided the SKIP_COST event-based rule to add any project-specific filter criteria for the initial load and online replication.

Initial Load and Replication of Cost Rates

Once the predefined load and replication objects are created and activated, you'll be able to load and replicate the activity rates from the source system to the Central Finance system. To start the load and replication, go to Transaction LTRC. Choose the **MT ID**, and click the **Data Provisioning** button located under **Participating Objects**. This will bring up the **Data Provisioning** popup screen, as shown in Figure 10.50. On the resulting screen ❶, enter "COST" in the **Table Name in Database** field. Next, choose the **Start Replication** radio button, and click the **OK** button. This action will load and replicate the data from source table COST to the target Central Finance system. This completes the initial load activity. To verify transfer of cost rates to the target system, navigate to interface **CC_AR** under **/FINCF** namespace on Transaction /AIF/ERR or /AIF/IFMON ❷.

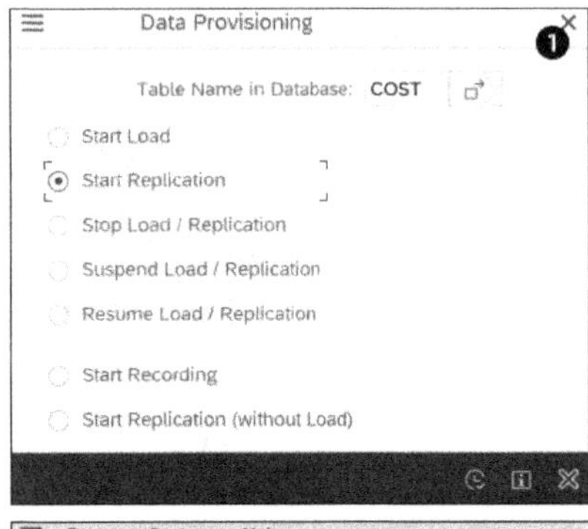

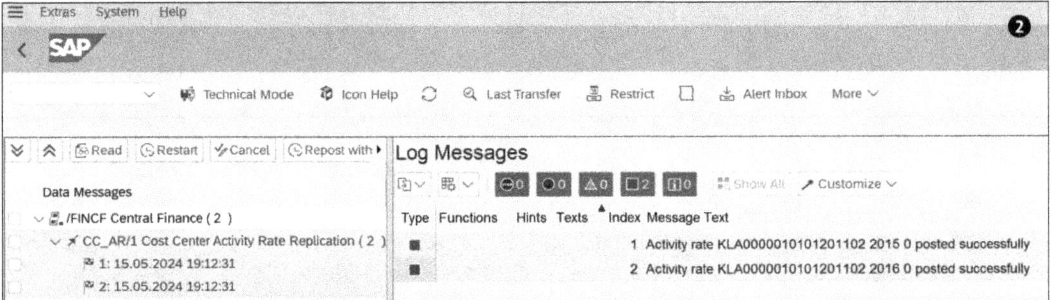

Figure 10.50 Start Replicate Activity Rate

10.5.2 Replicating Cost Rates Manually

SAP provides an option to manually replicate cost rates from the source system to the Central Finance system. The transaction is initiated from Central Finance by navigating from **SAP Easy Access** menu by choosing **Accounting • Central Finance • Manual Transfer Tools • Replicate Activities Rates Manually** or using Transaction FINS_CFIN_AR_REPL. After reaching the transaction, you'll arrive at the **Transfer Activity Rates from Source System to Central Finance** screen, as shown in Figure 10.51. On this screen, enter **Source System** under **Selection for Source System section** ❶. Next, enter **Parameters for**

Source System ❷ based on your requirements. In our example, we've entered values in the **Controlling Area**, **Cost Center**, **Activity Type**, and **Fiscal Year** fields. Optionally, based on your system, you can enter values in the fields located under **Settings for Parallel Processing** ❸. Click the **Execute** ❹ button. Upon execution of the transaction, the system checks activity rates against the transfer rule in the configuration prior to replication. On the resulting output screen, the system provides three options:

- **Replicate & Open AIF** ❺
  The system replicates cost rates and opens the SAP Application Interface Framework screen with the results for a successful replication. If no posting occurs, the system displays an error message.
- **Replicate** ❻
  The system just replicates cost rates from the source system to Central Finance. You'll navigate to the SAP Application Interface Framework screen manually.
- **AIF** ❼
  For the successfully posted cost rates, the system takes you to SAP Application Interface Framework directly for the selected row.

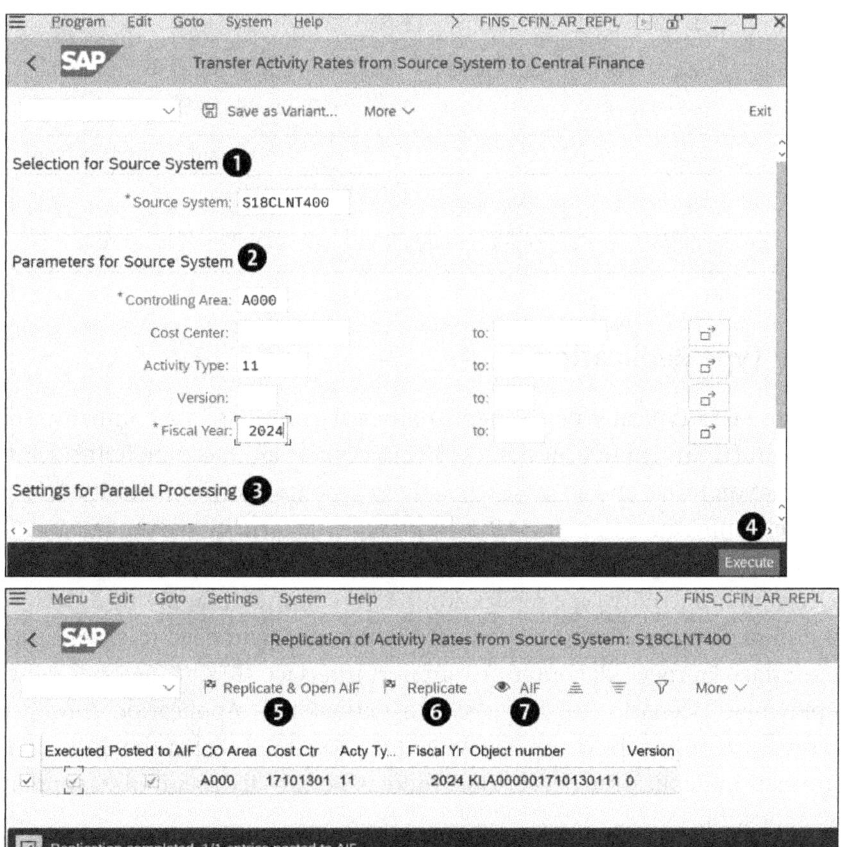

Figure 10.51 Manual Activity Rate Replication

Figure 10.52 displays the successfully replicated activity rate from the selected row in Central Finance.

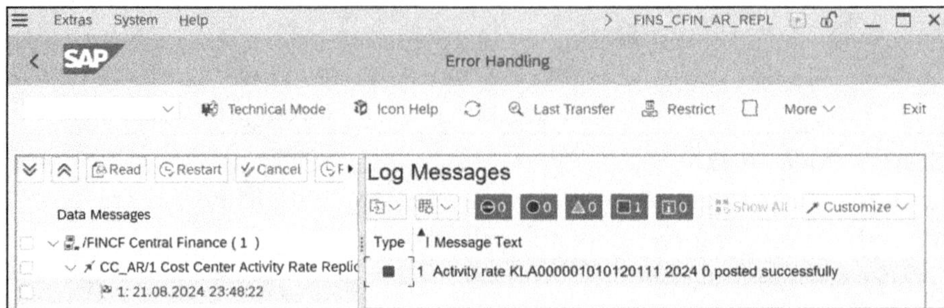

Figure 10.52 Manual Activity Rate Replication: SAP Application Interface Framework

To execute Transaction FINS_CFIN_AR_REPL, you need the authorization objects identified in Table 10.7.

Authorization Object	Activity	Purpose
F_CFIN_TRG	Application 16; Activity 16	Execute the transaction.
F_CFIN_TRG	Application 09; Activity 33	Read and fetch activity rate from table COST from the source system.
F_CFIN_SRC	Application 07; Activity 33	Fetch and display value help data from the source system.

Table 10.7 Authorization Objects for Transaction FINS_CFIN_AR_REPL

10.6 Activity Type Replication

Activity types are tasks typically performed in any environment in a company. For example, in a manufacturing environment, typical activities are machine hours, labor hours direct, direct material, and so on. Each activity type uses a specific unit of measure such as hours, units, and so on. In SAP, activity type is a form of master data in CO. Central Finance supports replicating activity type master from a source system to Central Finance. To support activity type replication, SAP recommends applying SAP Note 2935857 in the source system. From the Central Finance side, you need to extract SAP Application Interface Framework content to support interface ACTY_TP. SAP has delivered the deployment scenario SAP_AIF_00400 to extract SAP Application Interface Framework content using Transaction AIF/CONTENT_EXTRACT. From the SAP Landscape Transformation Replication Server side, there is no activity required to support activity type replication.

In this section, we'll discuss replication of activity types and relevant mapping scenarios.

10.6.1 Replicate Activity Types Manually

Activity types are replicated directly from the source system to Central Finance with Transaction CFIN_ACTY_TP_REPL. You can reach this transaction by navigating from **SAP Easy Access** Menu and going to menu path **Accounting • Central Finance • Manual Transfer Tools • Replicate Activity Type Manually** or using Transaction FINS_CFIN_TP_REPL. After reaching the transaction, you'll arrive at the **Transfer Activity Types from Source System to Central Finance** screen, as shown in Figure 10.53. On this screen, enter **Source System** under **Selection for Source System** ❶. Next, enter **Parameters for Source System** ❷ based on your requirements. In our example, we've entered values in the **Controlling Area** and **Activity Type** fields. Once all fields are filled, click the **Execute** ❸ button. This leads to the **Replication of Activity Types from Source System: S18CLNT400** screen where the system displays a list of activity types that can be replicated from the source system to the Central Finance system. Select activity type row **11**, and click the **Replicate** ❹ button. After clicking the **Replicate** button, the system replicates the selected activity type master data from the source system to the Central Finance target system. Optionally, you can choose the **AIF** ❺ button to navigate to SAP Application Interface Framework.

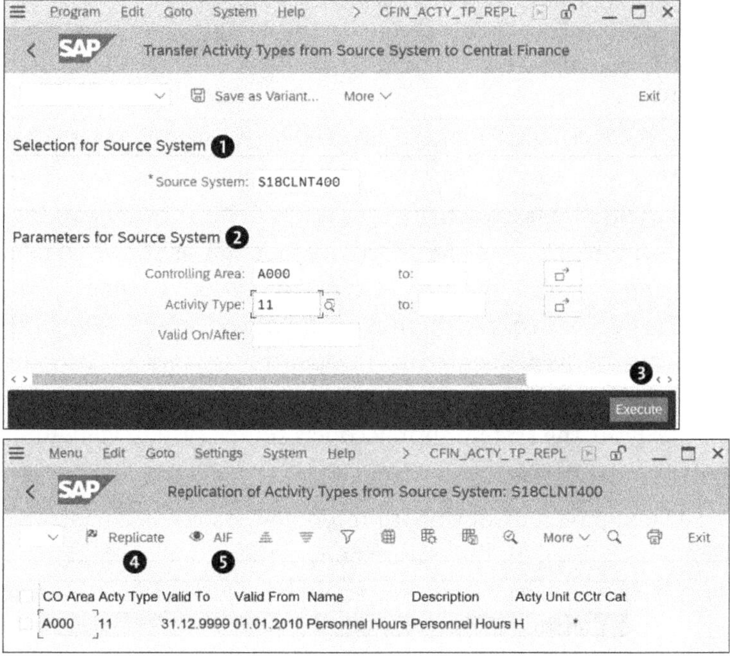

Figure 10.53 Transfer Activity Type

Figure 10.54 displays the SAP Application Interface Framework window showing activity type **11** replicated successfully and master data created in Central Finance. Ensure you have SAP Master Data Governance cost element mapping in the Central Finance system to prevent mapping errors.

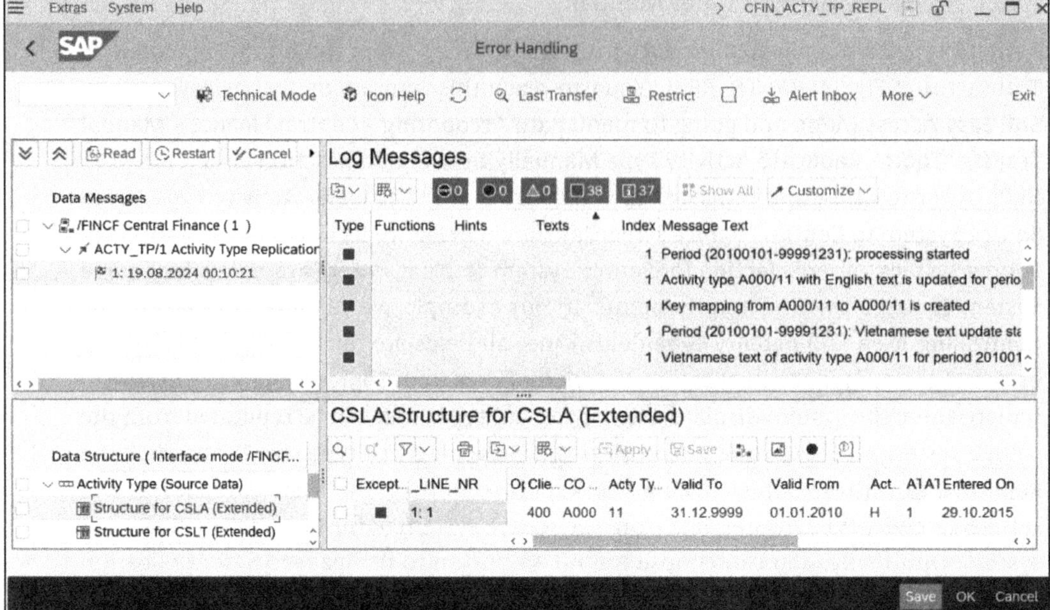

Figure 10.54 Activity Type Replication in SAP Application Interface Framework

10.6.2 Mapping Scenarios

If a mapping relationship doesn't exist for the source activity type in the Central Finance system for mapping entity **ACTIVITY_TYPE_ID**, then the system creates key mapping in Central Finance with the same activity type in the Central Finance system. Let's explore this scenario with replicating source activity type **102**, as shown in table `CSLA` in Figure 10.55 ❶. Upon replicating the activity type, the system creates activity type **102**, and key mapping is generated automatically. SAP provides BAdI `BADI_FINS_CFIN_ACTY_TP_REPL`, which can be implemented in the Central Finance system customer-specific logic to derive specific parameters on the activity type master data. Without implementing this BAdI, the system simply copies source system values into the target system activity type upon creation.

On the other hand, if a key mapping relationship is present in the Central Finance system for mapping entity `ACTIVITY_TYPE_ID`, the system derives the mapping of the activity type from SAP Master Data Governance mapping. Activity type **ZCFAT1** is mapped to **YCFAT** ❷ in key mapping in the Central Finance system.

Now execute Transaction CFIN_ACTY_TP_REPL to manually replicate the source activity type, as shown in left side of Figure 10.56. On the resulting **Transfer Activity Types from Source System to Central Finance** screen, enter values in **Source System** under the **Selection for Source System** ❶ section. Next, enter values in the **Controlling Area**, **Activity Type**, **Valid On/After** fields under **Parameters for Source System** ❷. Click the **Execute** ❸ button. Upon execution of the transaction, the system displays activity type **ZCFAT1** replicated successfully. The system displays the confirmation message **Activity type A000/YCFAT with English text is created for period 19940101-24001231** ❹ on the **Error Handling** screen. This means target activity type **YCFAT** is derived from the key mapping and created in the Central Finance system successfully.

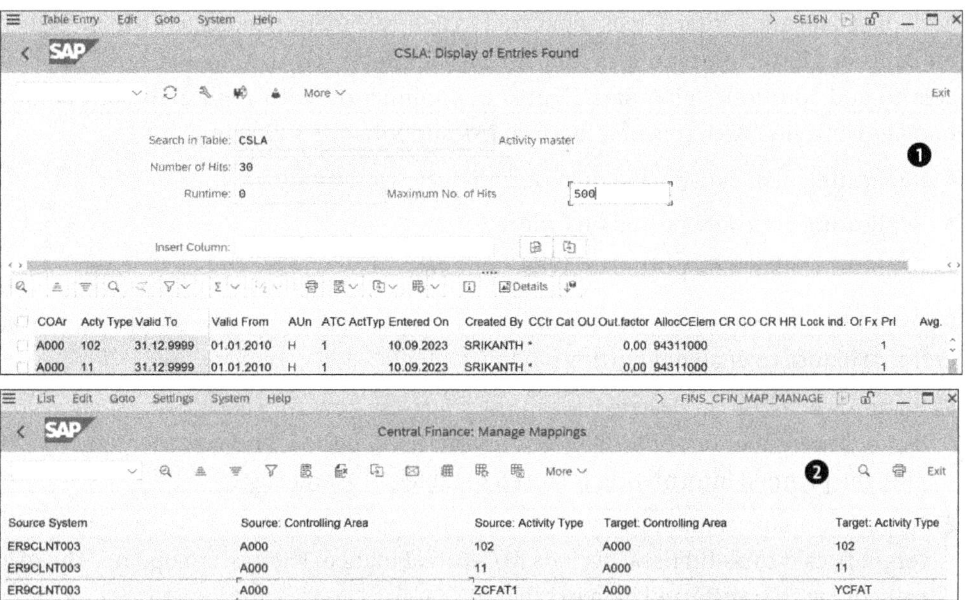

Figure 10.55 Activity Type Replication Mapping

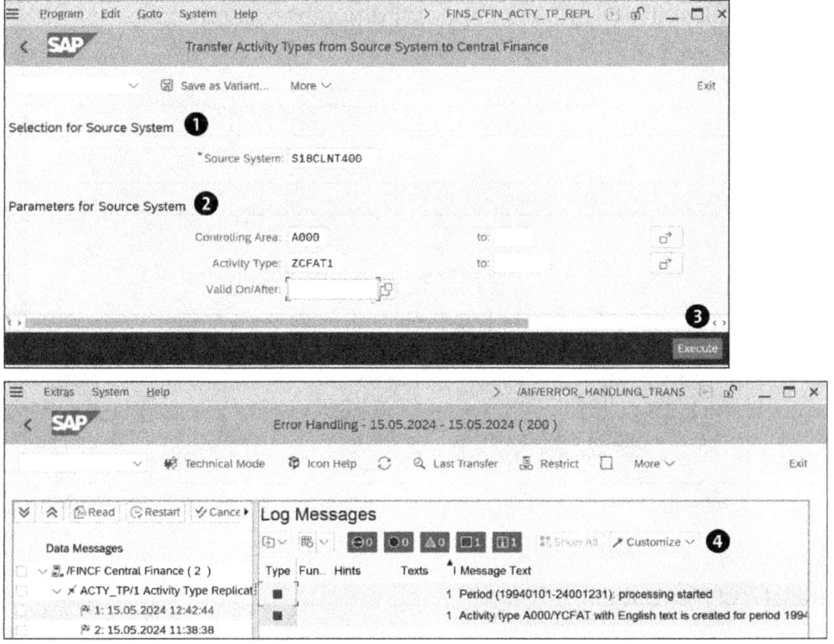

Figure 10.56 Activity Type Replication: Manual

10.7 Transfer of Plan Data

SAP supports transfer of plan values on cost objects from source systems to Central Finance with the SAP S/4HANA 2023 FPS01 release. Transferred plan data from the source system is updated in the Universal Journal table ACDOCP through an RFC. Once both plan data and actual data is replicated to Central Finance, managers can make better decisions based on plan vs. actual analysis in SAP S/4HANA. In addition, you'll be able to add additional plan data figures in conjunction with replicated plan data to manage centrally. With this new feature, SAP supports the following:

- Replicating plan data posted on cost centers
- Replicating plan data posted on orders
- Replicating plan data posted on WBS elements

Figure 10.57 illustrates how the system transfers plan data from the source system to Central Finance to enable reporting:

❶ Planning activities are carried out in the source system for cost objects such as cost center, internal order, or WBS by fiscal year, fiscal period, and cost element. In SAP ERP, the planned information is stored in tables COSP and COSS.

❷ As part of financial planning, you'll maintain a plan category for cost objects for different fiscal years and fiscal periods in Central Finance. The system updates the plan category with transferred data in table ACDOCP.

❸ With both plan data and actual data collected from the source system, you can carry out plan vs. actual analysis with SAP Analytics Cloud or any other ad hoc reporting tool. SAP also provides several SAP Fiori apps for different cost objects.

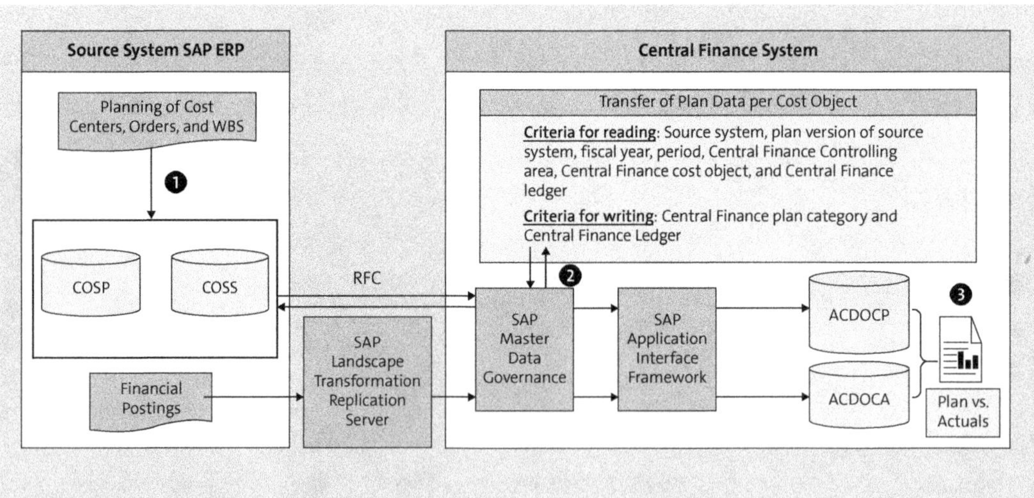

Figure 10.57 Transfer of Plan Data

In this section, we'll discuss the configuration requirements to support plan data transfer in the source system and Central Finance system, and then we'll highlight security requirements.

10.7.1 Configuration Requirements

In this section, we'll explore configuration requirements for transfer of plan data from source systems to the Central Finance system.

Source System Configuration

To support transfer of plan data from the source systems, SAP recommends implementing SAP Note 3379286, which activates the transfer of plan data functionality in source systems. In addition, the SAP Note is applicable for both SAP ERP and SAP S/4HANA source systems. Ensure SAP source systems are enabled with the financial planning functionality to support plan data transfer.

Central Finance System Configuration

Table 10.8 provides a list of configuration activities in the Central Finance system to support transfer of plan data from the source system to Central Finance. SAP recommends applying SAP Note 3392740, which enables the plan transfer replication functionality.

Customizing Activity	Transaction/Menu Path	Purpose
SAP Application Interface Framework	Transaction /AIF/CONTENT_EXTRACT	Extract deployment scenario SAP_AIF_0077 to enable interface **/FINCF/PD_C2A** Version **1** on Central Finance namespace **/FINCF** in Transaction /AIF/ERR or /AIF/IFMON.
Background RFC (bgRFC)	Navigate from IMG customizing menu path **Financial Accounting • Central Finance • Central Finance: Target System Settings • Set Up Systems • AIF Runtime Configuration • Settings for bgRFC • Assign Default bgRFC Subprocess to Replication Objects**.	Assign process and subprocesses to replication object 77 **Plan Data Postings (COSP/COSS - ACDOCP)**.

Table 10.8 Customizing Activities in Central Finance for Transfer of Plan Data

Customizing Activity	Transaction/Menu Path	Purpose
RFC destination	Navigate from Central Finance customizing (Transaction CFINIMG) menu path **Central Finance • Central Finance: Target System Settings • Set Up Systems • Set up RFC Destination for Source System**.	Use required connection to connect the source system via RFC.
RFC destination for RFC usage	Navigate from Central Finance customizing (Transaction CFINIMG) menu path **Central Finance • Central Finance: Target System Settings • Set Up Systems • Maintain RFC Assignments and Settings for Source System**.	Extract plan data from the source system.
Plan category	Navigate from IMG menu path **Controlling • General Controlling • Planning • Maintain Category for Planning**.	Activate **Importing Financial Plan Data** in the plan category.

Table 10.8 Customizing Activities in Central Finance for Transfer of Plan Data (Cont.)

10.7.2 Security Requirements

Table 10.9 provides a template of SAP roles that can be copied and used to read relevant objects to transfer plan data.

Customizing Activity	System/Transaction	Purpose
SAP_CFIN_RFC_USAGE_24	Source system via RFC Transaction FIN_CFIN_PLAN_DATA_READ	Read plan data tables COSP/COSS.
SAP_CFIN_PLANDATA_REPLICATOR	Central Finance system granting access to the following transactions: ▪ Transaction FINS_CFIN_PD_C2A_CC ▪ Transaction FINS_CFIN_PD_C2A_OR ▪ Transaction FINS_CFIN_PD_C2A_PR	Transfer plan data.
SAP_SFIN_CFIN_ADMIN	Transaction FINS_CFIN_PD_AIF_C2A	Delete. cancel, or reprocess SAP Application Interface Framework messages transfer of data.

Table 10.9 Security Requirements for Plan Data Transfer

Tip

Consult SAP Note 3403561 for an overview plan data transfer in Central Finance.

10.8 Margin Analysis

Margin analysis allows businesses to analyze the profitability of specific market segments such as products, customers, and material groups. The end goal is to provide management with contribution margin reporting capability to aid in the management decision-making process. With SAP S/4HANA, you can perform profitability analysis using both costing-based profitability analysis and margin analysis. SAP renamed *account-based profitability analysis* as *margin analysis* because it brings in more features and innovations. SAP recommends margin analysis as the go-to solution, but both margin analysis and costing-based profitability analysis can coexist. However, only margin analysis updates the Universal Journal (table ACDOCA), and all future innovations will happen only with margin analysis. Customers considering both approaches may find it difficult to justify why costing-based profitability analysis is required in SAP S/4HANA as all costing-based features are now available with margin analysis.

Central Finance supports replication of profitability analysis data from the source systems. Table 10.10 lists several possible scenarios where profitability data is replicated from the source system to the Central Finance system.

Source SAP System	Target System	Replication
Costing-based profitability analysis	Margin analysis	Supported, but no costing-based profitability analysis documents are replicated directly, and no value fields are replicated. Instead, the system reads cost-based profitability analysis data from staging table CFIN_ACCPA_CHAR and feeds it into the Universal Journal during the transfer of documents.
Accounting-based profitability analysis	Margin analysis	Supported.
Margin analysis	Margin analysis	Supported; SAP recommended approach.
Cost-based profitability analysis	Cost-based profitability analysis	Not supported.

Table 10.10 Profitability Analysis Supported Scenarios

Source SAP System	Target System	Replication
Costing-based profitability analysis and margin analysis	Costing-based profitability analysis and margin analysis	Costing-based profitability analysis isn't supported; accounting-based profitability analysis to margin analysis replication supported.
Attributed profitability segment not available in SAP ERP-based source systems	Attributed profitability segment available in the Central Finance system	Supported for non-profitability analysis objects.

Table 10.10 Profitability Analysis Supported Scenarios (Cont.)

Figure 10.58 illustrates how costing-based profitability analysis data is replicated from a source system to the Central Finance system where the source system is configured with costing-based profitability analysis and the Central Finance system is configured with margin analysis. Let's explore the flow of these values:

❶ FI documents and CO documents use profitability segments as the account assignment cost objects to collect cost and then relevant financial documents are replicated to Central Finance. In addition, FI and CO transactional line items are posted in reconciliation objects (object type = AO) with a profitability segment in the source system with full details.

❷ With costing-based profitability analysis active in the source system, FI documents posted in the source system are updated in table CE4XXX.

❸ Upon replication of the FI document, the system doesn't replicate costing-based profitability analysis directly; instead, costing-based profitability analysis characteristics are rederived based on the Central Finance costing-based profitability analysis mapping configuration. The system tries to match or sync the existing metadata structure to the actual metadata structure from table CE1* and table CE4*. In Central Finance, the system checks the cluster view FINS_CFINVC_COPA_MAPPING to locate whether any mapping rules for the characteristics and value fields exist. In the case of CO secondary document replication, tables CFIN_COPA and CFIN_COEP_ADD are built for costing-based profitability analysis segments based on PAOBJNR from table CE4*. In addition, the system replaces reconciliation objects with profitability analysis segments on the line items in Central Finance. If CO document line items are summarized and posted to one reconciliation object (object type = AO) with multiple profitability analysis line items, CO document line items are split into multiple line items based on profitability analysis segments in Central Finance. Central Finance supports replicating costing-based profitability analysis data from transaction types.

❹ The system regenerates a new profitability segment (PAOBJNR) again after going through field mapping in Central Finance. Documents are posted to the Universal Journal based on new costing-based profitability analysis characteristics derived from the Central Finance system.

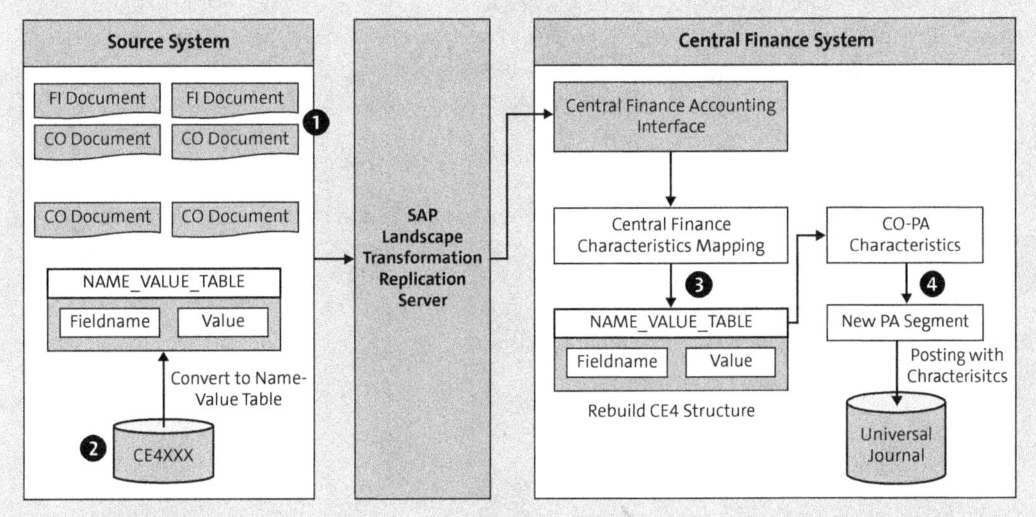

Figure 10.58 Margin Analysis Data Flow

Note

SAP provides BAdIs BADI_FINS_AC_INTERFACE and BADI_FINS_CO_INTERFACE to enhance costing-based profitability analysis data before posting financial data to the Central Finance system.

Now that you've learned the basic concepts behind margin analysis data flow in Central Finance, let's understand costing-based profitability analysis supported scenarios in Central Finance.

10.8.1 Replicating Costing-Based Profitability Analysis Data in Central Finance

In general, SAP ERP costing-based profitability analysis enables source systems implemented with costing-based profitability to connect to Central Finance. Both costing-based profitability analysis and margin analysis can coexist in Central Finance. However, margin analysis is the recommended approach in Central Finance as it's integrated with the Universal Journal. Margin analysis is built on an account-based profitability foundation. Table 10.11 provides the flows of actual values replicated from the source system to the Central Finance system.

Source System: Costing-Based Profitability Analysis	Target Central Finance System: Margin Analysis	Central Finance Profitability Analysis Replication Logic/ Recommendation
During direct FI posting, two documents are generated: ▪ FI-CO document ▪ Profitability analysis document	FI-CO document is posted in the Universal Journal with profitability analysis characteristics. The profitability analysis document isn't replicated.	Profitability segment on the line items of FI or CO documents in the source system will be posted to a profitability segment in the Central Finance system after mapping and rederiving values.
Goods issue FI-CO document	FI-CO document is posted in the Universal Journal with profitability analysis characteristics along with COGS split by account.	This brings a cost component split from the source system based on settings. SAP recommends the profitability segment should be defined as the default cost object for all COGS split accounts in automatic account assignment Transaction OKB9. In this case, source cost objects are cleared through a BAdI enhancement during replication.
During billing, the system generates the following: ▪ FI-CO document. ▪ Profitability analysis document with cost components and sales order conditions values. ▪ The billing document and subsequent FI document are posted with the same account key and general ledger account and may not contain details per condition type.	FI-CO document is posted in the Universal Journal with profitability analysis characteristics along with COGS split by account.	SAP recommends assigning different condition types to different general ledger accounts in customizing to analyze values in each condition type. If derivation rules differ between the source system and the Central Finance system, SAP recommends clearing the source profitability segment during replication.

Table 10.11 Flows of Costing-Based Profitability Analysis Actual Values When the Profitability Segment Is Defined as a Cost Object

Source System: Costing-Based Profitability Analysis	Target Central Finance System: Margin Analysis	Central Finance Profitability Analysis Replication Logic/ Recommendation
Reposting direct activity allocation in CO: ■ CO document ■ Profitability analysis document For costing-based profitability analysis, the reconciliation object is updated in table COSP with object type = AO.	Accounting document is posted in the Universal Journal with profitability analysis characteristics.	The system converts the reconciliation object to a profitability analysis segment in Central Finance.
Profitability analysis assessment in CO: ■ CO document ■ Profitability analysis document	Accounting document is posted with profitability analysis characteristics.	The system splits table COEP line items per profitability analysis segments (PAOBJNR).
CO settlement performed in CO: ■ CO document ■ Profitability analysis document	Accounting document is posted in the Universal Journal with profitability analysis characteristics.	The system converts reconciliation objects to profitability analysis segments (PAOBJNR).
Distribution and assessments in CO generates the following: ■ CO document ■ Profitability analysis document	Accounting document is posted in the Universal Journal with profitability analysis characteristics.	Profitability analysis segments are converted to margin analysis in Central Finance: ■ The system splits line items posted on reconciliation objects per profitability analysis document line items. ■ Reconciliation objects are replaced by profitability analysis segments.
Top-down distribution generates only a profitability analysis document in the source system.	Replication of top-down distribution isn't supported.	SAP recommends running the top-down distribution directly in Central Finance as it's not relevant for replication.

Table 10.11 Flows of Costing-Based Profitability Analysis Actual Values When the Profitability Segment Is Defined as a Cost Object (Cont.)

Source System: Costing-Based Profitability Analysis	Target Central Finance System: Margin Analysis	Central Finance Profitability Analysis Replication Logic/ Recommendation
Realignment.	Replication of realignment isn't supported.	SAP recommends running the realignment directly in Central Finance.
Attributed profitability segment isn't available in the source system.	The attributed profitability analysis functionality is available for cost objects based on customizing.	Attributed profitability analysis generates profitability analysis even if the profitability segment isn't a cost object.

Table 10.11 Flows of Costing-Based Profitability Analysis Actual Values When the Profitability Segment Is Defined as a Cost Object (Cont.)

Tip

For third-party source systems, SAP Landscape Transformation Replication Server provides staging table /1LT/CF_E_COPA, which can be filled with profitability analysis information to replicate profitability analysis data to the Central Finance system by using profitability analysis configuration, mapping tools, and SAP Application Interface Framework error handling.

10.8.2 Characteristics Mapping Tool

If costing-based profitability analysis is configured in the source system and margin analysis is implemented in Central Finance, characteristics and value fields are different between these systems. Due to the differences, Central Finance doesn't directly replicate costing-based profitability analysis documents from the source system. In earlier releases, often custom code was added to enrich characteristics of the profitability segment in methods MAP_VALUES or MAP_VALUES_FOR_COPA_SEGMENT of BAdI FINS_CFIN_AC_INTERFACE before generating the profitability segment number (PAOJBNR) to derive profitability characteristics in Central Finance. However, SAP provided a costing-based profitability analysis characteristics mapping tool in SAP S/4HANA 1709 to overcome the custom BAdI approach. With this tool, as shown in Figure 10.59, you can map characteristics and value fields from several operating concerns ❶ in source systems to different characteristics in operating concerns ❷ in the Central Finance system. In addition, the system can clear or derive unique characteristics during the transfer of data based on each referenced characteristics or value field per the configuration in this mapping tool.

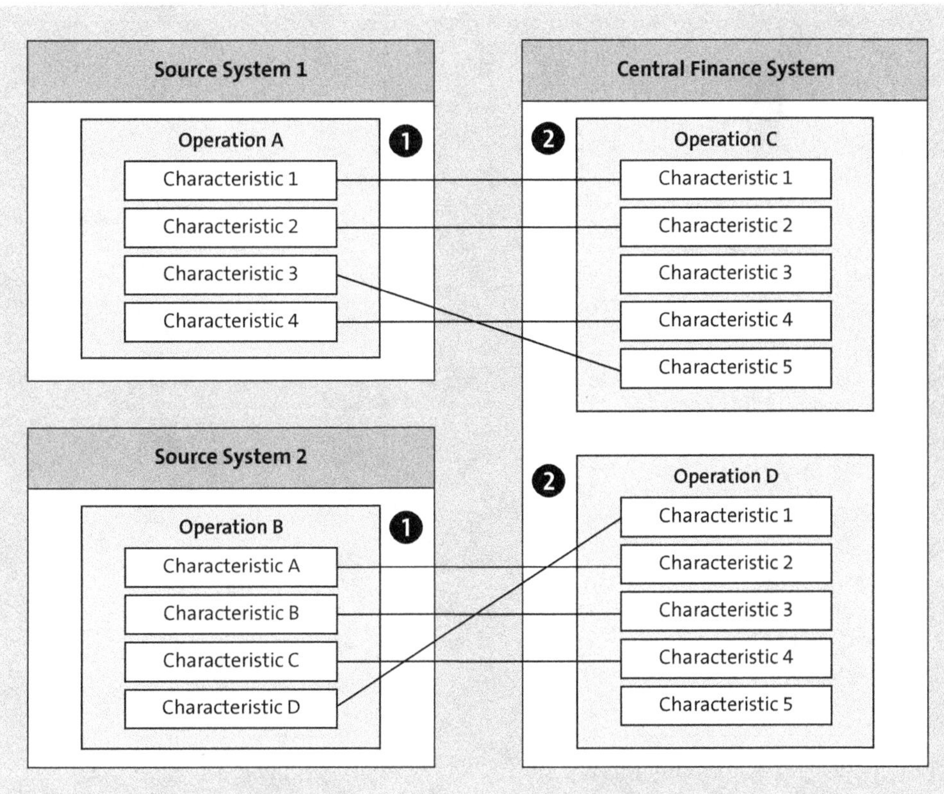

Figure 10.59 Mapping Characteristics and Value Fields

To map characteristics and value fields from the source to Central Finance, navigate customizing menu path **Financial Accounting • Central Finance • Central Finance – Target System Settings • Mapping • COPA Mapping • Define COPA Mapping**. You'll arrive at the **Display View "Maintain COPA Mapping": Overview** screen, as shown at the top of Figure 10.60. On this screen, choose the operating system row ❶, and double-click **Operating Concern in Source System** ❷ on the left panel under **Dialog Structure**. This leads to the **Change View "Operating Concern in Source System": Overview** screen. On this screen, maintain source logical system and operating concern combinations ❸. Select the row, and double-click **Characteristics Mapping** on the left panel.

Based on defined characteristics in your system, the system displays characteristics mapping, as shown in Figure 10.61 ❶. Next, double-click on the **Value Fields Mapping** ❷ under **Dialog Structure** on the left subscreen to maintain the required value mapping per business requirements. Click the **Save** ❸ button to save the entries.

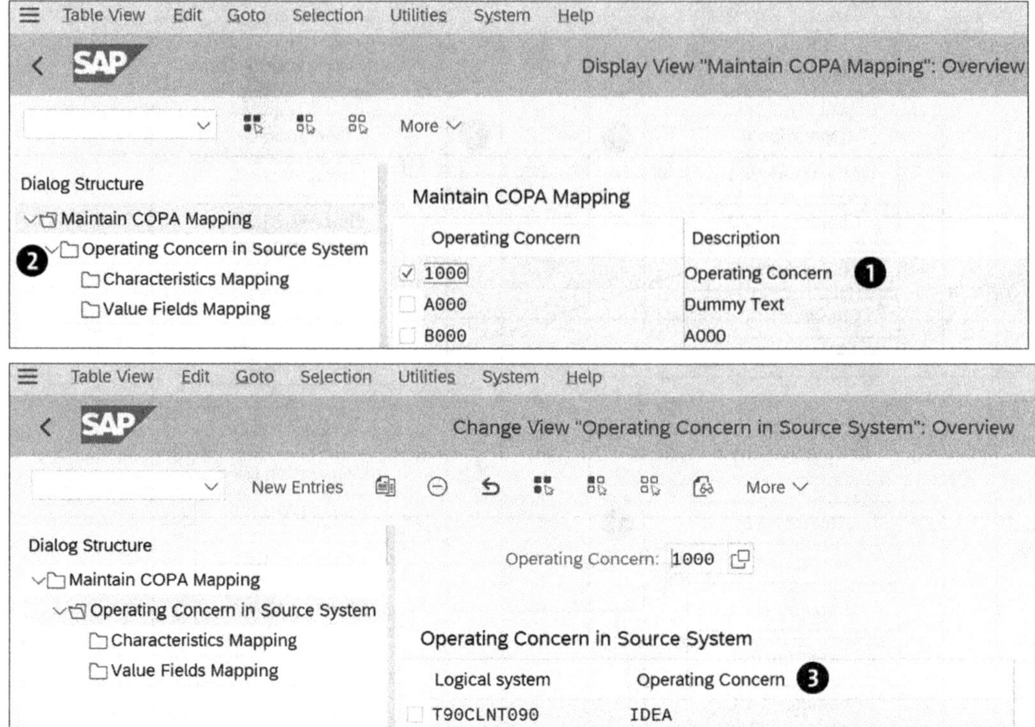

Figure 10.60 Maintain Costing-Based Profitability Analysis Mapping

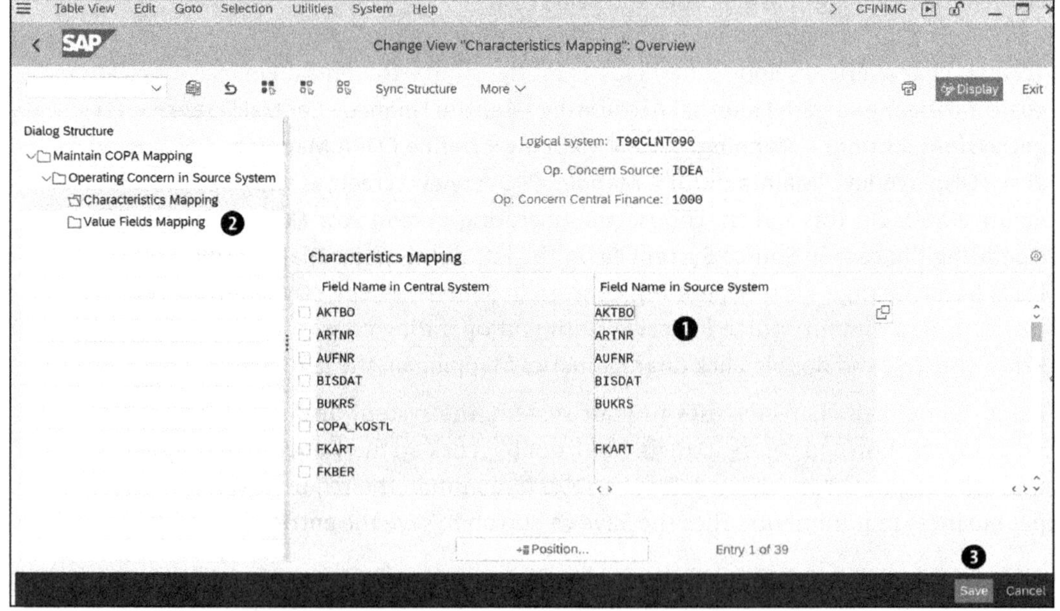

Figure 10.61 Maintain Costing-Based Profitability Analysis Characteristics Mapping

10.8.3 Additional Quantities Replication

Central Finance supports replicating billed and delivered quantities as part of financial postings. Upon a successful FI document replication with quantities, the system updates tables ACDOCA-MSL and BSEG-MENGE. The quantity field (quantity indicator in Central Finance structure ACCIT-XMFRW) is filled based on data received from upstream processes such as SD billing or other logistics processes. If your source system is classic general ledger or new general ledger, then how system fills the quantity (MENGE) field varies with each entry view and general ledger view. Consult SAP Note, 1063198 for further guidance on this.

The Universal Journal also comes with three additional quantities fields: QUNIT1, QUNIT2, and QUNIT3. By default, these fields aren't populated by financial postings replication from the source systems. Often you may need conversion of quantities or aggregate quantities by unit of measure upon transfer of data from the source system to the Central Finance system. You can populate these fields in two approaches. We'll review each approach in the following sections.

Assign a Dimension to the Additional Quantity Fields and Fill with a Business Add-In

In this approach, first you'll first assign dimensions to additional quantity fields (AC) via configuring **Define Additional Quantity Fields**. You can reach this configuration by navigating to it from IMG menu path **Controlling • General Controlling • Additional Amounts • Define Additional Quantity Fields**. Next, enter the quantity fields by implementing BAdI FCO_COEP_QUANTITY with method CALCULATE_QUANTITIES for each COEP line replicated. To support this, the transferred line item from the source system must contain material (COEP-MATNR) and quantity (COEP-MBGBTR). In addition, to support the conversion or aggregation, relevant master data is maintained in table MARA. This approach works for both replication and direct update from Central Finance scenarios. Once you finish the configuration and the BAdI is implemented, the Display Line Items - Margin Analysis (app ID F4818) app can be used to display additional quantities for each replicated line item.

Populate Additional Quantity Fields with Source System Business Add-In

In this alternate approach, Central Finance provides source system BAdI BADI_FIN_CFIN_DOCUMENT, which can be used to populate additional quantities. However, the additional quantities fields first must be added to standard structure FIN_CFIN_S_ACCIT_APP_CUS to transfer the quantity fields during replication. Likewise, the structure must be enhanced with the corresponding additional quantity fields in the structure FINS_CFIN_S_ACCIT_INT from the Central Finance side. The transfer of quantities fields through this approach is available in method PREPARE_INPUT_DATA of Central Finance BAdI BADI_FINS_CFIN_AC_INTERFACE.

10.8.4 Sales and Distribution Condition Types for Margin Analysis

Most SAP ERP systems activated with costing-based profitability analysis uses condition types to determine the price of a product (PR00), discount for a product (K004), freight charges (KF00), tax (MWST), and so on. In addition, you have statistical condition types created in sales and distribution (SD). As you've seen earlier, Central Finance supports replicating costing-based profitability analysis characteristics to margin analysis in Central Finance using the characteristics mapping tool. The system doesn't support replication of value fields. In costing-based profitability analysis, transfer of condition type values from billing documents to value fields are captured if relevant configuration is maintained in Transaction KE4I. Statistical and nonstatistical condition values are captured and posted during SD invoicing. From the Central Finance side, you need an extension ledger and extension ledger group setup to post SD condition types to the Universal Journal. In addition, if you want to analyze values in each condition type for margin analysis purposes, you must maintain the **Assign G/L Accounts to SD Condition Types** customizing activity in Central Finance.

In the customizing activity, you assign both statistical and nonstatistical condition types to the general ledger account. As a prerequisite, you must **Assign Ledger Group (Margin Analysis)** in the extension ledger. To get to this customizing activity, navigate from IMG customizing menu path **Financial Accounting • Central Finance • Central Finance: Target System Settings • Settings for Account Document Replication • SD Condition** Type **Mapping • Assign G/L Accounts to SD Condition Types**. You maintain this customizing activity for each company code. After reaching the configuration, you'll arrive at the **Change View "Account Determination for SD Conditions": Overview** screen. Click the **New Entries** button to reach the **New Entries: Overview of Added Entries** screen. On this screen, maintain values in the **CoCd**, **CnTy**, **G/L Account**, and **Offsetting Acct** fields, as shown in Figure 10.62 ❶. The **Offsetting Acct** field is populated with real statistical conditions to match the entries posted in the extension ledger. For nonstatistical conditions, leave this field empty. Click the **Save** ❷ button to save the entries.

If you want to override the default document type derived from the billing document with a different document type, you must maintain the document type defined in the **Define Document Type for Additional Document for SD Condition Type Split** configuration. This is especially useful if you want to isolate additional lines created for SD condition types for margin analysis. If you don't maintain an entry this configuration, the system uses the default document type from the billing document type for the additional lines created for condition types. To use this customizing activity, navigate from IMG customizing menu path **Financial Accounting • Central Finance • Central Finance: Target System Settings • Settings for Account Document Replication • SD Condition Type Mapping • Define Document Type for Additional Document for SD Condition Type Split**. After reaching the configuration, you'll arrive at the **Change View "Additional Document for SD Condition Type S": Overview** screen. Click the **New Entries** button to reach the **New Entries: Overview of Added Entries** screen. On this screen, maintain a document type in the **Doc Type for Add Doc** field as shown in Figure 10.62 ❸. Click the **Save** ❹ button to save the entries.

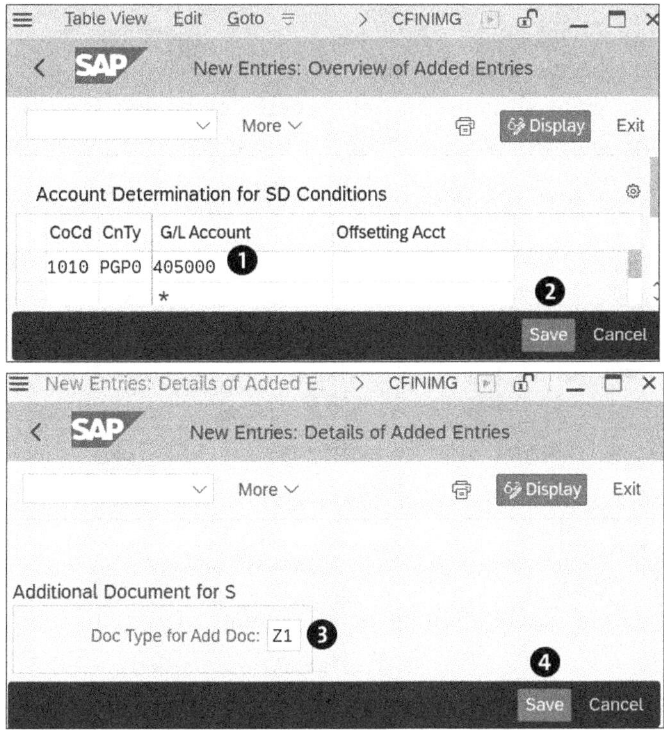

Figure 10.62 SD Conditions Configuration

10.9 Cost of Goods Sold Split

In SAP, the main goal of the COGS split is to obtain granular cost components to make better management decisions. When delivery documents are posted, cost components are posted based on the cost component structure definition of the material cost estimate. In costing-based profitability analysis, cost component split is updated to value fields based on the valuation strategy. On the other hand, the system posts cost components to different general ledger accounts in margin analysis.

In the Central Finance replication scenario, three use cases are possible based on where the COGS split occurs and the requirements of business. Figure 10.63 illustrates these three scenarios as explained here:

❶ **COGS split is carried out in the source system.**
COGS split is carried out and posted to a number of general ledger accounts in the source system. The system checks whether the **Define Accounts for Splitting the Cost of Goods Sold** configuration exists in the source system via customizing IMG

menu path **Financial Accounting • General Ledger Accounting • Periodic Processing • Integration • Materials Management**. If an entry exists, then the split is carried out in the source system. The split individual documents are then replicated to Central Finance. This means you won't maintain an entry in the **Define Accounts for Replication of Splitting the Cost of Goods Sold** configuration, located under Transaction CFINIMG. The system doesn't consider this configuration. As a prerequisite, you'll maintain a valid cost component structure in **Define Cost Component Structure** (Transaction OKTZ), release standard cost estimate for the material (Transaction CK11N or CK40N), and other relevant configuration in the source system.

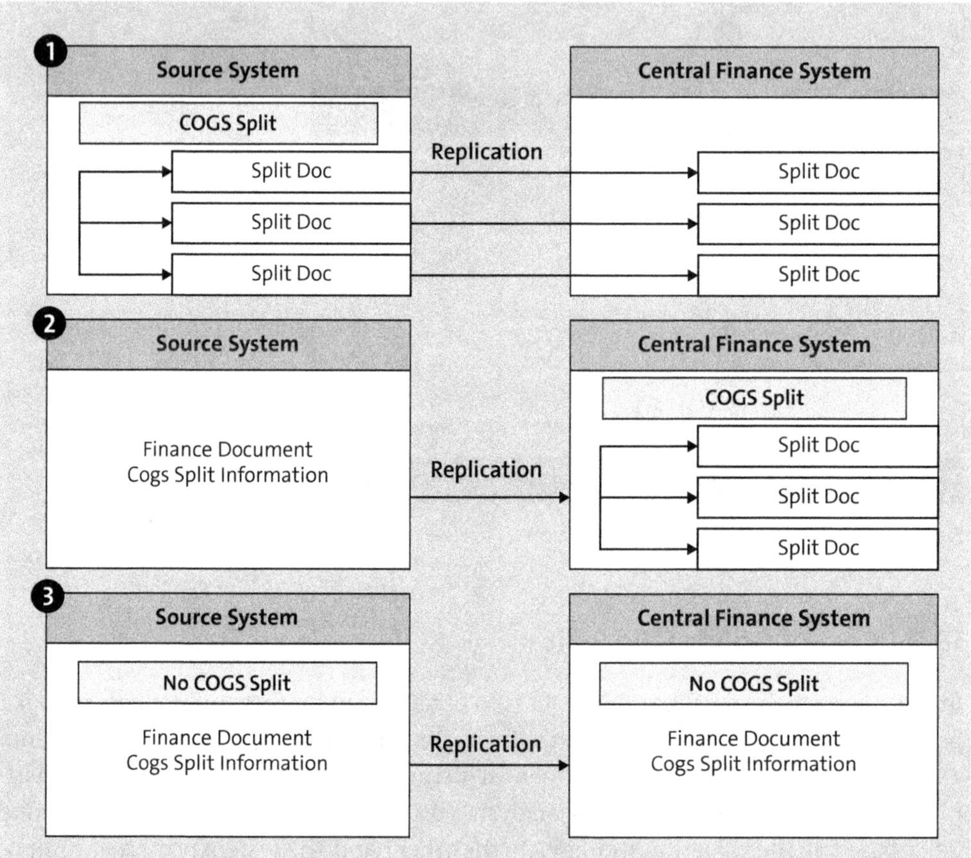

Figure 10.63 COGS Split Options

❷ **COGS split is carried out in the Central Finance system.**

If general ledger accounts are maintained in the **Define Accounts for Replication of Splitting the Cost of Goods Sold** configuration located under Transaction CFINIMG in Central Finance, the system transfers the COGS split information to Central Finance. The system is then split based on each company code, valuation view, and split general ledger account maintained in the Central Finance system. However, the system reads the COGS component structure defined in the source system. In addition, the system uses general ledger accounts and valuations configured in **Define Accounts for Replication of Splitting the Cost of Goods Sold** to transfer COGS split information from the source system to the Central Finance system. In Central Finance, the COGS split is possible at different valuation views, such as legal, group, and profit center.

❸ **COGS split is sent to the Central Finance system for goods movements based on sales orders.**
If no accounts are maintained in either **Define Accounts for Splitting the Cost of Goods Sold** configuration or **Define Accounts for Replication of Splitting the Cost of Goods Sold** configuration, the system simply transfers split information for goods movements for the legal view based on sales orders. In this case, no COGS split is carried out either in the source system or in the Central Finance system. The information is collected to track goods movements for legal valuation.

10

[!]

Warning
COGS split isn't supported in Central Finance initial loads. The functionality is available only for online replicated documents.

In the following sections, we'll discuss the configuration steps involved with defining accounts for COGS split in Central Finance for the scenario from Figure 10.63 ❷. Next, we'll cover how to enable simultaneous transfer of profitability analysis data and cost splits data in the source system.

10.9.1 Define Accounts for Replication of Splitting the Cost of Goods Sold

If the solution from Figure 10.63 ❷ is adopted, you'll define accounts for replication of splitting the COGS for combination of each company code, general ledger accounts, and valuation views in this configuration. Before you begin this configuration, ensure the following prerequisites are met in the source system and the Central Finance system:

Source System

To support cost of goods split, you'll need these three configurations from source system perspective:

- A valid cost component structure exists by navigating from customizing IMG menu path **Controlling • Product Cost Controlling • Product Cost Planning • Basic Settings for Material Costing • Define Cost Component Structure** or using Transaction OKTZ.
- Standard cost estimate is released for materials in scope (Transaction CK11N or Transaction CK40N).

- To transfer split information from the source system to the Central Finance system, ensure accounts are maintained for replication by navigating from customizing IMG menu path **Financial Accounting • Central Finance • Central Finance: Source System Settings) • Define Accounts for Replication of Splitting the Cost of Goods Sold**. The system considers this configuration to identify general ledger accounts and valuation views to replicate split and post to accounts maintained in Central Finance.

Central Finance System

In Central Finance, to define the accounts for replication of splitting COGS, navigate to IMG menu **SAP Customizing Implementation Guide • Financial Accounting • Central Finance: Source System Settings • Settings for Cost of Goods Sold/Profitability Analysis • Define Accounts for Replication of Splitting the Cost of Goods Sold** or navigate to it via Transaction CFINIMG. After reaching the configuration, you'll arrive at the **Change View "Account-based COGS split relevance": Overview** screen, as shown in Figure 10.64. On this screen, click the **New Entries** ❶ button. Next, on the resulting **New Entries: Overview of Added Entries** screen, maintain values for **Company Code**, **G/L Account**, and **Valuation View** ❷. If you leave the **Valuation View** column blank, the system considers all three valuation views for posting the COGS split. Click the **Save** ❸ button to save the configuration values.

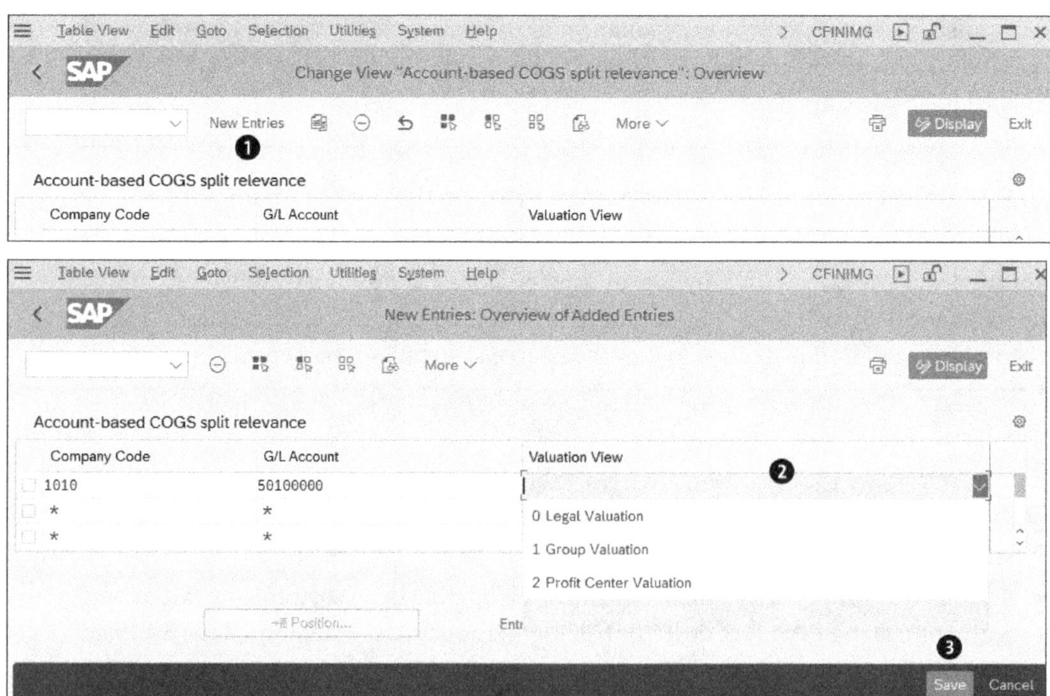

Figure 10.64 Define Accounts for Replication of Splitting the COGS

[«]

Note

SAP recommends applying SAP Note 2615469. Check whether this SAP Note is relevant for your source system.

10.9.2 Define Company Codes for Replication of Profitability Analysis Data

If the source system is activated with costing-based profitability analysis and Central Finance is configured with margin analysis, profitability analysis data is derived from the billing document and transferred to the Universal Journal. However, with this configuration, profitability analysis data can be replicated at the same time as COGS is split, and the split data is included as part of replication. This is called simultaneous replication of profitability analysis data and COGS split data. To achieve this functionality, the source system offers this configuration at the company code level. Having costing-based profitability active in the source system and having COGS split configured in the source system are the two prerequisites that must be met prior to enabling this functionality.

To reach configuration related to this functionality, navigate to IMG menu **SAP Customizing Implementation Guide • Financial Accounting • Central Finance: Source System Settings • Settings for Cost of Goods Sold/Profitability Analysis • Define Company Code for Replication of Profitability Analysis Data** or navigate to it via Transaction CFINIMG. After reaching the configuration, click **New Entries** on the **Change View "Define Company Code for Replication of Prof..": Overview** screen, as shown in Figure 10.65 ❶. In the resulting **New Entries: Overview of Added Entries** ❷ screen, maintain values for the **CoCd** field and then press [Enter] to populate **Company Name**. Click the **Save** ❸ icon to add the entry to the configuration table.

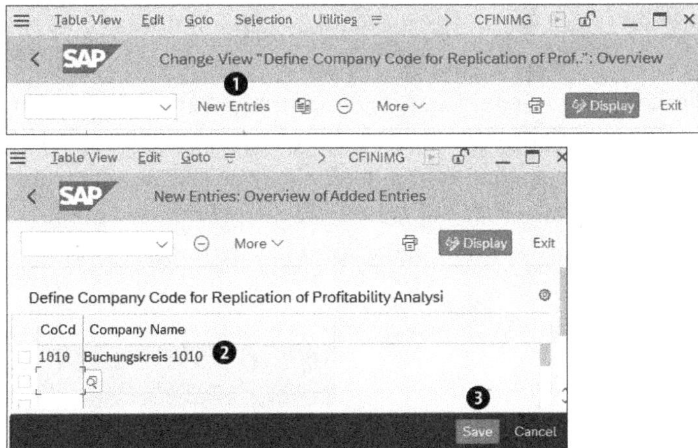

Figure 10.65 Define Company Code for Replication of Profitability Analysis Data

[»]

Note

A prerequisite for this functionality is that you've applied SAP Note 2962686.

10.10 Price Difference Split

In cost accounting, variances occur when comparing different cost buckets such as planned cost, budgeted cost, standard cost, and actual cost. In SAP cost object controlling, a standard cost estimate is prepared to set the standard price for a material or service item costing. This standard itemization cost is used to ascertain target costs for manufacturing orders. Several variance categories, for example, price, quantity, resource usage, and overhead variances, can be determined with the help of target cost and actual cost. Central Finance supports splitting of price differences based on the source system connected to the Central Finance system. The configuration requirements vary based on the source system. Figure 10.66 shows price difference splitting based on different source systems connected to Central Finance. Let's explore each source system scenario:

❶ **Price difference splitting in the Central Finance system**

In this case, the SAP ERP 6.0 source system enabled with costing-based profitability analysis is integrated with the Central Finance system. As you can see at the top of Figure 10.66, variances are calculated in costing-based profitability analysis in the source system prior to the actual settlement of the production order. The FI document of the production order settlement is then replicated to the Central Finance system. This FI document contains price split information based on variance categories. Upon replication of the FI document and variance information transferred from the source system, the system determines price differences splitting by applying variance categories from configuration **Define Accounts for Splitting Price Differences** in the Central Finance system. You can navigate to the IMG customizing activity from **Financial Accounting • General Ledger Accounting • Periodic Processing • Integration • Material Management • Define Accounts for Splitting Price Differences**. As a prerequisite, SAP recommends applying SAP Note 2633832 in the source system to support the price difference split. Table 10.12 displays supported manufacturing orders for the price difference split when the source system is SAP ERP.

Source System	Central Finance System Recommendation
Production order	Supported
Product cost collectors	Not supported

Table 10.12 Supported Scenario Orders for Price Difference Split

Source System	Central Finance System Recommendation
Production order mapped to the product cost collector in Central Finance	Supported

Table 10.12 Supported Scenario Orders for Price Difference Split (Cont.)

❷ **Price difference splitting in source system**
This scenario isn't common as SAP recommends margin analysis for SAP S/4HANA. In this case, SAP S/4HANA 1503 or higher release is set up as a source system, which is enabled with costing-based profitability analysis in some situations. As you can see from the bottom of Figure 10.66, the system determines price difference splitting by applying the configuration **Define Accounts for Splitting Price Differences** in the source system directly. The FI document of settlement of the production order already contains price difference split lines that are then replicated to the Central Finance system.

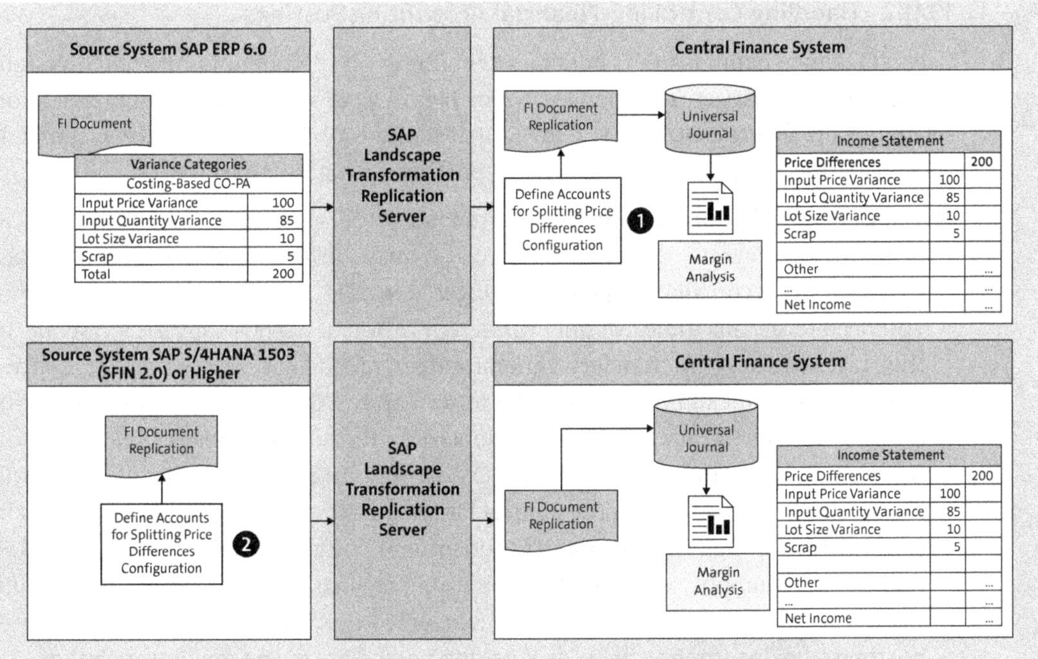

Figure 10.66 Pricing Difference Splitting in Central Finance

10.11 Additional Considerations

In this section, we'll discuss additional considerations that weren't covered in earlier sections.

10.11.1 Multiple Valuation Approaches

Central Finance doesn't support transfer pricing fully out of the box. However, transfer pricing requires additional evaluation such as how valuations and currencies are configured in source systems. Financial postings transfer all currency-related information from line items, for example, transaction currency (00), local currency (10), and group currency (30). If FI integrated currencies are configured along with currency types 12 and 32 for transfer pricing, these currencies are transferred via the initial load and online replication if they are configured similarly in Central Finance. In addition, runtime structure CFIN_ACCCR holds currency type 32, which is used to transfer currency type 32 to Central Finance. SAP recommends currency type 32 must also be configured in Central Finance just like the source system transfer pricing currency settings. You need to perform a proof of concept in a lower environment to confirm that the key processes aren't impacted. Consult SAP Note 2882025 on multiple valuations in SAP S/4HANA for additional guidance.

10.11.2 Handling Controlling-Financial Accounting Postings

Central Finance handles CO-FI postings depending on the type of SAP source system that is connected to Central Finance system. Figure 10.67 shows the general replication settings view **VCFIN_SOURCE_SET** for SAP ERP and SAP S/4HANA source systems. It shows how Central Finance handles CO-FI postings in both scenarios:

❶ The **GL Reconciliation Postings Transferred** indicator is shown in the general replication settings view VCFIN_SOURCE_SET, which is available only in SAP ERP. If you want to transfer FI reconciliation postings triggered in CO to the Central Finance system, you need to set this indicator only when replication of CO postings isn't in scope. In this case, the system transfers general ledger reconciliation postings to Central Finance via financial postings. On the other hand, if CO secondary postings are in scope, then general ledger reconciliations flow through CO postings. SAP recommends not activating this flag when CO postings are in scope. Otherwise, this will lead to duplicate postings in the Central Finance system. If CO-FI postings cleared in source which is triggered through a CO document, initial load of CO postings include this CO-FI document. However, the line items of such postings are not technically cleared in central finance. If a subsequent CO-FI posting tries to clear through FI postings, the document will fail as it couldn't associate to the previous postings.

❷ This screen shows the general replication settings view (VCFIN_SOURCE_SET-cofi_enabled is disabled from this view) for SAP S/4HANA as a source system. In this case, FI and CO are reconciled by the SAP S/4HANA architecture design, so reconciliation postings are no longer needed, and there is no such flag in the SAP S/4HANA system. In SAP S/4HANA, Transaction KALC is obsolete and the reconciliation ledger is no longer needed.

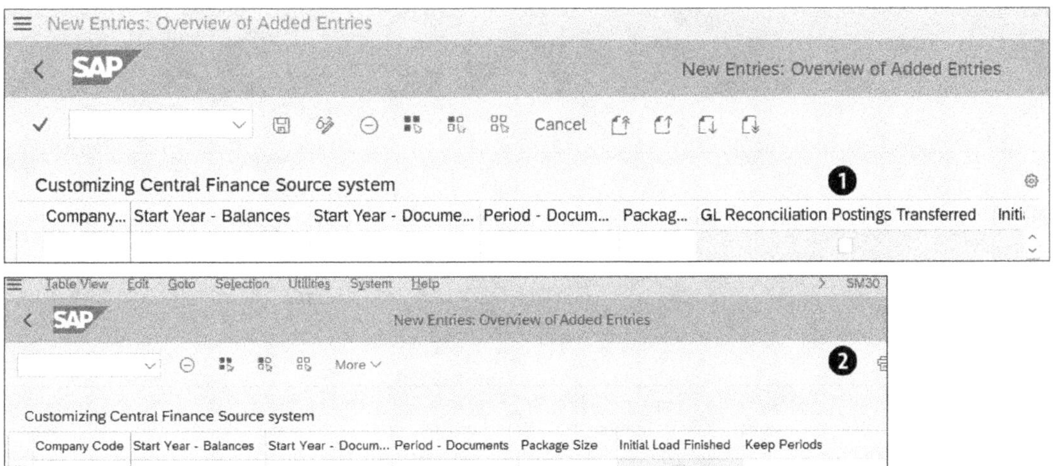

Figure 10.67 CO-FI Posting

10.12 Summary

When you explore SAP road maps, there are planned and future innovations for Central Finance in the CO space. SAP continuously introduces new features every major release cycle and through the release of SAP Notes periodically. As a result, you see more features in the CO space and other areas for Central Finance constantly. In this chapter, you've learned about several CO features supported in Central Finance. We began with project systems master data replication and then transitioned to other CO component replication functionalities. We reviewed supported functionality in margin analysis, which is often used in Central Finance implementations. Finally, we briefly looked into other important considerations in the CO space in the Central Finance context. In the next chapter, we'll discuss another important topic: accounting view of logistics (AVL).

Chapter 11
Accounting View of Logistics

Real-time visibility of both financial data and nonfinancial data is essential for improved reporting, quick decision analysis, and accelerated financial closing in Central Finance. With the accounting view of logistics (AVL) functionality offered in SAP Central Finance, you can bridge accounting data with logistics data to provide additional reporting capabilities that enhance decision-making processes.

Central Finance as a product evolved and expanded with additional innovations to meet the demands of complex landscapes connecting multiple source systems. The accounting view of logistics (AVL) solution offered with Central Finance replicates operational logistics data from the source systems to Central Finance. The replicated AVL data from operational tables in the source system resides on its own tables in Central Finance. Initially, Central Finance offered financial reporting capabilities by combining replicated data from multiple source enterprise resource planning (ERP) systems. Prior to the introduction of AVL in Central Finance, for any operational queries or reporting, you'd always go to the source systems. With the introduction of AVL in Central Finance from SAP S/4HANA 1809 FPS02 release onwards, SAP offered a subset of logistics data replicated to Central Finance without creating logistics-relevant configurations. In other words, AVL data replicated from the source system to Central Finance doesn't check logistics-related customizing in the Central Finance system. A subset of logistics data is merely transferred to Central Finance rather than reposting.

In this chapter, you'll learn the concepts and architecture of the AVL solution offered with Central Finance. Furthermore, we'll walk you through configuration requirements in the source system, SAP Landscape Transformation Replication Server, and the Central Finance system. You'll also learn about replicating AVL from a third-party system to Central Finance.

The most important use cases of AVL data in Central Finance are the following:

- Provide operational reporting capabilities in real time in Central Finance to reflect operational activities in the source system.
- Provide business users with additional reporting capabilities by combining both accounting data and logistics data to provide a comprehensive view of the Central Finance system.

- With availability of logistics data in Central Finance, you can perform better cost management by looking into granular details, enabling allocations to be done accurately.
- Support of accruals (supported from SAP S/4HANA 1909 onward) and deferrals (from SAP S/4HANA 2022 onward) is provided.
- Support is provided for the reporting needs of local statutory requirements especially to support tax compliance. The AVL solution, along with SAP Document and Reporting Compliance, may fulfill certain mandatory country-specific requirements.
- Support is provided for period-end reporting and fulfilling audit requirements.
- With combined data from financial accounting (FI) and logistics, you can perform predictive accounting in Central Finance.
- With logistics data in Central Finance, you can plan and forecast data accurately.
- Pricing data replication helps to perform profitability analysis and aids in better decision-making.
- Run ad hoc reporting for day-to-day business activities.
- Analyze and report logistics data from third-party systems.
- Accelerate financial closing by automating certain tasks based on the availability of logistics data. A faster period-end or year-end close process validation with additional logistics data is possible.
- Enhanced decision-making is a result of available comprehensive data from accounting and logistics.
- Leverage SAP S/4HANA reporting tools such as SAP Analytics Cloud to combine accounting and logistics data.
- Now that you understood the concept of AVL and use cases, let's understand how data flows between the source system and Central Finance next.

11.1 Getting Started with Accounting View of Logistics

Central Finance supports replicating a subset of logistics information from the source systems to Central Finance. The scope of logistics data includes data from the following:

- **Sales orders, including pricing document**
  Primarily used to support local and statutory reporting requirements. If costing-based profitability analysis data is used in sales documents, the data is included as part of AVL replication. On the other hand, with customer invoices, profitability analysis is included with financial (FI) and controlling (CO) replication.

- **Customer invoices, including pricing documents**
  In conjunction with sales order information, customer invoices are used to support local and statutory requirements. In addition, the replicated data used in SAP Document and Reporting Compliance to support reporting requirements in certain countries.
- **Purchasing documents**
  In addition to ad hoc reporting, the replicated subset of purchase orders is used in managing central accrual management in Central Finance.
- **Supplier invoices**
  The replicated supplier invoices are used in conjunction with SAP Document and Reporting Compliance to fulfill statutory and local reporting requirements.

Central Finance brings logistics data from different tables and field sets from the source systems. Table 11.1 lists replication objects, SAP Landscape Transformation Replication Server lead tables, dependent tables, and corresponding Central Finance tables.

AVL Replication Objects	SAP Landscape Transformation Replication Server Lead Table	Source-Dependent Tables	Corresponding Central Finance Table	Conditions
Sales Document	VBAK	ADRC	CFIN_AV_SO_ADRMP	One-time customer's address where VBPA-XCPDK = 'X'
		VBKD	CFIN_AV_SO_BIZ	
		VEDA	CFIN_AV_SO_CTR	
		VBAP	CFIN_AV_SO_ITEM	
		VBPA	CFIN_AV_SO_PART	
		VBAK		Where VBAK-VBTYP in : – service order C – Order H – Returns I – Order without change K – Credit memo request L – Debit memo request
		VBEP	CFIN_AV_SO_SLN	
		KONV/PRCD_ELEMENTS	CFIN_AV_PRC_ELEM	

Table 11.1 AVL Tables

AVL Replication Objects	SAP Landscape Transformation Replication Server Lead Table	Source-Dependent Tables	Corresponding Central Finance Table	Conditions
Customer Invoice	VBRK	ADRC	CFIN_AV_CI_ADRMP	One-time customer's address where VBPA-XCPDK = 'X'
		VBRP	CFIN_AV_CI_ITEM	
		VBPA	CFIN_AV_CI_PART	
		VBRK	CFIN_AV_CI_ROOT	
		KONV/PRCD_ELEMENTS	CFIN_AV_PRC_ELEM	
Purchase Document	EKKO	EKKN	CFIN_AV_PO_ACC	
		EKPO	CFIN_AV_PO_ITEM	
		EKBE	CFIN_AV_PO_RO	Where EKBE-VGABE in 1 – Goods Receipt 2 – Invoice Receipt 3 – Subseq. debit/credit '8' – Deliver (stock transfer) 'P' – Invoice parking
		EKBE_MA	CFIN_AV_PO_ROACC	Where EKBE_MA-VGABE in 1 – Goods receipt 2 – Invoice receipt 3 – Subseq. debit/credit 8 – Deliver (stock transfer) P – Invoice parking
		EKKO	CFIN_AV_PO_ROOT	
		EKET	CFIN_AV_PO_SCH	
Supplier Invoice	RBKP	RBCO	CFIN_AV_SI_ACCAS	Where RBCO-KOART = ' '
		RBCO	CFIN_AV_SI_GLACC	Where RBCO-KOART = 'S' – General ledger account
		RSEG	CFIN_AV_SI_ITEM	
		RBKP	CFIN_AV_SI_ROOT	

Table 11.1 AVL Tables (Cont.)

[+]

Tip

For a complete list of fields transferred, consult SAP Note 3220450, which provides a Microsoft Excel attachment with a list of fields for each AVL replication object.

11.1.1 Replicating Accounting View of Logistics Data

AVL data offers customers an option to perform operational reporting in Central Finance. In addition, with integration with SAP Document and Reporting Compliance, you can fulfill statutory and regulatory reporting requirements. To transfer a subset of logistics data from the source system to Central Finance, you'll need to understand data flow and components involved with AVL. Figure 11.1 illustrates the flow of logistics data replication from the source systems to the Central Finance system.

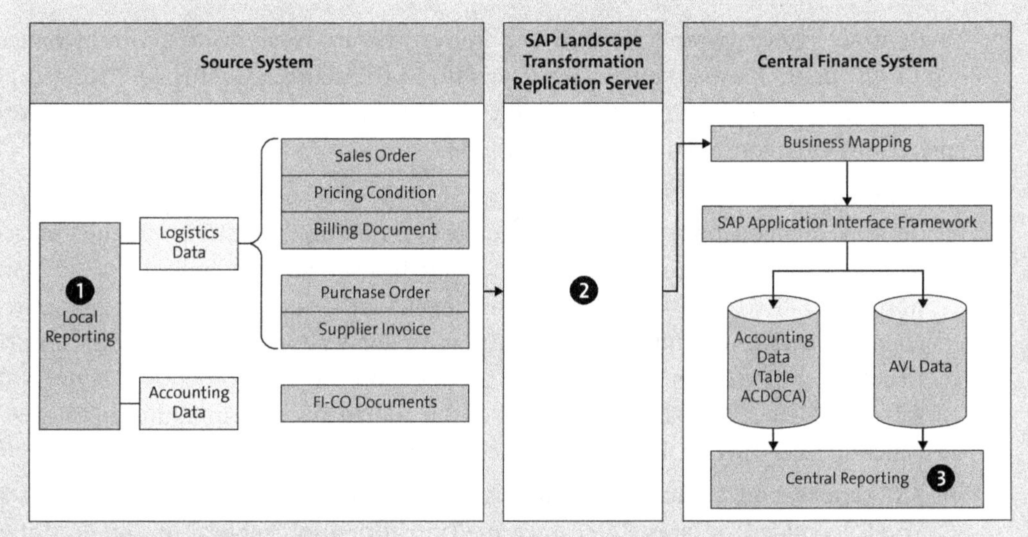

Figure 11.1 Accounting View of Logistics Overview

Let's explore each component next:

❶ **Source system reporting**
In a Central Finance landscape, day-to-day business activities are carried out in the source system. Both logistics data and accounting data are available to perform local reporting in the source systems.

❷ **SAP Landscape Transformation Replication Server**
This is used to carry out both the initial load and online replication of AVL data from the source system to the Central Finance system. SAP recommends downloading SAP Landscape Transformation Replication Server content to ensure you've got the most up-to-date content in SAP Landscape Transformation Replication Server for AVL.

❸ **Central reporting**
AVL data goes through business mapping and SAP Application Interface Framework for error correction and handling in Central Finance. Both accounting data (table ACDOCA) and logistics data (from AVL tables) are available in Central Finance for multiple use cases highlighted earlier.

[»]

Note

The architecture diagram displayed here assumes SAP as a source system. Central Finance supports replicating logistics data from a third-party system. SAP Landscape Transformation Replication Server comes with staging tables to support AVL data replication.

11.1.2 Prerequisites and Restrictions

In Central Finance, to replicate AVL data from the source systems to Central Finance, you must maintain the required prerequisites in the source systems, SAP Landscape Transformation Replication Server, and Central Finance. In this section, let's explore prerequisites to consider for AVL replication:

- **Source system**
  Central Finance supports AVL replication only on SAP ERP 6.0 EHP3 or higher release level as a source system. In addition, for purchase document replication, SAP supports SAP ERP 6.0 EHP4 or higher release level. Furthermore, SAP allows one source client for each SAP Landscape Transformation Replication Server configuration. To make the source systems ready for AVL replication, SAP recommends applying SAP Notes 2679506, 2647022, and 2735852 in the source systems In addition, business function LOG_ESOA_OPS_2 must be activated if applicable.
- **SAP Landscape Transformation Replication Server**
  You'll download the latest SAP Landscape Transformation Replication Server configuration content before configuration of SAP Landscape Transformation Replication Server configuration tables. Consult SAP Note 3013312 on SAP Landscape Transformation Replication Server configuration content.
- **Central Finance system**
  You'll maintain necessary business mapping using Transaction FINS_CFIN_MAP_MANAGE in Central Finance for cost objects, projects, and costing-based profitability analysis.

[»]

Note

Consult SAP Note 3013315 for a comprehensive overview on installation, prerequisite, and configuration guidance for AVL.

11.2 Configuring Accounting View of Logistics

In this section, you'll learn the configuration requirements for AVL with SAP's recommended sequence:

1. Source system configuration
2. Central Finance configuration
3. SAP Landscape Transformation Replication Server

Let's explore each configuration next in the order specified.

11.2.1 Source System

First, replication of changes made to sales orders requires business function LOG_ESOA_OPS_2 to be activated. This business function was released with SAP ERP 6.0 EHP3 (SAP_APPL_603). To activate the business function, navigate to **SAP Customizing Implementation Guide • Activate Business Functions** or use Transaction SFW5. After reaching the configuration, the system displays the **Security Information** popup warning screen, as shown at the top of Figure 11.2.

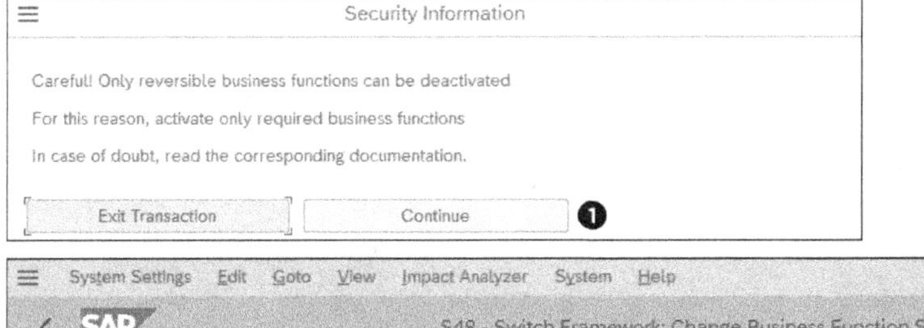

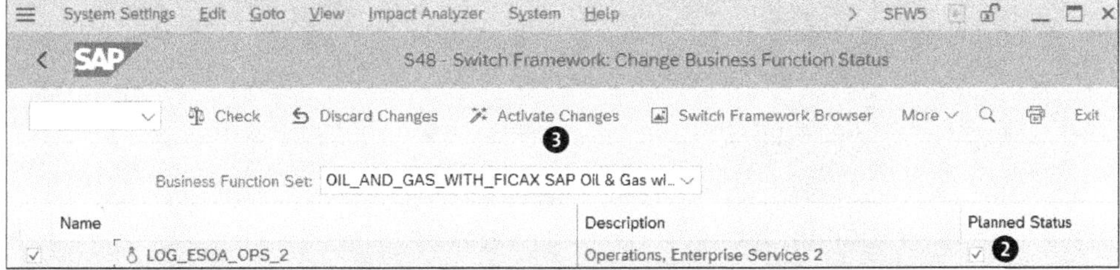

Figure 11.2 Activate Business Function LOG_ESOA_OPS_2

Click the **Continue** ❶ button, which brings up the **S48: Switch Framework: Change Business Function Status** screen. On this screen, locate the business function to check whether business function LOG_ESOA_OPS_2 is activated. If the business function is already active, no action is required. If the business function isn't activated, highlight the business, select the **Planned Status** ❷ checkbox, and then click the **Activate Changes** ❸ button. The system brings up the activation window to submit the task in the background. Upon submission, the system displays the message **Job SFW_ACTIVATE_SFOX was started in the background** in a popup window. After successful activation, the **Planned Status** column displays **Business func. will remain activated.** Note the business function ESOA_OPS01 is a prerequisite to activating business function LOG_ESOA_OPS_2.

Besides activating the business functions in the source system, SAP recommends implementing certain SAP Notes as highlighted in Table 11.2.

Source System SAP Notes	Purpose
2647022 - Central Finance (source) Accounting View - SAP Landscape Transformation Replication Server configuration "Application Select"	To support AVL features, corresponding SAP Landscape Transformation Replication Server configuration content makes use of SAP Landscape Transformation Replication Server configuration "Application Select" logic, which requires code changes in the source system. As a result, SAP released this Note, which needs to be implemented in both source and Central Finance system.
2679506 - Central Finance (source) Accounting View - WBS Element in Costing-Based Profitability Analysis No Conversion	If a work breakdown structure (WBS) is used as a characteristic in costing-based profitability analysis, SAP recommends this SAP Note be applied in both the source and Central Finance systems to bring AVL data for sales orders that contain costing-based profitability analysis characteristics.
2735582 - Accounting View - Replication of Purchase Order (Source System)	This SAP Note, along with several prerequisite SAP Notes, enables replication of a subset of purchase order information from the source systems to the Central Finance system.

Table 11.2 SAP Recommended Prerequisite SAP Notes in the Source System

11.2.2 Configuration in the Central Finance System

In Central Finance, you must set up the following configurations to support logistics data replication from the source system to the Central Finance system:

- **SAP Application Interface Framework content deployment**
  This is required to monitor AVL in Central Finance.
- **Business mapping**
  This is needed based on business requirements.

SAP Application Interface Framework Content Deployment

SAP Application Interface Framework is used to monitor and handle errors for replicating AVL data from the source system to the Central Finance system. As shown in Table

11.3, SAP provides predelivered SAP Application Interface Framework content to support AVL deployment scenarios for each AVL replication object.

SAP Application Interface Framework Interface	Deployment Scenario	Description
AV_SO	FINCF_AV_SO	Sales order (including pricing documents)
AV_CI	FINCF_AV_CI	Customer invoice (including pricing documents)
AV_SI	FINCF_AV_SI	Supplier invoice
AV_PO	FINCF_AV_PO	Purchasing document

Table 11.3 SAP Application Interface Framework Deployment Scenarios for AVL

You'll extract SAP Application Interface Framework content by executing Transaction /AIF/CONTENT_EXTRACT. Upon entering the transaction, on the **AIF Content Transport - Deploy** screen, for example, enter "FINCF_AV_SO" ❶ for the sales order **Deployment Scenario ID** field, as shown in Figure 11.3. Select further parameters ❷ to record registration and transport recording ❸, respectively. Click the **Execute** ❹ icon to extract the content. Once the content extraction is successful ❺, you'll see interface AV_SO available under namespace /FINCF in Transaction /AIF/ERR monitoring and error handling.

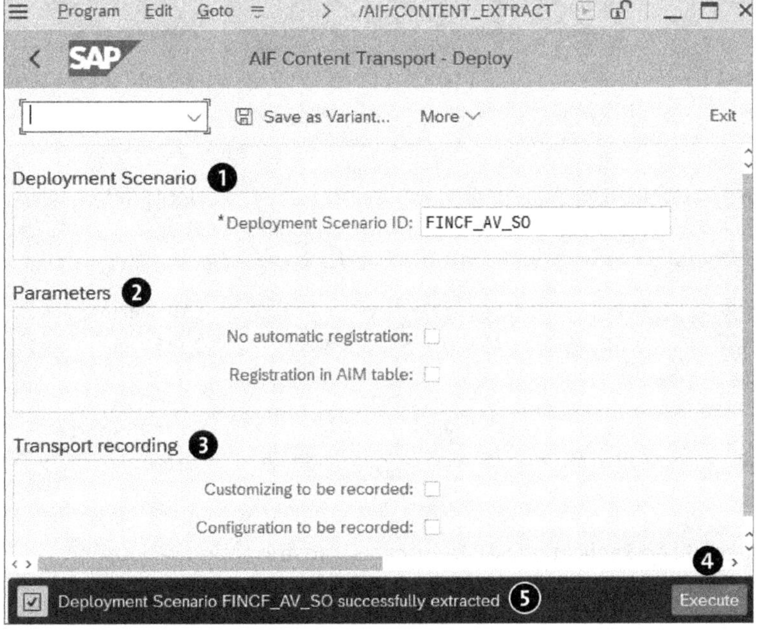

Figure 11.3 SAP Application Interface Framework Content Extraction for AVL

Repeat the SAP Application Interface Framework content activation steps for other AVL deployment scenarios listed in Table 11.3.

Business Mapping

Depending on your business requirements, you'll also maintain necessary mapping of master data in Central Finance. For example, if AVL data in the source system uses order number (AUFNR) or project/WBS, then you'll maintain cost object mapping and WBS (PS_PSP_PNR) mapping, respectively, in Central Finance. If a profitability segment is used in AVL data, then you'll define costing-based profitability analysis mapping as well.

11.2.3 SAP Landscape Transformation Replication Server Configuration

Prior to starting AVL replication, you must create the required configurations in SAP Landscape Transformation Replication Server configuration. You'll use an existing mass transfer ID (MT ID) you've previously created or create a new MT ID based on the project requirements. If only one source client is connected to the Central Finance system, you can use the existing MT ID. Again, this depends on your project requirements. However, SAP recommends creating a dedicated MT ID for each source client if multiple source clients are connected to the Central Finance system.

Table 11.4 shows a comprehensive list of predefined and replication SAP Landscape Transformation Replication Server configuration objects for AVL objects.

Replication Object	Table	Project	Subproject	Replication Object	Load Object	Source System Version
Sales document	VBAK	REPL_CFIN	REPL_CFIN_AS	CFI_VBAK_AS_R	CFI_VBAK_AS_L	SAP ERP source system
Sales document	VBAK	REPL_CFIN	REPL_CFIN_AS	CFI_VBAK_S4_AS_R	CFI_VBAK_S4_AS_L	Until SAP S/4HANA 2022
Sales document	VBAK	REPL_CFIN	REPL_CFIN_AS	CFI_VBAK_AS300_R	CFI_VBAK_AS300_L	SAP S/4HANA 2023
Customer invoice	VBRK	REPL_CFIN	REPL_CFIN_AS	CFI_VBRK_AS_R	CFI_VBRK_AS_L	SAP ERP source system

Table 11.4 Predefined Load and Replication SAP Landscape Transformation Replication Server Configuration Objects for AVL Objects

Replication Object	Table	Project	Subproject	Replication Object	Load Object	Source System Version
Customer invoice	VBRK	REPL_CFIN	REPL_CFIN_AS	CFI_VBRK_S4_AS_R	CFI_VBRK_S4_AS_L	Until SAP S/4HANA 2022
Customer invoice	VBRK	REPL_CFIN	REPL_CFIN_AS	CFI_VBRK_AS300_R	CFI_VBRK_AS300_L	SAP S/4HANA 2023
Purchasing document	EKKO	REPL_CFIN	REPL_CFIN_AS	CFI_EKKO_AS_R	CFI_EKKO_AS_L	SAP ERP and SAP S/4HANA source system
Supplier invoice	RBKP	REPL_CFIN	REPL_CFIN_AS	CFI_RBKP_AS_R	CFI_RBKP_AS_L	SAP ERP and SAP S/4HANA source system

Table 11.4 Predefined Load and Replication SAP Landscape Transformation Replication Server Configuration Objects for AVL Objects (Cont.)

[+]

Tip

Refer to SAP Note 3013315 for additional guidance and the latest information on predefined load/replication objects and release versions for various AVL deployment scenarios.

In this section, you'll learn important configuration for both sales- and purchasing-related logistics data. We'll begin with sales orders and customer invoices first, and then transition to the purchasing side.

Sales Document Initial Load and Replication

After creating the MT ID for loading/replicating AVL data from the source system to the Central Finance target system, you must set up the appropriate **Reading Type** under **Initial Load Options** for sales order table VBAK replication in Transaction LTRS. After opening the transaction, choose the MT ID you've created earlier, and double-click the MT ID. This will populate the **Mass Transfer** ID under **Advance Replication Settings** on the left pane. Right-click on **Performance Options**, and choose **Add Object** to add table **VBAK** under **Advanced Replication Settings**, as shown in Figure 11.4 ❶. Choose **1 Range Calculation** from the **Reading Type** ❷ dropdown. Click the **Save** ❸ button to save the configuration.

[+]

Tip

For AVL, SAP recommends choosing **Application Select** as the reading type to retrieve additional posting data during SAP Landscape Transformation Replication Server configuration. SAP doesn't recommend setting up reading type 4 or 5. Default reading type **5 Sender Queue** causes problems during the initial load. In general, reading type 5 is used for transparent table replication.

SAP instead recommends setting reading type **1 Range Calculation**, which can be found in Transaction LTRS under **Performance Options**, as it's resource-optimized.

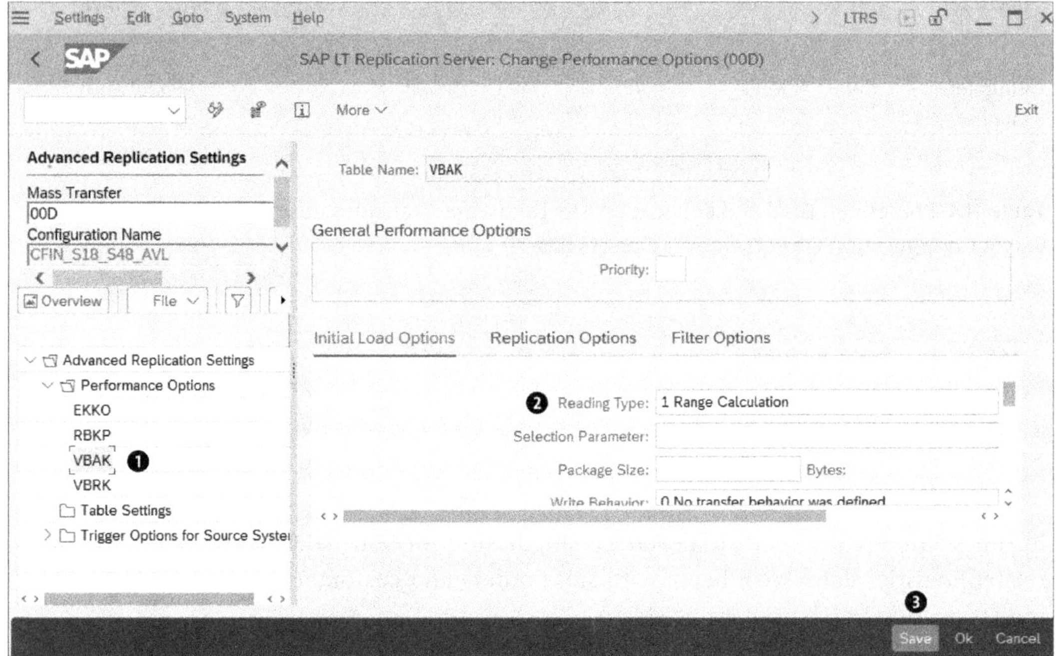

Figure 11.4 SAP Landscape Transformation Replication Server configuration Reading Type Setup for the Initial Load

Next, you'll create predefined load and replication objects for table VBAK from project REPL_CFIN and subproject REPL_CFIN_AS. To perform this task, choose **Copy Predefined Object** by starting program IUUC_REPL_PREDEF_OBJECTS by selecting the MT ID via Transaction SE38. After entering the program, on the resulting screen, enter values "REPL_CFIN" and "REPL_CFIN_AS" in the **Project** and **Subproject** fields, respectively. Choose **Predefined Object** for **Replication Object** and **Load Object** based on the source system version, as shown in Table 11.5.

Object	Table	Project	Subproject	Replication Object	Load Object	Source System Version
Sales document	VBAK	REPL_CFIN	REPL_CFIN_AS	CFI_VBAK_AS_R	CFI_VBAK_AS_L	SAP ERP source system
Sales document	VBAK	REPL_CFIN	REPL_CFIN_AS	CFI_VBAK_S4_AS_R	CFI_VBAK_S4_AS_L	Until SAP S/4HANA 2022
Sales document	VBAK	REPL_CFIN	REPL_CFIN_AS	CFI_VBAK_AS300_R	CFI_VBAK_AS300_L	SAP S/4HANA 2023

Table 11.5 Predefined Load and Replication Objects for Sales Documents

After successful creation of the predefined load and replication objects, you can modify or add additional filter criteria using Transaction MWB for AVL replication. Using this transaction, you'll create event-based and field-based rules. SAP-delivered SKIP_VBAK is an event-based rule to support and apply filters for AVL specifically for sales order replication. If you want to replicate/load sales documents with specific filter criteria such as sales organization or company code, then you can add the filter rules in the SKIP_VBAK event-based rule. As you've observed from Figure 11.5, in our example, we've added filter rule on the BUKRS_VK ❶ field to load and transfer sales documents from the source system to the Central Finance system for company code 1010. Once filter rules are added in the SKIP_VBAK event-based rule, click the **Save** ❷ button to save the rule.

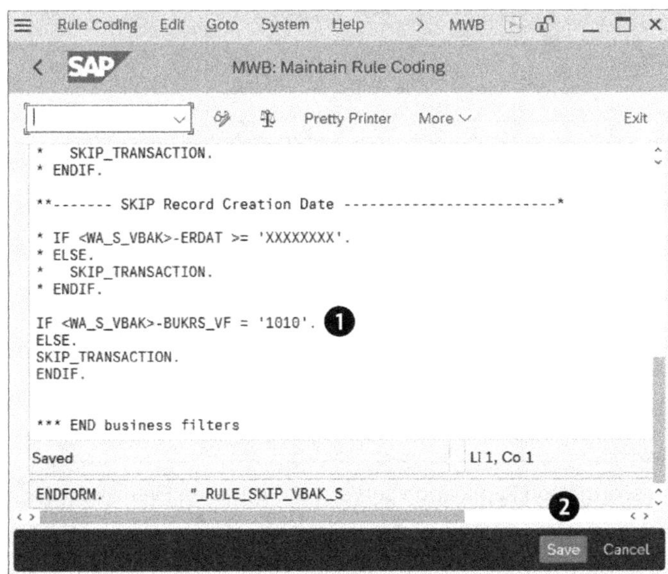

Figure 11.5 Transaction MWB Filter Rule: Company Code

[!]

Warning

Inconsistencies can happen if changeable fields are used as filters at event-based rules. If you set the filter using sales organization VBORG with a value XXXX, and then you change the sales organization value to YYYY, the system won't replicate the change to the Central Finance system. You must carefully evaluate the business requirements and add the required filter criteria. Changing the filter criteria could impact data consistency between the source system and the Central Finance system.

Next, within the event-based rule, SAP predelivered commented code. By default, SAP doesn't allow archived and deleted data to be replicated to the Central Finance system. If you have a requirement to transfer archived or deleted AVL data, you'll activate code to enable the transfer of archived/deleted data, as shown in Figure 11.6. If some documents don't replicate to the Central Finance system, then it can produce errors in the Central Finance system when deletion happens. Therefore, you must carefully review the requirements and update the code accordingly.

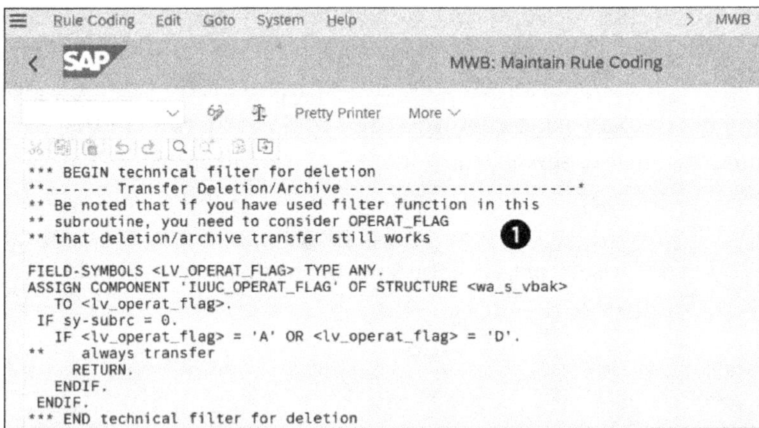

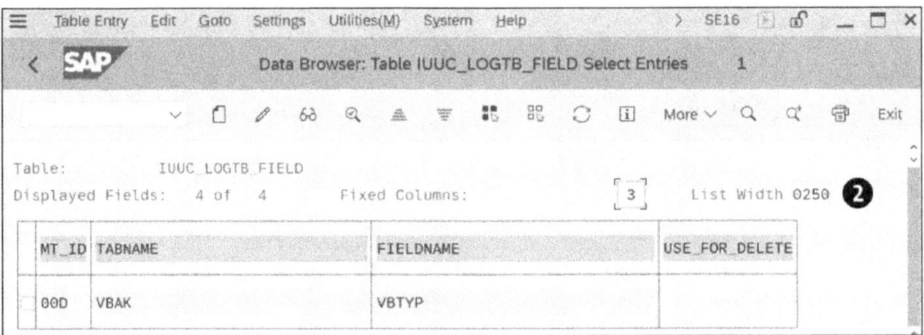

Figure 11.6 SAP Landscape Transformation Replication Server Configuration Event-Based Filter Rule for AVL

[+]

Tip

Activating this code will always bypass the filter and transfer deleted/archived data, whereas the downside of this approach is if some documents never reach the Central Finance system, producing errors in the Central Finance system when deletion happens.

Logistics data can be set with a deletion flag or archived in the source system. However, if there is a requirement to replicate deleted data or archived data from the source system to the Central Finance system, then you'll add the nonkey field VBTYP to the source logging table. To add the nonkey field to the logging table, you need to add a nonkey field for the table in table IUUC_LOGTB_FIELD in the SAP Landscape Transformation Replication Server, as shown in Figure 11.6 ❶.

Next, go to Transaction SE16, and create an entry in table IUUC_LOGTB_FIELD with **MT_ID**, **TABNAME**, and **FIELDNAME** ❷ for nonkey field **VBTYP**.

Figure 11.7 shows the successful addition of the nonkey field **VBTYP** to the source logging table created for SAP Landscape Transformation Replication Server configuration table VBAK.

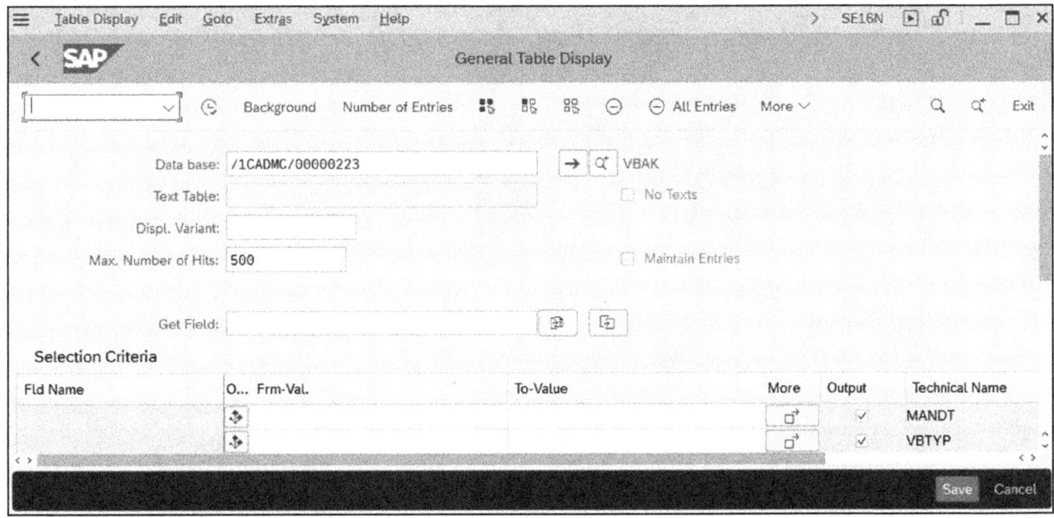

Figure 11.7 AVL Logging Table for Table VBTYP

[+]

Tip

In case of deletion, a record is deleted completely from source table VBAK, but nonkey fields still have the value of VBTYP, which can ensure that the sales document isn't filtered in the Central Finance system.

After performing the end-to-end setup at SAP Landscape Transformation Replication Server configuration to load/replicate the sales document from the source system to the target Central Finance system, you'll add table VBAK in Transaction LTRC under the previously created MT ID.

To start the initial load replication of sales documents from table VBAK, choose the mass transfer id by clicking on MT ID you've just created from Transaction LTRC. On the resulting screen,, on the **Participating Object** tab, choose **Data provisioning** to start the load. Populate all fields on this screen for the VBAK load and click **Continue**. Upon returning to the previous screen, click the **Refresh** button on the **Participating Object** list to see the replication status. Sales documents with valid document categories (VBTYP) can only be loaded/replicated to the Central Finance system. Allowed sales document categories (VBAK-VBTYP) are maintained in table CFIN_AV_SO_CAT in the Central Finance system. Furthermore, successfully loaded/replicated sales documents can be checked on SAP Application Interface Framework with interface **AV_SO** under namespace **/FINCF** in the Central Finance system, as shown in Figure 11.8.

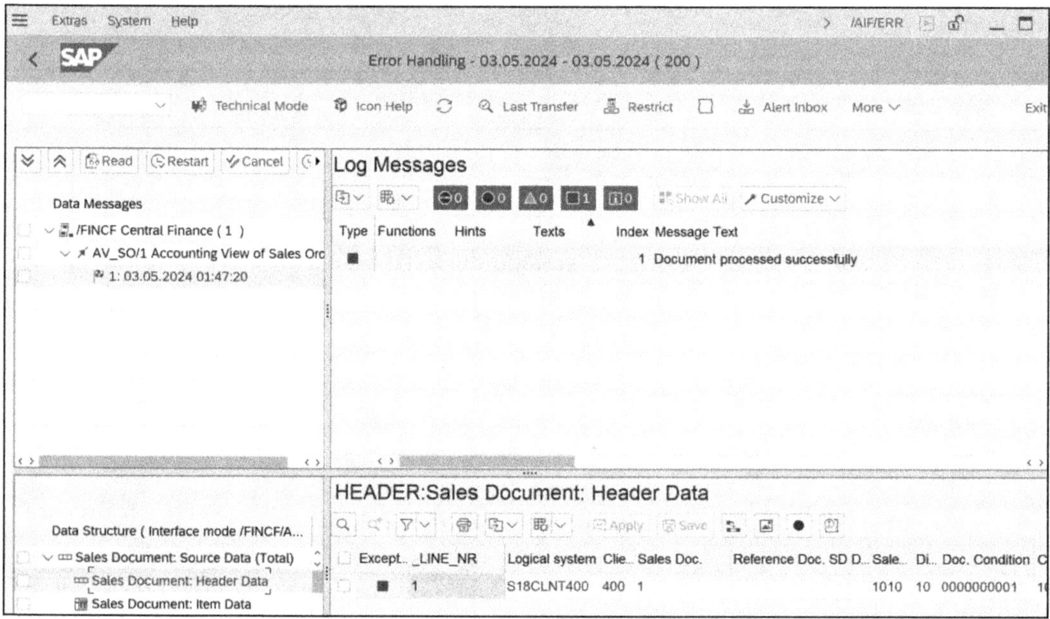

Figure 11.8 SAP Application Interface Framework for Sales Orders

Table 11.6 displays a list of tables updated for sales order tables in the Central Finance system.

Table	Description
CFIN_AV_SO_ROOT	Sales Document: Header Data
CFIN_AV_SO_ITEM	Sales Document: Item Data
CFIN_AV_SO_SLN	Sales Document: Schedule Line Data
CFIN_AV_SO_PART	Sales Document: Partner
CFIN_AV_SO_BIZ	Sales Document: Business Data

Table 11.6 Sales Order Tables for Sales Order Replication

Customer Invoice Initial Load and Replication

In this section, you'll learn about customer invoice replication from the source system to the Central Finance system. We'll use the same MT ID **00D** that you've used for sales document replication. Likewise, in sales document replication, we'll use a **1 Range Calculation** table for invoice table VBRK replication located in **Performance Options** under **Advanced Replication Settings** of Transaction LTRS. In addition, SAP predefined load and replication objects are generated for customer invoice table VBRK under project REPL_CFIN and subproject REPL_CFIN_AS. As shown in Table 11.7, you'll pick the load and replication objects based on your source system release.

Replication object	Table	Project	Subproject	Replication Object	Load Object	Source System
Customer invoice	VBRK	REPL_CFIN	REPL_CFIN_AS	CFI_VBRK_AS_R	CFI_VBRK_AS_L	SAP ERP source system
Customer invoice	VBRK	REPL_CFIN	REPL_CFIN_AS	CFI_VBRK_S4_AS_R	CFI_VBRK_S4_AS_L	Until SAP S/4HANA 2022
Customer invoice	VBRK	REPL_CFIN	REPL_CFIN_AS	CFI_VBRK_AS300_R	CFI_VBRK_AS300_L	SAP S/4HANA 2023

Table 11.7 Predefined Load and Replication Objects for Customer AVL Data

Like sales document replication filters, you can also apply filters to load and replicate invoice data to Central Finance by writing a rule in the SKIP_VBRK event-based rule in Transaction MWB.

After adding table VBRK in Transaction LTRC, you can choose **Start Replication** (including the initial load) on the **Participating Object** tab under the **Data Provisioning** option. Upon completion of the initial load, SAP Landscape Transformation Replication Server configuration will start reading the logging table in the source system and replicate the customer invoice to the target Central Finance system. Upon successfully transferring the customer invoice data, the system updates billing staging tables in Central Finance, as shown in Table 11.8.

Table	Description
CFIN_AV_CI_ROOT	Billing Document: Header Data
CFIN_AV_CI_ITEM	Billing Document: Item Data
CFIN_AV_CI_PART	Billing Document: Partner

Table 11.8 AVL Customer Invoice Tables

In addition, you can also check the replicated customer invoice documents in interface **AV_CI** under namespace **/FINCF** and other selection criteria in SAP Application Interface Framework, as shown in Figure 11.9.

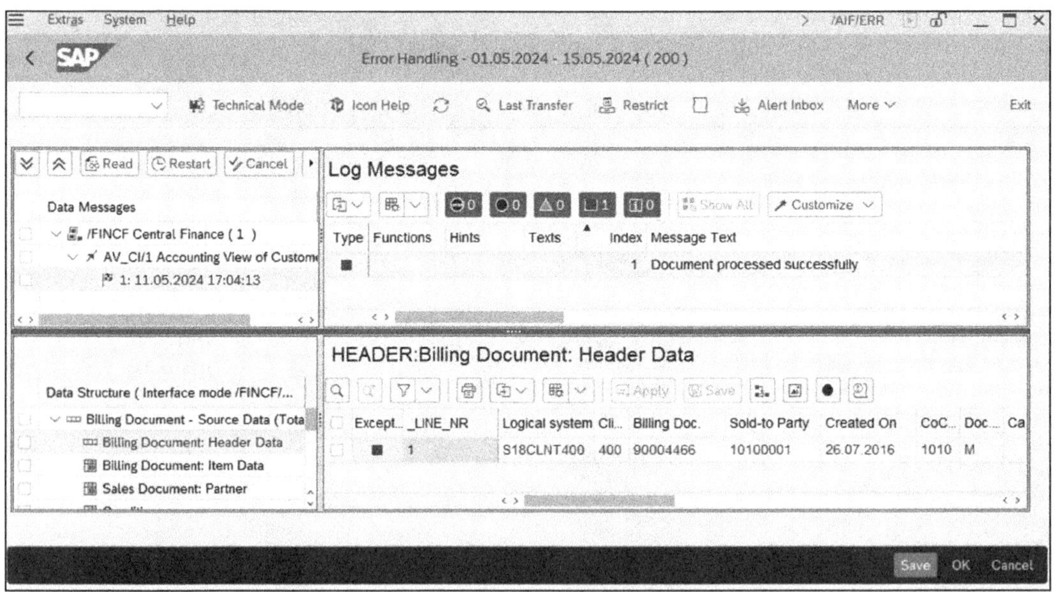

Figure 11.9 SAP Application Interface Framework for Customer Invoices

Purchasing Document Initial Load and Replication

In this section, you'll learn about purchase document replication as part of the Central Finance AVL functionality. Unlike sales documents, configuration and setup of purchase document replication has two parts:

- **Source to Central Finance**
  Transfers the purchase document data.
- **Source to source**
  Transfers any changes in purchase document line items or any other child nodes.

We'll discuss each configuration with an example next.

Source to Central Finance

In this approach, AVL objects are configured the same way you've configured other objects in SAP Landscape Transformation Replication Server configuration as discussed earlier. We'll use the same MT ID **00D** that you've used for sales documents and customer invoice replications. Likewise, in sales document replication, we'll use **1 Range Calculation** table for the purchase document table EKKO replication located in **Performance Options** under **Advanced Replication Settings** of Transaction LTRS. In addition, SAP delivered and predefined load and replication objects are generated for purchase document table EKKO under project REPL_CFIN and subproject REPL_CFIN_AS. The table displays the load and replication objects for purchase document replication. As shown in Table 11.9, purchase document replication uses the same load and replication objects for the SAP ERP and SAP S/4HANA source systems.

11

Object	Table	Project	Subproject	Replication Object	Load Object	Source System
Purchasing document	RBKP	REPL_CFIN	REPL_CFIN_AS	CFI_RBKP_AS_R	CFI_RBKP_AS_L	SAP ERP and SAP S/4HANA

Table 11.9 Predefined Load and Replication Objects for Purchasing Documents

Like any other replication objects, you can apply filters to load and replicate purchase documents to Central Finance by writing a rule in the SAP-delivered SKIP_EKKO event-based rule in Transaction MWB.

After adding table EKKO in Transaction LTRC, you can choose **Start Replication** (including the initial load) on the **Participating Object** tab under the **Data Provisioning** option. Upon completion of the initial load, SAP Landscape Transformation Replication Server configuration will start reading the logging table in the source system and replicate purchase documents to target central system. Figure 11.10 shows the **SAP LT Replication Server - Cockpit** screen after adding the **EKKO** object.

Purchase document history with valid document type can only be loaded/replicated to the Central Finance system. This is part of EKBE-VGABE, and valid document types are maintained in table CFIN_AV_RO_TYPE in the Central Finance system.

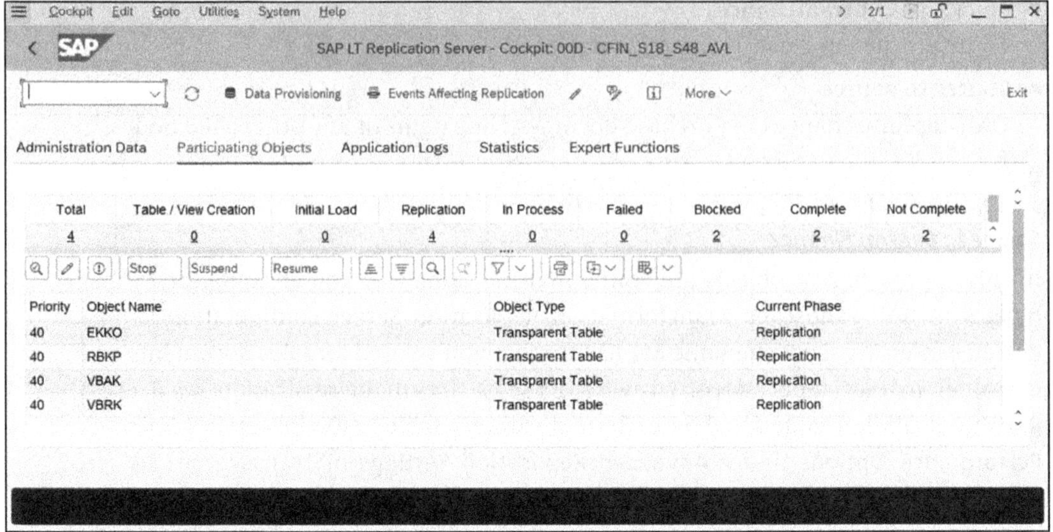

Figure 11.10 Mass Transfer ID: Table EKKO

You can also check the replicated purchase documents in interface **AV_PO** under namespace **/FINCF** along with other selection criteria in SAP Application Interface Framework. Figure 11.11 displays a successful replicated message in SAP Application Interface Framework for a purchase order document replication.

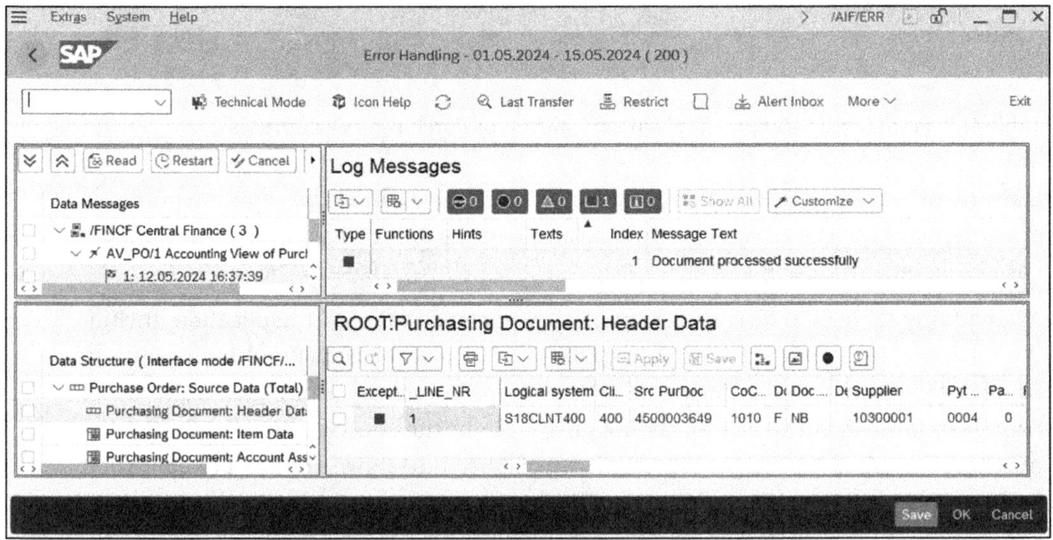

Figure 11.11 SAP Application Interface Framework for Purchase Orders

Table 11.10 displays a list of staging tables updated for purchase documents in Central Finance.

Table	Description
CFIN_AV_PO_ROOT	Purchasing Document Header
CFIN_AV_PO_ITEM	Purchasing Document Item
CFIN_AV_PO_SCH	Scheduling Agreement Schedule Lines
CFIN_AV_PO_ACC	Account Assignment in Purchasing Document
CFIN_AV_PO_RO	Related Object for Purchasing Document
CFIN_AV_PO_ROACC	Related of Purchasing Document at Account Assignment Level

Table 11.10 AVL Tables for Purchase Documents

Source to Source

Now that you understand the source-to-Central Finance configuration, let's dive deeper into the source-to-source configuration of replicating purchase document line-item changes or other child nodes. We'll first discuss the difference between the two configurations. Changes made at the purchasing document header level in the source systems trigger the transfer of change data through SAP Landscape Transformation Replication Server configuration to the target Central Finance system. However, changes made at the purchasing document line item or other nodes don't influence the purchasing document header. In other words, line-item level changes on purchase documents are done independently of header data. As a result, the purchase order line-item changes aren't transferred to the Central Finance system through the source-to-Central Finance setup we've demonstrated in the earlier section. With the source-to-source approach, any changes made at the subnodes or line-item level of a purchase document are first collected to the source logging table of purchasing document header table EKKO.

From the SAP Landscape Transformation Replication Server configuration side, SAP recommends creating a separate MT ID to replicate the changes made at the purchasing document subnodes level. In addition, the source system and target system RFC connections are always source RFC connected to SAP Landscape Transformation Replication Server configuration, for instance, the source RFC system created at the SAP Landscape Transformation Replication Server, as shown in Figure 11.12. Here, **S18CLNT400** is the source system from which the purchasing document data is loaded/replicated to the target Central Finance system.

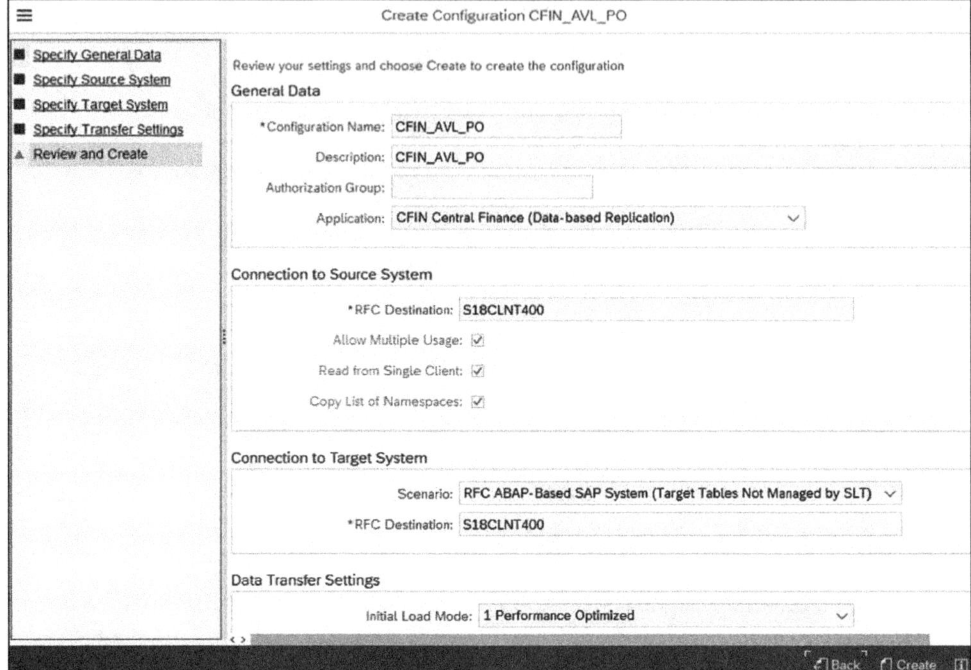

Figure 11.12 Creating the Mass Transfer ID: Source to Source

To demonstrate the source-to-source configuration, we've created MT ID 00E. Next, you'll create predefined load and replication objects for the subnode tables of purchase documents. Table 11.11 lists predefined replication objects for purchase order tables EKPO, EKET, EKKN, EKBE, and ESKN created under project REPL_CFIN and subproject REPL_CFIN_SRC.

Table	Project	Subproject	Replication Object	Load Object	Source System Version
EKPO	REPL_CFIN	REPL_CFIN_SRC	CFI_PO_EKPO_R	NA	SAP ERP and SAP S/4HANA source system
EKET	REPL_CFIN	REPL_CFIN_SRC	CFI_PO_EKET_R	NA	SAP ERP and SAP S/4HANA source system
EKKN	REPL_CFIN	REPL_CFIN_SRC	CFI_PO_EKKN_R	NA	SAP ERP and SAP S/4HANA source system

Table 11.11 Predefined Load and Replication Objects for Purchase Order Data

Table	Project	Subproject	Replication Object	Load Object	Source System Version
EKBE	REPL_CFIN	REPL_CFIN_SRC	CFI_PO_EKBE_R	NA	SAP ERP and SAP S/4HANA source system
ESKN	REPL_CFIN	REPL_CFIN_SRC	CFI_PO_ESKN_R	NA	SAP ERP and SAP S/4HANA source system

Table 11.11 Predefined Load and Replication Objects for Purchase Order Data (Cont.)

Create predefined replication objects for each purchasing document subtable for replication of purchase order line-item changes using the **Copy Predefined Object** button in program IUUC_REPL_PREDEF_OBJECTS by calling Transaction SE38. Choose **MTID** and continue. On the resulting screen, after choosing the **Copy Predefined Object** option, the system brings up the **Copy existing Predefined Object** screen, as shown in Figure 11.13.

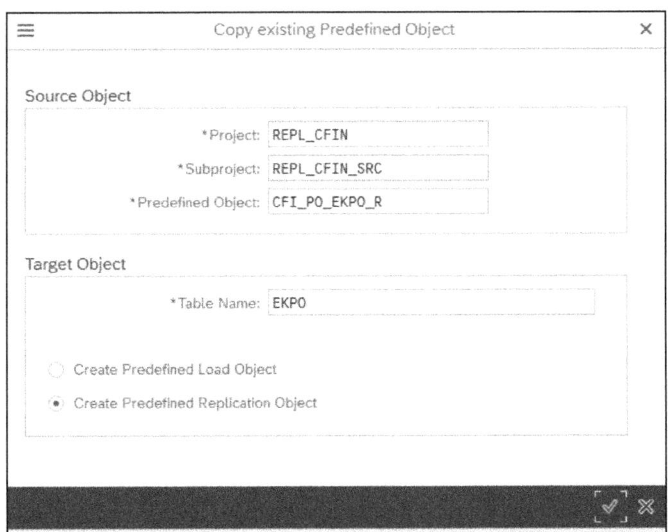

Figure 11.13 Create Predefined Replication Object EKPO

On this screen, enter values in the **Project**, **Subproject**, and **Predefined Object** fields under the **Source Object** section for predefined object CFI_PO_EKPO_R. Next, enter "EKPO" for the **Table** field, and then choose the **Create Predefined Replication Object** radio button under the **Target Object** section. Click the **Continue** icon to save the entry. Repeat the same exercise for all tables identified in Table 11.11.

After creating the predefined load and replication objects, set **Active** status from the **Repl. Object State** dropdown, as shown in Figure 11.14.

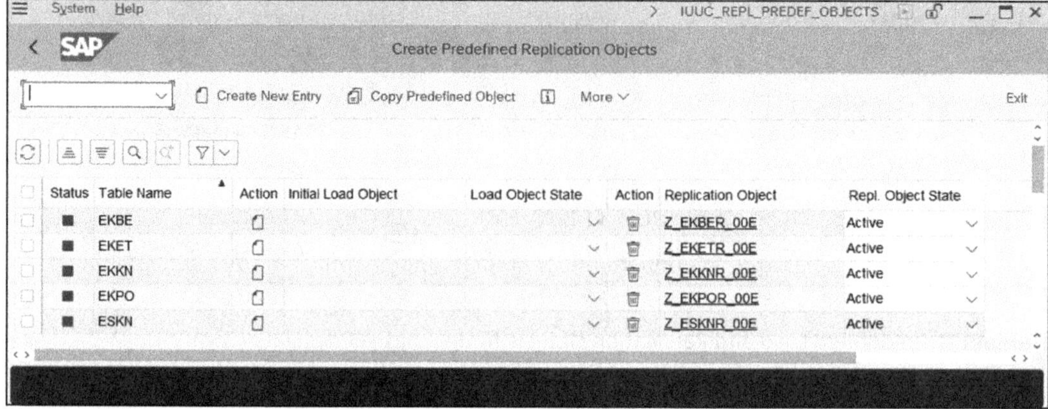

Figure 11.14 Replication Status for EKPO

Once you've created the load and replication objects for tables EKBE, EKET, EKKN, EKPO, and ESKN, go to Transaction MWB, and change the subproject. Click the **Change** button which leads you into the **DMC Tool: Change Subprojects** screen. On this screen choose the **Conversion Logic** button. On the resulting screen, click on the **GET_LOGTAB** field-based rule, as shown in Figure 11.15.

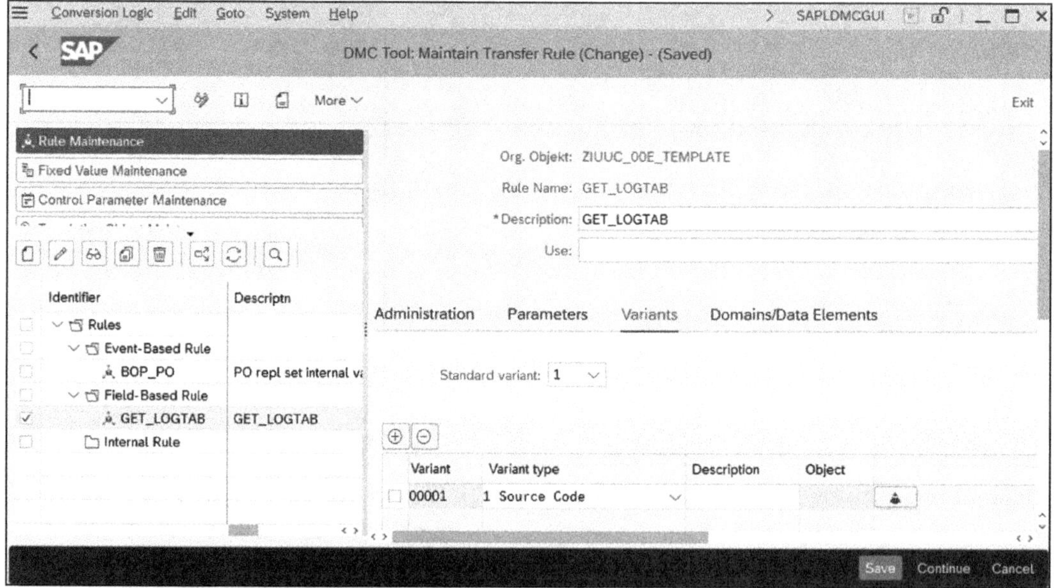

Figure 11.15 Maintain Transfer Rule GET_LOGTAB

On the **MWB: Maintain Rule Coding** screen, you need to uncomment the standard code and add the MT ID **00D** and **EKKO** for the **MT_ID** and **TABNAME** fields, as shown in Figure 11.6 ❶. Here, we're specifying the MT ID and table name to load/replicate the purchasing document data from the source system to the Central Finance system through SAP Landscape Transformation Replication Server. Click the **Save** ❷ button to save the code changes.

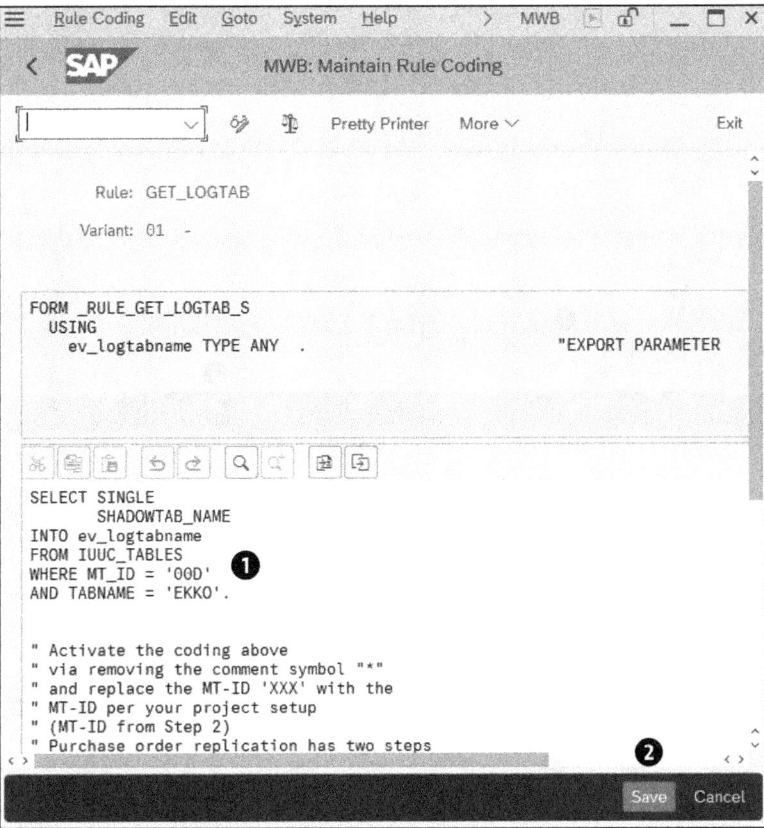

Figure 11.16 GET_LOGTAB Field-Based Rule for EKKO

Next, start the recording, and then turn on the real-time replication of purchasing document child tables EKPO, EKET, EKKN, EKBE, and ESKN in SAP Landscape Transformation Replication Server to transfer changes made at purchasing document items or other subnodes in the source system. To start recording in SAP Landscape Transformation Replication Server configuration, go to Transaction LTRC, and click on the MT ID **00E**. From there, navigate to the **Participating Objects** tab, and click the **Data Provisioning** option at the top of the screen. This action leads to the **Data Provisioning** popup, as shown in Figure 11.17. On this screen, enter "EKPO" for the **Table Name in Database** field. Next, choose the **Start Recording** ❶ radio button, and click the **Execute** ❷ icon to start recording. As a result, SAP Landscape Transformation Replication Server configuration creates a logging table and a trigger in the source system to capture changes in purchase order line-item information. SAP Landscape Transformation Replication Server configuration transfers changes registered in the source system logging table to the Central Finance system. The bottom of figure displays the **Current Phase** ❸ of recording.

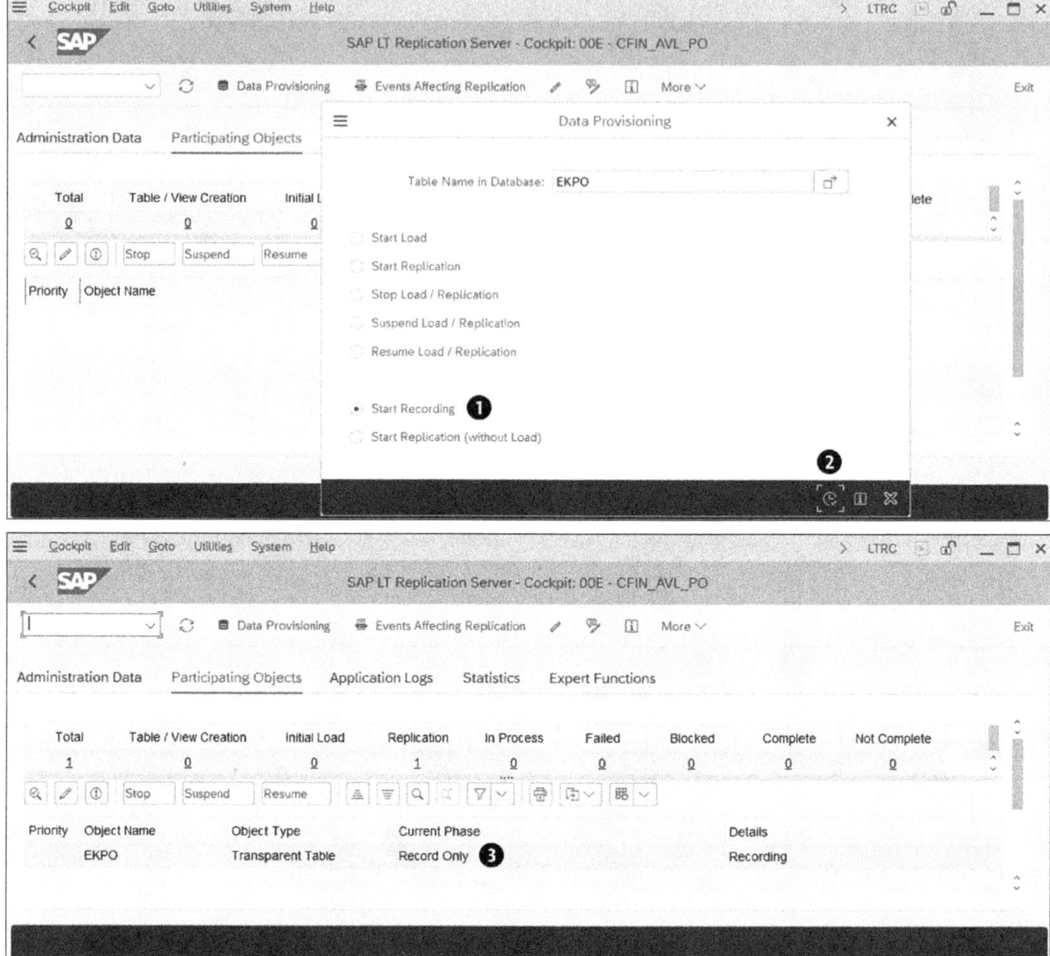

Figure 11.17 Start Recording EKPO

Now, the next step is to start the replication for table EKPO. Click on the **Data Provisioning** button, and a new popup will open with the **Data Provisioning** options, as shown in Figure 11.18. Enter the table "EKPO", and select **Start Replication (without Load)** ❶. Click the **Execute** ❷ icon to continue. You can view the replication status in Transaction LTRC ❸.

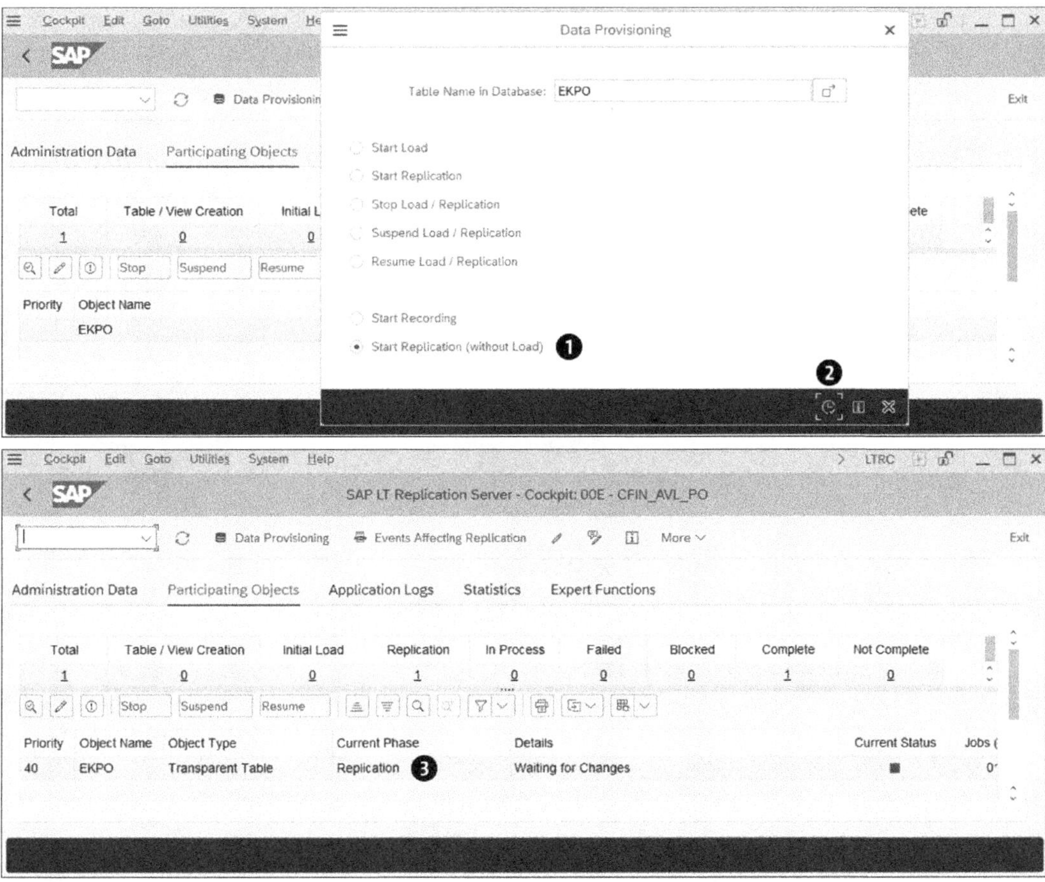

Figure 11.18 Start Replication EKPO

Supplier Invoice Initial Load and Replication

For supplier invoice replication, we'll use the same MT ID **00D** that we've used for purchase document replication. Like in purchase document replication, we'll use **1 Range Calculation** for the supplier invoice table RBKP replication located in **Performance Options** under **Advanced Replication Settings** of Transaction LTRS. In addition, SAP-delivered and predefined load and replication objects are generated for supplier invoice table RBKP under project REPL_CFIN and subproject REPL_CFIN_AS. Table 11.12 displays the load and replication objects for supplier invoice load and replication.

Object	Table	Project	Subproject	Replication Object	Load Object	Source System
Supplier invoice	RBKP	REPL_CFIN	REPL_CFIN_AS	CFI_RBKP_AS_R	CFI_RBKP_AS_L	SAP ERP and SAP S/4HANA

Table 11.12 Predefined Load and Replication Objects for AVL: Supplier Invoice

As we discussed earlier, you can also apply filter criteria in the SAP-delivered SKIP_RBKP event-based rule in Transaction MWB. Filter criteria requirements must be captured during the design phase of the project. Once all the SAP Landscape Transformation Replication Server configuration setup is completed, you'll add table RBKP in Transaction LTRC to **Start Replication** (including the initial load) on the **Participating Objects** tab under **Data Provisioning**. Upon completion of the initial load, SAP Landscape Transformation Replication Server configuration will start reading the logging table in the source system and replicate supplier invoices to the target Central Finance system. Replicated invoices can be viewed in SAP Application Interface Framework in interface **AV_SI** under Central Finance namespace **/FINCF**, as shown in Figure 11.19.

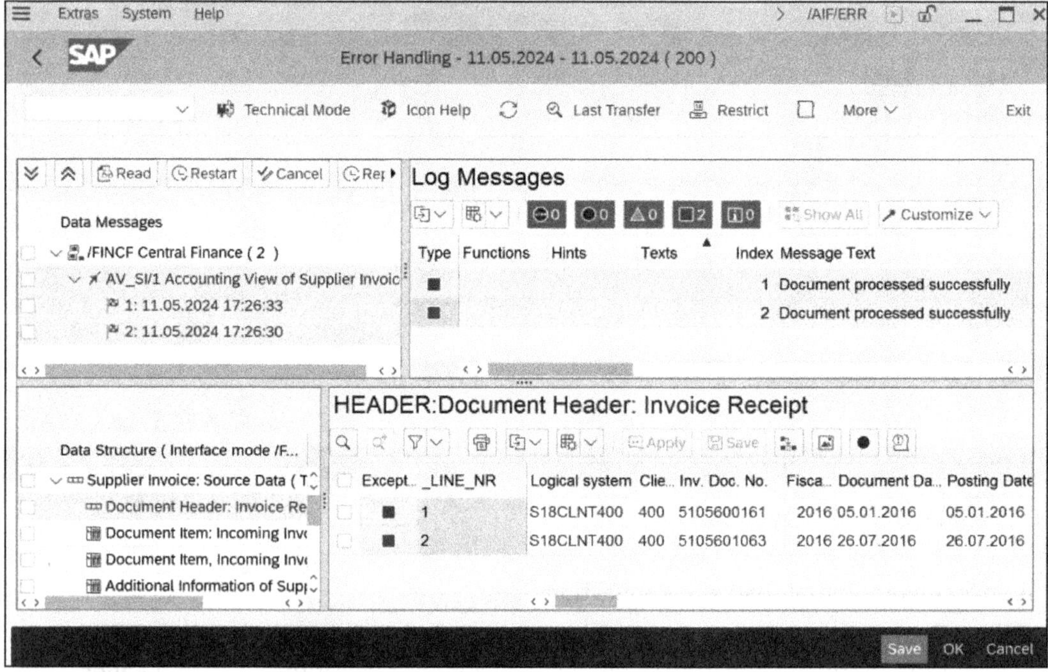

Figure 11.19 AVL Replication in SAP Application Interface Framework: Supplier Invoice

In addition, upon successful transfer of supplier invoice data from the source system to the Central Finance system, you can query the data used in AVL staging tables for supplier invoices in the Central Finance system. Table 11.13 lists supplier invoice staging tables updated in Central Finance.

Table	Description
CFIN_AV_SI_ROOT	Supplier Invoice: Header Data
CFIN_AV_SI_ITEM	Supplier Invoice: Item Data

Table 11.13 AVL Supplier Tables in Central Finance

Table	Description
CFIN_AV_SI_ACCAS	Supplier Invoice: Account Assignment-Related Data
CFIN_AV_SI_GLACC	Supplier Invoice: General Ledger Account-Related Data

Table 11.13 AVL Supplier Tables in Central Finance (Cont.)

11.2.4 Accounting View of Logistics Deletion Report

SAP provides Transaction FINS_CFIN_AVL_DEL to delete AVL staging table data in the Central Finance system. This report can be used when logistics data is no longer used, corresponding logistics data is archived, or data is set with the deletion flag. Upon reaching the transaction, you'll arrive at the **Accounting View of Logistics Information Deletion Report** screen, as shown in Figure 11.20. On this screen, enter filter information on **Generic selection** ❶, **Additional Parameters** ❷, and **Settings for Parallel Processing** ❸ sections as required. Click the **Execute** ❹ button.

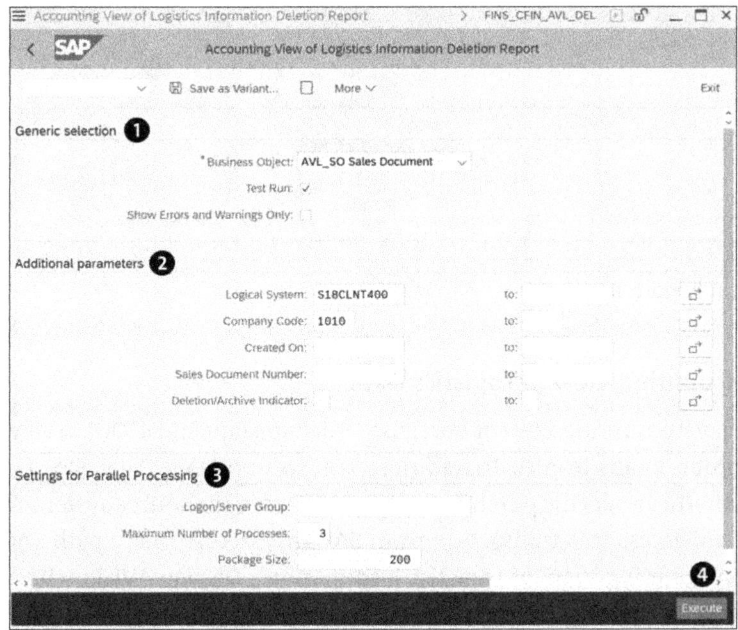

Figure 11.20 Accounting View of Logistics Deletion Report: Selection

Upon successful deletion of AVL staging table data, the system displays a summary of deleted information in the resulting screen, as shown in Figure 11.21 ❶. You can also navigate to the **Display logs** screen ❷ to view status of the confirmation.

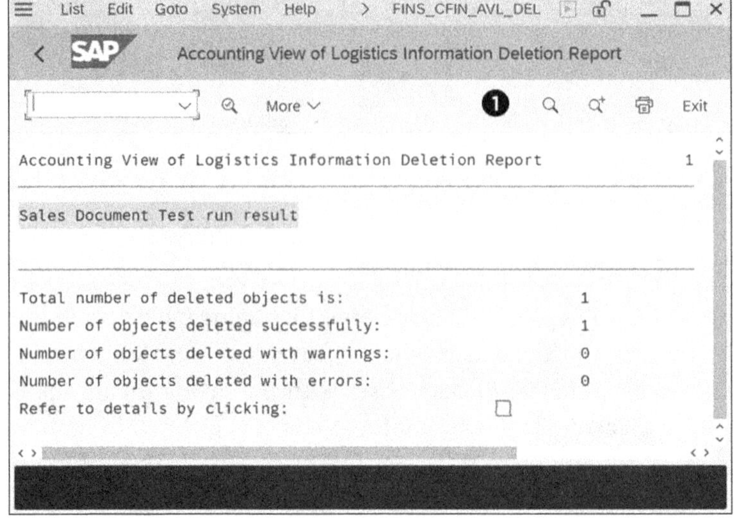

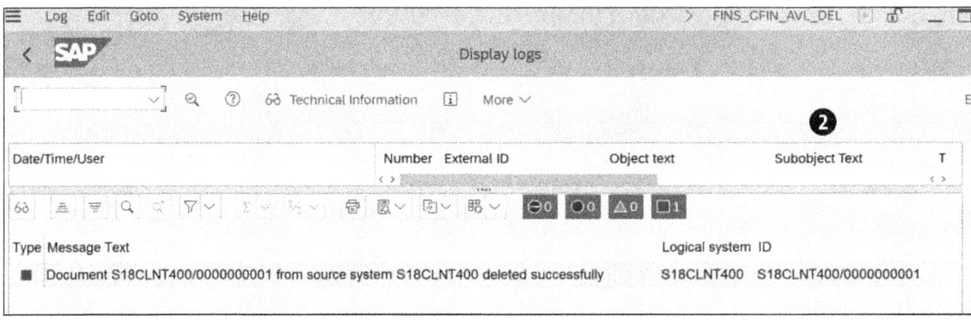

Figure 11.21 AVL Deletion Report: Results

11.2.5 Displaying Accounting View of Logistics Data

SAP provides the Display Accounting View of Purchase Order app (app ID F4700) to view replicated purchase order details in AVL. In addition, Transaction FINS_CFIN_DIS_PO can be used to check whether a specific purchase is replicated from the source system to Central Finance. You can access this transaction from **SAP Easy Access** menu path **SAP Menu • Accounting • Accounting View of Logistics Information • Display AVL Purchase Order**. After reaching the transaction, enter values in the **Logical System** and **Purchase Doc** fields on the **Display AVL Purchase Order** screen, as shown in Figure 11.22 ❶. Click the **OKAY** ❷ button. On the resulting screen, the system displays replicated purchase orders from the source system. In addition, you can navigate to purchase orders in the source system by clicking **Display in Source** ❸ button.

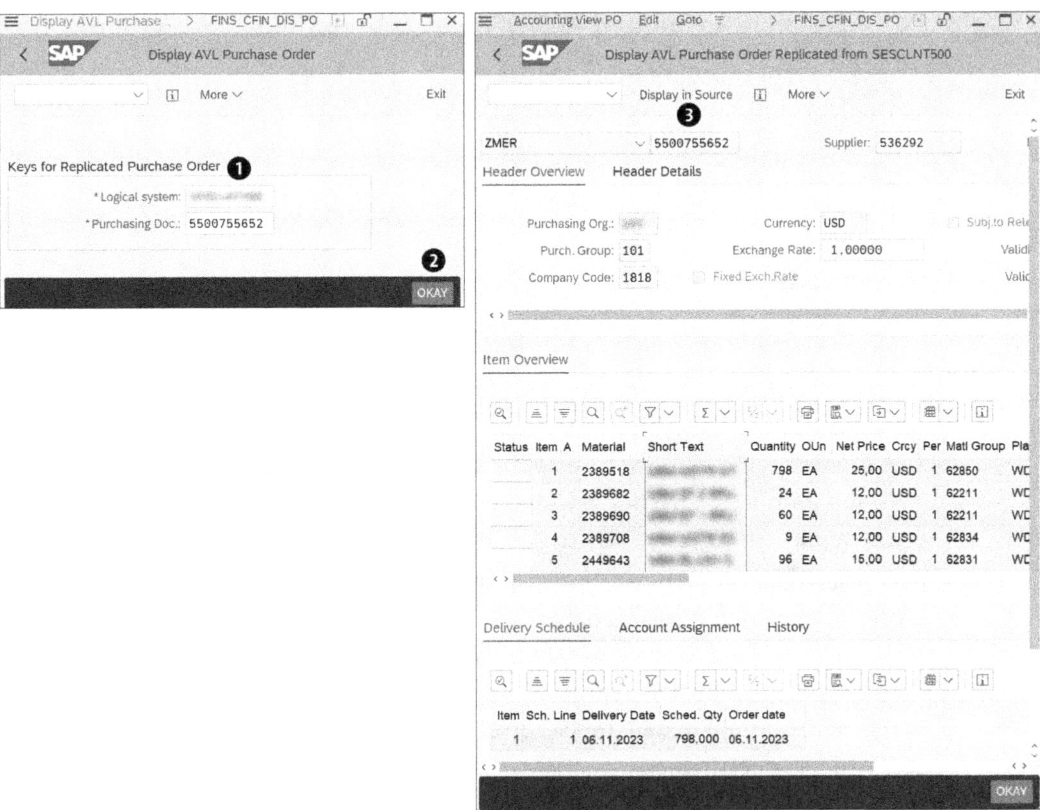

Figure 11.22 Display AVL Purchase Orders

Besides the app and GUI transaction, you can use comparison reports to compare transferred data from the source system to the Central Finance system. Table 11.14 lists the SAP-delivered comparison reports for AVL.

Transaction	Description of Transaction
FINS_CFIN_DFV_AV_SO	Sales Document
FINS_CFIN_DFV_AV_CI	Customer Invoice
FINS_CFIN_DFV_AV_PO	Purchasing Document
FINS_CFIN_DFV_AV_SI	Supplier Invoice

Table 11.14 AVL Comparison Reports

11.2.6 Replicating Accounting View of Logistics from Third-Party Systems

SAP supports replicating logistics data from a third-party source system. Logistics data supported for a third-party system includes the following:

- Sales documents (service orders, orders, returns, orders without charge, credit memo requests, debit memo requests)
- Customer invoices
- Purchase orders
- Supplier invoices

To support AVL replication from a third-party system to the Central Finance system, SAP recommends two approaches:

- **SAP Landscape Transformation Replication Server approach**
  With this approach, logistics data and its changes from a third-party source system can be replicated continuously to the Central Finance target system.
- **OData services V4 APIs**
  With this approach, logistics data from a third-party source system can be created, updated, or deleted in the Central Finance target system as needed using API calls. You'll be able to transfer a specific entity data or entire document from the third-party source system to update the corresponding data in the Central Finance system required for reporting.

Let's explore the replicating logistics data information using the SAP Landscape Transformation Replication Server configuration approach first.

Replicating Accounting View of Logistics Information Using SAP Landscape Transformation Replication Server configuration

In this approach, you'll be able replicate the AVL data from a third-party source system to the Central Finance system using the SAP Landscape Transformation Replication Server configuration approach. To achieve this, ensure that the following prerequisites are met:

- SAP recommends SAP Landscape Transformation Replication Server configuration 3.0 or SAP Landscape Transformation Replication Server configuration for SAP HANA for setting up the AVL structures and related configurations.
- Ensure the required AVL data structures are installed in the SAP Landscape Transformation Replication Server configuration client. Refer to Chapter 5, Section 5.1.6 for more information.

- The Central Finance system is configured to support AVL for a third-party system replication (refer to Chapter 7, Section 7.1.1).
- SAP Application Interface Framework is used to monitor error messages for AVL data. Therefore, you'll ensure that SAP Application Interface Framework content for third-party interfaces is extracted using Transaction /AIF/CONTENT_EXTRACT in the Central Finance system.
- If logistics data contains customer fields, implement the necessary extensions to support custom fields.
- Set up the required SAP Master Data Governance data mapping in Central Finance.

In the following sections, we'll discuss the steps involved in installing the AVL data structures in SAP Landscape Transformation Replication Server configuration.

Installing Accounting View of Logistics Data Structures in SAP Landscape Transformation Replication Server Configuration

SAP Landscape Transformation Replication Server configuration staging tables are used to transfer the AVL information from a third-party system to the Central Finance system. These staging tables are filled by extracting the AVL data from the third-party source systems. In addition, any changes made in the SAP Landscape Transformation Replication Server configuration staging tables automatically triggers AVL data replication to the Central Finance system.

[+]

Tip

SAP doesn't provide any tools to extract data from a third party and load it into SAP Landscape Transformation Replication Server configuration staging tables. It's the customer's responsibility to build a program or interface that extracts data from a third-party system and loads it into SAP Landscape Transformation Replication Server configuration staging tables.

First, you'll download the latest SAP Landscape Transformation Replication Server configuration content as referred to in SAP Note 3079411. Once downloaded, you can check the SAP Landscape Transformation Replication Server configuration content under project `REPL_CFIN` and subproject `REPL_CFIN_AE` in Transaction MWB. SAP Landscape Transformation Replication Server configuration content can be uploaded using report DMC_UPLOAD_OBJECT in SAP Landscape Transformation Replication Server, as shown in Figure 11.23.

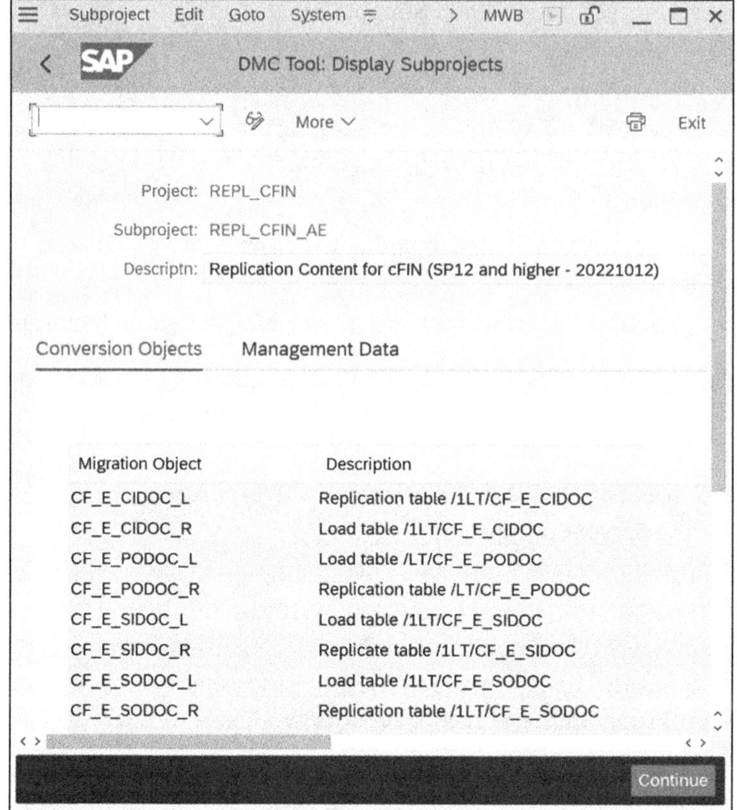

Figure 11.23 SAP Landscape Transformation Replication Server Content for AVL

Once SAP Landscape Transformation Replication Server configuration content is loaded, you can create SAP Landscape Transformation Replication Server configuration staging tables to transfer the logistics data from a third-party system to the Central Finance system. SAP delivered the necessary programs to generate SAP Landscape Transformation Replication Server configuration staging tables per guidance from SAP Note 3079411. Table 11.15 lists program names and file names for different AVL objects.

Object	Program Name	File Name
Sales document	ZFINS_CFIN_AV_GEN_TABLE_SO	*ZFINS_CFIN_AV_GEN_TABLE_SO.txt*
Customer invoice	ZFINS_CFIN_AV_GEN_TABLE_CI	*ZFINS_CFIN_AV_GEN_TABLE_CI.txt*
Purchase order	ZFINS_CFIN_AV_GEN_TABLE_PO	*ZFINS_CFIN_AV_GEN_TABLE_PO.txt*
Supplier invoice	ZFINS_CFIN_AV_GEN_TABLE_SI	*ZFINS_CFIN_AV_GEN_TABLE_SI.txt*

Table 11.15 Programs for SAP Landscape Transformation Replication Server Configuration Staging Tables Generation

Per SAP Note 3079411 guidance, you can create report ZFINS_CFIN_AV_GEN_TABLE_SO with an ABAPer's help. The SAP Note comes with a snippet of code to enable the availability of the report to create the sales documents for SAP Landscape Transformation Replication Server configuration staging tables to transfer third-party source system data to the Central Finance target system. Upon execution of the report, you'll arrive at the **ZFINS_CFIN_AV_GEN_TABLE_SO** screen shown in Figure 11.24. Enter a value in the **Central Finance System** field ❶. You can run the report in the **Test Run** to see if SAP Landscape Transformation Replication Server configuration tables are being created without any dump. Once this check is done, uncheck the **Test Run** checkbox and click the **Execute** ❷ button. The system prompts with the **Data** in staging table will be deleted. Make sure data in staging table has been replicated to central finance. To avoid inconsistency, please deactivate/suspend mass transfer popup screen. Click the **Confirm & Execute** button to continue. The system displays the output of the report in the following screen. As you can see in the figure, all four SAP Landscape Transformation Replication Server configuration staging tables required for the sales document data transfer are created in the SAP Landscape Transformation Replication Server.

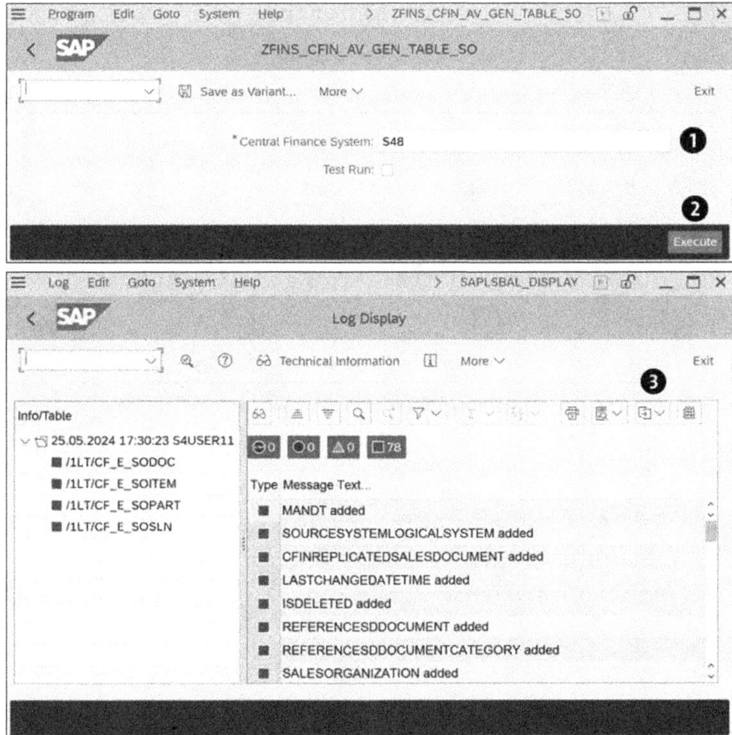

Figure 11.24 Generating SAP Landscape Transformation Replication Server Configuration AVL Tables for a Third-Party Interface

[+]

Tip

To avoid any inconsistencies during the program run, deactivate or suspend MT ID.

Once the report is executed, SAP Landscape Transformation Replication Server configuration staging tables are created in SAP Landscape Transformation Replication Server for sales document replication, as shown in Figure 11.24 ❸.

You'll repeat the same steps for other AVL objects referred to in Table 11.15. After execution of all reports, you'll see the SAP Landscape Transformation Replication Server configuration tables created, as shown in Table 11.16. You can verify these tables from Transaction SE11 or Transaction SE16N in SAP Landscape Transformation Replication Server.

Replication Object	SAP Landscape Transformation Replication Server configuration Staging Tables
Sales Document	/1LT/CF_E_SODOC: Sales Document: Header Data /1LT/CF_E_SOITEM: Sales Document: Item Data /1LT/CF_E_SOPART: Sales Document: Partner /1LT/CF_E_SOSLN: Sales Document: Schedule Line Data
Customer Invoice	/1LT/CF_E_CIDOC: Customer Invoice: Header Data /1LT/CF_E_CIITEM: Customer Invoice: Item Data
Purchase Order	/1LT/CF_E_PODOC: Purchase Order: Header Data /1LT/CF_E_POITEM: Purchase Order: Item Data /1LT/CF_E_POSCH: Purchase Order: Schedule Line Data /1LT/CF_E_POACC: Purchase Order: Account Assignment Data /1LT/CF_E_POMRD: Purchase Order: Material Document Data /1LT/CF_E_PORSI: Purchase Order: Supplier Invoice Data
Supplier Invoice	/1LT/CF_E_SIDOC: Supplier Invoice: Header Data /1LT/CF_E_SIITEM: Supplier Invoice: Item Data /1LT/CF_E_SIACC: Supplier Invoice: Account Assignment Data /1LT/CF_E_SIGLAC: Supplier Invoice: General Ledger Account Data

Table 11.16 SAP Landscape Transformation Replication Server Configuration Staging Tables for a Third-Party Interface

[+]

Tip

SAP recommends the following considerations when enabling SAP Landscape Transformation Replication Server configuration staging tables for third-party integration:

- Before executing the reports, it's always good practice to check if SAP Landscape Transformation Replication Server configuration staging tables for AVL exist. If tables exist, check if data exists as these programs overwrite the data on the tables.
- If you install a new service pack or upgrade to a new release, you need to re-create the SAP Landscape Transformation Replication Server configuration table. Verify that all data is transferred over to Central Finance prior to re-creating SAP Landscape Transformation Replication Server configuration staging tables
- Deactivate the MT ID to avoid data inconsistencies.

Consult SAP Note 3079411 for further guidance on this topic.

SAP Landscape Transformation Replication Server Configuration Setup to Transfer the Accounting View of Logistics Data

Now that you know how to upload SAP Landscape Transformation Replication Server configuration content and generate SAP Landscape Transformation Replication Server configuration staging tables, let's explore how to set up SAP Landscape Transformation Replication Server configuration to transfer the AVL data. First, you'll create an MT ID to transfer AVL data from a third-party system to the Central Finance system. Here, the source RFC destination is the SAP Landscape Transformation Replication Server configuration itself while creating the MT ID in the SAP Landscape Transformation Replication Server configuration. Next, you'll create predefined load and replication objects by copying existing predefined objects provided by SAP. Table 11.17 provides a list of predefined load and replication objects for each AVL replication object.

AVL Object	SAP Landscape Transformation Replication Server Configuration Staging Tables	Project	Subproject	Replication Object	Load Object
Sales Document	/1LT/CF_E_SODOC	REPL_CFIN	REPL_CFIN_AE	CF_E_SODOC_R	CF_E_SODOC_L
Customer Invoice	/1LT/CF_E_CIDOC	REPL_CFIN	REPL_CFIN_AE	CF_E_CIDOC_R	CF_E_CIDOC_L
Purchase Order	/1LT/CF_E_PODOC	REPL_CFIN	REPL_CFIN_AE	CF_E_PODOC_R	CF_E_PODOC_L
Supplier Invoice	/1LT/CF_E_SIDOC	REPL_CFIN	REPL_CFIN_AE	CF_E_SIDOC_R	CF_E_SIDOC_L

Table 11.17 Replication and Load Objects for AVL

In general, there is no event-based rule to filter the data to be transferred to the Central Finance system. Leading practice is that data scoping must be done before the extraction is started to load data into SAP Landscape Transformation staging tables using a custom extraction report. Filter logic must be added in the extraction report to load the selected data per the project scope.

Configuring Central Finance

For AVL replication, certain activities are required to be carried out in the Central Finance system under the Central Finance customizing menu path **Central Finance: Target System Settings**.

First, you must define the logical system for the third-party source system. To carry out this configuration, navigate to Central Finance customizing menu path **Financial Accounting • Central Finance • Central Finance – Target System Settings • Set Up Systems • Define Logical System for Source and Central Finance Systems**. After reaching the configuration, the system displays a warning popup window stating that the settings are cross-client. Click the **Continue** icon. This will bring up the **Change View "Logical Systems": Overview** screen. On this screen, click the **New Entries** button. On the resulting **New Entries: Overview Added Entries** screen, enter values in the **Log.System** and **Name of Logical System** fields, as shown in Figure 11.25. Click the **Save** button to save the entry.

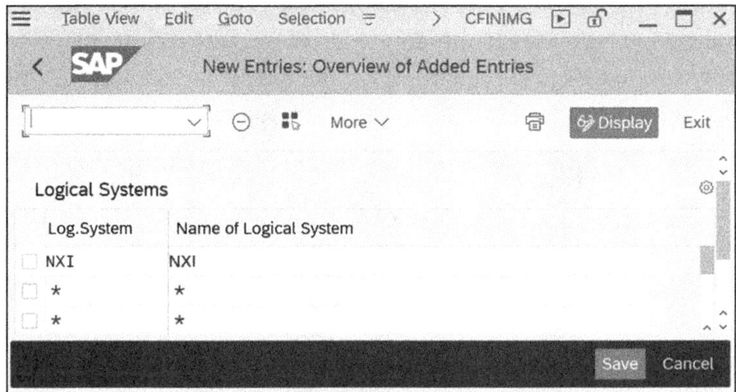

Figure 11.25 AVL Defined Logical Systems for a Third-Party System

Next, define a business system that links to the third-party logical system. To carry out this configuration, navigate to Central Finance customizing activities menu path **SAP Customizing Implementation Guide • Financial Accounting • Central Finance: Target System Settings • Mapping • Define Technical Settings for All Involved System**. After reaching the configuration, you'll land at the **Change View "Define Business Systems": Overview** screen, as shown at the top of Figure 11.26. On this screen, click the **New Entries** button. You'll arrive at the **New Entries: Overview of Added Entries** screen. Maintain values for the fields for the third-party system:

- **Business System**
- **Logical System**
- **RFC Destination**

Click the **Save** button to save the entries. Maintained values are reflected upon returning to the previous screen (see Figure 11.26 ❶).

Next, for each third-party logical system, RFC usage and RFC destinations must be maintained for AVL Central Finance scenarios. To maintain RFC assignments and settings for the Central Finance system, navigate to IMG menu path **SAP Customizing Implementation Guide • Financial Accounting • Central Finance: Source System Settings • Set Up Connection to Central Finance System • Maintain RFC Assignments and Settings for the Central Finance System** or navigate to it from Transaction CFINIMG. Once you reach the settings, click the **New Entries** button on the **Change View "Source System": Overview** screen, as shown at the bottom of Figure 11.26. You'll arrive at the **New Entries: Overview of Added Entries** screen. Here, maintain values for the **Logical System** and **Name of Logical System** fields for the third-party system. You must select the **Third-Party System checkbox** ❷ to designate the logical system as the third-party source system. Click the **Save** ❸ button to save the entry.

11

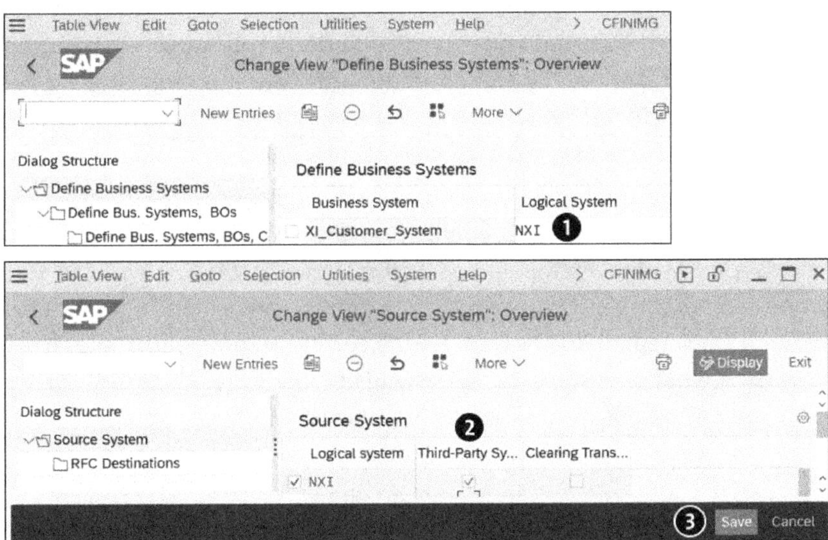

Figure 11.26 Define Technical Settings for All Involved Systems

Next, you'll set up SAP Application Interface Framework in the Central Finance for ALV data transfer from third-party source systems. SAP predelivered deployment scenarios should be implemented to install the SAP Application Interface Framework content using Transaction /AIF/CONTENT_EXTRACT for the third-party interfaces for AVL in the Central Finance system. Table 11.18 lists the SAP-delivered deployment scenarios for AVL replication in a third-party system.

Object	Deployment Scenario	SAP Application Interface Framework Interface	Central Finance System Version
Sales document	SAP_AIF_0041	AV_SO_EX (version 1)	Since SAP S/4HANA 2021 FPS0
Customer invoice	SAP_AIF_0042	AV_CI_EX (version 1)	Since SAP S/4HANA 2021 FPS0
Purchase order	SAP_AIF_0043	AV_PO_EX (version 1)	Since SAP S/4HANA 2021 FPS0
Supplier invoice	SAP_AIF_0044	AV_SI_EX (version 1)	Since SAP S/4HANA 2021 FPS0

Table 11.18 SAP Application Interface Framework Deployment Scenarios for Third-Party Systems

Accounting View of Logistics Data Transfer from Third-Party Source System to Central Finance via SAP Landscape Transformation Replication Server Configuration

In this section, we'll explore how AVL data transfer happens from a third-party source system to the Central Finance target system. Replication of AVL is divided into a two-step process:

1. Extract data from the third-party source system using a custom extraction program, and load data into the AVL SAP Landscape Transformation Replication Server configuration staging tables created in the earlier section, "Installing Accounting View of Logistics Data Structures in SAP Landscape Transformation Replication Server Configuration."
2. Once data is loaded in the SAP Landscape Transformation Replication Server configuration staging tables, you'll create predefined load and replication objects after copying an existing SAP-delivered load and replication object. Then, you can start the replication of the corresponding SAP Landscape Transformation Replication Server configuration staging table to transfer the corresponding logistics information to the Central Finance system. As a prerequisite, the Central Finance system configuration and SAP Application Interface Framework interface content must be installed before the AVL data is transferred to the target Central Finance system.

Transferred AVL data to Central Finance can be checked and monitored in the corresponding SAP Application Interface Framework interface and versions with Transaction /AIF/ERR in the Central Finance target system.

Replicating Accounting View of Logistics Data Using OData Services V4 Application Program Interfaces

With this approach, AVL data from a third-party source system can be created, updated, or deleted in the Central Finance target system as needed using API calls. You can transfer the specific entity or the entire document from the third-party source system to update the corresponding data in the Central Finance system as required for reporting.

To use OData Services V4 APIs, you must do the following to transfer the AVL data to the Central Finance system:

- Perform Central Finance configuration.
- Install SAP Application Interface Framework content at the Central Finance system.
- Use SAP Master Data Governance data mapping.

As shown in Table 11.19, SAP provides APIs in the Central Finance system to replicate the AVL data from a third-party system.

Object	OData Service Name	Type
Sales document	API_CFinRpldSalesDocument	OData V4
Customer invoice	API_CFinRpldBillingDocument	OData V4
Purchase order	API_CFinRpldPurchaseOrder	OData V4
Supplier invoice	API_CFinRpldSupplierInvoice	OData V4

Table 11.19 OData Services APIs

Note these OData services enable you to replicate AVL data from a third-party source system to the Central Finance target system. If the corresponding replicated AVL object data doesn't exist in the Central Finance system, then the system creates the AVL object. If the AVL object already exists, the corresponding AVL object data can be updated or deleted depending on what operation is executed in the third-party system. The services are developed in the SAP backend system and registered with Transaction /IWBEP/V4_ADMIN at the SAP backend. In addition, you can get the model and data provider class information for the corresponding service group under Transaction / IWBEP/V4_ADMIN.

11.3 Summary

AVL provides logistics data replication capabilities from source systems to the Central Finance system. This data is helpful to stakeholders to perform management reporting, local reporting, and statutory reporting. In addition, replicated purchase orders can be used for accrual processes in conjunction with the accrual engine. In this chapter, you've learned configuration prerequisites for the source and Central Finance systems. Then, we discussed step-by-step details on how to configure AVL in SAP Landscape Transformation Replication Server configuration. Next, the chapter walked you through steps involved with third-party AVL data replication. In the next chapter, you'll learn about another challenging and exciting area: reconciliation and reporting in Central Finance.

Chapter 12
Reconciliation and Reporting in Central Finance

Successful Central Finance implementation relies on reconciliation and reporting of financial data at various stages during implementation and after go-live. Reconciliation and reporting are carried out at transactional level, individual account level, and entity or enterprise level. In addition, if group reporting is in scope for the project, the consolidated data from various source systems, integration of analytics, and harmonized master data facilitate reporting accurately in Central Finance. These activities aid in faster period closing, continuous accounting, and consolidation anytime.

Financial closing involves a series of tasks performed to ensure the company's book of records reflects events that have a financial impact during a period. Financial closing involves processes and resources to execute both pre-closing and periodic closing activities, which are essential for accurate financial reporting. The primary goal of financial closing is to ensure companies record transactions accurately in the system of record to meet the following:

- Financial objectives of both external and internal stakeholders
- Regulatory compliance (e.g., generate financial statements per GAAP)
- Operational objectives to reduce cost
- Increased transparency and efficiency in processes
- Effectiveness of internal controls

With Central Finance connected to different enterprise resource planning (ERP) systems, you have the same primary goals as the general financial closing goals mentioned earlier. In addition, the complexity of financial closing in Central Finance may increase due to real-time replication of data and integration with multiple source systems. For financial closing in Central Finance, you'll align with key stakeholders on process definition, process steps, and leading practices to be followed. During the explore phase, you'll review and discuss key questions with stakeholders:

- What are the critical reporting requirements and the reporting strategy?
- What critical reports are required: financial statements, flexible hierarchies, cost center, profit center, profitability analysis, and so on?

- What master data objects require extensive harmonization, so the reporting reflects accordingly?
- What custom attributes or dimensions are required?
- Which fields are required when transactional data is posted from the source system or pass through?
- Which fields are required to be derived in Central Finance?
- Which fields are required to be cleared during replication that are no longer necessary in Central Finance or don't add any value?
- Which fields are optional?
- What tax reporting requirements are required for entities/countries?
- What processes are included in closing steps?
- What's the sequence of activities you'll perform during the initial load and real-time replication?
- Who is responsible to execute the closing steps and monitor the progress of the steps in Central Finance?
- Who is responsible for validating and generating reports necessary to support GAAP and management reporting?
- What reporting is required for scenarios such as group close, local close, intercompany eliminations, and so on?
- Are entry and exit points determined clearly as a reconciliation strategy for both SAP-based and non-SAP-based systems?

Based on requirements and answers to the questions, you'll document key requirements and log all open issues for follow-up with relevant teams. The initial design leads to defining the global closing process across entities. However, variations to global closing design templates are adapted per local requirements. Leading practices are followed wherever applicable. A well-defined closing process should lead to achieving a common reporting information model (CRIM). This helps organizations with several benefits as follows:

- A common reporting framework for financial reporting that defines the structure of financial data, attributes, and standards to follow for reporting
- Faster financial close
- Increased transparency and traceability
- Improved quality of data
- Quicker insights into financial data using SAP S/4HANA reporting tools
- Consolidated reporting in Central Finance
- Adaptable for local and statutory reporting requirements with improved compliance
- Better decision making

In this chapter, you'll learn the reconciliation approach in Central Finance, with step-by-step instructions on how to run SAP-delivered comparison reports. You'll then transition into SAP-delivered comparison reports. Finally, the chapter discusses other SAP S/4HANA finance reporting tools that can be leveraged during Central Finance implementation.

12.1 Reconciliation Approach

Before replicated data can be reconciled, master data mapping and transformation of data accurately from source systems to Central Finance are critical steps. In this section, we'll discuss the reconciliation approach to follow in Central Finance.

12.1.1 Reconciliation Process

From a technical perspective, the end goal of Central Finance is to replicate financial data from an SAP source system or a third-party system to Central Finance in real time. SAP provides the necessary tools to enable real-time replication in Central Finance. To reach consolidated reporting in Central Finance, you'll go through several steps, starting from the source system. Figure 12.1 illustrates the data transformation process, which starts from the source system to Central Finance to achieve consolidated reporting at a high level. Following are the process steps, which are organized into six major topics as follows:

❶ **Pre-closing and closing activities in the source system**
Day-to-day business processes are executed in the source system, and relevant financial postings are captured in source systems. This includes both SAP and non-SAP systems. As part of Central Finance initial load preparation, you'll perform pre-closing and closing activities such as running depreciation, posting any adjustments in the source system.

❷ **Data extraction**
The system prepares and extracts data for the initial load from the source systems. The extraction can be limited to selected company codes or all company codes in scope.

❸ **Data validations and transformation**
Once data is extracted from the source systems and brought into Central Finance, you'll carry out validation through simulation tools provided with Central Finance. In addition, the system uses SAP Master Data Governance mapping to undergo transformation from the source system to Central Finance.

❹ **Reconciliation**
Once the initial load is completed, you'll run comparison reports and reconciliation reports. Running comparison reports and other finance reports identify discrepan-

cies in transferred data. Once the initial load is reconciled, you'll initiate online replication via SAP Landscape Transformation Replication Server. You'll then run comparison reports to verify replicated data.

❺ **Consolidation and reporting**
Once the reconciliation activities are completed, you'll generate consolidated financial reports and other ad hoc reports per business requirements.

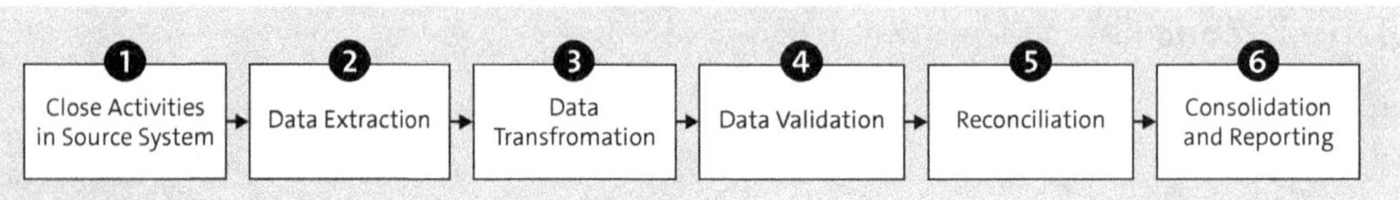

Figure 12.1 Financial Close in Central Finance

12.1.2 Migration Clearing and Substitution Accounts

Central Finance requires you to maintain one migration clearing account for each company code and substitution accounts, which are primarily used during the initial load postings. When the initial load data is extracted and transferred to Central Finance, the system uses the migration clearing account as an offset account to general ledger account balances. Likewise, for open items, you'll define substitution accounts, which act as offsetting accounts for balances transferred from reconciliation accounts and open-item managed accounts. Once all balances and open item balances are loaded, both the migration and substitution account balances become zero automatically. When there are balances in any of these accounts, you'll analyze mapping of these accounts. You've learned configuration steps in Chapter 7, Section 7.4.6. Figure 12.2 illustrates how the system uses migration clearing and substitution accounts during FI-CO initial load postings for accounts payable (AP) and accounts receivable (AR) reconciliation accounts. Let's explore this further:

❶ **Reconciliation account balances**
As you can see in Figure 12.2, the system transfers reconciliation accounts balances to relevant substitution accounts configured in the system. For example, in the case of AP, balances from AP from reconciliation are posted to corresponding substation clearing accounts as a credit entry.

❷ **Transferring open items**
During the initial load, all open items are transferred regardless of the year. The system transfers open items from the source system reconciliation account to the corresponding reconciliation account in the target system. Meanwhile, the offsetting entry is posted to the corresponding substitution account configured in the system. At the end, this makes the balances on the substitution account zero.

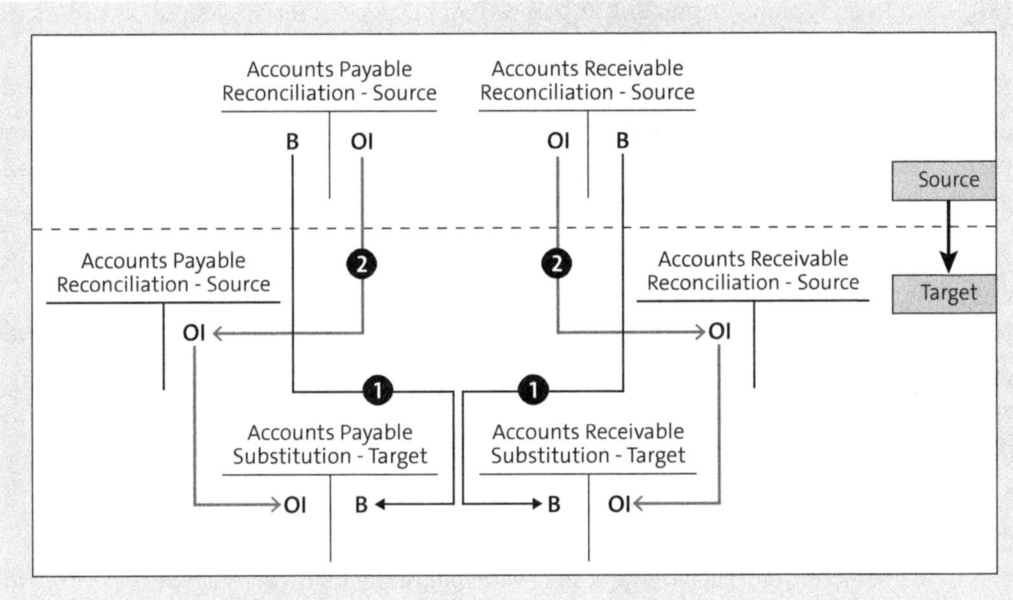

Figure 12.2 Initial Load Reconciliation Overview

12.2 Checks and Reconciliations at Various Stages

In Central Finance, you'll reconcile data and take snapshots at various stages during the transfer of data from the source system to Central Finance. Figure 12.3 displays different stages when you'll reconcile data and run reports to validate data:

❶ **Before the initial load**
As a leading practice, SAP recommends reducing open items for customers, vendors, and open-item-managed general ledger accounts in the source systems. As a result of this activity, both performance and time duration greatly improved to reconcile data. SAP also recommends running consistency check reports to identify and fix inconsistent data. In addition, you'll execute period-end closing tasks such as the asset accounting depreciation run. Finally, taking snapshots of data prior to the initial load helps for comparative purposes.

❷ **After the initial load**
Once you've posted the initial load, you'll carry out validating balances and line items. You'll run SAP-delivered comparison reports and other financial reports to validate data and tie balances between the source system and the Central Finance system.

❸ **After activating real-time replication**
Once the initial load is reconciled, real-time replication is activated to replicate data from the source system to the Central Finance system. Depending on how your

cutover is planned, you'll clear bulk SAP Application Interface Framework messages before going into normal replication mode.

❹ **Subsequent period-end closing**
After the initial load and enabling real-time replication, subsequent period-end activities can be carried out in Central Finance. Depending on the business processes in Central Finance, period-end activities are carried out in Central Finance.

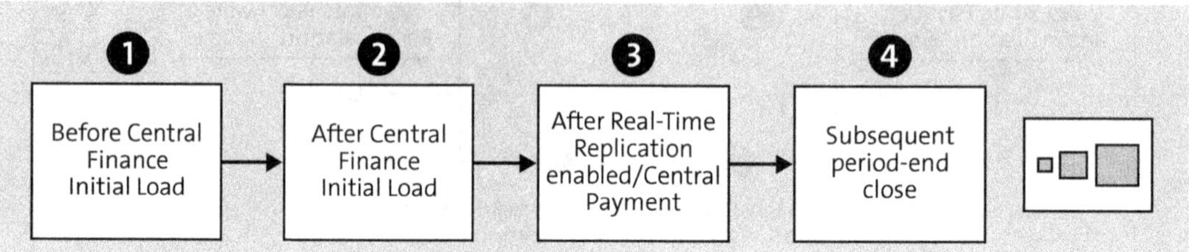

Figure 12.3 Reconciliation Approach

We'll walk through each of these stages further in the following sections.

12.2.1 Before Initial Load

From a functional system readiness perspective, prior to beginning the initial load, the source system preparatory tasks must be completed for FI documents replication. The preparatory tasks are divided into the following categories:

- **Initial load data volume**
  Assess the source system data volumes for FI initial load purposes based on general replication settings (VCFIN_SOURCE_SET).
- **Data cleansing**
  During the preparatory phase, carry out cleansing activities in the source system for the following open items:
  - Open item general ledger accounts
  - AR open items
  - AP open items
- **Consistency checks**
  Run consistency checks to repair any inconsistencies in document indexes, transactions figures, and so on.
- **Run reconciliation reports**
  Before the initial load extraction, all subledgers are reconciled to avoid any errors detected during the initial load process. You'll run reconciliation reports to capture a snapshot of the system before the initial load.

- **Timeout parameter**
  During the initial load, the system packages open items, balances, and detail documents into different packages. Depending on the volume of data extraction, you'll specify the timeout parameter for the initial load program to prevent job failure during the initial load extract.
- **General replication settings**
  Maintain general replication settings in view VCFIN_SOURCE_SET in the source system.

Table 12.1 summarizes key activities you'll typically perform before the initial load extraction process. Perhaps all the steps mentioned and any other project-specific tasks such as the requesting process must be documented in the cutover plan.

Activities Carried Out Prior to Initial Load Extraction		
Cleansing Activities	**Consistency Checks**	**Taking Snapshots before Initial Load Extraction**
▪ Clear Customer Open Items (Transaction F.32) ▪ Clear Vendor Open Items (Transaction F.44) ▪ Clear General Ledger Open Items (Transaction F-03) ▪ Clear General Ledger Account for Ledger Group (Transaction FB1SL) ▪ Automatic Clearing Using Transaction (Transaction F.13)	▪ Run consistency check report RFINDEX to check the document against indexes, documents against transaction figures, and indexes against transaction figures ▪ Reconcile the general ledger with AP and AR subledgers using Transaction FAGLF03 or running report TFC_COMPARE_VZ. ▪ Reconcile ledgers with report RGUCOMP4 or Transaction GCAC. ▪ Run reports RM07MBST and RM07MMFI to reconcile materials management (MM) with general ledger.	▪ Financial statements using Transaction F.01 or report RFBILA00 ▪ Totals report for cost centers Transaction S_ALR_87013611 ▪ General ledger account balance using program RFSSLD00 ▪ Compact document journal ▪ Other specific reports per project requirements

Table 12.1 Source System Initial Load Preparation

Tip

Table 12.1 provides general guidance based on standard transactions or reports you would run prior to the Central Finance initial load extraction. This reconciliation activity varies for each customer, so use the list as a starting point and adapt it to your business requirements.

12.2.2 After Initial Load

After the FI initial load is completed in Central Finance, based on project requirements, several reconciliation activities are performed. Similarly, this list varies with each customer's requirements. Nevertheless, the following list provides a starting point for reconciliation activities:

- Count documents transferred from the source system to Central Finance.
- Perform technical reconciliation of migration/substitution clearing account balances to zero.
- Run Central Finance comparison reports to ensure all documents are accounted for and that balances tie at the individual general ledger account or group of general ledger accounts.
- Perform open item reconciliation for customer and vendor balances.
- Ensure balances tie for cost center, profit center, and any other characteristics between the source system and Central Finance.
- Run intercompany reconciliation reports to ensure data is accurately transferred.
- Complete asset accounting reconciliation with general ledger manual accounts used in Central Finance.
- Carry out year-end carryforward programs to bring general ledger balances to following year if necessary.
- Run any other custom reports based on your project requirements.

12.2.3 During Ongoing Replication

Once the FI initial load and SAP Application Interface Framework bulk load are completed, during online replication, you'll run comparison reports and other reconciliation reports based on your project requirements. The SAP Application Interface Framework is an interface monitoring tool to continuously monitor replicated data from the source system to Central Finance. You'll also carry out typical tasks:

- Periodically run comparison reports to compare replicated data between source systems and Central Finance systems.
- Regularly analyze mapping of new master data, and perform consistency checks.
- Review SAP Application Interface Framework error logs, and perform error resolution.
- Run other ad hoc reports as necessary based on SAP Fiori, SAP Analytics Cloud, SAP Analysis for Microsoft Office, and so on.

12.2.4 At Subsequent Period End

In Central Finance, depending on the business processes in scope, you'll carry out period-end closing activities. Central closing often involves closing business activities for multiple entities and the source system in Central Finance. The following closing task list is not exhaustive as the closing activities vary with every customer and required reporting requirements:

- Perform pre-close activities that often involve data validation and completing incomplete transactions.
- Open financial periods.
- Perform AP and AR clearing activities; close AP and AR books.
- Run intercompany reconciliations.
- Perform treasury- and risk-related functions.
- Execute asset under construction (AuC) periodic settlement, final settlement, and asset closing processes.
- Perform asset accounting close activities.
- Perform universal allocations.
- Carry out accruals and deferrals activities.
- Perform foreign currency valuation.
- Close subledgers and the general ledger.
- Close financial periods.

12.3 Running Central Finance Comparison Reports

SAP provides comparison reports to analyze different levels of transferred data from the source system to Central Finance. All Central Finance comparison reports are executed from the Central Finance system. Table 12.2 (in the next section) provides SAP-delivered Central Finance comparison reports. When reconciling data from the source system to Central Finance, the leading practice is to begin with level 1 reports, which provide an overall count of documents transferred from the source system to Central Finance as expected. For example, for FI comparison reports, if level 1 reports tie, then you'll begin reconciling level 2 and level 3 reports, which provide further granular level details. Level 2 reports provide debit or credit balances such as a general ledger account or cost element account, whereas level 3 reports offer line items comparison details for a given selection criterion. Generally, you'll run these reports at various stages during the Central Finance implementation such as the following:

- After a financial initial load is carried out in Central Finance
- After clearing an SAP Application Interface Framework bulk load or clearing SAP Application Interface Framework messages

- After real-time replication is turned on either during or after completing the initial load or after a downtime of SAP Landscape Transformation Replication Server
- During period-close activities
- During year-end closing activities
- On an ad hoc basis as necessary

In this section, you'll learn about SAP-delivered Central Finance comparison reports by each application area.

12.3.1 Standard Comparison Reports

Table 12.2 shows a list of the SAP-provided comparison reports in Central Finance for different application areas. In conjunction with these reports, you'll run other relevant reports wherever necessary. Later in this chapter, Section 12.4.1 provides a list of SAP Fiori apps that might be useful for reconciliation.

Area	Report Name	Transaction	Report Usage
Financial Accounting	Count of Journal Entries	FINS_CFIN_DFV_FI_NUM	Displays count of journal entries transferred to Central Finance.
	General Ledger Account Line Items	FINS_CFIN_DFV_FI_DOC	Shows FI line-item details for a given selection criteria such as document number and general ledger account.
	General Ledger Account Balances	FINS_CFIN_DFV_FI_BAL	Displays debit or credit balances for the selected general ledger account between source and Central Finance systems.
	FI Clearing Status of Document Line Items	FINS_CFIN_DFV_CLR	Displays clearing status of FI documents transferred from the source system to the Central Finance system.
Management Accounting	Count of CO Documents	FINS_CFIN_DFV_CO_NUM	Provides a count of transferred CO secondary posting in Central Finance.
	Line Items for Cost Elements	FINS_CFIN_DFV_CO_DOC	Displays CO line items for selected CO documents and cost elements transferred to Central Finance.

Table 12.2 Central Finance Comparison Reports

Area	Report Name	Transaction	Report Usage
Management Accounting (Cont.)	Balances for Cost Elements	FINS_CFIN_DFV_CO_BAL	Compares balances of cost elements between the source system and the Central Finance system.
	Count of Line Items for Commitments in CO	FINS_CFIN_DFV_CMTNUM	Provides count of commitment line items transferred from the source system to the Central Finance system.
	Line Items for Commitments in CO	FINS_CFIN_DFV_CMTDOC	Displays commitment line items in the Central Finance system for the selected cost elements in the source system.
	Count of Profit Center Documents	FINS_CFIN_DFV_PCA_CN	Provides a count of enterprise controlling profit center accounting (EC-PCA) documents replicated from the source system to the Central Finance system.
	Balances for Profit Centers	FINS_CFIN_DFV_PCA_BA	Provides EC-PCA balances between the source system and the Central Finance system.
Logistics	AVL Sales Document	FINS_CFIN_DFV_AV_SO	Displays the count of sales documents and their details replicated from the source system to the Central Finance system.
	AVL Customer Invoice	FINS_CFIN_DFV_AV_CI	Displays the count of customer invoices transferred from the source system to the Central Finance system. In addition, you can view invoice details.
	AVL Purchasing Document	FINS_CFIN_DFV_AV_PO	Displays the count of purchase documents and their details replicated from the source system to the Central Finance system.
	AVL Purchase Order	FINS_CFIN_DIS_PO	Displays a specific purchase order, which is replicated from the source system to the Central Finance system.

Table 12.2 Central Finance Comparison Reports (Cont.)

Area	Report Name	Transaction	Report Usage
Logistics (Cont.)	AVL Supplier Invoice	FINS_CFIN_DFV_AV_SI	Displays the count of customer invoices transferred from the source system to the Central Finance system. In addition, you can view invoice details.
SEPA Mandate	SEPA Mandates	CFIN_CPAY_SEPA_REC	Displays SEPA mandates transferred from the source system to Central Finance.
Configuration	Configuration Consistency Check Report	FINS_CFIN_CC	Displays configuration inconsistencies between the source system and the Central Finance system.

Table 12.2 Central Finance Comparison Reports (Cont.)

Note

To execute these reports in Central Finance, you need the appropriate authorization to run the relevant transaction codes. In addition, you'll need to display authorization for the source company code (authorization object F_BKPF_BUK) for the drill-back functionality for finance reports.

In case of any missing comparison report functionality, consult SAP Note 2352483, which provides an overview of SAP Notes required to run comparison reports. If you see any performance issues, refer to source SAP Note 2747107.

12.3.2 Financial Accounting

In this section, you'll learn how to execute comparison reports for FI. As we've mentioned, you'll run these reports at various stages of the Central Finance implementation.

Count of Journal Entries

As the report name implies, the count of journal entries comparison provides a count of journal entries or documents transferred from the source system to Central Finance. In addition, it combines the total number of journal entries with errors. You'll execute them after an initial load is carried out or anytime during the online replication activity. For a given selection criterion, the system displays a count of the number of documents transferred to the Central Finance system. In addition, the report also displays journal entries stuck in SAP Application Interface Framework with error status and number of documents that didn't make it to Central Finance.

To execute the count of journal entries report, navigate to the report from **SAP Easy Access** menu **SAP Menu • Accounting • Central Finance • Comparison Reports • Financial Accounting (FI) • Count of Journal Entries** or use the Transaction FINS_CFIN_DFV_FI_NUM. After executing the transaction, you'll arrive at the **Central Finance: Comparison of FI Document Headers** screen, as shown at the top of Figure 12.4. On this screen, enter the source system under **Selection for Source System** ❶. Next, enter the relevant parameters to apply filters under the **Parameters for Source System** ❷ section as follows:

- **Company Code**
  At minimum, enter a company code or list of company codes.
- **Document Number**
  Enter a specific document or list of documents to check whether the documents were transferred to the Central Finance system.
- **Posting Period**
  Enter the posting period or a range of posting periods in which documents are posted in the source system.
- **Fiscal Year**
  Enter the fiscal year of documents posted in the source system. For the initial load, enter the fiscal year maintained in view `VCFIN_SOURCE_SET`.
- **Posting Date**
  Enter a specific posting date or a range of posting dates.
- **Entry Date**
  Enter the entry date of the document. For the initial load, enter the date when the initial load was extracted.
- **Document Type**
  Optionally, enter a document type or list of document types by which you want to get a journal entries count.

Once the selection parameters are entered, click the **Execute** ❸ button or use F8 to continue. Next, the system displays journal entries categories as follows ❹:

- **In Source System**
  Displays the total count of all journal entries in the source system based on filter criteria.
- **In Central Finance System**
  Displays the count of financial documents transferred to Central Finance.
- **In AIF: With Errors**
  Displays any journal entries stuck in SAP Application Interface Framework with error status.

- **In AIF: Without Errors**
  Displays documents or journal entries that were transferred, for example, foreign currency valuation postings, but weren't posted due to missing setups.
- **Document not in Central Finance**
  Displays only documents that didn't make it to Central Finance.

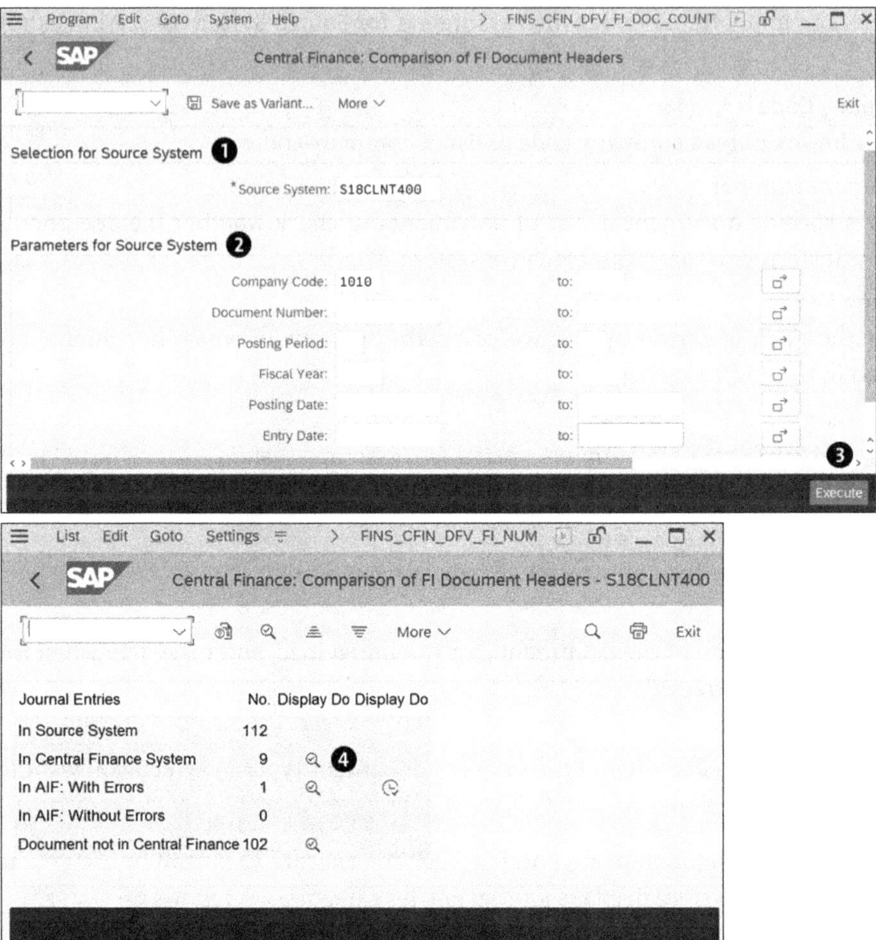

Figure 12.4 Comparison of FI Document Headers

[»]

> **Note**
>
> Depending on the volume, the report might take several minutes to complete. Generally, it's recommended to restrict the selection criteria as much as possible. Coordinate with the Basis team to allow the report to run much longer, if necessary, especially during period-end closing.

By double-clicking any count ❹, the system shows you the detail documents, as shown in Figure 12.5 ❶. In addition, you can navigate to SAP Application Interface Framework by clicking the **Execute** ❷ icon shown under the **Status Display Documents in AIF** column.

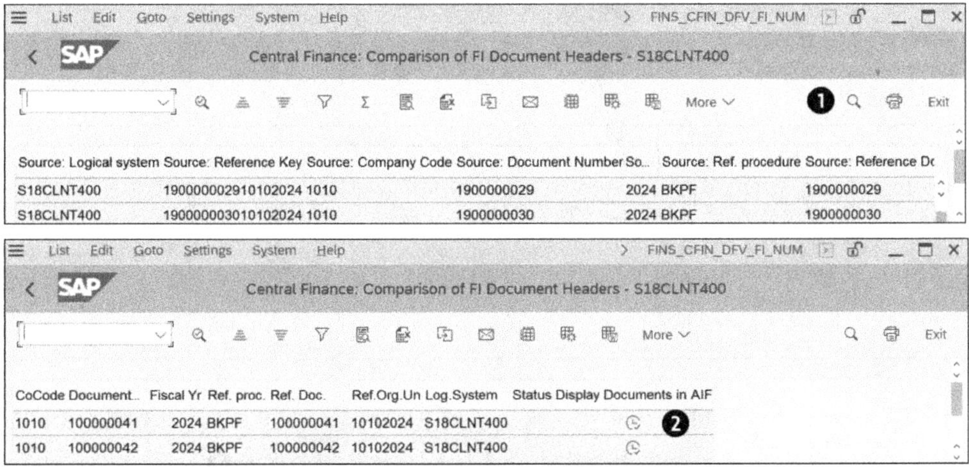

Figure 12.5 Comparison of FI Document Headers: View Documents

> **Tip**
>
> Transaction FINS_CFIN_DFV_FI_NUM relies on source system view VCFIN_SORUCE_SET to determine the posting date after the financial initial load is carried out.

General Ledger Account Balances

The report is useful to identify which general ledger accounts have different balances when comparing the source system and the Central Finance system. For a given general ledger account selection, the system brings in balances from the Universal Journal (table ACDOCA) for the Central Finance system balances. Conversely, a similar search is done in the source system tables, such as accounting document table BSEG, to show relevant credit or debit general ledger balances from the source system. The report compares both the source system and Central Finance data and then displays report output for a given selection criterion:

- Debit balance amount
- Credit balance amount
- Account balance

To execute the general ledger account balances report, navigate to the report from **SAP Easy Access** menu **SAP Menu • Accounting • Central Finance • Comparison Reports •**

Financial Accounting (FI) • G/L Account Balances or use Transaction FINS_CFIN_DFV_FI_BAL. After executing the transaction, you'll arrive at the **Central Finance: Comparison of FI Balances** screen, as shown in Figure 12.6.

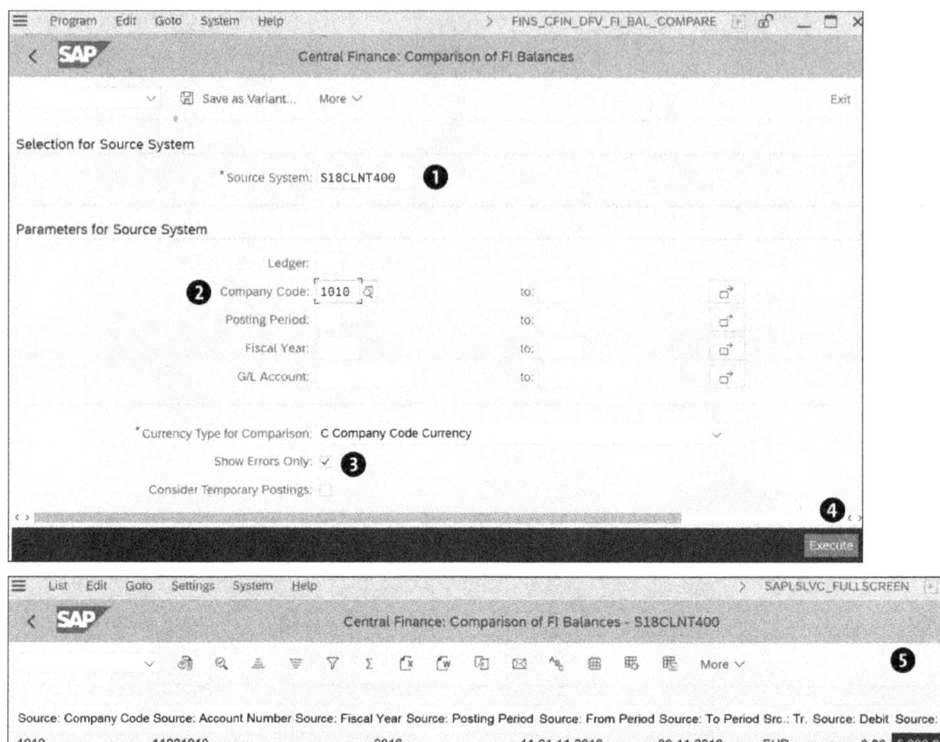

List Edit Goto Settings System Help | SAPLSLVC_FULLSCREEN

Central Finance: Comparison of FI Balances - S18CLNT400

More | Exit

Source: Company Code	Source: Account Number	Source: Fiscal Year	Source: Posting Period	Source: From Period	Source: To Period	Src.: Tr.	Source: Debit	Source: Credit	Source: B
1010	11001010	2016	11	01.11.2016	30.11.2016	EUR	0,00	5.000.000,00-	5.000.0
1010	12100000	2016	7	01.07.2016	31.07.2016	EUR	169,00	0,00	1
1010	12600000	2016	7	01.07.2016	31.07.2016	EUR	23,47	0,00	
1010	13100000	2016	12	01.12.2016	31.12.2016	EUR	130,00	0,00	1
1010	13600000	2016	4	01.04.2016	30.04.2016	EUR	127,40	0,00	1
1010	13600000	2016	5	01.05.2016	31.05.2016	EUR	0,00	12,74-	
1010	21100000	2016	1	01.01.2016	31.01.2016	EUR	0,00	1.000,00-	1.0
1010	21100000	2016	7	01.07.2016	31.07.2016	EUR	0,00	146,97-	1
1010	21120000	2016	4	01.04.2016	30.04.2016	EUR	0,00	127,40-	1
1010	21300000	2016	1	01.01.2016	31.01.2016	EUR	0,00	197.000,00-	197.0
1010	21300000	2016	2	01.02.2016	29.02.2016	EUR	0,00	18.000,00-	18.0
1010	21300000	2016	3	01.03.2016	31.03.2016	EUR	0,00	23.000,00-	23.0
1010	21300000	2016	4	01.04.2016	30.04.2016	EUR	0,00	103.000,00-	103.0
1010	29500100	2016	5	01.05.2016	31.05.2016	EUR	12,74	12,74-	

Figure 12.6 Comparison of FI Balances

On this screen, enter the source system under **Selection for Source System** ❶. Next, enter relevant parameters to apply filters under the **Parameters for Source System** ❷ section as follows:

- **Ledger**
  Select the ledger only if you want to compare a specific parallel ledger. Otherwise, the system uses the leading ledger by default.

- **Company Code**
  At minimum, enter a company code or list of company codes.
- **Posting Period**
  Enter a posting period or a range of posting periods in which you want to compare balances for a given general ledger account.
- **Fiscal Year**
  Enter the fiscal year of documents posted in the source system. For the initial load, enter the fiscal year maintained in view VCFIN_SOURCE_SET.
- **G/L Account**
  Enter a specific general ledger account, range of general ledger accounts, or list of general ledger accounts for balance comparison. If you've mapped multiple general ledger accounts and one general ledger account in Central Finance, the system displays differences with respect to each general ledger account. This must be considered for reconciliation purposes.
- **Currency Type for Comparison**
  Defaults to company code currency. You can choose a different currency type.

To improve performance of the report and only show reconciliation errors, choose **Show Errors Only** ❸ to restrict the selection data. Choose the **Consider Temporary Postings** checkbox if you want to include temporary postings for comparison.

Once the selection parameters are entered, click the **Execute** ❹ button or use F8 to continue. Next, the system displays any differences in the debit and credit balances format for the selected source general ledger account ❺.

At the bottom of Figure 12.6, the differences are highlighted in each row in red for the selected accounts.

[«]

Note

If balances don't tie although the system transferred all documents from the source system to the Central Finance system, there are several reasons for the differences. The most important reasons are listed here:

- Many general ledger accounts are mapped to a single general ledger account. This should be considered as part of the reconciliation process.
- For reversal documents replicated to Central Finance, the reversal reason is selected as normal reversal in the source system, whereas negative posting is selected in Central Finance. The system posts documents, but the totals of general ledger accounts may not tie correctly.
- If CO/FI replication is posted as an FI posting for the CO posting in the source SAP ERP system. This happens when you've selected the **G/L Reconciliation Postings Transfer** checkbox from VCFIN_SOURCE_SET when CO replication isn't in scope. Note that this flag isn't relevant in SAP S/4HANA because as a source system, the system is reconciled by design.

- If classic general ledger is connected as a source system, then the system supports only transaction and company code currencies during the initial load. Any hard currencies defined in the source system aren't transferred to Central Finance. You need to keep this in mind when you're reconciling balances for classic general ledger.

General Ledger Account Line Items

You'll run general ledger account line-item reports to display details at the line-item level information from the source system and the Central Finance system. The report provides various parameters to filter data on and displays details at the line-item level.

To execute the general ledger account line-items comparison report, navigate to the report from SAP Easy Access menu **SAP Menu • Accounting • Central Finance • Comparison Reports • Financial Accounting (FI) • G/L Account Line Items** or use Transaction FINS_CFIN_DFV_FI_DOC. After executing the transaction, you'll arrive at the **Central Finance: Comparison of FI Line Items** screen, as shown at the top of Figure 12.7. On this screen, like other reports you've seen earlier, enter the source system under **Selection for Source System** ❶. Next, enter relevant parameters to apply filters under the **Parameters for Source System** ❷ section as follows:

- **Company Code**
  At minimum, enter a company code or list of company codes.
- **Document Number**
  Enter a specific document, range of documents, or list of documents to check whether those documents were transferred to the Central Finance system.
- **Posting Period**
  Enter the posting period or a range of posting periods in which you want to compare balances for a given general ledger account.
- **Fiscal Year**
  Enter the fiscal year of documents posted in the source system. For the initial load, enter the fiscal year maintained in VCFIN_SOURCE_SET.
- **G/L account**
  Enter a specific general ledger account, range of general ledger accounts, or list of general ledger accounts for balance comparison. If you've mapped multiple general ledger accounts and one general ledger account in Central Finance, the system displays differences with respect to each general ledger account. This must be considered for reconciliation purposes.
- **Posting Date**
  Enter a specific posting date or a range of posting dates.

- **Entry Date**
  Enter the entry date of the document. For the initial load, enter the date when the initial load was extracted.
- **Document Type**
  Optionally, enter a document type or list of document types by which you want to get a journal entries count.
- **Currency Type for Comparison**
  The system defaults to the company code currency. You can choose a different currency type.

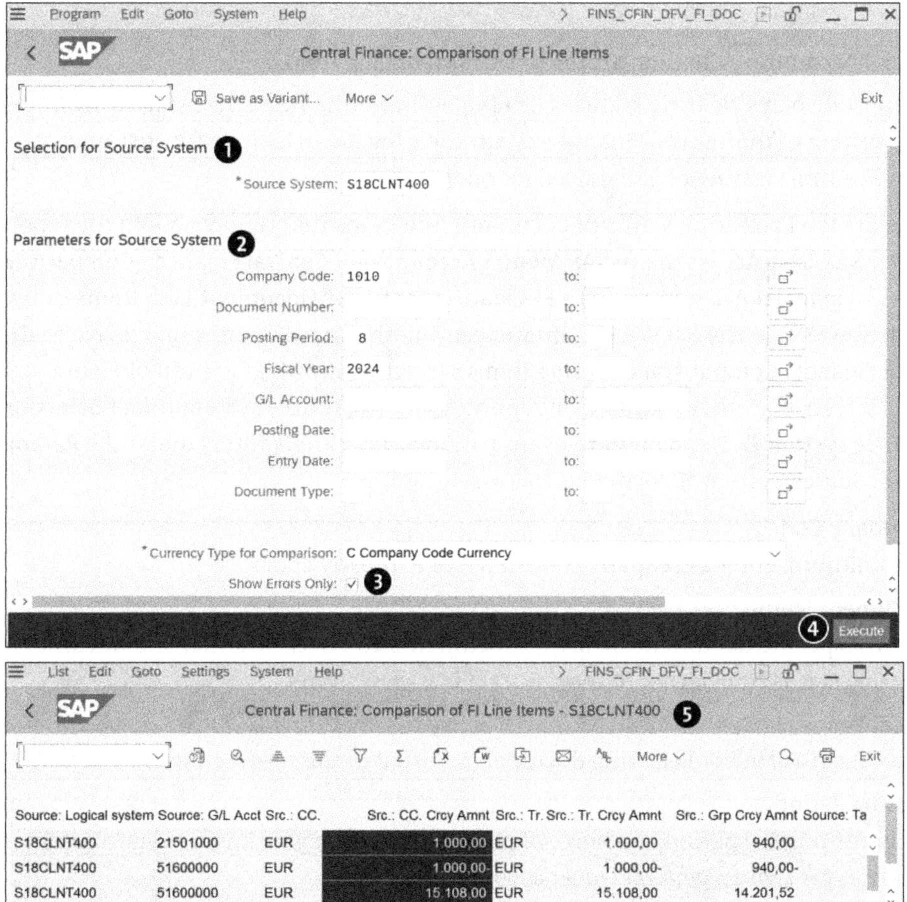

Figure 12.7 Comparison of FI Line Items

To improve the performance of the report and only show reconciliation errors, choose the **Show Errors Only** ❸ flag to restrict selection criteria. Choose the **Consider Temporary Postings** checkbox if you want to include temporary postings for comparison.

Once the selection parameters are entered, click the **Execute** ❹ button or use [F8] to continue. Next, the system displays any differences in the debit and credit balances format for the selected source general ledger account ❺.

[»]

Note

When reconciling at the line-item level for clearing items, pay attention to open items cleared in foreign currency or local currency in the source system. The system is posted in the company code currency in Central Finance. Consult SAP Note 2549601 for further information on this topic.

Financial Accounting Clearing Status of Document Line Items

The report displays clearing statuses of FI open item line items transferred from the source system to the Central Finance system. Clearing items include for customer, vendor, and central clearing of general ledger open items.

To execute the FI clearing status of document line items report, navigate to the report from the SAP Easy Access menu **SAP Menu • Accounting • Central Finance • Comparison Reports • Financial Accounting (FI) • FI Clearing Status of Document Line Items** or use Transaction FINS_CFIN_DFV_CLR. After executing the transaction, you'll arrive at the **Central Finance: Comparison of FI Line Items** screen, as shown at the top of Figure 12.8. On this screen, like other comparison reports, enter the source system under **Selection for Source system** ❶. Next, enter relevant parameters to apply filters under the **Parameters for Source System** ❷ section as follows:

- **Company Code**
  At minimum, enter a company code or list of company codes.
- **Document Number**
  Enter a specific document or list of documents to check whether the documents clearing status transferred to the Central Finance system.
- **Fiscal Year**
  Enter the fiscal year of clearing documents posted in the source system.
- **Posting Period**
  Enter the posting period or a range of posting periods in which you want to compare balances for a given general ledger account.
- **Posting Date**
  Enter a specific posting date or a range of posting dates.
- **Vendor**
  Enter a specific vendor, range of vendors, or list of vendors.
- **Customer**
  Enter a specific customer, range of customers, or list of customers.

- **G/L account**
  Enter a specific general ledger account, range of general ledger accounts, or list of general ledger accounts.
- **Document Type**
  Optionally, enter a document type or list of document types by which you want to get a journal entries count.
- **Currency Type for Comparison**
  The system defaults to company code currency. You can choose a different currency type.

Likewise, to improve performance of the report and only show reconciliation errors, choose the **Show Errors Only** flag to restrict selection data. Choose the **Consider Temporary Postings** checkbox if you want to include temporary postings for comparison.

Once the selection parameters are entered, click the **Execute** ❸ button or use [F8] to continue. Next, the system displays any differences in debit and credit balances format for the selected source general ledger account ❹.

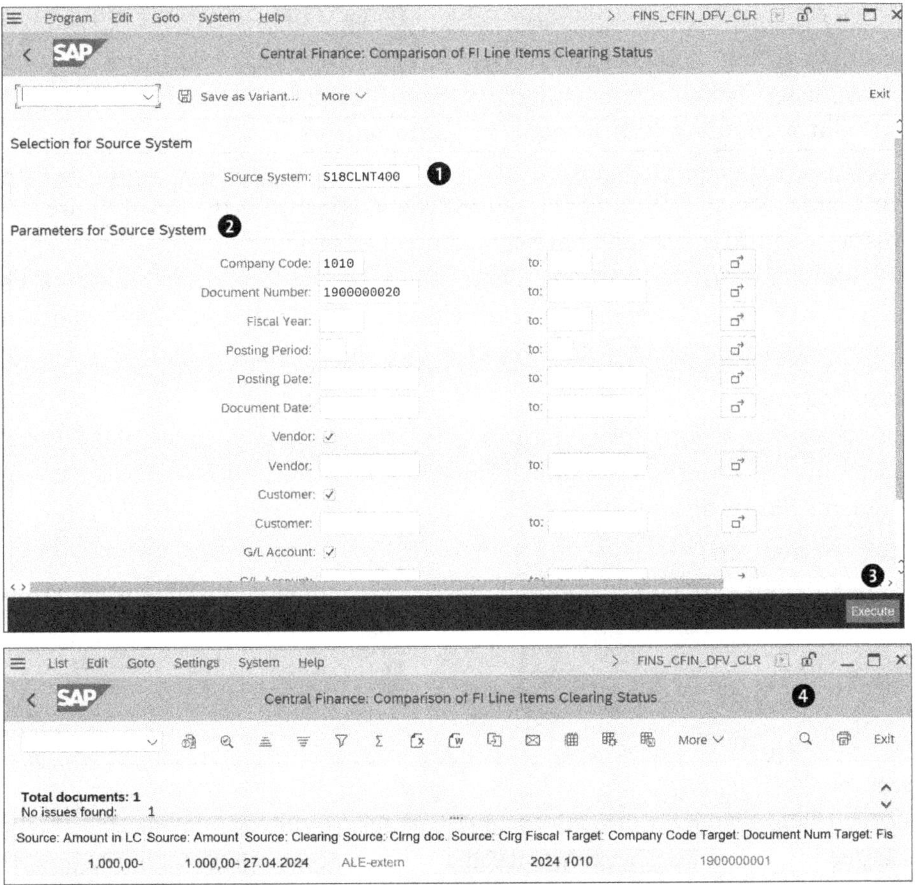

Figure 12.8 Comparison of FI Line Items Clearing Status

[»]

Note

If the clearing status comparison report is missing in your system, refer to SAP Note 3408990 to implement missing SAP Notes.

12.3.3 Management Accounting

SAP provides comparison reports for management accounting–related reporting activities. In this section, first you'll learn Controlling (CO) secondary postings reports and then we'll transition into other comparison reports such as commitment comparison reports in management accounting.

Count of Controlling Documents

The first level 1 report in report controlling is the count of CO secondary posting documents. Using this report, you can get a count of CO secondary documents replicated from the source system to the Central Finance system with a specific source system selection criterion such as controlling area, fiscal year, posting date, and so on.

To execute the count of CO documents report, navigate to the report from **SAP Easy Access** menu **SAP Menu • Accounting • Central Finance • Comparison Reports • Management Accounting (CO) • Count of CO documents** or use Transaction FINS_CFIN_DFV_CO_NUM. After executing the transaction, you'll arrive at the **Central Finance: Comparison of FI Document Headers** screen, as shown at the top of Figure 12.9. Note the selection screen doesn't have company code as a selection because the report is based on table COBK. On this screen, enter the source system under **Selection for Source system** ❶. Next, enter relevant parameters to apply filters under the **Parameters for Source System** ❷ section as follows:

- **Controlling Area**
  Enter the source system controlling area, which represents an organizational unit in cost accounting. If you enter a nonexistent controlling area, the report excludes any documents from that controlling area.
- **Document Number**
  Enter a specific CO document number or list of documents to check whether the documents were transferred to the Central Finance system.
- **Posting Period**
  Enter a posting period or a range of posting periods in which documents are posted in the source system.
- **Fiscal Year**
  Enter the fiscal year of documents posted in the source system. For example, for the initial load, enter the fiscal year maintained in VCFIN_SOURCE_SET.

- **Posting Date**
  Enter a specific posting date or a range of posting dates.
- **Entry Date**
  Enter the entry date of document. For the initial load, enter the date when the initial load was extracted.

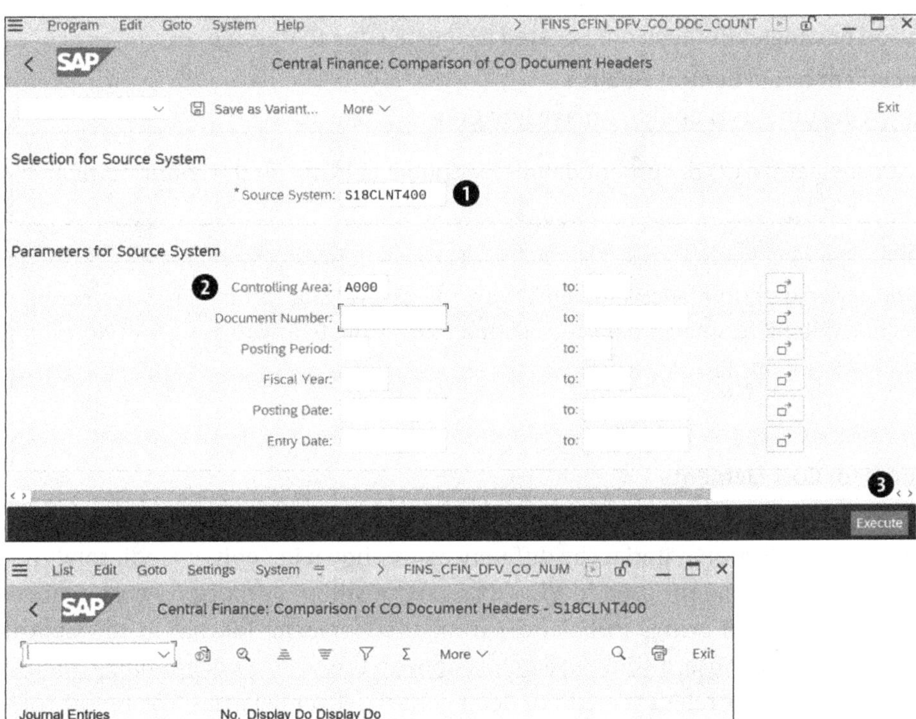

Figure 12.9 Comparison of CO Document Headers

Once the selection parameters are entered, click the **Execute** ❸ button or use F8 to continue. Next, as shown in Figure 12.9, the system displays journal entries categories:

- **In Source System**
  Displays the total count of all CO documents in the source system based on filter criteria.
- **In Central Finance System**
  Displays the count of CO secondary documents transferred to Central Finance.

- **In AIF: With Errors**
  Displays the count of CO documents stuck in SAP Application Interface Framework with error status. When you click on the magnifying glass ❹, the system navigates to the SAP Application Interface Framework correction tool to review further results.
- **In AIF: Without Errors**
  Shows documents or journal entries transferred, for example, foreign currency valuation postings, but they aren't posted for posted due to missing setup.
- **Document not in Central Finance**
  Shows documents that didn't make it to Central Finance.

You can navigate to CO document details by double-clicking on the count number.

Note

If you observe performance problems with the report, SAP recommends restricting selection criteria to smaller data sets. In addition, evaluate any missing SAP Notes in the source system.

Balances for Cost Elements

With this level 2 report, the system displays differences for the cost elements you've selected in the selection criteria. The differences are shown in debit or credit totals format for each cost element used in CO secondary postings. For a given cost element selection, the system brings in balances from the Universal Journal in the Central Finance system. Likewise, a similar search is done in the source system tables such as table COEP to show the relevant credit or debit general ledger balances. The report compares both the source system and the Central Finance system and displays balances.

To execute the cost element balances comparison report, navigate to the report from **SAP Easy Access** menu **SAP Menu • Accounting • Central Finance • Comparison Reports • Management Accounting (CO) • Balances for Cost Elements** or use Transaction FINS_CFIN_DFV_CO_BAL. After executing the transaction, you'll arrive at the **Central Finance: Comparison of CO Balances** screen, as shown in Figure 12.10. On this screen, enter the source system under the **Selection for Source System** ❶. Next, enter relevant parameters to apply filters under the **Parameters for Source System** ❷ section as follows:

- **Controlling Area**
  Enter the source system controlling area, which represents an organizational unit in cost accounting. If you enter a nonexistent controlling area, the report excludes any documents from that controlling area.
- **Company Code**
  Enter a company code or list of company codes.

- **Posting Period**
  Enter a posting period or a range of posting periods in which you want to compare balances for a given general ledger account.
- **Fiscal Year**
  Enter a fiscal year of documents posted in the source system. For the initial load, enter a fiscal year maintained in VCFIN_SOURCE_SET.
- **Cost Element**
  Enter a specific cost element, range of cost elements, or list of cost elements for balance comparison. If you've mapped multiple cost elements to one cost element in Central Finance, the system displays differences with respect to each cost element. This must be considered for reconciliation purposes.
- **Order**
  Enter a specific order, range of orders, or list of orders used in secondary posting for comparison purpose.
- **Order Type**
  The system provides the order type as the selection option rather than entering an individual order.
- **Order Group**
  Enter an order group or list of order groups rather than selecting individual orders.
- **Cost Center**
  Cost centers represent an organization unit within SAP such as a department or where cost is collected for further analysis. Enter the cost center used in secondary postings.
- **Cost Center Group**
  Cost center group can be entered rather than entering individual cost centers.
- **Currency Type for Comparison**
  The system defaults to the controlling area currency. You can choose a different currency type such as transaction currency. Only controlling area currency and transaction currency are available. Object currency varies with CO object and hence doesn't provide consistent data.

To improve the performance of the report and only show reconciliation errors, choose the **Show Errors Only** ❸ flag to restrict selection data.

Once the selection parameters are entered, click the **Execute** ❹ button or use F8 to continue. Next, the system displays any differences in debit and credit balances format for the selected criteria ❺.

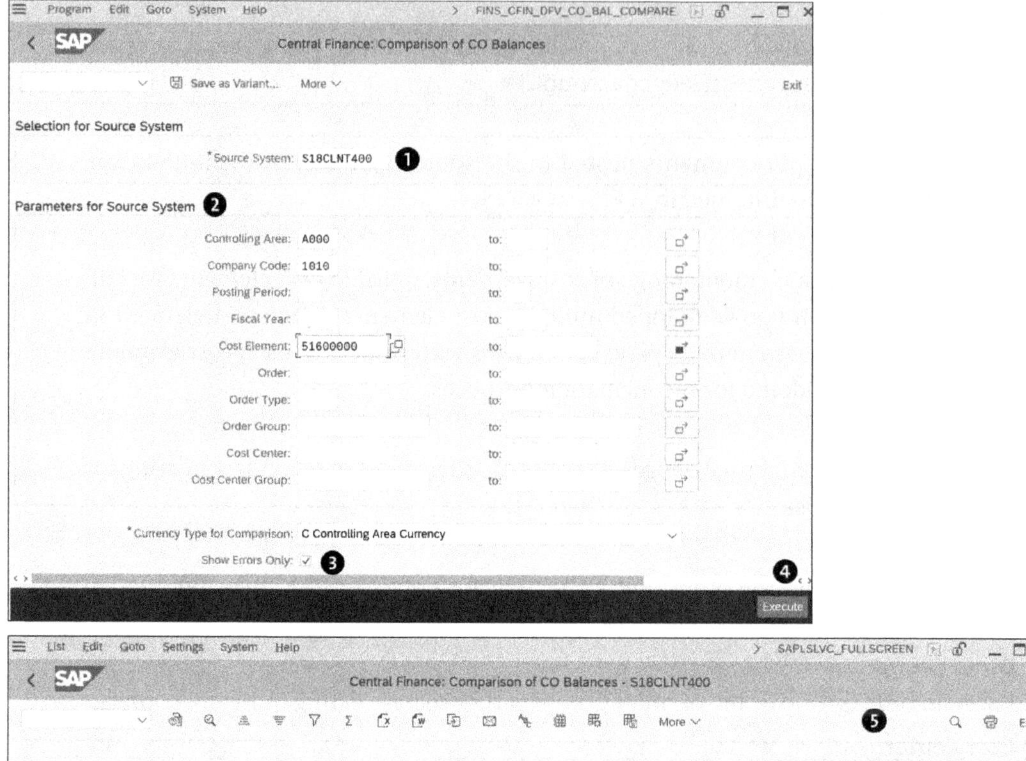

Figure 12.10 Comparison of CO Balances

Note

If balances don't tie although the system transferred all CO secondary documents, check for reasons such as missing SAP Notes, nonexistent controlling areas, or many-to-one mapping of cost elements. In addition, check transaction currency and the document entry date in the selection criteria, which may impact the outcome of the report.

Line Items for Cost Elements

With this level 3 CO line-item comparison report, you can compare CO lines transferred from the source system to the Central Finance system by document number, fiscal year, cost element, currency, and other selection parameters.

To execute the CO line-item comparison report, navigate to the report from **SAP Easy Access** menu **SAP Menu • Accounting • Central Finance • Comparison Reports • Consistency Checks • Line Items for Cost Elements** or use Transaction FINS_CFIN_DFV_CO_

DOC. After executing the transaction, you'll arrive at the **Central Finance: Comparison of CO Line Items** screen, as shown at the top of Figure 12.11. On this screen, like any other comparison report, enter the source system under **Selection for Source System** ❶. Next, enter relevant parameters to apply filters under the **Parameters for Source System** ❷ section as follows:

- **Controlling Area**
  Enter the source system controlling area, which represents an organizational unit in cost accounting. If you enter a nonexistent controlling area, the report excludes any documents from that controlling area.
- **Company Code**
  Enter a company code or list of company codes for which you want to perform CO documents transferred to the Central Finance system.
- **Document Number**
  Enter a specific CO document number or list of documents to check whether the documents were transferred to the Central Finance system.
- **Posting Period**
  Enter a posting period or a range of posting periods in which documents were posted in the source system.
- **Fiscal Year**
  Enter a fiscal year of documents posted in the source system. For example, for the initial load, enter a fiscal year maintained in `VCFIN_SOURCE_SET`.
- **Posting Date**
  Enter a specific posting date or a range of posting dates.
- **Entry Date**
  Enter the entry date of the document. For the initial load, enter the date when the initial load was extracted.
- **Cost Element**
  Enter a specific cost element, range of cost elements, or list of cost elements for balance comparison. If you've mapped multiple cost elements to one cost element in Central Finance, the system displays differences with respect to each cost element. This must be considered for reconciliation purposes.
- **Order**
  Enter a specific order, range of orders, or list of orders used in secondary posting for comparison purposes.
- **Order Type**
  The system provides the order type as a selection option rather than entering individual orders.
- **Order Group**
  Enter an order group or list of order groups rather than selecting individual orders.

- **Cost Center**
  Enter cost centers used in secondary postings. Cost centers represent organization units within SAP such as a department or where cost is collected for further analysis.
- **Cost Center Group**
  Enter a cost center group rather than individual cost centers.
- **Currency Type for Comparison**
  The system defaults to controlling area currency. You can choose a different currency type such as transaction currency. Only controlling area currency and transaction currency are available. Object currency varies with each CO object and hence doesn't provide consistent data.

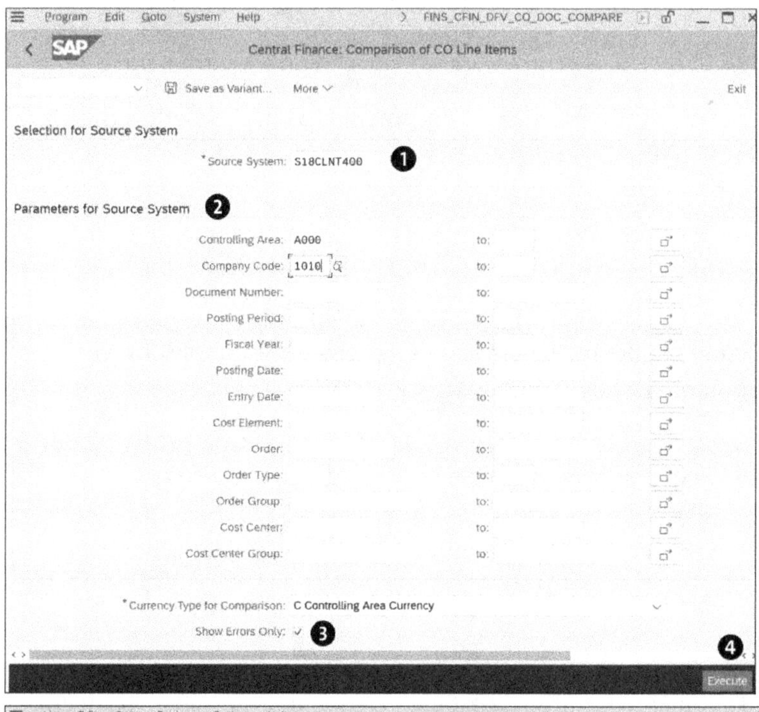

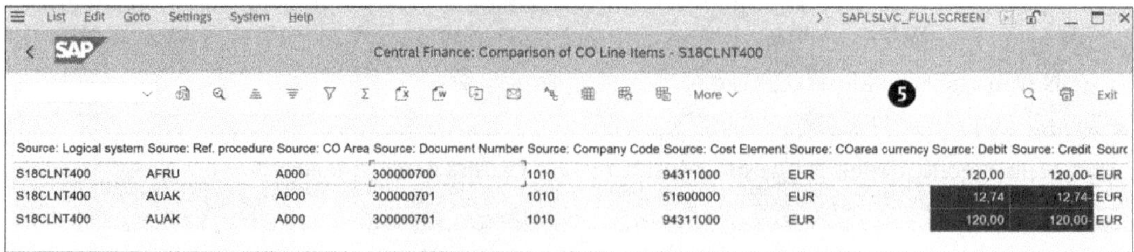

Source: Logical system	Source: Ref. procedure	Source: CO Area	Source: Document Number	Source: Company Code	Source: Cost Element	Source: COarea currency	Source: Debit	Source: Credit	Sourc
S18CLNT400	AFRU	A000	300000700	1010	94311000	EUR	120,00	120,00-	EUR
S18CLNT400	AUAK	A000	300000701	1010	51600000	EUR	12,74	12,74-	EUR
S18CLNT400	AUAK	A000	300000701	1010	94311000	EUR	120,00	120,00-	EUR

Figure 12.11 Comparison of CO Line Items

To improve performance of the report and only show reconciliation errors, choose the **Show Errors Only** ❸ flag to restrict selection data

Once the selection parameters are entered, click the **Execute** ❹ button or use [F8] to continue. Next, the system displays any differences in debit and credit balances format for the selected criteria ❺.

12.3.4 Commitments

Commitments can be replicated in real time from the source system to the Central Finance system. In this section, you'll learn how to run SAP-delivered commitment comparison reports.

Count of Line Items for Commitments in Controlling

The count of line items for commitments in CO comparison report provides users with a count of commitment line items transferred from the source system to the Central Finance system. You'll execute the report after an initial load is carried out or anytime during the online replication activity. For a given selection criterion, the system displays a count of the number of commitment documents transferred to the Central Finance system.

To execute the count of line items for commitments in CO comparison report, navigate to the report from **SAP Easy Access** menu **SAP Menu • Accounting • Central Finance • Comparison Reports • Commitments • Count of Line Items for Commitments in CO** or use Transaction FINS_CFIN_DFV_CMTNUM. After executing the transaction, you'll arrive at the **Central Finance: Comparison of FI Document Headers** screen, as shown at the top of Figure 12.12. On this screen, enter the source system under **Selection for Source System** ❶. Next, enter relevant parameters to apply filters under the **Parameters for Source System** ❷ section as follows:

- **Controlling Area**
  Enter the source system controlling area, which represents an organizational unit in cost accounting. If you enter a nonexistent controlling area, the report excludes any documents from that controlling area.
- **Company Code**
  Enter a company code or list of company codes.
- **Document Number**
  Enter a specific document or list of documents to check whether the documents were transferred to the Central Finance system.
- **Fiscal Year**
  Enter a fiscal year of documents posted in the source system. For the initial load, enter a fiscal year maintained in `VCFIN_SOURCE_SET`.
- **Period**
  Enter a period or a range of posting periods in which commitments documents are posted in the source system.

The report provides options to choose either the **Old Commitment** or **New Commitment** ❸ radio button under the **Search Type** section. With the **Old Commitment** option, the system compares entries from table COOI table in Central Finance and the source system. On the other hand, for the **New Commitment** option, the system compares commitment values stored from the Universal Journal (table ACDOCA) to commitment table entries in the source system.

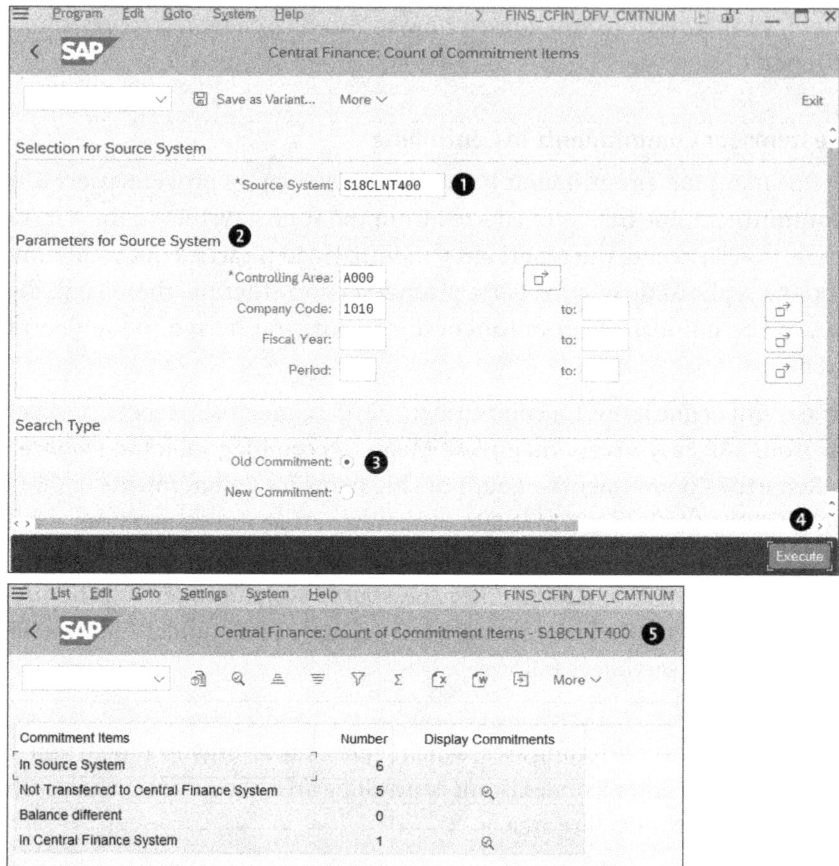

Figure 12.12 Count of Line Items for Commitments in CO

Once the selection parameters are entered, click the **Execute** ❹ button or use F8 to continue. Next, the system presents the output of the count of commitments with the following categories ❺:

- **In Source System**
  Displays the total count of all commitments in the source system based on applied filter criteria.
- **Not transferred to Central Finance System**
  Displays a count of commitment items not transferred to the Central Finance system.

- **Balance different**
  Displays a count of commitments items that are different in balances between the source system and the Central Finance system.
- **In Central Finance**
  Displays only commitment documents that made it to Central Finance.

From Figure 12.12, when you click the magnifying glass under **Display Commitments**, the system navigates to the **Central Finance: Count of Commitment Items - S18CLNT400** screen, as shown in Figure 12.13.

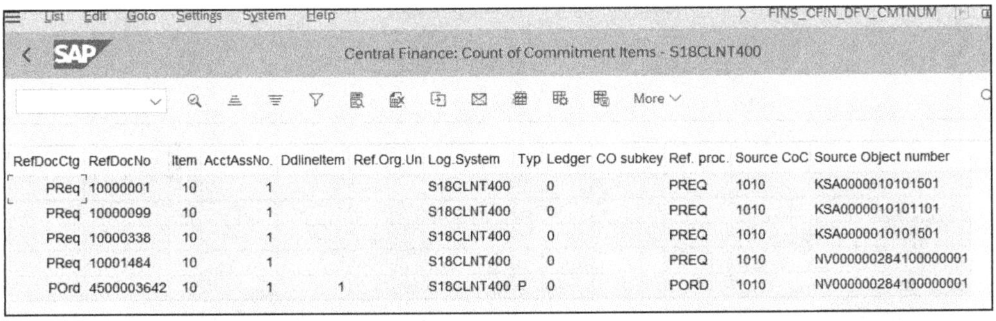

List Edit Goto Settings System Help FINS_CFIN_DFV_CMTNUM

Central Finance: Count of Commitment Items - S18CLNT400

More

RefDocCtg	RefDocNo	Item	AcctAssNo.	DdlineItem	Ref.Org.Un	Log.System	Typ	Ledger	CO subkey	Ref. proc.	Source CoC	Source Object number
PReq	10000001	10	1			S18CLNT400		0		PREQ	1010	KSA0000010101501
PReq	10000099	10	1			S18CLNT400		0		PREQ	1010	KSA0000010101101
PReq	10000338	10	1			S18CLNT400		0		PREQ	1010	KSA0000010101501
PReq	10001484	10	1			S18CLNT400		0		PREQ	1010	NV000000284100000001
POrd	4500003642	10	1	1		S18CLNT400	P	0		PORD	1010	NV000000284100000001

Figure 12.13 Count of Line Items for Commitments in CO Details

Line Items for Commitments in Controlling

With this level 3 line-item for commitments in CO comparison report, you can compare commitments items transferred from the source system to the Central Finance system by document number, fiscal year, cost element, currency, and other selection parameters.

To execute the line-item for commitment in CO comparison report, navigate to the report from **SAP Easy Access** menu **SAP Menu • Accounting • Central Finance • Comparison Reports • Commitments • Line Items for Commitments in CO** or use Transaction FINS_CFIN_DFV_CMTDOC. After executing the transaction, you'll arrive at the **Central Finance: Comparison of Commitment items** screen, as shown at the top of Figure 12.14. On this screen, like any other comparison reports we've covered, enter the source system under **Selection for Source System** ❶. Next, enter relevant parameters to apply filters under the **Parameters for Source System** ❷ section as follows:

- **Controlling Area**
  Enter the source system controlling area, which represents an organizational unit in cost accounting. If you enter a nonexistent controlling area, the report excludes any commitments from that controlling area.
- **Company Code**
  At minimum, enter a company code or list of company codes for which you want to perform commitments count transferred from the source system to the Central Finance system.

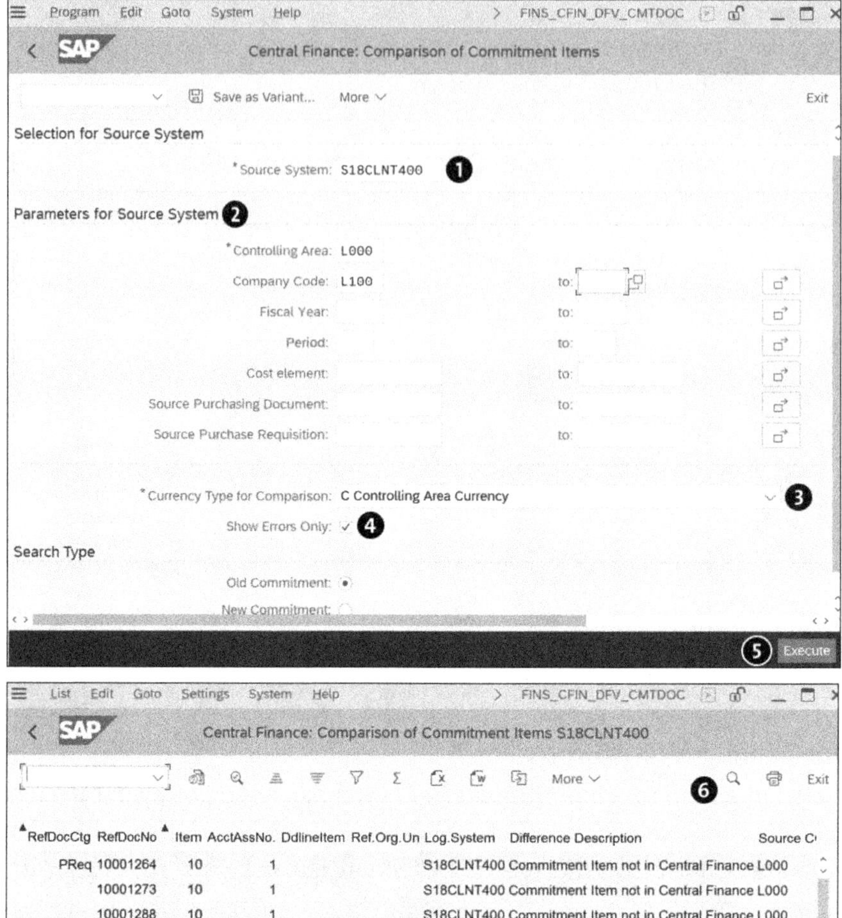

Figure 12.14 Comparison of Commitments Items

- **Fiscal Year**
  Enter a fiscal year of commitment documents posted in the source system.
- **Period**
  Enter a posting period or a range of posting periods in which documents are posted in the source system.
- **Cost element**
  Enter a specific cost element or list of cost element to search based on cost elements.
- **Source Purchasing Document**
  Enter a specific purchase document, a range of purchasing documents, or list of purchase orders.
- **Source Purchase requisition**
  Enter a specific purchase requisition, a range of purchase requisitions, or a list of purchase requisitions.

- **Currency Type for Comparison ❸**
  The system defaults to the controlling area currency. You can choose a different currency type such as object currency.
- **Show Errors Only ❹**
  Select to restrict selection data to improve the performance of the report and only show reconciliation errors.
- **Old Commitment** or **New Commitment**
  Choose the appropriate radio button based on activation of the new commitment functionality.

Once the selection parameters are entered, click the **Execute ❺** button or use F8 to continue. Next, the system displays any differences in debit and credit balances format for the selected criteria ❻.

12.3.5 Profit Center Accounting

Central Finance provides comparison reports to compare PCA data transferred between the source systems and the Central Finance systems. To run these reports, ensure you have access to authorization objects F_BKPF_BUK and F_CFIN_SRC. As you recall, PCA supports only three types of replication comparison. Table 12.3 provides PCA-supported business transaction types for Central Finance replication.

Business Transaction Type	Description
PRC5	Direct PCA document entry
PCAA	PCA assessments
PCAD	PCA distributions

Table 12.3 PCA Business Transaction Types for Comparison Reports

Let's explore each PCA comparison report next:

- **Central Finance: Comparison of PCA Document Counts (Transaction FINS_CFIN_DFV_PCA_CN)**
  This provides a count PCA document replicated from the source system to Central Finance. Once the report is executed with the **Ongoing Replication** radio button selection, the comparison report reads data from tables GLPCT or GLPCA in the source system and table BKPF in Central Finance to get the count. In addition, the report provides document details and balances by drilling down from the output screen. You can look up document details and balances by various selection criteria such as fiscal year, posting period, transfer status, and so on. If you select the **Initial Load** radio button, the system reads data from tables GLPCT or GLPCA based on selection criteria,

displays the count of PCA documents, total number entries in Central Finance, and total number of documents not in Central Finance.

- **Central Finance: Comparison of PCA Balances (Transaction FINS_CFIN_DFV_PCA_BA)**
  You'll use this report to compare the initial load balances transferred for PCA. One limitation of this report is that it can't display details of PCA document. The report compares data from table GLPCT or table GLPCA source system to tables BKPF and ACDOCA from the Central Finance system and displays comparative data for the initial load or ongoing replication.

Figure 12.15 displays the selection screens of Transaction FINS_CFIN_DFV_PCA_CN ❶ and FINS_CFIN_DFV_PCA_BA ❷.

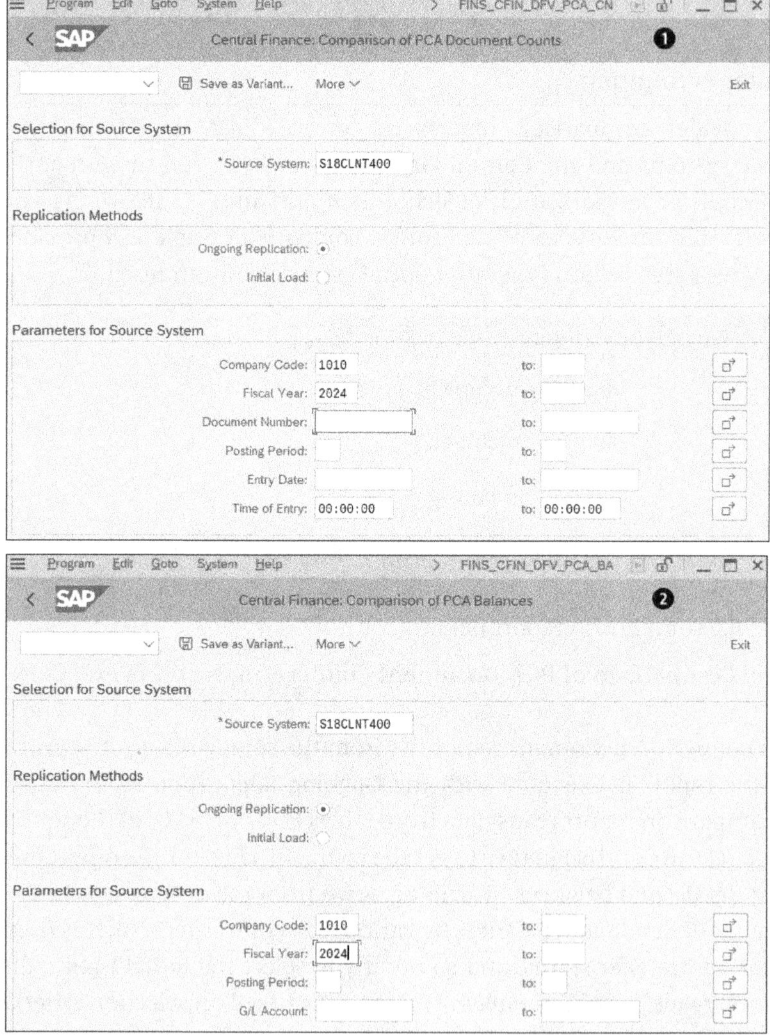

Figure 12.15 Comparison Reports for PCA

12.3.6 SEPA Mandates Reconciliation

Central Finance provides SEPA mandates reconciliation Transaction CFIN_CPAY_SEPA_REC (program FINS_CFIN_APAR_SEPA_REC). You can navigate to this transaction from **SAP Easy Access** menu **SAP Menu • Accounting • Central Finance • Comparison Reports • SEPA Mandates • SEPA Mandate.** With this transaction, you can do the following:

1. Perform a comparison of replicated mandate details between the source system and the Central Finance system.
2. Dive into the comparison results for each field of the mandate replicated by double-clicking the **Display** icon.
3. Change a mandate through the **Change Mandate** screen.
4. Match SEPA mandates between the source system and Central Finance. While synchronizing mandates, if errors occur, you'll use Transaction /AIF/ERR monitoring and error handling or Transaction WE02 to display the IDoc transaction to troubleshoot the errors.
5. Refresh the migrated results.

Now, let's run the SEPA reconciliation tool with detailed instructions. After reaching the transaction, the system takes you to the **Reconciliation Tool for SEPA Mandate** screen, as shown in Figure 12.16. On this screen, enter the search criteria values in the sections as follows:

❶ **SEPA Mandate Common Data**
Enter values in any of the following fields: **Mandate ID**, **Receiver ID**, **Receiver Creditor ID**, or **Sender ID**.

❷ **Administrative Date**
Enter values in this section if you want to search by user ID or timestamp when SEPA mandates were created.

❸ **System Data**
Enter the source logical system in the **Source System** field.

❹ **Search Options**
From the options **Search by Source Data** or **Search by Central Data**, choose **Search by Source Data** for this example.

❺ **Status Selection**
Choose the **Correct Data**, **Lost in Source or Central**, and **Not Identical Data** checkboxes as required by your requirements.

Once all fields are completed, click the **Execute** ❻ button. The system displays SEPA mandate comparison results on the resulting **SEPA Mandate Reconciliation Statistics** ❼ screen.

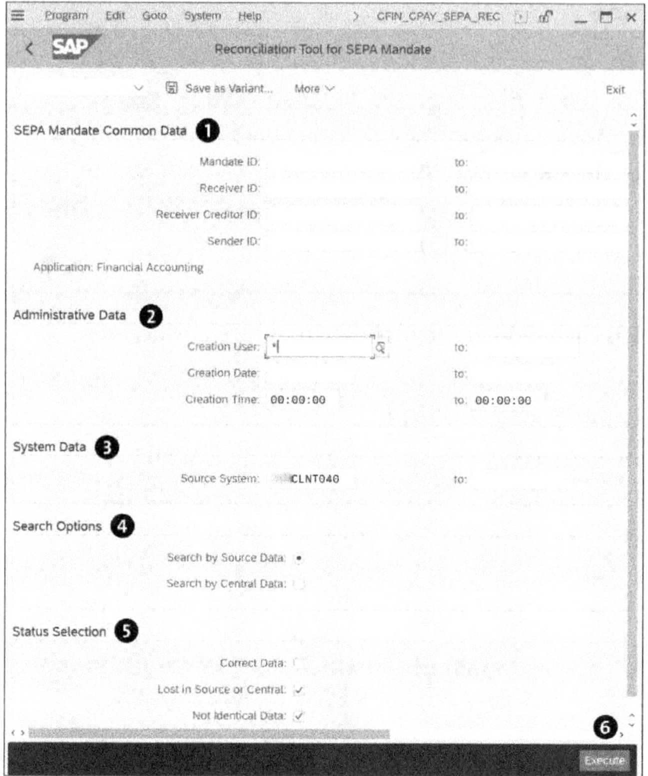

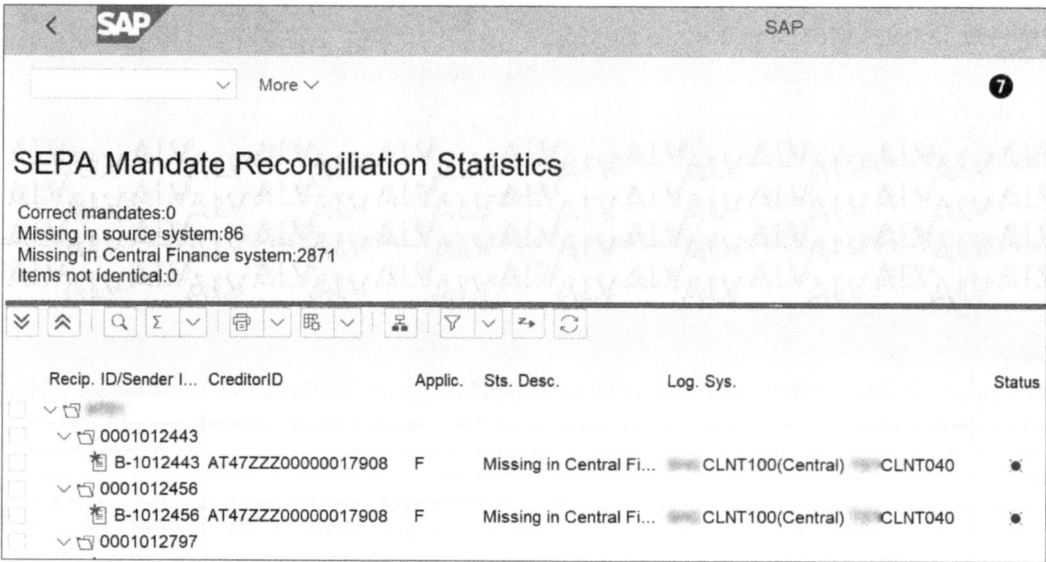

Figure 12.16 SEPA Mandates Reconciliation Tool

12.4 Running SAP S/4HANA Reporting Tools

In this section, we'll explore other reconciliation tools that can be used in conjunction with comparison reports in Central Finance for reconciliation.

12.4.1 SAP Fiori Apps

In SAP S/4HANA, with SAP Fiori apps, users have a new and enhanced experience for interacting with the system. The main benefit of SAP Fiori apps is that users can personalize and contextualize their content. In addition, information is available to analyze and make decisions quickly. For example, users get answers to questions such as these:

- What's my profitability during close?
- What are my overview receivables?
- Do I have enough cash on hand?
- What are my days payable outstanding?

SAP Fiori represents the future of user interfaces (UIs) to perform day-to-day activities. In this section, you'll learn the fundamentals of SAP Fiori. Then, we'll transition into the other SAP tools—core data services (CDS) views, SAP Analysis for Microsoft Office, and SAP Analytics Cloud.

SAP Fiori Overview

There are four types of SAP Fiori apps to perform Central Finance activities such as executing day-to-day operational tasks, workflow approvals, self-service tasks, and running reports:

- **Transactional apps**
  Transactional apps support users in their day-to-day operational activities in an enterprise. Typical tasks include creating, changing, or approving business processes. Security access is given to users based on tasks and the type of functions they perform. For example, a general ledger accountant reverses a manual journal entry using the Manage Journal Entries app (app ID F0717A).
- **Fact sheets**
  As the name implies, fact sheets provide users with factual business information to make critical decisions. An example of a fact sheet is the Customer Master fact sheets for Master Data Specialist - Business Partner Data (app ID F0046A), which provide users with contextual information on customer data quickly. From the fact sheet, you can navigate to other relevant transactional apps as necessary.
- **Analytical apps**
  Analytical apps provide real-time business insights based on assigned roles. For example, the Profit Centers - Plan/Actual app (app ID W0079) enables users to report

profit center actuals and plan data for a given selection criterion. An analytical app can be either launched as a tile or as a key performance indicator (KPI). In addition, drilldown capabilities are available with analytical apps to navigate to details.

- **Smart business**
  This combines multiple app types coupled with real-time analytics and KPIs to take necessary actions based on the information available.

Tip

SAP provides the SAP Fiori apps reference library to find an app and its details. You can search for apps by different classifications, such as line of business, industry, role, application component, product version, backend product, and best practice categories. The SAP Fiori apps library can be found at *http://s-prs.co/v590103*.

In SAP S/4HANA, the SAP Fiori launchpad is the recommended frontend interface accessed by business users. Compared to traditional SAP GUI programs and reports, in SAP S/4HANA, SAP Fiori apps can be widely used to perform day-to-day operational activities and period-closing activities in Central Finance. SAP Fiori has a friendly UI that runs in a web browser, tablet, or mobile phone. It's built with the latest SAPUI5 framework (HTML5) technology. In addition, SAP Fiori apps are built with the role-based model, which means apps are decomposed or separated for each role. Another important goal of SAP Fiori apps is to provide users with the ability to perform day-to-day tasks from a smartphone or tablet. Table 12.4 shows the important differences between traditional SAP GUI transactions and SAP Fiori apps.

SAP GUI	SAP Fiori
Transactions are added to multiple roles.	Apps are decomposed for each role, yet each app can be interlinked.
Separate transactions are required for add, change, and display functions.	A single app or tile combines multiple functionalities.
Complex transactions have inconsistent navigation.	Navigation is intuitive and consistent.
SAP GUI software access is required to log on to the system.	Browser-based apps can be accessed from mobile phones or tablets using multiple operating systems.
SAP GUI is based on the ABAP programming language.	The SAP Fiori architecture is based on HTML5, and representational state transfer (RESTful) application programming interfaces (APIs) enable the integration with cloud technologies and technologies of the future.

Table 12.4 SAP GUI versus SAP Fiori Apps

Tip

Consult SAP Note 2881803 for SAP Fiori best practices.

As shown in Figure 12.17, the SAP Fiori apps layout is intuitive, built with dynamic pages and flexible column layouts. The SAP Fiori launchpad is the frontend interface for business users and power users, serving as an entry point for the users. Based on security access, the SAP Fiori launchpad initial screen appears, which features tiles ❶, tile groups ❷, the **App Finder** ❸, and **Settings** ❹ for personalizing the homepage.

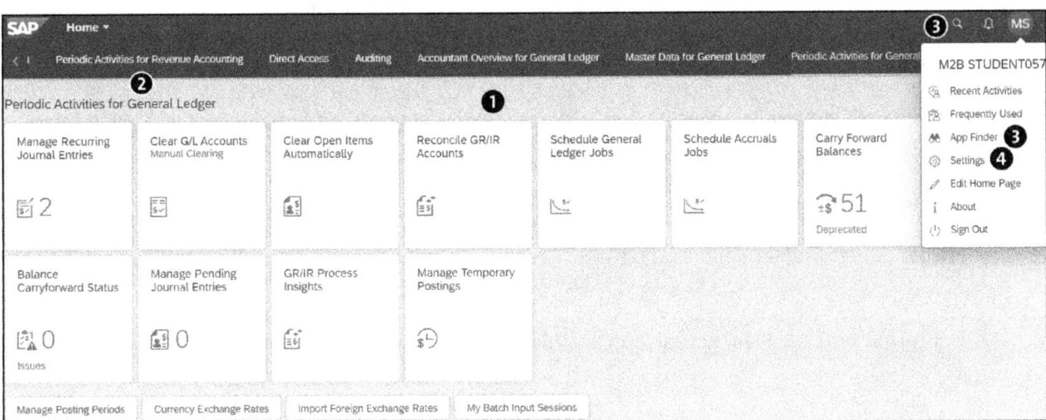

Figure 12.17 SAP Fiori Launchpad

For Central Finance reconciliation, in conjunction with Central Finance delivered comparison reports, several operational SAP Fiori apps can be used for reconciliation and reporting. Table 12.5 describes the most important reports that can be used to review and reconcile the initial load and online replicated data in Central Finance. Although this isn't an exhaustive list as the requirements vary with every project, you can use it as a starting point to build your own list.

SAP Fiori App	Purpose
Trial Balance (app ID F0996)	Display debit and credit balances based on a time interval
Balance Sheet/Income Statement (app ID F0708)	Display balance sheets and profit and loss (P&L) statements for selected selection criteria using operational, local, and global charts of accounts
Display Journal Entries (app ID F7697)	Analyze and view journal entries

Table 12.5 SAP Fiori Apps for Initial Load and Online Replication Validation

SAP Fiori App	Purpose
Display Financial Statement (app ID FSE3_XBRL)	Display financial statement in HTML-based classic financial statement version
Display G/L Account Line Items (app ID F2217)	Display general ledger account line items
Display Line items – Cost Accounting (app ID F4023)	Display line items from cost accounting documents
Display Account Balances (app ID W0130)	Display general ledger account balances
Display Supplier Balances (Version 2) (app ID F0701A)	Display vendor balances for a given financial period
Display Customer Balances (Version 2) (app ID F0703A)	Display customer balances for a given financial period
Manage Customer Line Items (app ID F0711)	Display customer line items
Cost Center Budget Report (app ID F3871)	Display cost center budget actual costs, commitments
Internal Orders – Plan/Actual (app ID W0076)	Display actual and plan costs for the given selection criteria
Market Segment – Actual (app ID F0960)	Actual and plan data for profitability analysis
Profit and Loss – Plan/Actual (app ID F1710)	Display actual and plan data for P&L accounts
Projects – Actual (app ID F0961)	Display actual and plan data for project/work breakdown structure (WBS) elements
My Spend (app ID F0393)	Track budget and actuals for a department or project
My Unusual Items (app ID F0368)	Display usual expense items for a department or project

Table 12.5 SAP Fiori Apps for Initial Load and Online Replication Validation (Cont.)

Ensure that you test all SAP Fiori apps as many apps which contain both finance and logistics data as you mayn't find logistics data in Central Finance.

Core Data Services Views

In Central Finance, you can leverage SAP-delivered CDS views, which form the basis to perform ad hoc reporting. SAP S/4HANA reporting tools consume CDS views to display reporting data. CDS views are used by technical teams to create virtual data models (VDMs). SAP offers three types of VDMs:

- **Basic views**
  These views are created on top of ABAP Data Dictionary (DDIC) tables or views with databases to retrieve transactional and master data.
- **Composite views**
  These views are created on top of basic views, as they don't access database tables directly. Composite views are considered intermediate views.
- **Consumption views**
  These views are built on top of composite views, and they don't interact either with basic views or DDIC tables directly. Consumption views are consumed by frontend tools. Like analytical queries, consumption views can be consumed from various reporting tools, such as the Query Browser app or the View Browser app, as discussed next.

Query Browser and View Browser

SAP delivers several CDS views for reporting both in Central Finance and SAP S/4HANA implementation projects. Users can access these CDS views via the Query Browser app or the View Browser app.

In SAP Fiori, the View Browser (app ID F2170) and Query Browser (app ID F1068) apps are available to explore CDS views. Users can search for views and queries related to relevant tasks using these browsers. To access the View Browser app and the Query Browser app, users should be assigned the `SAP_BR_ANLYTICS_SPECIALIST` role and the `SAP_BR_EMPLOYEE` role, respectively. The key difference between the two apps is that the View Browser app provides detailed information on the field definitions, annotations, and cross-references of the selected views. In addition, the View Browser app is used to display basic, composite, and consumption CDS view types. On the other hand, the Query Browser app is used to display standard SAP Fiori content and only shows consumption views that have been annotated as analytical queries. Therefore, end users are generally given access to the Query Browser app to perform ad hoc reporting, whereas access to the View Browser app should be limited to advanced users and super users.

The Query Browser app can be accessed by clicking on the **Query Browser** tile, as shown in Figure 12.18 ❶, from the SAP Fiori launchpad. Once the tile is selected, the tile displays a count of available views ❷. This count changes as SAP adds new CDS views. On this screen, you can search ❸ for a specific report, for example, "profit center plan" ❹. In our example, the SAP-delivered analytical query **C_PROFITCENTERPLANACTQ2701** ❺ is displayed.

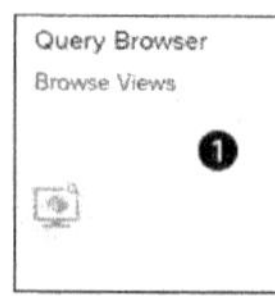

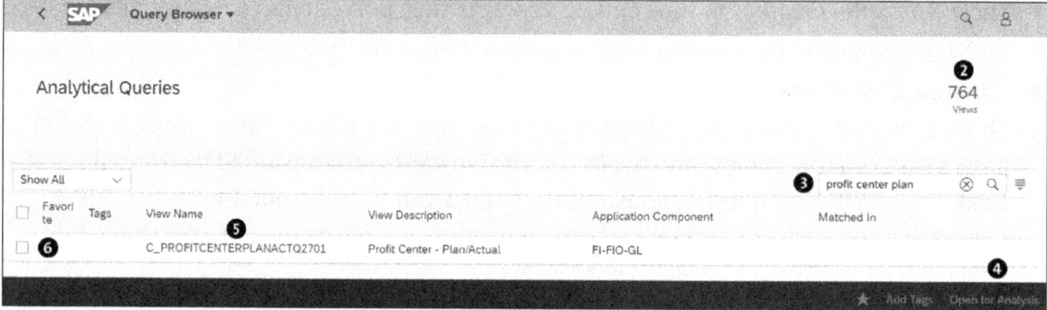

Figure 12.18 Query Browser

The system provides two ways to display query results. From Figure 12.18, you can select the checkbox for a view ❻ and then click on **Open for Analysis** ❹. The other way is to double-click on the entry under **View Name** ❺.

As shown in Figure 12.19, all field definitions will be displayed ❶. Click on **Open for Analysis** ❷, and a popup window will open to select parameters ❸. After filling in the required filter criteria, click **OK** ❹ to continue.

Figure 12.19 Query Browser Selection

As shown in Figure 12.19, the **Annotation** tab ❺ provides additional metadata relevant to CDS views. The annotations express the properties and functionality of the CDS view such as whether the CDS view is modifiable for inserts, updates, or deletes. In addition, it describes whether you can add filters.

The resulting screen, as shown in Figure 12.20, displays **DIMENSIONS** ❶, **COLUMNS** ❷, and **ROWS** ❸ on the left. Users can simply drag and drop the required fields as required. In addition, the output report provides additional tools, such as the ability to apply additional filter criteria ❹ and to export data to a Microsoft Excel spreadsheet ❺.

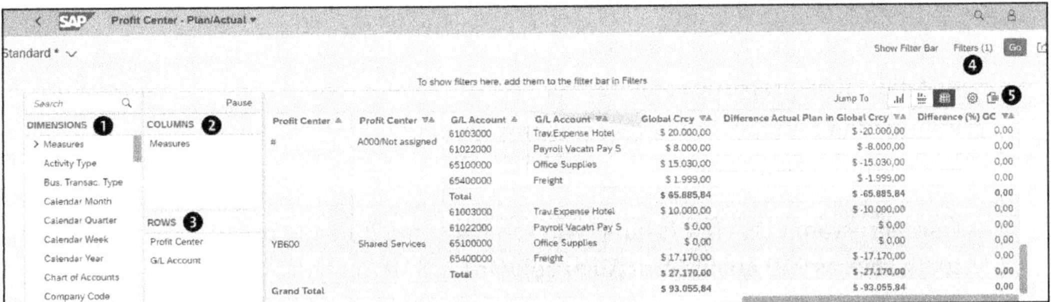

Figure 12.20 Query Browser Output

To access the view browser, log on to the SAP Fiori launchpad. On the homepage, search for "View Browser", and then click on the **View Browser** tile, as shown in Figure 12.21 ❶. On the resulting screen, notice the different categories available in the header: **Basic**, **Composite**, **Consumption**, **Extension**, and **Undefined** ❷. As you recall, these options aren't available in the query browser. Like the Query Browser app, you can search ❸, for example, for "Profit Center Plan" views. In our example, the system displayed the **C_ProfitCenterPlanActQ2701** ❹ view. An understanding of the View Browser app requires technical knowledge, so we recommend consulting with technical teams.

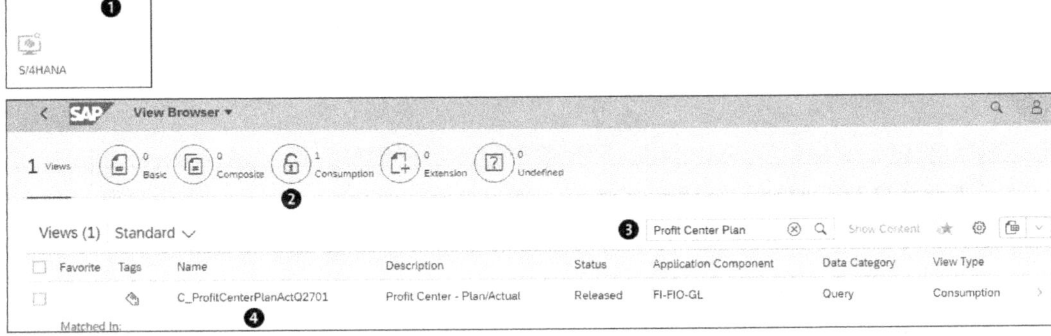

Figure 12.21 View Browser App

Custom Core Data Services Views

CDS views are foundational elements for the SAP S/4HANA VDM. You can use the Custom CDS Views app (app ID F1866A) to build customer-specific CDS views. The main advantage of using this app is that you can copy an existing CDS view and build your own CDS view to meet your business requirements. Central Finance generally leverages this app to build custom reports. With the Custom CDS View app, you can perform the following tasks:

- Add additional custom fields to the CDS, which means you can join other tables, depending on your requirements.
- Define custom calculated fields.
- Adjust field properties as required.
- Add filter criteria to get an accurate results set.
- Maintain parameters.

Once a custom CDS view is built, you can consume it in any of the available frontend tools, such as SAP Analysis for Microsoft Office, SAP Analytics Cloud, or SAP Fiori apps. Consult with your technical team on implementing custom CDS views. As shown in Figure 12.22, the app provides step-by-step instructions to define custom CDS views with customer-relevant field definitions and filter criteria.

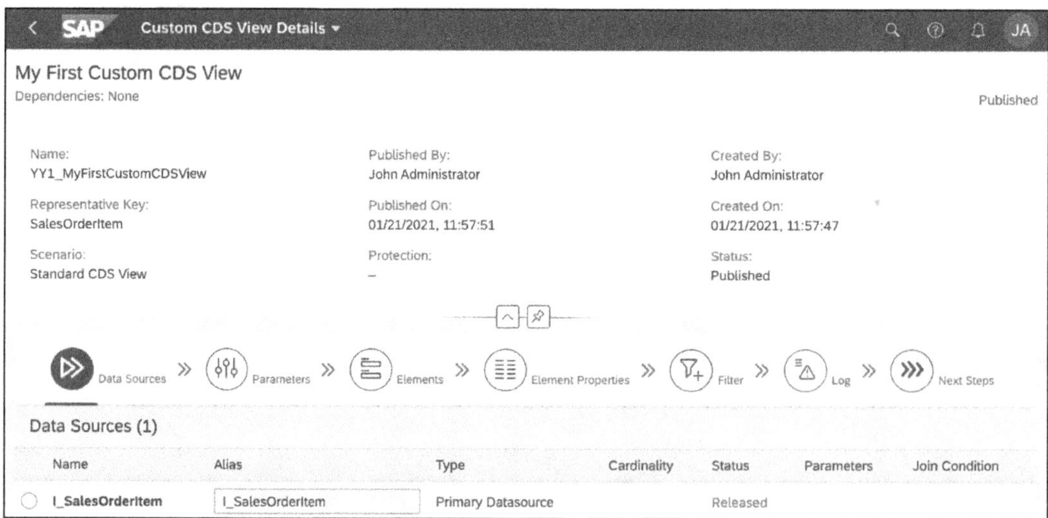

Figure 12.22 Custom CDS Views

12.4.2 SAP Analysis for Microsoft Office

SAP Analysis for Microsoft Office is a Microsoft Excel add-on tool offered by SAP. SAP Analysis for Microsoft Office provides users with the ability to perform multidimensional analysis using SAP data sources such as reading off from the Universal Journal (table ACDOCA) or SAP BW. Once you've installed and enabled SAP Analysis for Microsoft Office, as shown in Figure 12.23, Microsoft Excel shows the **Analysis** ❶ and **Analysis Design** ❷ features. The **Analysis** screen provides additional filtering of data by **ROWS** and **COLUMNS** ❸.

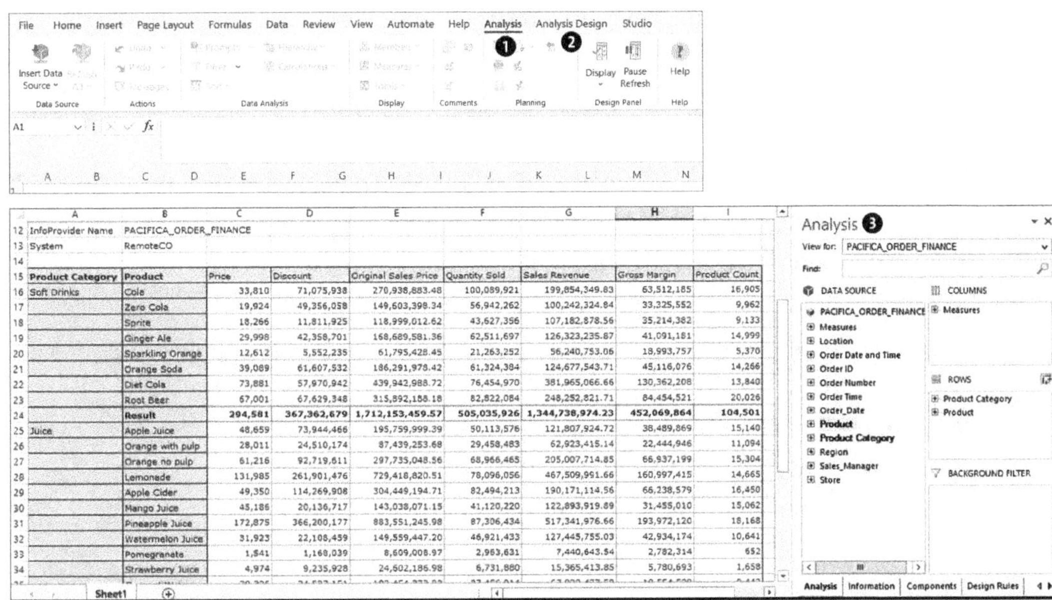

Figure 12.23 SAP Analysis for Microsoft Office: Excel Add-On

SAP Analysis for Microsoft Office is used in the Central Finance environment for a wide variety of reporting requirements, and it provides the following benefits:

- Perform multidimensional reporting along with Microsoft Excel built-in functions.
- Use Microsoft VBA features.
- Generate dynamic reports and interactive dashboards.
- Connect to multiple data sources such as SAP HANA, SAP Business Warehouse (SAP BW), and SAP Analytics Cloud.
- Perform enterprise management activities as an SAP Business Planning and Consolidation (SAP BPC) plug-in or Enterprise Performance Management OnDemand add-in.
- Enable the use of different filters and display capabilities.
- Generate SAP Analysis for Microsoft Office reports, and download workbooks for offline analysis.

12.4.3 SAP Analytics Cloud

SAP Analytics Cloud is a cloud-based tool used to visualize data, perform planning, and analyze data with user-friendly dashboards. As it's built in the cloud, you have the flexibility of accessing the tool from anywhere either using a PC or mobile phone. SAP Analytics Cloud combines multiple features that were previously done separately with different tools. SAP Analytics Cloud is a combination of three components—business intelligence (BI), predictive analysis, and business planning—to provide features such as the following:

- Create interactive dashboards to get a quick visual snapshot of data and trends.
- Get real-time insights to make informed decisions.
- Improve performance with access to real-time information.
- Perform predictive analytics to forecast future events.
- Perform enterprise-wide planning by leveraging predictive analytics and forecasting features.
- Leverage flexibility in integrating with multiple data sources. SAP Analytics Cloud can connect to SAP HANA, SAP BW, and other data sources.
- Generate interactive charts and tables.

At a high level, SAP Analytics Cloud has built-in modeling tools, stories, and catalogs, which we'll discuss next.

Models

In SAP Analytics Cloud, models include dimensions and measures that represent a subset of transactional data and are used to structure and analyze that data. Dimensions refer to a category of data that provides a perspective and contains properties to differentiate from the other dimensions. For example, vendor is a dimension that holds several attributes, such as address book data (street address, phone number, email ID, etc.). On the other hand, a measure is a numeric representation of a quantitative value. For example, you want to see sales revenue per customer or geographic dimension. SAP Analytics Cloud stories (defined in the next section) use models as the primary source of input.

In SAP Analytics Cloud, the system provides two types of models—*analytic model* and *planning model*. The Analytic model is used for displaying data in a read-only format. Conversely, as the name implies, the Planning model is used for planning purposes as it comes with date and version predelivered. SAP provides the modeler tool to create and maintain models. Figure 12.24 shows the initial screen of the SAP Analytics Cloud modeler.

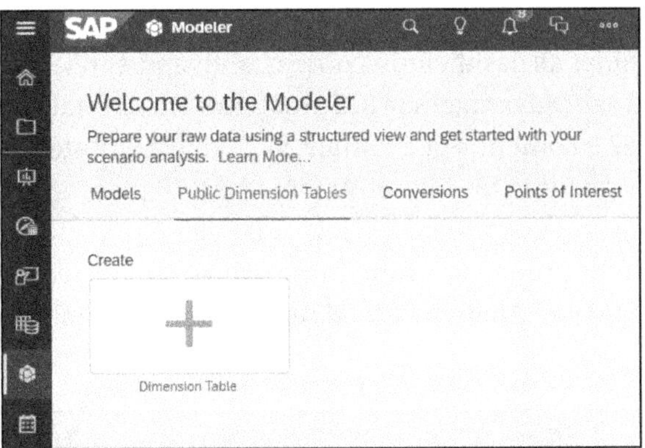

Figure 12.24 SAP Analytics Cloud: Data Modeling

Stories

In SAP Analytics Cloud, a *story* is a dashboard that provides business data insight using data visualization tools such as graphs, tables, and interactive elements. To create a story in SAP Analytics Cloud, the system uses a connection to SAP or another source to supply data. Based on business use cases, you can create stories either via real-time or scheduled data imports. Users of SAP Analytics Cloud can use stories to gain insights to identify patterns, trends, and outliers in the data. SAP Analytics Cloud stories can be created by super users or IT personnel. With appropriate authorization, you can create graphs or tables using SAP Analytics Cloud delivered widgets, for example, as shown in Figure 12.25.

12

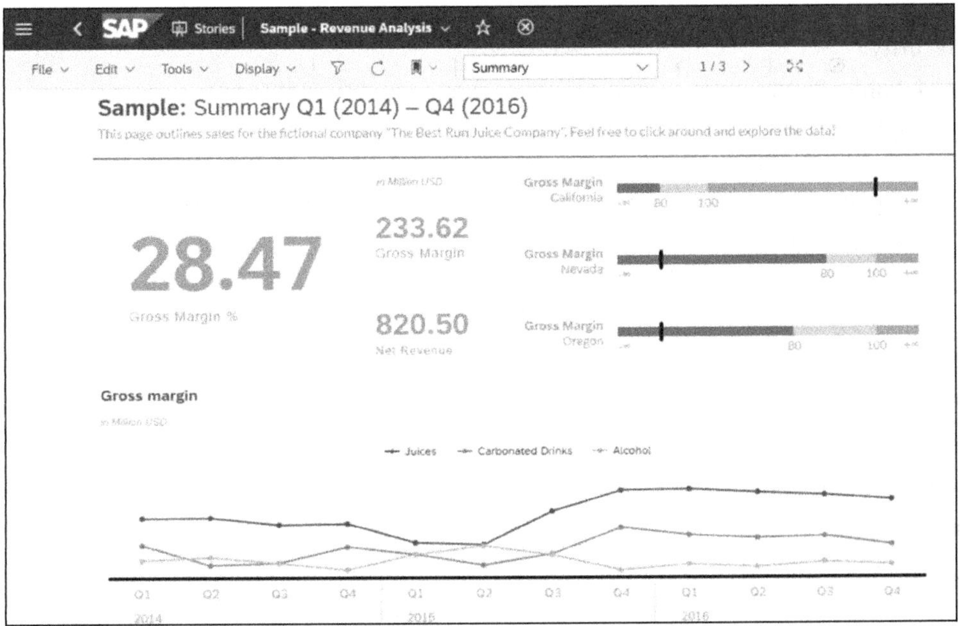

Figure 12.25 SAP Analytics Cloud: Story

Catalog

In SAP Analytics Cloud, a *catalog* includes all dashboards, content, and reports relevant to your business process. One of the main advantages of SAP Analytics Cloud catalog is that it can house both SAP and non-SAP content. For example, you'll have SAP stories, Microsoft Power BI dashboards, SAP Analysis for Microsoft Office, ABAP objects, and SAP Fiori reports. To view the catalog contents, you'll need SAP Analytics Cloud catalog access. Security access to catalogs can be grouped by functions such as general ledger, AP, or AR. Figure 12.26 displays a sample SAP Analytics Cloud catalog showing multiple SAP Fiori apps.

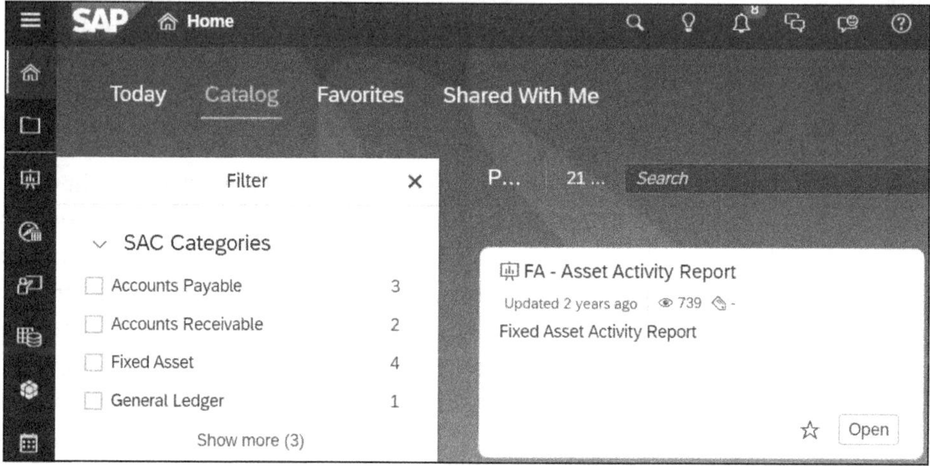

Figure 12.26 SAP Analytics Cloud: Catalog

Security

Based on SAP roles, the security team can provide access to different types of users. For example, you can have three types of users—end users, super users or key users, or functional team. The end user can view most stories in read-only mode. Super users can be granted the opportunity to create their own dashboard reports and have access to manage all reports. On the other hand, a functional team member can have edit access. Your security team can design authorizations based on your business requirements. The following list provides the types of access given to different user groups:

- View a story
- Export a story
- Share a story
- Create a story
- Edit a story
- Publish a story to a catalog

- Create a content link
- Access and manage analytical reports

12.4.4 Group Reporting

One of the main use cases of Central Finance is to perform group reporting. SAP recommends group reporting to achieve consolidated reporting to maximize Central Finance investments. Central Finance can use the full potential of SAP S/4HANA features to represent a financial platform that offers seamless integration with multiple systems, including both SAP and non-SAP source systems, SAP Analytics Cloud, embedded analytics, and systems outside of SAP S/4HANA.

Figure 12.27 illustrates how group reporting integrates with Central Finance where both SAP-based and non-SAP based (third-party) systems are used. Central Finance collects real-time data from these source systems in the Universal Journal (table ACDOCA) ❶, which provides a single source of truth. Group reporting consumes a significant amount of data from the Universal Journal. Similarly, financial plan data (table ACDOCP) can be replicated from the source systems using the **Plan Data Transfer** functionality, directly imported using the **Import Financial Plan Data** app (App ID F1711) or received from SAP's analytics and planning platform ❷. SAP recommends using solutions from the SAP analytics platform to perform financial planning activities in Central Finance. With real-time transactional data in Central Finance, you can carry out accounting close ❸. In addition, you can perform group reporting as the system has direct access to data in the Universal Journal and analytical platforms. Group reporting can drill through line items and access data from central processes directly in Central Finance. In addition, data can be transferred from the Universal Journal (table ACDOCA) to the consolidation data table (table ACDOCU) in close to real time. This facilitates central consolidated reporting using SAP group reporting ❹, which goes through *data collection*, *preparation*, *consolidation*, and *reporting* activities. With group reporting in Central Finance, you can perform the following:

- Group closing using Central Finance master data and consolidated data in the Universal Journal
- Intercompany eliminations using trading partners and partner units
- Consolidated investments
- Company code mapping to consolidation units
- Perform consolidation process
- Consolidate using partner profit center or profit segment
- Document type classification for group reporting
- Foreign currency translations
- Generating financial statements with financial statement hierarchy

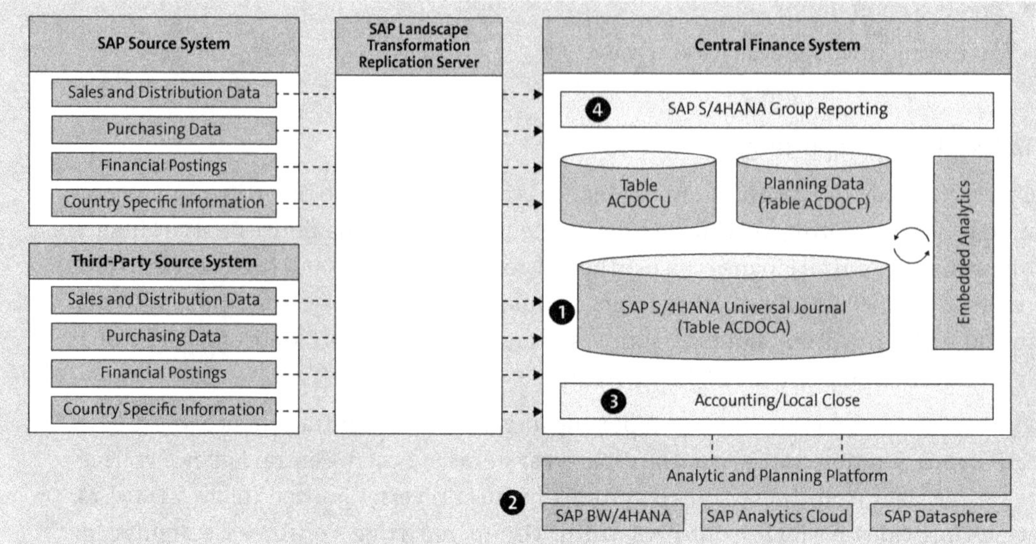

Figure 12.27 Group Reporting in Central Finance

Tip

SAP recommends leveraging SAP Best Practice content for group reporting and SAP Analytics Cloud for planning activities in Central Finance.

12.5 Statutory and Local Reporting in Central Finance

Central Finance supports statutory and local reporting requirements for countries for both SAP and non-SAP source systems. As shown in Figure 12.28, to support local statutory reporting, both logistics data ❶ and finance postings ❷ are transferred to Central Finance from source systems. In addition, if country-specific data is included with FI postings in the source systems, the same information is replicated to the Central Finance system ❸. Once the source system data is available in Central Finance, SAP Document and Reporting Compliance ❹ can leverage the collected data from the Universal Journal to support country-specific statutory reporting in Central Finance ❺. The SAP Document and Reporting Compliance component, called advanced compliance reporting (ACR) earlier, is a separately licensed product. SAP merged ACR with SAP Document and Reporting Compliance. However, existing ACR isn't impacted, and support is still available. On the other hand, SAP Document and Reporting Compliance comes with the latest set of statutory reports with the SAP Document and Reporting Compliance framework to comply with recent legal changes.

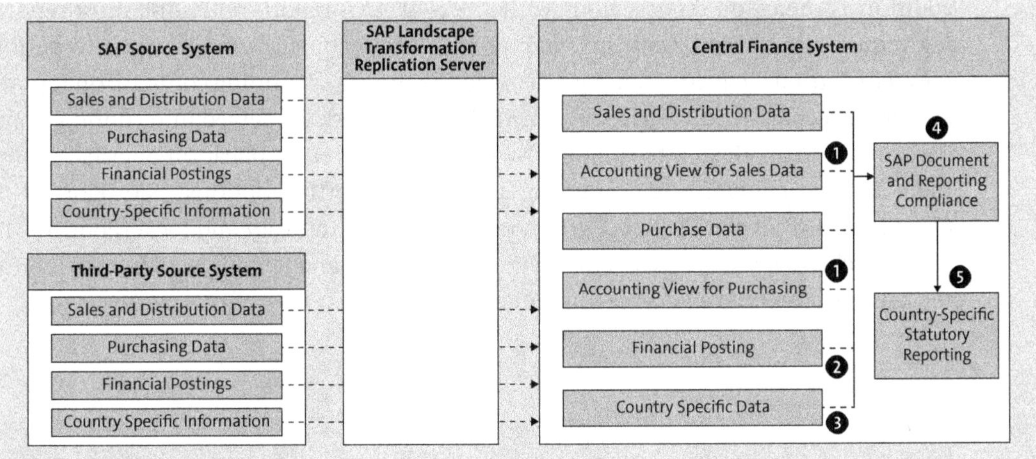

Figure 12.28 Statutory Reporting in Central Finance

12

If central payment is activated in Central Finance, you can perform central tax reporting in Central Finance. Refer to Chapter 2, Section 2.1.2, for a list of supported tax reports to perform tax reporting in Central Finance. These reports are included during period-close activities to get accurate tax reporting in Central Finance. If central payment isn't active, you'll perform tax reporting in the source system to support comprehensive tax calculation procedures based on country-specific requirements. During the explore phase, tax harmonization must be considered, which directly impacts central tax reporting in Central Finance.

[+]

Tip

Review the following SAP Notes for additional guidance on SAP Document and Reporting Compliance and ACR:

- SAP Note 2480067: Provides guidance on new reports in SAP Document and Reporting Compliance.
- SAP Note 2620211: SAP Notes required to implement SAP Document and Reporting Compliance for different SAP S/4HANA releases.
- SAP Note 3302971: Summarizes changes in SAP Fiori app names.
- SAP Note 2584083: Documentation for ACR in SAP S/4HANA, which provides administration and extensibility guides.

12.6 Summary

Comparison reports in Central Finance play a crucial role in achieving reconciliation of transferred data at different points of time during Central Finance implementation. In addition, comparison reports along with SAP S/4HANA reports will fulfill most reporting requirements. We began this chapter with the reconciliation approach to be followed in Central Finance. Then, we discussed the timing of running comparison reports and other reconciliation reports in Central Finance. Next, we delved into running comparison reports with step-by-step instructions for different replication objects. Finally, the chapter concluded with exploring additional reporting options—SAP Fiori apps, CDS views, SAP Analytics Cloud, group reporting, and statutory and local reporting. In the next chapter, you'll learn Central Finance project management, which is critical for the success of a Central Finance implementation.

Chapter 13
Central Finance Project Management

In this chapter, you'll learn how the SAP Activate methodology and SAP-delivered adoption content helps customers manage Central Finance projects effectively. The chapter walks you through change management aspects to be considered during a Central Finance implementation.

For a successful Central Finance implementation, you need to identify the right resources and tools to execute project tasks. For example, project management plays a key role in setting project goals, setting realistic timelines, and tracking project activities. Project management glues project stakeholders together with clear communication throughout the project. In addition, project managers face challenges such as scope creep and user acceptance, and understanding project requirements is no different from any other enterprise resource planning (ERP) implementations. Mitigating risks with appropriate actions and follow-up actions is critical for a successful implementation.

To implement Central Finance without interrupting existing processes, you'll start with an implementation strategy composed of four broad levels, substrategies, or approaches. The end goal is to reach SAP S/4HANA innovations quickly without interrupting existing processes, especially to leverage the single source of truth with the Universal Journal. Figure 13.1 displays a typical Central Finance implementation strategy that can be adopted for a complex heterogenous landscape. The strategy comprises the following:

❶ First, a *high-level assessment* is required to understand Central Finance capabilities and how Central Finance addresses business requirements. The assessment helps stakeholders achieve key strategic decisions. During evaluation of the high-level assessment strategy, you'll gain an understanding of the following:

- Overview of Central Finance and how it addresses key business issues
- Deployment options of SAP Landscape Transformation, SAP Master Data Governance tools, and Central Finance
- As-is processes in both financial and logistics operations
- Layout of current and future road map
- Buy-in from management to support the strategic decision

❷ *Proof of concepts/prototype/pilot* may be required to buy-in from stakeholders on the new processes or solutions offered with Central Finance. With proof of concept, you'll gain an understanding of the following:
 - Central Finance features
 - Central Finance use cases validated at the pilot phase
 - Blueprint or design of use cases, including key and value mappings
 - Financial data replication from a source system
 - Central Finance design elements such as parallel ledgers, document splitting, and reporting needs
 - Organizational changes due to the introduction of new processes

❸ *Technical implementation* lays out planned activities for Central Finance. With the technical implementation strategy, you'll understand the following:
 - Setting up an SAP S/4HANA system with the Central Finance solution
 - Performing real-time replication for selected objects considered for proof of concept
 - Ensuring no disruptions to existing processes in the source system
 - Using the end-to-end Central Finance process for the selected scoped items

❹ *Value realization* happens at the end when stakeholders observe the Central Finance features, thus providing opportunities to expand the capabilities to additional systems. Customers realize value realization with the following:
 - Incremental approach of starting reporting first
 - Expanding capabilities enterprise-wide
 - Exploring local and statutory reporting

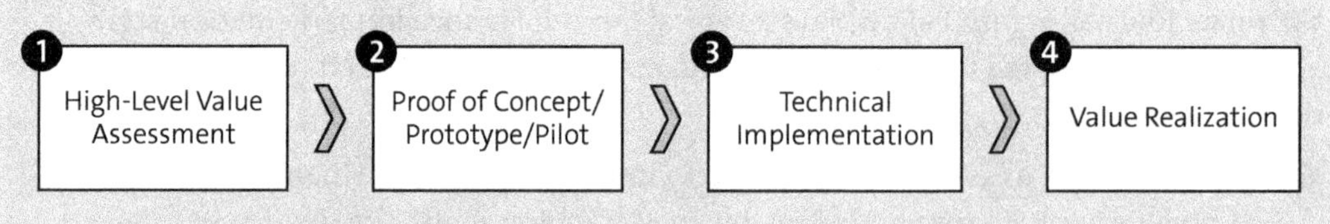

Figure 13.1 Central Finance Approach

In this chapter, we begin learning how the SAP Activate methodology fits in Central Finance implementations. Next, we'll discuss the Central Finance implementation change impact in the context of the organization, technology, and process changes. The chapter concludes with Central Finance leading practices.

13.1 SAP Activate Methodology

SAP Activate is an agile methodology that helps customers use SAP best practices and migration tools to leverage SAP S/4HANA innovations quickly. The goal of the methodology is to provide customers with ready-to-run proven processes as a starting point rather than starting from ground zero. The SAP Activate methodology provides the following benefits to customers:

- Best road map to implement Central Finance
- Identifies custom code analysis with requirements adjustments to adopt SAP S/4HANA features
- Identifies efforts involved with various teams
- Provides risk and mitigation approaches

The SAP Activate methodology is equipped with three main pillars that support implementation teams with best practices, guided configuration tools, and methodology accelerators:

- **SAP Best Practices**
  SAP offers best practices packages for different SAP product versions that have scope items relevant for the product offering. Often, you'll install best practices packages on a sandbox environment to demonstrate the standard functionality without going through extensive configuration.
- **Guided configuration**
  SAP offers adoption content for Central Finance with assisted guided configuration that comes with initial implementation and continuous adoption tools and accelerators to speed up the Central Finance implementation. With guided configuration tools, you'll jump-start activities immediately.
- **Methodology**
  The SAP Activate methodology is divided into six phases: *discover*, *prepare*, *explore*, *realize*, *deploy*, and *run*.

Table 13.1 displays the key benefits of the three pillars of the SAP Activate methodology framework we just discussed.

> **Note**
> The SAP Activate methodology elements include phases, workstreams, deliverables/tasks, and artifacts. The workstream referred to as an activity can be, for example, customer team enablement or analytics. A project plan generally contains deliverables and tasks with assigned dates and responsible persons. Artifacts are referred to as input and output variables used to communicate and deliver results when managing projects.

SAP Activate Pillar	Benefit	SAP Deployment Scenario
SAP Best Practices	▪ Ready-to-run business processes to support SAP S/4HANA migration ▪ Best practices scenarios: process flows, roles, and sample test scripts ▪ Predefined load templates and tools with Central Finance adoption content	SAP S/4HANA Cloud, on-premise SAP S/4HANA, hybrid options, and Central Finance implementation scenarios
Guided configuration	▪ Guided implementation steps to configure the new system, including initial configuration and continuous changes to support business processes, migration engine with monitoring cockpit, and end user training	
Methodology	▪ Using best practices to deliver project delivery quickly ▪ Agile project methodology with iterative approach to continuously improve processes ▪ Reduced lifecycle – prepare, explore, realize, and deploy ▪ Scalable methodology with varying size of projects	

Table 13.1 Three Pillars of SAP Activate

In general, complex SAP implementations are adopting the SAP Activate methodology. This applies to any Central Finance project implementation processes to enhance the overall execution of a project. The methodology is divided into six major phases for easy management of project activities, as shown in Figure 13.2. Let's explore each phase in detail and see how it can be applied in Central Finance project implementation scenarios:

❶ **Discover phase**

In the preparation phase of a Central Finance project, you'll kick off the project. In this phase, the project team establishes a project governance structure, which includes defining the project team's roles and responsibilities. The project team also starts discussing the landscape environment, setting up a dedicated sandbox environment for proof of concepts, and exploring new features offered with Central Finance to become familiar with them. Typical activities you'll perform during this phase include the following:

– First, you'll get familiar with the SAP S/4HANA and Central Finance ecosystem and its terminology. Focus must also be given to new innovations that come with the Universal Journal, SAP Fiori, reporting tools, choice of technology, and the SAP cloud platform offering.

- Next, you'll explore Central Finance features and become familiar with the current road map, planned and future innovations, use cases, and training opportunities.
- As a final step in this phase, you'll begin with validation workshops or proof of concept workshops with key stakeholders. Information gathered at this phase will eventually serve as an input to subsequent phases. You'll determine the overall Central Finance strategy, vision, business, IT objectives, and use cases relevant for Central Finance.

❷ **Prepare phase**
In the preparation phase, you kick off the project with the overall strategy and vision of Central Finance. Following are typical tasks performed during the prepare phase:

- Enable the teams for the Central Finance project.
- Establish priorities based on the understanding of use cases from the previous phase.
- SAP Solution Manager is leveraged to support project execution and functional design work.
- The project team establishes a project governance structure that includes defining the project team's roles and responsibilities.
- The project team also starts discussing the landscape environment and setting up a dedicated sandbox environment with best practices content installed.
- The project team gets familiar with the new Central Finance functionality to be designed and implemented in the subsequent phases.

A first quality gate is achieved as the preparatory phase comes to an end.

❸ **Explore phase**
After completing the preparatory phase, you'll transition into the explore phase, assuming you've conducted detailed planning sessions for both functional and technical teams. This phase is typically done with three unique components: define scope, conduct workshops, and create strategy.

In this phase, you'll carry out tasks such as these:

- As part of the scope definition, evaluate the master data scope for both SAP-based and non-SAP-source systems.
- The project team also conducts requirements gathering workshops with functionality demos to identify fit gaps on all business processes. The workshops focus on identifying design elements such as enterprise structure, master data, and SAP Landscape Transformation aspects.
- The project team prepares blueprints or documents, delta configuration documents, test planning, and training strategy documents.

- Create key decision documents on major design challenges that are created and approved by project stakeholders.
- All requirements are captured with user stories in the backlog list to plan the work to be performed in different sprint cycles.
- Develop a Central Finance test strategy considering the design elements aligning with the business requirements. You'll develop a strategy for both the initial load and real-time replication.
- Finally, align on the strategy from the preparation phase to refine topics requiring additional attention. For example, master data and test strategy can be refined based on inputs from both the prepare and explore phases.

At end of this phase, you'll end with a second quality gate that ensures the project deliverables are meeting predefined goals prior to transitioning to the next phase.

❹ **Realize phase**

In the realization phase, the project teams set up systems; complete baseline configurations, master data, and readiness checks; and run multiple mock and test cycles. The project team builds solutions by configuring the system and developing custom objects based on fit-gap analysis and then shows the functionality to business users with walkthrough sessions to validate the solutions. Several end-to-end testing scenarios are carried out in an iterative process until the business signs off on the final solution. The project team also prepares an operational runbook for production cutover tasks. Activities typically carried at this phase include the following:

- Set up source systems, SAP Landscape Transformation, SAP Master Data Governance, any third-party software, and Central Finance systems. This includes applying SAP Notes, establishing connectivity, and carrying out any other important Basis-related tasks.
- Complete baseline configuration with Central Finance design elements based on the design document from the explore phase.
- After configuration and master data setup, execute SAP-delivered consistency check programs and readiness programs to ensure no inconsistencies are found. Any inconsistencies must be resolved prior to testing Central Finance features.
- Next, move to the testing phase. Project teams and stakeholders carry out different types testing—unit testing, integration testing or system integration testing, and user acceptance testing—at various stages. In each test cycle, you'll use entry and exit criteria defined per the test strategy, which was documented in the explore phase earlier.
- For the initial load and real-time replication, run through the mock test cycle to ensure data is transferred from the source system and reconciled with Central Finance comparison reports. You'll run multiple mock runs per business requirement.

- The cutover plan is updated in parallel as testing is in progress to reflect the time taken to complete each mock cycle and production dry run. In addition, you'll create additional test cycles to refine the test plan and refine entry and exit criteria.
- Begin creating troubleshooting guides and user documentation that will be carried to the deployment phase.
- If required, complete performance testing by copying production data into a separate client full volume.
- Finally, conduct user acceptance testing sessions with project stakeholders to sign off on the end-to-end solutions.

At the end of this phase, you'll have a third quality that ensures the project deliverables are met and solution built is meeting the expectations.

❺ Deploy phase

In the deploy phase, the project team assesses business readiness for go-live. After the stakeholders' approval, the project team executes cutover tasks in the production system as part of go-live preparation. Sometimes, a cutover rehearsal is done in a different client before executing final cutover steps in the production system. The deploy phase activities include the following:

- Execute the cutover plan with project teams and stakeholders.
- Carry out data load/migration activities per the cutover plan.
- Review all open critical issues, and determine showstoppers. Assess if you require an additional testing cycle.
- Kick off organizational change management (OCM) plans, and engage stakeholders' communication.
- Conduct user training as planned from the earlier phases.
- Document critical issues impacting processes as a result of the new solutions introduced.
- Perform smoke tests in relevant areas, and monitor business processes.
- Establish a center of excellence or support team to provide support wherever needed.
- Track delivery of the solution, and assess the feedback.

❻ Run phase

In the run phase, the project team provides hyper care support for a short duration. Then, the project team provides necessary knowledge to support the team in transitioning to the run phase. You'll also create run books for the production support teams. Project teams or support teams continue to enhance the system by introducing incremental changes through change requests or service ticket requests. In addition, as SAP releases new features through feature packs or service packs, you'll continue to enhance the system.

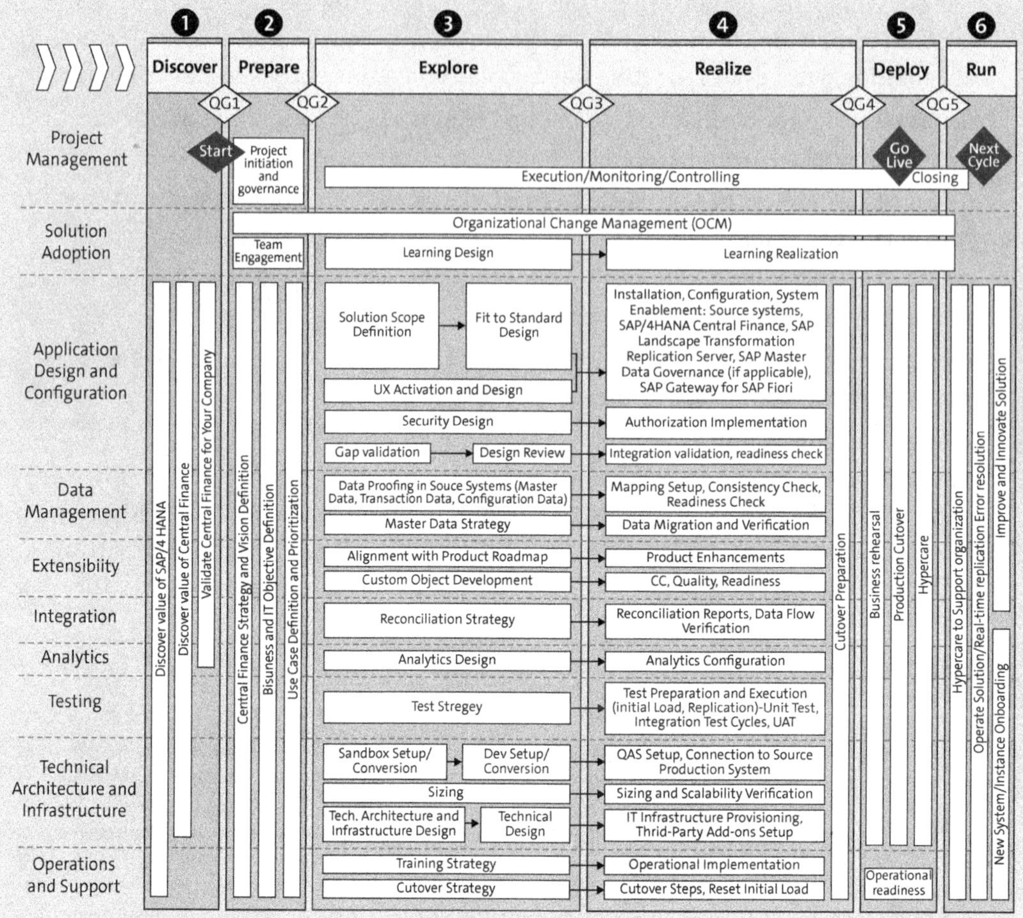

Figure 13.2 Central Finance Activate Methodology

[»]

Note

SAP Activate methodology for SAP S/4HANA is based on the agile approach. The SAP Activate framework offers projects with automated tools, ready-to-use content, and ready-to-run tools to make Central Finance project delivery faster. SAP Activate isn't just limited to Central Finance, but it can be used in any financial transformation projects involving greenfield, system conversion, bluefield, or Central Finance scenarios either in a cloud or on-premise implementation.

13.2 Predelivered Adoption Content

SAP offers predelivered adoption content for Central Finance customers and implementation partners to accelerate the project implementation without starting from ground up. The content is catered for different project stakeholders to quickly adopt reference materials to support the project needs. SAP provides adoption content in different categories as reference material collected from the past implementations. You'll need to adjust the adoption content per project needs as the example screenshots and references cited may not be applicable for specific versions or landscapes. The adoption content is available for project managers, functional consultants, and technical consultants. Table 13.2 shows an example usage of adoption content.

Category	Project Stakeholders	Central Finance Adoption Content
1 – Documentation assets	Project and project managers	▪ Project management template ▪ Sample project plan ▪ Scoping and effort estimation templates
	Functional consultants	▪ Configuration and how-to guides ▪ Configuration tracker template ▪ General ledger account harmonization guide ▪ Deferred taxes guide ▪ Cross-company posting handling ▪ Document splitting design considerations
	Technical consultants	▪ Configuration and how-to guides ▪ SAP Landscape Transformation and SAP Application Interface Framework configuration guides ▪ Relevant SAP Notes ▪ BAdI Guide ▪ Data flow architecture diagrams for different replication scenarios
2 – Sample programs or sample enhancement	Functional consultants	▪ FI initial load simulation report ▪ Third-party ERP integration via file upload report
	Technical consultants	▪ Business process cost object enhancement ▪ Project systems network WBS mapping enhancement ▪ WBS status handling initial load enhancement ▪ Additional quantities handling enhancement
3 – Delivered as services	Functional/technical consultants	▪ Tools with specific functionality

Table 13.2 Central Finance Adoption Content Categories

Note

Consult SAP Note 2827145 to find additional information on adoption content (*http://s-prs.co/v590104*).

Now, that you've learned about Central Finance adoption content, let's look at a sample project activity in a Central Finance implementation in the following sections.

13.2.1 Central Finance Project Plan

Figure 13.3 shows a sample project plan for Central Finance highlighting key activities. Project plans for Central Finance vary with every project and scope of the implementations. This isn't an exhaustive list of Central Finance activities, but it provides you with an idea of typical central project activities. You'll have to build a project plan adopting project-specific requirements.

Central Finance Activities	T1	T2	T3	T4	T5	T6	T7
Planning sessions prioritize user stories and backlogs							
Requirements gathering workshops							
Proof of concept sessions							
Complete business requirements document							
System landscape build							
Assess source system requirements							
SAP Landscape Transformation Replication Server build/DMIS installation							
Central Finance build							
Determine master data requiements and landscape strategy							
Detailed requirements workshops							
Central Finance foundational design elements							
Document key design decisions							
Confirm mapping requirements							
Determine reporting requirements							
Configure Central Finance							
SAP Fiori setup							
SAP Landscape Transformation Replication Server setup for tax tables and withholding tax							
Install source system notes or upgrade to latest service pack							
Source system configuration							
Configuration consistency checks							
Perform Central Finance initial load activities							
Perform reconciliation and run comparison reports							
Perform SAP Application Interface Framework bulk load							

Figure 13.3 Central Finance High-Level Project Plan

13.2.2 Security Deployment

SAP is an integrated system that comes with complex security role requirements. During the design phase, you'll carefully design security roles for various business requirements and at the same time comply with the company's security policies and internal controls. Following are typical requirements you'll gather during security requirements assessments:

- Identifying business roles
- Identifying correct authorization objects

- Organization-level security requirements
- Field-level security requirements
- Developing a transaction security matrix
- Display roles
- General user roles
- Temporary roles
- Derived business roles and composite role requirements
- New role request process
- Audit roles
- Security monitoring

13.3 Understanding Change Impact

For any ERP implementation projects, change management is vital for the success of the project. The primary goal of change management is to prepare organizations to adopt Central Finance as a new way of working with transformed processes. This often includes changes in work culture, streamlined processes, and landscapes driven by new technology tools. With the project end goals in mind, leadership engages and inspires employees with the changes brought in by Central Finance. By embracing change management early in the project, companies can avoid risks of failures. Change management must be considered in three aspects especially:

- **Organizational**
  The change management team ensures that all members of the organization are prepared for the change and agree with all project stakeholders. Change management is also responsible for adopting strategies early in the project to help the business reduce risks or failures.
- **Technology**
  Any finance transformation brings a plethora of technological changes on business processes. For example, although introducing the SAP Fiori user interface or similar tools may be easier to use and adopt, you'll still develop appropriate training and job aids to close any potential gaps. SAP S/4HANA also brings new intuitive advanced reporting tools such as SAP Analytics Cloud, SAP Analysis for Microsoft Office, and other tools. You'll ensure employees receive adequate training and the resources needed to perform their job.
- **Process**
  Central Finance also brings centralization of business processes across the enterprise. As a result, users must learn new ways of performing their day-to-day tasks. Centralized processes bring advanced automation capabilities that may simplify

execution of those tasks. However, comprehensive training is required for employees to learn the new processes, so at go-live, they are prepared to function with new processes.

Let's understand the impact on each of these areas further in the following sections.

13.3.1 Organizational

Unlike greenfield and brownfield approaches, the Central Finance transformation approach gives customers a quick way to adopt SAP S/4HANA innovations without interrupting current business processes. Central Finance still requires an organizational assessment to understand the changes and business buy-in. Following are typical questions you'll ask, especially when it comes to changes involving people:

- Who performs the processes in scope today? Do the processes different across business units or regions?
- What's the impact of new processes on current roles? How would users adapt to the new way of working?
- Does the process affect a specific single group or cross functional teams?
- Do users need to obtain skills to perform the new processes? What extent of job aids and user documentation are required?
- Does the process of change affect people's motivation? Do managers consider new objectives or performance goals and incentives to achieve target goals?

Besides organizational change impact, you'll identify stakeholders involved in Central Finance implementation. At a high level, management leaders engage key stakeholders to derive and govern projects. The leadership also provides necessary support to achieve project goals. Figure 13.4 shows important stakeholders involved in a Central Finance implementation project. Let's consider each stakeholder's role and typical tasks they carry out when executing Central Finance projects:

- **Executive sponsors**
  They set the vision and goals of the Central Finance project. They'll get necessary funding and approvals from relevant stakeholders. In addition, they exhibit leadership to ensure project teams get the necessary resources and support.
- **CFO/CIO leadership**
  They set the tone and provides inspiration, motivation, and support throughout the project duration and its success. They also foster purpose and commitment from project stakeholders.
- **Project management office (PMO)**
  This office plays an important role in planning, coordinating, and executing project plans. Project managers develop a detailed project plan to identify milestones and deliverables.

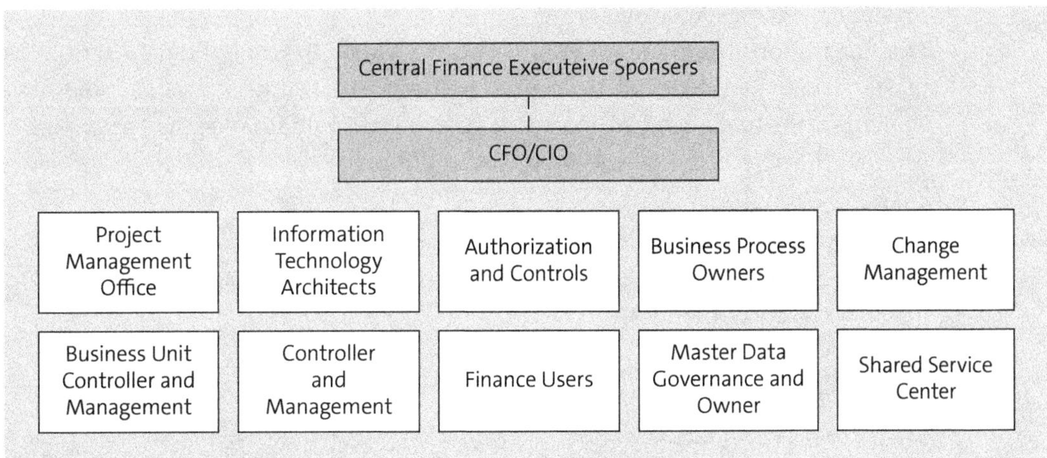

Figure 13.4 Key Stakeholders

- **Information technology architects**
  They work with business counterparts to design and develop Central Finance capabilities with minimal interruption to infrastructure and ongoing business processes.
- **Authorization and controls**
  This group needs to be aligned with new processes and functions. Security teams involved in designing and implementing security roles per new processes design.
- **Business process owners**
  They are involved with the help of IT architects to design blueprints and functional documents. In addition, they help stakeholders validate end-to-end solutions.
- **Change management team**
  This team assess changes and help stakeholders adopt transitions to new processes with minimal interruptions. They also develop a communication plan and engage stakeholders.
- **Business unit controller and management**
  They are involved in key design decisions as the changes impact their department and how they function going forward. Project managers engage business unit controller and management teams per the project plan.
- **Corporate controllers**
  They are involved in understanding financial aspects of key decisions impacting finances and the budget of the project. They need to understand major risks and any potential issues that will disrupt financials and budgets.
- **Finance users**
  They are involved in system validations, executing test cases for user acceptance testing (UAT) reporting defects, and retesting the fixes. They'll also be involved in user training to understand new processes.

- **Master data governance owners**
  They are responsible for defining which master data is in scope and help harmonize master data in Central Finance. They work closely with business process owners and IT architecture to identify the master data scope and mapping requirements, that is, whether the objects are set with 1:1 or N:1 condition.
- **Shared service center owners**
  They are involved in executing centralized processes in Central Finance. They're involved during different phases of projects to identify central processes to adopt new ways of doing their job.

Note

Organization structure and project teams vary with every organization. The example chart shows a sample chart, and it needs to be adapted to project requirements.

13.3.2 Technology

Although technology has no impact or minimal impact on source systems, the technology needs on SAP Landscape Transformation Replication Server and Central Finance could be significant and can't be underestimated. Companies should assess technology for various factors such as scalability, adopting future innovations, and cloud computing. During project phases, you'll find answers to these questions:

- What systems are replaced by new systems?
- What technology tools are considered for process automation?
- What analytics and reporting are considered? Are these tools capable of predictive analysis, machine learning, and artificial intelligence features?
- Does the technology support security and compliance aspects?
- How is data maintained during interim periods?
- What are the archiving goal and solutions adopted?
- Are performance aspects considering volume?

13.3.3 Process Changes

By implementing Central Finance process changes, stakeholders get an opportunity to adopt new unified financial and operational processes across the enterprise. This significantly improves the current processes using the Universal Journal in SAP S/4HANA. The process opportunities include the following:

- In a heterogenous landscape, you can use disparate financial processes across the enterprise to bring unified processes via the Universal Journal as the single source of truth.

- Standardize processes.
- Harmonize master data by reducing master data maintenance in disparate source systems.
- Consolidate reporting using the Universal Journal.

As part of process change, you'll bring up typical process changes:

- Is the new process adopted from standardization or industry leading practices?
- How much deviation exists between current and new global processes?
- How difficult is it for business users to adopt new processes? Does the new process change leave business users to find an alternative workaround? Does the workaround increase or decrease overall process time? Are there improved processes available or redesign opportunities for the new process?
- Is the documentation prepared to fulfill users' requirements?
- Are the processes standardized across business units or regions?
- What local process deviation is required? What additional or special processes are required for local or regional requirements?

13.3.4 Challenges and Mitigation

Like any other SAP transformation project, Central Finance faces challenges when integrating multiple source systems to ensure consistency and that data is replicated accurately. You'll address general challenges by following agile processes:

- Understanding objectives and aligning with project goals
- Sprint planning to identify user stories and creating a backlog of tasks
- Following activate agile methodologies throughout the project cycle
- Conducting workshops for data gathering in iterative cycles
- Engaging stakeholders early in the project cycle for alignment and buy-in
- Understanding reporting requirements early in the project cycle
- Engaging stakeholders in every sprint cycle to validate solution development and approval
- Documenting a detailed cutover plan and performing dress rehearsal cycles prior to production deployment

In addition, most of the Central Finance specific challenges are mitigated through following Central Finance leading practices explained later in Section 13.5.

13.4 SAP Focused Run

SAP Focused Run, a licensed solution, is provided to Central Finance customers and partners with high-volume data and systems to monitor applications, perform alerting, and provide analytics. The current version of SAP Focused Run is 4.0 FP03, and the service offers customers the following:

- Centralized real-time SAP Application Interface Framework monitoring for different application areas in Central Finance
- Analytics and tracking using message ID or key fields
- Real-time replication, portioning, predictive analytics, and compression with the advent of the SAP HANA platform
- SAP Application Interface Framework monitoring and automating email alerts to users about exceptions or errors
- SAP Landscape Transformation monitoring and system monitoring at different levels
- Root-cause analysis

Note

For additional information on SAP Focused Run, visit the SAP Support Portal: *http://s-prs.co/v590105*

13.5 Central Finance Leading Practices

We recommend following leading practices for a successful Central Finance implementation. Table 13.3 provides leading practices generally followed in Central Finance by each application area.

Area	Subtopic	Central Finance Leading Practice
Master data	Data cleansing	Open items transferred during the initial load may cause performance bottlenecks and delays in completing the initial load validation. SAP strongly recommends converting unwanted general ledger open items to normal general ledger accounts. In addition, you must clear open items from both current and prior years to the extent possible as it directly impacts performance of the initial load.
		Review important master data elements such as general ledger accounts, profit centers, cost centers, customers, and vendors in the source system. Lock inactive, redundant, and duplicate objects.

Table 13.3 Central Finance Leading Practices

Area	Subtopic	Central Finance Leading Practice
Master data (Cont.)	Data cleansing (Cont.)	SAP recommends the MCDelta tool for master data profiling. The tool offers customers accelerate the implementation with Central Finance adoption content.
	Business mapping rules	Ensure you've followed data harmonization rules for all master data, especially the general ledger harmonization guide to avoid mistakes during the mapping exercise. General ledger attributes must be aligned with corresponding general ledger master in the source system. Master data is the most time-consuming activity in Central Finance projects.
		■ Data mapping also derives final reporting requirements. You must keep reporting requirements while performing business mapping rules. ■ Avoid generic or dummy mapping. ■ Remove the blocking indicator of master data, especially during the initial load.
	Data inconsistencies and configuration inconsistencies	Be sure to run configuration and master data consistency reports to avoid data inconsistencies.
		SAP provided the permanent constancy check to alert stakeholders to check configuration consistency and resolve errors quickly.
	Master data load and mapping	Perform data profiling prior to the initial load and real-time replication to avoid master data errors.
		Both financial and nonfinancial master data should be loaded prior to beginning the initial load and online replication activity.
		Both SAP Master Data Governance key mapping and value mapping must be completed.
	Data governance	Enforce the right level of data governance on creation and change of master data to prevent duplicate records.
	Third-party system	Be sure adequate scope is considered for third-party system master data harmonization as different ERP systems store data differently and often terminology is different. This makes more time to achieve data harmonization.

Table 13.3 Central Finance Leading Practices (Cont.)

Area	Subtopic	Central Finance Leading Practice
SAP Landscape Transformation Replication Server	Installation	Have a separate SAP Landscape Transformation box to connect the source system and Central Finance system as it brings multiple benefits, especially during monthly maintenance and upgrades.
		Determine the Data Migration Server (DMIS) version of SAP Landscape Transformation ahead of time based on the source system and Central Finance service pack versions.
	Performance	Data volume must be assessed prior to configuring SAP Landscape Transformation. The landscape sizing and memory should support a high volume of background processing.
		Configure SAP Landscape Transformation in an optimum way, so the parallelization of the calculation jobs transfers data without any performance bottlenecks.
		Maintain SAP Landscape Transformation data transfer jobs by considering data volume.
SAP Application Interface Framework	Runtime configuration object	Before you begin replication, define SAP Application Interface Framework runtime configuration groups (Transaction /AIF/PERS_CGR) under the `/FINCF` namespace for each replication object. This will help on serialization of data being posted.
		Assign the runtime configuration to the appropriate replication objects.
	SAP Application Interface Framework message archiving	For companies with large volumes of transaction data, SAP Application Interface Framework messages fills up quickly and thereby impact performance. Archive SAP Application Interface Framework messages on a periodic basis to improve performance of the interface. SAP recommends the following: ▪ Archive the SAP Application Interface Framework XML messages relating to documents that have been already posted successfully or posted with warning. To implement archiving, Transaction SARA can be used. `/AIF/PERSX` is the archiving object. ▪ For SAP Application Interface Framework XML messages that are in-process or with error status, the system doesn't allow you to archive or delete.

Table 13.3 Central Finance Leading Practices (Cont.)

Area	Subtopic	Central Finance Leading Practice
SAP Application Interface Framework (Cont.)	SAP Application Interface Framework security	Assign SAP Application Interface Framework security roles as necessary to prevent last-minute errors.
FI-CO Initial Load	Initial load groups	Have a separate initial load group for company codes with intercompany transaction postings.
		Configure the initial load groups for the source company codes with a smaller data set to avoid performance issues.
	Initial load extraction	Work with Basis to ensure you have enough dialog work processes in the source system and background processes in the target Central Finance system to extract the FI documents, open items, and balances.
		Ensure you have enough memory to process the initial load extraction. Otherwise, you'll see cancelled jobs that require further efforts to delete and restart.
		During cutover, extract data during minimal postings or off business hours to save time on the delta run.
	Initial load mapping and simulation	Run both mapping and posting simulations to identify and resolve potential errors.
		Increase memory parameters to allow the job to run in multiple parallel processes.
		If errors exceed greater than two million, you won't be able to download error message to a Microsoft Excel worksheet. In such cases, you'll fix the errors and rerun the simulation jobs.
	Initial load posting	You must post all documents in the initial load to avoid any manual journal entry adjustments. This will also help in reconciliation of balances, open items, and documents. Furthermore, a full posting of the initial load also helps in posting subsequent follow-on documents through SAP Application Interface Framework. For example, you'll avoid **Reference document isn't posted** errors.

Table 13.3 Central Finance Leading Practices (Cont.)

Area	Subtopic	Central Finance Leading Practice
FI Accounting Online Replication	SAP Application Interface Framework bulk load	Immediately after the initial load extraction is triggered in the source system, any new postings start being captured into source staging tables (table CFIN_ACC*). For example, there will be invoices, payments, and reset and reversals postings in the source system captured under the source staging tables. These documents will be transferred to the Central Finance system through SAP Landscape Transformation once SAP Landscape Transformation initial load/ replication is triggered for SAP Landscape Transformation table CFIN_ACCHD. When all captured data transfers to the Central Finance system and once SAP Landscape Transformation triggers the SAP Landscape Transformation initial load, it doesn't load in the sequence into the Central Finance system. There could be chances of clearing documents transferring before the invoice or reset/reversal, which will lead to serialization errors such as **Reference document isn't posted** or **Reference open item is already cleared** in SAP Application Interface Framework. Take care of SAP Application Interface Framework serialization before transferring the FI accounting documents through the SAP Landscape Transformation initial load. Consult SAP Note 2679070 on addressing serialization and steps to be followed.
Source Staging Table Cleanup	Staging table cleanup	In the source system, Central Finance staging tables (table CFIN_AC*) filled up after the initial load extraction is triggered. Similar staging tables exist for controlling secondary postings (tables CFIN_CO_ADD and CFIN_COPA). As the table data grows, this may have performance problems. SAP delivered two reports to clean up the staging tables: RFIN_CFIN_CLEANUP for FI-CO staging tables and FIN_CFIN_CO_CLEANUP for CO staging tables. Determine the appropriate strategy and timing to clean up these tables to avoid performance bottlenecks.

Table 13.3 Central Finance Leading Practices (Cont.)

Area	Subtopic	Central Finance Leading Practice
Reporting	Reporting requirements	■ Understanding the vision of reporting and the right level of data with appropriate granularity is replicated from the source systems, especially for consolidated central reporting. ■ Assess which reports needs to be carried out in the source system versus the Central Finance system. ■ Execute Central Finance comparison reports and other reconciliation reports regularly to ensure data replicated is accurate and consistent. ■ Enable SAP Fiori apps, CDS views, SAP Analytics Cloud, and other reporting tools to ensure they fit into your solution. Tools such as SAP Analytics Cloud or SAP Analysis for Microsoft Office requires additional licenses. ■ Ensure roles and authorization objects are aligned with stakeholders. ■ Group reporting and document and reporting compliance (DRC) teams are aligned on the solution.
Maintenance		You must have a maintenance plan for the target SAP S/4HANA Central Finance, source SAP ERP/SAP S/4HANA, and SAP Landscape Transformation Replication Server. This isn't new as planned technical maintenance of the SAP system is common in any project. As part of planned maintenance, suspend the replication. If the Central Finance system or SAP Landscape Transformation Replication Server technical maintenance is planned and SAP Landscape Transformation Replication Server is suspended, postings in the source system can keep accumulating in the source staging table along with the SAP Landscape Transformation Replication Server logging table. As we know, accumulated postings in the SAP Landscape Transformation Replication Server logging table could be invoice, clearing, or reset/reversal documents. SAP recommends following the best practices in SAP Note 2679070.

Table 13.3 Central Finance Leading Practices (Cont.)

13

13.6 Summary

In this chapter, you've learned how the SAP Activate methodology fits with Central Finance implementations. We've explored six phases of the SAP Activate methodology and typical activities you'll carry out in each phase. Then, we transitioned to discuss the predelivered adoption content and how it helps customers start the Central Finance project with predelivered templates and help content. Then, we reviewed the change management topics, especially organizational, technology, and process changes in the context of Central Finance. We briefly touched on how customers benefit from the SAP Focused Run offering for Central Finance. Finally, you learned all about the Central Finance leading practices.

The Author

Anand Seetharaju is an SAP financial and management accounting certified consultant, a tax certified consultant, a CPA, and a CMA with more than 25 years of hands-on experience in financial, managerial, tax and accounting applications. He specializes in SAP finance transformation projects, including Central Finance, central payments, credit, collections, and dispute management, cash operations, margin analysis, project systems, revenue accounting and reporting, contract accounting, capital asset management, and leasing/flexible real estate management. His expertise also includes in global tax and legal reporting, integration of tax compliance software such as Thomson Reuters ONESOURCE and Vertex with ERP systems, and statutory reporting with SAP Document and Reporting Compliance.

Index

C

D

N

O

P

Q

R

S

T

U

V

W

Y

- Set up financial accounting and controlling processes in SAP S/4HANA with step-by-step instructions
- Plan your SAP S/4HANA Finance project
- Prepare for testing, go-live, and production support

Stoil Jotev

Configuring SAP S/4HANA Finance

Get your SAP S/4HANA Finance configuration right the first time! Whether you're running a new implementation or transitioning from SAP ERP, this comprehensive guide walks you through each project task. Start by setting up an organizational structure and defining global master data. Next, follow step-by-step instructions organized by functional area: general ledger, AP, AR, controlling, margin analysis, predictive accounting, and more. Your new system awaits!

738 pages, 2nd edition, pub. 08/2021
E-Book: $84.99 | **Print:** $89.95 | **Bundle:** $99.99

www.sap-press.com/5361

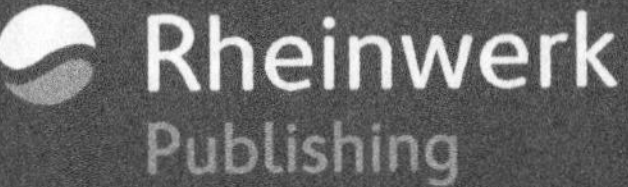